Trace Metals and other Contaminants in the Environment 10

Lead and Public Health

Trace Metals and other Contaminants in the Environment 10

Series Editor: Jerome O. Nriagu
Department of Environmental and Industrial Health
School of Public Health
University of Michigan
Ann Arbor, Michigan 48109-2029
USA

Other volumes in this series:

Volume 1: Heavy Metals in the Environment, edited by J.P. Vernet

Volume 2: Impact of Heavy Metals on the Environment, edited by J.-P. Vernet

Volume 3: Photocatalytic Purification and Treatment of Water and Air, edited by D.F. Ollis and H. Al-Ekabi

Volume 4: Trace Elements – Their Distribution and Effects in the Environment, edited by B. Markert and K. Friese

Volume 5: Metals, Metalloids and Radionuclides in the Baltic Sea Ecosystem, edited by P. Szefer

Volume 6: Bioindicators and Biomonitors: Principles, Concepts and Applications, edited by B.A. Markert, A.M. Breure and H.G. Zechmeister

Volume 7: Long-term Performance of Permeable Reactive Barriers, edited by K.E. Roehl, T. Meggyes, F.-G. Simon and D.I. Stewart

Volume 8: Hazardous Substances and Human Health, Till M. Bachmann

Volume 9: Arsenic in Soil and Groundwater Environment, edited by P. Bhattacharya, A.B. Mukherjee, J. Bundschuh, R. Zevenhoven and R.H. Loeppert

Trace Metals and other Contaminants in the Environment 10

Lead and Public Health

Science, Risk and Regulation

Paul Mushak

ELSEVIER

AMSTERDAM • BOSTON • HEIDELBERG • LONDON • NEW YORK • OXFORD
PARIS • SAN DIEGO • SAN FRANCISCO • SINGAPORE • SYDNEY • TOKYO

Elsevier
Radarweg 29, PO Box 211, 1000 AE Amsterdam, The Netherlands
The Boulevard, Langford Lane, Kidlington, Oxford, OX5 1GB, UK

British Library Cataloguing-in-Publication Data
A catalogue record for this book is available from the British Library

Library of Congress Cataloging-in-Publication Data
A catalog record for this book is available from the Library of Congress

ISBN: 978-0-444-51554-4
ISSN: 1875-1121

Printed and bound in Great Britain
11 12 13 14 10 9 8 7 6 5 4 3 2 1

Contents

Part 3
Lead Toxicity in Humans

Part 4
Human Health Risk Assessment

Part 5
Regulatory Approaches to Control

Acknowledgments

Numerous clinicians, scientists, and public health practitioners have contributed to the huge scientific and medical literature that underpins the topic of lead and public health. I have acknowledged their contributions as used in this book by citing their efforts in the respective chapter bibliographies.

A number of these clinicians and scientists merit added acknowledgment for their contributions to the broader clinical, regulatory, and societal aspects of lead beyond their many scientific contributions. Of particular note for such recognition are the late Clair Patterson and Herbert L. Needleman. Also to be recognized are the multiple contributions of the late Julian J. Chisolm, Jr., Lester D. Grant, Bruce Lanphear, the late Kathryn R. Mahaffey, John F. Rosen, Joel Schwartz, and Ellen K. Silbergeld.

SPECIAL THANKS

I want to express a special thanks to my wife, Betty, for her many contributions to this book's preparation, particularly assistance in its organizing and editing. I also greatly appreciated her many words of encouragement during the preparation of this complex and involved writing effort. Short fuses and long subordinate clauses gave way to more patience and shorter grammatical constructions in the course of this interesting experience.

Part 1

Lead in the Human Environment

Chapter 1

Introduction

Why a book that combines the science of lead with its risk assessment and its regulation?

The answer is partly contained in the question itself. The status of environmental lead as a factor in public health, particularly in heavily contaminated industrial societies such as those of the United States and Europe, has evolved to the point that it is no longer sufficient to merely describe the problems of lead contamination and its adverse impacts by enumerating examples of lead's potency as a toxic substance or its propensity for pervasive contamination. The current interest in the relationship of lead contamination to human health spans more dimensions than toxicology or epidemiology in isolation. These added dimensions came about through major developments in parallel areas such as human health risk assessment as an applied discipline. A significant body of information and activity that basically covers and combines regulatory science, regulatory policy, and public health policy for environmental lead has also evolved.

Earlier monographs and other texts dealing with lead have typically consisted of multiauthor tracts summarizing topic-specific research presented at conferences and symposia or summations representing state-of-the-art progress in some relatively narrow technical aspect of lead. Those works have also been the product of scientists and others working within some organizational framework, such as reports of committees of the National Research Council and National Academy of Sciences. Other works have consisted of collections of lead-related topics by different authors writing in their specialty area, under the aegis of an editor commissioned by some commercial publishing house. Few have been written and presented in a way that permits ready use by those who are not narrowly specialized scientists, but scientists charged with translating the technical underpinnings of our knowledge of lead to the needs of the larger regulatory and public health communities.

Trace Metals and other Contaminants in the Environment, Volume 10
ISSN: 1875-1121 DOI: 10.1016/B978-0-444-51554-4.00001-8

The disciplinary diversity and multifaceted complexity of the database for lead has seemingly limited efforts at producing and presenting comprehensive and integrated assessments of lead in its scientific and biomedical totality.

The basic rationales for this book are several. First, the current volume of information on lead relevant to human and ecological health impacts is sufficiently large and diverse to need a broad integration of diverse components of the topic. Such a broad, integrating assessment for lead has not appeared before. Second, any analysis requires sorting through two contending realities. While lead has played a major role in the cultural, technical, and economic developments of cultures and civilizations, it is simultaneously a multisource, multiexposure pathway contaminant that pervasively produces multiple adverse health effects across a quite broad segment of its dose–response spectrum of adverse health effects.

Third, the scientific and societal "maturation" of lead as a public health issue has increasingly compelled both the scientist and the physician to view the element in ways qualitatively and quantitatively different from what has traditionally been the case. Lead has long been known to produce severe adverse effects in identifiable poisoned subjects, and those impacts required clinician and physician involvement within the framework of a one-on-one medical intervention model. However, we are increasingly made aware that lead also produces a variety of subtle but important adverse effects on the developing central nervous system and other systems in humans. These are not only irreversible in some cases but can exact significant economic and public health tolls when those effects are metered or scaled on a "macro" or population basis.

Fourth, the area of lead and health needs a central source of information that is descriptively and prescriptively useful to both the scientist and the health practitioner. Virtually, all state-of-the-science and proceedings monographs on lead are largely descriptive in nature, while published sources of prescriptive information have largely been confined to such material as the periodic statements of the U.S. Centers for Disease Control on lead poisoning in children. The latter have traditionally been a guidance document for medical identification and subsequent medical management. Descriptive information has typically been devoid of any efforts at prescriptive guidance based on the data presented, while prescriptive sources have not been very broad based in terms of where much of the interest in lead currently lies, i.e., dealing with those elevated lead exposures that may produce subtle effects.

Lead is an environmental contaminant that produces both reportable human disease and risks of human disease, depending on where along the dose–response curve actual prevalence and incidence of significant systemic exposures have occurred or are occurring. As such, this dose–response duality of lead's behavior is unlike many pollutants of public health and regulatory interest whose potentials for adverse health impacts on humans typically and often only exist in such illustrative forms as species-extrapolated mathematical estimates of lifetime cancer risk or estimates of relative freedom from adverse effect risk, e.g., the oral reference dose (RfD) or the inhalation reference concentration (RfC) methodology of the online Integrated Risk Information System of the U.S. Environmental Protection Agency (U.S. EPA). A detailed discussion of these two facets for lead and human health risk assessment is included in a report by the author to the U.S. EPA that dealt with a historical and technical comparative perspective on lead versus mercury (Mushak and Mushak, 2000).

This book is intended to not only present a coherent body of health science dealing with lead but also present the most reliable portions of that information in the context of human health risk assessment and regulatory science and policy. It has been written by one author, a multidisciplinary scientific practitioner with over four decades of research and advisory experience in all three areas germane to the book. As such, this book serves as a bridge between the most relevant, credible science of lead and society's uses of that information to quantify the hazards to human health from lead and to effectively reduce or eliminate that risk.

This book is not intended to serve as an exhaustive compendium of the entire published literature on lead in the human environment. The volume of that literature is enormous and continuously growing; it is not even apparent what an uncritical recapitulation of this volume of information on lead would achieve for any public health goal or other specific purpose. Most immediately germane to this book, this aggregate volume of lead data is comprised of published findings that range greatly in research quality and vary greatly as to relative importance to useful assessments of the major features of environmental lead as a public health issue in human populations. This book, in particular, required the selection and use of scientific data sets for Pb judged to be of sufficient scientific quality and importance to address the trio of topics covered here: lead science and research, risk assessments for Pb exposures and toxic effects, and regulatory approaches to effective lead controls in the human environment.

1.1 THE SCIENCE OF LEAD AND HUMAN HEALTH

Environmental lead poisoning is still considered to be the principal environmental health threat to preschool children in industrialized countries, particularly for American infants and toddlers. A second risk population equally vulnerable to lead exposure and associated toxicity is the fetus, arising from maternal lead exposure and ready transplacental movement of lead from early in gestation until birth. The categorical and quantitative criteria by which we can establish this premier hazard ranking for lead in the world of environmental health are presented in later chapters.

Lead's unrivaled status persists in the face of recent attempts at reducing or regulating lead emissions into the human environment. This status persists despite information indicating that for certain segments of the affected population in the United States and other developed countries, the levels of lead in the body have declined compared to previous decades. One reason for persistence is that these desirable declines in systemic exposures in some individuals have not occurred to the same extent in those individuals who are simultaneously at risk for other threats to health, e.g., poor nutrition, exposures to numerous environmental contaminants other than lead, and a low probability of escaping such exposures. Children in this category can be demographically characterized as those from low-income or impoverished families residing in deteriorated neighborhoods of older urban centers, where lead exposures from paint, atmospheric emissions from industrial operations, and other emission sources such as incinerators largely have still not yielded to remediation.

The cumulative global literature on lead as a public health problem over recorded history is enormous and continuously growing. What is more, this information base has been expanding at a much more rapid rate over the last several decades. Among environmental contaminants, more is probably known about lead pollution and its various adverse impacts than is known about any other substance. For just 1 year, 1991, the author reported that about 1,500 published papers dealing with health-related lead appeared in scientific journals and other media abstracted by the Chemical Abstracts Service of the American Chemical Society (Mushak, 1992a). These citations ranged from results of basic research on lead's chemical characteristics relevant to *in vivo* behavior to the outcome of various environmental epidemiological studies of subtle neurobehavioral effects in infants and older children.

1.2 LEAD AND HUMAN HEALTH RISK ASSESSMENT

Risk assessment in human populations is broadly defined within the vocabulary of the National Academy of Science/National Research Council (NAS/NRC) (1983) monograph on the topic as [quantitative] characterization of the potential adverse health effects from quantified exposures of humans to environmental hazards. In keeping with the NRC report's use of the term, both qualitative and quantitative risk assessments are considered when dealing with lead contamination and adverse effects in this text, although the most useful information emerged when empirical determination or modeling estimations produced a numerical outcome. Part 4 (Chapters 20–24) of this monograph presents the topic of human health risk assessment in detail.

As discussed later, the relatively recent emergence of the theory and practice of risk assessment on the regulatory and public health policy scene, with its associated vocabulary and diverse methodologies, did not occur with universal acclaim. Responses in the scientific, public health, and economic worlds have been quite mixed. For example, one response has been that of a technological and arguably simplistic conviction among certain regulated stakeholders that risk assessment removes significant guesswork from regulatory actions, hopefully, thereby, bringing the level of adverse economic impacts on those interests below what might occur in its absence. Risk assessment methodologies and their claimed accommodation of variability and uncertainty and impacts on regulatory policy are also simplistically viewed as a considerable improvement over purely judgmental decision making.

Other segments of the scientific and health communities take the more conservative or even opposite view that using health risk assessment methods necessarily entails its own cluster of uncertainties, i.e., one set of uncertainties is replaced by a second, perhaps more complex set of uncertainties. In these cases, risk assessment can be *risky* assessment. Consequently, persisting uncertainties in the risk assessment process, whatever the specifics of methodology, still require the extensive overlay of professional judgments. These judgments are, ideally, informed and guided by the intervention of strong, multidisciplinary backgrounds. That is, the traditional biological and toxicological view that informed judgment should ultimately dominate regulatory policy and/or action would ideally persist.

The antithesis of formulaïc human health risk assessments for populations exposed to environmental contaminants, i.e., a *post hoc* process, is the "precautionary principle," an *ante hoc* rationale. This approach appears to some as relatively draconian in its prescriptive formulation, i.e., no substances should be released into the human environment unless the likelihood of harm to health is determined to be acceptably small.

Whatever the range of scientific perceptions about the relative utility of risk assessment methodologies in pollutant regulation, this methodological tool rarely offers a quick, simple, and ready solution to finalizing judgments about real potential or actual harm to individuals and populations. This limit applies when considering all the settings where environmental exposures to a particular substance might occur or in instances with a mixture of substances occurring as multiple contaminants at a particular site. Sufficient uncertainty requiring scientific judgment calls will invariably remain. It is also the case that the extent to which one can do quantitative risk assessments for a contaminant hinges largely on the available database subsumed in any of the four generic components of a typical risk assessment: (1) hazard characterization; (2) dose–response relationships; (3) site- or case-specific exposure characterization; and (4) the ultimate product, overall risk characterization. The more complete the information base, the more extensive and quantitative the risk assessment can be. Hazard characterization and dose–response relationships are general for a particular substance, while exposure assessment and final risk characterization require specific information as input.

The size and quality of the available database for an environmental pollutant will vary greatly across substances and will also vary within the four components of the typical risk assessment. The variety of adverse health risks of exposure may be qualitatively known but dose–response relationships may be poorly quantifiable, either because of such limits as inadequate biomarkers of exposure or early effect or absence of information on the full span of the dose–response curve.

Risk assessment is an increasingly contentious issue for those outside the immediate health and environmental science areas. The notion of using a tool called risk assessment to underpin legislative, policy, and regulatory actions, a tool which seemingly exploits quantitative science, appears to be particularly appealing to legislators, health officials, and regulatory staff who traditionally have been forced to wrestle with the conundrums and charges of arbitrariness from some quarters that often accompany categorical or qualitative guidelines.

One difficulty with this "end user" view, especially among those who are charged with developing or implementing environmental health policy, is that these policy players are likely to underestimate the extent to which technical judgments on risk assessments' uncertainties are required. This is especially so for specific environmental contaminants whose economic and political dimensions are considerable. In practice, resolution of questions about how risk assessment is applied and questions of how results of such assessments are interpreted are left to others. These others may, as illustration, be experts retained by specifically affected economic or public advocacy interests who have the ear of elected representatives. A common charge from elected officials and stakeholder lobbyists has been that regulatory agencies need to be doing more risk assessments on the nature and extent of potential ecological and human effects of environmental contaminants. In the case of the U.S. EPA and other governmental agencies, risk assessments have increasingly been part of the regulatory scene in recent years.

In a number of significant ways, lead's current status as a public health issue makes the substance ideally suited for use as a model for quantitative application of the elements of human and ecological risk assessment. Lead's status also makes it a good model for examining health risk assessment interactively with the supporting science and regulatory initiatives which may result. One is faced with looking at a contaminant like lead either across a wide range of exposures and associated effects or where the types of risk assessments are specific for either a particular type of lead exposure or a particular form of lead in the environment.

First, we now have a rather detailed grasp of the many and varied adverse impacts of lead exposure, i.e., lead's hazard characterization, based on the enormous information base for that hazard characterization. Lead produces a range of adverse effects in humans, beginning with very subtle but still significant effects in "asymptomatic" children at blood lead (PbB) levels beginning at <10 μg/dl. These include decreases in IQ and a range of neurobehavioral problems including attention deficits, growth decrements, neuroelectrophysiological deficits such as reduced hearing acuity, alterations in hematological indices, and early effects on the cardiovascular, renal, and immune systems. Higher levels of exposure produce overt signs and symptoms of lead poisoning such as anemia, chronic, and acute lead encephalopathy and, in certain cases, adverse effects on the kidney in both adults and children. At PbB values of 80–100 μg/dl and above, severe lead-induced brain damage, e.g., coma and convulsions, and eventually death, become increasingly likely. Hazard characterization for lead, therefore, spans a range from changes in subtle neurodevelopmental indices and biochemical measures at the low end to death at very high lead exposures. A full expression of lead's toxicity in humans is therefore realized within one to two orders of magnitude. This markedly narrow range for hazard expressions has considerable public health and regulatory implications.

Dose–response relationships for lead, in terms of PbB as the dose/exposure biomarker or administered doses in experimental systems versus some adverse effect, are probably better known for lead than for virtually any other environmental contaminant and certainly for any other metal or metal-related pollutant. Furthermore, these relationships are buttressed by a vast scientific and public health literature derived from clinical and epidemiological studies, experimental animal testings, and a large variety of sophisticated mechanistic toxicological studies *in vivo* and *in vitro*.

Exposure assessments for lead in its various dimensions for both specific and general cases are associated with an equally huge database. Exposure assessments of environmental contamination at various hazardous waste sites are increasingly important for lead given that lead occurs at a high frequency at these sites, much more so than most other substances.

In addition to a large database on lead exposures assembled empirically, a number of biokinetic models to ascertain exposure biomarkers and body lead burdens exist in the more recent lead literature, mainly in the form of such biomarkers of exposure as PbB. These predictive models of systemic lead exposure are of differing complexity and utility in diverse exposure settings. Historically, they can be defined as classical compartment models, a hybrid of the compartmental and physiologically based pharmacokinetic (PB-PK) models or the PB-PK model type.

One can also view the lead picture as being an excellent model for integrated risk assessment across the dose spectrum for substances that not only produce risk of toxic harm but have been shown to actually cause disease. Integrating risk for a global contaminant and toxicant like lead can also be considered with respect to those risks and associated risk populations that exist across population and system boundaries. This includes ecological risk assessment as well as human risk assessments from the standpoint of how the outcomes of the former affect the latter and vice versa.

While this book is confined to lead and human health effects, its message is not to suggest any affected species bias or to hold that anthropocentric lead exposure and health issues are more important than those involving ecological systems, in which humans occupy but one niche of toxic impact. It may well be the case that some ecological systems and populations are comparatively more toxicologically sensitive to lead than human high risk populations.

The various links of lead and human health with the growing area of risk assessment methodologies are limited by the fact that risk assessment efforts for lead exposure and poisoning risk have been described rather piecemeal in the recent literature. Much of the information covers such areas as extractive industry wastes in the U.S. Inter-Mountain West and industrial or process waste risk assessments across America.

1.3 REGULATION OF LEAD IN THE HUMAN ENVIRONMENT

It is not unexpected that a widely used but toxic substance in cultures and commerce would eventually provoke interest in regulating and controlling that substance. This writer provided historical and technical discussions of rarely successful early attempts at controlling human lead exposures from the very beginning of a recognition that lead poisoning was a significant health threat to whole human populations and subsets within these populations (Mushak, 1992b; Mushak and Mushak, 2000).

Any controls on lead production, use, or dispersal into the human environment were, first, historically linked to the level of development of the society wherein attempts at lead control were made. In the Greco-Roman era, little or no regulatory controls on lead accompanied the metal's use; the producer or consumer of the substance basically determined its presence in commercial channels.

In Medieval Europe, crude attempts at lead control are discernible, producing a record of numerous examples where adulteration of foods and beverages, especially wines, was so prevalent that edicts proscribing the practice were posted on an *ad hoc* (albeit largely ineffectual) basis.

A second factor was the relative power of the economic interests producing or using lead versus those concerned about the element's adverse health impacts. This writer's description of the issue via a comparative perspective on the relative significance of lead as a public health issue (Mushak, 1992b) versus its value in commerce plus the analysis of Graebner (1987) show that lead's utilitarian popularity and its prominent role in the industrial and economic mainstream overrode any public health concerns for lead's health hazards. This dominance persisted from antiquity through much of the twentieth century.

Public awareness about health hazards that would logically feed a demand for regulatory measures would have arguably been spotty, rarely shared, and largely ineffectual in the period from the Greco-Roman era up to about the seventeenth century. The means for engendering public awareness after this period improved considerably, but it is of critical importance to point out that public awareness on any widespread or collective scale is of quite recent vintage. The reasons for this are complex and multifactorial and are provided elsewhere (Mushak, 1992b; Mushak and Mushak, 2000).

1.3.1 Interactive Feedback Relationships Among Lead Research, Risk Assessment, and Regulation

Much of the history of lead as a public health issue over the millennia unfolded as lead use and resulting adverse health impacts occurred in a fragmented and relatively isolated manner over time (see, for example, Nriagu, 1983, 1985; Mushak, 1992b; Mushak and Mushak, 2000; Waldron, 1973). This led to little awareness within the larger cultures or societies of any need for preventive or protective responses to adverse health impacts of lead use. Recent decades, however, have witnessed a burgeoning growth of scientific research on the topic of lead pollution in industrialized and industrializing societies. Robust growth in the scientific database for lead has been accompanied by marked growth in the development and use of health risk methodologies to quantify the likelihood of toxic effects of contaminants such as lead in human populations and individuals. The growth of the latter occurred in tandem with steady growth in legislative initiatives mandating lead environmental regulation and regulations for control of lead emissions and dispersal into the human environment.

This abrupt turnabout from erratic, fragmented, and problematic lead exposure and toxicity information to a systematic growth in the multiple and related databases noted above appears to spring from the recent evolution of interactive, positive feedback relationships among the areas of lead science, risk analysis, and lead regulation. These interactive relationships are positive feedback in mechanistic form in that they uniformly prompt more activity. An early depiction of an aspect of this interactive cycle was discussed by Mushak (1991) and later in a report on lead exposures in sensitive populations by the NAS/NRC (1993).

The science sector produces research data on lead. The health risk and policy communities employ that scientific data to quantify risks of human health hazards. Both the scientific and the risk analysis communities feed information on lead to lawmakers who enact laws that mandate regulatory controls on lead. Regulators then take steps to issue rules and regulations to control lead. Subsequently, users of risk methodologies for lead—individuals involved in the legislative process and the regulatory apparatus for lead—identify gaps in the existing database in terms of their needs and, therefore, the need to support the acquisition of more data. Providing such research support prompts more research and more research data, sustaining this interactive cycle. This interactive feedback cycle is depicted in Figure 1.1.

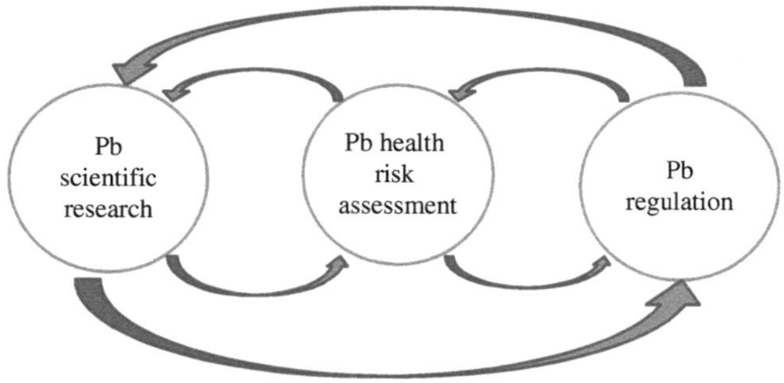

FIGURE 1.1 Interactive, feedback-loop relationships for lead science, health risk assessment and regulation.

The catalysts prompting the evolution of such an interactive cycle for lead in the human environment were, first, the establishment of a sustainable scientific research apparatus through competitive research grants created by various public agencies in the 1960s and 1970s. This public competitive, arm's-length research granting system in turn resulted from growing public and legislative awareness of hazards from lead in the environment and the need to provide a neutral, nonconflicted alternative to lead industry-supported research (Graebner, 1987; Mushak, 1992b). Further assistance in the creation and maintenance of this lead-based interactive science and control cycle came from legislative actions, particularly in the United States and in Europe, mandating that various public agencies control environmental pollutants (discussed in Part 5 of this book).

Operation of such an interactive, positive feedback cycle for environmental lead, while helpful for effective societal responses to lead's hazards, is nevertheless challenged by some of the environmental and toxicological characteristics of human lead exposures.

First, environmental lead is a multimedia pollutant. It travels various pathways in various environmental media (e.g., air, water, dusts) to impart potentially toxic lead exposures in humans. This characteristic requires that specific lead exposure settings be characterized and quantified, along with the identification of principal and likely sources of lead. By contrast, lead contamination has typically been addressed legislatively and through regulatory actions on a medium-specific basis by some particular agency with the mandated jurisdiction and responsibility for the medium.

In 1980, Mushak and Schroeder authored the first report to a U.S. Federal agency that focused exclusively on the science of lead within a multimedia pollutant framework. It was combined with the legislative history of lead and other multimedia pollutants (Mushak and Schroeder, 1980) with reference to effectiveness. This report to the U.S. National Commission on Air Quality was summarized and cited in the Commission's report, *To Breathe*

Clean Air, released in March 1981 (National Commission on Air Quality, 1981). Recommendations included the need for additional methodologies for further regulation of lead within the context of multimedia controls.

Second, lead is a pollutant and toxicant that produces both demonstrable, diagnosable disease and disease risks depicted probabilistically. That is, risk assessment methodologies are multifactorial across dose or exposure ranges. Environmental epidemiologists and other researchers investigate the impact of lead exposures and lead's health effects on a human study group or population subset with regard to parameters such as prevalence or incidence of some adverse outcome measure studied by various statistical designs, e.g., studied prospectively. Clinicians informed in the area of childhood lead poisoning will proffer diagnoses and management of individual patients, relying on available epidemiological data on the propensity of lead to produce relevant adverse health impacts in humans. This reliance is particularly significant in cases of asymptomatic lead poisoning.

Epidemiologists and clinicians issuing diagnoses occasionally speak to this issue of the individual versus the group as a toxicological target of lead using somewhat different vocabularies. The lead epidemiologist may be principally concerned about effect likelihood in groups exposed to lead versus individuals exposed to lead, but epidemiologists do not always recognize that clinical diagnostic practice is partly built on epidemiological data to inform the reasonableness and rigor of a diagnosis of lead poisoning in a specific individual. Epidemiological data are typically one of the prerequisites for settling on a particular diagnosis or identification of a given etiological factor in specific causality. The reverse is less commonly done, i.e., clinical and diagnostic practice results in identification of an etiological factor in some disease process in the total absence of any published, peer-reviewed epidemiological data showing links between a suspected causative factor and any relevant outcomes. The case series reports of pediatric morbidity and mortality in the 1910s–1920s early epidemiological efforts promoted more epidemiological efforts with lead.

Third, lead's toxicological potency in terms of dose–response relationships has so far been expressed empirically without a threshold, and certainly without a threshold for certain developmental neurotoxicological effects. A mechanistic threshold may theoretically exist, but it would have to be expressed in the very low and narrow range of 0 to 2–5 μg/dl PbB. A no-threshold or very low-threshold risk model for lead toxicity then poses the problem of how the element should be regulated appropriately. Conventional risk assessment methodologies for nonthreshold substances such as documented or probable human carcinogens differ from those involving threshold substances. Nonetheless, lead remains regulated as a threshold agent.

The difficulties engendered by this risk status for lead can be illustrated by the recent decisions in 2010 by several expert committees advising member states of the European Union to basically void the existing FAO/WHO provisional tolerable limit for lead intake (a threshold metric) in children.

The reason for this, as discussed in Chapter 29, was concern over whether the current Pb limit or any limit would be adequate, given that no empirical threshold for lead's toxic effects in children has been determined.

Finally, the reach of lead contamination around the globe in both developed and developing areas has been pervasive and at quite elevated concentrations above natural inputs to the human environment. The twin threats of pervasive contamination by lead in the face of documented toxicity of lead in humans without any apparent threshold clearly signal that lead reduction and remediation in the human environment will be a complex and expensive process.

This book provides a broad set of data-based elements with which one can develop guidelines for addressing these challenges. The regulatory aspects for lead most germane to public health policies and initiatives comprise the last portion of this monograph, Part 5.

1.4 ORGANIZATION OF THE BOOK

This book is organized into five coherent and interactively related parts containing a total of 29 chapters. Each of the five parts, in turn, is subdivided into chapters that are sequenced to develop topics and their rationales and their contributions to the overall purposes of both the part and the entire book. Parts 1–3 address the science of lead as a pollutant in the human environment and as a potent human toxicant. Part 1 presents multiple chapters on lead in the human environment. Part 2 deals with relevant aspects of defining human lead exposures. Part 3 presents an examination of the nature and extent of human lead toxicity in subsets of the population using criteria for importance and for reliable data availability to inform evaluations done on a body system or organ/organ grouping basis.

Part 4 concerns itself with the use and purposes of various human health risk assessment approaches by which one attempts to quantify and translate the large amount of winnowed scientific data into quantitative statements of human health risk in various human risk populations. That is, the 19 chapters on lead science in Parts 1–3 are integrated into the five chapters on human health risk assessment.

Part 5 addresses the broad topic of regulatory approaches to control in the United States and elsewhere of lead in the human environment. Its five chapters include an introductory treatment of some of the challenges facing the effective regulation of environmental lead, followed by environmental medium-specific discussions of regulations, for example, for lead in ambient air.

The principal focus of the book is the array of lead's human health effects from nonoccupational exposures. However, occupational exposure and effect epidemiology data are presented where they are useful for informing the overall lead and health picture as it exists for the upper end of dose–response relationships in human risk populations. Experimental results in terms of various animal model surrogates of human lead exposure and toxicology are included where they are useful for rationalizing dose–response, mechanistic, or other relationships of lead across exposed species. Finally, as noted earlier, human populations are the focus of this book, not those nonhuman species distributed across the global total environment which are now known to be adversely impacted by environmental lead.

There are six chapters in Part 1. Chapter 1 provides an introduction to all chapters in the book. Chapter 2 offers a historical and complementary context for lead in the environment, in terms of contamination and lead exposure. The historical context is a summary of topics including early lead production, uses of the element, the evolutionary chronology of Pb dispersal and human exposures, and the interplay between lead economics and public concerns.

Chapter 3 provides much of the quantitative data defining Pb production, consumption, dispersal and related topics, data mainly involving the U.S. picture. Tabulations in Chapter 3 cover the years 1720–2009 as summarized by such U.S. Federal agencies as the U.S. Geological Survey and the U.S. Bureau of Mines. Tabular data in this chapter include such categories as annual United States and World lead production, consumption by use, etc. Trend analysis is important here, since formerly major lead sources have declined in terms of ongoing inputs to environmental compartments. However, a cumulative contamination legacy remains. Other sources continue to be significant or actually promise to increase as significant sources in future years.

Chapter 4 presents data for and discussion of various inputs of lead into environmental compartments. The chapter includes typical forms of lead emissions, emission rates as a function of source type, and temporal trends in lead emissions. Of particular interest are comparisons of the broad categories of stationary versus mobile lead source emission rates. There are many lead sources and emissions, but they differ as to potential impacts on human exposures and associated toxicity risks. Of concern for this discussion is the fact that economic development in developing nations often presages the appearance of lead in commerce and then in waste streams linked to lead production and use.

A lead source is a qualitative term that identifies some environmental origin of lead. A lead emission for our purposes is mainly quantitative in nature, providing the amounts of lead being dispersed from some source of lead. Lead sources can be defined by their type. In the case of lead emissions to the atmosphere, we have typically defined these as being from either mobile or stationary sources. Such a duality in definition has both technical and regulatory rationales. Because there are technical differences, there have been differences in regulatory approaches. Some origins of lead in the human environment are technically classified as pathways of exposure by regulatory agencies such as the U.S. EPA. A pathway within this meaning is simply the interfacing environmental medium between an originating source and the exposed subject or population. Within this definition lie lead in dusts and soils and lead in foods.

Chapter 5 comprises fate and transport processes for environmental lead. These processes enter into and underpin the environmental cycling of the element and the contaminant's appearance in various media encountered by human populations.

Chapter 6 explores lead as a multimedia contaminant in the human environment. It discusses the qualitative and quantitative aspects of lead in ambient air, in paint, in diet, in drinking water, as well as lead in soils and dusts from some contributing source or mixture of sources. Some limited discussion is provided of various idiosyncratic sources of lead, i.e., sources that impact certain clusters of individuals or limited exposure circumstances for large numbers of individuals.

Part 2 consists of four chapters that address the topics associated with human lead exposures. These topics are quantitative in nature, with quantification scaled in terms of amounts, rates, time periods, etc. The chapters appear in the order of their toxicokinetic and toxicological sequences for overall expressions of adverse health risks.

Chapter 7 presents the levels of lead intakes in various subsets of affected human populations. The term "intake" as employed in the book describes amounts of media-specific Pb inhaled or ingested per unit time, typically daily, that enter the chief receiving body compartments: the gastrointestinal (GI) and respiratory tracts. Chapter 8 describes the biokinetics of Pb, specifically the toxicokinetics of Pb in human populations and the toxicokinetic basis of lead exposure biomarkers. It deals with the absorption (uptake) rate of Pb, subsequent distribution of the element into the body post-uptake, the rate of retention over the short and long term, and the rate of short- or long-term excretion of the substance.

Chapter 9 sets forth predictive modeling approaches for estimating Pb exposures in humans. It discusses the various predictive models of human Pb exposure with respect to parameters of complexity, utility, limitations, comparative advantages versus competing approaches, and scope. A number of the predictive, simulation approaches for exposure modeling entail the use of computerized components while others are relatively simple, *ad hoc* statistical forms.

Chapter 10 closes Part 2 with a discussion of the most reliable and useful environmental epidemiological data gathered for Pb exposure studies in subsets of human populations. Chapter 10 is confined to Pb exposure epidemiology and differs from the various epidemiological studies addressing dose–adverse response relationships appearing in Part 3 as part of the chapters dealing with human lead toxicity.

Parts 2 and 3, Pb toxicity, are focused on general human populations exposed to Pb in nonoccupational settings. Occupational Pb exposure data in both parts, however, are included where appropriate to inform discussions on topics including the nature of high Pb exposures as part of the full spectrum of human exposures to Pb with regard to such factors as relative sensitivity of subsets of human populations to Pb and frequency of toxic endpoints.

Part 3, dealing with human Pb toxicity in human populations, consists of nine chapters. Chapter 11 provides a historical perspective and context on human lead toxicity over the centuries and millennia. The remaining eight chapters examine particular human systems and organs adversely affected by Pb exposure. Various criteria were employed to prioritize and organize the discussions on human Pb toxicity in Part 3. One criterion was the amount of reliable, valid data for the particular systemic targets. A second was the relative usefulness of the available information on toxicity in systems and organs to derivation of risk assessments and conclusions from the effort, and a third criterion was the utility of the toxicity data in providing guidance for regulatory science and public health policies.

While some of the scientific literature has addressed organolead compounds as environmental contaminants and hazards, the predominant concern is with the inorganic, divalent form and that is the form covered in this book.

The chapters in Part 3 are principally confined to studies of subsets of human populations, and human data appear first. Experimental data for *in vivo* animal, *in vitro*, or mechanistic studies which provide support or refutation of human population findings are then presented, as are data which permit developing mechanistic understandings of toxicity processes observed for Pb exposures in humans. Overall, there is an extensive literature on experimental animal and mechanistic toxicological data for Pb that lends support to the various interpretations of the nature and extent of the element's toxic impacts in humans. Lead is arguably unique among environmental contaminants in the extent to which there is very close adherence among human toxicity, experimental toxicity, and the elucidation of mechanisms of toxicity in both. The sizes of the latter databases for environmental lead confer a high level of confidence to the parallel results found in humans and support a weight-of-the-evidence approach.

The chapters in Part 3 are subdivided into sections that first present the picture for lead toxicity in organs and systems that produces some level of adverse effects in various subsets of human populations especially sensitive to Pb exposures. Two major impact groups in such human populations are young children and older children or adults. These are followed by data from experimental studies.

Chapter 12 presents a critical discussion of Pb neurotoxicity in human populations. This is broadly held to be the most significant toxic expression of Pb exposures in sensitive human populations, notably developmental neurotoxicity *in utero* and in the early postnatal developmental window in infants, toddlers, and even older children. Two factors defining effect centrality are the seeming absence of neurotoxicity empirical thresholds indexed as Pb exposure biomarkers and the widely held scientific notion that neurotoxicity is persistent and probably irreversible. Human Pb neurotoxicity is represented by a large and relatively reliable database and includes a suite of adverse health impacts with major individual and macroscale socioeconomic and other societal implications. Some of these have yielded to econometric quantification in the form of cost–benefit analyses.

Cardiovascular effects of Pb in humans are the subject of Chapter 13, particularly with respect to effect potency in older exposure subjects but with inclusion of other risk groups. Cardiovascular effects, while inconsistently quantified across human populations, have been identified in multiple epidemiological studies, supported by a number of experimental data sets appearing in the global literature. Chapter 14 on human reproductive and developmental impacts of lead exposures presents the more useful data across several risk groups within human populations. Chapter 15 discusses adverse effects of lead on the renal system as nephrotoxicity is considered to occur across both occupational and nonoccupational subsets of human populations and subsets within nonoccupational categories. Discussions in Chapter 15, much like those in Chapter 18 on immunotoxicity, have benefited from quite recent findings.

Chapter 16 covers what has historically been one of the oldest somatic categories of Pb toxicity in humans. This arose in part from the utility of hematological indices of Pb toxicity in defining both early effect biomarkers of the element's toxicity and functional toxicity in outcomes such as anemia. Carcinogenic and genotoxic effects of Pb, Chapter 17, while more firmly rooted in occupational Pb exposure data, have also been elucidated to some extent by mixed population studies. Chapter 18 discusses a relatively newer category of Pb's adverse health impacts, immunotoxicity. The topic has gained in relative importance through informing dose–toxic response relationships at quite low levels of exposures, low levels around or not greatly above empirical relationships noted for developmental neurotoxicity. Part 3 closes with Chapter 19 on various effects in other organs and systems, effects more commonly seen at higher Pb exposures.

Part 4, human health risk assessment approaches for Pb exposures, consists of five chapters. This part begins with a broad look at the evolution of health risk assessment as an applied scientific discipline (Chapter 20). The currently formalized concepts and methods of human (and ecological) health risk assessment for lead, as they are for other pollutants and toxicants, are of relatively recent vintage and remain in flux as to appearance of new findings. However, in the case of lead, earlier elements of risk determinations for lead were in place in terms of diagnostic frameworks for clinicians identifying lead poisoning.

An added complication of applying risk assessment methods for contaminants is their dependence for form and use on the political and economic matrices in which they operate. As noted elsewhere, some governments employ risk assessment methodologies that are categorical and deterministic or probabilistic in nature. Those methods are *post hoc* as to contaminant life cycles in that they are implemented postuse and postdispersal. Others, such as European Union member states, proceed through predispersal, thresholds for official approval and permissible use models. The latter approach has been subsumed under labels such as the "Precautionary Principle."

Part 4 continues with lead-specific discussions of the four components of a human health risk assessment as structurally articulated in 1983 by the NAS/NRC (1983): Chapter 21, human health hazard characterization for lead and diverse human populations; Chapter 22, dose–toxic response relationships for lead in humans; Chapter 23, illustrative uses of case- or setting-specific lead exposure characterizations; and, Chapter 24, the last part of health risk assessment, the overall final and most quantitative step in actualizing (in a relative sense) the estimates of risk outcomes.

Human health risk assessment as used in this book refers to risk assessment as an applied scientific discipline in the area of medical and environmental science. This applied discipline partakes of activities and backgrounds of scientists and engineers with extensive formal and advanced academic expertise in relevant topics. It does not refer to, nor should it be confused with, the more narrowly defined regulatory label of "risk assessment" in the United States that addresses the evaluation of lead paint hazards in lead-painted residential units for mainly children within the statutory definitions and requirements of risk assessors set forth in the 1992 Housing and Community Development Act Reauthorization, commonly referred to as "Title X." Residential lead paint risk assessors confine their activities under Title X, after completion of a narrowly structured training regimen, to field evaluations of residential housing, typically public, Indian, or Federally assisted housing.

Noted earlier in passing, lead is different from many toxic substances subjected to human health risk assessment methods in that the toxicant not only poses risk of disease but causes actual disease as well. What is more, this disease-producing substance works to do so in human populations as well as in experimental animals. This propensity of lead is amply characterized in a huge database that permits us to discover this duality and to integrate each form with the other. There are many hundreds and thousands of environmental contaminants which are not well characterized, so that we are left with quantifying hazards to human health in the form of calculated probabilities for harm using formulaic methodologies.

Part 5 covers the issue of lead control and regulation in the human environment. It consists of five comprehensive chapters. Chapter 25 deals with the general aspects of regulatory initiatives and regulation of environmental lead in the United States and elsewhere. The broad topic of lead regulation is complex at multiple levels. There is the jurisdictional/governmental level across nations and areas within nations. Within countries, furthermore, there are divisions of regulatory labor which operate in temporal and legal sequences. In the United States, legislative initiatives via the U.S. Congress mandate regulatory action. Regulatory agencies within the Executive Branch, and often under the scrutiny of the Federal courts, then respond with rule making. In the European Union, multinational bodies such as the European Economic Community or the European Commission issue distinct directives or regulations to compel regulatory actions.

While Part 5 deals with the regulation of lead in the human environment, it is not meant to be a statutory and governmental executive history of the element in various environments. The focus is on the processes by which regulatory actions for lead have occurred and are occurring within interactive, positive feedback loop cycles with the other two main topics in this book—scientific research and human health risk assessment. The extent to which these interactive relationships will continue to feed overall protections for public health is also discussed.

Lead acts as a multimedia pollutant in the human environment. Multimedia pollutants are complex to regulate because of some larger realities. For example, what are the major sources and pathways for the contaminant in terms of human exposures on either a national or case-specific basis? How well are levels in the major media controlled relative to concentrations in the minor media and *vice versa*?

Another complicating feature of pollutant regulation in general is that of intermedia regulatory integration. That is, are certain sources and pathways of exposure more effectively controlled than others? If so, what is the nature and history of this disparity in regulatory approaches?

A second issue for Pb regulation goes to the question of how well pollutant regulation tracks the scientific and biomedical understanding of these contaminants. That is, how does what one sees in Part 5 represent a harmonization with the previously presented four parts? Put simply, one has to know something is a problem before one can begin to deal with that problem. This, in turn, begs the further question: Who is expected to be the first to be cognizant of any adverse societal consequences of environmental contamination?

The regulatory control of lead in the human environment has not only a scientific component but, necessarily, legislative and political dimensions as well. Here, the interactive feedback loop cycle operates fully. The loop portion, from public concerns and demands back to scientific research and definition of the problem, works through laws and rules at various Federal, state, and local elected bodies.

The regulatory picture for environmental lead in the United States and other industrialized and developed societies is the most advanced and extensive. Lead is most heavily regulated in these areas of the world for the simple reason that they have also had the largest problems with lead contamination as a consequence of lead's role in industry and commerce that accompanied industrial development, e.g., the Industrial Revolution.

The remaining four chapters in Part 5 are confined to medium-specific lead regulation: Chapter 26, lead in paint; Chapter 27, lead in ambient air; Chapter 28, lead in drinking water; and Chapter 29, lead in the human diet.

REFERENCES

Graebner, W., 1987. Hegemony through science: information engineering and lead toxicology, 1925–1965. In: Rosner, D., Markowitz, G. (Eds.), Dying for Work: Workers' Safety and Health in Twentieth-Century America. Indiana University Press, Indianapolis, IN, pp. 140–159.

Mushak, P., 1991. U.S. Agency for Toxic Substances and Disease Registry's report to Congress on childhood lead poisoning in America: review and update. In: Proc. First National Conference on Laboratory Issues in Childhood Lead Poisoning Prevention, Columbia, MD, pp. 79–104.

Mushak, P., 1992a. Commentary. The landmark Needleman study of childhood lead poisoning: scientific and social aftermath. PSR Quart. 2, 165–170.

Mushak, P., 1992b. Perspective: defining lead as the premiere environmental health issue for children in America: criteria and their quantitative application. Environ. Res. 59, 281–309.

Mushak, P., Mushak, E.W., 2000. A comparative analysis of the evolution of lead and mercury as public health hazards. In: Nriagu, J. (Ed.), 11th International Conference on Heavy Metals in the Environment. CD-ROM: Manuscript No. 1445, 6–10 August 2000, University of Michigan, Ann Arbor, MI.

Mushak, P., Schroeder, C., 1980. Multiple Media Pollutant Exposures and Their Regulation. Report to the National Commission on Air Quality. Contract No. 23a-AQ-6981. Discussed and cited in the National Commission on Air Quality report: To Breathe Clean Air, pp. 2.1.9, 2.2–4, 3.1–30, 3.1–34.

National Academy of Sciences: National Research Council, 1983. Risk Assessment in the Federal Government: Managing the Process. National Academy Press, Washington, DC.

National Academy of Sciences: National Research Council, 1993. Measuring Lead Exposure in Infants, Children, and Other Sensitive Populations. National Academy Press, Washington, DC.

National Commission on Air Quality, 1981. Report: To Breathe Clean Air, Washington, DC, March.

Nriagu, J.O., 1983. Lead and Lead Poisoning in Antiquity. John Wiley & Sons, New York.

Nriagu, J.O., 1985. Historical perspective on the contamination of food and beverages with lead. In: Mahaffey, K.R. (Ed.), Dietary and Environmental Lead: Human Health Effects. Elsevier, New York, pp. 1–41.

Waldron, H.A., 1973. Lead poisoning in the ancient world. Med. Hist. 17, 391–399.

Chapter 2

A Brief Early History of Lead as an Evolving Global Pollutant and Toxicant

Principal interest in the early industrial and public health history of lead for this book is in those aspects of lead that allowed lead production and consumption to evolve as a useful and desirable economic enterprise while overshadowing societal concerns about lead's adverse human health and safety aspects in the ensuing centuries and millennia. This intertwined history of lead as an economic commodity of major importance and lead's potential as a widely dispersed environmental contaminant in early and later, industrializing societies guaranteed that the element and its compounds would become a major player in human environmental health in both the workplace and the ambient environment of nonoccupational populations. This chapter deals with lead as a contaminant; Chapter 11 considers its early history as a toxic substance.

The earlier history of lead has been described in the scientific and science history literature and has been the subject of a number of treatises and individual reviews. These reports include such individual contemporary writings on various aspects of this history as those of Aitchison (1960), Drasch (1982), Eisinger (1982), Felton (1965), Gilfillan (1965), Hodge (1981), Hunter (1978), Lessler (1988), Major (1945), McCord (1953a,b,c; 1954a,b), Nriagu (1983a,b, 1985), Patterson (1965), Patterson et al. (1987), Retief and Cilliers (2006), Tylecote (1992), Waldron (1973), and Woolley (1984).

Various writings included discussions of primary historical media, in translation from the Greek, Latin, French, German, etc. Original tracts in translation by primary writers of the early periods included those of Agricola (Hoover and Hoover, 1950), Columella (Foster and Heffner, 1979), Vitruvius (Granger, 1931), Tanquerel des Planches (Dana, 1848), Apicius (Fowler and Rosenbaum, 1958), Tronchin (Schomberg, 1764), Hippocrates as several volumes with several editors–translators (Adams, 1849), Pliny the Elder (Rackham et al., 1938–1962, 10 volumes), Paulus Aeginata (Paul of Aegina, Major, 1945), and Nikander (Major, 1945).

Trace Metals and other Contaminants in the Environment, Volume 10
ISSN: 1875-1121 DOI: 10.1016/B978-0-444-51554-4.00002-X

Multiauthor compendia discussing aspects of lead history include expert consensus treatises such as those of the U.S. EPA (1986), the U.S. ATSDR (1988), the various statements on childhood lead poisoning by the U.S. CDC (1985, 1991, 2005), the U.S. NAS/NRC (1972, 1980, 1993), and the WHO (1995). Two comparative historical and scientific analyses by Mushak have also appeared (Mushak, 1992; Mushak and Mushak, 2000), ranking or otherwise comparing lead as an economic and public health issue with environmental contaminants such as other metals, notably comparisons of lead to mercury as the highly neurotoxic methylmercury form (Mushak and Mushak, 2000).

Reports of lead's early metallurgical, utilitarian, and human health history differ considerably in scope and depth and in the dimensions of time and geography. Some included discussions dating to the beginning of historical records and the introduction of early advances in technology that benefited lead's production, commerce, and use, e.g., the invention of cupellation at least 5500 years Before Present (BP) or even much earlier. These include the writings of Nriagu (1983a), Waldron (1973), and Aitchison (1960). Others have confined their historical explorations to Greco-Roman or Medieval times or somewhat later (Nriagu, 1983b, 1985; Eisinger, 1982) while the efforts of others have been confined to evaluation of more recent archival information, typified by the history of lead in early America presented as a series of papers authored by McCord (McCord, 1953a,b,c, 1954a,b) spanning the period of ~1600–1850 in America.

Some reports of lead's early history focused on a particular aspect of lead and public health, whatever their chronological reach. Examples of these are the claims of various human health hazards of lead contamination of water, foods, and beverages from the Greco-Roman era forward (Gilfillan, 1965; Needleman and Needleman, 1985; Nriagu, 1983b, 1985; Scarborough, 1984; Wedeen, 1984).

Other historical reports on lead and human health impacts were restricted to more recent source-specific lead exposures and/or lead poisoning. They are of particular usefulness in addressing various contemporary lead and health issues. This group, obviously, benefited from the availability of more data, and this data accumulation allowing characterization of sources, exposures, and effects. Leaded paint was employed on interior and exterior surfaces of structures for centuries in Europe and Colonial America, continuing through much of the twentieth century. Lead paint may have been a poorly understood topic for public health prior to the mid- to late nineteenth century when the Industrial Revolution in both Europe and North America fostered growing middle-class and upper-class populations. From this time onward, however, the evidence for lead poisoning associated with lead paint and pigment production and use grew substantially (Gibson et al., 1892; Hamilton, 1913, 1919, 1924; McKhann and Vogt, 1926; Thomas and Blackfan, 1914).

Newly affluent population segments in the United States and Europe produced a commensurate growth in residences and public buildings where lead paint was considered a desirable surface coating. This demand translated to a workforce producing lead, lead pigments, and lead paints, and users of such products, particularly house painters. These developments, in turn, translated into a lead-exposed and lead-poisoned workforce. Occupational health consequences of lead pigment and paint uses had already been distilled and evaluated in the classic nineteenth century work of Tanquerel des Planches (1839) in French.

Leaded gasoline appeared on the industrialized scene in the United States and eventually elsewhere in the 1920s and reviews of this topic cover this period to the present (Graebner, 1987; Mushak and Mushak, 2000; Nriagu, 1990; Rosner and Markowitz, 1985). Lead as a drinking water contaminant became a public health issue with the growth of cities in Europe and America, catalyzed by the Industrial Revolution and its aftermath (Dana, 1848; McCord, 1954a; multiple citations in later chapters).

Some works have dealt with the interplay of some of the more recent roles of lead in commercial and industrial history with social and public health issues and their respective proponents. Rosner and Markowitz (1985) detailed the enormous industrial and political pressures brought to bear to ease entry of the gasoline lead additive, tetraethyllead (TEL), into commerce as an effective antiknock compound for high compression engines. Graebner (1987) has detailed the history of the lead industry's participation in lead research with an eye to controlling the flow of such research information.

2.1 EARLIEST RECORDS AND USES OF LEAD

It is likely that the use of lead-bearing ores by early peoples predated isolation and use of the metal itself (Aitchison, 1960; Waldron, 1973). Aitchison (1960) refers to the use of galena (lead sulfide) for decorative purposes and body paints, particularly for the eyes. Such early uses are known with ores of other metals. For example, powdered cinnabar (mercuric sulfide) had a number of decorative uses before it was processed to yield elemental mercury (Goldwater, 1972).

The earliest reported use of lead as the metal itself, based on a surviving artifact of early civilizations, is that of a lead statue unearthed in Turkey (Hunter, 1978; NAS/NRC, 1993) and dated to 6500 Before the Common Era (BCE) or 8500 BP. This age estimate pushes back the introduction and early adoption of the metallurgical process of cupellation for ore smelting. It is likely, or at least reasonable to assume, that any systematic form of smelting, such as cupellation, would be preceded by inadvertent or experimental contact with early tribe campfires and lead-bearing geomaterials such as silver ores. Gowland (1906) stated that copper was isolated from malachite ore in this way, and Waldron (1973) speculatively extended this metallurgical serendipity to lead. This crude process would have produced impure lump lead and lead isolation via this crude means would have become known over time.

The available records indicate that early lead metallurgy included the use of crude smelting processes to liberate lead but was mainly used to separate lead from more desired metals, particularly silver, well before the Common Era (CE) (Nriagu, 1983a; Tylecote, 1992; Waldron, 1973). It is plausible that pre-Greco-Roman societies processed lead ores for the small silver content. Waldron (1973) argues that the sole interest of earlier peoples in galena was the fact that it was the principal source of silver. According to Aitchison (1960), the Laurion lead mine that fueled the wealth of Athens yielded 130 ounces of silver per ton of galena ore.

Early human communities would have had varying uses for smelted lead on a limited scale, followed by large-scale uses beginning with the Greco-Roman era. Production and use of lead water pipe and other plumbing components in the Greco-Roman era, a well-populated historical technological record, was extensive by that period and lead piping may have predated this era by many hundreds of years (Aitchison, 1960; Nriagu, 1983a). Most lead uses of any quantitative consequence were linked to the development and growth of towns and cities within the Greco-Roman imperial orbits. Based on the aggregated historical reports, demand for the metal within the communal infrastructure was for such items of everyday use as water delivery systems, construction and architectural items, currency and tokens, pigments and paints, and even medicine (Aitchison, 1960; Nriagu, 1983a; Waldron, 1973).

2.2 THE INTERPLAY OF LEAD PRODUCTION, USE, AND ENVIRONMENTAL CONTAMINATION

2.2.1 Highlights of the Early Period

It was probably not until the founding and growth of the Roman Empire that lead production and use first took on an economic and industrial momentum that cemented its utilitarian role well into the future. It was also during this historical interval that early writings drew attention to the potential toxicity of lead in the forms being widely used then. By the time of the Greco-Roman era, cupellation smelting of lead ores to recover both their lead and silver contents apparently was a widely exploited process. Equally important, the holdings of the Empire encompassed a number of lead mines, including the particularly rich deposits in Britain (Aitchison, 1960; Nriagu, 1983a; Waldron, 1973). Early smelting technology was crude and inefficient, so only ores of relatively high lead content like those in Britain were of interest or value. An early technology for fabricating lead products had also accrued by the time of the Roman imperial era.

The largest use category for lead in early Rome and elsewhere in the Roman Empire was for water distribution systems (Aitchison, 1960; Hodge, 1981; Nriagu, 1985), including lead pipes and lead siphon systems. The amounts of lead consumed for these uses apparently were enormous. Aitchison (1960) notes that the fabrication of just one water siphon unit for one aqueduct, that at Lyon, required 12,000 tons of lead. The techniques for fabricating and using lead pipe probably did not originate with the Romans, although they were the first to put lead to wide-scale use in plumbing (Aitchison, 1960).

The popularity of metallic lead that compelled uses such as construction of potable water systems certainly derived in part from the characteristics of this substance such as its inertness, resistance to rapid environmental deterioration, and malleability. Nriagu (1985) noted such characteristics as: (1) resistance to corrosion and deterioration; (2) malleability that responded to and avoided the freeze-thaw cycle, eliminating burst pipes during hard freezes; (3) malleability that made it easy to work or fit into a myriad of plumbing configurations; (4) making and adapting changes in pipe length and configurations was quite easy to do, which meant immediate responses to needs; and (5) generating scale from the interior lead surface itself would be minimal, extending the lifetime of the pipe.

Other uses of lead on a broad scale during the Greco-Roman era figured significantly in terms of potential health hazards and similarly had their roots in more limited form in earlier cultures of China, Egypt, and elsewhere. These uses included utensils, cups, and vessels made of lead, lead alloyed with other metals such as copper for leaded bronzes and zinc for pewter, and a variety of vessels lined on the inside with lead (tinning) to minimize the leaching of elements such as copper which would spoil the flavor of the contents (Nriagu, 1985). Lead compounds were also deliberately used as adulterants to disguise poor quality in wines and even for coloring various food items. Widespread use of lead in these ways, like the wide use of lead plumbing, persisted for many centuries.

The mechanisms by which using these everyday items resulted in significant lead exposures were several. For example, the concentrated grape syrup called sapa (defrutum), used as the principal sweetening agent for foods and as a preservative agent for fruits, other foods, and wines during this time, was prepared by cooking down (a net 30–50% concentrating) grape juice in lead-lined pots. Lead-lined pots were specifically prescribed as part of various recipes which survived to be recorded by various writers. Sapa preparation methods meant that relatively large amounts of lead leached from the leaded surface of the pots during cooking. Hofmann (1885, cited in Eisinger, 1982), following the usual recipe for sapa preparation and using Austrian white grapes in a lead-lined vessel, found lead leaching was extensive, 237 mg lead/l grape concentrate. In a similar test conducted by Eisinger (1982), using the recorded recipe of Columella, the resulting lead content of the sapa was highly elevated, 1,000 mg/l.

The high lead content of sapa accounted for its potent preservative efficacy, since fermentation enzymology would have been strongly affected (Eisinger, 1982). That is, key fermentation enzymes would have been strongly inhibited. Sapa lead would also serve as a strong fungicide to prevent mold formation in stored materials such as stored fruits. Sapa's high lead content also assured that human lead exposures through its various uses during the Greco-Roman era and long after would have been serious enough to induce some level of lead poisoning. The levels experimentally measured in the sapa food and wine additive simulating recorded ancient recipes, up to about 1,000 mg/l, would certainly have been adequate in dose–toxic response terms to produce chronic and even episodic acute Pb poisoning bouts. One of the most common toxic manifestations of very high lead consumption through leaded food and beverage adulterants would have been the disorder largely characterized in the sixteenth to eighteenth centuries as colica Pictonum (historically and toxicologically synonymous with the colic of Poitou, the West Indies dry gripes, and the Devonshire colic), a set of signs and symptoms attending severe gastrointestinal injury and accompanied over time by peripheral and central neuropathy (see Chapter 11).

Lead or lead-alloyed metal vessels leached lead in other ways. Storing foods for preservation allowed extensive lead leaching over time. Lead drinking cups likewise permitted leaching high amounts of lead, followed by ingestion. Widespread use of lead-glazed earthenware, often poorly fired ware, during this period and long afterward has also been held to be a significant source of lead exposure (Nriagu, 1985). Lead exposure through using that ware would be especially amplified when these containers were used for storage of acidic beverages and foods.

The variety of ways that lead was extensively used in the Greco-Roman era (and afterward as well) would appear to increase the likelihood that all levels of the society at the time would have had significant lead exposure and potentially debilitating, if not always lethal, lead poisoning.

2.2.2 Lead Contamination Causing Pb Exposures in Early Human Populations

Lead contamination of the early human environment would appear to have been adequate to produce lead exposures at toxic levels, the extent of which is discussed in toxicity terms in Chapter 11. Other data relevant to early exposures are presented here.

Lead exposures in the Roman Imperial era sufficient for chronic, multigenerational reproductive and developmental toxicity are supported by data from Nriagu (1983a,b, 1985) and Eisinger (1982) for estimates of lead exposures and likely body lead burdens of Romans of the period, stratified by class. That is, the most severe toxic responses would have been expressed mainly among the aristocracy, with lower exposure and perhaps lower toxic responses among the plebeians and slaves.

An aggregate reading of bone lead levels in and around relevant time periods and from widely different areas (Drasch, 1982; Ericson et al., 1979, 1991; Grandjean et al., 1979; Mackey et al., 1975) provide data that are ambiguous in some cases but consistent with elevated exposures in other cases. The hypothesis of class-stratified exposures of the Roman aristocracy is supported by analogous results of bone sample Pb levels stratified by class and gathered in the American Colonial era (Aufderheide et al., 1981). It is supported by the modern database (later chapters in this book) for the toxicokinetic behavior of lead in humans during development and metabolic stresses, such as bone lead releases during pregnancy. It is consistent with the modern database for dose–toxic response relationships among multisystem toxic effects (later chapters in this book) at clinical and subclinical exposures.

Rebuttals to the high-Pb-exposures theory invoke factors such as the existence of complex familial and dynastic dynamics within and among families and groups of the Roman aristocracy (Needleman and Needleman, 1985; Scarborough, 1984). Other factors may well have been in play, but these do not rule out a cofactorial role for lead. Nothing in the elements of this debate compels an either–or judgment. Hodge (1981) and Forbes (1964) argue that lead exposures in the Roman Imperial era from Rome's water supplies would have been much less than suggested, noting that the drinking water in modern city supplies is relatively hard. Nriagu (1985) responds that there were also soft (corrosive) water supplies in the Roman Imperial era and such supplies were actually preferred.

Opponents of Pb etiology for Imperial decay also argue (Needleman and Needleman, 1985) that, whatever the role of ancient water lead in Roman population exposures, the use of the sweetener *sapa* (the principal sweetening agent for foods and as a wine additive) was also exaggerated in terms of overestimated toxic lead intake rates from this source. This argument is thinly drawn. Comparatively, sapa lead levels based on the experiments of Eisinger (1982) and others would have clearly been the more likely toxic lead source compared to drinking water, whatever the relative corrosivity of the water distribution system. Eisinger (1982), using a specifically detailed recipe described in his paper for the sweetener's preparation from a widely followed source in those ancient times, obtained a level of 1,000 mg Pb/l sapa (1 mg/ml, ~1,000 ppm). Eisinger also noted that the dilution volume factor for sapa to wine explicitly described in the ancient recipes typically was 1:48 v/v.

One readily estimates high amounts of lead ingested with even moderate volumes of wine intake. Use of sapa as a sweetener in various foods was an equally potent source of toxic lead exposures. Of further concern, use as a food sweetener rather than as a wine additive would include exposures of very young children as well as adults in the aristocratic families, young children being the highest risk group for multiple lead effects.

Retief and Cilliers (2006) question whether widespread reproductive toxicity arising from major Pb contamination routes could have occurred without the presence of severe maternal lead poisoning. Such co-occurring maternal poisonings, they argue, would have been widely noted and recorded. This concern, however, is contradicted on several counts. First, and as noted by the authors themselves, the signs and symptoms of lead poisoning were largely unknown and difficult to recognize as an identifiable disease and toxicological entity until the much later seventh century CE writings of Paul of Aegina. He described early chronic lead poisoning epidemics presenting typically as lead colic with neuropathy.

Second, the requirement of severe Pb exposures in mothers as necessary for lead reproductive and developmental toxicity is not a tenable premise. As noted in later chapters, embryo- and fetotoxicity is quite a sensitive endpoint in humans and occurs at relatively low systemic lead levels owing to ready transplacental movement of lead early in pregnancy (NAS/NRC, 1993; U.S. ATSDR, 1988; U.S. CDC, 1985, 1991, 2005; U.S. EPA, 1986, 2006). The occupational history of lead in the late nineteenth and early twentieth centuries establishes various adverse reproductive outcomes in pregnant women not themselves demonstrably debilitated by manifest lead poisoning (Legge, 1901; Oliver, 1911; U.S. EPA, 1986, 2006).

A third factor in this connection is the low likelihood that information would be widely dispersed within and among socially stratified groups and populations in the Greco-Roman era, information which would have led to wide recognition of co-occurring lead poisoning and pregnancies, at least among the aristocracy and imperial leadership. Mushak and Mushak (2000), for example, noted the extreme difficulty of widespread, rapid dispersal of information about public health hazards in antiquity in the face of little information about the hazards.

2.2.3 Occupational Lead Exposures in the Greco-Roman Era

An obviously significant source of lead exposure and poisoning during the early periods of lead production and use would have been workplace exposures during lead mining and smelting. The available primary information on occupational lead poisoning *per se* is lost to recorded history, with the exception of some general comments by early writers. Vitruvius (cited in NAS/NRC, 1993) made somewhat sparse reference to workers suffering from lead poisoning. Whatever the completeness of the early occupational health record for lead, we would fully expect that occupational lead poisoning among lead workers was extremely common. Early smelting technology, in terms of engineering controls and worker protection, was hazardous to workers, with workplace air and dust lead levels extremely high, in all likelihood.

Also, laborers in lead production would have been considered relatively expendable, i.e., prisoners of war, criminals and, in some cases, slaves. These groups were also doomed to spend their short working lives in Rome's lead mines.

Early records likewise make little reference to lead poisoning of children as a specific health hazard; this may be attributable to a view that did not recognize the higher susceptibility of young children. Child laborers working side by side with adults would also have blurred any heightened sense of toxicological vulnerability of the former. Certainly the opportunities for lead exposures of children, via contaminated drinking water and contaminated confections and other foods using sapa, were present.

2.3 LEAD PRODUCTION, USE, AND EXPOSURES: THE MIDDLE AGES TO THE INDUSTRIAL REVOLUTION

The extent to which those lead sources of concern as health hazards in antiquity remained as economically valuable and utilitarian commodities that persisted through the 18 centuries from the beginning of the CE to the Industrial Revolution in Europe, and in most cases into the present, is quite striking. Nriagu (1983a) has provided a historical overview of lead uses and exposure sources that were widespread in Greco-Roman times and that continued through recent times. Lead pipe and other plumbing components continued to be popular in the Middle Ages and later, and many European cities assembled and installed extensive lead pipe water distribution networks as the cities grew during this period and later. Records of the time also make it clear that other drinking water-related lead sources could be especially hazardous to health. Lead used for roofing, guttering, and other rainwater dispersal components was a hazard when drinking water was mainly obtained by collecting rain runoff from those surfaces. Related to such Pb exposures was roofing painted with lead paint. As such paint weathered, it was borne away with runoff to cisterns or other containers. The problems were exacerbated when rainwater running over leaded roofing and guttering was collected in lead-lined cisterns. Recurring episodes of widespread lead poisoning from lead plumbing or collected rainwater consumption are dispersed through the medical and public health literature.

Typical of these poisoning outbreaks were those that occurred in Amsterdam and Haarlem, Holland, described by several writers (Tronchin, 1764). Tronchin's description of the Amsterdam lead poisoning epidemic showed an astute recognition of the role of acidification of water in enhancing lead leaching from leaded surfaces. He noted that accumulated, rotting leaves on roofs of Amsterdam buildings created an acidic environment that figured significantly in the poisoning epidemic. A similar, later epidemic of lead poisoning occurred in Tunbridge, England, and was described by Dana (1848) as an appendix to his English translation of Tanquerel des Planches' classical nineteenth century work on occupational lead exposures. Dana also appreciated that large amounts of lead in drinking water were not necessary to eventually induce the classical symptoms of clinical lead poisoning, noting that water samples conveyed in lead plumbing failed to show detectable lead levels, using the relatively crude analytical methods of the time.

McCord (1954a) has provided a rather extensive historical survey of lead plumbing in early America, including the Colonial period. By the late 1700s, lead piping and other components of plumbing appeared to be widely used in the major settled areas of the original colonies, particularly in the Middle Atlantic and Northeastern areas. However, these were not the longer-distance systems that gained wider use in the nineteenth century. Rather, short lengths of lead conduit were used to direct rainwater or nearby pond water to dwelling units. As in Europe, lead roofing and guttering as well as lead-lined collection cisterns posed additional avenues for lead contamination of the predictably aggressive, soft rainfall waters. Dana's research efforts on lead plumbing for drinking water, described in the Appendix to his translation, showed that water lead exposures were high enough to induce widespread lead poisoning in Lowell, MA, in the first half of the nineteenth century. His work was supplemented by information from a number of physicians practicing in that city at the time showing that patients were commonly presenting with symptoms attributed by the physicians to lead poisoning. From that period to the present, New England has been at particularly high risk for lead poisoning from lead plumbing, owing in part to the extent of that use and the corrosive nature of the water sources available.

Lead in foods and beverages, whether inadvertent or via deliberate use of lead adulterants, similarly persisted from Greco-Roman to more recent times. This certainly was the case for lead-contaminated wines. If anything, it appears that the occurrence of lead in low-quality wines became more of a public health issue from the Middle Ages onward than it was in ancient times. Nriagu (1985) and Eisinger (1982) have provided more recent critical discussions of the lead and wine problem, especially in Europe from the Middle Ages through the nineteenth century.

The persisting problem of lead contamination of wines, particularly deliberate addition of palatable lead compounds to disguise spoiled or mediocre vintage wines, was an enormous public health problem for centuries in Europe. Municipalities in France, Germany, and elsewhere, reacting to epidemics of wine-related lead poisoning, repeatedly enacted ordinances proscribing wine adulteration with lead salts under threat of severe punishment (Eisinger, 1982). Nonetheless, widespread adulterative contamination continued, and factors in this continued practice likely included the difficulty in demonstrating at what stage adulteration occurred and the absence of analytical tests sensitive enough to show that adulteration had, in fact, occurred.

Measurement methods of the time would have been insensitive for measuring even highly toxic quantities of lead.

Adulterating wines with lead salts was accompanied over the centuries by other forms of lead contamination. The use of lead-fitted wine presses and storage vats lined (tinned) with lead alloy likely would have produced extensive lead contamination, whether recognized or not. Using stone presses jointed with lead for grape and apple pressings was shown by Baker (1767) to be a significant source of lead in the resulting liquids in his classic research on the cause of the Devonshire Colic.

Contamination of other alcoholic beverages, especially those produced in distillation units containing lead condensers and condenser heads, was a particular problem in Europe and early America. Benjamin Franklin (cited in McCord, 1953a) referred to the widespread lead poisoning induced by imbibing lead-contaminated rum from lead unit distillation, and to reaction by the Massachusetts Bay Colony to such rum production within its borders in the form of an ordinance banning the practice in 1723.

The use of the sweetening agent sapa, described earlier as a potent source of lead contamination of numerous foods in the more elegant Greco-Roman cuisines, persisted for many centuries and Eisinger (1982) noted its use as part of classic recipes in French and Italian cook books as late as the nineteenth century.

Pewter ware containing high fractions of lead—up to 50%—was popular for about 600 years, until the early nineteenth century, because of durability and attractive appearance, and the wide variety of dining and food-related objects made of this alloy (Hatcher and Barker, 1974) produced ample opportunities for lead leaching into foods and beverages. Lead-free pewter emerged on the scene rather late, and it appears that the principal reason for abandoning pewter ware had little to do with any perceived hazards in the ware and much to do with the growing fashion of stoneware. Similarly, the production and use of earthenware with poorly fired lead glazes increased greatly over the centuries and produced more widespread threats of lead poisonings.

The Industrial Revolution created an enormously increased industrial and middle-class consumer demand for commodities that entailed lead in some form. These demands in turn produced a greatly expanded population of lead workers who were clearly lead poisoned at the workplace, given the relatively primitive working conditions. The nature and extent of the occupational lead problem was documented in the classical 1839 work of the French physician Tanquerel des Planches. As noted in the 1848 English translation from the French of this opus, all classes of occupational uses of lead were accompanied by recorded lead poisoning hazards.

Of particular interest to Tanquerel des Planches in his work was the widespread occupational poisonings of painters, particularly those using white lead for repainting or new painting and those exposed during removal of the old leaded paints. Lead paint, especially the basic lead carbonate pigment commonly known as white lead, was widely used in Europe in the eighteenth and nineteenth centuries for various categories of construction. This use continued until it was proscribed by the White Lead Convention of 1921, subsequently signed by many European nations. This period is discussed in a later chapter. The Convention advocated the use of substitutes, taking a page from Tanquerel des Planches, who urged painters to abandon white lead use and instead use zinc white (zinc oxide).

2.4 LEAD PRODUCTION AND USES IN THE TWENTIETH CENTURY

The more recent or current uses of lead by developed or industrialized societies represent an interesting and somewhat complex mix of those uses with ancient utilitarian pedigrees—lead plumbing, lead-contaminated foodstuffs, lead roofing, and related, lead-glazed earthenware—and new utilitarian forms: residential and commercial lead paint use, lead-seamed food and beverage cans, lead antiknock additives such as TEL for gasoline, lead acid storage batteries, and more pervasive workplace lead exposures arising from expanding lead production and use in the first half of the twentieth century and afterward.

Especially troublesome at this stage in lead's history are the myriad ways in which lead emissions from lead production and use became significant exposure sources for large numbers of individuals in various risk populations. Lead milling, smelting, and refining produced significant atmospheric emissions which then affected nearby communities by various pathways. Added to these stationary sources of lead for much of the twentith century were mobile source emissions in the form of auto exhausts from leaded gasoline combustion. The enormous production of lead-seamed canned foods and beverages in the first three-quarters of the twentieth century posed lead exposure risks for all segments of the population, but especially for infants and toddlers consuming condensed milk and canned fruit juices. Many of these lead emission sources led to continuing contamination of large tracts of various environmental compartments, notably soil, water, and sediments. Current lead sources and exposure pathways are detailed in later chapters of this book.

2.4.1 The Corporate Lead Industry and Lead Pollution in the Twentieth Century

Small-scale lead production in the United States and other lead-producing countries gave way to industrial corporation production in the last quarter of the nineteenth century. This transition included production of lead in the course of silver and zinc production (Knerr, 1992; Nriagu, 1983a). Unlike the small lead producer, the evolving extractive industry corporation produced a relatively large volume of the metal and various metal compounds as well as equally enormous amounts of production-associated leaded waste streams. These included mining, milling, and smelting wastes. The worst environmental impacts of this production were relatively localized to the mining and/or smelting operations. They took the form of ecological damage to land, waterways, and wildlife, and lead exposure for workers and residents of adjacent communities. In time, however, with large increases in production volumes and national-scale marketing, that orbit of adverse impacts was expanded to embrace contamination and exposures associated with a myriad of lead's uses remote from lead production.

The corporate lead industry in its evolution and growth understandably had a great deal of economic and social influence because of size and financial leverage at the community, regional, and national levels. This leverage could be applied to preserve the company's economic well-being and lessen constraints on its operations, either through efforts of its own or through the aegis of the trade association (Derickson, 1987; Graebner, 1987). In time, the continued growth of large lead production operations required similar development of markets for lead on a large scale. Expanded use of lead and its products meant greater potential dispersion of these substances. Marketing promotion encountered a receptive consumer population. An expanding middle class and the number of commercial and industrial operations using lead during this period produced increased demand for lead paint pigments, lead paint, lead plumbing, a myriad of household products containing lead, and for uses in the urban infrastructure. An example of infrastructure demand would be a huge requirement for red lead paint for numerous public structures such as bridges.

A quantitative and comparative look at the impact of large later versus small earlier lead producers on overall United States and world lead production and consumption is readily discernible from the tabulations for lead production and uses in Chapter 3.

Development of the lead corporation in the waning years of the nineteenth century and into the first quarter of the twentieth century was especially rapid in the Midwest and, somewhat later, the Intermountain West and Southwest. A detailed description of this development is included in the corporate profile of the Eagle-Picher Industries by Knerr (1992). The analysis and review of Derickson (1987) made it clear that a well-capitalized metals corporation was required for: (1) the acquisition and settling of numerous contested mineral claims; (2) the hiring and retention of a relatively skilled mining, milling, and smelter workforce; (3) the acquisition of the necessary equipment; and (4) the production and transport of large amounts of lead and other economic metals with efficiencies of scale. Consolidation of numerous small metal producing operations into large multistate enterprises was common around the turn of the century and was typified by the creation of the American Smelting and Refining Co. (ASARCO) in 1895 from many smaller operations.

The ultimate outcome of attempts at consolidating industrial production of many commodities into fewer hands in the late nineteenth century was the monopolistic trust and its successors in interest, arising from regulatory dismemberment. Lead production and marketing followed the trend. Knerr (1992) points out that the National Lead Trust, established in 1887, managed to control, by 1890, 80% of America's white lead production and 70% of the nation's red lead output. Its eventual dissolution, in turn, resulted in the National Lead Company, which managed to maintain control of 70% of white lead production into the twentieth century.

REFERENCES

Adams, F. (Ed. and Trans.), 1849. Hippocrates, The Genuine Works of Hippocrates, 2 vols. Sydenham Society, London, UK.

Aitchison, L., 1960. A History of Metals, vol. 1. Interscience, New York.

Aufderheide, A.C., Angel, J.L., Kelly, J.O., Outlaw, A.C., 1981. Lead in bone. III. Prediction of social correlates from skeletal lead content in four colonial American populations. Am. J. Phys. Anthropol. 66, 353–361.

Baker, G., 1767. An essay concerning the cause of the endemical colic of Devonshire. [Read before the College of Physicians, Royal Society, London.] Med. 1768. Trans. Coll. Phys., Lond. 1, 257–318.

Dana, S.L. (Ed. and Trans.), 1848. Lead Diseases: A Treatise from the French of L. Tanqueral des Planches. With Notes and Additions on the Use of Lead Pipe and its substitutes. Daniel Bixby and Company, Lowell, MA.

Derickson, A., 1987. "To be his own benefactor": The founding of the Coeur d' Alene Miners' Union Hospital, 1891. In: Rosner, D., Markowitz, G. (Eds.), Dying for Work. Workers' Safety and Health in Twentieth-Century America. Indiana University Press, Bloomington, IN, pp. 3–18.

Drasch, G.A., 1982. Lead burden in prehistorical, historical, and modern human bodies. Sci. Total Environ. 24, 199–231.

Eisinger, J., 1982. Lead and wine: Eberhard Gockel and the *colica Pictonum*. Med. Hist. 26, 279–302.

Ericson, J.E., Shirahata, H., Patterson, C.C., 1979. Skeletal concentrations of lead in ancient Peruvians. N. Engl. J. Med. 300, 946–951.

Ericson, J.E., Smith, D.R., Flegal, A.R., 1991. Skeletal concentrations of lead, cadmium, zinc, and silver in ancient North American Pacos Indians. Environ. Health Perspect. 93, 217–223.

Felton, J.S., 1965. Man, medicine and work in America: an historical series. II. Lead, liquor and legislation. J. Occup. Med. 7, 572–579.

Forbes, R.J., 1964. Studies in Ancient Technology, vol. 7–9. Brill, Leiden, The Netherlands.

Foster, E.S., Heffner, E.H. (Ed. and Trans.), 1979. Columella. On Agriculture and Trees, vol. 3. Harvard University Press, Cambridge, MA.

Fowler, B., Rosenbaum, E. (Ed. and Trans.), 1958. Apicius. The Art of Cooking. George G. Harrapa and Co., London, UK.

Gibson, J.L, Love, W., Hardine, D., Bancroft, P., Turner, A.J., 1892. In: Huxtable, L.R. (Ed.), Transactions of the Third Intercolonial Medical Congress of Australasia. Charles Potter, Sydney, pp. 76–83.

Gilfillan, S.C., 1965. Lead poisoning and the fall of Rome. J. Occup. Med. 7, 53–60.

Goldwater, L.J., 1972. Mercury. A History of Quicksilver. York Press, Baltimore, MD.

Gowland, W., 1906. Copper and its alloys in prehistoric times. JAI [J. Anthrop. Inst.] XXXVI, 17.

Graebner, W., 1987. Hegemony through science: information engineering and lead toxicology, 1925–1965. In: Rosner, D., Markowitz, G. (Eds.), Dying for Work: Workers' Safety and Health in Twentieth-Century America. Indiana University Press, Indianapolis, IN, pp. 140–159.

Grandjean, P., Nielsen, O.V., Shapiro, I.M., 1979. Lead retention in ancient Nubian and contemporary populations. J. Environ. Pathol. Toxicol. 2, 781–787.

Granger, F. (Ed. and Trans.), 1931. Vitruvius. On architecture. Harvard University Press, Cambridge, MA. Reprinted 1962.

Hamilton, A., 1913. Hygiene of the Painters' Trade. Bull. U.S. Bur. Lab. Stat., No. 120. U.S. Government Printing Office, Washington, DC.

Hamilton, A., 1919. Lead poisoning in American industry. J. Indust. Hyg. 8–21.

Hamilton, A., 1924. The prevalence and distribution of industrial lead poisoning. JAMA. 83, 583–588.

Hatcher, J., Barker, T.C., 1974. History of British Pewter, first ed. Prentice Hall Press, New York.

Hodge, A.T., 1981. Vitruvius, lead pipes and lead poisoning. Am. J. Archaeol. 85, 486–491.

Hofmann, K.B., 1885. Das Blei bei den Völkern des Altertums. In: Dentike, F. (Ed.), Beiträge aus der Geschichte der Chemie, Leipzig, Germany, Vienna, Austria.

Hoover, H.C., Hoover, L.C. (Ed. and Trans.), 1950. G. Agricola. 1556. De Re Metallica. Dover Publications, New York, NY.

Hunter, D., 1978. The ancient metals. The Diseases of Occupations. sixth ed. Hodder and Stoughton, London, pp. 248–297.

Knerr, D., 1992. Eagle-Picher Industries. Strategies for Survival in the Industrial Marketplace, 1840–1980. Ohio State University Press, Columbus, OH (Chapters 1 and 2).

Legge, T.M., 1901. Industrial lead poisoning. J. Hyg. 1, 96–108.

Lessler, M.A., 1988. Lead and lead poisoning from antiquity to modern times. Ohio J. Sci. 88, 78–84.

Mackey, A.C., Townsend, A., Waldron, H.A., 1975. Lead concentrations in the bones of Roman York. J. Archaeol. Soc. 2, 235–237.

Major, R.H., 1945. Classic Descriptions of Disease. III. Lead Poisoning. third ed. C.C. Thomas, Springfield, IL, pp. 312–325.

McCord, C.P., 1953a. Lead and lead poisoning in early America: Benjamin Franklin and lead poisoning. Ind. Med. Surg. 22, 393–399.

McCord, C.P., 1953b. Lead and lead poisoning in early America: lead mines and lead poisoning. Ind. Med. Surg. 22, 534–539.

McCord, C.P., 1953c. Lead and lead poisoning in early America: the pewter era. Ind. Med. Surg. 22, 573–577.

McCord, C.P., 1954a. Lead and lead poisoning in early America: the lead pipe period. Ind. Med. Surg. 23, 27–31.

McCord, C.P., 1954b. Lead and lead poisoning in early America: clinical lead poisoning in the Colonies. Ind. Med. Surg. 23, 120–125.

McKhann, C.F., Vogt, E.C., 1926. Lead poisoning in children, with notes on therapy. Am. J. Dis. Child. 32, 386–392.

Mushak, P., 1992. Perspective: defining lead as the premiere environmental health issue for children in America: criteria and their quantitative application. Environ Res. 59, 281–309.

Mushak, P., Mushak, E.W., 2000. Lessons Learned: a comparative analysis of the evolution of lead and mercury as public health hazards. In: Nriagu, J. (Ed.), 11th International Conference on Heavy Metals in the Environment, CD-ROM: Manuscript No. 1445, 6–10, August 2000, University of Michigan, Ann Arbor, MI.

National Academy of Sciences: National Research Council, 1972. Lead—Airborne Lead in Perspective. National Academy Press, Washington, DC.

National Academy of Sciences: National Research Council, 1980. Lead in the Human Environment. National Academy Press, Washington, DC.

National Academy of Sciences: National Research Council, 1993. Measuring Lead Exposure in Infants, Children, and Other Sensitive Populations. National Academy Press, Washington, DC.

Needleman, L., Needleman, D., 1985. Lead poisoning and the decline of the Roman aristocracy. Class. Views 4, 63–94.

Nriagu, J.O., 1983a. Lead and Lead Poisoning in Antiquity. John Wiley & Sons, New York.

Nriagu, J.O., 1983b. Saturnine gout among Roman aristocrats: Did lead poisoning contribute to the fall of the Empire? N. Engl. J. Med. 308, 660–663.

Nriagu, J.O., 1985. Historical perspective on the contamination of food and beverages with lead. In: Mahaffey, K.R. (Ed.), Dietary and Environmental Lead: Human Health Effects. Elsevier, New York, pp. 1–41.

Nriagu, J.O., 1990. The rise and fall of leaded gasoline. Sci. Total Environ. 92, 13–28.

Oliver, T., 1911. Lead poisoning and the race. BMJ. 1, 1096–1098.

Patterson, C.C., 1965. Contaminated and natural lead environments of man. Arch. Environ. Health 11, 344–363.

Patterson, C.C., Ericson, J., Shirahata, H., 1987. Lead in ancient human bones and its relevance to historical developments of social problems with lead. Sci. Total Environ. 61, 167–200.

Rackham, H. (Ed. and Trans.), 1962. Pliny the Elder. Natural History, 10 vols. Harvard University Press, Cambridge, MA.

Retief, F.P., Cilliers, L., 2006. Lead poisoning in ancient Rome. Acta Theol. 26 (Suppl. 7), 147–164.

Rosner, D., Markowitz, G., 1985. A "gift of God"? The public health controversy over leaded gasoline during the 1920s. Am. J. Publ. Health 75, 344–352.

Scarborough, J., 1984. The myth of lead poisoning among the Romans: an essay review. J. Hist. Med. 39, 469–475.

Schomberg, R. (Ed. and Trans.), 1764. T. Tronchin. A Treatise on the Colica Pictonum or the Dry Belly-Ache (Original Edition, 1757). London, UK, Geneva, Switzerland.

Tanquerel des Planches, 1839. Lead Diseases. (S.L. Dana, Trans. 1848). Daniel Bixby and Co., Lowell, MA.

Thomas, H., Blackfan, K., 1914. Recurrent meningitis due to lead, in a child of five years. Am. J. Dis. Child. 8, 377–380.

Tronchin T., 1764. A Treatise on the Colica Pictonum; or the Dry-Belly-Ache. Schomberg, R. (Ed. and Trans.). London, U.K. (1764 Edition), Geneva, Switzerland (1757 Original Edition).

Tylecote, R.F., 1992. A History of Metallurgy. The Institute of Materials, London.

U.S. Agency for Toxic Substances and Disease Registry, 1988. The Nature and Extent of Lead Poisoning in Children in the United States: A Report to Congress. Atlanta, GA: U.S. Department of Health and Human Services.

U.S. Centers for Disease Control, 1985. Preventing Lead Poisoning in Young Children. A Statement by the Centers for Disease Control. U.S. Department of Health and Human Services, Atlanta, GA.

U.S. Centers for Disease Control, 1991. Preventing Lead Poisoning in Children: A Statement by the Centers for Disease Control. U.S. Department of Health and Human Services, Atlanta, GA.

U.S. Centers for Disease Control, 2005. Preventing Lead Poisoning in Young Children: A Statement by the Centers for Disease Control. Department of Health and Human Services, Atlanta, GA.

U.S. Environmental Protection Agency, 1986. Air Quality Criteria for Lead, 4 Vols. Report No. EPA/600/8-83/028bF. Environmental Criteria and Assessment Office, Washington, DC.

U.S. Environmental Protection Agency, 2006. Air Quality Criteria for Lead, Vol. 1. Report No. EPA/600/R-05/144 aF (Chapters 4–6).

Waldron, H.A., 1973. Lead poisoning in the ancient world. Med. Hist. 17, 391–399.

Wedeen, R.P., 1984. Poison in the Pot: The Legacy of Lead. Southern Illinois University Press, Carbondale, IL.

Woolley, D.E., 1984. A perspective of lead poisoning in antiquity and the present. Neurotoxicology 5, 353–362.

World Health Organization, 1995. Environmental Health Criteria 165: Inorganic Lead. International Programme on Chemical Safety, Geneva, Switzerland.

Chapter 3

Lead in the Human Environment: Production, Uses, Trends

3.1 INTRODUCTION

This chapter presents the information for a key question relevant to this book: How much lead has entered the environment of human populations? Data for such topics as the rates of lead entry over time, the critical timelines for major increases in lead's dispersal into the human environment, the past and present uses of the element, and discernible trends in the production, consumption, and dispersal of lead are included.

Available data to address this question vary as a function of factors including period of production and use and regions of the globe for which figures are sought. Lead production is quantified with respect to both primary and secondary production. Primary production totals and rates over the long term provide a cumulative picture of lead entry into the environment, while secondary production, e.g., recycled lead amounts, provide a measure of what is being recycled versus what is being dispersed and not retrieved. The historical and technical record for lead production and use indicates variability over time and also indicates the international reach of production and use, especially in the modern era. That is, lead production in ancient times is difficult to quantify, but the amounts at issue are relatively modest compared to those produced in, say, the era of the Industrial Revolution and later. Lead production and use also do not necessarily track each other in any close, geographic manner, so that lead produced in one country or region will enter international trade and be consumed elsewhere. This economic reality in recent decades complicates the search for complete data.

Lead production—mining, milling, smelting, refining—in the United States has historically occurred in areas relatively remote from points of lead consumption (NAS/NRC, 1993; U.S. EPA, 1986, 2006; Nriagu and Pacyna, 1988; McCord, 1953; Murray, 1926). This particularly applies to consumption of lead in urbanized, heavily populated areas for diverse purposes such as lead for paint pigments, lead used in the production of

Trace Metals and other Contaminants in the Environment, Volume 10
ISSN: 1875-1121 DOI: 10.1016/B978-0-444-51554-4.00003-1

lead-acid batteries, and lead for the production of gasoline antiknock additives. Communities proximate to lead production facilities sustain their own lead exposures in the form of contact by risk groups with various waste streams (McCord, 1953; NAS/NRC, 1993, 2005; U.S. EPA, 1986, 2006).

It would be difficult to get a quantitative sense of the overall adverse impact of environmental lead on human populations without comprehending the magnitude of the multimillennial lead releases into the culture and commerce of various peoples. A highly toxic substance that has an obscure and highly restricted commercial and industrial pedigree poses less cumulative risk to human health than a second equally toxic substance which has a long history of widespread and long-term utility and associated widespread exposures in human populations. Lead releases to the environment over time typify the latter category.

3.2 LEAD PRODUCTION

It has been said that, in the fullness of time, all lead used by human populations eventually becomes lead dispersed and lead contamination in the form of future "background" with the rise and fall of cultures and civilizations (NAS/NRC, 1993, p. 18). There are numerous examples of this inexorable movement of lead from use to waste stream, a movement aided by complex entries and exits from numerous and diverse environmental compartments. Examples include lead that becomes the material of discrete artifacts as well as lead that is dispersed uniformly with its production into environmental compartments as constituents of the atmosphere, leaded dusts from atmospheric fallout, lead in riverine and marine sediments, lead in biota, etc. The close relationship of lead production to eventual lead dispersal and environmental contamination over both the near and long term means that one cannot get a handle on the latter without some quantitative assessment of the former.

In quantifying lead's entry and persistence in cultural and commercial channels, it is necessary to consider a rather broad time frame both for setting some form of baseline to contamination (the "natural" baseline) and for the steady additions of anthropogenic lead to this baseline. The extended period for lead contamination and linking expansion of that contamination with codevelopment and industrial expansion of peoples requires both an evaluation of the early days of lead's first human uses in the ancient world and a quantitative determination of the massive amounts of lead production and dispersal accompanying the development of industrialized societies. This is not to say that the amounts of lead entering the human environment were relatively stable over the millennia. In particular, the past century or so witnessed the most intensive growth in lead production, use and environmental dispersal.

3.2.1 Lead in the Ancient World

The qualitative archeological record for types of lead use was noted in Chapter 2. Earliest records that might produce some inkling of the quantitative picture for amounts of lead production or use in early recorded history are quite spotty and provide an incomplete picture of lead production and use. Investigators have used various metrics for evaluating either cumulative lead releases or the rates of such releases (Flegal and Smith, 1992; Nriagu, 1983; Settle and Patterson, 1980; Woolley, 1984). The earliest timeline for any widespread production of lead in the then known world is likely the introduction and use of cupellation, i.e., 5000 years BP, (ca. 3000 years BCE; Settle and Patterson, 1980). Cupellation (L., *cuppa*, cup) is the crude if relatively effective metallurgical process of roasting metalliferous ores in air. The principal interest in this metallurgical method was in roasting gold and silver ores to recover these precious metals. Base metals such as lead are typically present in those ores.

The available historical materials describe to some extent the variety of uses for lead by early peoples, notably Greeks and Romans and early Asian cultures but do not permit quantifying lead consumption via any of these uses. Some uses would have likely involved modest lead consumption rates, e.g., certain beverage adulterants, but others likely entailed more significant consumption. The writings of the Roman architect Vitruvius, for example, make note of the apparently common use of lead in potable water delivery systems, to include cisterns and piping (Forbes, 1964; Hodge, 1981; Nriagu, 1983). Worker exposures in lead production and uses suggest that the amounts of the element involved in some production settings were not trivial, even if difficult to quantify. Lead glazes for pottery and vitreous ware, in the meanwhile, came into common use in China during the Chou Dynasty (1122–256 BCE).

Flegal and Smith (1992) have estimated total global lead production since the advent of cupellation to the late twentieth century at about 300 million MT (Figure 3.1). This cumulative tally, as we would expect, was not produced in a smooth monotonic fashion with a stable growth rate in production and consumption. Rather, the record for accumulated lead production (most if not all being the amounts dispersed into the environment) is one of abrupt rise and occasional decline. The reason for this is simply that the nonlinear consumption rates tracked closely the leaps in growth of civilizations and empires.

Settle and Patterson (1980; see also Figure 3.1 in NAS/NRC, 1993) reported the estimates of annual global lead production rates from the invention of cupellation to the present. Moderate increase in annual production is estimated until around the beginning of the Greco-Roman era, where the first large boosts in lead production accompanied both demands for lead in the Roman Empire and the introduction of coinage and requirements for coinage

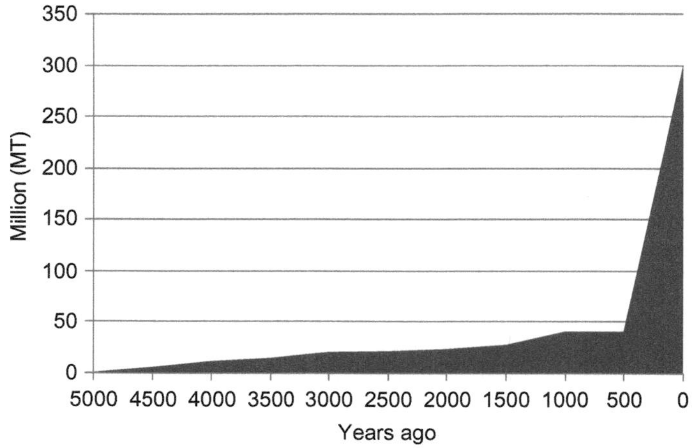

FIGURE 3.1 Cumulative production of lead over historic time. *Source: Flegal and Smith (1992). © Elsevier Inc. Reprint with permission.*

currency in trade, particularly by the Greeks and Romans in the period beginning about 2700 BP. Lead was coproduced metallurgically with silver in the course of the precious metal's mining and smelting. This development logically led to renewed mining activity and silver production in Europe and also to the exploitation of new lands. The latter is illustrated by Spanish mining of silver ores and their processing for silver in the new world. Earlier production of lead, i.e., anthropogenic inputs of lead into global environmental compartments, should not be considered as only of historical and academic interest. One critical reason, noted in the 1993 NAS/NRC report on lead exposure, is that anthropogenic lead inputs to the environment over past centuries and millennia become future "background" fractions of total global lead content in various environmental media. "Background lead" defined as lead present in the environment before, say, the twentieth century is a mixture of early anthropogenic and natural lead.

3.2.2 Modern U.S. and Global Lead Production

It is modern lead production that is of particular interest, given the vastly higher production rates of lead over the last century-and-a-half or so and the associated higher rates of environmental dispersal and contamination. Emphasis is on the period of time with available estimates of U.S. production and use, 1825 to the present. However, lead production and consumption figures dating to 1900 are the most detailed. This section deals with this modern lead production in its diverse aspects. Various official or otherwise reliable sources for national and global lead production were employed to provide quantitative characterizations of this modern production.

Several sources report lead production figures for the United States and the rest of the world as used in this chapter. Principal sources are those from the U.S. Bureau of Mines and the U.S. Geological Survey (USGS). Other sources are as footnoted in various tables. The U.S. domestic lead production figures for three periods—1720–1824, 1825–1899, and 1900–2009—are presented in this section. The periods 1720–1824 and 1825–1899 are covered in Table 3.1, while Table 3.2 presents primary annual production rates and diverse other categorical statistics for the period 1900–2009.

The U.S. colonial and early Federal lead production data for the period 1720–1824 are lumped as one estimate in footnote (b) of Table 3.1. This total was estimated at 50,636 MT. For the 105-year interval, the average production can be estimated as about 500 MT/year. Figures for 1830–1899 were retrieved as Salient Lead Statistics and were available online at http://usgs.gov/minerals/pubs/commodity/lead/stat/tbl1.txt. Figures for 1825–1829 were gathered from annual reports of retrospective estimates in Mineral Resources of the United States. Primary production refers to lead chiefly produced from ore concentrates as opposed to recycling existing lead from scrap materials. The latter would be included in the figures estimated for secondary production.

As seen in Table 3.1, and beginning in 1830, primary production rates increased in stages. The 1830 figure rises to 7,260 MT with further steady increases afterward, reflecting the early industrialization of the Eastern U.S. and early Midwestern industrial growth along with the opening of Midwestern lead belts. Notable in this regard for the mid- and later nineteenth century was the opening of the Tri-State lead/zinc belt, centered in Ottawa County, OK; Cherokee County, KS; and Jasper County, MO (Knerr, 1992).

Primary U.S. production stabilized in the period 1845 and shortly thereafter and declined in the decade or so leading up to the Civil War (1850–1860), with a marked decline during the Civil War. It is a puzzle that the Civil War years were not associated with increased lead production to feed military ordnance and ammunition needs of Union forces. As noted, figures for lead production during the war years were lower than prewar figures. For example, the primary domestic production in the war year of 1864 was 13,880 MT, a figure half that for 1845, a tonnage of 27,220 MT, 19 years earlier. It should be kept in mind that lead production figures in the Civil War period would have reflected only those production levels from the Union states and perhaps quite imperfectly that, given the security requirements. Lead production or even ore or bullion availability to the Confederate states was unknown. Lead primary production did not return to the levels of the 1840s until the early 1870s. These changes are consistent with the larger economic and political picture for the United States in those decades.

The second of the nineteenth century spurts in primary lead production began in the post-Civil War period, the 1870s. For example, the 1874

TABLE 3.1 Production of Refined Lead in the United States (MT), 1720–1899[a,b]

Year	Primary Production[c]	Consumption[d]
1720–1824	50,636	NA
1825	1,364	NA
1830	7,260	NA
1831	6,800	NA
1832	9,070	NA
1833	9,980	NA
1834	10,880	NA
1835	11,800	NA
1836	13,600	NA
1837	12,250	NA
1838	13,600	NA
1839	15,880	NA
1840	15,420	NA
1841	18,600	NA
1842	21,770	NA
1843	22,680	NA
1844	23,590	NA
1845	27,220	NA
1846	22,680	NA
1847	25,400	NA
1848	22,680	NA
1849	21,320	NA
1850	19,960	NA
1851	16,780	NA
1852	14,240	NA
1853	15,240	NA
1854	14,970	NA
1855	14,330	NA

(*Continued*)

TABLE 3.1 Production of Refined Lead in the United States (MT), 1720–1899[a,b]—(cont.)

Year	Primary Production[c]	Consumption[d]
1856	14,520	NA
1857	14,330	NA
1858	13,880	NA
1859	14,880	NA
1860	14,150	NA
1861	12,790	NA
1862	12,880	NA
1863	13,420	NA
1864	13,880	NA
1865	13,330	NA
1866	14,600	NA
1867	13,790	NA
1868	14,880	NA
1869	15,880	NA
1870	16,175	NA
1871	18,145	NA
1872	23,480	NA
1873	38,590	NA
1874	47,245	NA
1875	54,105	NA
1876	58,125	NA
1877	74,300	NA
1878	82,610	NA
1879	84,170	NA
1880	88,745	NA
1881	106,218	NA
1882	120,556	NA
1883	130,596	NA

(Continued)

TABLE 3.1 Production of Refined Lead in the United States (MT), 1720–1899[a,b]–(cont.)

Year	Primary Production[c]	Consumption[d]
1884	126,912	NA
1885	117,401	NA
1886	123,041	NA
1887	132,177	145,785
1888	137,819	136,581
1889	141,881	165,985
1890	130,299	146,741
1891	161,982	181,170
1892	157,220	181,789
1893	148,762	176,296
1894	147,586	174,516
1895	154,220	218,672
1896	170,550	194,048
1897	192,320	198,898
1898	201,400	206,341
1899	190,960	205,310

[a] *Adapted from USGS; Annual Data:* Mineral Resources of the United States; no secondary production figures found through 1899 *(http://minerals.usgs.gov/minerals/pubs/commodity/lead/stat/tbl1.txt).*
[b] *Lumped figures for 1720–1824 from USGS* Mineral Resources of the United States—1914, *U.S. Government Printing Office, 1916; original figure in short tons.*
[c] *Recovered from domestic ores; metric tons, original figures expressed in short tons.*
[d] *Data for 1887–1899 is primary consumption, metric tons.*
Total primary production for the period 1720–1899 is 4,054,800 MT.
Total recorded domestic consumption for the period 1887–1899 is 2,132,132 MT.
NA, not available.

production estimate was already twice that for 1872, at 47,245 versus 23,480 MT, respectively. A sharp rise in production continued for the balance of the century. By 1881, the annual lead production estimate was 106,218 MT, rising to 161,982 MT in 1891. The century ended with an annual production of 190,960 MT.

TABLE 3.2 Lead Production and Use[a,b]

Year	U.S. Primary Production	U.S. Secondary Production	U.S. Imports	U.S. Exports	Apparent U.S. Consumption[c]	World Production
1900	343,000	NA	410	90,000	253,000	749,000
1901	341,000	NA	490	91,000	260,000	NA
1902	342,000	NA	1,400	74,000	260,000	NA
1903	343,000	NA	1,600	87,000	289,000	NA
1904	367,000	NA	630	77,000	313,000	NA
1905	366,000	NA	890	55,000	311,000	NA
1906	378,000	NA	11,000	300	387,000	1,040,000
1907	410,000	23,400	8,400	59,000	390,000	993,000
1908	422,000	16,800	2,500	90,000	339,000	1,280,000
1909	426,000	37,300	3,200	90,000	378,000	1,060,000
1910	449,000	50,200	32,000	78,000	436,000	1,100,000
1911	465,000	49,200	2,400	120,000	425,000	1,110,000
1912	457,000	61,000	250	74,000	439,000	1,160,000
1913	455,000	66,100	38	50,000	476,000	1,150,000
1914	524,000	55,400	150	81,000	496,000	NA

(Continued)

TABLE 3.2 Lead Production and Use[a,b]—(cont.)

Year	U.S. Primary Production	U.S. Secondary Production	U.S. Imports	U.S. Exports	Apparent U.S. Consumption[c]	World Production
1915	544,000	71,600	410	120,000	492,000	NA
1916	517,000	87,400	5,100	110,000	499,000	NA
1917	597,000	84,800	5,300	90,000	606,000	NA
1918	611,000	88,100	4,800	95,000	611,000	NA
1919	452,000	111,000	4,600	58,000	510,000	764,000
1920	506,000	113,000	32,000	24,000	626,000	804,000
1921	412,000	94,100	28,000	33,000	502,000	783,000
1922	501,000	145,000	3,200	36,000	613,000	972,000
1923	581,000	176,000	19,000	52,000	724,000	1,080,000
1924	657,000	186,000	74,000	75,000	840,000	1,220,000
1925	711,000	206,000	9,500	95,000	829,000	1,410,000
1926	744,000	252,000	74,000	65,000	1,000,000	NA
1927	736,000	250,000	0	110,000	877,000	1,540,000
1928	735,000	280,000	0	110,000	904,000	1,680,000
1929	724,000	282,000	0	67,000	939,000	1,610,000
1930	603,000	232,000	0	44,000	791,000	1,520,000

1931	419,000	213,000	9	20,000	558,000	1,260,000
1932	278,000	180,000	40	21,000	396,000	1,050,000
1933	263,000	204,000	99	21,000	417,000	1,040,000
1934	295,000	189,000	257	5,400	462,000	1,200,000
1935	308,000	245,000	1,200	6,300	547,000	1,380,000
1936	371,000	239,000	2,350	18,000	654,000	1,470,000
1937	447,000	250,000	4,450	18,000	713,000	NA
1938	367,000	204,000	2,940	42,000	528,000	1,700,000
1939	457,000	191,000	6,480	68,000	577,000	1,740,000
1940	508,000	206,000	137,000	68,000	764,000	NA
1941	551,000	345,000	249,000	13,000	1,170,000	NA
1942	558,000	280,000	332,000	8,800	1,170,000	NA
1943	481,000	282,000	222,000	22,000	973,000	NA
1944	470,000	296,000	202,000	14,000	1,076,000	NA
1945	450,000	281,000	206,000	2,000	915,000	1,250,000
1946	350,000	313,000	105,000	640	816,000	1,030,000
1947	474,000	403,000	145,000	1,400	1,090,000	1,310,000
1948	455,000	393,000	224,000	370	1,050,000	1,380,000

(Continued)

TABLE 3.2 Lead Production and Use[a,b]—(cont.)

Year	U.S. Primary Production	U.S. Secondary Production	U.S. Imports	U.S. Exports	Apparent U.S. Consumption[c]	World Production
1949	515,000	330,000	250,000	880	1,070,000	1,370,000
1950	510,000	388,000	401,000	2,500	1,330,000	1,640,000
1951	434,000	401,000	162,000	1,200	1,050,000	1,600,000
1952	478,000	374,000	463,000	1,600	1,290,000	1,810,000
1953	477,000	389,000	349,000	730	1,460,000	1,870,000
1954	493,000	386,000	251,000	540	901,000	2,000,000
1955	490,000	408,000	240,000	370	1,390,000	2,010,000
1956	504,000	404,000	238,000	4,200	1,060,000	2,400,000
1957	502,000	392,000	294,000	4,500	1,280,000	2,380,000
1958	441,000	311,000	334,000	2,800	975,000	2,350,000
1959	321,000	356,000	239,000	3,600	937,000	2,320,000
1960	210,000	370,000	187,000	5,400	720,000	2,390,000
1961	430,000	354,000	233,000	11,000	1,000,000	2,390,000
1962	366,000	357,000	233,000	6,400	1,030,000	2,510,000
1963	366,000	424,000	206,000	3,600	1,040,000	2,560,000
1964	408,000	426,000	189,000	21,000	997,000	2,530,000

1965	385,000	450,000	202,000	11,000	1,000,000	2,700,000
1966	410,000	440,000	259,000	5,400	1,100,000	2,850,000
1967	386,000	433,000	330,000	5,900	1,130,000	2,870,000
1968	487,000	428,000	307,000	8,200	1,200,000	3,010,000
1969	594,000	468,000	253,000	6,400	1,320,000	3,240,000
1970	626,000	459,000	222,000	11,000	1,360,000	3,390,000
1971	604,000	444,000	177,000	21,000	1,210,000	3,490,000
1972	617,000	452,000	220,000	40,000	1,260,000	3,450,000
1973	624,000	489,000	170,000	110,000	1,420,000	3,490,000
1974	620,000	545,000	107,000	56,000	1,430,000	3,490,000
1975	579,000	597,000	90,000	19,000	1,040,000	3,440,000
1976	596,000	659,000	129,000	5,000	1,280,000	3,690,000
1977	552,000	637,000	230,000	9,000	1,180,000	3,410,000
1978	568,000	650,000	230,000	8,000	1,220,000	3,460,000
1979	578,000	673,000	190,000	11,000	1,130,000	3,510,000
1980	548,000	581,000	85,000	164,000	906,000	3,520,000
1981	498,000	578,000	100,000	23,300	977,000	3,350,000
1982	512,000	521,000	95,000	55,600	1,110,000	3,450,000

(Continued)

TABLE 3.2 Lead Production and Use[a,b]—(cont.)

Year	U.S. Primary Production	U.S. Secondary Production	U.S. Imports	U.S. Exports	Apparent U.S. Consumption[c]	World Production
1983	515,000	452,000	180,000	24,400	1,140,000	3,350,000
1984	396,000	586,000	162,000	16,600	1,140,000	3,200,000
1985	487,000	570,000	134,000	27,300	1,130,000	3,390,000
1986	366,000	575,000	140,000	12,600	1,150,000	3,240,000
1987	374,000	658,000	190,000	10,000	1,200,000	3,430,000
1988	392,000	691,000	150,000	14,000	1,230,000	3,420,000
1989	396,000	842,000	120,000	34,000	1,330,000	3,400,000
1990	404,000	874,000	90,900	64,000	1,300,000	3,370,000
1991	346,000	830,000	117,000	102,000	1,230,000	3,260,000
1992	305,000	861,000	191,000	70,000	1,270,000	3,200,000
1993	335,000	843,000	196,000	58,500	1,340,000	2,900,000
1994	351,000	858,000	231,000	53,500	1,490,000	2,800,000
1995	374,000	963,000	264,000	57,000	1,570,000	2,710,000
1996	326,000	1,030,000	268,000	61,000	1,630,000	2,920,000
1997	343,000	1,040,000	265,000	53,000	1,610,000	3,100,000
1998	337,000	1,060,000	310,000	40,000	1,690,000	3,060,000

1999	350,000	1,060,000	323,000	37,000	1,760,000	3,080,000
2000	341,000	1,080,000	365,000	48,600	1,740,000	3,200,000
2001	290,000	1,090,000	284,000	38,100	1,690,000	3,120,000
2002	262,000	1,100,000	218,000	43,400	1,530,000	2,850,000
2003	245,000	1,140,000	183,000	123,000	1,490,000	3,200,000
2004	148,000	1,130,000	208,000	82,600	1,470,000	3,150,000
2005	143,000	1,150,000	310,000	64,600	1,480,000	3,480,000
2006	153,000	1,160,000	343,000	68,500	1,580,000	3,600,000
2007	123,000	1,180,000	267,000	56,400	1,540,000	3,680,000
2008	135,000	1,150,000	314,000	75,000	1,500,000	3,840,000
2009[d]	115,000	1,120,000	275,000	85,000	1,420,000	3,900,000
Total	48,203,000	48,399,400	15,928,993	4,710,430	108,144,000	220,625,000

[a]*Lead statistics. USGS. Last modified November 5, 2009, for U.S. and world figures.*
[b]*Values in metric tons lead unless otherwise noted for U.S. and world figures.*
[c]*Apparent consumption = production + imports − exports ± changes in stocks ± government shipments. The two categories of "stocks" and "government shipments" are in the USGS tabulations but omitted here.*
[d]*USGS, Mineral Commodity Summaries, January 2010.*
NA, not available.

The basis of this rise was both demand via consumption and newly discovered Western U.S. lead belts, particularly those in the Inter-Rocky Mountain West. Lead consumption grew across a number of categories, e.g., lead paint and lead plumbing, while the entry of Western lead became economically feasible and highly competitive for commodity production and marketing through the growth of the U.S. railroad system, particularly west of the Mississippi River.

U.S. consumption rates in the postbellum nineteenth century were difficult to quantify prior to the late 1880s and only then in terms of aggregated estimates. Comparison of U.S. primary production versus aggregated consumption in Table 3.1 for 1887–1899 indicates that domestic primary production could barely keep up with demand and in the late years fell short of consumption, requiring some modest levels of lead importing.

Primary U.S. and world lead production levels, along with statistics for secondary production, U.S. imports and exports, and estimates of apparent U.S. consumption are set forth in Table 3.2. From 1900 to 1923, the USGS provided production rates via the series, Mineral Resources of the United States. From 1927 to 1934, the production figures were produced by the U.S. Bureau of Mines, which became the sponsor of Mineral Resources. From 1932 to 1994, the Bureau became the principal source of U.S. lead production data as the annual Mineral Yearbook series. This continued until 1994, the last year in which the Bureau of Mines existed as a discrete governmental agency. Post-1994, the sources include figures of the USGS, which took over the metal production estimation tasks of the former Bureau in 1995.

The data in Table 3.2 for U.S. domestic primary and secondary lead production and estimates of global production for the period 1900–2009 require clarifying comments. The most useful index of lead tonnage that would eventually figure in environmental pollution is that of primary U.S. and world production. Primary production refers to newly produced, refined lead. Secondary production refers to recyclable lead sources, such as recycled lead-acid batteries. Secondary production, although quantified annually and useful for evaluating annual consumption figures, has within those estimates percentages of prior primary production estimates that now are part of the recycled numbers. Table 3.2 also includes the tonnage of lead imported and exported. Imports add to the overall amounts of lead available in the United States for consumption, while exports deduct from the amounts available for domestic consumption.

Finally, Table 3.2 presents annual world production, but not every year has a figure available for world production. Total tallies of global production for the entire period should be taken as a lower bound to the actual sums, since lead production continued regardless of availability of information on that production.

The level of domestic lead production varied considerably with factors such as national economic growth, growth in domestic lead consumption, disruptions in such growth, and population growth. Overall primary and secondary production increased throughout the twentieth century (Table 3.2). Early years of the century saw primary production dominate that from lead recycling and other secondary production, but the latter became increasingly higher as the century unfolded. Secondary production began to approach primary production by the early 1960s in the United States and eclipsed primary production by the 1980s.

U.S. domestic primary lead production in 1900 was 343,000 MT, with no figure available for secondary production until 1907. Of this figure for 1900, 90,000 MT were exported. By 1907, primary domestic production rose to 410,000 MT, while recorded secondary production was 23,400 MT. Domestic primary and secondary production of lead increased significantly during the robust industrial growth years of the 1920s, reaching a peak primary domestic production figure of 744,000 MT in 1926. The corresponding secondary production figure for that year was 252,000. Of the combined tonnage, 65,000 MT were exported.

During the Depression years, the 1930s, domestic primary lead production tumbled to as low as about one-third that of the 1926 peak, with some rebound in the late 1930s. Primary production in 1933 fell to 263,000 MT, the second lowest figure recorded for the twentieth century U.S. domestic primary production. The lowest primary lead production tally in these figures was 210,000 MT for 1960. By contrast, lead scrap recycling and secondary production were much less affected, sustaining a secondary lead production figure of 204,000 MT in 1993.

The war years of the 1940s saw a rebound in both primary and secondary production. For example, primary and secondary domestic lead production in 1942 was 558,000 and 280,000 MT, respectively.

The recorded domestic primary lead production peak for the twentieth century in 1926 of 744,000 MT was surrounded by primary production in excess of 700,000 MT for 5 years during the 1920s, a performance not to be repeated during the balance of the century. While primary production declined or remained relatively static after the 1920s, secondary production continued to increase. Secondary production exceeded 500,000 MT in 1974, exceeded 800,000 MT by 1989, and exceeded 1,000,000 MT by 1996. In 1996, the figure for primary production was only 326,000 MT, but the import level was 280,000 MT.

The peak in the combined twentieth century primary and secondary domestic U.S. lead production occurred quite recently, excluding the level of lead importing. The 1999 combined production figure was 1,410,000 MT. For the first 10 years of the twenty-first century, the peak occurred in 2000: 1,421,000 MT.

The growth of secondary lead production in the face of declining primary production reflected at least two factors. The closing or exhaustion of U.S. lead mines and the associated closing of lead smelters and refineries, and the growth of a huge inventory of lead scrap sources by at least the 1950s accounted for this later divergence. Scrap in the secondary stream included lead from recycling lead-acid batteries.

The relative amounts of exports and imports across the years affected the total amounts of lead available for domestic consumption, and both categories have varied considerably. Imports for approximately the first 40 years of the twentieth century were only a fraction of exports, indicating that domestic production for these four decades commonly exceeded domestic consumption requirements. This direction in lead's international trade reversed itself around the outbreak of World War II (WWII). At that point, domestic production began to fall short of annual lead consumption requirements and the shortfall was made up by imports.

World production levels remained under 2 million MT until 1954 (Table 3.2) and increased rapidly thereafter. These increases, proportionately, were less reflective of U.S. production increases than they were in earlier decades. For example, global production increased ~1 million MT from 1954 (2.00 million MT) to 1968 (3.01 million MT), while U.S. production in this same interval remained relatively unchanged.

One obvious factor influencing the chronology of these global production figures was the extent to which the ravages of WWII and to some extent WWI had disrupted European and Asian economies. This is clearly reflected in the marked increase beginning in the 1950s and continuing, in lockstep with the economic recovery and expansion of various countries after WWII.

Table 3.2 also presents the totals for each of the categories indicated. The total U.S. primary production of lead for the period 1900–2009 was 48,200,000 MT (rounding). This figure is ~12-fold higher than that for the combined eighteenth and nineteenth centuries. The total for secondary production for the period 1900–2009 was 48,400,000 MT (rounding). Secondary production figures for the period 1900–2009 approximate primary production figures. The total for primary and secondary U.S. lead production is 96,600,000 MT. Lead imports exceed exports by ~3:1, i.e., 14,700,000 MT (rounding) versus 4,800,000 MT, respectively.

3.3 MODERN ERA LEAD CONSUMPTION IN THE UNITED STATES AND ELSEWHERE

Limited data exist for U.S. lead consumption during the nineteenth century or before. Specifically, annual U.S. domestic lead consumption data are not available for years prior to 1887. Table 3.1 provided estimates of domestic lead consumption late in the last quarter of the nineteenth century, 1887–1899. Consumption of lead domestically in this short span ranged from 137,000 to 219,000 MT (rounding). Consumption was higher in the 1890s. These annual figures total 2,300,000 MT (rounding).

World lead consumption figures for the period 1891–1900, figures which include those of the United States, were estimates provided by The Metallgesellschaft, Frankfurt-am-Main, Germany and tabulated in the USGS Mineral Resources of the United States annual document for the calendar year 1900. The world lead consumption rates (MT) per year were:

Year	World Pb Consumption (MT)
1891	617,030
1892	631,295
1893	635,532
1894	630,072
1895	660,000
1896	677,734
1897	713,321
1898	786,372
1899	778,773
1900	813,470

The total world lead consumption in these 10 years was 6,944,000 MT (rounding). Worldwide lead consumption increased for all countries between 1891 and 1900. However, the largest nation-specific proportional increases in overall increased consumption were those of Germany and the United States. Germany's lead consumption, from 1891 to 1900, almost doubled (88,268–172,940 MT), while consumption for the United States increased from 181,842 to 239,407 MT in this period (this figure includes adjustments for changes in stocks and government shipments, while the entry in Table 3.2 does not).

Annual consumption figures for combined primary and secondary domestic production, along with amounts imported for consumption, are not reported for years prior to 1900. Table 3.2 includes such data sets for 1900–2009. Annual figures for Table 3.2 are derived from sources indicated in footnotes to the table.

A U.S. domestic consumption tally is provided in Table 3.2 for the years 1900–2009 based on the USGS annual tabulations using the original data sets from either the U.S. Bureau of Mines or the USGS. For the first several decades, apparent consumption was in the approximate range of 300,000–600,000 MT. U.S. consumption greatly increased in the 1920s in tandem with primary and secondary production. In 1926, for example, U.S. consumption reached 1 million MT.

A marked decline followed with the Great Depression, succeeded by a marked increase during and after WWII. From 1941 to 1967, annual U.S.

consumption hovered around or slightly above 1 million MT. A marked spurt in consumption prevailed in the first half of the 1970s. From the 1970s forward, annual U.S. consumption never reached 2 million MT. The highest annual rate occurred for 1999 and is 1.8 million MT (rounding). The total apparent U.S. lead consumption amounted to 108 million MT (rounding) for the period 1900–2009. The bulk of this cumulative amount was achieved post-WWII.

Table 3.3 reports the annual U.S. apparent consumption by selected categories of product or uses that also comprise the major uses. Table 3.3 is not an exhaustive compilation of all uses, but a compilation whose stratified annual use and totals represent a great majority of domestic U.S. lead production over the decades and whose ultimate dispersals to the human environment were significant. Estimates for indicated annual category-specific U.S. consumption only date to 1920 in terms of relatively complete data for most of the use categories. Prior years combine a number of these into the total apparent consumption figures (see Table 3.2.).

Some clarifying comments are required in considering the data presented in Table 3.3. For example, the category "Building" included different consumption subcategories prior to 1941 than from 1941 onward. Different groups of years were presented with tonnages expressed as short tons or metric tons as indicated in relevant footnotes. For tabulation of subtotals, all estimates were converted to metric tons to allow direct comparisons of different data sets. Some categories appear *de novo*, reflecting new technological developments. The antiknock additive TEL was only recorded as a discrete category of use beginning in 1941, although it entered the U.S. and international markets earlier.

Totals for the various consumption categories bear critical data and implications of the data. For example, lead-acid batteries made up the category consuming most lead by 1920 and afterward. The white lead category of basic white lead carbonate and sulfate forms, the highest fraction of which was employed for lead paint pigment production, was higher in the 1920s through 1940s than later, but still comprised a significant use grouping into the 1950s and 1960s. The white lead/pigment category, specifically for basic white lead carbonate, had some production data available as annual figures predating 1920, from 1915 to 1919. For the period 1915–1919, Pb production as the basic white lead carbonate for this category amounted to:

Year	Pigment Production (MT)[a]
1915	141,979
1916	117,371
1917	104,832
1918	102,888
1919	126,572
Total	593,642

[a]*Converted from short tons, USGS data.*

TABLE 3.3 United States Lead Consumption by Selected Major Categories of Use[a,b,c]

Year	Storage Batteries[d]	White Lead	Cable Covering	Building	Ammunition	Bearing Metal	Solder	Red Lead and Litharge[e,f]	Antiknock Additive
1920	108,000	116,500	79,000	71,000	33,600	26,000	15,000	45,000	NA
1921	87,000	136,000	67,000	48,000	26,500	21,000	12,000	23,500	NA
1922	130,000	156,000	93,000	71,000	38,000	26,000	20,000	30,000	NA
1923	143,000	130,000	131,000	75,000	40,000	28,000	30,000	46,000	NA
1924	170,000	150,000	138,000	83,100	27,000	32,000	30,000	34,000	NA
1925	180,000	131,000	156,000	88,400	31,500	34,000	35,000	42,000	NA
1926	190,000	120,000	185,000	93,700	32,000	36,000	37,000	36,000	NA
1927	175,000	126,000	161,000	88,000	34,000	31,000	35,000	38,000	NA
1928	220,000	123,000	180,000	96,000	39,600	32,000	37,000	31,000	NA
1929	210,000	119,700	206,000	96,000	41,100	33,000	37,000	30,000	NA
1930	163,000	83,900	195,000	67,000	33,300	20,000	27,000	32,000	NA
1931	157,000	77,500	117,000	40,000	29,700	12,000	20,500	18,000	NA
1932	138,000	54,500	55,000	22,000	23,300	10,000	14,000	15,800	NA
1933	147,000	59,100	31,000	26,000	32,300	16,000	16,000	19,000	NA
1934	163,000	65,500	34,100	30,000	34,800	16,000	16,000	28,000	NA

(Continued)

TABLE 3.3 United States Lead Consumption by Selected Major Categories of Use[a,b,c]—(cont.)

Year	Storage Batteries[d]	White Lead	Cable Covering	Building	Ammunition	Bearing Metal	Solder	Red Lead and Litharge[e,f]	Antiknock Additive
1935	175,000	80,000	37,800	32,000	29,200	20,000	20,000	47,500	NA
1936	191,000	85,500	61,400	40,000	32,500	16,500	22,000	54,000	NA
1937	192,000	86,000	90,000	45,000	39,500	15,000	22,000	57,000	NA
1938	167,000	71,000	60,000	36,000	31,200	9,000	15,000	43,000	NA
1939	198,000	75,000	74,400	50,000	42,300	12,800	20,000	57,200	NA
1940	220,200	65,500	107,400	65,000	56,000	14,000	24,000	59,400	NA
1941	73,982	83,230	141,161	NA[g]	32,959	10,160	35,630	143,496	48,327
1942	62,604	56,476	128,535	NA[g]	48,025	8,466	13,371	100,563	50,152
1943	68,239	36,809	117,802	NA[g]	64,023	10,189	15,472	124,715	65,320
1944	68,769	54,333	110,417	NA[g]	31,479	15,941	22,390	157,080	83,067
1945	60,179	35,611	86,158	NA[g]	29,315	14,104	27,475	157,171	75,890
1946	56,726	43,294	60,004	NA[g]	16,857	11,012	33,279	128,513	47,965
1947	103,896	51,347	91,826	NA[g]	18,471	8,487	22,402	166,763	78,280
1948	354,405	30,970	171,654	NA[g]	49,635	42,594	71,025	80,356	83,809
1949	313,718	18,400	144,340	NA[g]	24,111	29,189	62,104	70,832	94,644
1950	393,409	36,181	131,989	NA[g]	38,438	38,241	94,606	101,974	113,846

1951	375,384	25,578	131,863	NA[g]	40,242	35,410	82,465	88,031	128,407
1952	350,930	22,943	142,571	NA[g]	36,182	36,545	72,664	76,742	146,723
1953	367,575	17,775	146,565	NA[g]	45,147	38,591	78,743	88,649	162,443
1954	337,272	17,704	127,939	NA[g]	40,206	27,166	71,122	76,472	160,436
1955	380,033	18,549	121,165	NA[g]	46,816	34,567	88,749	87,503	165,133
1956	370,771	16,951	134,339	NA[g]	44,438	28,321	75,290	79,199	191,990
1957	361,015	15,701	108,225	NA[g]	42,509	26,997	70,684	78,323	177,001
1958	312,725	18,360	74,981	NA[g]	40,215	18,980	59,653	64,892	178,444
1959	380,732	10,955	61,626	NA[g]	45,328	23,298	68,871	74,116	160,020
1960	353,196	8,432	60,350	NA[g]	43,577	20,717	60,013	74,901	163,826
1961	367,998	7,615	57,458	NA[g]	45,837	17,757	54,838	72,022	169,802
1962	419,906	11,091	56,676	NA[g]	47,779	16,472	66,873	76,325	168,926
1963	439,081	8,846	57,707	NA[g]	49,894	21,713	67,945	70,649	192,811
1964	429,348	8,802	56,225	NA[g]	56,493	22,754	71,186	74,802	223,461
1965	455,347	8,414	59,645	NA[g]	57,322	21,600	77,819	79,853	225,203
1966	472,492	8,131	66,491	NA[g]	78,435	21,588	78,898	89,500	246,879
1967	466,665	8,087	63,037	NA[g]	78,766	19,561	68,833	76,589	247,170
1968	453,178	5,857	53,456	NA[g]	82,193	18,441	74,074	86,480	261,897

(Continued)

TABLE 3.3 United States Lead Consumption by Selected Major Categories of Use[a,b,c]—(cont.)

Year	Storage Batteries[d]	White Lead	Cable Covering	Building	Ammunition	Bearing Metal	Solder	Red Lead and Litharge[e,f]	Antiknock Additive
1969	582,546	6,617	54,203	NA[g]	79,233	17,406	72,626	79,898	271,128
1970	593,453	5,936	50,772	NA[g]	72,726	16,328	69,707	77,215	278,505
1971	679,803	4,731	52,920	NA[g]	87,567	16,285	70,013	61,838	264,240
1972	726,592	2,814	45,930	NA[g]	84,699	15,915	71,289	69,799	278,340
1973	769,447	1,749	43,005	NA[g]	81,479	15,657	71,770	89,577	274,410
1974	851,881	1,996	43,426	NA[g]	87,090	14,609	66,280	96,163	250,502
1975	699,414	2,498	22,099	NA[g]	75,081	12,184	57,344	65,457	208,605
1976	822,404	2,993	15,930	NA[g]	73,478	13,063	63,324	85,403	239,758
1977	945,876	6,612	15,107	NA[g]	68,390	11,985	64,286	78,071	232,910
1978	879,274	1,225	13,851	NA[g]	55,776	9,510	68,390	113,769	178,331
1979	813,932	1,167	16,393	NA[g]	53,235	9,631	54,278	101,675	186,945
1980	645,357	889	13,468	NA[g]	48,682	7,808	41,304	49,909	127,903
1981	770,132	818	12,072	NA[g]	49,514	6,922	29,705	56,873	111,367
1982	705,323	1,174	15,181	NA[g]	44,287	6,123	28,500	60,590	119,254
1983	806,899	841	10,505	NA[g]	43,697	5,844	28,490	68,814	89,118
1984	865,542	964	12,270	NA[g]	47,828	4,677	24,411	66,511	78,933

1985	840,940	382	15,501	NA[g]	50,233	5,392	21,374	91,628	45,694
1986	853,810	376	17,061	NA[g]	44,382	5,525	21,302	67,507	28,541
1987	953,598	NA	20,140	NA[g]	46,835	5,260	19,758	73,414	NA
1988	955,263	NA	16,170	NA[g]	52,708	6,034	19,064	77,747	NA
1989	1,012,155	NA	22,605	NA[g]	57,310	2,586	17,009	73,000	NA
1990	1,019,637	NA	18,253	NA[g]	58,210	2,878	16,490	79,717	NA
1991	1,007,117	NA	17,472	NA[g]	58,477	3,669	14,750	80,614	NA
1992	998,184	NA	15,993	NA[g]	64,845	4,785	13,518	70,355	NA
1993	1,050,000	NA	17,200	NA[g]	65,100	4,830	14,400	68,800	NA
1994	1,220,000	NA	16,000	NA[g]	62,400	5,560	12,400	70,100	NA
1995	1,330,000	NA	5,640	NA[g]	70,900	6,490	16,200	14,601	NA
1996	1,340,000	NA	NA	NA[g]	52,100	4,350	9,020	14,415	NA
1997	1,390,000	NA	4,930	NA[g]	52,400	2,490	9,580	21,018	NA
1998	1,430,000	NA	4,630	NA[g]	52,800	2,210	10,900	17,391	NA
1999	1,470,000	NA	2,400	NA[g]	58,300	1,570	13,100	NA	NA
2000	1,490,000	NA	NA	NA[g]	63,500	1,480	11,500	NA	NA
2001	1,350,000	NA	NA	NA[g]	53,600	837	6,120	19,995	NA
2002	1,190,000	NA	NA	NA[g]	57,600	406	6,450	18,972	NA

(Continued)

TABLE 3.3 United States Lead Consumption by Selected Major Categories of Use[a,b,c]—(cont.)

Year	Storage Batteries[d]	White Lead	Cable Covering	Building	Ammunition	Bearing Metal	Solder	Red Lead and Litharge[e,f]	Antiknock Additive
2003	1,170,000	NA	NA	NA[g]	48,800	406	6,310	13,764	NA
2004	1,170,000	NA	4270	NA[g]	51,000	NA	1,460	NA	NA
2005	1,280,000	NA	NA[h]	NA[g]	61,300	1180	8,370	NA	NA
2006	1,300,000	NA	NA[h]	NA[g]	65,700	1240	7,280	NA	NA
2007	1,380,000	NA	NA[h]	NA[g]	69,400	1410	7,220	NA	NA
2008	1,290,000	NA	NA[h]	NA[g]	71,900	1250	6,880	NA	NA
2009	1,210,000	NA	NA[h]	NA[g]	70,500	NA	7,270	NA	NA
Total (MT)	52,336,287	2,586,856	5,134,113[h]	1,158,612[g]	4,207.089	1,244,013	3,060,131	5,114,292	6,799,432

[a]*Short tons, 1920–1977; metric tons, 1978 and later.*
[b]*Various sources:* Mineral Resources of the United States, *various annual compendia, and* Minerals Yearbook, *U.S. Bureau of Mines, annual compendia.*
[c]*White lead is the sum of dry white lead and white lead in oil.*
[d]*Sum of battery plate lead plus battery oxides in 1948 and later.*
[e]*In 1948 and later, red lead and litharge tallies not inclusive of battery posts fraction.*
[f]*For 1979 and later, lead consumed in red lead or litharge estimated as either production (1979–1994) or shipments (1995–later).*
[g]*Building consumption category after 1940 subsumed under other categories.*
[h]*For years 2005–2009, Pb cable cover tonnage not tallied as a separate product from totals nor included with other product categories.*

The increased use in the red lead/litharge combined group reflected increased use for structural metal surface coatings and increased production of lead-acid batteries. The antiknock additive increased from a figure of 48,327 MT in 1941 to a peak of 278,340 MT in 1970, associated with the peak use of leaded gasoline, the principal end use for the product.

The highest U.S. domestic lead use category throughout virtually all of the period analyzed, 1920–2009, was for lead-acid storage batteries, mainly for automobile and truck use. In 1923, storage batteries consumed 143,000 short tons (130,000 MT, rounding) of lead, followed by cable covering at 131,000 tons, and white lead for lead paint pigment at 130,000 tons.

By 1940, growth in this category reached ca. 220,000 tons, compared to 75,000 tons (rounding) for white lead and 107,000 tons (rounding) for cable covering. The war years and shortly afterward saw a steep decline in battery lead use. The steepest decline was to 60,000 tons (rounding) in 1945. White lead declined to 36,000 tons in 1945, while cable covering lead consumption hit its lowest point in this period in 1946. Lead consumption for battery use quickly recovered in close lockstep with the post-WWII prosperity of the nation. By contrast, recoveries in a number of other categories were short-lived. White lead continued to decline, as did bearing metal and cable covering. Uses in building construction and maintenance declined and then ceased being reported as a separate category in the 1940s.

Table 3.4 provides additional lead production and consumption statistics for the period 1960–1988 as compiled by the metals trade organization based in Frankfurt-am-Main, Germany, *Metallgesellschaft AG*, as contained in various issues of *Metallstatistic*. They are presented and discussed in *World Metal Demand. Trends and Prospects* (Tilton, 1990). Annual consumption rates in this monograph are calculated for the world and various economic and political groupings of nations. For the years 1960–1988, these data augment the figures of the USGS. Table 3.4 includes consumption levels for the world and the aggregate Organisation for Economic Co-operation and Development (OECD) rates. The OECD roster of consumption in this monograph includes the individual nations having the largest lead consumption: United States, Japan, United Kingdom, France, and the former West Germany (FRG). Included in this grouping are the nations of the European Economic Community, designated as the EEC-9: Belgium-Luxembourg, Denmark, Federal Republic of Germany, France, Greece, Ireland, Italy, The Netherlands, and United Kingdom.

Table 3.4 also tallies consumption rates for most of the less-developed countries (LDCs), defined as such in Tilton (1990) as all countries except the industrialized nations and those categorized as having centrally planned economies.

In Table 3.4, the world consumption figures appear to include uses drawn from both primary and secondary lead production, and as such are considerably higher than the world production figures given in Table 3.2 for the overlapping years 1960–1988. The latter numbers are primary production data.

TABLE 3.4 Global and Regional Consumption of Refined Lead, 1960–1988 ('000 MT)[a,b]

Consuming Entity	1960	1961	1962	1963	1964	1965	1966	1967
1. World	2621.4	2706.3	2818.9	2968.7	3149.5	3182.2	3332.9	3323.0
2. Total OECD[c]	1890.7	1916.6	1980.6	2063.0	2174.1	2198.1	2283.2	2236.7
3. United States	646.6	640.8	689.6	718.3	727.9	753.5	821.8	2476.4
4. LDC[d]	156.2	196.6	203.3	259.8	240.8	244.2	262.4	272.3
	1968	**1969**	**1970**	**1971**	**1972**	**1973**	**1974**	**1975**
1. World	3668.2	3837.4	3914.2	4024.4	4168.6	4472.9	4419.8	4758.8
2. Total OECD	2476.4	2575.2	2623.0	2626.8	2732.5	2943.2	2790.3	2942.2
3. United States	912.0	911.3	943.3	941.6	1009.6	1093.2	1055.1	1122.7
4. LDC	292.3	300.8	308.6	341.2	334.0	378.2	427.0	505.1
	1976	**1977**	**1978**	**1979**	**1980**	**1981**	**1982**	**1983**
1. World	5178.9	5491.7	5522.4	5644.8	5396.1	5282.0	5223.9	5278.3
2. Total OECD	3340.6	3568.4	3540.2	3547.4	3293.0	3208.7	3146.7	3149.9
3. United States	1272.3	1417.9	1404.5	1345.4	1094.0	1127.8	1106.1	1134.2
4. LDC	515.8	577.6	560.4	660.2	639.7	616.8	609.2	671.5
	1984	**1985**	**1986**	**1987**	**1988**			
1. World	5434.9	5420.9	5502.9	5622.5	–			
2. Total OECD	3323.8	3226.1	3265.0	3323.2	3398.6			
3. United States	1142.7	1099.1	1118.5	1202.8	1184.0			
4. LDC	664.8	721.1	802.5	858.8	–			

[a]*Adapted from Tilton (1990).*
[b]*Data from Metallgesellschaft Aktiengesellschaft, Metallstatistic, Frankfurt, Germany.*
[c]*OECD: Australia, Austria, Belgium, Canada, Denmark, Finland, France, Greece, Iceland, Italy, Japan, Luxembourg, The Netherlands, New Zealand, Norway, Portugal, Spain, Sweden, Switzerland, Turkey, United Kingdom, United States.*
[d]*LDCs: All countries except OECD nations and Centrally Planned Economies (CPE countries), as listed in Tilton (1990).*

3.4 TRENDS IN U.S. AND GLOBAL Pb CONSUMPTION

Table 3.3 shows that U.S. consumption trended upward from 1900 through the early postwar years, followed by considerable fluctuation afterward. Annual consumption rates were around 1.0–1.3 million MT from the 1950s until the early 1990s. From the 1990s, consumption increased to 1.5–1.7 million MT.

Table 3.4, based on data presented by Tilton (1990), provides a mixed picture. The Tilton monograph offers a glimpse of trends in production and use as well as tabulations of production and consumption within various economic geographic entities (see Appendix B-4 in Tilton, 1990).

The trend in the world and regional lead consumption in industrialized countries and regions over the long term generally shows an upswing, but within that general direction are highly variable short-term data. The world and industrialized regional figures generally increased from 1960 to the late 1970s, followed by relatively stable consumption out to 1987–1988.

The trend in LDCs is clearly upward, because earlier periods of economic activity in these countries were relatively modest. For example, the LDC category's total consumption in 1960 was 156,000 MT, rising to 859,000 MT by 1987. This represented a greater than fivefold growth in that period.

Table 3.3 shows continuing trends upward for some consumption categories and declines in other categories. The continuing upward trend for lead use in storage battery manufacture is notable.

The recent economic growth in certain countries which formerly had modest lead consumption levels appears to have propelled world lead production and consumption upward, which is evident in statistics for the most recent years indicated in Table 3.3. This particularly appears to reflect the rapid economic growth and commodity demands in China for lead for such uses as lead storage batteries.

3.5 LINKAGES AMONG LEAD PRODUCTION, CONSUMPTION, DISPERSAL, AND POTENTIAL EXPOSURES

The preceding sections make it clear that huge amounts of lead entered commercial streams in the United States and the rest of the world beginning in the nineteenth century and continuing to the present. These lead use and dispersal burdens added to the more slowly accumulating but still significant amounts of lead already in environmental compartments from releases over millennia. Primary lead production initially and both primary and secondary lead production subsequently have provided us a quantitative look at the potential in industrialized countries (and recently in LDCs) for dispersal of lead into environmental compartments providing contact for human and ecological populations.

Figures for lead production and releases to the early human environment are mainly determinable over large blocks of time. The period 5000 to ca. 3000 BP, a period spanning 2,000 years of recorded history, was typified by very modest growth in cumulative lead production, on the order of

8–9 million MT. This was an annual average global production within that period of ~4,500 MT.

The post-Roman Empire demands for lead increased the cumulative lead production total to about 20 million MT by 2500 BP (Figure 3.1), followed by a modest increase up to about 1200 BP (800 in the CE). This lag corresponded to the depletion of deposits in Roman lead mines, the decline of the Roman Empire and the onset of the Dark Ages.

The next burst in lead production was, again, one associated with interest in and production of silver in Europe. This was followed by early exploitation of silver deposits by Spain in its New World colonies (Settle and Patterson, 1980). This produced a cumulative figure of about 50 million MT in the early stages with a significant increase over time.

The largest pulse of additional lead production was in the period from 1500 CE to the present (Figure 3.1). This increment of 250 million MT traced in large part to the industrialization of societies and the myriad of new uses for lead. That is, the vast majority of environmental lead dispersal has been of recent vintage. The most recent contribution was 85% of the global total produced over the last 5,000 years. For this 200-year period, an annual global average of about 1.3 million MT was produced, with current primary production of lead of ~3.4 million MT (NAS/NRC, 1993, p. 102; also Tables 3.1–3.3). Of the total of ~250 million MT, the great majority was produced in the twentieth century. That is, the period of most interest in quantitative terms has been approximately the last 100 years.

The cumulative U.S. primary lead production noted in Table 3.1 for the period 1720–1899 is not inconsequential at 4,054,800 MT, most of that tonnage representing the output from the 1870s onward. It is also not inconsequential when taking account of the consumption of this total within confined population and socioeconomic strata of the United States at the time. However, this tally is dwarfed by the U.S. twentieth century production and consumption. The total recorded U.S. consumption of lead in Table 3.1 is 2,132,132 MT. This figure cannot be taken to reliably reflect actual consumption, given the absence of consumption data prior to 1887.

Global production of lead increased considerably during the twentieth century, reflecting industrial growth of Europe and parts of Asia. Such growth was more modest for the first half of the twentieth century, compared to U.S. production. U.S. tallies accounted for sizeable fractions of global production up to about 1950. Global production in 1900 amounted to 749,000 MT, while the first year for production of 1 million MT or more (1,040,000 MT) occurred in 1906. Corresponding U.S. figures were 343,000 and 378,000 MT, respectively. Gaps in tallies include the years of global wars, WWI and WWII. Lead production continued and would have been much higher than in peacetime, but associated production figures were not available. One should therefore consider any total global lead production level for the twentieth century discernible from the indicated sources as providing a lower-bound estimate.

World production for the period 1900–2009 totaled 220,625,000 MT. Of this total, U.S. combined primary and secondary production comprised 22%. Apparent U.S. consumption across categories amounted to 108,144,000 MT and was ~49% of total world production. This percentage takes account of primary and secondary production, imports and exports, but not the two small categories of "stocks" and "government shipments" as noted in footnote (c) in Table 3.2.

The first section of this chapter noted that the accumulated global lead production in recorded history from the invention of cupellation to the late twentieth century was ~300,000,000 MT. Table 3.2 indicates that over two-thirds of this accumulated global production estimate, 220,625,000 MT, occurred in the period 1900–2009. Of this ~220 million MT figure, 180,420,000 MT or ~82% of global production for 1900–2009 took place from 1950 to 2009, a period of 59 years. Even more significant, global lead production of 180,420,000 MT in this relatively brief period of recorded history accounted for over 60% of the entire estimated lead production over the previous five millennia of 300 million MT.

The various annual use categories in Table 3.3 differ greatly in the fractions of production and use which eventually were dispersed to various compartments of the human environment. The highest category of lead-acid batteries represents a lead use with a relatively high recycling profile for its life cycle. By contrast, lead paint and antiknock TEL added to leaded gasoline are categories with high fractions if not total original consumption volumes of lead that eventually enter the environment. Of lead paint and leaded gasoline, emissions of the latter with combustion range farther spatially but with decreases over time and distance. Lead paint is known to be a persisting lead exposure factor in numerous U.S. areas with old housing stock (Jacobs et al., 2002; U.S. Department of Housing and Urban Development, 2001).

Given the above, the use-specific totals for the United States provide interesting guidance to the issue of cumulative lead pollution in the United States and likely in other developed and industrial nations. The two categories of use in Table 3.3 and additional sums given above with the highest potential for subsequent conversion to environmental contamination in various forms were the white lead pigment and antiknock TEL uses. The white lead pigment tally, determined as use statistics and production statistics for 1915–1919, was 3.2 million MT (rounding). The TEL lead consumption total over the period 1941–1986 was 6.8 million MT (rounding).

REFERENCES

Flegal, A.R., Smith, D.R., 1992. Current needs for increased accuracy and precision in measurements of low levels of lead in blood. Environ. Res. 58, 125–133.

Forbes, R.J., 1964. Studies in Ancient Technology, vol. 7–9. Brill, Leiden, The Netherlands.

Hodge, A.T., 1981. Vitruvius, lead pipes and lead poisoning. Am. J. Archaeol. 85, 486–491.

Jacobs, D.E., Clickner, R., Zhou, J., Viet, S.M., Marker, D.A., Rogers, J.W., et al., 2002. The prevalence of lead-based paint hazards in U.S. housing. Environ. Health Perspect. 110, A599–A606.

Knerr, D., 1992. Eagle-Picher Industries. Strategies for Survival in the Industrial Marketplace, 1840–1980. Ohio State University Press, Columbus, OH (Chapters 1 and 2).

McCord, C.P., 1953. Lead and lead poisoning in early America: lead mines and lead poisoning. Ind. Med. Surg. 22, 534–539.

Murray, A.L., 1926. Lead poisoning in the mining of lead in Utah. Technical Paper No. 389. U. S. Government Printing Office, Washington, DC.

National Academy of Sciences: National Research Council, 1993. Measuring Lead Exposure in Infants, Children, and Other Sensitive Populations. National Academy Press, Washington, DC.

National Academy of Sciences: National Research Council, 2005. Superfund and Mining Megasites. Lessons from the Coeur D' Alene River Basin. National Academy Press, Washington, DC.

Nriagu, J.O., 1983. Lead and Lead Poisoning in Antiquity. John Wiley & Sons, New York.

Nriagu, J.O., Pacyna, J.M., 1988. Quantitative assessment of worldwide contamination of air, water and soils by trace metals. Nature 333, 134–139.

Settle, D.M., Patterson, C.C., 1980. Lead in albacore: guide to lead pollution in the Americas. Science 207, 1167–1176.

Tilton J.E. (Ed.), 1990. World Metal Demand. Trends and Prospects. Washington, DC: Resources for the Future (Appendix B: Table B-4).

U.S. Department of Housing and Urban Development, 2001. National Survey of Lead and Allergens in Housing. R.B. Clickner et al., Final Report. Westat, Inc., Rockville, MD.

U.S. Environmental Protection Agency, 1986. Air Quality Criteria for Lead, 4 Vols. Report No. EPA/600/8-83/028bF. Environmental Criteria and Assessment Office, Washington, DC.

U.S. Environmental Protection Agency, 2006. Air Quality Criteria for Lead, vol. 1. Report No. EPA/600/R-05/144aF. National Center for Environmental Assessment, Washington, DC (Chapters 4–6).

Woolley, D.E., 1984. A perspective of lead poisoning in antiquity and the present. Neurotoxicology 5, 353–362.

Chapter 4

Lead in the Human Environment: Lead Emissions and Emission Trends

4.1 INTRODUCTION

This chapter provides the connection between the annual and cumulative U.S. and global lead production and consumption figures presented in Chapter 3 and figures in the next two chapters, those for fate and transport of emitted lead through environmental compartments (Chapter 5) and levels of lead in environmental media relevant to human lead exposures (Chapter 6). These three chapters collectively provide a quantitative sense of what amounts of lead entering the broad channels of the U.S. and global commerce have been translated downstream of production or use into levels of lead contamination in the human environment and associated exposures in, at least, the near term.

The basic types of lead emissions into some given environmental compartment differ with where along the life cycle for lead production, use, and dispersal emissions occur. Lead can figure in emissions at the production step, such as lead smelter stack emissions to the atmosphere. The element can also be emitted at the use stage. Production of elemental lead, production of lead compounds as substrates, conversion of lead or compounds into the antiknock additive TEL, addition of TEL to form leaded gasoline, and combustion of leaded gasoline as a vehicular fuel are examples. Tailpipe emissions of lead to the atmosphere were then available for localized and long-range transport.

Lead emission rates in terms of annualized tonnages are described here for mainly contemporary industrialized/industrializing societies emission rates. Data for earlier centuries or millennia are not available. Investigators examining such matters instead are confined to comparative contamination chronologies to evaluate current levels of media-specific Pb and relative changes in these concentrations over time as discerned from media that reflect such changes. Levels of lead in environmental media that contribute to current human exposures are the main subject of Chapter 6.

Trace Metals and other Contaminants in the Environment, Volume 10
ISSN: 1875-1121 DOI: 10.1016/B978-0-444-51554-4.00004-3

Within recent decades we can evaluate measured and modeled short-term lead emission rates and trends in these near-term rates, especially for uses which have changed greatly with technological changes. Those efforts are the focus of this chapter.

It is critical to point out that cumulative Pb emissions and rates of emissions to the various environmental compartments are not equivalent to lead production and consumption, at least in the near term. Over the long term, i.e., over the lifetime of cultures and societies, all lead present in human consumption channels can be viewed as potential lead contamination. Overall, and by estimating lead release dispersal across categories of lead production and consumption, a report of the National Academy of Sciences (NAS/NRC, 1980) estimated that at least 50% of cumulative lead production eventually becomes lead emissions. This fraction in quantitative terms translates to 150 million MT of a total global estimate of 300 million MT produced since the invention of successful smelting (Flegal and Smith, 1992).

A major reason for the lack of close tracking by lead production and consumption figures with source-specific lead emission rates in at least the near term is that the different classes of lead consumption involve lead products that differ greatly as to the U.S. and global dispersal rates to the human environment and what fractions of the original production and consumption amounts enter the human and ecological populations' environment.

For example, a major use of lead in the United States and globally is the manufacturing of lead-acid batteries for vehicular and other energy uses. For the foreseeable future, lead consumption for this category will not only continue but also increase globally as large nations such as China and India increase their private and commercial vehicular fleets. Overall, lead-acid battery production is the one consumption category that has continued to increase, save for some breaks during the Great Depression and then WWII. This trend is clear from the production and consumption figures described in Chapter 3.

Lead-acid batteries, after consumer use, do not typically release all of their lead contents into environmental dispersion channels. Instead, the lead content of much lead-acid battery production is recycled. This is not to say that the cruder forms of battery recycling are or have not been associated with waste streams, particularly at the breaking phase where lead components are first recovered for eventual secondary smelting. Unsecured disposal on land of battery acid containing lead provides not only plumes of the metal in toxic bioavailable forms but also mobility for it as the acid retards soil binding of lead. That recycling, often classified as part of scrap lead inventories, is a significant contributor to secondary lead production. As seen in various tables in Chapter 3, secondary lead production over recent decades has become a major part of total production.

Other consumption categories, by contrast, are linked to total or near-total environmental dispersal, although the rate of dispersal can be incremental.

Virtually all lead consumed in leaded gasoline production becomes lead dispersed to the atmosphere as part of vehicular exhaust. That dispersal does not allow for any economic reclamation and the dispersed lead remains in the atmosphere or is transported into and out of other environmental compartments (next chapter). In the case of lead consumption in lead fuel additives, there is close overlap in the near term between lead consumed for additive production and lead dispersed to the atmospheric environment from actual use in gasoline. Lead fallout through dry and wet precipitation processes transports lead to other media. This topic is addressed in Chapters 5 and 6.

A second case of near-total overlap of categorical lead use and environmental dispersal is in lead-based pigments for assorted paints used in residences, vehicular coatings, public buildings, other public structures such as bridges, etc. Once lead paint has been applied to surfaces, there is little or no reclamation. Removing lead paint from surfaces is not done to reclaim the lead but for hazard removal or reduction. This would be the case with lead paint abatement, where any removed lead paint is supposed to be disposed of under regulatory constraints and guidelines. The current inventory of old U.S. housing stock that contains millions of tons of old leaded paints is still huge (Jacobs et al., 2002; U.S. HUD, 2001). See, also, Table 3.3 for total Pb tonnages consumed for lead paint pigments, 1920–2009.

A separate case is that of near-total dispersal of a lead consumption category into the environment in the near term, with potential recapture for recycling in the long term. An example here is the use of lead piping in plumbing. Lead piping in water distribution systems may remain for decades until its removal is required, as is occurring in the United States subsequent to requirements of the Safe Drinking Water Act (Chapter 29). The exceptions in the lead plumbing category are soldered joints in copper household plumbing, where the older soldered joints (50:50 lead–tin) may or may not be recycled with salvaged copper plumbing.

Numerous examples exist of incremental lead dispersal to the environment with disposable consumer products that contain lead as plasticizers in plastics, lead solder in countless inexpensive electronic devices, lead-painted imported objects, and others. The lead content of an individual item in such cases may be small, but collectively such items contribute to enormous quantities of municipal solid and hazardous wastes that are either landfilled or incinerated.

4.2 HISTORICAL PERSPECTIVE ON LEAD SOURCES/ EMISSIONS

Chapter 2 provided a historical perspective on the long-term relationship of lead via contamination to the exposure of populations and cultures.

The elements of the lead exposure history for humans are largely from the general record and are qualitative in nature. There are technical approaches to quantifying the history of lead emissions and dispersals, but such approaches are principally in the form of concentrations of lead relative to corresponding time periods. For example, various investigators have examined the buildup of lead over time in accumulating media from largely the global atmospheric lead burdens at time points and by links to anthropogenic milestones for activity. It has been more difficult to reach back one step to quantify the lead tonnages released to early global environmental compartments beyond deriving estimates of tonnages of "natural" emissions.

One can use data that allow quantifying the relative increases of large-area lead emission rates over time, if not allowing precise quantification of actual tonnages of Pb emissions per unit time. Such scientific approaches have been applied to chronological assessment of relative lead deposition rates in stable accumulating media such as polar snow strata, marine water and freshwater sediments, coral skeletal media, and lead content in tree rings. A distinguishing feature among these records is the overall time frame that they report. Table 4.1 sets forth the various methods used by investigators and their reported results.

TABLE 4.1 Comparative Historical Pb Emission Inputs to Environmental Compartments[a,b]

Site and Test Media	Time Period	Ratio Early Level to 1975 Level	References
Polar ice strata	~1750 CE to 1975 CE	0.03	Murozumi et al. (1969), Ng and Patterson (1981)
Freshwater sediments	~1780 CE to ~1975 CE	0.22	Edgington and Robbins (1976)
Pond sediments	~1860 CE to ~1975 CE	0.02	Shirahata et al. (1980)
Marine sediments	~1940 CE to ~1975 CE	0.60	Ng and Patterson (1982)
Tree rings	~1920 CE to ~1965 CE	~0.25	Rolfe (1974)
Sediment Pb in Michigan lakes ($N = 12$)	1975 to 1995 CE	~0.2–0.4	Yohn et al. (2004)

[a]*Specific time periods adapted from U.S. EPA (1986), Ch. 5; U.S. EPA, (2006), Ch. 2.*
[b]*Various chronological ranges and tested media for Pb.*

Relative increases in Pb emissions from the relatively short historical interval of the eighteenth century or later to about 1975 indicate increases of 50-fold or more. For more remote periods over several millennia, increases in emission rates are two to three orders of magnitude, based on polar ice Pb levels stratified chronologically.

Attenuation of relative Pb emissions to various receiving compartments from the post-leaded gasoline and other source-based Pb inputs from ca. 1970–1975 to more recent years has also been documented. For example, Yohn et al. (2004), using sediment samples from 12 Michigan lakes, reported data indicating that sediment lead content had declined fivefold from the peak of leaded gasoline use.

Polar ice strata analysis conducted by Murozumi et al. (1969) showed a record for a period of about three millennia of increasing global lead contamination, a finding deemed valid in later investigations by Wolff and Peel (1985) and Ng and Patterson (1981). Notable is the fact that Antarctic snow and ice lead levels are an order of magnitude lower than comparable samplings for polar regions in the northern hemisphere. EPA concluded (U.S. EPA, 1986, Ch. 5) that (1) prehistoric levels of lead in snow and ice of Greenland and Antarctica were a maximum of 1.4 and 1.2 pg Pb/g; (2) then-current atmospheric lead emissions had increased 2,000-fold since the pre-Roman era; and (3) global atmospheric levels of lead had increased "dramatically" since the 1920s.

Pond and other freshwater sediments expand the recording interval considerably. Shirahata et al. (1980) showed a record of lead deposition amounting to increases up to 50-fold in the western United States, while the magnitude of increase was consistent with other findings from freshwater sediments (Christensen and Chien, 1981; Dominik et al., 1984; Wong et al., 1984).

The marine sediment Pb data of Ng and Patterson (1982) showed that time intervals as short as several decades permitted one to estimate that from about the 1940s to the 1970s, Pb accumulation continued to increase by some 50%.

Tree rings appear to capture the shortest time frame, several decades, for indexing the increases of anthropogenic lead emissions over background levels. The reliability of tree ring lead depositions across tree species and areas where growth appears to be irregular needs to be taken into account (Baes and Ragsdale, 1981; Rolfe, 1974; Symeonides, 1979; U.S. EPA, 1986, Ch. 5; U.S. EPA, 2006, Chs 2 and 3).

4.3 CURRENT LEAD SOURCES AND EMISSIONS

There are two general classes of lead emissions to the environment. These are natural emissions and anthropogenic releases.

4.3.1 Natural Pb Emissions

Natural emissions include lead released from volcanoes, seawater sprays, forest fires, and wind-borne soil particles in remote areas. These releases are typically to the atmosphere and are set forth in Table 4.2. Cumulative atmospheric releases in the natural source category average ~19,000 MT/year and a median of 12,000 MT/year (Nriagu, 1989). It should be noted that one has to distinguish between true natural sources and emissions and those which are more inclusive, i.e., "background" lead estimates. Such background levels can represent releases to the atmosphere and subsequent deposition of reentrained dusts contaminated with lead from past anthropogenic activities.

Table 4.2 depicts the five highest categories or category groups of natural lead emissions, the range and the medians of annual emitted amounts (10^3 MT/year). Wind-borne soil lead and volcano emission lead are the two largest categories indexed as median figures followed by forest wild fires and sea salt spray. Three types of biogenic natural emissions have been quantified (Nriagu, 1989): continental biogenic volatiles and particulates, and marine biogenic emissions.

Anthropogenic activities involved in the production, distribution, consumption, and disposal of lead and lead products produce estimates of lead emissions that are many times the amount assigned to natural emission rates. The rates in prior years would have been higher and the ratios of anthropogenic to natural emissions are much higher, but the retention of those past emissions in soil and sediment environmental compartments means that the amounts of anthropogenic lead in the human and ecological environments

TABLE 4.2 Natural[a,b] Lead Emission Rates to Atmospheric Compartments

Source	Range (1,000 MT/year)	Median (1,000 MT/year)
Wind-borne soils	0.3–7.5	3.9
Volcanoes	0.54–6.0	3.3
Wild fires	0.06–3.8	1.9
Sea spray	0.02–2.8	1.4
Continental biogenic particles	0.02–2.5	1.3
Marine biogenic emissions	0.02–0.45	0.24
Continental biogenic volatiles	0.01–0.04	0.20
Total	0.97–23	12

[a]*Nonanthropogenic emissions differ from "background" or "remote" levels.*
[b]*From Nriagu (1989).*

still dwarf those arising from naturally derived lead sources. Nriagu (1989) estimated that the ratio of anthropogenic to natural lead emission rates were on the order of 700:1 for earlier times. Current anthropogenic lead releases still remain at least an order of magnitude over natural releases (USEPA, 2006, Ch. 2).

4.3.2 Current Anthropogenic Pb Emissions

Current anthropogenic Pb emissions differ with the source producing the emissions, the environmental compartments receiving various source emissions, the relative mobility of lead in and out of environmental compartments, and the relative propensity for providing human exposures within the receiving compartments. Two of the principal receiving compartments are the atmosphere and soils.

Lead Emissions to the Atmosphere

There are two broad categories of anthropogenic lead emissions to the atmosphere, mobile and stationary sources, sources largely defined in terms of their regulatory identity. The first category included large amounts of lead emitted in vehicular exhaust from previous leaded gasoline consumption (mobile sources). In this case, emissions are not confined to one site or locale and consequently require regulatory controls different from those for nonmobile emitters (see chapters in Part 5).

Stationary sources, in turn, are of two types. A point stationary source is a specific facility such as a lead smelter, lead ore milling operation, waste incinerator, and the like. A nonpoint stationary source is a lead emitter arrayed collectively across more than one specific site. One example would be residential heating that is not inventoried at a specific location, but aggregated at a municipality, county, or other jurisdictional level.

The relative significance of these two categories of lead emissions to overall environmental impact has varied markedly over the decades and in industrialized and industrializing societies. These changes are noted below for trends in Pb emissions.

Table 4.3 sets forth global annual emissions to the atmosphere as estimated by Nriagu and Pacyna (1988) for 1983. These were on the order of 330,000 MT as a median (range 289,000–376,000) and the totals combined both stationary and mobile source emitter contributions. The major component of that total was leaded gasoline combustion, ca. 248,000 MT or 75% of the annual total for 1983. Other major sources of global atmospheric input in the table included mining, coal combustion, cement production, manufacturing, wood and oil combustion, and accumulated miscellaneous categories.

TABLE 4.3 Global Emissions of Pb to the Atmospheric Environment[a,b]

Emission Source[c]	Range (MT/year)
Leaded gasoline	248,030
Mining	30,060–69,640
Coal combustion	1,765–14,550
Cement production	18–14,240
Manufacturing	1,065–14,200
Miscellaneous	3,900–5,100
Wood combustion	1,200–3,000
Incineration	1,640–3,100
Oil combustion	948–3,890
Fertilizer	55–274
Total	288,700–376,000 (median 332,000)

[a]Source: *Nriagu and Pacyna (1988).*
[b]*For year 1983 CE, MT/year.*
[c]*In descending order of Pb quantities emitted.*

The atmosphere is the short-term, global conduit for transport of Pb from source to eventual repositories, so that the quantities set forth in Table 4.3 are eventually dispersed to the terrestrial and aquatic compartments. Concurrently, however, atmospheric Pb that impacts human population exposures is more accessible and biologically active than lead in other forms.

Table 4.4 offers an update of global emissions to the mid-1990s (Pacyna and Pacyna, 2001; UNEP, 2008) and includes summary air emission data for illustrative countries with available emission data. Relative to the 1983 global air Pb inputs, emissions by the mid-1990s had declined to about 120,000 MT/year. Of this figure, alkyl Pb additives for gasoline still comprised a significant fraction, ~75%, of total global air Pb inputs. This figure reflects persisting leaded gasoline use outside of the United States and Canada in the 1990s.

Tables 4.5 and 4.6 provide data for the U.S. emissions of Pb to the atmosphere as a function of various source categories, noted by the U.S. EPA (2006). Two time points for atmospheric Pb inputs are estimated in Table 4.5: 1990 and 2002. EPA's figures are for categories involving Pb emissions of >20 MT/year for 1990 or 2002, ranked by 2002 Pb emission rates.

Marked changes are seen in both rankings as percentages of total emissions, 1990–2002. This reflects, among other things, the phase down and

TABLE 4.4 Current Global and Selected Non-U.S. Atmospheric Lead Emissions[a,b,c,d]

		Selected Non-U.S. Pb Emissions				
Emission Category	Global Emissions	Canada (2004)	Australia (2003–2004)	Japan (2003)	Slovak Republic (2003)	Denmark (2000)
Power and heat production	11,690	13.8	8.4	–	4.3	0.20–0.59
Nonferrous metal production	14,815	231.3	330	29.8	2.8	–
Lead fuel additives	88,739	–	85	–	2.0	1.6–2.0
Iron and steel	2,926	17.9	4.9	2.0	1.3	0.51
Cement and related	268	1.1	1.4	–	–	0.13
Waste disposal	821	0.4	–	<0.3–7.2	10.8	1.2–3.8
Glass and glass products	–	–	1.6	–	14.2	0.05–0.4
Total (rounding)	119,259	289	1,022	51–59	61	5–19

[a] *Adapted from UNEP (2008).*
[b] *MT/year.*
[c] *Non-U.S. national figures derived from various inventory and registry data sets for year indicated.*
[d] *Global data for mid-1990s.*

TABLE 4.5 Current U.S. Annual Emissions of Pb from Various Sources to the Atmosphere[a,b,c,d]

Source Category	1990 TPY	Category % of 1990 Total	2002 TPY	Category % of 2002 Total
Industrial/commercial/institutional boilers and heaters	34	1.0	247	17.2
Utility coal boilers	72	2.0	165	11.5
Mobile sources	1198	36.7	143	10.0
Iron/steel foundries	14	0.40	110	7.7
Hazardous waste incineration	97	3.0	70	4.9
Primary Pb smelting	220	6.7	59	4.1
Electric services	[e]	–	53	3.7
National security	[e]	–	34	2.4
Municipal waste incineration	80	2.4	33	2.3
Total of source emissions with <2% as each contribution in 2002 (iron and steel production, glass/glassware making, secondary metals, Pb/Zn ores, acid battery production, stainless steel and related production, primary copper production, cement making, metal products)	425	13.0	229	16.0
Grand total—all sources	3,270[f]		1,435[f]	

[a]*Adapted from U.S. EPA (2006), rounding from original.*
[b]*Emission values for 1990 based on EPA estimates within CAA Sec. 112(k) mandates.*
[c]*Emission values for 2002 based on U.S. EPA's (2002) National Emissions Inventory.*
[d]*Sources emitting >20 tons/year (TPY) for 1990 or 2002, ranked by 2002 emissions.*
[e]*Not included or not quantified.*
[f]*Includes sources <20 TPY.*

phase out of leaded gasoline in the United States. While mobile sources (mainly leaded gasoline use) comprised 37% (rounding) of all emissions in 1990 and accounted for the highest fractional contribution, this category declined to third place (10%) by 2002. Table 4.5 outlines the source contributions and percents of totals for 1990 and 2002.

A second source of the U.S. atmospheric Pb emissions data is found in the annual Toxics Release Inventory (TRI) figures gathered by the U.S. EPA

TABLE 4.6 U.S. Lead and Lead Compound Releases Reported for the USEPA Toxic Release Inventory (TRI)[a,b,c,d]

TRI Category	Year Covered in Reporting for TRI		
	2002	2005	2008
Total releases: on-site plus off-site, to all compartments	200,856	213,182	220,909
Releases on-site	183,411	198,636	210,455
On-site % of all releases	91%	93%	95%
Releases off-site	17,425	14,923	10,455
Off-site % of all releases	9	7	5
Releases to the atmosphere	663	449	500
On-site disposal to land	104,259	117,273	133,182
On-site disposal to impoundments	57,656	64,091	60,910
Off-site land disposal (no RCRA landfills included)	–	5,909	5,455
Off-site disposal with solidification or stabilization	–	4,091	2,273

[a]*Converted to metric tons from pounds in original reports.*
[b]*Figures combine Pb releases as the core element and as various lead compounds.*
[c]*Figures prior to 2001 employed a reporting Pb/Pb compound threshold of 25,000 pounds, reduced to 100 pounds in 2001.*
[d]*2002 figures released June 2004; 2005 figures released March 2007; 2008 figures released December 2009.*

from the U.S. operations and facilities required to provide such data (U.S. EPA, 2002, 2007, 2009). TRI data for lead and lead compounds also quantify dispersal to other environmental compartments.

Table 4.6 indicates air Pb emissions for the years 2002, 2005, and 2008 as 663, 449, and 500 MT, respectively. These amounts represent relatively modest fractions of the corresponding total on-site and off-site releases: 200,856, 213,182, and 220,909 MT for the same years. These years' emissions are estimated in years long after leaded gasoline was essentially banned in the United States in the 1990s and thus reflect Pb inputs from various remaining Pb waste streams.

The U.S. picture for major point or nonpoint stationary sources for lead emissions has changed in the general direction of significant decline. However, both stationary and mobile sources emitted lead in past decades that remains in receiving environmental compartments. Today, there is only one active primary smelter in the United States, no existing lead additive facility, and a number of the mines have closed. Secondary lead battery smelters and scrap battery recycling facilities still exist. Major changes particularly occurred in 1990 and 2002 for reasons other than leaded gasoline phase-out. The categories of commercial/industrial/institutional boilers and utility coal boilers both showed marked increases in the 12 years as did emissions from iron and steel foundries. Hazardous waste and municipal waste incineration, as well as primary Pb smelting emissions, declined. Collectively, various small emitters (each having <2% of total contributions for 2002) amounted to 16% of the total and a tonnage figure of 229 MT/year for 2002.

Tonnages of Pb emitted to the U.S. atmosphere from combusted leaded gasoline in years prior to 1990, i.e., before the effects of regulations against such use, were extremely high and repeatedly comprised the highest contributor to total U.S. atmospheric emissions. Table 4.7 tabulates figures for annual amounts of lead consumption for leaded gasoline as either total TEL consumption rates for the period or amounts of Pb additive actually blended into gasoline in the indicated year. Figures for the former, for the period 1941–1970, are from Table 3.3. Figures for the latter are from U.S. EPA (1986), Ch. 5.

In the first year of reported TEL Pb consumption by the U.S. government, 1941, this figure was 48,000 MT. Succeeding years witnessed greatly increased consumption rates, at 165,000 MT in 1955 and at 279,000 MT in 1970, the peak of such consumption. By 1981, the U.S. EPA estimate of lead consumed in combusted leaded gasoline was 61,000 MT and the figure in Table 3.3 for this same year was 111,000 MT. The dramatic declines in 1987 and 1989 reflected the mandated phasedown of lead in gasoline for the main U.S. vehicular fleet.

Deposition of atmospheric lead to land via dry and wet precipitation is lower on an annual basis than noted above for direct dispersal of materials such as mine tailings and smelter slags, but the many decades of lead fallout onto soils, especially urban soils adjacent to major road corridors, can be sizeable. It has been estimated that more than 3,000,000 MT has been deposited on urban and suburban terrestrial surfaces from leaded gasoline combustion (NAS/NRC, 1993). This figure is derived from summing lead additive consumption figures (see Chapter 3) and assuming rapid consumption of leaded gasoline produced with these additives. Sizeable quantities have also been deposited from point and nonpoint sources although quantitation of these deposition masses is difficult.

Global Lead Emissions to Land

Table 4.8 sets forth global lead emissions to the terrestrial compartment as estimated by Nriagu and Pacyna (1988) for the year 1983. The total Pb dispersals to the terrestrial compartment are in the range 803,000–1,819,000 MT/year (rounding). In descending order of emitted Pb quantities, lead

TABLE 4.7 U.S. Consumption of Lead in Gasoline (10^3 MT/year): 1941–1989[a,b]

	Gasoline Lead Emitted via		
Year	Table 3.3, this book[c]	Table 5.5, USEPA (1986)[d]	Ch. 2, USEPA (2006)
1941	48	–	–
1945	76	–	–
1950	114	–	–
1955	165	–	–
1960	164	–	–
1965	225	–	–
1970	279	–	–
1975	209	167	–
1977	233	169	–
1979	187	129	–
1981	111	61	–
1983	89	52	–
1985	46	26	–
1987	–	3	–
1989	–	2	–
1990[e]	–	–	1.2
2002[e]	–	–	0.11

[a]*Adapted from U.S. EPA (1986), Ch. 5 for 1975–1989.*
[b]*Data 1985–1990 estimates are projections from U.S. EPA (1986) or are derived from Ch. 3, this book; provides additive consumption.*
[c]*Table 3.3; total TEL consumption.*
[d]*EPA figures for Pb additive blended into gasoline.*
[e]*Data for 1990 and 2002 are from U.S. EPA (2006), Ch. 2.*

releases from extractive metal industries top the list in the form of mine tailings and smelter wastes, followed by commercial product wastes and Pb transport to land surfaces from atmospheric Pb fallout. Coal fly and bottom ash, mainly from electrical utilities and urban municipal waste combustion, follow. Other emission sources are animal wastes, agricultural and food wastes, metal manufacturing and miscellaneous sources expressed in Table 4.4 as a total.

The quantities of Pb emitted to the terrestrial compartment are clearly much above those recorded in Table 4.3 as entering the global atmosphere, up to about fivefold greater. Annual atmospheric Pb burdens, on the other

TABLE 4.8 Global Emissions of Lead (MT/year) to the Terrestrial Environment[a,b,c]

Category of Pb Emissions	Pb Emission Rate (MT/year)
Commercial product wastes	195,000–390,000
Smelter wastes	194,000–390,000
Mine tailings	130,000–390,000
Atmospheric fallout	202,000–263,000
Coal fly ash/bottom ash	45,000–242,000
Urban refuse	18,000–62,000
Animal wastes	3,200–20,000
Agricultural/food wastes	1,500–27,000
Solid wastes: metal manufacturing	4,100–11,000
Miscellaneous sources total: logging, sewage sludge, organic wastes, fertilizer, peat	10,290–23,800
Total (rounding)	803,000–1,819,000

[a]Source: *Adapted from Nriagu and Pacyna (1988).*
[b]*For year 1983 CE.*
[c]*In descending order of Pb quantities emitted.*

hand, are much more mobile in terms of human exposures and have a potentially greater spatial reach. In contrast, land inputs of Pb from diverse inputs are temporally much more stable over the long term, so that these receiving compartments continue to contain elevated Pb levels for many years.

It is important to distinguish between Pb releases to the land compartment, *per se*, and those to surface soils as a readily accessible component of the land compartment. The latter are principally impacted by atmospheric lead fallout to the topmost portions of soils across large soil expanses, while the former include huge tonnages of tailing and smelter waste (slag) piles confined to specific areas. Nriagu and Pacyna (1988) appropriately differentiated the two fractions in their analyses.

Table 4.6, discussed earlier for the U.S. air Pb inputs, depicts the huge amounts and large fractions of total environmental Pb releases to land and impoundments. Total lead releases on-site from the U.S. data for 2002, 2005, and 2008 are 183,411, 198,636, and 210,455 MT, respectively, comprising 91%, 93%, and 95% of all on-site and off-site Pb/Pb compound releases.

Lead Emissions to the Aquatic Environment

Lead can enter rivers, lakes, estuaries, etc. by direct runoff from a source or through effluent discharges. As in the case of leaded waste dispersed to land, lead entering aquatic systems comes heavily from extractive industry operations, such as the disposal of tailing wastes from ore mills processing lead, zinc, copper, and other ores. Lead smelters and refineries, for example, were typically sited along waterways for access to both a processing water source and subsequent disposal path. Waterway disposal of metallurgical wastes in rivers has the potential to distribute toxic substances marked distances from discharge points and has the potential to produce toxicity at multiple trophic levels in ecological systems.

Nriagu and Pacyna (1988) reported that while direct dispersal of lead to the aquatic environment was considerably smaller than for land, the figures were still significant. For the period of the estimations, these authors reported a range of such lead discharges of 25,000–50,000 MT/year. This range was produced by contributions from manufacturing (14,000 MT/year), sewage sludge (9,000 MT/year), domestic wastewater (7,000 MT/year), smelting and refining (6,000 MT/year), and mining (1,000 MT/year).

The U.S. EPA and others have used atmospheric lead emission factors to estimate lead entry to air in more recent years. An emission factor is a mathematical constant which links estimates for volumes of lead production or amounts of feed stocks to amounts of lead exiting the air lead waste stream as a complex function of type of feed material and the engineering characteristics of lead metallurgical and other pyrogenic processes, such as municipal and other waste combustion.

4.4 TRENDS IN Pb EMISSIONS TO GLOBAL AND U.S. ENVIRONMENTAL COMPARTMENTS

Globally, there was a major decline in lead emissions to the atmosphere in the period 1983 to ca. 1995 (compare Tables 4.3 and 4.4). A 64% decline was seen, despite the lingering and significant quantity of alkyl Pb gasoline additives still being blended into part of the global gasoline pool.

Individual countries such as the United States and Canada have seen the ban on leaded gasoline reduce estimated levels to 110 MT in 2002 from the peak 1970 figure of 279,000 MT, a decline of 99.96%. All major lead emission categories collectively show a marked decline over time from the 1970s to 2002. Much of this decline was mandated by law and regulatory rulemaking. The most drastic reduction was due to the mandated removal of lead from gasoline. In 1970, the U.S. atmospheric lead emissions rate was about 280,000 MT, falling to 35,000 MT in 1984 and to barely measurable amounts by 2002. In 1975, leaded gasoline consumption accounted for 167,000 MT, which in turn was 95% of 1975 total U.S. air emissions, or 176,000 MT (rounding). Using the 2002 total lead atmospheric emissions figure of 1,435 MT and the estimated total 1975 U.S. EPA (1986, Table 5.5) figure of 176,000 MT, the former value comprises 0.8% of the latter. For the United States, total emission rates per year have declined 99.2%.

Trends in the U.S. lead emissions or those emissions in other industrialized or postindustrialized countries can differ from the trends in less developed or slowly developing nations. Overall, however, global declines in Pb emissions have been traced to declining leaded gasoline use. In the 1970s, global consumption of Pb for gasoline blending was ~400,000 MT, declining about sixfold by 1993, to 70,000 MT (Socolow and Thomas, 1997). In other countries, leaded gasoline is still used to some extent (see Chapter 28).

In the United State, the 2002 estimate of atmospheric lead emissions on the order of <1,500 MT will likely further decline only slowly. We would not expect the great decline seen in decades prior to 2000 simply because the cause of that decline, removing lead additives from gasoline, has largely been accomplished.

A second factor may be emergence or existence of a floor to further atmospheric Pb emission declines in certain locales. For example, modeled estimates indicate that 90% of the atmospheric Pb particulate loading in California is attributable to reentrainment of leaded roadway dusts from roadways contaminated over past decades by leaded gasoline combustion and fallout (Harris and Davidson, 2005). Lankey et al. (1998) noted that resuspension accounts for ~40% of Pb emissions in the South Coast Air Basin of CA, while a much higher fraction of 75% was reported for lead resuspension in an Armenian city (Kurkjian et al., 2002).

Stationary point source and nonpoint source inputs to environmental compartments in the coming years will also decline as part of regulatory requirements or simply the size of the remaining lead emission generators in coming years versus their historical impact. For example, only one major primary lead smelter remains in the United States, the facility in Herculaneum, MO. In 1985, there were five major primary lead smelters and refineries (ILZRO, 1982): East Helena, MT; Omaha, NE; and three in the Missouri Lead Belt.

Emissions from waste combustion of various types will likely decline as well, given the mix of new rules on municipal solid waste (MSW) and hazardous waste combustion. Walsh et al. (2001) reported that in past decades, the late 1960s, MSW burning in the New York City area emitted up to ca. 800 MT Pb/year.

The trend in U.S. releases of lead and lead compounds noted in the U.S. TRI reveals some salient features:

- In 2002 and 2005, lead and lead compounds accounted for 98% of total disposal of persistent, bioaccumulative and toxic (PBT) chemicals; relative to 2001, there was an increase of 3% in 2002 and 6% in 2005 from 2004.
- From 2001 to 2005, Pb/Pb compounds disposal increased 12% (+49 million pounds).

- From 2007 to 2008, Pb/Pb compounds total disposal decreased 2% (−12 million pounds).
- From 2001 to 2008, total disposal increased by 15% (+64 million pounds).
- A longer comparison interval is not possible, because of a major lowering of the reporting threshold for Pb/Pb compounds prior to 2001.

The expected annualized inputs of lead via atmospheric emissions into other environmental compartments (see Chapter 5) must be balanced against the enormous quantities of postdeposition lead sequestered in soils, dusts, sediments, etc. over previous centuries, especially after the introduction of lead additives for fuels and the use of lead pigments in paints. Current lead inputs to environmental compartments will, through accumulation, be added to the historical lead deposition amounts.

REFERENCES

Baes, C.F., Ragsdale, H.L., 1981. Age-specific lead distribution in xylem rings of three tree genera in Atlanta, Georgia. Environ. Pollut. Ser. B 2, 21–35.

Christensen, E.R., Chien, N., 1981. Fluxes of arsenic, lead, zinc, and cadmium to Green Bay and Lake Michigan sediments. Environ. Sci. Technol. 15, 553–558.

Dominik, J., Mangini, A., Prosi, F., 1984. Sedimentation rate variations and anthropogenic metal fluxes into Lake Constance sediments. Environ. Geol. 5, 151–157.

Edgington, D.N., Robbins, J.A., 1976. Records of lead deposition in Lake Michigan sediments since 1800. Environ. Sci. Technol. 10, 266–274.

Flegal, A.R., Smith, D.R., 1992. Current needs for increased accuracy and precision in measurements of low levels of lead in blood. Environ. Res. 58, 125–133.

Harris, A.R., Davidson, C.I., 2005. The role of resuspended soil in lead flows in the California South Coast Air Basin. Environ. Sci. Technol. 39, 7410–7415.

ILZRO, 1982. Locations of major lead operations in the United States [map]. As cited in USEPA, 1986, Ch. 5. International Lead Zinc Research Organization, New York.

Jacobs, D.E., Clickner, R., Zhou, J., Viet, S.M., Marker, D.A., Rogers, J.W., et al., 2002. The prevalence of lead-based paint hazards in U.S. housing. Environ. Health Perspect. 110, A599–A606.

Kurkjian, R., Dunlap, C., Flegal, A.R., 2002. Lead isotope tracking of atmospheric response to post-industrial conditions in Yerevan, Armenia. Atmos. Environ. 36, 1421–1429.

Lankey, R.L., Davidson, C.I., McMichael, F.C., 1998. Mass balance for lead in the California south coast air basin: an update. Environ. Res. 78, 86–93.

Murozumi, M., Chow, T.J., Patterson, C., 1969. Chemical concentrations of pollutant lead aerosols, terrestrial dusts and sea salts in Greenland and Antarctic snow strata. Geochim. Cosmochim. Acta. 33, 1247–1294.

National Academy of Sciences: National Research Council, 1980. Lead in the Human Environment. National Academy Press, Washington, DC.

National Academy of Sciences: National Research Council, 1993. Measuring Lead Exposure in Infants, Children, and Other Sensitive Populations. National Academy Press, Washington, DC.

Ng, A., Patterson, C., 1981. Natural concentrations of lead in ancient Arctic and Antarctic ice. Geochim. Cosmochim. Acta 45, 2109–2121.

Ng, A., Patterson, C., 1982. Changes of lead and barium with time in California off-shore basin sediments. Geochim. Cosmochim. Acta 46, 2307–2321.

Nriagu, J.O., 1989. A global assessment of natural sources of atmospheric trace metals. Nature 338, 47–49.

Nriagu, J.O., Pacyna, J.M., 1988. Quantitative assessment of worldwide contamination of air, water and soils by trace metals. Nature 333, 134–139.

Pacyna, J.M., Pacyna, E.G., 2001. An assessment of global and regional emissions of trace metals to the atmosphere from anthropogenic sources worldwide. Environ. Rev. 9, 269–298.

Rolfe, G.L., 1974. Lead distribution in tree rings. For. Sci. 20, 283–286.

Shirahata, H., Elias, R.W., Patterson, C.C., Koide, M., 1980. Chronological variations in concentrations and isotopic compositions of anthropogenic atmospheric lead in sediments of a remote subalpine pond. Geochim. Cosmochim. Acta 44, 149–162.

Socolow, R., Thomas, V., 1997. The industrial ecology of lead and electric vehicles. J. Ind. Ecol. 1, 13–36.

Symeonides, C., 1979. Tree-ring analysis for tracing the history of pollution: application to a study in northern Sweden. J. Environ. Qual. 8, 482–486.

United Nations Environmental Programme, 2008. Draft Final Review of Scientific Information on Lead. UNEP Governing Council: GC Decision 23/9 III, November. Chemical Branch, DTIE, Geneva, Switzerland.

U.S. Department of Housing and Urban Development, 2001. National Survey of Lead and Allergens in Housing. R.B. Clickner et al., Final Report. Westat, Inc. Rockville, MD.

U.S. Environmental Protection Agency, 1986. Air Quality Criteria for Lead. 4 vols. Report No. EPA/600/8-83/028bF. Environmental Criteria and Assessment Office, Washington, DC.

U.S. Environmental Protection Agency, 2002 [Released June 2004]. Toxics Release Inventory (TRI) Public Release Report. Report No. EPA 260-R-04-003. Data Table: TRI Disposal or other Releases, PBT Chemicals, 2002. www.epa.gov/tri (accessed 22.10.10).

U.S. Environmental Protection Agency, 2006. Air Quality Criteria for Lead, Vol. 1. Report No. EPA/600/R-05/144aF (Chapters 2 and 3).

U.S. Environmental Protection Agency, 2007 [Released March 2007]. U.S. EPA Toxics Release Inventory Reporting Year 2005 Public Data Release. Summary of key findings. Lead and Lead Compounds, pp. 12–14. www.epa.gov/tri (accessed 23.10.10).

U.S. Environmental Protection Agency, 2009 [Released December 2009]. U.S. EPA Toxics Release Inventory Reporting Year 2008 National Analyses. Summary of key findings. Lead and Lead Compounds, pp. 17–19. www.epa.gov/tri (accessed 24.10.10).

Walsh, D.C., Chillrud, S.N., Simpson, H.J., Bopp, R.F., 2001. Refuse incinerator particulate emissions and combustion residues for New York City during the 20th century. Environ. Sci. Technol. 35, 2441–2447.

Wolff, E.W., Peel, D.A., 1985. The record of global pollution in polar snow and ice. Nature 313, 535–540.

Wong, H.K.T., Nriagu, J.O., Coker, R.D., 1984. Atmospheric input of heavy metals chronicled in lake sediments of the Algonquin Provincial Park, Ontario, Canada. Chem. Geol. 44, 187–201.

Yohn, S., Long, D., Fett, J., Patino, L., 2004. Regional versus local influences on lead and cadmium loading to the Great Lakes region. Appl. Geochem. 19, 1157–1175.

Chapter 5

Lead in the Human Environment: Fate and Transport Processes

5.1 INTRODUCTION

Lead fate and transport refer to those environmental, geochemical, and biogeochemical processes that move lead from emission sources to environmental compartments. These intermediary roles for lead fate and transport are brought about, first, by lead as some chemical or physical species passing from an industrial, mobile, or other emission source to an environmental compartment for transport, followed by deposition in or movement out of one or more environmental compartments. Lead transport and transformations in the human and ecological environment are necessary preconditions for eventual lead exposures of various populations. For example, lead emissions to the atmosphere from a lead primary or secondary smelter require the operation of a source (pyrometallurgy), followed by entry of source emissions into a transport and transformation series of processes. Many other examples can be cited. Second, fate and transport processes figure in the overall quantitative risk assessments that will eventually lead to the overall health and regulatory policies for this toxicant.

This chapter also serves as an environmental road map, laying out the physical, physicochemical and chemical means by which lead is transported and mobilized in and out of environmental compartments that serve as lead contact points for human and ecological populations. Environmental compartment cycling of lead until recently was a largely unknown cluster of phenomena, and the multimedia impact of lead emissions on the larger biosphere was little understood and even misunderstood. This chapter is not intended to be encyclopedic, but it focuses on data eventually useful for human health risk assessment and regulatory initiatives.

Figure 5.1 presents a comprehensive schematic for the environmental cycling of lead, movement presently understood in terms of compartments

Trace Metals and other Contaminants in the Environment, Volume 10
ISSN: 1875-1121 DOI: 10.1016/B978-0-444-51554-4.00005-5

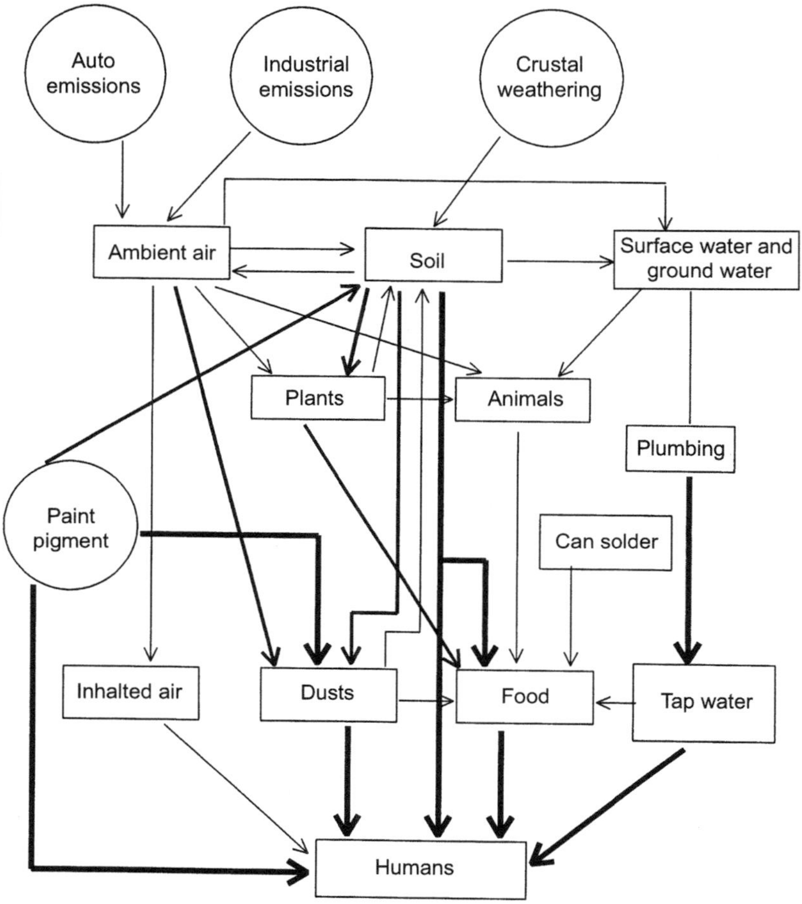

FIGURE 5.1 Environmental cycling of lead in the human environment. Line thickness varies with pathway importance for children. *Source: Mushak 1992. © Elsevier Inc. Reprinted with permission.*

within which the representative media are defined as pathways to human population lead exposures (Mushak, 1992). The interconnections in Figure 5.1 are made by either heavy or light arrows, the former representing the proportionately more important movements among lead pathways.

Figure 5.1 represents deposition into and movement among compartments over very different time frames (Mushak, 1992). For example, the largest particles containing lead emitted from stationary source stacks have a short life in the atmosphere, deposited over a matter of days. Lead in soils, by contrast, is relatively immobile and has a retention half-life on the order of decades.

Some illustrative salient features of the depiction of lead's fate and transport can be summarized as follows:

- Ambient air is the main conduit for transport of lead from emission sources to other compartments: soils, dusts, plants, animals, surface water and groundwater.
- Ambient air lead receives the element directly from source emissions or through reentrainment of dust lead particles after deposition onto soils or through mobilization of industrial process dusts.
- Predominant pathways for lead from air to humans are soil, plants, as deposited dusts, and direct exposures of humans through inhalation.
- Soil lead, from deposition of atmospheric lead to soil or by means of direct deposition of solid wastes or effluents containing lead onto soils, can undergo further movement to dusts, plants, surface water and groundwater.
- Soil lead is a major contributor to dusts, as are exterior or interior leaded paints in deteriorated or deteriorating condition.
- Lead in dust, especially interior dusts, is a major pathway for lead exposure of risk populations such as very young children through hand–mouth activity and ingestion or by direct inhalation. Industrial dusts provide exposures for lead workers on the job.
- Dust lead can then readily enter the food supply at various levels: during crop growth through foliar deposition, contamination of crop surfaces during transport, etc.
- Lead enters surface water or eventually groundwater from ambient air or from soil via runoff, and eventually provides tap water lead exposures either directly through drinking water or through use of lead-contaminated water for cooking.
- Lead contaminates plant crops through surface deposition or root uptake to some extent. Human lead exposures from crops involve direct consumption of food crops by humans or exposure through lead intakes of livestock that become meat sources.

This figure depicts fate and transport among environmental compartments for a chemical element. Lead, being an element, cannot be destroyed or created in its environmental lifetime, and its relative mobility has little effect on its overall environmental survival. Consequently, significant movements of atmospheric lead to receiving compartments such as soils during past years are still relevant for today's lead exposure assessments regardless of marked declines in current air lead emissions and lead deposition to soils.

5.2 ATMOSPHERIC LEAD TRANSPORT

Ambient air is the major conduit for lead movement from industrial, commercial, and institutional combustion sources (Reuer and Weiss, 2002; U.S.

EPA, 2006). Reentrainment of lead in particulate back into the atmosphere from dusts and soils after initial deposition of lead from the atmosphere also occurs due to atmospheric mobilization (Cowherd et al., 1985; U.S. EPA, 2006). The range of lead transport via the atmosphere can be many miles for the smallest size particles to as short as meters for reentrained dust particles at roadways for heavy traffic or for large lead paint particles, as occurs with weathering and flaking of lead-painted surfaces in older buildings where particles are mobilized to nearby soil surfaces, around the "drip line."

One can stratify atmospheric air lead transport into two categories, depending on the emissions source: mobile sources—largely vehicular emissions; and stationary or point sources—smelters, refineries, incinerators, industrial boilers, power plants, etc. This division is important to understanding the relative spatial reach of lead movement through the atmosphere and the types of chemical and physical species being transformed there. These two categories are incorporated into three subheadings for lead atmospheric transport: dispersion, transformation, and removal through precipitation processes.

5.2.1 Atmospheric Dispersion of Lead Emissions from Mobile and Fixed Sources

The current level of lead emissions from mobile sources such as autos, trucks, farm equipment, and aircraft, is a small fraction of the total lead releases nationally and globally that occurred up to the early and mid-1970s. Chapters 3 and 4 detailed and quantified this dramatic reduction. However, lead from mobile sources largely settled in the environmental compartments of soils and dusts, where residence time is on the order of decades. This cumulative tally has been estimated for the United States at over 3 million MT. Furthermore, current lead inputs to existing lead loadings in soils and as dusts even at reduced values occur because of the ready remobilization of small dust particles from leaded soil surfaces. In a number of geographic areas with heavily used traffic arteries, roadside soil lead releases as fine particle dusts can produce significant localized elevations in air lead content.

Lead emissions at the tailpipe as recorded in the 1970s, when leaded gasoline had relatively high average lead content of 1.5–1.6 g Pb/gal, were on the order of 20–30 mg Pb/m^3 (Dzubay et al., 1979; U.S. EPA, 1986, Ch. 6). Comparison of concentrations near or at exhaust pipes with those at increasing distances from traffic indicates most deposition and impact occurs at or near roadside.

The magnitude of reduction in air lead *vis-a-vis* tailpipe exhaust air concentrations depends on a number of parameters: distance from the source to where air lead is measured, the physical and physicochemical characteristics of the exhaust lead particulate, the presence or absence of atmospheric confinement, and the extent of lead removal from air by wet and dry deposition processes. With distance, there is increasing dilution of the amount of lead in the atmosphere through mixing with clean air.

A major factor in the relative rate and extent of atmospheric dilution of emitted lead is the presence or absence of confinement. The confinement factor, in turn, figures in lead exposure risks to individuals within the confinement zones. Ingalls and Garbe (1982) reported estimates for relative levels of air lead in garages, tunnels, and street canyons. Low air volumes and little ventilation are particularly robust enhancers for relative persistence in high air lead values, with concentrations up to about three orders of magnitude higher than levels found in open air. These enhancements have implications for workers' lead exposures in garages and roadway tunnels.

Table 6.1 of U.S. EPA (1986, vol. II) illustrates both typical and severe air lead concentrations for confined spaces such as residential and parking garages and roadway tunnels. Typical concentrations of air lead in garages range from 40 to 80 $\mu g/m^3$, and severe condition concentrations for garages range up to 670 $\mu g/m^3$, the latter arising from extended idling times. Tunnel air lead values for typical and severe scenarios were noted to be 11 and 29 $\mu g/m^3$, respectively. Street canyons, where sidewalk readings are taken, show typical and severe scenario air lead levels of up to 0.9 and 2.8 $\mu g/m^3$, respectively.

Open air lead levels attributable to mobile sources vary with meteorological conditions, topography, and distance from major traffic arteries. This is particularly the case for major urban areas with central business districts. Using various measurement methods (Burton and Suggs, 1984; Clift et al., 1983), atmospheric lead levels were shown to decrease exponentially with distance from heavily traveled roadways. U.S. EPA (1986, pp. 6–8) estimated that United States air lead levels in the 1980s declined two- to threefold in going from central city areas to the suburbs, with a further twofold decline to the outer perimeters of suburban zones.

Air lead concentrations for purposes of determining dispersion data can be found by actual measurement over some unit of selected time, e.g., 24-hour high-volume sampling or with various dispersion models. A major factor in either air lead measurement or modeling is the type of lead emission source. Mobile sources or mixes of mobile and point lead sources when modeled within relatively broad geographic areas such as urban zones require quite different quantification approaches than modeling particular point source emissions, i.e., from a primary or secondary lead smelter.

In the case of point sources, available dispersion models are typically based on assumption of a Gaussian distribution. The relative complexity and reliability of Gaussian distribution models are affected by parameters such as turbulence although modifications can be made to account for complex atmospherics. Reliability of the Gaussian approach is relatively satisfactory for distances up to about 100 km from the stationary source. Seinfeld and Pandis (1998) have published a comprehensive treatment of Gaussian dispersion models. The particulars include meteorological conditions, the rates of stack lead emissions, stack height, and presence or absence of nearby obstructing buildings.

5.2.2 Resuspension of Leaded Soils and Dusts in the Atmosphere

The decades of leaded gasoline use and the emissions of lead from numerous industrial and other pyrogenic sources created enormous inventories of lead in soils and as dusts. These pathways for direct and indirect lead interactions with human and ecological populations are discussed separately below. The role of these lead-containing media in generating localized atmospheric lead burdens is presented in this section. The phase-down and phase-out of leaded gasoline and the marked reduction in lead-emitting point sources in the United States and globally has served to significantly reduce inputs from traditionally dominant sources of lead to the atmosphere.

However, lead inputs to a more localized atmosphere can occur by reentrainment of lead from depositional sites such as roadway soils and dusts to the contiguous atmosphere. Other sources of reentrained dust lead are fugitive dusts mobilized from point source waste storage or persisting surface contamination. The long half-life of lead in these media assures that resuspended lead levels in the atmosphere will pose exposure problems for future decades.

The physical and chemical characteristics of this post-depositional phenomenon obviously differ significantly from those describing lead species from primary exhaust pipe or stack emissions. The physics of resuspension operates through two mechanisms, wind-based mobilization and mechanical disturbances. The latter arise via vehicular traffic. Sehmel (1980) and Nicholson (1988a) reported that mechanical stresses are more effective than wind-derived mobilization, but that arguably would depend on the source of mobilizable leaded dusts. Clearly, the miles of contaminated roadsides in the United States and other developed countries would be especially impacted by mechanical agitation arising from continuous traffic flow, while dusts present on stationary lead site surfaces probably would be more affected by wind speeds and frequencies.

Resuspended lead-containing dusts are not inconsequential for public health. One health issue with reentrained dust lead is that of exposure risks to adults by inhalation of particles or risks to young children from both inhalation and ingestion of resuspended and redeposited dust lead. Cowherd et al. (1985) calculated respirable (PM_{10}) levels of resuspended dust lead through the equation below:

$$R_{10} = \alpha E_{10} A$$

where R_{10} is emission rate of respirable particles as PM_{10} (mass/time), alpha (α) is the fraction (mass/mass) of contaminant in this respirable size range,

E_{10} is the PM_{10} emission factor (mass/source extent), and A is the source extent (area or volume).

Soil and dust particle mobilization by winds entails physical parameters such as saltation, surface creep, and suspension (Nicholson, 1988a; Sehmel, 1980). Saltation, relevant to particles in the size range of 100–500 μm, can move 50–75% of surface particles. Larger particles, 500–1000 μm in size, move by surface creep, up to 25% of all particles. Suspension, the process by which particles enter the ambient air, applies to a size <100 μm and involves up to 40% of soil by weight.

Vehicular resuspension arises from either shearing stress or vehicle-generated turbulence as a function of vehicle size, speed, moisture, and particle size.

5.2.3 Transformations Affecting Atmospheric Lead

Three areas of the atmospheric lead transformation process as they relate to lead dispersion are physical and chemical in nature. They include particle size distribution and chemical transformation processes for both the uncombusted organic air lead and inorganic lead from the exhaust stream in the case of mobile (vehicular) sources, and inorganic transformation of emissions from point sources. Whitby et al. (1975) characterized atmospheric lead particles as being of three size types based on modes of generation at sources or afterward: (1) the nucleation mode; (2) the accumulation mode; and (3) the large particle mode. The particle sizes, i.e., mass median diameters (MMDs) for these three types are <0.1, 0.1–2, and >2 μm, respectively. Both the nucleation and coarse particle types are generated proximate to the emissions point, but then there is a divergence. Large particles are deposited close to the emission source, while the nucleation type undergoes agglomeration to form the accumulation type. It is the last type that experiences long-distance dispersion. Numerous studies, summarized in U.S. EPA (1986, Ch. 6), show that both urban and rural ambient air contain lead predominantly in the fine particle size.

The fractional distribution of lead as a function of particle size has been reported and the data cluster around a relatively small range of values. Stevens et al. (1978) reported the ratios of fine to coarse particle lead measured for six U.S. urban areas: Glendora, CA; New York, NY; Philadelphia, PA; Portland, OR; South Charleston, WV; and St. Louis, MO. The corresponding ratios of lead distributions (measured in $\mu g/m^3$) into fine and coarse particles (F/C ratio) were 6.0, 5.6, 4.6, 3.4, 5.0, and 6.7. These ratios are consistent with the widely reported removal of large particles relatively soon after emission and a distribution skewed to small particle sizes (Pierson and Brachaczek, 1983). These authors reported data showing gutter debris lead particulate size ranges of 40–1,000 μm, supporting the assumption of early removal of the larger particles. Available evidence also supports the conclusion that the size ranges in urban and remote areas are quite similar, consistent with an early stabilization after emission and dispersion of lead particulate.

It has also been documented that much of the lead involved in long-range transport is in the submicron aerosol form. Lin et al. (1993) reported that in 1991, when leaded gasoline combustion was no longer the major source of atmospheric lead, the fine fraction of lead particulate had an MMD of 0.38 μm versus a coarse fraction average of 8.3. Earlier, Milford and Davidson (1985) reported an average MMD of 0.55 μm. It is the fine fraction that remains in atmospheric suspension and is borne great distances, including as far as the Polar Regions.

Dusts resuspended in localized atmospheres reflect both the distribution of lead species and particulate sizes in the dusts or soils serving as the source and the main mechanisms producing the resuspension. Street dusts have a bimodal distribution, the fine particle sizes arising from the tailpipes directly and the coarse materials from such mechanisms as smaller particle distribution. Al-Chalabi and Hawker (1997) reported that with increasing resuspension, the lead levels decreased, consistent with either coalescing small and coarser particles or increasing dispersion from the source. Similarly, Abu-Allaban et al. (2003) reported that road dust had most of its lead content in the coarse fraction.

Resuspended dust and soil entering the atmosphere produce varying lead concentration, especially in the respirable fraction. Lead bound in the PM_{10} fraction ranged from 0.012 to 1.2 ppm of bulk suspended soils, two orders of magnitude difference (Young et al., 2002). Airborne particulate lead level was up to 30-fold higher than the levels in road dust serving as the source, suggesting lead is easily mobilized into air (Tsai and Wu, 1995). Fine fraction lead in resuspended dusts was shown in the study of Thurston and Spengler (1985) to be up to 0.02 μg Pb/m^3, while an Armenian study (Kurkjian et al., 2002) credited resuspension for 75% of all atmospheric lead. In the South Coast Air Basin of California, Lankey et al. (1998) estimated that resuspension accounted for about 40% of total lead emissions to ambient air.

Table 5.1 presents some illustrative data gathered for California sites for lead content of resuspended materials in the respirable particulate fraction as a function of type of dust source. The levels range from 0.002% to 0.3% with agricultural soils presenting with the lowest lead concentrations, as might be expected. An urban (Fresno, CA) test site showed a level of 0.3%, consistent with the relationship of dust and soil lead content determining resuspended particle lead content.

Chemical transformations occur with lead species on entering the atmosphere or during transport within ambient air. They consist in the case of mobile (vehicular) sources mainly of (1) conversion of organic forms of lead to inorganic forms and (2) changes among inorganic forms. Organolead

TABLE 5.1 Percent Pb in Resuspended Particles of Differing Size in Various California and Nevada Locales[a,b]

Source	Locale	% Pb in PM_{10}	% Pb in $PM_{2.5}$	Reference
Paved road dust	Urban San Joaquin Valley, CA	0.016	–	Chow et al. (2003)
Paved road dust	Urban Fresno, CA	0.3	0.4	Chow et al. (1994)
Paved road dust	Urban Reno, Sparks, NV	–	0.01	Gillies et al. (1999)
Paved road dust	Rural San Joaquin Valley, CA	0.006	–	Chow et al. (2003)
Unpaved road dust	Rural San Joaquin Valley, CA	0.006	–	Chow et al. (2003)
Unpaved road dust	Residential San Joaquin Valley, CA	0.020	–	Chow et al. (2003)
Agricultural soil	San Joaquin Valley	0.002	–	Chow et al. (2003)
Playa dust	Rural Reno, Sparks, NV	0.006	0.001	Gillies et al. (1999)
Construction site	Urban Reno, Sparks, NV	–	0.001	Gillies et al. (1999)

[a] *Resuspended matter from diverse dusts and soils.*
[b] *Pb in particulate mass.*

compounds in gasoline, largely represented in past years by the ethyl and methyl alkyl lead additives, survived to some small extent during vehicular operation but in all cases were minor amounts compared to total air lead, ca. 1–6% (Harrison et al., 1979; Huntzicker et al., 1975; Purdue et al., 1973). Organolead antiknock additives such as tetraethyl and tetramethyl lead rapidly undergo photochemical decomposition when emitted with engine exhaust (Huntzicker et al., 1975). Cold-started engine emissions produce the highest levels of organolead additive.

The chemical species of inorganic lead emitted to the atmosphere varies with the type of lead emission. Auto exhaust emits lead as the halides or as double salts (Biggins and Harrison, 1978, 1979), in particular, PbBrCl and the salt with ammonium chloride, $PbBrCl \cdot 2NH_4Cl$. Initial chemical species in emissions from point sources such as smelters yield lead sulfate, the double salt $PbO \cdot PbSO_4$, and lead sulfide. Whatever the emission source, however, the dominant species in ambient air is lead sulfate.

The chemical composition of emitted forms provides some clues to source. For example, Pb:Br ratios can indicate auto exhaust lead but these halide forms do not persist (Dietzmann et al., 1981). Loss of halide over time to yield the sulfate may implicate acidic gas replacement or photochemical decomposition (Chang et al., 1977). Another factor in chemical composition is particle size. The smallest particles have lead as the $2PbBrCl \cdot NH_4Cl$ salt, while particles of 2–10 μm in size contain PbBrCl. Illustrative studies addressing chemical and physical forms of lead in air emissions are in Table 5.2.

5.2.4 Lead Removal from the Atmosphere

Lead exits the atmosphere through dry and wet deposition processes. Each mechanism for lead removal from ambient air has its own set of characteristics and differs in relative importance for impact on receiving environmental compartments and lead-exposed populations. The removal processes are reasonably well understood, particularly in terms of the physics of dry deposition (Friedlander, 1977; U.S. EPA, 1986). There are three zonal or layer elements in the dry precipitation process for lead removal: the main airstream, the boundary surface, and the receiving surface. Each of these zones is viewed in terms of aerodynamic resistance, boundary layer resistance, and surface resistance.

Dry deposition of lead is quantifiable by means of the downward flux, F, which is expressed in the equation (Davidson and Wu, 1990; Wu et al., 1992):

$$F = V_d C$$

where C is the airborne lead concentration in μg/m^3, and V_d is the deposition velocity. Deposition velocity is in units of meters per second, m/s. Particle size is the dominant factor in determining deposition rates for some given vertical height reference. Brownian motion governs deposition of very small particles at the surface border layer, while the largest particles are removed by sedimentation. The intermediate particle range deposition is mainly governed by impaction and interception (Nicholson, 1988b).

Dry deposition appears to mainly affect the largest fraction of particles. Lin et al. (1993) reported that only about 0.5% of particles undergoing dry removal are in the most respirable range, i.e., <2.5 μm in diameter. This percentage is much less than the fraction of this particle size in ambient air.

Various investigators over the years have developed dry deposition models with special regard to lead particle deposition velocities involving diverse

TABLE 5.2 Chemical and Physical Forms of Lead in Atmospheric Emissions

Emission Source	Pb Chemical/ Physical Form(s)	Testing Specifics	Reference
Ambient and confined roadway atmospheres	Particle size distributions differ with nature of atmospheric compartment: larger particles deposited roadside; smaller particles transported farther	Aerodynamic diameters measured in ambient versus vehicle tunnel	Pierson and Brachaczek (1983)
Ambient air	Pb in submicron vapor aerosols	Mass median diameters for fine and coarse fractions	Lin et al. (1993)
Ambient urban air	Organolead species comprise 1–6% of total ambient lead	Gas absorbers behind filters in air samples	Harrison et al. (1979)
Auto exhaust emissions to air and ambient air in England	Inorganic Pb emitted as double salts, e.g., $PbBrCl \cdot 2NH_4Cl$, conversion to Pb sulfate in air	X-ray diffractometry	Biggins and Harrison (1978, 1979)
Pb releases to the atmosphere from lead ore, sintering, and smelting in Missouri	Air lead from sintering and furnace emissions $PbSO_4$, $PbO \cdot PbSO_4$, PbS, some elemental Pb	X-ray diffractometry	Foster and Lott (1980); Spear et al. (1998)
Pb releases to air from burning waste oil	Inorganic Pb; emitted as $PbCl_2$, PbO, and elemental Pb	–	Nerin et al. (1999)
Tetraalkyl Pb in ambient air	Half-life of degradation to trialkyls differs with alkyl group: ~6 hours, TEL; 65 hours, TML	–	Nielsen et al. (1991)

surfaces, particularly vegetative surfaces (Davidson et al., 1982; Slinn, 1982).

The modeling of Davidson et al. (1985), using multiple vegetation measurements and wind speeds, indicates a deposition velocity in the range of 0.05–1.0 cm/s. Slinn (1982) reported a lower range of velocities, 0.01–0.1 cm/s, but the latter range was not adjusted for actual field data, as it was with the Davidson et al. (1985) model.

The relationship of dry deposition velocities to overall particle size appears to be logarithmic and V-shaped, showing slowing rates to a trough in velocity at a particle diameter of about 0.1–0.2 μm with increases either side of this trough (Main and Friedlander, 1990).

Wet precipitation covers the processes of removing lead through either rainout or washout. The former describes lead in particles already present in developing clouds, which also serve to promote formation of droplets and eventually raindrops. Washout, as the term implies, collects lead in particles by impaction and diffuses the particles by a rainfall event. The rate of removal through wet deposition, flux, is expressed by an equation broadly analogous to that for dry removal (Miller and Friedland, 1994):

$$F = V_p C_p$$

where V_p is the rate of precipitation in cm/s, and C_p is the concentration of the chemical species in the precipitation in μg/l.

As with dry precipitation, a number of factors govern the rate of wet deposition lead removal rates. These include particle size, seasonality, altitude, amount of precipitation, and thermal inversion (Conko et al., 2004; Davidson and Rabinowitz, 1992; Miller and Friedland, 1994). Input rates of lead to the atmosphere affect rainwater lead content. Table 5.4 presents water lead levels for various areas over a broad time frame. There is a clear trend to lower levels with lower anthropogenic impact, and a clear trend with time. Compared to the 1960s and 1970s when leaded gasoline consumption was at a maximum in the United States, later years showed dramatic drops in lead content.

Gatz and Chu (1986) reported that the great majority of lead in rainwater samples is in soluble form, about 83%, an observation of importance to the subsequent environmental fate of lead in receiving environmental compartments. Hicks (1986) reported that the amounts of lead removed by dry versus wet deposition phenomena differed greatly, ranging in ratio from 0.4 to 1.8.

5.3 FATE AND TRANSPORT OF LEAD IN SOILS

5.3.1 Transport of Lead to Soils

The lead content of soils will vary depending on factors such as the impact of anthropogenic activity, presence or absence of mineralized soils, the physical and geochemical characteristics of the soils, including pH, binding capacity, and the presence or absence of organic detritus. The role of anthropogenic activity, in turn, hinges on population density, proximity to mobile and point sources, and transport mechanisms bringing lead to soils. Anthropogenic contamination occurs by a number of mechanisms: (1) atmospheric lead transfer to soils by dry and wet deposition; (2) transfer of lead from urban and industrial surfaces through suspension or dissolution in

TABLE 5.3 Chemical and Physical Forms of Lead in Soils

Soil Type(s)	Physical/Chemical Forms	Testing Specifics	Reference
Soils with inorganic or organic ligands	Lead complexes with organic and inorganic ligands in solution; precipitates with insoluble inorganic or organic ligands; relative distribution of complexes (pH dependent)	Organic ligands—fulvic, humic acids; inorganic ligands Fe- or Mn-hydrous oxides; speciation based on charge separation	Camerlynck and Kiekens (1982)
Various soil treatments	99.8% of total Pb in soils not liberated by shaking	Dispersion tests for Pb mobility using various metals and phosphorus	Dong et al. (1985)
Contaminated industrial site soils	77% of total Pb in exchangeable or organic forms; 0% in aqueous solution	Organic content: 7–16%; Pb level, 64–360 ppm	Miller and McFee (1983)
Soils tested for Pb immobilization	Air Pb enters soils as sulfate and is then immobilized by precipitation via ion exchange with hydrous oxides and clays or organic acid chelation	Various binding systems	Olson and Skogerboe (1975); Zimdahl and Skogerboe (1977)
Soil types tested for Pb species	All Pb released to soils—$PbSO_4$, $PbCO_3$, PbS, PbClBr, etc.—is connected to adsorbed complexed forms; binding sites are clays, organic acids, hydrous iron oxide	Various binding systems	Chaney et al. (1988)
Soil Pb test systems	Soil Pb solubility dependent on soil pH; high solubility at pH ~ 5	Various chemical testing systems	Chuan et al. (1996)

runoff water; (3) transfer of lead to arable soils by lead-contaminated effluents or other waters used for irrigation; and (4) direct contamination by disposal of wastes containing lead. Illustrative studies for chemical forms of lead in soils are given in Table 5.3.

Soils comprise the most significant repository of emitted lead. First, the receiving soils have enormous capacity and a huge surface area available for emitted lead. Second, lead entering soil in most chemical forms undergoes relatively strong binding with only slow release. The half-life of lead in soils has been estimated to range from several decades to centuries (U.S. EPA, 1986, Ch. 2). In some cases, lead-enriched soils remain at former mining and smelting sites from ancient times.

It has long been understood that deposition of airborne lead from leaded gasoline combustion or from diverse point sources was a significant pathway for soil lead contamination (U.S. EPA, 1986; World Health Organization, 1995). More recently, the added role of lead transfer to receiving soils from storm water runoff generated by heavy rain flowing over and through heavily contaminated urban and industrial surfaces has been defined and partially quantified. Besides soil contamination, runoff water lead can be a factor in waterway and catchment basin contamination.

The loading rate of lead into soils is a complex function of air lead concentration and the interplay of dry and wet deposition processes for atmospheric lead and those fractions augmented by lead in runoff water. Rural soils appear to have taken up lead at an average annual rate of 0.26 ppm (Ewing and Pearson 1974), but this figure does not fully account for the high leaded gasoline combustion years (Shirahata et al., 1980). Where soil lead testings were done proximate to high traffic density, a much higher annual accumulation rate, 0.83 ppm Pb, was calculated.

5.3.2 Transport and Transformation of Lead Within Soils

Numerous studies have shown that lead entering the soil matrix becomes rapidly and tightly bound, with binding typically confined to the top several centimeters (see U.S. EPA, 1986, Chs 6 and 7 for illustrative citations) for soils which have been undisturbed. This propensity for surface binding has implications for lead exposure risk assessment as well as sampling protocols and sampling statistics. Children, for example, will come into contact with the surfaces of soils while playing, yet soil sampling that extends past vertical coring of several centimeters will produce a lower bias to the measured levels. That is, soil below the top several centimeters will not add to lead content but will add to mass, producing lower concentrations.

Lead within soils is distributed between solid and liquid phases, with the latter of major importance to the issue of lead bioavailability, for example, to plant roots where uptake can occur. Studies of lead species in this liquid mobile phase indicate that they exist as both complexed and ionic forms although the latter as simple ions are present in very low concentrations. The extent to which lead can move through soils, in turn, is the extent to which lead binds to insoluble organic and mineralogical inorganic species. The former are typically humic and fulvic acid derivatives, and the latter are surfaces of clays and metal oxides (U.S. EPA, 1986). The factors most important for lead movement within soils are pH, cation exchange capacity of soils, and the type and amounts of lead complexing agents, both organic and inorganic. Across the pH range, pH levels of 4–6 are most conducive to leaching lead from soil. A measure of the tight binding of lead to soil matter is the study of Dong et al. (1985), who showed that only about 0.2% of total soil lead content is readily released to a simple aqueous medium on shaking.

It is generally assumed that atmospheric lead (as opposed to runoff lead) enters soils as lead sulfate since atmospheric lead ages to this form (Olson and Skogerboe, 1975). Lead sulfate is soon converted to complexed forms entailing both inorganic and organic complexation. Available data indicate that inorganic and organic binding occur in about equal amounts, 43% versus 39%, respectively. One distinction that can be made here is binding of lead of natural origin in soils versus lead of anthropogenic origin (Elzahabi and Yong, 2001; Erel et al., 1997). Lead of industrial origin entering soils was shown not to be easily leachable, retaining 50–60% of the element regardless of extraction method (Cajuste et al., 2000). The remaining 40–50% was found to be present as carbonate and Fe- and Mn-bound fractions, along with lead in organic complexes.

Lead in soils can undergo transport through several physical and chemical forms. Karathanasis (2000) and Denaix et al. (2001) reported that lead transport on soil columns can occur through both dissolution and as colloidal suspensions, the latter in turn being variably affected by surface charge, pH, and organic carbon. About half of lead transport occurs through colloidal forms (Denaix et al., 2001). Lead leaching from soils is principally governed by pH of the pore water portion, lower pH being associated with increased leaching rates, especially of such lead forms as carbonates and hydroxides (Badawy et al., 2002). This propensity for leaching has implications at waste-site impacted soils, where, for example, the codischarge of lead residues with battery acid produces rapid movement of the lead plume through soils and potentially into groundwaters (Jensen et al., 2000).

The rate of lead movement through soils is of considerable importance in human health risk assessments for groundwater contamination and movement of lead into plant root zones. Migration rates, based on the previous discussions, are strongly influenced by factors such as pH, cation exchange capacity, and populations of adsorption sites. Sheppard and Sheppard (1991) reported a migration rate of 0.3 cm/year, while Miller and Friedland (1994) calculated rates of 0.11 and 0.29 cm/year for lead movement through forest organic horizons. Other studies estimated that anthropogenic lead will take 60–150 years to move through the organic horizon in such areas (Kaste et al., 2003).

5.4 FATE AND TRANSPORT OF LEAD IN DUSTS

Lead in dusts was discussed earlier as a factor in producing localized increases in ambient air lead from resuspension. Numerous studies have shown that dust lead generation itself arises from five major sources of input to environments likely to be of significance to human exposures: ambient air lead deposition to interior and exterior surfaces; interior and exterior paint lead deterioration; generation from soils and subsequent movement to interior or exterior surfaces; industrial activity such as the mining and milling of lead–zinc ores and generation of tailing piles and ponds; and inadvertent transport by workers of workplace dusts to their homes.

Dusts are unlike soils in that they have much finer particulate size distributions and differing chemical composition and are also likely to be considerably more mobile into and out of various environmental compartments. Friedlander (1977) describes dusts as products from disintegration of other materials, and this certainly would be the case with generation from soil particles or lead paint deterioration from exterior and interior lead-painted surfaces. Atmospheric or extractive industry dusts from tailings and dusts brought home from the workplace fit this characterization less well. In the former case, movement into the atmosphere or attachment to surfaces hinges on particle size. Workplace dusts brought home on workers' shoes, clothing, etc. would directly reflect the finer particles of these lead materials.

Leaded dusts can occur in soils, in the household, on hard exterior surfaces such as streets, and as occupational dusts in the specific context noted above. The ability of lead in dusts to rapidly accumulate on hard surfaces produces both a problem for quantification and a potent potential factor in human lead exposure assessments (U.S. EPA, 1986, Ch. 7). The physical and chemical compositions of dusts as well as the levels of lead are determined in large measure by their origin. It is also typical of dusts that they can deposit onto an almost infinite array of hard surfaces, and as a consequence can accumulate to quite high levels over time. This accumulation can be depicted quantitatively as either concentration or as dust lead loadings.

Concentrations of lead in outside dusts vary with their location and inputs. A summary of studies in U.S. EPA (1986, Ch. 7) shows that in rural environments at a time when leaded gasoline was in use, rural roadside dusts were on the order of 80–130 ppm, and urban roadways had dust levels in the range of 1,000–20,000 ppm. Interior household dusts in the United States can present an upper range of 5,600–16,000 depending on source inputs to these dusts such as deteriorating lead paint (Angle and McIntire, 1979; Clark et al., 1985) or lead smelter emissions (Landrigan et al., 1975).

5.5 FATE AND TRANSPORT OF LEAD IN PLANTS

The focus in this section is principally on terrestrial botanica, particularly those plants that comprise economically important agricultural commodities and are an important part of the human food supply. Attention here is on lead entry and deposition during plant growth rather than lead contamination occurring post-harvest or during processing and distribution. The latter aspects are dealt with in Chapter 6.

Plants may take up lead through foliar surfaces or through their roots. In areas of high atmospheric lead levels, foliar lead deposition can be the principal pathway for contamination that is also difficult to remove by ordinary household washing (Arvik and Zimdahl, 1974; Lagerwerff et al., 1973). As a general rule, plants grown in lead-contaminated soils have higher concentrations of the toxicant (Finster et al., 2004).

Soil pH appears to be the main factor affecting plant uptake of lead through the root system. Other parameters are the species of plant, the chemical form of lead, and diverse soil conditions (Dudka and Miller, 1999). Root and leafy vegetables appear to be the two types of food crop forms most prone to lead uptake (Dudka and Miller, 1999; Finster et al., 2004), while fruits and grains take up much less.

5.6 FATE AND TRANSPORT OF LEAD IN SURFACE AND GROUNDWATERS

5.6.1 Surface Waters

Lead enters surface water, freshwater, and salt water by direct deposition onto surfaces of lakes, impoundments such as dams and rivers or suspension in runoff waters, or discharges of lead-containing effluents. Runoff water lead can come from urban runoff or from soil erosion (e.g., Dolske and Sievering, 1979; Peltier et al., 2003). These runoff waters can enter soils, while lead in both stable water bodies and runoff waters can subsequently be transported to sediments. Nationally, proportionately more lead enters waters through atmospheric deposition than by other pathways although waterways with high lead loadings from industrial point discharges and urban runoff will show much higher lead concentrations close to the discharge points (Peltier et al., 2003).

The extent to which runoff water moves contaminating surface lead is strongly affected by a number of factors: (1) rainfall intensity and maximum inflow; (2) rainfall duration; and (3) the preceding dry period length. Rainfall intensity and the persistence of the rain event govern the volume of runoff, while the duration of the preceding dry period would be directly proportional to the accumulation of lead in diverse forms on surfaces.

The question of particulate versus dissolved lead content in runoff water was addressed by various researchers. Hewitt and Rashed (1992) noted that more than 90% of lead contained in highway runoff water was in the particulate phase. Davis and Burns (1999) and Davis et al. (2001) reported water runoff from lead-painted surfaces. Up to 93% of lead was in the particulate form. Table 5.4 summarizes data for lead species measured in various water samples.

TABLE 5.4 Chemical and Physical Forms of Lead in Waters

Water Type[a]	Chemical Forms	Test Specifics	Reference
Natural waters	Pb solubility in natural waters depends on pH and salt (carbonate, sulfate, phosphate) content	Equilibrium tests and calculations	Davies and Everhart (1973); Hem and Durum (1973)
River, stream waters	Undissolved Pb can be a significant fraction of total river Pb: colloidal Pb; Pb carbonate, oxide, etc.; Pb also sorbed to sediment particles	Cation exchange and other methods	Hem (1976); Lovering (1976)
Urban or rural streams	Major portion of stream-borne Pb in urban or rural waterways found in suspended solids: ~85% urban; 75% rural streams; remainder in filtrates	Filtration studies of bulk water samples	Getz et al. (1977)
Laboratory water systems	Fulvic acid has significant effect on Pb dissolution in waters, as does pH; fulvates binds Pb maximally at pH ~7.0 or ~10	Laboratory tests simulating natural waters	Lovering (1976)
Aquatic systems	Tetraalkyl Pb compounds, notably the methylated Pb unstable in aquatic systems, degrading to the trialkyl Pb ion; overall *in situ* formation of alkyl Pb species is minor	Simulated alkylating systems and alkyl measurements	Bëijer and Jernelöv (1984); Jarvie et al. (1981)
Urban surface runoff waters	Large amounts of Pb removed in rainwater runoff from roadways; only 5–10% is soluble	Roadway runoff water Pb transport studies	Harrison and Laxen (1981)
Laboratory tests of water systems	Pb forms especially strong complexes in water with humic acid and other organics	Binding studies	Denaix et al. (2001); Gao et al. (1999)

[a]*As term is used in reference(s).*

Up to 50% of roadway lead was removed during rainwater runoff in the Japanese study of Shinya et al. (2000). How much was removed depended on the amounts of lead on the surfaces prior to the rains and the length of the rain periods.

The fate and transport of lead within surface waters, as in soils, is a complex function of chemical and physical processes. Lead entering surface waters is diluted in concentration. Dilution occurs relatively quickly, its extent governed by the nature of the body of water (Kurkjian et al., 2004; Peltier et al., 2003) and the nature of the lead source. Schell and Barnes (1986) showed overall residence times for lead in lakes and reservoirs to be on the order of 77–250 days.

Schell and Barnes (1986) and Arakel and Hongjun (1992) described the modes of lead transport in aquatic systems. The predominant forms of lead that are bioactive in aquatic systems are lead oxide and lead carbonate, the element being adsorbed and transportable in colloidal species of clay minerals, Fe–Mn oxides and hydroxides, and suspended organic compounds such as humates and fulvates. Other transportable forms are soluble chelates and ions, or as particulates.

Lead solubility in waters tends to be under various control mechanisms. Two factors that control aquatic solubility are water pH and dissolved salt content. Davies and Everhart (1973) noted that, at pH above 5.4, solubility of lead is about 30 μg/l in hard waters and 300 μg/l in soft waters. Overall, soluble lead in rivers is in proportionately lesser amounts than the insoluble (i.e., colloidal) forms.

Lead is removed from the water column through transport to sediments, with sediment levels correlated with lead levels in the water (Kurkjian et al., 2004). Anthropogenic lead is removed from aquatic systems through adsorption, which is preferentially onto smaller rather than larger particles owing to the larger surface area for contact in the former. A second mechanism for lead removal to sediments is through binding with organic matters such as the humates (Rhoads and Cahill, 1999). Other removal pathways are through binding with sulfides and Fe–Mn oxides.

5.6.2 Groundwaters

Municipal and private wells account for a large fraction of drinking water sources. The vulnerability of groundwaters to lead contamination is relatively lower than for surface waters because of the terrestrial barriers existing for the former. The overall extent to which lead can migrate through soils to groundwater is contingent on various factors discussed in the section on soil lead. Groundwaters typically have neutral pH. Groundwater is also principally reflective of the surrounding bedrock in which the aquifer exists (U.S. EPA, 1986, Ch. 7).

5.7 FATE AND TRANSPORT OF LEAD IN SEDIMENTS

Sediments, like soils, serve as a major repository for lead entering sediments. This characterization resides in tight binding of lead as the carbonate, sulfate, and sulfide or as insoluble complexes adsorbed to particulate matter (Prosi, 1989).

The National Water Quality Assessment (NWQA) program of the U.S. Geological Survey has been examining lead and other metal contents of aquatic systems including sediments since about 1991 (see U.S. EPA, 2006, vol. II, Appendix AX-8). Sediment samples were gathered for the surficial layer and were sieved to $<63\ \mu m$ size. A total of 1,466 sediment samples were analyzed, showing an ambient mean lead level of 120 ppm, a median of 28 ppm, and a 95th percentile figure of 200 ppm. In terms of U.S. geographic trend data, sediment Pb increased from West to East, with "hot spots" identified in western states. Those localized elevations were in Idaho, Utah, and Colorado, with the majority associated with mining and mining-related activities.

REFERENCES

Abu-Allaban, M., Gillies, J.A., Gertler, A.W., Clayton, R., Proffitt, D., 2003. Tailpipe, resuspended road dust, and brake-wear emission factors from on-road vehicles. Atmos. Environ. 37, 5283–5293.

Al-Chalabi, A.S., Hawker, D., 1997. Response of vehicular lead to the presence of street dust in the atmospheric environment of major roads. Sci. Total Environ. 206, 195–202.

Angle, C.R., McIntire, M.S., 1979. Environmental lead and children: the Omaha study. J. Toxicol. Environ. Health 5, 855–870.

Arakel, A.V., Hongjun, T., 1992. Heavy-metal geochemistry and dispersion pattern in coastal sediments, soil, and water of Kedron Brook floodplain area, Brisbane, Australia. Environ. Geol. Water Sci. 20, 219–231.

Arvik, J.H., Zimdahl, R.L., 1974. Barriers to the foliar uptake of lead. J. Environ. Qual. 3, 369–373.

Badawy, S.H., Helal, M.I.D., Chaudri, A.M., Lawlor, K., McGrath, S.P., 2002. Soil solid-phase controls lead activity in soil solution. J. Environ. Qual. 31, 162–167.

Bëijer, K., Jernelöv, A., 1984. Microbial methylation of lead. In: Grandjean, P. (Ed.), Biological Effects of Organolead Compounds. CRC Press, Boca Raton, FL, pp. 13–19.

Biggins, P.D.E., Harrison, R.M., 1978. Identification of lead compounds in urban air. Nature 272, 531–532.

Biggins, P.D.E., Harrison, R.M., 1979. Atmospheric chemistry of automotive lead. Environ. Sci. Technol. 13, 558–565.

Burton, R.M., Suggs, J.C., 1984. Philadelphia Roadway Study. EPA Report No. 600/4-84-070. U.S. Environmental Protection Agency: Environmental Monitoring Systems Laboratory, Research Triangle Park, NC.

Cajuste, L.J., Cruz-Diaz, J., Garcia-Osorio, C., 2000. Extraction of heavy metals from contaminated soils: I. Sequential extraction in surface soils and their relationship to DTPA extractable metals and metal plant uptake. J. Environ. Sci. Health A A35, 1141–1152.

Camerlynck, R., Kiekens, L., 1982. Speciation of heavy metals in soils based on charge separation. Plant Soil 68, 331–339.

Chaney, R.L., Mielke, H.W., Sterret, S.B., 1988. Speciation, mobility and bioavailability of soil lead. Environ. Geochem. Health 9, 105–129.

Chang, S.-G., Dod, R.L., Giauque, R.D., Novakov, T., 1977. Photodecomposition of lead bromochloride. In: Rosen, H. (Ed.), Annual Report 1976-77. Lawrence Berkeley Laboratory, Berkeley, CA, pp. 74–82. As cited in U.S. EPA (1986, Ch. 6).

Chow, J.C., Watson, J.G., Houck, J.E., Pritchett, L.C., Rogers, C.F., Frazier, C.A., et al., 1994. A laboratory resuspension chamber to measure fugitive dust size distributions and chemical compositions. Atmos. Environ. 28, 3463–3481.

Chow, J.C., Watson, J.G., Ashbaugh, L.L., Magliano, K.L., 2003. Similarities and differences in PM_{10} chemical source profiles for geological dust from the San Joaquin Valley, California. Atmos. Environ. 37, 1317–1340.

Chuan, M.C., Shu, G.Y., Liu, J.C., 1996. Solubility of heavy metals in a contaminated soil: effects of redox potential and pH. Water Air Soil Pollut. 90, 543–556.

Clark, C.S., Bornschein, R.L., Succop, P., Que Hee, S.S., Hammond, P.B., Peace, B., 1985. Condition and type of housing as an indicator of potential environmental lead exposure and pediatric blood lead levels. Environ. Res. 38, 46–53.

Clift, D., Dickson, I.E., Ross, T., Collins, P., Jolly, M., Klindworth, A., 1983. Accumulation of lead beside the Mulgrave Freeway, Victoria. Search 14, 155–157.

Conko, K.M., Rice, K.C., Kennedy, M.M., 2004. Atmospheric wet deposition of trace elements to a suburban environment, Reston, Virginia, USA. Atmos. Environ. 38, 4025–4033.

Cowherd, C., Muleski, G.E., Englehart, P.J., Gillette, D.A., 1985. Rapid Assessment of Exposure to Particulate Emissions from Surface Contamination Sites. U.S. EPA Report No. EPA/600/8-85/002. Office of Health and Environmental Assessment, US Environmental Protection Agency, Washington, DC.

Davidson, C.I., Rabinowitz, M., 1992. Lead in the environment: from sources to human receptors. In: Needleman, H.L. (Ed.), Human Lead Exposure. CRC Press, Ann Arbor, MI, pp. 65–86.

Davidson, C.I., Wu, Y.-L., 1990. Dry deposition of particles and vapors. In: Lindberg, S.E., Page, A.L., Norton, S.A. (Eds.), Acidic Precipitation. Vol. 3, Sources, Deposition, and Canopy Interactions. Springer-Verlag, New York, pp. 103–216.

Davidson, C.I., Miller, J.M., Pleskow, M.A., 1982. The influence of surface structure on predicted particle dry deposition to natural grass canopies. Water Air Soil Pollut. 18, 25–43.

Davidson, C.I., Goold, W.D., Mathison, T.P., Wiersma, G.B., Brown, K.W., Reilly, M.T., 1985. Airborne trace elements in Great Smoky Mountains, Olympic and Glacier National Parks. Environ. Sci. Technol. 19, 27–35.

Davies, P.H., Everhart, W.H., 1973. Effects of Chemical Variations in Aquatic Environments: Lead Toxicity to Rainbow Trout and Testing Application Factor Concept. Report No. EPAR373011C. U.S. Environmental Protection Agency, Washington, DC. As cited in U.S. EPA (1986, Ch. 6).

Davis, A.P., Burns, M., 1999. Evaluation of lead concentration in runoff from painted structures. Water Res. 33, 2949–2958.

Davis, A.P., Shokouhian, M., Ni, S., 2001. Loading estimates of lead, copper, cadmium, and zinc in urban runoff from specific sources. Chemosphere 44, 997–1009.

Denaix, L., Semlali, R.M., Douay, F., 2001. Dissolved and colloidal transport of Cd, Pb, and Zn in a silt loam soil affected by atmospheric industrial deposition. Environ. Pollut. 114, 29–38.

Dietzmann H.E., Parness M.A., Bradow R.L., 1981. Emissions from gasoline and diesel delivery trucks by chassis transient cycle. In: Energy-Sources Technology Conference and Exhibition, Houston, TX, January. ASME Paper No. 81-DPG-6. American Society of Mechanical Engineers, New York.

Dolske, D.A., Sievering, H., 1979. Trace element loading of southern Lake Michigan by dry deposition of atmospheric aerosol. Water Air Soil Pollut. 12, 485–502.

Dong, A., Simsiman, G.V., Chesters, G., 1985. Release of phosphorus and metals from soils and sediments during dispersion. Soil Sci. 139, 97–99.

Dudka, S., Miller, W.P., 1999. Accumulation of potentially toxic elements in plants and their transfer to human food chain. J. Environ. Sci. Health B 34, 681–708.

Dzubay, T.G., Stevens, R.K., Richards, L.W., 1979. Composition of aerosols over Los Angeles freeways. Atmos. Environ. 13, 653–659.

Elzahabi, M., Yong, R.N., 2001. pH influence on sorption characteristics in the vadose zone. Eng. Geol. 60, 61–68.

Erel, Y., Veron, A., Halicz, L., 1997. Tracing the transport of anthropogenic lead in the atmosphere and in soils using isotopic ratios. Geochim. Cosmochim. Acta 61, 4495–4505.

Ewing, B.B., Pearson, J.E., 1974. Lead in the environment. In: Pilt, J.N., Metcalfe, R.L., Lloyd, A. (Eds.), Advances in Environmental Science and Technology, vol. 3. John Wiley & Sons, New York, pp. 1–126.

Finster, M.E., Gray, K.A., Binns, H.J., 2004. Lead levels of edibles grown in contaminated residential soils: a field survey. Sci. Total Environ. 320, 245–257.

Foster, R.L., Lott, P.F., 1980. X-ray diffractometry examination of air filters for compounds emitted by lead smelting operations. Environ. Sci. Technol. 14, 1240–1244.

Friedlander, S.K., 1977. Smoke, Dust and Haze: Fundamentals of Aerosol Behavior. John Wiley & Sons, New York, pp. 1–23.

Gao, K., Pearce, J., Jones, J., Taylor, C., 1999. Interaction between peat, humic acid and aqueous metal ions. Environ. Geochem. Health 21, 13–26.

Gatz, D.F., Chu, L.-C., 1986. Metal solubility in atmospheric deposition. In: Nriagu, J.O., Davidson, C.I. (Eds.), Toxic Metals in the Atmosphere. John Wiley & Sons, Inc., New York, pp. 391–408.

Getz, L.L., Haney, A.W., Larimore, R.W., McNurney, J.W., Leland, H.V., Price, P.W., et al., 1977. Transport and distribution in a watershed ecosystem. In: Boggess, W.R. (Ed.), Lead in the Environment. NSF Report No. NSF/RA-770214. National Science Foundation, Washington, DC, pp. 105–133.

Gillies, J.A., O'Connor, C.M., Mamane, Y., Gertler, A.W., 1999. Chemical profiles for characterizing dust sources in an urban area, western Nevada, USA. In: Livingston, I. (Ed.), Aeolian Geomorphology: Papers from the 4th International Conference on Aeolian Research, Oxford, UK, 1998. Z. Geomorphol. 116 (Suppl.), 19–44.

Harrison, R.M., Laxen, D.P., 1981. Measurements of gaseous lead alkyls in polluted atmospheres. Atmos. Environ. 15, 422–424.

Harrison, R.M., Laxen, D.P., Birch, J., 1979. Tetraalkyl lead in air: sources, sinks and concentrations. In: International Conference: Management and Control of Heavy Metals in the Environment. CEP Consultants, Ltd., Edinburgh, UK, pp. 257–261.

Hem, J.D., 1976. Geochemical controls on lead concentrations in stream water and sediments. Geochim. Cosmochim. Acta 40, 599–609.

Hem, J.U.D., Durum, W.H., 1973. Solubility and occurrence of lead in surface water. J. Am. Water Works Assoc. 65, 562–568.

Hewitt, C.N., Rashed, M.B., 1992. Removal rates of selected pollutants in the runoff waters from a major rural highway. Water Res. 26, 311–319.

Hicks, B.B., 1986. Measuring dry deposition: a re-assessment of the state of the art. Water Air Soil Pollut. 30, 75–90.

Huntzicker, J.J., Friedlander, S.K., Davidson, C.I., 1975. Material balance for automobile-emitted lead in Los Angeles basin. Environ. Sci. Technol. 9, 448–457.

Ingalls, M.N., Garbe, R.J., 1982. Ambient pollution concentrations from mobile sources in microscale situations. In: 1982 Passenger Car Meeting, Troy, MI, June. SAE Paper No. 820787. Society of Automotive Engineers, Warrendale, PA.

Jarvie, A.W.P., Markall, R.N., Potter, H.R., 1981. Decomposition of organolead compounds in aqueous systems. Environ. Res. 25, 241–249.

Jensen, D.L., Holm, P.E., Christensen, T.H., 2000. Leachability of heavy metals from scrap dirt sampled at two scrap iron and metal recycling facilities. Waste Manage. Res. 18, 367–379.

Karathanasis, A.D., 2000. Colloid-mediated transport of Pb through soil porous media. Int. J. Environ. Studies 57, 579–596.

Kaste, J., Friedland, A., Stürup, S., 2003. Using stable and radioactive isotopes to trace atmospherically deposited Pb in montane forest soils. Environ. Sci. Technol. 37, 3560–3567.

Kurkjian, R., Dunlap, C., Flegal, A.R., 2002. Lead isotope tracking of atmospheric response to post-industrial conditions in Yerevan, Armenia. Atmos. Environ. 36, 1421–1429.

Kurkjian, R., Dunlap, C., Flegal, A.R., 2004. Long-range downstream effects of urban runoff and acid mine drainage in the Debed River, Armenia: insights from lead isotope modeling. Appl. Geochem. 19, 1567–1580.

Lagerwerff, J.V., Armiger, W.H., Specht, A.W., 1973. Uptake of lead by alfalfa and corn from soil and air. Soil Sci. 115, 455–460.

Landrigan, P.J., Gehlbach, S.H., Rosenblum, B.F., Shoults, J.M., Candelaria, R.M., Barthel, W. F., et al., 1975. Epidemic lead absorption near an ore smelter: the role of particulate lead. N. Engl. J. Med. 292, 123–129.

Lankey, R.L., Davidson, C.I., McMichael, F.C., 1998. Mass balance for lead in the California south coast air basin: an update. Environ. Res. 78, 86–93.

Lin, J.-M., Fang, J.-C., Holsen, T.M., Noll, K.E., 1993. A comparison of dry deposition modeled from size distribution data and measured with a smooth surface for total particle mass, lead, and calcium in Chicago. Atmos. Environ. A 27, 1131–1138.

Lovering, T.G. (Ed.), 1976. Lead in the Environment. Professional Paper No. 957. U.S. Geological Survey, U.S. Department of the Interior, Washington, DC.

Main, H.H., Friedlander, S.K., 1990. Dry deposition of atmospheric aerosols by dual tracer method—I. Area source. Atmos. Environ. 24 A, 103–108.

Milford, J.B., Davidson, C.I., 1985. The sizes of particulate trace elements in the atmosphere-A review. J. Air Pollut. Control Assoc. 35, 1249–1260.

Miller, E.K., Friedland, A.J., 1994. Lead migration in forest soils: response to changing atmospheric inputs. Environ. Sci. Technol. 28, 662–669.

Miller, W.P., McFee, W.W., 1983. Distribution of cadmium, zinc, copper and lead in soils of industrial northwestern Indiana. J. Environ. Qual. 12, 29–33.

Mushak, P., 1992. PERSPECTIVE. Defining lead as the premiere environmental health issue for children in America: criteria and their quantitative application. Environ. Res. 59, 281–309.

Nerin, C., Domeno, C., Garcia, J.L., Alamo, A., 1999. Distribution of Pb, V, Cr, Ni, Cd, Cu and Fe in particles formed from the combustion of waste oils. Chemosphere 38, 1533–1540.

Nicholson, K.W., 1988a. A review of particle resuspension. Atmos. Environ. 22, 2639–2651.

Nicholson, K.W., 1988b. The dry deposition of small particles: a review of experimental measurements. Atmos. Environ. 22, 2653–2666.

Nielsen, O.J., O'Farrell, D.J., Treacy, J.J., Sidebottom, H.W., 1991. Rate constants for the gas-phase reactions of hydroxyl radicals with tetramethyllead and tetraethyllead. Environ. Sci. Technol. 25, 1098–1103.

Olson, K.W., Skogerboe, R.K., 1975. Identification of soil lead compounds from automotive sources. Environ. Sci. Technol. 9, 227–230.

Peltier, E.F., Webb, S.M., Gaillard, J.-F., 2003. Zinc and lead sequestration in an impacted wetland system. Adv. Environ. Res. 8, 103–112.

Pierson, W.R., Brachaczek, W.W., 1983. Emissions of ammonia and amines from vehicles on the road. Environ. Sci. Technol. 17, 757–760.

Prosi, F., 1989. Factors controlling biological availability and toxic effects of lead in aquatic organisms. Sci. Total Environ. 79, 157–169.

Purdue, L.J., Enrione, R.E., Thompson, R.J., Bonfield, B.A., 1973. Determination of organic and total lead in the atmosphere by atomic absorption spectrometry. Anal. Chem. 45, 527–530.

Reuer, M.K., Weiss, D.J., 2002. Anthropogenic lead dynamics in the terrestrial and marine environment. Philos. Trans. R. Soc. Lond. A 360, 2889–2904.

Rhoads, B.L., Cahill, R.A., 1999. Geomorphological assessment of sediment contamination in an urban stream system. Appl. Geochem. 14, 459–483.

Schell, W.R., Barnes, R.S., 1986. Environmental isotope and anthropogenic tracers of recent lake sedimentation. In: Fritz, P., Fontes, J.C. (Eds.), Handbook of Environmental Isotope Geochemistry. The Terrestrial Environment. B., vol. 2. Elsevier Science Publishers, New York, pp. 169–206.

Sehmel, G.A., 1980. Particle and gas dry deposition: a review. Atmos. Environ. 14, 983–1011.

Seinfeld, J.H., Pandis, S.N., 1998. Atmospheric Chemistry and Physics: From Air Pollution to Climate Change. John Wiley and Sons, New York.

Sheppard, S.C., Sheppard, M.I., 1991. Lead in boreal soils and food plants. In: Adriano, D.C. (Ed.), Metals in Soils, Waters, Plants and Animals: Proceedings of an International Conference, Orlando, FL, April, 1990. Water Air Soil Pollut. 57–58, 79–81.

Shinya, M., Tsuchinaga, T., Kitano, M., Yamada, Y., Ishikawa, M., 2000. Characterization of heavy metals and polycyclic aromatic hydrocarbons in urban highway runoff. Water Sci. Technol. 42, 201–208.

Shirahata, H., Elias, R.W., Patterson, C.C., Koide, M., 1980. Chronological variations in concentrations and isotopic compositions of anthropogenic atmospheric lead in sediments of a remote subalpine pond. Geochim. Cosmochim. Acta 44, 149–162.

Slinn, W.G.N., 1982. Predictions for particle deposition to vegetative canopies. Atmos. Environ. 16, 1785–1794.

Spear, T.M., Svee, W., Vincent, J.H., Stanisch, N., 1998. Chemical speciation of lead dust associated with primary lead smelting. Environ. Health Perspect. 106, 565–571.

Stevens, R.K., Dzubay, T.G., Russwurm, G., Rickel, D., 1978. Sampling and analysis of atmospheric sulfates and other species. Atmos. Environ. 12, 55–68.

Thurston, G.D., Spengler, J.D., 1985. A quantitative assessment of source contributions to inhalable particulate matter pollution in metropolitan Boston. Atmos. Environ. 19, 9–25.

Tsai, J.-H., Wu, Y.-L., 1995. Contributions of road dust resuspensions to the airborne particle concentrations in Taipei. Part. Sci. Technol. 13, 55–67.

U.S. Environmental Protection Agency, 1986. Air Quality Criteria for Lead. 4 vols. Report No. EPA/600/8-83/028bF. Environmental Criteria and Assessment Office, Washington, DC (Chapter 6).

U.S. Environmental Protection Agency, 2006. Air Quality Criteria for Lead, vol. 1. Report No. EPA/600/R-05/144 aF. National Center for Environmental Assessment, Washington, DC (Chapters 2 and 3).

Whitby, K.T., Clark, W.E., Marple, V.A., Sverdrup, G.M., Sem, G.J., Willeke, K., et al., 1975. Characterization of California aerosols—I. Size distributions of freeway aerosol. Atmos. Environ. 9, 463–482.

World Health Organization, 1995. Environmental Health Criteria 165: Inorganic Lead. International Programme on Chemical Safety, Geneva Switzerland.

Wu, Y.-L., Davidson, C.I., Dolske, D.A., Sherwood, S.A., 1992. Dry deposition of atmospheric contaminants: the relative importance of aerodynamic, boundary layer and surface resistances. Aerosol. Sci. Technol. 16, 65–81.

Young, T.M., Heeraman, D.A., Sirin, G., Ashbaugh, L.L., 2002. Resuspension of soil as a source of airborne lead near industrial facilities and highways. Environ. Sci. Technol. 36, 2484–2490.

Zimdahl, R.L., Skogerboe, R.K., 1977. Behavior of lead in soil. Environ. Sci. Technol. 11, 1202–1207.

Chapter 6

Lead Concentrations in Environmental Media Relevant to Human Lead Exposures

6.1 LEAD AS MULTIMEDIA POLLUTANT

Figure 5.1 in the previous chapter presented a graphic depiction of lead as a substance which enters and departs multiple environmental compartments, multiple environmental media, with relative physical and chemical ease. This recognition and graphic characterization of lead as a multimedia pollutant is of recent vintage, dating to the 1977 U.S. EPA Air Quality Criteria for Lead report (U.S. EPA, 1977), the first of a series of Federal lead documents, and the 1980 report to the National Commission on Air Quality by Mushak and Schroeder (see also Chapter 25). Subsequent expert consensus treatises such as EPA's later Air Quality Criteria Documents for Lead (U.S. EPA, 1986a, 2006), the U.S. CDC Statements on Childhood Lead Poisoning (U.S. CDC, 1985, 1991, 2005), the NAS/NRC (1980, 1993), and the WHO (1995) helped establish the nature and extent of lead's multimedia behavior.

Lead's behavior as a multimedia pollutant poses problems for human health risk assessment and regulatory science at several levels, a number of which are presented in detail in later chapters. First, it is important to establish the full extent of lead exposures of human populations, especially those subsets of the population at elevated risk for exposure and/or harm. Establishing the extent of exposure mainly includes identifying and quantifying lead intakes and uptakes into the human body.

A second important factor for lead as a multimedia pollutant is the recognized toxicokinetic and toxicological fact that all sources of lead intake and uptake by human populations contribute to a single, integrated internal toxicological exposure or "dose." It is not biologically or toxicologically required that a single source of lead provides all of the potentially toxic lead exposure, merely that it provides a measurable contribution to the total. The body, furthermore, does not typically preserve the chemical form of the lead entering the receiving compartments, for example, the gastrointestinal and

Trace Metals and other Contaminants in the Environment, Volume 10
ISSN: 1875-1121 DOI: 10.1016/B978-0-444-51554-4.00006-7

the respiratory tracts. Once absorbed, a given quantity of lead is toxic to the same extent, regardless of its environmental chemical or physical form.

A third factor in assessing lead in multiple contact media and those associated lead exposures in human populations is that lead is biologically cumulative and is significantly cumulative over time in human populations. Lead is cumulative in humans principally because of its propensity to lodge in skeletal mineral tissue with time, growth, and age. The accumulation begins in childhood and continues well into adulthood (Mushak, 1993, 1998; U.S. EPA, 2006). However, lead is unlike methyl mercury, in that it does not biomagnify across species or trophic orders as one goes up the food web.

There are two consequences for risk assessment of this potential for accumulation in bone. First, lead lodged in bone can be mobilized under diverse physiological and other stresses, e.g., pregnancy, lactation, menopause, physical immobilization (Gulson et al., 1995, 1997; Markowitz and Weinberger, 1990; Silbergeld et al., 1988), and it subsequently enters the bloodstream, producing an "endogenous" pool of lead for inducing toxic effects. Secondly, lead's systemic accumulation places a premium on quantification of lead intakes and uptakes over much of the lifetime of human populations, since the main bone depository for lead, cortical bone, has a half-life for lead release of up to several decades from that compartment (Rabinowitz et al., 1976).

This chapter examines the levels of lead in environmental media with which human populations interact by such means as ingestion and inhalation. Quantification of lead levels in media not only applies to the ongoing and near-term case, but also requires a depiction of the environmental lead picture over population lifetimes. If lead deposited in bone 30 years ago can be released and produce harm, one needs to evaluate as best as possible the magnitude of lead intakes via various media 30 years ago. This would particularly be the case in attempting predictive modeling of lead exposures that occurred decades ago.

For measuring lead in environmental media providing potential human lead exposures, this chapter includes older published data for lead concentrations in media, data which are old enough to encompass the full lifetimes of living populations. This is because of long-term Pb storage in bone. One concern with any appraisal of older lead measurement data in media is that of analytical and statistical data reliability versus that of methods employed with more recent accepted techniques. Sensitivity is of particular concern. A potent toxicant such as environmental lead requires methods for quantification of concentrations of lead at ultra-trace levels in order to permit estimates of the full range of Pb exposures.

Evolution of methodologies for measuring lead in media and associated statistical analysis require a brief discussion of current methods in perspective. Methods have greatly improved over the years, so that one must confine reliance on older data to those likely to be most reliable.

6.2 SUMMARY OF SAMPLING AND LABORATORY ANALYTICAL METHODS FOR ENVIRONMENTAL LEAD

Environmental media of interest in this section on measurement are the same as those producing potential human lead exposures: ambient air, lead paints, diet, drinking water, soils and dusts, and some of the more problematic idiosyncratic sources. Sampling and laboratory measurement techniques now widely used are emphasized with comparative statements for older methods provided mainly to offer perspective. Biomarker sampling and measurement methodologies, i.e., procedures for lead in biological media directly relevant to human lead exposures, are presented in a later chapter.

Lead has been and continues to be so pervasive in human environments that it poses special challenges for sampling and analysis of either environmental or biological lead when present in extremely small amounts. Because of the contamination problem, no serious attempt at reliable lead measurements in environmental media can be done without rigorous quality assurance and quality control (QA/QC) protocols. For regulatory compliance with enforceable U.S. standards, such QA/QC steps are explicitly prescribed in order to have legal meaning (Code of Federal Regulations, CFR, 1982, 40:§58).

6.2.1 Analysis of Lead in Ambient Air

Sampling for ambient air lead measurements is quite complex in implementation and interpretation. Furthermore, that sampling complexity is arguably greater than it is for most other lead-containing environmental media. This arises partly from the nature of how human populations encounter lead in air and lead's fate and transport from points of emission. This section also summarizes some changes in sampling and methodology that have accompanied changes in lead emission sources. Specifically, lead from mobile sources and air lead analyses to quantify this source contribution have declined significantly in the last 15–20 years while stationary or point lead sources have increased in relative significance. The magnitude of these changes was noted earlier.

Methodologies for air lead sampling and laboratory analysis are limited to specific reference methods officially prescribed to accommodate the fact that air lead analysis in the United States is directed to, among other purposes, compliance with an ambient air lead standard within the framework of State Implementation Plans (SIPs). SIPs are the regulatory and legal means by which the various states implement the nationally enacted ambient air lead standard. The details are discussed in such treatises as Chapter 4 of the 1986 U.S. EPA Air Quality Criteria for Lead document (U.S. EPA, 1986a) and Chapters 2 and 3 of U.S. EPA, 2006. The required reference methods for legal enforcement are codified in the CFRs (CFR, 1982, 40:§58).

Three factors govern air lead sampling: (1) site selection; (2) appliances used for air sampling and form of lead being sampled; and (3) sample preservation prior to laboratory analysis. For any area's site sampling for airborne lead, some minimum number of sampling stations are required, depending on both population category and ranges of total suspended particulate (TSP). Air lead quantification reports, from earlier times through the present, typically reported air lead concentrations based on TSP. Other criteria pollutants have adopted size-selective sampling techniques with an eye to focusing on the most readily respirable particulate forms of the pollutants. These requirements mainly applied to those earlier years when the predominant contributor, on the order of 90–95% of total air lead, was leaded gasoline combustion.

Up to 6 to 8 air-monitoring stations were spelled out for areas with populations exceeding 500,000 and where prior testings showed particulate air levels within the "high" range, that is, when the TSP level exceeded 20% of the TSP standard. At the other extreme, areas with populations of only 50,000–100,000 and where the TSP level is less than the ambient air standard require no monitoring sites (Appendix D, CFR, 1982, 40:§58).

Ambient air lead partitions into vertical gradients, especially near mobile lead sources, e.g., vehicular exhaust from leaded gasoline combustion. There is special emphasis on those heights above source relevant to human lead exposures. Other specific monitor locating requirements include specifications for distances from roadways (5–100 m) and distances from obstacles between the monitor and the emitting source (Appendix E, CFR, 1982, 40:§58).

Ambient air sampling within typical regulatory and other contexts uses a high-volume ("hi-vol") aerosol sampler. Other collection devices may include filters, impactors, and impingers (U.S. EPA, 1971). This overall approach is based on the existence of lead in ambient air as largely inorganic particulate matter rather than in predominantly vapor form. Quantification using this sampler is as micrograms per cubic meter of air ($\mu g\ Pb/m^3$).

The dichotomous and impact samplers are other devices for air lead monitoring. The former collects particulate and segregates it into two size ranges: 0–2.5 μm and 2.5 to the maximum opening of the intake port, typically 10 μm (Loo et al., 1979). These ranges, generally conforming to fine and coarse particle categories, respectively, provide a close approximation to those particulate sizes that are relevant to respirable lead fractions and that fraction entering the deep pulmonary compartment.

Impactor samplers, which permit a broader range of measurable particulate size distributions, are constructed so that air entering at high flow rates encounters a series of trapping surfaces and particles are trapped based on size. First trapped are the coarsest particles followed sequentially by smaller and smaller fractions. The cascade samplers typically allow fractionations within narrow ranges, such as 0.01–0.1 μm (Dzubay et al., 1976).

Laboratory analysis of air lead contained on various sampler trapping materials typically uses reference methods, i.e., methods taken to be particularly reliable, that have relatively reliable track records and which have been codified for use to ascertain compliance with existing enforceable standards in the case of regulatory legal compliance testings (CFR, 1982, 40:§58).

The approved reference method for enforcement purposes (CFR, 1982, 40:§58) uses hi-vol samplers and measures lead by atomic absorption spectrometry (AAS). This laboratory method, which has been available in various analytical configurations for several decades, has been shown to be particularly reliable and sensitive for measuring lead quantitatively in a large range of environmental media. Flameless AAS is a more sensitive variation of this technique than conventional flame methods and has been the choice for many years. As with any lead measurement method, sample handling must minimize both contamination with lead and loss of lead from the sample. Comparatively, the contamination problem is still the more problematic and this is certainly the case for analyses in U.S. urban areas and in other industrialized nations (NAS/NRC, 1993; Patterson, 1983; Settle and Patterson, 1980). For air sample analyses, the codified reference method using AAS is quite adequate for a wide range of air lead concentrations.

Other methodologies fall into the category of definitive or alternative reference methodologies. The definitive method for lead, against which other methods are qualified for reference use, and the one employed for standard sample lead certification by the National Institute of Science and Technology (NIST), is isotope-dilution mass spectrometry (IDMS). Among its cardinal virtues, in addition to accuracy and precision, are sensitivity and applicability to many lead-containing environmental matrices.

6.2.2 Analysis of Lead in Paint

Lead in paint remains a major source of lead in human environments in terms of the U.S. national picture, and paint lead remains the dominant general lead source for humans residing in inner city, densely populated neighborhoods having a high fraction of deteriorated housing. Lead in paint, like lead in ambient air, is defined as a source of lead that works through pathways to provide human lead exposures. Shared pathways from these originating lead sources include interior and exterior dusts and yard soils. It is therefore appropriate to discuss this lead source with air lead and prior to discussions of lead analysis in pathway media such as soils, dusts, plants, water, and food. Paint lead also shares the characteristic of sampling complexity with air lead and, equally important, much of paint lead testing and quantification of lead in paint lies within regulatory and legal constraints that prescribe certain sampling and analysis protocols. Lead paint testings and assessments as part of risk assessment for U.S. housing and in other frameworks lie within the jurisdictions of two U.S. agencies. The principal agency having oversight of lead paint-containing U.S. housing units is the U.S. Department of Housing and Urban Development (U.S. HUD, 1995), with the U.S. EPA sharing statutory mandated responsibility (U.S. EPA, 2001).

Sampling lead-painted surfaces is prescribed for the purpose of "... a surface-by-surface investigation to determine the presence of lead-based paint ... ," the latter being present when any measured lead paint content in terms of lead loading is $\geq$1.0 mg/cm^2 or has a concentration $\geq$0.5% lead by weight (CFR, 2001, 40: Part 745; CFR, 1996, 24: Part 35). Two types of residential units are covered in these regulations, single family units and multifamily units. In the latter case, statistical formulae are used to randomly select a fraction of all the units broken into two categories of multiunit age ranges, pre-1960 or 1960–1977. This is necessary since it usually would not be feasible to test the entire tally of units at, for example, an apartment complex. Table 7.3 of U.S. HUD (1995, Ch. 7) sets forth how many units are to be tested at a multiunit complex as a function of complex age.

Testings of lead-painted surfaces typically begin with *in situ* examination using portable X-ray fluorescence (XRF) spectrometers that have been calibrated and are of an acceptable type as prescribed by U.S. HUD regulations. There are statistical protocols that employ random sampling techniques that apply for either single residence or multifamily residences (U.S. HUD, 1995). Documenting surface samplings typically includes testing forms that spell out both the area of a residential unit tested, such as a living room, and components within that area, e.g., baseboards, window sills/sashes/wells.

A reading will produce one of three results: (1) a level below which a reading is considered negative; (2) an intermediate range where the result is deemed inconclusive; and (3) a reading which is positive, i.e., exceeds the inconclusive reading. Surfaces with inconclusive readings can be also examined by collecting paint chips at the same surface to determine whether the 0.5% lead chip standard has been exceeded.

Surfaces to be sampled are those often found to either be higher in paint lead content and/or within easy reach of those most likely to have exposure, i.e., infants and toddlers. They especially include window components—sills, frames/jambs, wells—easily accessible surfaces that have been commonly associated with lead exposures of very young children. Surfaces for testing besides painted areas are those that are varnished, stained, shellacked, or painted surfaces under wallpaper.

Frequency or other criteria for lead paint surface testing depends on the purpose of the effort. U.S. HUD (1995, Ch. 7, revised 1997) spells out two types of lead paint measurement protocols for paint surface testing. These are lead paint inspection testings and risk assessments, the latter to identify what HUD terms lead paint "hazards." The term "risk assessment" carries a regulatory definition (CFR 40: Part 745, 2001) focused on lead paint hazards. Lead paint inspections deal with determining the presence of lead paint and usually test many more areas than lead paint hazard risk assessments. Both types of surface testing have gathering paint chips and how to collect the paint chips in their protocols, but the number of chips required is fewer with the risk assessment. This is because the risk assessment protocol is directed to chip sampling only in areas of deteriorated paint surfaces. In addition, as noted in more detail later, risk assessment of lead paint hazards entails lead paint dust testings and lead paint-impacted soil lead measurements.

There are other circumstances for which one or the other of the two types of surface testings is done. For example, collection of paint chip samples and laboratory testing are recommended for inaccessible or irregular surfaces that provide problems for XRF surface testings. *In situ* testing, on the other hand, is nondestructive and does not entail paint removal from portions of surfaces. Sampling for lead paint using chips, however, requires that the samples represent all layers of lead paint that could be present. Chips that are mainly surficial represent more recent, and usually lower lead or nonlead paint films. Surface delamination (broad surface peeling from poorly prepared surfaces), for example, produces chips that do not capture the earlier history of the entire paint depth on the residential surfaces. Such sampling would potentially result in understatements of likely lead exposure of child residents or visitors. The most common metric for paint chip lead measurement is percent by weight. To express lead content by area rather than percent by weight, a measured area of paint surface is typically obtained.

Portable XRF testing is the conventional method for *in situ* lead paint testings on older painted surfaces. The XRF method records lead present in all layers in the painted surface via full depth penetration of the X-rays. Irradiated lead from X- or gamma rays produces X-rays at a characteristic frequency or energy and at an intensity proportional to the integrated total amount of lead present throughout the irradiated layers. Legally prescribed performance by acceptable portable XRF analyzers is specifically spelled out by both the U.S. HUD and U.S. EPA with the inspector's use of an XRF Performance Characteristic Sheet. This form describes results in the context of analytical findings of negative, inconclusive, or positive readings. Where readings are less than either the threshold or the lower boundary of "inconclusive" the result is taken as negative, i.e., legally defined lead paint is absent. Similarly, the readings that exceed the upper boundary of the inconclusive range are taken as positive. The XRF Performance Characteristic Sheet also sets forth guidelines for calibration, substrate issues and corrections, etc.

Laboratory analysis of lead in paint chips should only be carried out by those laboratories certified for such analyses by the U.S. EPA National Lead Laboratory Accreditation Program (NLLAP), using approved laboratory methods. Use of AAS methodology, particularly in the ICP-AAS configuration, is the common approach. Another routine method is anodic stripping voltammetry (ASV).

The American Society for Testing Materials (ASTM) has provided validated procedures for quantification of lead in various types of paint samples. ASTM E 1645 prescribes preparation of paint lead chips for analysis while methods ASTM E 1613, ASTM E 1775, and ASTM PS 88 are directed to measurements of lead. The NIST has made paint standard reference materials (SRM) available for calibration and validation of lead paint measurement methodologies. SRM 2579 is available as a paint film while SRM 2589 consists of paint samples collected from interiors of various homes and prepared as a homogeneous powder, $\sim$100,000 ppm (10%) lead, with >99% of the particles being <100 μm in size.

6.2.3 Analysis of Lead in Soils

The variety of sampling approaches for soil lead analyses is as complex as for lead paint. The type of sample taking for soils depends on the purpose of the analyses. Residential, industrial, and public area soils are often analyzed for regulatory, legal, risk assessment, or scientific research purposes. Soils may also be tested to quantify the amounts of lead present, to determine the source(s) of lead in the soils, and, in certain cases, to ascertain the chemical and physicochemical species of lead in the soils to ascertain both likely source and relative bioavailability of lead forms present.

A number of methodological and statistical criteria govern the actual physical nature of the soil sampling. There is the matter of spatial sampling, where one or more surface area soil samples are collected. Collection may entail grab sampling, where one sample is collected, or composite sampling, where subsamples are gathered and combined prior to analysis. Subsamples may also be analyzed individually before combining to provide a composite value. Compositing is typically done where there is some information available about the likely nature of the lead source providing the contamination and its distribution. That is, is the soil lead relatively uniform in distribution or heterogeneous, with one or more potentially troublesome "hot spots" that can produce human lead exposures (U.S. EPA, 1989; U.S. EPA 2008a, Child-Specific Exposure Factors Handbook)?

Soil sampling may also entail spatially variable collection in area or vertical depth. For example, if one is interested in whether exterior lead paint has weathered and released to adjacent soils, one would collect foundation perimeter, i.e., "drip-line" samples. If one is interested in whether roadways are affecting property front soils, samples can be collected at the curb. A critical factor in soil sampling is depth of sampling. Soil lead arising from atmospheric dry and wet deposition will typically deposit on the top 1–2 cm of soil surface. Absent other lead sources, deeper soil strata will have little lead content. Sample cores which are gathered well below this 1–2 cm depth have the potential to statistically dilute the soil lead loading with reference to lead exposure of young children who play in such soils and only come in contact with the surface soil through typical child hand–mouth activity. Agricultural soils may require greater sampling depth or include deeper cores in multidepth sampling if one wishes to ascertain the subhorizon lead concentrations in crop root zones. Such deeper samplings, in tandem with surface testing, may also be required if Pb-impacted soils are tilled and turned over.

A critical factor in soil sampling is the question of lead content as a function of soil particle size distribution. Anthropogenic lead contamination of soils often entails deposition of lead particulate or suspended lead from runoff entering soils (see preceding chapter). The lead content of resulting soil particulate forms is often inversely proportional to particle size, owing to increasing surface to volume ratio with decreasing particle size. Larger surface areas permit more lead binding for a given mass. A second factor is the variability with particle size of lead exposures for children contacting leaded soils (Mushak, 1991). The smaller the lead-containing soil particle, the higher the propensity for adhering to hands and the higher the likelihood of sticking to clothing, shoes, etc. for transport indoors and later potential ingestion (see later chapters).

Bulk soil samples which have not been sieved into different size fractions have the potential to underestimate lead content relative to those smaller particle fractions which have the higher lead exposure potential. That is, larger particles will have lower lead content but may contribute substantially to sample mass. Other mass determination concerns include the need to exclude large-sized detritus and organic matter, since the latter contributes to sample mass but has questionable relevance as a lead exposure medium.

There are a variety of field and laboratory analytical methods for soil lead measurement, depending on the type of analysis and its purposes in a given evaluation. Bulk soil lead measurement refers to measurement of the total lead content of the soil sample. Chemical speciation and micromineralogical studies in the context of human lead exposure variability refer to amounts of specific chemical forms of lead and their geochemical states. These studies are sometimes done in tandem with relative bioavailability testings, i.e., amounts of lead being absorbed under *in vivo* or *in vitro* simulation of *in vivo* conditions (Casteel et al., 2006) with respect to Pb source attribution. Stable isotopic analysis studies deal with the quantitative stratification of lead's stable isotopic composition into the four main stable isotopes: lead-204, lead-206, lead-207, and lead-208 (Gulson et al., 1995, 1997).

Field measurement of bulk soil lead by XRF instruments will typically require confirmation analysis through some randomly selected subset of further testing by some reference technique in the laboratory: AAS, inductively coupled plasma-atomic emission spectroscopy (ICP-AES), or ICP-mass spectrometry (ICP-MS). Other methods are electrochemical in nature, such as ASV and differential pulse polarography. Many soil samples are processed and analyzed directly in the laboratory.

Detection limits for lead in soil matrices have markedly improved over the last several decades owing to improvements in the signal detection systems, e.g., charge-coupled devices in place of photomultiplier tubes.

U.S. EPA (2001) holds ICP-AES or ICP-MS to be the soil lead measurement methods of choice. Detection limits are on the order of 40 parts per billion (ppb) which are quite adequate for an environmental medium where even background, i.e., uncontaminated soil lead, concentrations are on the order of 10–20 ppm.

A number of soil-derived SRMs are available for QA/QC use in the quantitative analysis of lead in soils. The SRM numbers, their corresponding matrix type and lead level (mg/kg, ppm) are 2709, soil, 18.9; 2710, soil, 5532; 2711, soil, 1162; 2586, soil (paint), 432; 2587, and soil (paint) 3242.

Analytical speciation methods, referred to earlier, generally differ from the methods adequate for bulk quantification of lead. As with bulk methods, nonetheless, contaminating levels should not be permitted in interfering amounts. Several lead speciation approaches are X-ray absorption spectrometry, X-ray diffraction, and electron-microprobe microanalyses (U.S. EPA, 2006; Welter et al., 1999).

6.2.4 Analysis of Lead in Dusts

Dust lead arises through transport mechanisms for lead in originating sources, typically lead in paint (Lanphear et al., 1996) and/or ambient air lead deposited onto various surfaces (Adgate et al., 1998). Ambient air lead from both mobile and stationary sources produces interior dust lead (U.S. EPA, 1986a, Ch. 7). Interior dust lead from these input media is of particular concern for childhood lead exposures. Exterior dust lead may also elevate lead exposures from outside toys, play sets, etc. Dust lead can arise from soil lead, which is an environmental medium pathway that receives lead via wet and dry deposition from atmospheric lead or exterior lead paints weathering onto adjacent soils (von Lindern et al., 2003a). A fourth generator of interior dust lead is clothing, shoes, etc. of family members or others who work in leaded settings including battery production and lead secondary smelting and wear their work clothes home. This is termed secondary or "take-home" lead exposure.

Dust sampling for lead uses two different metrics: lead concentration or lead loadings. Each method has its advantages and drawbacks. Determining lead concentrations in dusts typically requires a gravimetric approach, where the mass of lead in collected dusts is quantified from the mass of total dust amounts. Dust lead loadings are done by collecting leaded dusts within a predetermined surface area, e.g., per square meter or per square foot. The lead concentration is influenced by the presence of nonlead materials and housekeeping efficiency differences from site to site. Lead loading per unit area is not affected in this manner (see, e.g., Milar and Mushak, 1982; U.S. EPA, 2001). In addition, leaded dust regulatory standards are currently expressed as loadings per square foot. However, expressing lead content as a loading per unit area makes it difficult to draw comparisons with lead in diverse other media where concentration is expressed, and it is relatively difficult to use lead dust loading for lead exposure modeling of human exposure populations (Leggett, 1993; Mushak, 1998; O'Flaherty, 1998).

Surface or media sampling for dust lead is as complex as it is for lead paint, owing to the number of sampling variables. Dust by its nature is relatively variable mass-wise as a lead-bearing environmental medium compared to lead in paint or lead in soil. For example, the amount of dust accumulation or lead accumulation in dusts is significantly affected by the residence times of the dusts. Housekeeping practices will affect dust sampling measurements within a given home and between homes for samplings of dusts in active contact areas. Surfaces that are relatively undisturbed over extended periods of time—in attics, basements, inaccessible living area surfaces—provide a longer lead accumulation record than areas routinely cleaned.

Other parameters have to do with the purposes of the dust sampling, especially of residential unit exteriors. Dust wipe sampling procedures as part of lead-abatement guidelines or requirements of the U.S. EPA (2001) and U.S. HUD (1995, 1999) are explicitly prescribed. Regulatory lead dust testing to indicate current levels of lead contamination for purposes of human health risk assessment or for setting cleanup levels at Superfund and other contaminated sites are similarly prescribed (von Lindern et al., 2003b). Research projects dealing with dust lead measurements, by contrast, are not as constrained by regulation as by requirements of the study, for example, determination of dust lead loadings versus dust lead concentrations.

Interior paint lead dusts to be sampled in the context of regulatory requirements for either carrying out lead paint hazard risk assessment or for determining the efficacy of lead paint abatement or interim controls are described in such treatises as the 1995 U.S. HUD guidelines for lead paint. Dust lead sampling inside residences with leaded paint surfaces is best done in areas commonly contacted by young children, especially infants and toddlers. These include play areas within rooms, high-traffic commons areas such as hallways, or, particularly important, surfaces beneath windows and window components with leaded paint. Especially problematic for lead exposures are those windows that are frequently opened and closed, and that young children frequently contact.

As set forth in U.S. HUD (1995), dust sampling by use of surface wipes is preferred on the basis of ease of use, relative cost, and a generally favorable performance record for routine samplings. For example, children's blood lead levels and dust wipe lead content in the same units are well correlated (Farfel et al., 1994; Lanphear et al., 1995). Certain vacuum collection approaches can also be used with trained operators in research studies. Measuring dust lead loading from hard surfaces is more reliable than from complex surfaces such as upholstery or carpeting (Ewers et al., 1994).

One can collect dust samples as grab samples or through compositing. In the latter case, at least three subsamples for a composite are recommended where surfaces are relatively similar. Single, grab wipe sampling should be done where a particular surface is not similar to others or where surface deterioration is quite pronounced. That is, it is a surface producing a "hot spot" dust lead location. Compositing, furthermore, should be component- and area-specific. Bare floor samplings should not be combined with carpeted surface collections. Baseboard wipes should not be combined with window sill/trough wipes, etc. Multiple surface wipings with the same wipe should never be done. Each surface should use a new wipe each time. Recommended testing areas in residences occupied by children would include principal playroom, kitchen, bedroom of the youngest child, and bedroom of the second youngest. Dust sampling for multiunit residences expands on the above protocol for single units mainly in adding several common area samples.

Dust sampling for evaluation of residences as part of hazardous waste site evaluations as occur in Superfund activities entails determination of dust lead concentration, usually by vacuum collection through use of vacuum cleaners (von Lindern et al., 2003b) or, preferably, filter collection units attached to vacuum sources.

Laboratory analysis of dust samples employs methods similar to those for measuring lead in soils. However, the total amounts of dust mass sampled for analysis are typically much lower than for soil lead, where soil typically is available in large amounts and sample mass is not problematic. Dust lead levels, on the other hand, tend to track higher at a given testing site than associated soil lead levels. On balance, a more sensitive measurement method is desirable. Flameless AAS and ICP-MS provide both the sensitivity and specificity for acceptable dust lead bulk measurements. Speciation of lead in dusts generally employs methodology similar to that for lead speciation in soils (see earlier).

There are three SRMs that are available from NIST for QA/QC assessment in the laboratory for dust lead measurements (SRM #, matrix, and lead level, ppm): 1649a, urban dust, 12,400; 2583, indoor dust, 85.9; 2584, indoor dust, 9,761.

6.2.5 Analysis of Lead in Diets

Sampling and analysis of lead in diets of human populations require taking account of some characteristics of lead in human diets. First, any given dietary item will have relatively low amounts of lead but the amounts consumed daily can lead to relatively high total lead intakes. Second, while human populations all receive some fraction of their lead exposures from their diets, the size of the fractional intakes and uptakes of lead from diet will vary with such factors as the subset of the population, the amount of the centralized versus local food supplies, dietary lead content over time, etc. Adults have a different dietary pattern than school-age children, and school-age children have a different dietary mix than early infants.

Third, populations in industrialized countries, particularly the United States, receive food through a largely centralized food distribution system which simplifies somewhat food staple sampling protocols on a population or national basis. However, there are exceptions such as those who grow and consume their own garden crops or purchase from local farmers' markets. Rural residents are more apt to have a higher fraction of local foods than residents of large cities.

Lead in diets presents potential exposures for humans in a myriad of ways and through complex pathways. Individuals in the United States largely rely on a centralized food distribution system but the huge diversity of components of that food system and diversity of food consumption patterns across communities and individuals make quantification of dietary lead exposure a special challenge. An added factor is time. The relative contribution of dietary lead to total lead intakes for human populations, especially in the United States, has markedly changed over time. This means that tabulations of dietary lead levels given later in this chapter have to take account of this overall change. Lead intakes also differ over time as a result of changes in dietary patterns for reasons including concerns over relative health value of particular dietary components, e.g., avoiding obesity and associated morbidity.

Sampling methods for determining dietary lead levels range from specific approaches—determining dietary lead intakes of individuals at the one extreme—to national or international evaluations of the central food supply. Galal-Gorchev (1991) described the international dietary lead surveys done in the 1980–1988 period under the auspices of the United Nations Environmental Program's (UNEP's) Global Environmental Monitoring System (GEMS-Food, 1991). Galal-Gorchev reported lead level results for 25 countries in the GEMS-Food effort, including those for the United States. Such samplings are mainly focused on adults in the population, with less attention paid to children's dietary lead intakes across countries. However, Galal-Gorchev does tabulate the GEMS-Food network's results for lead intakes of children up to 12 years of age for 13 countries.

The U.S. dietary lead picture has been evaluated by the U.S. Food and Drug Administration (FDA) with varying intensity since the early 1960s, largely in the form of the Total Diet Study (TDS). The TDS evaluates nutrients, pesticides, and other environmental contaminants in the U.S. food supply. For a history of the TDS, see Adams (1991). Since the early 1980s, lead in diet has been reported annually through the market-basket approach under the aegis of the TDS program (Bolger et al., 1991; Pennington, 1983; Pennington and Gunderson, 1987). Currently, the TDS approach categorizes representative U.S. diets through eight age-sex groups gathered from four geographic areas of the country and for 12 commodity groups. The 234 subgroups within the 12 commodity groups in turn represent up to 5,000 different foods (Adams, 1991). The TDS approach also relies on data from the U.S. Department of Agriculture's (USDA's) National Food Consumption Survey (NFCS) for both selection of food types and translation of data for lead in diet components into daily dietary lead intakes (Adams, 1991).

Determining lead intakes and exposures for individuals as part of some research purpose will require not general but specific lead intakes. Pao (1989) described four sampling methods for individual dietary assessments, whatever the substances being measured. Two were retrospective in nature and include the recall of past food intake over the last 24 hours or some other time interval and the recall of past usual intake as might be recapitulated in a dietary history. The remaining two were prospective, involving an ongoing record of dietary intake of food components or using duplicate diet food intakes. Of these, the 24-hour food recall is probably the approach which serves for current intakes of dietary lead in either large groups being studied or those for whom the other approaches would not be feasible.

Lead in diet occurs at lower concentrations than in a number of other media, e.g., soils and dusts. Measurement methods therefore require high sensitivity for detection. Lead levels were originally measured by conventional AAS, but now there is more use of bulk sample ashing and Flameless AAS. Capar (1991) described a food lead quantitation limit of 20 ppb.

6.2.6 Analysis of Lead in Drinking Water

Transport of deposited ambient air and soil lead to surface water and groundwater was described in the previous chapter. This chapter is principally concerned with these waters to the extent they serve as drinking water sources for human populations. Human populations typically get their drinking water from individual wells in rural areas or through public water supply distribution systems in suburban and urban locales. Neither well water aquifers nor surface water supplies, e.g., reservoirs, are significant sources of those amounts of lead that enter residential or public tap water. Rather, the distribution system and/or residential/public site tap water plumbing systems are the main sources adding lead to the water. Of these two components of the tap water chain of distribution, household plumbing with leaded connections or fittings is more often the culprit.

Tap water lead is regulated by the U.S. EPA through an action level for the part of the system water in household plumbing and a regulatory maximum contaminant level (MCL) that applies to water leaving water treatment plants. The former is a level of 15 ppb for no more than 10% of cumulatively sampled water testings in a community and the enforceable MCL is 5 ppb. The 1991 EPA lead and copper drinking water rule spells out details for tap water sampling (42 USC §300f; 40 CFR Part 141).

In general, two types of water samples may be gathered, standing tap water and flushed samples. The former allows evaluation of the lead level in tap water that comes from both household plumbing leaching and lead in water from the distribution system. The latter provides both the distribution contribution to tap water lead and, by difference from the stagnant sampling protocol, individual household-specific plumbing contributions.

Lead in tap water can occur in either fully soluble form or a mixture of soluble and particle-bound, suspended lead. If filtered samples are used, the lead in the insoluble fraction is not accounted for although it can contribute to lead exposures. The latter is of relatively more significance for individual or community wells than for water being distributed from public treatment plants in relatively high-population communities. However, fine particles of lead soldering in household plumbing in suburban and urban areas can contribute to water lead intakes and need attention. In regulatory assessments, such as Superfund risk assessment, total tap water lead, i.e., use of unfiltered samples, is recommended (U.S. EPA, 1989).

6.3 CONCENTRATIONS OF LEAD IN ENVIRONMENTAL MEDIA PRODUCING EXPOSURES

Lead-contaminated environmental media differ in their relative significance for human lead exposures. The first difference is who among human populations are exposed to the Pb in the matrix. Dust and soil lead concentrations are highly significant for lead exposures of young children, especially infants and toddlers as discussed in the following chapter. Nonoccupational adult exposures mainly involve lead in air and diet. In some cases, both children and adults may sustain drinking water lead exposures. Workers are exposed to workplace lead in facilities such as lead smelters and lead-acid battery operations through both inhalation and ingestion, the latter through contact with settled lead dusts. Occupational exposures are not addressed *per se* in this monograph and are mainly noted in the context of secondary or "take-home" lead exposure of workers' families, especially their young children. This type of lead contact occurs through workplace dusts brought home on clothing, shoes, work items, etc.

A second significant factor concerns both the relative amounts of environmental media contamination from lead and the relative amounts of the medium contacted by exposed populations. Both lead intake scenarios can be important, as seen in the next chapter on human exposures. A third broad characteristic has to do with the physical, chemical, physicochemical, and geochemical nature of the element present in some medium. These characteristics affect the extent to which lead levels in some medium are absorbed into the bloodstream.

Lead-containing environmental media and lead concentrations in the media are mainly presented and tabulated in this chapter for ambient air lead, soil lead, interior and exterior dust lead, dietary lead, and lead in tap water. Lead levels in other, idiosyncratic sources of lead are provided only in brief summary.

Concentrations of lead in the various environmental media described in this section are presented for extended periods. The available data that meet minimal statistical and measurement criteria generally only extend from the late 1960s/early 1970s to the present. The purposes of a wide temporal look at environmental lead concentrations are several. First, the nature of lead as an accumulating contaminant in the bodies of human populations requires an appreciation of the amounts of environmental lead that existed in past decades. As noted earlier, lead levels in media have been changing, mainly downward, so that current human body lead burdens are only partially quantifiable from current lead intakes into body compartments. Secondly, the use of predictive, biokinetic models of human lead exposures for simulating lifetime lead exposures requires knowledge of lead intakes from the earliest periods of life.

6.3.1 Concentrations of Lead in Ambient Air

Prehistoric/Natural Levels in Ambient Air

Prehistoric or "natural" global levels of lead in ambient air are not currently measureable. Only projected estimates of early concentrations can be done. One approach has been to assume that prehistoric levels approximate lead levels measured in the most remote areas of the world using scrupulously rigorous methodologies for both sampling and measurement (see Table 6.7). Maenhaut et al. (1979) measured a value of 7.6×10^{-5} μg/m^3 at the South Pole, while Settle and Patterson (1982) recorded an Eniwetok atoll level of 1.7×10^{-4} μg/m^3 and a similar value of 1.5×10^{-4} μg/m^3 was recorded by Davidson et al. (1981) at a Greenland site. A second approach, that of Nriagu (1979) and Settle and Patterson (1980), entailed calculations of air Pb levels derived from estimating natural global emissions and dispersion of the emissions into selected tropospheric volumes, and assuming a residence time of 10 days. The two corresponding estimates are 2.6×10^{-4} μg Pb/m^3 (Nriagu, 1979) and 2.1×10^{-5} μg Pb/m^3.

The U.S. EPA, in its 1986 Air Quality Criteria for Lead document, selected the figure of 5×10^{-5} μg Pb/m^3 as the most reasonable depiction of the various estimating outcomes for purposes of subsequently calculating a background air lead value.

Ambient Air Pb Levels in the Modern Era

Ambient air Pb levels in the United States or elsewhere were not recorded or reported in any systematic way until the late 1950s. The establishment of the National Air Surveillance Network (NASN) in the United States was typical of such efforts. By the late 1970s, the NASN sites numbered 300 urban and 30 nonurban monitoring units. Since then, the numbers have declined with the phasing out of lead additives in gasoline (see later). However, other U.S. surveillance networks have been established and remain operational mainly through the U.S. EPA. These current networks measure different characteristics of ambient air Pb.

The total suspended particulate (TSP) measuring networks, managed by local and state agencies, currently numbering about 250 sites, measure lead via official Federal methods (40 CFR: Part 40, Appendix G) in particulate matter sized up to 30 μm. The retention of the TSP networks is driven by the regulatory structure of the previous and current ambient air lead primary and secondary standards. Locations of these sites are depicted in U.S. EPA (2006, Ch. 3).

Beginning in 2000, the Speciation Trends Network (STN) operated to quantify the lead content of particulate matter 2.5 μm or less ($PM_{2.5}$). Active $PM_{2.5}$ sites, as of this writing, number about 50 and look at long-term trends in the United States. These are augmented with 150 additional sites (U.S. EPA, 2006, Ch. 3). These STN sites are principally oriented toward urban areas.

The remaining two U.S. networks are the Interagency Monitoring of Protected Invisible Environments (IMPROVE) system and the National Air Toxics Trends Stations (NATTS). The IMPROVE network, as the label implies, deals with lead and other contaminants in rural areas measured in the $PM_{2.5}$ fraction. The National Park Service manages the principal 110 rural IMPROVE sites with shared management through other parties for 80 more sites that are a mix of urban and rural. The NATTS network is the newest of the group of networks, coming on stream in 2004, and is a set of 23 sites for urban and some rural areas. Lead measurement is confined to particle sizes of 10 μm or less.

The earliest U.S. data gathering, before 1966, had problematic results according to the U.S. EPA (1977). Because of that finding, data recorded in this chapter begin with reports for 1966. Concerns about earlier data included factors such as analytical inadequacy, statistical design problems, and sampling issues. Similar concerns about reliability of earlier air lead data elsewhere in the world limited international monitoring in general. As described earlier, adequate and sensitive air particulate sampling and laboratory measurement methods were still in development.

Three periods are covered in tabulating U.S. ambient air lead levels: 1966–1974, 1975–1984, and 1985 to the present. The first interval captures the increase and peak in air lead from mobile (leaded gasoline use) and stationary (e.g., industrial, fossil fuel, and waste combustion) facilities. The second period overlaps the regulatory phasing down period for lead emissions to the air, while the final period of 1985 to the present represents the current picture for air lead levels. As noted earlier, inclusion of earlier periods with their much higher atmospheric lead burdens is intended to help quantify

TABLE 6.1 Percentage of NASN Stations Reporting Urban Air Lead Data at Indicated Air Lead (μg Pb/m³) Intervals[a,b]

	Air Pb Interval (μg/m³)				
Year	<0.5	0.5–0.99	1.0–1.9	2.0–3.9	4.0–5.3
1966	9	42	42	6	–
1967	3	32	55	7	–
1968	9	45	36	6	1
1969	2	25	57	12	1
1970	5	33	50	9	1
1971	–	21	58	19	1
1972	9	37	47	7	0
1973	15	55	26	3	1
1974	15	53	29	3	0
1966–1974 (Mean)	8	38	45	8	1

[a]*Adapted from U.S. EPA (1977, Ch. 7).*
[b]*Percentages of reporting stations within the air Pb intervals indicated sum to approximately 100%.*

estimates for atmospheric lead inputs to long-lived reservoirs of anthropogenic lead—soils, dusts, sediments—and to assist in modeling lifetime lead exposures in older populations exposed to those earlier levels. The topic of cumulative lead exposures of human populations in developed, industrialized societies is developed later in this text (Mushak and Mushak, 2000).

Table 6.1 provides a useful distribution tally of urban NASN stations reporting U.S. urban air lead ranges (μg/m³) within the indicated intervals and over the period 1966–1974. The figures are adapted from U.S. EPA (1977, Ch. 7).

As expected in this period of peak leaded gasoline use and heavy emissions from stationary sources such as smelters, the highest air lead readings occur for the intervals representing traffic densities and/or stationary site emissions. The two highest air lead ranges largely capture areas of point source lead emissions, with the highest air lead range covering one emission site.

Table 6.1 shows that the highest percentage of stations reporting within the air lead range 0.50–0.99 μg/m³, 55%, occurred in 1973. The highest percentage, 57%, of all NASN monitoring sites for the next highest air lead interval of 1.0–1.9 μg/m³ was reported in 1969. For the next highest air lead

TABLE 6.2 Percentage of Nonurban NASN Stations Reporting Air Lead Data at Indicated Air Lead (μg Pb/m³) Intervals[a,b]

Year	Air Lead Interval (μg/m³)			
	<0.03	0.03–0.099	0.10–0.19	0.20–0.45
1966	–	52	32	16
1967	–	35	50	10
1968	5	75	20	–
1969	–	52	43	5
1970–1971	–	–	70	30
1972	29	12	26	33
1973	39	31	26	4
1974	19	31	38	12
1966–1974	15	36	35	14

[a]*Adapted from U.S. EPA (1977, Ch. 7).*
[b]*Number of reporting NASN stations varies across the air Pb intervals.*

range, 2.0–3.9 μg/m^3, the highest percentage, 19%, occurred in 1971. For the time period 1966–1974, the average percentage distribution of reporting sites for the ranges <0.5, 0.5–0.99, 1.0–1.9, 2.0–3.9, and 4.0–5.3 μg/m^3 were 8%, 38%, 45%, 8%, and 1% respectively. The highest percentage, 45%, was reported in the air lead range of 1.0–1.9 μg/m^3, followed closely by 38% for the range 0.5–0.99 μg/m^3. Combined, 83% of all stations reported air Pb in the range 0.5–1.9 μg/m^3.

Analogous percentage distribution calculations for reporting NASN stations in nonurban areas are presented in Table 6.2. Air Pb levels in the two higher ranges of 0.03–0.099 and 0.10–0.19 μg/m^3 account for the majority of the reporting site levels up to about 1971, followed by a decline in later years. The highest percentage of reporting stations were generally for air Pb values 0.030–0.099 μg/m^3 followed by the range of 0.10–0.19 μg/m^3. For the years 1966–1974, the average percentages of sites reporting various air Pb ranges showed 36% and 35% for the intervals 0.030–0.099 and 0.10–0.19 μg/m^3.

Table 6.3 presents additional data on the descriptive statistics for U.S. urban air lead measured quarterly in the 1970–1974 time frame in the form of means, maxima, and percentile distributions of air lead concentrations. Both arithmetic (1.19, 1.23, 1.13 μg/m^3) and geometric (0.99, 1.00, 0.03 μg/m^3) means were similarly elevated from 1970 to 1972, respectively, followed

TABLE 6.3 Mean, Percentile Distribution, and Maximum Urban Air Lead Quarterly Measurements from Urban Stations, 1970–1974[a,b]

Year	Arithmetic Mean (μg/m³)	Geometric Mean (μg/m³)	Maximum (μg/m³)	Percentile Distribution 10	50	70	95	99
1970	1.19	0.99	5.83	0.47	1.05	1.37	2.59	4.14
1971	1.23	1.00	6.31	0.42	1.01	1.42	2.86	4.38
1972	1.13	0.93	6.88	0.46	0.97	1.25	2.57	3.69
1973	0.92	0.76	5.83	0.35	0.77	1.05	2.08	3.03
1974	0.89	0.75	4.09	0.36	0.75	1.00	1.97	3.16

[a]*Adapted from U.S. EPA (1977, Ch. 7).*
[b]*Number of quarterly composites varies.*

TABLE 6.4 Mean, Maximum, and Percentile Distribution of Nonurban Air Lead Quarterly Measurements from Nonurban Stations, 1970–1974[a,b]

Year	Arithmetic Mean (μg/m³)	Geometric Mean (μg/m³)	Maximum (μg/m³)	Percentile Distribution 10	50	70	95	99
1970	0.09	0.04	1.47	0.00	0.00	0.00	0.38	0.63
1971	0.05	0.01	1.13	0.00	0.00	0.00	0.20	0.78
1972	0.14	0.09	1.05	0.01	0.11	0.17	0.39	0.95
1973	0.11	0.07	0.94	0.02	0.06	0.13	0.39	0.70
1974	0.11	0.08	0.53	0.01	0.09	0.14	0.32	0.50

[a]*Adapted from U.S. EPA (1977).*
[b]*Number of quarterly composites varies.*

by onset of declines in 1973 and 1974. Median (50th percentile) air lead levels showed a similar comparative relationship, at 1.05, 1.01, and 0.97 μg/m^3 for 1970–1972, respectively, followed by declines to 0.77 and 0.75 μg/m^3 for the later 2 years. Maxima for these years were 5.83, 6.31, 6.88, 5.83, and 4.09 μg/m^3.

Descriptive statistics are similarly provided for nonurban air lead values in the 1970–1974 time period, in Table 6.4. Here, the picture is less clear across the 5 years of measurement, giving no definitive trend. Results also reflect the lower number of monitoring stations and variability in the number

TABLE 6.5 Average Air Pb and Particle Size Distributions for Six Major U.S. Cities in 1970[a]

City	Average Annual Pb Level (μg/m^3)	Average Particle Size MMD (μm)[b]	% Particles ≤1 μm
Chicago, IL	3.2	0.68	59
Cincinnati, OH	1.8	0.48	72
Denver, CO	1.8	0.50	70
Philadelphia, PA	1.6	0.47	62
St. Louis, MO	1.8	0.69	62
Washington, DC	1.3	0.42	74

[a]*Adapted from U.S. EPA (1977, Ch. 7).*
[b]*MMD = mass median diameter.*

of quarterly composites. During the period 1970–1974, the arithmetic mean values were 0.09, 0.05, 0.14, 0.11, and 0.11 μg/m^3, respectively. Corresponding geometric mean values were 0.04, 0.01, 0.09, 0.07, and 0.08 μg/m^3, respectively, while maxima for the 5 years were 1.47, 1.13, 1.05, 0.94, and 0.53 μg/m^3, respectively. Unlike the measures of central tendency, the maxima show a downward trend from 1970 to 1974.

Mean values of air lead concentration were on the order of 10- to 20-fold lower for nonurban versus urban monitoring reports, but maxima for the two categories varied less, about fourfold lower. The latter likely reflects the principal contributor to nonurban air lead, point source contributions such as lead smelters. The nonurban sites are typically located in more remote areas, in the Midwest and West.

Table 6.5 presents mean air lead values in 1970 for six major U.S. cities as well as data for lead-bearing particle sizes. The mean air lead concentrations for Chicago, Cincinnati, Denver, Philadelphia, St. Louis, and Washington, DC, were 3.2, 1.8, 1.8, 1.6, 1.8, and 1.3 μg/m^3, respectively. Average sizes of measured air particulate ranged from 0.42 to 0.69 μm average MMD, while the percentages of particles ≤1 μm ranged from 59% to 74%. The fraction of lead in the small particle fraction indicates both a size with longer geographic reach and a size that is respirable by human populations.

Table 6.6 depicts ambient air lead statistics gathered by the U.S. EPA for 1973, the Seven-City Study of U.S. air lead for commercial, industrial, and residential monitoring sites. The U.S. cities were Chicago, Cincinnati, Houston, Los Angeles, New York, Philadelphia, and Washington, DC. Data reported are broken into monthly mean, minimum, and maximum values.

TABLE 6.6 U.S. Ambient Air Lead Statistics[a] for the Seven-City Study: Chicago, Cincinnati, Houston, Los Angeles, New York, Philadelphia, and Washington, DC[b]

		Monthly Level ($\mu g/m^3$)[d]		
City	**Site Type[c]**	**Mean**	**Minimum**	**Maximum**
Chicago, IL	C	4.5	2.5	6.7
	I	3.7	1.7	7.0
	R	3.3	2.1	4.7
Cincinnati, OH	C	2.0	1.3	3.1
	I	2.2	1.2	2.8
	R	1.5	0.8	2.6
Houston, TX	C	1.9	1.1	3.1
	I	–	–	–
	R	1.0	0.6	1.8
Los Angeles, CA	C	0.2	0.1	0.3
	I	–	–	–
	R	0.2	0.1	0.3
New York, NY	C	–	–	–
	I	–	–	–
	R	1.4	1.0	1.8
Philadelphia, PA	C	2.8	1.9	3.9
	I	2.2	1.5	3.0
	R	1.3	0.8	1.9
Washington, DC	C	1.4	3.1	2.0
	I	–	–	–
	R	1.5	1.1	1.8

[a]*From Tepper and Levin 1975. As discussed and cited in U.S. EPA (1977, Appendix C).*
[b]*Adapted from U.S. EPA (1977, Appendix C).*
[c]*C = commercial; I = industrial; R = residential.*
[d]*Mean of multiple values for site types, mainly 12 months of data.*

Some comparisons cannot be made for a site category across cities because data were not available. Nonetheless, all stations for these cities reported residential area monitoring measurements, while all but one city reported commercial district air lead concentrations.

The city reporting the highest residential and commercial district values was Chicago. Mean, minimum, and maximum residential air Pb concentrations were 3.3, 2.1, and 4.7 $\mu g/m^3$, respectively. Los Angeles provided the lowest air lead results, with the mean, minimum, and maximum concentrations being 0.2, 0.1, and 0.3 $\mu g/m^3$, respectively. Mean data for the remaining five cities in descending order were 1.5, 1.5, 1.4, 1.3, and 1.0 $\mu g/m^3$ for Washington, DC, Cincinnati, New York, Philadelphia, and Houston, respectively. The air lead values reflect mobile lead emissions from vehicular consumption of leaded gasoline but clearly other factors are operative. The relative urban character and population densities of each city's residential areas arguably play a role but multiple emission sources of airborne lead contribute. For example, residential air lead values for Chicago are several-fold higher than those of New York, despite similar population density or even greater multiple-unit residential density in New York. Chicago data may reflect both vehicular and industrial emissions, a more heterogeneous mix of multiple lead emission sources to the atmosphere.

The U.S. air lead levels are presented here for later periods as well. These later periods generally reflect decline in consumption of lead and consequently production of lead (U.S. EPA, 1986a, 2006). There was the major decline in leaded gasoline consumption due to regulatory restrictions leading to, first, a phasedown of the amounts of antiknock additive permitted in fuels in the 1980s and early 1990s, followed by an eventual phaseout. The numbers of point source lead emissions from primary and secondary lead smelters markedly declined as well.

Such changes are captured in Table 6.7, which depicts urban air lead levels for downtown sites in nine major U.S. cities from 1975 to 1984. Most of these cities showed air lead quarterly averages on the order of 1.0–2.0 $\mu g/m^3$ for the period 1978–1979, followed by declines of 50–70% or more from 1980 to 1984. Marked declines in some cities to 15–20% of 1975 figures were reported by 1984.

The tabulated air lead figures for U.S. areas were approximated globally, especially in major cities outside the United States. In a number of cases, cities outside the United States presented figures greater than those in the United States for the period 1975–1984. This difference arose in some significant part from the relatively slower speed of change in lead content of gasoline for non-U.S. vehicular fleets and other lead uses.

Table 6.8 tabulates air lead data for non-U.S. urban areas and rural around the globe. Locations, year(s) of measurement, and reported air lead concentration summaries are given. The earliest of the years monitored for each national locale in Table 6.8 show quite high air lead, compared to U.S. data shown in the earlier tables for about the same time period. Sites in Italy, Saudi Arabia, and Greece had particularly elevated air lead, ranging from 3.2 to 5.5 $\mu g/m^3$. In most cases where multiple years of measurement were noted, declines in air lead are seen post-1979. Table 6.8 also shows that size of urban area or

TABLE 6.7 Annualized Air Lead Levels[a,b,c] ($\mu g/m^3$) for Major U.S. Metropolitan Areas, 1975–1984

	U.S. Urban Areas Pb ($\mu g/m^3$)								
Year	Boston, MA	New York, NY	Philadelphia, PA	Washington, DC	Detroit, MI	Chicago, IL	Houston, TX	Dallas, TX	Los Angeles, CA
1975	0.93	0.93	–	1.1	0.98	–	2.1	2.8	2.1
1976	–	–	–	1.0	–	–	0.87	0.83	–
1977	0.65	–	1.4	1.4	1.0	–	0.80	1.5	2.1
1978	1.0	1.3	1.3	1.9	–	–	1.1	1.5	1.9
1979	0.70	0.95	1.1	1.7	–	0.60	0.73	0.85	1.0
1980	0.67	–	0.63	–	0.33	0.65	0.45	0.40	0.90
1981	0.35	–	0.43	–	0.30	0.33	0.55	0.45	1.0
1982	1.0	0.53	–	0.23	–	0.33	–	0.55	0.80
1983	0.50	0.35	–	0.23	–	0.38	0.30	0.70	0.70
1984	0.45	0.35	–	0.20	0.15	0.25	0.30	0.38	0.40

[a] *Adapted from U.S. EPA (1986a, Ch. 7).*
[b] *Average of reported quarterly averages given in U.S. EPA (1986a, Ch. 7).*
[c] *Four different types of reporting downtown commercial sites used.*

TABLE 6.8 Air Lead Level Reports for Locations Outside the United States: Urban and Rural Measurements for 1975–1984

Location			
Urban	Year(s) of Measurement	Air Pb Concentration ($\mu g/m^3$)	References
Brussels, Belgium	1978	0.50	Roels et al. (1980)
Brussels, Belgium	1979	1.05	Ducoffre et al. (1990)
Other Belgian sites	1983	0.66	
Ottawa, Canada	1975	1.30	NAPS (1971–1976)
Toronto, Canada	1975	1.30	NAPS (1971–1976)
Toronto, Canada	1984	0.45	O'Heany et al. (1988)
Montreal, Canada	1975	2.0	NAPS (1971–1976)
Helsinki, Finland	1983	0.33	Ponka et al. (1991)
Athens, Greece	1979	3.20	Chartsias et al. (1986); Kapaki et al. (1998)
	1982	1.76	
	1984	0.91	
Turin, Italy	1974–1979	4.5	Facchetti and Geiss (1982)
	1980	3.0	Facchetti (1989)
Riyadh, Saudi Arabia	1983	5.5	El-Shobokshy (1985)
Stockholm, Sweden	1980	1.20	Elinder et al. (1986)
	1983	0.50	
Sites in Wales			
Queensferry	1984	0.55	Page et al. (1988)
Wrexham		0.21	
Bedwas		0.22	
Port Talbot		0.14	
Cities in France			
Paris	1984–1985	0.44	Del Delumyea and Kalivretenos (1987)
Strasbourg		0.07	
Clermont		0.05	
Orleans		0.11	
Senonches		0.01	

(Continued)

TABLE 6.8 Air Lead Level Reports for Locations Outside the United States: Urban and Rural Measurements for 1975–1984—(cont.)

Location			
Rural	**Year(s) of Measurement**	**Air Pb Concentration ($\mu g/m^3$)**	**References**
Canada	1984	0.10	O'Heany et al. (1988)
Italy	1976–1980	0.33	Facchetti and Geiss (1982)
Belgium	1978	0.37	Roels et al. (1980)

level of industrialization is directly related to air lead content. Paris showed the highest air lead level for French cities, with the lowest value cited for the small town of Senonches. A similar relationship was noted for Wales.

Table 6.9 depicts peak air lead levels (quarterly maximum average, $\mu g/m^3$) reported for 1994 and for all major urban areas of the United States with a population of 1 million or more compiled in the 1990 U.S. Census and arranged alphabetically. These 48 urban areas are the metropolitan statistical areas (MSAs), comprising an identified central city and, where indicated, contiguous urban populations. Four did not provide air lead values, leaving a net of 44 sites. Table 6.9, adapted from U.S. EPA (1995), has data for monitoring sites which are directed toward the greatest populations even though EPA records air lead values for nonpopulation monitors as well. This adjustment avoids the use of anomalous values that capture large point source air lead emissions that have minimal impact in terms of numbers of individuals affected.

Of the 44 entries for the largest U.S. MSAs with air lead levels reported in 1994, over half ($N = 25$) reported a peak air lead $\leq 0.05\ \mu g/m^3$. A total of eight MSAs reported air lead in the range of >0.05–0.10 $\mu g/m^3$, while eight MSAs had peak air lead values of 0.11–0.30 $\mu g/m^3$. Three MSAs exceeded peak air lead of 0.30 $\mu g/m^3$, with the highest level for all MSAs being 0.89 $\mu g/m^3$. A majority of the locations (57%) had peak air lead of 0.05 $\mu g/m^3$ or less. Overall, the 1994 MSAs show peak air lead values that are but a small fraction of typical U.S. values in the 1970s and 1980s.

The U.S. air lead levels largely showed their maximal decline from peak values in the 1960s and early 1970s by 1994, with relatively more modest declines afterwards. However, relative to the picture in 1994, the proportional decline into the 2000–2005 time frame was still measurable.

TABLE 6.9 Peak Air Lead Levels[a] in Major U.S. Urban Areas[b,c,d], 1994[e]

Location	1990 Population	Air Lead ($\mu g/m^3$)
Atlanta, GA	2,833,511	≤0.05
Baltimore, MD	2,382,172	0.03
Bergen–Passaic, NJ	1,278,440	0.04
Boston, MA–NH	2,870,669	0.01
Buffalo–Niagara Falls, NY	1,189,288	0.05
Charlotte, NC–Gastonia, NC–Rock Hill, SC	1,162,093	0.03
Chicago, IL[f]	6,069,974	0.10
Cincinnati, OH–KY–IN	1,452,645	0.04
Cleveland–Lorain–Elyria, OH[g]	2,202,069	0.12
Columbus, OH	1,377,419	0.11
Dallas, TX[h]	2,553,362	0.08
Denver, CO	1,622,980	0.07
Detroit, MI	4,382,299	0.07
Fort Worth–Arlington, TX	1,332,053	0.03
Houston, TX	3,301,937	0.01
Indianapolis, IN[i]	1,249,822	0.20
Kansas City, MO–KS	1,566,280	0.03
Los Angeles–Long Beach, CA	8,863,164	0.08
Miami, FL	3,192,582	0.01
Middlesex–Somerset–Hunterdon, NJ	1,019,835	0.12
Milwaukee–Waukesha, WI	1,432,149	0.03
Minneapolis–St. Paul, MN–WI	2,464,124	Not available
Nassau–Suffolk, NY	2,609,212	Not available
Newark, NJ	1,824,321	0.30
New Orleans, LA	1,238,816	0.12
New York, NY	8,546,846	0.11
Norfolk–Virginia Beach–Newport News, VA	1,396,107	0.02
Oakland, CA	2,082,914	0.02

(Continued)

TABLE 6.9 Peak Air Lead Levels[a] in Major U.S. Urban Areas[b,c,d], 1994[e]—(cont.)

Location	1990 Population	Air Lead ($\mu g/m^3$)
Orange County, CA	2,410,556	0.04
Orlando, FL	1,072,748	0.00
Philadelphia, PA–NJ[j]	4,856,881	0.49
Phoenix–Mesa, AZ	2,122,101	0.04
Pittsburgh, PA	2,056,705	0.07
Portland–Vancouver, OR–WA[k]	1,239,842	0.27
Providence–Fall River–Warwick, RI–MA	1,141,501	Not available
Riverside–San Bernardino, CA	2,588,793	0.04
Rochester, NY	1,002,410	0.04
Sacramento, CA	1,481,102	0.02
St. Louis, MO–IL[l]	2,444,099	0.06
Salt Lake City–Ogden, UT	1,072,227	0.05
San Antonio, TX	1,302,099	0.03
San Diego, CA	2,498,016	0.02
San Francisco, CA	1,603,678	0.02
San Jose, CA	1,497,577	0.02
San Juan-Bayamon, PR	1,086,376	Not available
Seattle-Bellevue-Everett, WA	1,972,961	0.61
Tampa–St. Petersburg–Clearwater, FL[m]	2,067,959	0.89
Washington, DC–MD–VA–WV	3,000,504	0.04

[a] *Based on site air lead monitors for MSAs arranged alphabetically.*
[b] *Quarterly maximum average, $\mu g/m^3$.*
[c] *Defined as MSA with population ≥1 million.*
[d] *Population from 1990 census.*
[e] *Adapted from U.S. EPA (1995, Table A-12).*
[f] *Value for population-directed monitoring.*
[g] *Level for Cleveland itself.*
[h] *Value for population-oriented monitoring.*
[i] *Value for population-oriented monitoring.*
[j] *Highest population-oriented monitor result.*
[k] *Impact from an industrial source.*
[l] *Original value impacted by an industrial source; highest reading from population-oriented source used.*
[m] *Highest population-oriented monitor result.*

For example, air lead averages in the Los Angeles area noted by Hui (2002) were 0.015–0.019 μg/m^3 compared to the figure of 0.08 μg/m^3 in 1994, a decline of about 75%.

The most detailed data sets for U.S. national air lead statistics as of this writing are those contained in the Air Toxics Data Archive (2003–2005), a joint effort of the U.S. EPA, state/territorial air pollution officials, and local air control officials (U.S. EPA, 2007a). EPA (2007a) reported air lead in TSP measured at 189 monitoring sites, 140 of which were urban areas. The same material reported the national figures for these 3 years using four different statistical metrics: annual mean, maximum quarterly mean, maximum monthly mean, and second maximum monthly mean. For example, the national average annual mean for 2003–2005 was 0.09 μg/m^3, the corresponding national maximum quarterly mean was 0.17 μg/m^3, the national maximum monthly mean was 0.31 μg/m^3, and the national second maximum monthly mean was 0.21 μg/m^3.

Similar declines were seen for other countries over this period. Table 6.10 shows non-U.S. global air lead values reported from 1997 to

TABLE 6.10 Non-U.S. Global Urban Air Lead Levels, 1997–2003

Location	Air Pb (μg/m^3)	References
Yerevan, Armenia	<0.04	Kurkjian et al. (2002)
Australia roadsides	0.40–1.00	Al-Chalabi and Hawker (1997)
Vancouver, Canada	0.05	Brewer and Belzer (2001)
Eleusis, Greece	0.11	Torfs and Van Grieken (1997)
Hong Kong roadsides	0.13–0.17	Chan et al. (2000)
Caesarea, Israel	0.004–0.44	Erel et al. (1997)
Jerusalem–Tel Aviv Freeway, Israel	0.76	Erel et al. (1997)
Jerusalem, Israel	0.02	Erel et al. (2002)
Bari, Italy	0.01	Torfs and Van Grieken (1997)
Malta	0.06	Torfs and Van Grieken (1997)
Cadiz, Spain	0.01	Torfs and Van Grieken (1997)
Gothenberg, Sweden roadsides	0.05–0.11	Sternbeck et al. (2002)
Geneva, Switzerland	0.05	Chiaradia and Cuppelin (2000)
Birmingham, U.K. roadsides	0.03	Harrison et al. (2003)

2003. Table 6.10 reports few levels of air lead higher than 0.20 $\mu g/m^3$, and the latter are for roadside air lead monitoring. The highest measurements were for Australian roadsides, in the range of 0.40–1.00 $\mu g/m^3$ (Al-Chalabi and Hawker, 1997). Hong Kong roadsides showed values 0.13–0.17 $\mu g/m^3$ (Chan et al., 2000), with a value of 0.76 for an expressway between Jerusalem and Tel Aviv, Israel (Erel et al., 1997). Most reports recorded in Table 6.10 were for air lead levels around or below 0.10 $\mu g/m^3$.

Air lead levels in the proximity of stationary sources such as primary and secondary lead smelters, battery operations, etc. will be quite elevated regardless of the co-occurrence of high lead levels in past years from dense traffic. Most major U.S. smelters had ceased operations as of this writing. There is one remaining primary lead smelter operating in the U.S., in Herculaneum, MO, and there are 15 U.S. secondary smelters (U.S. EPA, 2007b). Table 6.11 depicts representative air lead levels near U.S. stationary lead emissions sites from the 1970s to the present. The listing is skewed to earlier years because the number of these facilities has dropped, as have their emissions of lead. Nonetheless, recording air lead levels for these operations regardless of period provides a measure of the extent of atmospheric lead deposition on impacted soils and other surfaces.

Measurements of Indoor Air Pb Levels

Human exposures to air lead by direct inhalation are a combination of outside ambient and interior air lead. Yocum (1982) reported that U.S. homes and other buildings without air conditioning have indoor/outdoor air Pb ratios higher than those with air conditioning or those that are otherwise better sealed. A typical range for the former is 0.6–0.8, while for the latter it is 0.3–0.5. Davies et al. (1987) reported a ratio of 0.6 for U.K. dwellings with young children, a figure similar to that of Diemel et al. (1981) for tested residences in Arnhem, a Dutch community with a secondary lead smelter. The U.S. EPA, in its guidance material for its Integrated Exposure-Uptake Biokinetic (IEUBK) model for predicting lead exposures in children up to 84 months of age (U.S. EPA, 1994), employs an indoor air lead fraction of 0.3 that of outdoor atmospheric levels.

6.3.2 Lead-Based Paint

Lead in paint has long been recognized as an environmental source that provides the most serious levels of lead exposure for human risk populations and associated lead poisoning (see Chapters 7 and 8). Children are at particular risk, especially infants and toddlers. The severity of lead paint poisoning differs geographically around the globe and geographically within the United States, the country most severely impacted by production, sale, and use of

TABLE 6.11 Illustrative Global Air Lead Levels near Major Stationary Pb Emission Sources, 1971–2005

Source	Location	Air Pb (μg/m^3)	References
Primary lead smelter	Silver Valley, ID	10.3 (<1 mi)	Yankel et al. (1977)
		8.6 (1–1.5 mi)	
		4.9 (2.5–6 mi)	
		2.5 (6–15 mi)	
Primary Pb smelter, refinery, battery plants	Omaha, NE	5.0	McIntire and Angle (1973)
		May–November 1970 composite	
Primary Pb smelter	El Paso, TX	2.7 (4.8 km)	U.S. EPA (1977)
		2.4 (2.4 km)	
Primary Pb smelter	East Helena, MT	3.9 (0.5 mi)	U.S. EPA (1977)
Secondary Pb smelter battery plant	Southern Ontario, Canada	3.8	Linzon et al. (1976)
		2.1	
Primary Pb smelter	Meza Valley, former Yugoslavia	24.2–38.4	U.S. EPA (1977); Fugas et al. (1973)
Secondary Pb smelter	California	1.7–4.0	Kimbrough and Suffet (1995)
Primary Pb smelter	Herculaneum, MO	2001: 1.3[a]	U.S. EPA (2007a, Attachment B-22)
		2002: 0.4[a]	
		2003: 0.4[a]	
Zinc and copper smelters	Ajo, AZ	0.3[b]	Hartwell et al. (1983)
	Anaconda, MT	0.3[b]	
	Bartlesville, OK	0.4[b]	
	Palmerton, PA	0.3[b]	
Secondary Pb smelter	Arnhem, the Netherlands	0.3–0.5	Diemel et al. (1981)

[a] *Average annual values for nine EPA Air Quality System (AQS) monitors, U.S. EPA (2007a, Table B-2).*
[b] *Air Pb values closest to the smelter, reference distances differing among smelters.*

lead-based paints (LBPs) in prior decades. However, it is likely no area of the globe has been fully free of lead paint risk.

Paint as a lead-bearing environmental medium differs qualitatively and quantitatively from other sources and pathways of lead for human contact.

LBP applied decades ago still exists on residential surfaces in millions of residences and other facilities occupied by young children. As such, it will continue to provide lead exposures to young children until these millions of units are rendered either lead-free or lead-safe for many years.

We characterize lead paint reservoirs in terms of decades, in contrast to lead emissions to the atmosphere where the emissions are associated with air lead changes that are on the order of hours or days in duration. A second contrasting element is that of the nature and extent of lead dispersal via various pathways. While the atmosphere (scaled as to specific areas or regions) is the common conduit for lead from emission sources, lead paint in each of the millions of U.S. residential units that contain it provides microscale risk across a macroscale of distribution.

LBP in the United States is an environmental medium broadly confined to the nation's older housing stock and its contiguous environmental compartments such as lead in building perimeter, i.e., "drip-line" soils, where lead comes mainly from exterior lead paint weathering and deposition onto contiguous surfaces. It is therefore appropriate to attempt to quantify the magnitude of the U.S. lead-paint problem by looking at various housing surveys. In particular, these are (1) the 2001 U.S. national housing survey for lead based on a representative sampling; (2) the 1997 American Housing Survey as a view of the national distribution of lead-painted housing; and (3) another national distribution survey called the Current Population Survey presented in 1999. Collectively, they provide the most reliable current picture for the U.S. lead paint problem. Of these, the 2001 National Survey of Lead and Allergens in Housing (NSLAH) is the most useful for purposes of this chapter and is a principal focus of this section (Jacobs et al., 2002; NSLAH, 2001).

Lead-containing paint as a medium for potential lead exposures for children and other risk groups is particularly complex for characterizing the nature and extent of human contact. First, lead paint's hazard as a solid surface on different residential areas and interior or exterior building components varies with the lead level (concentration of lead in paint samples) or loading (amounts of lead per unit area of measurement). These aspects were presented in the analytical measurements portion of this chapter. Surface condition, with deteriorating surfaces posing more of an overall risk to residents than surfaces in good condition at some specific instant in time, is another factor. This is not to imply that intact surfaces pose no risk. The regulatory scientific literature for LBP with regard to lead content, hazards to children's health from lead paint as a function of condition, etc., is a significant body of data. These topics are presented and discussed in more detail in a later part of this monograph. Here, these factors are included to the extent they are required to quantitatively describe the overall lead paint exposure picture for risk groups such as young children.

Both current and historic U.S. nationwide lead paint statistics are presented here for lead paint as a discrete lead source *per se*. Environmental pathways for lead from paint released to other media, such as lead paint-derived dusts and leaded soils, are presented later. Lead paint statistics presented in such compendia as the NSLAH (2001) survey classified the various categories of lead paint within those definitions established by the U.S. HUD and the U.S. EPA. The U.S. HUD defines categories of lead paint characterization for this survey according to the definitions set forth in the 1999 U.S. HUD lead-safe regulations. These regulations differ in some of their particulars from companion regulations issued by the U.S. EPA in its final lead hazard rule issued in 2001 (U.S. EPA, 2001). Both agencies, overall, have had quite similar regulations for lead paint hazards as mandated by the U.S. Congress in its 1992 "Title X" legislation.

LBP in U.S. housing and other structural stock is defined in regulatory and statutory terms as lead paint present at a loading of 1.0 mg Pb/cm^2 or higher and/or a lead concentration of 0.5% (5,000 ppm) or more. Two broad parameters characterizing variation in the amounts of lead paint in U.S. housing stock are region of the country and age of the housing units. These two variables are closely related, however. Lead paint as an environmental problem traces to the vintage of the lead-painted surfaces; the older the residential housing unit or the public building, the higher the likelihood of one or more painted surfaces meeting the definition of LBP. Likewise, the older the area of the country in terms of settlement history, the higher the likelihood of LBP surfaces in buildings.

Table 6.12 tabulates the prevalence of LBP in U.S. housing stock as a function of four geographic regions of the nation—Northeast, Midwest, South, and West—and four housing age bands—pre-1940, 1940–1959, 1960–1977, and 1978–1998—within each of those four geographic regions. The figures are estimates from the NSLAH (2001).

Table 6.12 makes it clear that the older the housing units in any of the four major regions, the higher the fraction of units with LBP. For pre-1940 housing, the percentages of LBP units for the regions range from 71% for units in the West to 91% in the Midwest. Corresponding figures for the Northeast and South were 86% and 91%, respectively. Table 6.12 also shows that the total number of units with LBP is highest in the two oldest areas of the country, the Northeast and Midwest, with a lower count for the South and the lowest number of LBP units found in the West.

Housing constructed in the 1978–1998 period, following the total U.S. ban on use of lead paint for residential interiors, showed the lowest percentages of LBP among housing units within each of the geographic areas. The persistence of some measurable percentage in the face of the lead paint ban likely is attributable to some combination of remaining lead paint in commercial channels after the ban and other factors. The persistence of significant percentages of LBP surfaces in residential units built from 1960 to 1977 is noteworthy. The fractions of housing units with LBP from this age band (%) and the corresponding regions are: 39, northeast; 28, Midwest; 17, South; 22, West.

TABLE 6.12 U.S. Prevalence Rates[a] of LBP in Housing[b] by Region and Ages of Housing

Region	Years Built	Estimated Number of Units with LBP (000)	Estimated % of Units with LBP
Northeast	Pre-1940	5,957	86
	1940–1959	3,089	73
	1960–1977	1,478	39
	1978–1998	76	2
Regional subtotal/% LBP[c]		10,300	27
Midwest	Pre-1940	4,658	91
	1940–1959	4,785	81
	1960–1977	1,771	28
	1978–1998	533	11
Regional subtotal/% LBP[c]		11,747	31
South	Pre-1940	3,065	89
	1940–1959	3,431	54
	1960–1977	1,914	17
	1978–1998	1,197	8
Regional subtotal/% LBP[c]		9,607	25
West	Pre-1940	1,437	71
	1940–1959	2,866	69
	1960–1977	1,414	22
	1978–1998	225	4
Regional subtotal/% LBP[c]		5,942	16

Total U.S. housing units: 95,688,000; total U.S. LBP-containing units: 37,897,000; % of total with LBP: 40.

[a] *Adapted from NSLAH (2001).*

[b] *As defined by U.S. EPA (2001) and U.S. HUD (1999); Pb loading, XRF: $\geq$1.0 mg/cm ^{2}, Pb concentration $\geq$0.5%.*

[c] *Regional percent of total 37,897,000 U.S. LBP units.*

Persistence of sizeable percentages of LBP in housing in all four regions from 1940 to 1959 is evident in Table 6.12. Percentages for this period seen in the Northeast, Midwest and West regions are generally not greatly less than those for pre-1940 housing, at 73%, 81%, and 69%, respectively.

TABLE 6.13 U.S. Prevalence Rates[a] of LBP[b] from Housing with Children <6 Years Old[c] by Age of Housing

Housing Ages	Estimated Number of Units (000) with LBP	Estimated % with LBP Units
Pre-1940	2,253	94
1940–1959	1,997	65
1960–1977	876	17
1978–1998	202	3

Total residential units with young children: 16,402,000; total LBP-containing units with young children: 5,328,000; % with LBP: 32.
[a]*Adapted from NSLAH (2001).*
[b]*As defined by U.S. EPA (2001) and U.S. HUD (1999): Pb loading, XRF ≥1.0 mg/cm^2, Pb concentration ≥0.5%.*
[c]*Defined by U.S. CDC as high-risk group.*

One area of concern about the persisting presence of LBP in U.S. housing is the equally persisting potential exposure threat to residents, particularly young children. The U.S. CDC in its 1991 and 2005 Statements on childhood lead poisoning defined young children less than six years of age as that subset of the population at highest risk because of various behavioral and developmental vulnerability factors (U.S. CDC, 1991, 2005).

Table 6.13 presents estimates of the number of U.S. housing units with both LBP and one or more children <6 years of age as a function of housing age. The oldest housing, pre-1940, was estimated in the 2002 NSLAH survey to have 2,253,000 housing units with both LBP and one or more young children. This comprises 94% of all units built pre-1940. The survey estimated 1,997,000 residential units had both one or more children and LBP for the 1940–1959 period, 65% of all units in that housing age group. For the 1960–1977 and 1978–1998 periods for housing construction, the counts of units with both LBP and one or more young children were 876,000 and 202,000, respectively. The corresponding percentages with LBP were 17% and 3%, respectively.

Overall, using the survey figures, there were 16,402,000 U.S. residential units with young children, and of these, 5,328,000 or 32% had LBP.

Urbanization as a factor in the relative likelihood of U.S. housing units having LBP was presented and quantified in this survey, as summarized in Table 6.14. Urbanization referred to (1) presence in or outside an MSA as defined by the U.S. Bureau of the Census; and/or (2) presence in one of two MSA categories differing in population size as also defined by the U.S. Bureau of the Census. Those U.S. MSAs with a population of 2,000,000 or higher had 24,967,000 residential units, of which 8,963,000 or 36%

TABLE 6.14 U.S. Prevalence Rates[a] of LBP[b] in Housing by Extent of Urbanization[c]

Degree of Urbanization	Estimated Number of Units with LBP (000)	Estimated % Units with LBP
Total number of units (000) in MSAs ≥2 million population 24,967	8,963	36
Total number of units (000) in MSAs ≤2 million population 42,782	16,250	38
Total number of units (000) not in an MSA 21,808	10,046	46

Total units in MSAs = 67,749,000; total units with LBP in MSAs = 25,213,000; % all MSA units with LBP = 37.
[a] *Adapted from NSLAH (2001).*
[b] *As defined by U.S. EPA (2001) and U.S. HUD (1999). Pb loading, XRF: ≥1.0 mg/cm^2; Pb concentration: ≥0.5%.*
[c] *Based on data from U.S. Bureau of the Census.*

contained LBP. MSAs with <2,000,000 population had 42,782,000 units, of which 16,250,000 or 38% had LBP. There were a total of 21,808,000 units outside of these MSAs, of which 10,046,000 units or 46 % contained LBP. The non-MSA tally excluded units not characterized or assigned to some metropolitan grouping. Overall, level of urbanization was not as significant a variable in prevalence of LBP in housing units as other variables such as age of housing and geographic location.

In the late nineteenth and twentieth centuries in U.S. households, LBP was used for both interior and exterior surfaces. Each type of surface presented different potential risks for contact and potential human exposures. Interior LBP posed the more significant risk in terms of duration of typical contact for young children, especially infants and toddlers. However, exterior lead paints produced a broader reach for lead contact through higher deterioration rates and wider dissemination of deteriorated paint residues. For example, exterior lead paint readily weathered over time and shed fine, high-hazard particles onto contiguous soils and as exterior dusts.

Table 6.15 presents U.S. prevalence rates for LBP-containing surfaces in housing units as a function of location within the housing structures: interior surfaces, exterior surfaces, or both interior and exterior surfaces. The survey estimated that 8,609,000 U.S. units or 9% of all units with LBP had interior LBP on some surface. By contrast, the U.S. prevalence rate for exterior surfaces was more than double the interior rate, 20,260,000 or 21% of LBP units. An estimate of 9,028,000 units, or 9% of all LBP units, is recorded in this table for those units having LBP on both interior and exterior surfaces.

TABLE 6.15 U.S. Prevalence Rates[a] of LBP[b] in Housing by Location in the Building

LBP Location	Estimated Number of Units with LBP (000)	Estimated % with LBP
Some interior surface	8,609	9
Some exterior surface	20,260	21
Both exterior and interior surfaces	9,028	9

No LBP in building: 57,791,000; % with LBP somewhere in building: 40.
[a]*Adapted from NSLAH (2001).*
[b]*As defined by U.S. EPA (2001) and U.S. HUD (1999): Pb loading, XRF: ≥1.0 mg/cm^2; Pb concentration ≥0.5% total units with LBP somewhere in building: 37,897,000.*

TABLE 6.16 U.S. Housing Distributions[a] of Highest Paint Lead Loading on Unit Interiors as a Function of Housing Age

Maximum Loading[b] by Increasing Pb Amount (mg/cm^2)	% of Units ≥ Indicated Highest Pb Loading by Housing Age			
	Pre-1940	1940–1959	1960–1977	1978–1998
≥0.6	83	59	21	9
≥1.0	79	46	16	4
≥1.3	72	41	12	3
≥4.0	60	19	6	1
≥10.0	38	7	2	1

[a]*Adapted from NSLAH (2001).*
[b]*Loading as maximum XRF reading (mg/cm^2) in the housing unit.*

Comparatively, about three-quarters of all LBP units—29,288,000 units—had exterior surfaces covered with LBP with or without interior LBP.

A critical factor in the level of toxicity risk posed by LBP is the level of lead loading or lead concentration. The former is indexed in units of mg Pb/cm^2 on painted surfaces using XRF detection (see Section 6.1) while the latter employs content of Pb per unit mass, typically as % Pb. All other factors being equal, the magnitude of Pb exposure hazard increases with the increase in Pb loading or concentration. Table 6.16 tabulates the relationship of the U.S. prevalence rates for selected threshold maximum lead loadings as a function of housing age. In general, the older the housing,

TABLE 6.17 U.S. Housing Distributions[a] of Highest Paint Lead Loadings[b] on Unit Exteriors as a Function of Housing Age

Maximum Pb Loading in Paint by Increasing Pb Amount (mg/cm^2)	% of Units with Indicated Highest Pb Loadings by Housing Age			
	Pre-1940	1940–1959	1960–1977	1978–1998
≥0.6	76	64	18	7
≥1.0	72	59	13	3
≥1.3	71	56	11	3
≥4.0	56	28	6	0
≥10.0	41	10	2	0

[a]*Adapted from NSLAH (2001).*
[b]*Loading as a maximum XRF reading (mg/cm^2) in the housing unit.*

the higher the percentage of maximum Pb loadings found in housing in the age band. This relationship prevails for interior surfaces, exterior surfaces, or LBP surfaces anywhere in the residential units.

Table 6.16 shows that, in pre-1940 housing, the rates at any threshold for the maximum measured Pb loading in housing interiors are much higher than those in any other housing age group. For example, maximum threshold Pb loadings ≥1.3, 4.0, and 10.0 mg/cm^2 in interiors occurred in pre-1940 housing at rates of 72%, 60%, and 38%, respectively. For the 1940–1959 age band, the corresponding Pb loading rates were 41%, 19%, and 7%. Measurable lead loadings above the definition of LBP were still estimated in the most recent age of housing, 1978–1998.

Table 6.17 similarly shows that in pre-1940 housing, the rates at any threshold for the maximum measured Pb loading on housing exterior surfaces were much higher than those in any other housing age group. For example, maximum threshold Pb loadings ≥1.3, 4.0, and 10.0 mg/cm^2 occurred in pre-1940 housing exterior surfaces at rates of 71%, 56%, and 41%, respectively. For the 1940–1959 age band, the corresponding Pb loadings in exteriors were 56%, 28%, and 10%. Measurable lead loading above the definition of a positive LBP reading at the ≥1.3 mg Pb/cm^2 threshold was also estimated for exteriors in the most recent housing studied, 1978–1998.

Table 6.18 shows that the percentage rates at any loading threshold for the maximum measured Pb loading in housing with LBP anywhere in the structures were highest in the oldest pre-1940 subset of housing stock in the United States. Prevalence rates for loadings ≥1.3, 4.0, and 10.0 mg Pb/cm^2 pre-1940 were 84%, 73%, and 55%, respectively. Moving through younger

TABLE 6.18 U.S. Housing Distributions[a] of Highest Paint Lead Loading[b] on Units Anywhere in Building as a Function of Housing Age

Maximum Pb Loading in Paint By Increasing Pb Amount (mg/cm^2)	Pre-1940	1940–1959	1960–1977	1978–1998
≥0.6	89	80	31	15
≥1.0	87	69	24	7
≥1.3	84	65	18	5
≥4.0	73	34	10	1
≥10.0	55	14	3	1

[a]*Adapted from NSLAH (2001).*
[b]*Loading as maximum XRF reading in the unit, mg/cm^2.*

housing stock, the percentage of units meeting those thresholds in Pb loadings declined significantly. A comparison of the two highest Pb loading thresholds in Table 6.18 demonstrates marked declines in percent occurrences across the four housing age groups, ranging from 55% for the highest loading pre-1940 to 1% in housing built between 1978 and 1998.

Deterioration of lead-painted surfaces in the form of peeling, chipping, abrasion, and chewing can provide added risk of lead exposure over that provided by LBP surfaces assumed to be relatively intact. Central to any discussion of the role of deterioration in defining exposure risk is the need to keep in mind that all intact and accessible LBP surfaces can readily transition to surfaces in some stage of deterioration for a variety of environmental and socioeconomic reasons.

Table 6.19 depicts the rates of deteriorated and significantly deteriorated LBP surfaces in U.S. housing units as a function of age of the housing unit. Refer to U.S. EPA (2001) and U.S. HUD (1999) for regulatory characterization of degrees of deterioration. Of 37,897,000 U.S. units with LBP, 17,425,000 or 46% had one or more deteriorated LBP surfaces while 13,635,000 or 36% had one or more significantly deteriorated LBP surfaces. Of these overall tallies, pre-1940 housing units were 9,866,000 or 26% of all U.S. LBP housing units. Similarly, pre-1940 units with significantly deteriorated LBP surfaces comprised 7,752,000 units, 20%, of U.S. LBP units. Housing units with deteriorated LBP built from 1940 through 1959 amounted to 6,510,000 or 17% of all units with LBP, while 5,190,000 or 14% of the total LBP group were estimated to have significantly deteriorated surfaces. These figures are consistent with the data showing that older housing has higher rates of deteriorated LBP surfaces and significantly deteriorated

TABLE 6.19 U.S. Prevalence Rates[a] of Deteriorated and Significantly Deteriorated LBP Surfaces by Age[b] of Housing Stock Versus All LBP Housing

Housing Age Band	Number of Units with Deteriorated LBP (000)	% of Total LBP Units	Number of Units with Significantly Deteriorated LBP (000)	% of Total with Significantly Deteriorated LBP
Pre-1940	9,866	26	7,752	20
1940–1959	6,510	17	5,190	14
1960–1977	910	2	610	1.6
1978–1998	139	0.4	83	0.2
Total units/% all LBP	17,425	46	13,635	36

Total number of units with LBP = 37,897,000; total number of units with deteriorated LBP = 17,425,000; total number of units with significantly deteriorated LBP = 13,635,000.
[a] *Adapted from NSLAH (2001, Table 4.2).*
[b] *As defined by U.S. EPA (2001) and U.S. HUD (1999).*

surfaces. Pre-1940 housing was that block of U.S. housing stock constructed in the period with highest use and consumption of LBP for home construction and repainting. At the other extreme, units painted in 1978 or later were those built after the Federal ban on interior LBP and most uses of exterior residential LBP.

Table 6.20 presents prevalence rates for deteriorated and significantly deteriorated units as a function of location in their structures.

Distributions of national total and individual unit average surface areas with LBP are given for interiors and exteriors of U.S. housing stock in Tables 6.21 and 6.22, respectively. Table 6.21 shows the national total and individual average LBP surface areas for interiors to be 7,448 million square feet and 259 square feet, respectively. Of the total national and typical LBP individual values, the highest contributor is from the wall–floor–ceiling category. Table 6.22 shows that the total national square footage for exterior surfaces is 29,159 million square feet while the average unit exterior LBP is 996 square feet. For both categories, the "wall" portion is the principal contributor.

It is important to keep in mind that there is no direct relationship between the area of any given LBP residential unit and the relative hazards to children for lead exposures. The lead exposure hazard is more a function of likelihood of contact, where deteriorated LBP on interior window sills and

TABLE 6.20 Estimated Prevalence Rates[a] of Deteriorated and Significantly Deteriorated[b] LBP in U.S. Housing Units with LBP

Location in Building	Number of Units with Deteriorated LBP (000)	Deteriorated LBP Units as % Total Housing Units	Number of Units with Significantly Deteriorated LBP (000)	Significantly Deteriorated LBP Units as % Total Housing Units
Interior surface	4,180	4	2,629	3
Exterior surface	7,009	7	3,487	4
Both interior and exterior surfaces	6,236	7	7,518	8
Total	17,425	18	13,634	14[c]

Total housing units = 95,688,000.
[a]*Adapted from NSLAH (2001, Table 4.3).*
[b]*As defined by U.S. EPA (2001) and U.S. HUD (1999).*
[c]*Rounding.*

TABLE 6.21 National and Single-Unit Amounts of LBP Surface Areas (ft^2) for Interior Components[a,b]

Component	National LBP Square Footage (000,000)	Average LBP Area (ft^2)/Housing Unit
Wall, floor, ceiling	4,993	173
Window	687	24
Door	911	32
Trim	499	17
Cabinets, beams, chimney	388	13
Total	7,448	259

[a]*Adapted from NSLAH (2001).*
[b]*LBP as defined by U.S. EPA (2001) and U.S. HUD (1999).*

troughs (wells) may pose more immediate exposure threats, as discussed in a later chapter, than LBP-covered ceilings with evidence of peeling or chipping paint. This is not at all to say that peeling and chipping LBP from poor-condition ceilings and walls pose little health hazard.

TABLE 6.22 National and Single-Unit Average LBP Surface Area (ft^2) for Exterior Components[a,b]

Component	National LBP Square Footage (000,000)	Average Unit LBP Surface Area
Wall	26,706	912
Window	365	12
Door	446	15
Trim	556	19
Porch	1,086	37
Total	29,159	996

[a]*Adapted from NSLAH (2001).*
[b]*LBP as defined by U.S. EPA (2001) and U.S. HUD (1999).*

6.3.3 Lead in Interior and Exterior Dusts

Dusts are technically defined as solids consisting of small particulate materials that in turn derive from a variety of materials through diverse processes of deterioration. While dusts can consist solid particulates of any size, those which both contain lead and arise from environmental transformations of lead sources are generally considered to have an average diameter of 50 μm.

Interior and exterior lead-containing dusts, especially the former, are known to be one of the most important, if not the most important, pathways by which lead from original sources, in ambient air and/or in exterior or interior paints, enters human environments and then enters the bodies of exposed populations, notably young children. Pathways for generation of dust lead can occur directly from these sources or can occur indirectly. For example, interior dust lead levels or loadings can occur via deposition of particulate lead from the atmosphere directly to interior surfaces or through initial loading onto exterior surfaces. The indirect path can involve interim Pb deposition on soils and then lead migration from soils to interiors as dusts. Another indirect pathway for interior dust lead is "take-home" or occupational secondary exposure lead, where workplace leaded dusts in a leaded work environment adhere to workers' clothing, shoes, etc. and are carried home in the absence of washing or clothes-changing facilities at work.

For purposes of categorizing environmental mobility, one can subdivide the topic of dust lead by origin: exterior leaded dusts from atmospheric deposition and/or weathering lead paints and interior dusts from atmospheric lead deposition and/or leaded paints. Atmospheric lead deposition to interior or exterior surfaces, in turn, can readily arise from either or both mobile and stationary atmospheric lead sources.

Little in the way of quantitative information on lead content of dusts impacting human populations appears in the literature before the late 1960s and 1970s, for many of the same reasons that air lead data were little reported: absence of both acceptable analytical Pb methodology and statistically standardized sampling methods. By the 1970s, measurements of dust lead tracing to mobile sources, i.e., vehicular emissions of leaded gasoline exhaust, or from stationary source air lead emissions like those from smelters, were appearing in the scientific literature.

Lead in Exterior Dusts

Significant exterior (outside) leaded dust deposition occurred near vehicular traffic, notably in the decades of leaded gasoline use for private and commercial vehicles (U.S. EPA, 2008b). Table 6.23 tabulates dust lead measurement results expressed in units of concentration for some illustrative U.S. vehicular traffic areas recorded in the late 1960s and 1970s. Measurements typically ranged up to about 10,000 ppm, and in the case of tunnels, up to 20,000 ppm. Table 6.23 generally depicts the high dust lead concentrations measured at U.S. roadsides close to motor lanes in those years of high leaded gasoline consumption, consisting of heavier fractions of auto exhaust particulate that settled close to the roadways soon after emission.

Table 6.24 presents U.S. roadside dust lead figures as dust lead loading rates onto receiving surfaces from atmospheric lead deposition in this same period of high-leaded fuel use. This table depicts lead loading results for outside dusts sampled in 77 U.S. Midwestern cities. Three cities had the highest lead loading rates indexed by type of outside lead deposition: residential, South Bend IN; commercial, Nashville, TN; industrial, Omaha, NE. Cities in this particular survey overall showed the geometric mean of deposition rates highest for industrial locales, intermediate for commercial sites, and lowest for residential areas. The highest comparative finding for industrial zones is to be expected based on contributions from both stationary and mobile lead emission sources. Residential areas were lowest in lead deposition, showing a clean auto traffic contribution and lower traffic densities than for commercial zones.

These data for U.S. locations with elevated roadway/outside dust lead residues in the era of high leaded gasoline consumption were largely replicated for non-U.S. findings in developed areas around the world with relatively large vehicular fleets and densely populated urban areas. Table 6.25 shows the case for locales in Europe, Asia, Canada, etc.

Roadway and other outside dust lead levels declined with the phasedown and phaseout of leaded gasoline but the rate of decline has been understandably slow given the long-term retention of deposited dusts in roadside soils

TABLE 6.23 Illustrative U.S. Roadside Dust Lead Concentrations (μg/g) in the Period of High Leaded Gasoline Consumption[a,b]

U.S. Location	Road Site(s)	Dust Lead Concentration (μg/g)
Washington, DC	a. Busy road intersection	12,800
	b. Multiple sites	4000–8000
Chicago, IL	Proximity to expressway	6600
Philadelphia, PA	Proximity to expressway	3000–8000
New York City, NY	Proximity to expressway	2000[c]
Detroit, MI	"Street dust"	966–1213[d]
Various U.S. Cities	Highways and tunnels	10,000–20,000
Several U.S. Cities	"Street dusts"	300–18,000[e]

[a]*Adapted from U.S. EPA (1977).*
[b]*Figures for early 1970s, U.S. EPA (1977).*
[c]*Pinkerton et al. (1973).*
[d]*Ter Haar and Aranow (1974).*
[e]*Nriagu (1978).*

TABLE 6.24 Illustrative U.S. Dust Lead Deposition Rates ($mg/m^2/mo$) in the Period of High Leaded Gasoline Consumption[a,b]

Location of Highest Deposition Rate	Deposition Rate	Geometric Mean Deposition, 77 Cities, by Category
South Bend, IN	80 $mg/m^2/mo$	Residential 5.2
Nashville, TN	346 $mg/m^2/mo$	Commercial 9.8
Omaha, NE	137 $mg/m^2/mo$	Industrial 12.8

[a]*Adapted from Hunt et al. (1971) and U.S. EPA (1977).*
[b]*For 77 Midwestern U.S. cities, gathered for the fourth quarter 1968.*

and organic litter covers. Some illustrative values are presented in Table 6.26 for U.S., European, and Asian locales. While the dust lead levels are elevated well above background amounts, they are simultaneously well below comparative figures from past decades of high leaded gasoline use.

TABLE 6.25 Illustrative Non-U.S., International Studies of Outside Dust Lead Levels (μg/g) in the Period of Elevated Gasoline Pb Consumption

Area(s)	Testing Sites	Dust Lead Level (μg/g)	References
Canada, Jamaica, New Zealand	Small urban areas	700–2,000	Fergusson and Ryan (1984)
Belgium	Busy roadways, street dust	500–2,500	Deroanne-Bauvin et al. (1987)
Hong Kong	Street dust	960–7,400	Lau and Wong (1982)
		130–3,900	
United Kingdom	Urban versus Rural street dusts	Urban, 970 mean	Day et al. (1975)
		Rural, 85 mean	
Lancaster, U.K.	Different test sites	Car parks (a) 46,300 mean	Harrison (1979)
		Car parks (b) 4,560 mean	
		Town center: 2,130 mean	
		Main roads: 1,890 mean	
		Residential areas: 850 mean	
London, U.K.	14 road areas in the city	1,360–3,400	Jensen and Laxen (1985)
The Netherlands	Roadways with high traffic density	Mean = 5,000	Rameau (1973)
Christchurch, New Zealand	Street dusts in selected neighborhoods	Mean = 1,160	Fergusson and Schroeder (1985)

Lead in Interior/Household Dusts

Relative to exterior leaded dusts, household and other interior dust lead reservoirs pose a higher lead exposure risk to vulnerable human populations such as young children. More so than roadside or other exterior leaded dusts, interior dusts generally reflect more lead source inputs, the relative size of

TABLE 6.26 Illustrative International Studies of Road Dust Pb Levels in Recent, Post-Gasoline Lead Use Years

Locale	Road Type	Pb Level (ppm)	References
Honolulu, HI	Three paved road areas	Means: 588	Sutherland et al. (2003)
		470	
		151	
Reno-Sparks, NV	Urban paved road	~100	Gillies et al. (1999)
San Joaquin Valley, CA	Six test areas of different road types	Means: Urban paved road, 161	Chow et al. (2003)
		Rural paved road, 57	
		Composite paved road, 109	
		Agricultural unpaved road, 58	
		Residential unpaved road, 203	
		Unpaved composite road, 101	
Oslo, Norway	Urban paved road	180	de Miguel et al. (1997)
Madrid, Spain	Urban paved road	1,927	de Miguel et al. (1997)
Hong Kong	Two urban paved roads	1,061	Ho et al., 2003
		1,209	

these inputs depending on specific circumstances. For example, old, deteriorated housing in United States' older, inner cities often have significant inputs from deteriorating leaded paints to household dusts. Interior/household dusts can also arise from factors other than purely environmental ones. Family members working in lead dust-generating industries and businesses pose the risk of secondary, indirect lead exposures of young children in the family when workplace dusts are brought home. The literature for household/interior dusts typically does not assign origin of their lead content, but some extensive databases exist for source-specific contributors including

TABLE 6.27 Illustrative International Household/Indoor Dust Levels (μg/g) or Loadings (μg/m^2)

Location	Dust Lead Concentration (μg/g)	References
Omaha, NE	18–5,600	Angle and McIntire (1979)
Cincinnati, OH (Various housing categories)	70–16,000	Clark et al. (1985)
Lancaster, U.K.	Mean: 720 Range: 510–970	Harrison 1979
New York City (Residential sites)	610–740	Pinkerton et al. (1973)
Philadelphia, PA (Industrial neighborhoods)	930–16,000	Needleman et al. (1974)
Edinburgh, Scotland (Floor dusts)	43–13,600 Median: 308	Laxen et al. (1987)
Jersey City, NJ (Floor dusts)	1,133 (PM_{10})	Adgate et al. (1998)
Midwest United States (Pb Loadings)	5,140 μg/m^2	Clayton et al. (1999)

those concerned with the relationship of deteriorating interior (and exterior) LBPs to interior dust lead health hazards. These data are included in the following tables.

Illustrative household dust lead amounts, as concentrations or loadings, reported in the United States and the United Kingdom are contained in Table 6.27. Levels by both metrics are quite high, ranging in time from the 1970s to the late 1990s. This persistence of interior dust lead content extends into later years after leaded gasoline was removed from vehicular fuel.

Table 6.28 presents some typical reports of interior dust lead levels near lead smelters and other point sources situated around the world, covering several decades. There is a considerable range of levels and relatively high mean values. In terms of source rankings, highest levels are associated with primary and secondary lead smelters, followed by residences impacted by mining and milling operations. Hartwell et al. (1983) reported median leaded dust values for lead–zinc and copper smelters of between 116 and 441 ppm, but other data in Table 6.28 cite levels of several thousand ppm or even

TABLE 6.28 Illustrative International Interior Dust Pb Levels Around Smelters and Other Point Sources

Location	Testing Area	Dust Pb Level (ppm)	References
Bartlesville, OK	Zone closest to Zn smelter	441 (median)	Hartwell et al. (1983)
Palmerton, PA	Zone closest to Zn smelter	438 (median)	Hartwell et al. (1983)
Ajo, AZ	Zone closest to copper smelter	116 (median)	Hartwell et al. (1983)
Anaconda, MT	Zone closest to copper smelter	398 (median)	Hartwell et al. (1983)
Belgium	<1 km from primary Pb smelter	2,517 (estimated from original figure)	Roels et al. (1980)
Trail, BC	Area closest to primary Pb–Zn smelter	4,676 (floor) 9,014 (window)	Hertzman et al. (1991)
Arnhem, the Netherlands	Proximity to secondary Pb smelter	457–8,100 Mean = 1,140 (total Pb) Mean = 1,050 (fine fraction)	Diemel et al. (1981)
Derbyshire, U.K.	a. Mining area, high soil Pb	1,050–28,000	Barltrop et al. (1975)
	b. Low soil Pb	130–3,000	
Telluride, CO	Former mining and milling area	281	Succop et al. (1998)
Midvale, UT	Former milling and smelter site	438	Succop et al. (1998)
Butte, MT	Former mining and milling site	451	Succop et al. (1998)
Leadville, CO	Former mining, milling, and smelter site	638	Succop et al. (1998)
Pribram, Czech Republic	Proximity to Pb smelter	1,984	Rieuwerts and Farago (1996)
Shipham, U.K.	Proximity to Pb mining site	786	Thornton (1988)
Port Pirie, Australia	Proximity to Pb smelter	1,407–4,590 (PM_{250}) 1,693–6,799 (PM_{53})	Oliver et al. (1999)

(*Continued*)

TABLE 6.28 Illustrative International Interior Dust Pb Levels Around Smelters and Other Point Sources—(cont.)

Location	Testing Area	Dust Pb Level (ppm)	References
Helena, MT	Proximity of primary Pb smelters	1,598	Schilling and Bain (1988)
Kellogg, ID			
El Paso, TX (Smeltertown)	Various distances from smelter	0–1 mi: 36,853	Landrigan et al. (1975)
		1–2 mi: 2,726	
		2–3 mi: 2,234	
		>4 mi: 2,151	

higher. Mining and milling sites, as summarized by Succop et al. (1998), show mean values of 281–451 ppm Pb.

The 2001 U.S. national housing survey described for the case of lead paint also presented details of the relationships of interior dust lead levels with LBP. Tables 6.29–6.31 show interior dust lead data for U.S. residential units with respect to lead paint-related housing variables: dust lead loadings (μg/ft^2) versus interior surfaces at different threshold values for dust Pb (Table 6.29); interior dust lead loadings versus housing age (years of construction; Table 6.30) at different threshold values for dust Pb; and the absence or presence of dust Pb hazards with respect to absence or presence of LBP (Table 6.31). Table 6.29 shows, for uncarpeted floors, that 8,512,000 U.S. housing units, or 9% of all such units, had interior dust lead at a loading of 10 μg/ft^2, while the tally and percent fraction of units ≥40 μg/ft^2, the current U.S. HUD threshold for defining a floor lead hazard, was 2,449,000 or 3% of all U.S. housing.

Window sill troughs showed dust lead loadings above the testing detection limit for 78,936,000 units, 83% of all U.S. units. A total of 8,287,000 units or 9% of the total tally had window sill dust lead loadings at or above 250 μg/ft^2, the dust hazard value for window sills. Window troughs (window wells) in 16,395,000 or 17% of U.S. housing showed surface Pb loading ≥800 μg/ft^2, the previous Pb hazard threshold value for troughs. The current window well/trough Pb level, defined as "a clearance" rather than a "hazard" standard, is 500 μg/ft^2. The clearance value is a figure abatement contractors must not exceed for satisfactory hazard reduction or control.

Table 6.30 presents the distribution of dust lead loadings on bare floors, window sills, and window troughs as a function of housing age, i.e., years when the buildings were constructed. For all dust lead loading threshold values for all three surfaces, the older the housing, the higher the number of

TABLE 6.29 Distribution of Average Loadings (μg/ft^2), All Interior Dust Lead, in U.S. Housing[a,b]

Threshold Value, Pb-Dust (μg/ft^2)	Estimated National Tally (000)	Percent of National Tally
Uncarpeted floors		
≥10	8,512	9
≥20	4,843	5
≥40	2,449	3
≥100	966	1
Window sills		
≥Detection limit	78,936	83
≥125	13,875	15
≥250	8,287	9
≥500	4,900	5
Window troughs		
≥Detection limit	72,349	76
≥800	16,395	17

[a]*Adapted from NSLAH (2001).*
[b]*Detection limits vary with tested surface, NSLAH (2001, Ch. 5).*

units in that category and thereby the higher the fractional percentage. This is especially so for the threshold values corresponding to the U.S. HUD definitions of the various surface dusts lead hazard loadings: 40, floors; 250, window sills; 800, window troughs. For uncarpeted floors, pre-1940 units accounted for 2,843,000 or 16% of dust lead measurements at or above the floor standard but 0% in units built between 1978 and 1998. The pre-1940 versus 1978–1998 categories offer particularly striking comparisons for the dust Pb hazard values for window sills and troughs.

The critical role of LBP in the distributions of dust Pb loadings on interior surfaces recorded in Tables 6.29 and 6.30 can be seen in the data depicted in Table 6.31. When LBP is present and in "good" condition as defined in the national survey, 33% of units had a dust Pb hazard as defined by the U.S. HUD (1999), but in cases of significant deterioration of the LBP present, the corresponding percentage was roughly double, 61%. The absence of LBP in the unit resulted in only 6% having a dust lead hazard of some type on some surface.

TABLE 6.30 Distribution of Maximum Surface Dust Pb Loadings (μg/m^2) in All U.S. Housing as a Function of Construction Years[a,b]

Threshold Value Pb-Dust (μg/ft^2)	Pre-1940 Number (000)	Pre-1940 %	1940–1959 Number (000)	1940–1959 %	1960–1977 Number (000)	1960–1977 %	1978–1998 Number (000)	1978–1998 %
A. Floor Pb loadings								
≥10	7,386	42	4,938	24	2,488	9	1,153	4
≥20	4,996	29	2,784	14	1,112	4	97	0
≥40	2,843	16	1,967	10	588	4	97	0
≥100	1,114	6	935	5	280	1	97	0
B. Window sill Pb loadings								
≥Detection limit	16,803	96	18,779	91	24,729	89	21,823	73
≥125	9,028	52	5,407	26	4,097	15	1,806	6
≥250	6,943	40	3,712	18	1,755	6	1,029	4
≥500	4,980	29	2,869	14	747	3	447	2
C. Window trough Pb loadings								
≥Detection limit	14,143	86	16,406	80	20,319	73	20,969	70
≥800	8,883	51	6,286	31	3,788	14	2,252	8

[a]*Adapted from NSLAH (2001).*
[b]*Detection limits vary with tested surface, NSLAH (2001, Ch. 5).*

TABLE 6.31 Relationship of U.S. Housing Units[a] with or without Interior Dust Lead Hazards[b] as a Function of Interior LBP[c]

Interior Dust Lead Hazards	No Interior LBP Number (000)	No Interior LBP %	LBP in Good Condition Number (000)	LBP in Good Condition %	Significant LBP Number (000)	Significant LBP %
A. Absent	62,752	94	15,244	67	2,389	39
B. Present	4,068	6	7,508	33	3,727	61
Total units	66,820	100	22,752	100	6,116	100

[a]*Adapted from NSLAH (2001).*
[b]*Dust lead hazard as defined by U.S. HUD (1999).*
[c]*LBP as defined by U.S. HUD (1999).*

6.3.4 Lead in Soils

Natural or background levels of lead in soils are quite low and largely reflect the low levels of lead in the parent rocks giving rise to the uncontaminated soils. One exception would be elevated lead levels in those limited metalliferous soils in regions of the world with that surface mineralogy and accompanying extractive industrial activities. Background levels of lead have typically been determined by sampling done in two ways. The first is soil core testings in rural or remote areas assumed to be free of contaminating, i.e., anthropogenic, activity. The second approach consists of deeper soil core measurements and analysis of soil segments below any depth assumed to be reflective of anthropogenic activity. Both strategies require supporting data ruling out any artifactual but undetected anthropogenic contributions that result in higher lead contents than expected for the unaltered soil. Extremely low air lead values would indicate no or little atmospheric anthropogenic depositions of lead onto these soils that could artificially elevate levels over true background. Vertical strata samplings of soils would require some evidence that soil layers have not been altered through, e.g., farming, grading during property development, etc. Vertical invariance in serial deep-core soil testings would indicate the testing region is below depth-variable additions from anthropogenic activities.

Measurements of soil Pb are critically dependent on the sampling methods for data gathering. As an environmental medium, soil is quite complex in its chemical, physical, and physicochemical matrix characteristics and dependence of Pb content on distributions in the soil matrix. For example, ambient air Pb depositions on soil or depositions from lead-painted surfaces weathering, or reentrained dusts from waste sites distribute the element in soil with vertical and/or horizontal soil area variability. Air Pb typically deposits onto soils from the atmosphere in the top 2–5 cm, owing to binding of Pb by bioorganic or geological material such as vegetative litter. Soil Pb also varies with distance from point or mobile sources in direct reflection of ambient air Pb. Lead releases from weathering exterior lead paint surfaces produce depositions proportional to distances of soil areas from the building foundation or "drip line." Increased Pb levels near the drip line may also occur to some extent from airborne Pb striking sides of buildings and fallout. However, this likelihood is ruled out when fronts and backs of structures show similar Pb levels. Anthropogenic Pb differs inversely with soil particulate size with coarse fractions of soil having lower Pb content while Pb content increases with decreased soil particle size (Young et al., 2002). This particular relationship is significant for measuring soil Pb relevant to human exposures because of the propensity for small particles to adhere to children's hands and then be ingested.

Shacklette et al. (1971) reported an arithmetic mean of 20 μg/g (ppm) and geometric mean of 16 ppm for 863 background soil lead samples at various U.S. sites, most of which were <30 ppm. These means were quite close to the arithmetic mean of 20 ppm reported for Canadian nonagricultural soils by McKeague and Wolynetz (1980). Rural soils in the United Kingdom were reported to have geometric means of 42 ppm (Davies, 1983) and 48 ppm (McGrath, 1986).

There is a global literature on the topic of lead contamination of soils, particularly for industrialized countries. A notable feature of Pb in soils is its persistence there, with studies showing a residence time or half-life estimated to range from decades to centuries (U.S. EPA, 2006). For example, Freitas et al. (2004) reported a mean soil Pb of 2,694 ppm for samples gathered at a long-defunct Portuguese copper mining site worked in pre-Roman and Roman times.

Retention of Pb in soils translates to concentrations of the element which do not vary widely over time, in contrast to data for atmospheric Pb emissions which show considerably more rapid changes in response to source emission rates. As one consequence, alterations in metallurgical operations such as primary lead smelters in the form of more efficient Pb emission controls lead to reductions in atmospheric air Pb and dustfall Pb but not reductions in lead content of nearby soils (Hilts, 2003).

Table 6.32 presents a range of largely urban soil Pb measurements gathered using diverse sampling methods and for various purposes from around the globe. Soil Pb content for urban properties in the United States typically range over 1,000 ppm expressed in various metrics, and in some cases, well above this figure. Central New Orleans, LA, provided measurements ranging up to 69,000 ppm. Yard soils were typically well above garden soil samples, expected from tilling soils when planting. The deeper the tilling, the greater the vertical mixing of higher Pb levels in upper strata with lower concentrations deeper in the soils. Soils less impacted by typical Pb input sources show lower values.

The impact of exterior LBP weathering over time on adjacent, i.e., dripline and other perimeter soil, zones is apparent in Table 6.32. Ter Haar and Aranow (1974) found that soil Pb adjacent to LBP-containing wood frame buildings was much higher in Pb level than soil Pb levels 10 or 20 ft from the foundation. Urban Detroit, MI, soils showed Pb at 2,349 ppm at 2 ft away, versus a much lower 447 ppm at 10 ft. Similar differences with increasing distance were observed for rural buildings. The levels of Pb in exterior paints used in U.S. housing differed with housing age in both frequency of use and Pb concentrations or loadings used, so that the older the housing the higher the Pb content and the higher likelihood lead paint was used, especially for exterior surfaces. Subsequent outside weathering of age-variable painted surfaces, in turn, would result in differing amounts of Pb in contiguous soils. Stark et al. (1982) showed that yard soils on lots with housing built in 1920–1929 were almost 10-fold higher in Pb content than were yards with younger units, built in 1970–1977.

TABLE 6.32 Illustrative Lead Levels (μg/g, ppm) in Global Soils[a,b,c]

Locale	Soil Sites	Pb Concentration (μg/g)	References
Various U.S. cities in a U.S. HUD program	Urban soil collections	1,043 ppm (GM)[d]	Clark et al. (2004)
Baltimore, MD, St. Paul, MN soils	Gardens or surface sampling	Baltimore, MD: Median 100	Chaney and Mielke (1986)
		St. Paul, MN: Median 228	
Three CA cities: Oakland, Los Angeles, Sacramento	Yard soils	Oakland: GN 897	Sutton et al. (1995)
		Los Angeles: GM 188	
		Sacramento: GM 227	
Central New Orleans, LA	Soils	Up to 69,000	Ter Haar and Aranow (1974)
Detroit, MI	Urban soils versus frame house distance	2 ft, front: 2,349	Ter Haar and Aranow (1974)
		10 ft, front: 447	
Rural area, MI	Soils versus frame house distance	2 ft: 2,529	Ter Haar and Aranow (1974)
		10 ft: 609	
		20 ft: 209	
Miami, FL	Urban soils	93	Chirinje et al. (2004)
Cincinnati, OH	Roadside soils	59–1,980	Turer et al. (2001)
New Haven, CT	Yard soils, variable housing ages	131, 1970–1977	Stark et al. (1982)
		1,273, 1920–1929	
Charleston, SC	Household soils	9–7,890, depending on traffic, LBP	Galke et al. (1975)
Chelsea, MA	Soils around a painted bridge being remediated	8,127: under bridge	Landrigan et al. (1982)
		<30 m: 3,272	
		30–80 m: 457	
		100 m: 197	

(*Continued*)

TABLE 6.32 Illustrative Lead Levels (μg/g, ppm) in Global Soils[a,b,c]—(cont.)

Locale	Soil Sites	Pb Concentration (μg/g)	References
Omaha, NE	Urban soils	16–4,792	Angle and McIntire (1982)
Boston, MA	Urban soils in Boston area	7–13,240 Mean = 702	Rabinowitz and Bellinger (1988)
Egypt	Multiple cities urban soils	23–200	Badawy et al. (2002)
United Kingdom	Urban soils	Up to 14,100	Culbard et al. (1988)
New Zealand	Wellington urban soils near old housing	GM: 21–1,890, increasing with housing age	Bates et al. (1995)
The Netherlands cities	Urban area garden soils	GMs: 43–336	Brunekreef et al. (1983)
Toronto, Canada	Urban soils in Toronto, Windsor	48–54	O'Heany et al. (1988)

[a]*Mainly urban soils not affected by smelting, mining, and milling.*
[b]*Reflects periods of leaded gasoline use or recent phaseout data.*
[c]*Various soil-sampling methods used.*
[d]*GM = geometric mean.*

Urban soil Pb values outside the United States show similar marked elevations over background values. Culbard et al. (1988), for example, reported U.K. urban soils ranging up to 14,100 ppm. Urban soils adjacent to LBP-containing housing in older sections of Wellington, New Zealand, were found to increase in Pb content with increasing housing age (Bates et al., 1995).

A large database has developed with reference to soil Pb values derived largely from Pb point sources, particularly smelting, milling, and mining sites. These point sources are often in rural areas, especially primary metal smelters, where extractive industrial activity will typically be the dominant contributor of Pb emissions to site and nearby community soils. By contrast, secondary Pb smelters or lead-acid battery manufacturing plants may be found in either rural or urban/suburban areas. Soil Pb contamination from extractive industrial operations can occur through several pathways: deposition of point source-generated atmospheric Pb, fugitive dusts windblown from on-site storage to soils offsite, or by deposition of windborne Pb dusts from tailing waste piles on community yards.

One of the largest extractive industry sites in the world was the Bunker Hill lead and zinc smelting, milling, and mining complex in the Silver Valley of Idaho. This site, closed in 1981 after more than a century of operation, soon became a U.S. EPA Superfund hazardous waste site. In the early 1970s, while the complex was still operating, an extensive series of environmental contamination studies began over concerns about childhood lead poisonings within nearby communities (Walter et al., 1980; Yankel et al., 1977) and was followed by multiple environmental regulatory characterizations of the Bunker Hill complex and its contiguous communities. These characterizations included the nature and extent of the Bunker Hill Pb waste stream within and downstream of the complex, including the levels of Pb and other geochemical contaminants in soils, dusts, air, and plant life as well as exposure levels in humans and ecological populations.

Table 6.33 is a tally of soil Pb levels measured within two communities significantly impacted by the Bunker Hill operations, Kellogg and Smelterville, ID, for the period 1974–1993, a period bracketing many of the

TABLE 6.33 Average Soil Pb Levels (μg/g, ppm) Versus Calendar Years of Intervention at the Silver Valley, ID, Smelter Site[a,b,c,d]

	Kellogg, ID[e]		Smelterville, ID[e]	
Year	AM[f]	GM[f]	AM[f]	GM[f]
1974	3,073	2,255	7,386	5,770
1975	3,918	2,658	5,581	3,907
1983	–	–	6,231	4,188
1988	3,195	2,609	2,932	2,198
1989	2,880	2,376	2,900	1,725
1990	1,572	608	1,906	719
1991	846	226	1,180	292
1992	994	276	1,264	292
1993	772	223	1,639	339

[a]*Adapted from TerraGraphics Environmental Engineering (2000) as prepared for Idaho Department of Health and Welfare.*
[b]*For the years 1974–1993.*
[c]*Smelting operation ended in 1981.*
[d]*Designated a Superfund site by the EPA.*
[e]*Community highly affected by Pb emissions.*
[f]*AM, arithmetic mean; GM, geometric mean.*

studies of communities within the Bunker Hill impact zone. Clearly evident are the declines in soil Pb levels in both communities, indexed as either arithmetic or geometric mean values. Soil Pb levels at the time of little soil Pb remediation, 1974 and 1975, indicated lead values in the thousands of ppm. For example, the arithmetic means for Smelterville, ID, soil Pb values in 1974 and 1975 were 7,386 and 5,581 ppm, respectively. Associated figures for the geometric means were 5,770 and 3,907 ppm, respectively. The soil Pb arithmetic mean in Smelterville in 1993, well after removals of residential soils exceeding the Pb action level were under way, was 1,666 ppm while the geometric mean was 300 ppm.

Evidence of increased Pb contamination from other extractive industry point sources of Pb has been seen in and around neighboring communities. Table 6.34 summarizes some illustrative examples. Leadville, CO, the site of many decades of smelting, milling, and mining activities dating to the nineteenth century, had a soil Pb mean (geometric) of 1,763 ppm (Colorado Department of Health, 1990). A large lead–zinc smelter in Trail, British Columbia, Canada, produced group means of soil Pb of 225–1,800 ppm (Neri et al., 1978), values not materially different from later testings in this community with changes in emission controls (Hilts, 2003). Derbyshire, United Kingdom, where lead mining occurred over several hundred years, produced group means in soil Pb ranging from 420 to 13,969 ppm.

One typical feature of extractive industry primary and secondary smelters as contaminating point sources is the inverse relationship of soil Pb levels to distances from the emitting sources to the soil being tested. The quantitative features of this inverse relationship are affected by height of stack, topographical features of the areas, wind patterns, and other meteorological features such as rainfall, etc. Godin et al. (1985) examined soil Pb versus distance at several smelter sites. Liu (2003) reported the soil Pb values at increasing distance (m) from several smelters: 313 ppm, 100 m; 217 ppm, 5,000 m; 110 ppm, 10,000 m; 57 ppm, 20,000 m; 33 ppm, 30,000 m.

The 2001 U.S. NSLAH included data for residential soil Pb levels associated with U.S. housing units characterized in various ways: number and percentage distributions by soil Pb thresholds, distribution by soil Pb thresholds and housing age, and distribution by absence or presence of significantly deteriorated exterior LBP surfaces. Collectively, these data define the relationship of lead paint to soil lead values in and around LBP residential units. Table 6.35 indicates that 15,299,000 U.S. housing units, 16% of the total, were associated with soil Pb levels of 200 ppm or higher. The 400 ppm or higher soil Pb level, which for bare, play area soils defines a regulatory soil Pb hazard level, was projected to be found at almost 10 million housing units, 10% of all residences. Soil Pb values of 5,000 ppm or higher were projected to be found at 1,580,000 U.S. units, or 2% of all units.

The age of U.S. housing, i.e., the year of construction, has a significant effect on the distribution of soil Pb values ranked by differing soil Pb

TABLE 6.34 Other Illustrative Soil Pb Levels in Proximity to Smelting, Milling, and Mining Point Sources[a,b,c]

Locale	Pb Concentration (μg/g, ppm)	References
Jasper County, MO, U.S.A.	574	Murgueytio et al. (1998)
Leadville, CO, U.S.A.	1,110[d]	Colorado Department of Health (1990)
	1,763[e]	
Ottawa County, OK, U.S.A.	Median = 148, white children's homes	Malcoe et al. (2002)
	Median = 103, Native American children's homes	
Midvale, UT, U.S.A.	295[f]	Bornschein et al. (1991)
Sandy, UT, U.S.A.	362[f]	Succop et al. (1998)
Bartlesville, OK, U.S.A.	821[f]	Hartwell et al. (1983)
Palmerton, PA, U.S.A.	331[g]	Hartwell et al. (1983)
Anaconda, MT, U.S.A.	424[g]	Hartwell et al. (1983)
Derbyshire, U.K.	420–13,969, group means	Barltrop et al. (1975)
Trail, BC, Canada	225–1,800, group means	Neri et al. (1978)
Trail, BC, Canada	GMs: 559–2,002, areas closest to smelter	Hertzman et al. (1991)
Arnhem, the Netherlands	240[f]	Diemel et al. (1981)

[a] *Urban or rural soils affected by smelting, milling, or mining activities.*
[b] *Includes past and present industry history.*
[c] *Various soil-sampling methods.*
[d] *Front yard soil, geometric mean.*
[e] *Front yard soil, arithmetic mean.*
[f] *Garden soil.*
[g] *Results closest to smelter.*

thresholds. In Table 6.36, for pre-1940 housing, the oldest age group, the regulatory bare soil Pb hazard level of 400 ppm or higher was encountered in 11,613,000 U.S. residences, or 67% of all U.S. housing built before 1940. The very high soil Pb threshold of 5,000 ppm or higher was seen in almost 2 million units, or 11%. By contrast, the newest housing, built in 1978–1998, showed no discernible percentage at or above 400 ppm.

TABLE 6.35 Distribution of Maximum Soil Pb Levels in U.S. Housing for Indicated Thresholds: Unit Counts and Percentages[a,b,c]

Bare Soil Threshold Pb	Number (000)	% of All U.S. Units
≥200	15,299	16
≥400	9,996	10
≥1,200	6,271	7
≥2,000	3,124	3
≥5,000	1,580	2

[a]*Adapted from NSLAH (2001).*
[b]*Total U.S. units = 95,688,000.*
[c]*Soil-sampling methods in original report.*

TABLE 6.36 Maximum Soil Pb Distributions in U.S. Housing for Indicated Soil Pb Thresholds as a Function of Construction Years[a,b,c]

	Years of Construction							
	Pre-1940		1940–1959		1960–1977		1978–1998	
Threshold soil Pb (ppm)	Number (000)	%[d]	Number (000)	%[d]	Number (000)	%[d]	Number (000)	%[d]
≥200	13,314	76	9,950	48	4,495	16	1,476	5
≥400	11,613	67	6,283	31	2,410	9	84	0
≥1,200	6,536	37	3,922	19	686	3	–	0
≥2,000	3,929	23	2,194	11	686	3	–	0
≥5,000	1,891	11	865	4	231	1	–	0

[a]*Adapted from NSLAH (2001).*
[b]*Bare and covered soils.*
[c]*Sampling methods described in original report.*
[d]*Percentage of all units in the indicated age period.*

Table 6.37 indicates that the absence or presence of significantly deteriorated exterior LBP is a major factor in soil Pb concentration. At the 400 ppm or higher bare soil hazard level, the presence of significantly deteriorated exterior LBP increases the Pb concentration almost fourfold (30% versus 8%). The ratios are even more telling at the upper end of the soil Pb thresholds. With significant LBP deterioration, soils with Pb ≥5,000 ppm are about eightfold higher than test sites without LBP.

TABLE 6.37 Bare Soil Pb Percentage Distributions in U.S. Housing for Indicated Soil Pb Thresholds in the Absence or Presence of Significant Exterior LBP Deterioration[a,b,c]

Threshold Bare Soil Pb (ppm)	Percent Versus Significant Exterior LBP Deterioration Status	
	Not Deteriorated	Deteriorated
≥200	13	39
≥400	8	30
≥1,200	4	24
≥2,000	2	13
≥5,000	1	8

[a]*Adapted from NSLAH (2001).*
[b]*All samplings.*
[c]*Sampling methods described in original report.*

6.3.5 Lead in Human Diets

Naturally occurring levels of Pb in diets consumed by early peoples have not been extensively studied. Settle and Patterson (1980), using careful laboratory techniques, reported that fresh albacore muscle in prehistoric times was one-tenth that of modern catches. Anchovies, similarly, were enriched in Pb content by 10-fold, from 2.1 to 21 ppb. Patterson (1982) found that edible plant Pb content averaged up to 2.0 ppb in prehistoric times.

Lead in the diet has long been known to be a significant source of ingested lead in human populations, with records of lead ingestion at toxic levels tracing back to at least the Greco-Roman and Roman eras (Nriagu, 1983a,b, 1985 and other citations, Ch. 2). The intervening centuries have recorded considerable evidence of dietary lead exposures. The very early record does not distinguish vulnerability to effects of dietary lead across subsets of human populations, but by the late nineteenth century episodes of childhood poisonings from lead ingestion were recorded. Stewart (1887, 1895) reported the poisoning of children in Philadelphia, PA, from bakery product coloring adulterated with lead chromate and possibly mixed with other lead compounds, producing yellow and highly toxic icings. A number of children died.

Dietary lead differs from other media-specific lead sources and pathways for human populations in a number of ways. Lead in soils and dusts typically affects young children more than adults, while food Pb can produce more Pb intakes and exposures in adults or, equally, children and adults. Dietary components acquire Pb via various, complex mechanisms. For example, leafy vegetables can acquire Pb by foliar deposition of airborne lead or by growing in contaminated soils. Canned foods are further contaminated through processing methods and lead-seamed cans, a common food container until relatively recently. Lead can bind to foods cooked in Pb-contaminated cooking water through Pb transfer from water.

Food components differ greatly in lead content, reflecting in part differences in production and distribution. Food Pb intakes differ as a function of, e.g., food consumption habits. This has required attention not only to concentrations of lead in classes of dietary components but also to quantities of these components consumed on a typical or daily basis. Various national or international surveys, for that reason, have determined and reported both levels of Pb in diet categories and population consumption patterns to permit quantification of net exposures to lead in diet.

Some reports on dietary Pb measure lead content in diet distinct from any Pb intakes in water used for preparation of beverages or food. This chapter discusses drinking water Pb separately, and attempts to address the additive nature of water Pb intakes from beverages and prepared foods when such data can be separately identified.

National food supplies in industrialized or developed societies are often centralized, so that national food Pb surveys in, e.g., the United States and/or Europe, provide relevant information about typical exposures across regions and demographic subsets of the population. There are idiosyncratic exceptions to this rule, however, in the case of ethnic foods such as canned goods in lead-seamed containers imported for traditional local or selective consumption.

The nature and extent of Pb intakes from diet are quite dependent on human host factors of age, gender, socioeconomic status, lifestyle practices, etc. Infants consuming infant formula reconstituted from tap water differ in their Pb intakes from older children consuming baby foods, and the latter differ from still older children consuming foods in patterns resembling those of adults.

Dietary Pb content is affected by contamination of dietary components on a much different time scale than either ambient air Pb at one temporal extreme (shortest) or soil and roadside dust Pb on the other. Food grown in areas with air Pb deposition on those crops will reflect deposition rates from air across the entire growing season, regardless of whether atmospheric emissions vary markedly over that interval. Root crops will sustain some uptake through the root system and by root surface contamination. Canned goods with an acidic matrix and stored in lead-seamed cans may have lead leaching over time of storage before use.

Lead in the diet comes from dietary components whose concentrations of the element are relatively low, compared to levels of the substance in, say, dusts and soils. However, unlike these other environmental media, intakes of

Pb in diet on a daily or other time scale entail much larger amounts in terms of relative mass. For example, adults typically will consume only small amounts of lead in dusts and soils, 25–150 mg for various age bands, while ingesting food at a daily rate of 1 kg or more. Consequently, total daily intake of 100 mg of dust having a concentration of 1,000 ppm Pb yields the same Pb intake as 0.1 ppm Pb in 1.0 kg of diet. Other factors differentiating diet Pb from Pb in other media can potentially include a higher relative bioavailability of food Pb once ingested.

Pb in dietary components at low concentrations versus Pb in other media dictates the need for quite sensitive measurement and sampling methods, methods that only became available for routine use in the mid–late 1970s. One should therefore be cautious in evaluating earlier Pb in human diets measurement data.

It can generally be said that, at least for food supplies consumed in developed or industrialized nations, dietary lead has declined over the past several decades, owing to declines in Pb released to food components, e.g., major reductions in lead-seamed can use to virtually zero, and reduced ambient air Pb levels. For example, evaporated milk in lead-seamed cans was heavily contaminated by Pb leaching from the seams, with Mitchell and Aldous (1974) reporting an average evaporated milk value of 202 μg/l and a range up to 820 μg/l. This section presents data for several blocks of time, from the 1960s–1970s to the present. This is done for the same reason a longer temporal reach was used for air and other media Pb emissions over time. Body Pb accumulations in older segments of human populations reflect Pb intakes and bone Pb deposition in past years. These reservoirs of body Pb can be sources of toxic exposures through Pb resorption to blood.

National and International Dietary Pb Surveys

A number of dietary Pb surveys have been carried out in the United States and around the world. Summaries of these surveys are presented in this section. Some surveys have simply reported descriptive statistics for lead concentrations in dietary groups and levels of Pb in individual food components within those groups, e.g., measured Pb levels in cereals as a group. Other surveys have reported Pb levels in dietary components and coupled these with consumption patterns to provide intakes of food Pb in some time frame, typically as daily total intakes. Some survey reports have mainly concerned themselves with total dietary Pb intakes. This chapter confines itself to Pb levels in foods and presents intakes and uptakes in the context of human exposures in later sections.

Early attempts at quantifying Pb in human diet were plagued by absence of appropriate statistical survey methods, limited reach of any survey, and the problematic nature of measurement methods in terms of laboratory sensitivity and specificity, clean-laboratory techniques, etc. The question of sensitivity is a critical one, given that even in contaminated environments dietary Pb occurs at much lower levels than in dusts and soils. The 1961 study by Schroeder and Balassa of U.S. unprocessed food components found Pb content of 0–1.3 ppm, vegetables; 0–1.4 ppm, grains; 0–3.7 ppm for meats and eggs; 0.2–2.5 ppm, seafoods; 0–1.5 ppm, condiments.

The U.S. Food and Drug Administration (U.S. FDA) has been doing surveys of Pb and other contaminants in the U.S. food supply since the 1970s. The structures of these surveys and their results have varied over the years, with combinations of food-specific Pb content, Pb content of food groups, and population subset-specific dietary Pb consumption rates. The U.S. FDA Compliance Program Evaluation for Fiscal Year 1974 (summarized in U.S. EPA, 1977) included the Heavy Metals in Food Survey. Throughout the U.S. FDA survey efforts, special focus has been on dietary lead ingestion rates in infants and toddlers, as well as older children. Besides greater toxicological vulnerability, the very young consume food at an age- and weight-specific rate that is two- to threefold higher than the rates for adults.

This early U.S. FDA survey offers some general conclusions. Most notably, canned foods typically had much higher Pb content than fresh food groups, all other Pb-contributing factors being more or less equal. For example, with adult food consumptions, the average Pb content for canned foods, 0.38 ppm, was more than twice that for noncanned samples, 0.16 ppm. Infant food lead contents included 0.33 ppm for juices and 0.09 ppm for food in jars.

Several food Pb surveys were carried out in the 1970s and 1980s. The most comprehensive U.S. survey was done by the U.S. FDA using various blocks of data, including groupings of U.S. foods for total diet profiles and Pb contents of diet components within these groupings (Beloian 1985; Pennington 1983; U.S. EPA, 1986a; U.S. FDA, 1985).

Table 6.38 lists illustrative food Pb concentrations that equal or exceed 0.05 ppm Pb fresh weight for food groups and individual foods within the groups for the U.S. population via market-basket surveys and across eight age/sex groups. Items were drawn from the total diet list of Pennington (1983) and presented in U.S. EPA (1986a), and U.S. FDA (1985). Overall, food Pb content at that time was significantly below 1.00 ppm and most items are <0.100 ppm. Canned foods were the major category of higher Pb-content foods. Of the canned products, the more acidic food items were higher overall. Comparing canned tomatoes, tomato sauce, and sauerkraut with other canned fruit and vegetables, the Pb levels were one or more orders of magnitude above natural values and represented increments of Pb contamination in the U.S. food supply during growing, processing, and marketing. Notable sources for this contamination were airborne Pb deposition and lead-seamed cans.

In the early 1980s, a second source of food Pb survey data was the 1982 Nutrition Foundation Report on the safety of Pb and Pb salts. A portion of that report tabulated U.S. and Canadian food supply lead levels. Table 6.39

TABLE 6.38 Lead Content of U.S. Dietary Items Having Pb Levels ≥0.05 μg/g, 1970s–Early 1980s[a,b,c]

Category Number	Food Item	Mean (μg/g)
8	Evaporated milk	0.083
27	Beef liver	0.083
32	Canned tuna	0.159
39	Canned pork and beans	0.130
45	Canned green peas	0.136
55	Canned corn	0.111
56	Canned creamed corn	0.102
57	White dinner rolls	0.084
82	Canned peaches	0.223
84	Canned applesauce	0.094
87	Canned fruit cocktail	0.221
93	Canned pineapple	0.093
101	Canned grape juice	0.053
104	Canned orange juice	0.053
106	Canned spinach	0.649
108	Frozen collards	0.074
112	Canned sauerkraut	0.524
118	Canned tomato juice	0.084
119	Canned tomato sauce	0.258
120	Canned tomatoes	0.218
122	Canned green beans	0.099
129	Canned mixed vegetables	0.081
130	Canned mushrooms	0.255
131	Canned beets	0.103
142	Spaghetti with meat sauce	0.136
145	Chili, beef, and beans	0.102
148	Meat loaf	0.093
150	Lasagna	0.070

(*Continued*)

TABLE 6.38 Lead Content of U.S. Dietary Items Having Pb Levels ≥0.05 μg/g, 1970s–Early 1980s[a,b,c]—(cont.)

Category Number	Food Item	Mean (μg/g)
153	Pork chow mein	0.076
157	Canned vegetable soup	0.073
176	Ice cream sandwich	0.058
187	Milk chocolate candy	0.073
189	Chocolate powder	0.055

[a]*Adapted from U.S. EPA (1986a, Ch. 7, Appendix D); U.S. FDA (1985).*
[b]*Means across four market-basket surveys, 1970s–early 1980s, and across eight age/sex groups.*
[c]*Total diet food list of Pennington (1983).*

depicts the Pb content data for illustrative U.S. foods within 12 food groups, including sugar items, desserts, and various types of beverages. Lead levels were recorded for uncanned foods and canned foods where indicated. In the baby foods group (canned), evaporated milk at 0.05–0.06 ppm Pb and canned juices and drinks ranked highest. Uncanned baby juices and drinks had Pb levels about half that of canned product. The highest food Pb value in the dairy group not in cans was butter at 0.07 ppm, while canned milk was highest overall, at 0.10–0.13 ppm.

The meat/poultry/fish group had highest Pb levels overall, in terms of fresh and canned forms. Canned foods Pb in this group was in the range 0.21–1.50 ppm. Cereals, potatoes, and leafy vegetables were intermediate in lead content and canned foods in these groups were two- to threefold higher than fresh/uncanned items. Root vegetables such as carrots and beets were relatively high in Pb, canned forms being 6- to 10-fold higher than uncanned examples.

Items in the garden vegetables, fruits, oils/fats, sugar items and desserts, and beverages generally were relatively lower in Pb content. Again, canned forms of these food groups were higher in Pb than uncanned examples, up to 10-fold higher in some cases.

Table 6.40 illustrates those Pb values in U.S. food items recorded from several sources during the 1990s, limited to reported concentrations at or above indicated Pb concentrations. The data from Dudka and Miller (1999) for foods ≥0.03 ppm are presented as concentrations per food mass, while the FDA Total Diet Survey records Pb content ≥2.0 μg/individual serving. Root and leafy vegetables were significantly elevated compared to other items in the Dudka and Miller report while the FDA survey noted, as was the case in

TABLE 6.39 Lead Levels in Selected U.S. Dietary Items from the 1982 Nutrition Foundation Report on Safety of Lead and Lead Salts in Food[a,b,c]

Food Group	Food Item	Lead Levels (μg/g, ppm)	
		Uncanned	Canned
Baby foods	Milk substitute	0.02	
	Mother's milk	0.03	
	Evaporated milk		0.05–0.06
	Meat/meat dinners	0.03–0.04	
	Vegetables	0.04	
	Juices and drinks	0.03	0.06
Dairy products	Milk	0.02	0.10–0.13
	Cheese	0.05	
	Butter	0.07	
Meat, poultry, fish	Beef, pork, lamb-veal	0.06	0.24
	Poultry	0.12	0.24
	Fish, excludes sardines	0.04–0.08	0.21–0.51
	Sardines	0.65	1.5
	Eggs	0.17	
	Cold cuts	0.06	0.24
Cereals	Breads	0.08	
	Flours	0.05	
	Cakes, cookies	0.03–0.05	
	Hot cereals	0.02	
	Ready-to-eat cereals	0.11	
	Rice, pasta	0.06	0.11
	Corn	0.01	
Potatoes	White cooked	0.05	0.12
	Sweet potatoes	0.08	0.25

(*Continued*)

TABLE 6.39 **Lead Levels in Selected U.S. Dietary Items from the 1982 Nutrition Foundation Report on Safety of Lead and Lead Salts in Food[a,b,c]—(cont.)**

Food Group	Food Item	Lead Levels (μg/g, ppm)	
		Uncanned	Canned
Leafy vegetables	Cabbage, broccoli	0.01–0.04	0.08
	Lettuce, other greens	0.12–0.15	0.39
	Salads	0.03–0.08	0.17
Legumes	Beans	0.01–0.04	0.16–0.?2
	Peas	0.03	0.27
	Soups	0.04	0.10–0.13
Root vegetables	Carrots	0.14	0.13
	Onions	0.18	0.32
	Beets	0.01	0.10–0.11
	Parsnips, turnips	0.05	0.32
Garden vegetables	Tomatoes	0.05–0.08	0.30–0.37
	Tomato paste	0.03	0.19–0.47
	Tomato juice	0.05	0.22
	Squash, pumpkin	0.03	0.36
	Peppers	0.02	0.32
	Juices	0.01–0.02	0.12–0.13
Fruits	Citrus	0.01	0.39
	Apples	0.02	0.22
	Apple juice	0.02	0.13
	Cherries	0.02	0.39
	Cranberries	0.05	0.25
	Grapes	0.01	0.28
	Peaches	0.01–0.03	0.19–0.39
	Pears	0.02	0.18–0.19

(*Continued*)

TABLE 6.39 **Lead Levels in Selected U.S. Dietary Items from the 1982 Nutrition Foundation Report on Safety of Lead and Lead Salts in Food[a,b,c]—(cont.)**

Food Group	Food Item	Lead Levels (μg/g, ppm) Uncanned	Canned
Oils and Fats	Salad dressing	0.01–0.02	
	Cooking oils	0.02	0.10
	Nuts	0.02	0.20
Sugar, desserts, jellies, etc.	Sugar	0.03	
	Chocolate	0.08	
	Candy	0.03	
	Puddings	0.01–0.03	
	Jellies	0.01	
	Pickles, olives	0.10	0.70
Beverages	Soft drinks	0.01–0.02	0.14–0.21
	Coffee	0.01	
	Tea	0.01	0.07
	Alcoholic beverages	0.01	0.02

[a]*Adapted from combined U.S. surveys in Nutrition Foundation (1982) report.*
[b]*Twelve food groups; includes beverages.*
[c]*Uncanned, plus canned levels where available.*

earlier surveys, that canned foods were significantly higher in Pb content than uncanned items. Among beverages, wine was highest at 7.7 μg/serving.

The most recent year for FDA survey efforts was 2004. The raw data set had all classes of contaminants and hundreds of food items containing each of those substances. Some food items shown in earlier surveys are not listed in the 2004 compilation, while some foods in this latest survey do not appear in earlier surveys. For purposes of this chapter, FDA 2004 dietary Pb data were first isolated from the full data set. Selected food items were then tabulated. Table 6.41 provides the illustrative 2004 data, along with Pb detection and quantitation measurement limits. It is readily apparent from Table 6.41 that virtually all foods in the U.S. FDA (2004) diet survey were very low in Pb, most items presenting as not measurable, or at trace (ppb) values. This marked decline for lead content in the U.S. diet is generally consistent with the removal of lead in gasoline, the abandonment of any lead-seamed

TABLE 6.40 Illustrative Pb Levels in the U.S. Diet Reported for 1990s Sources[a,b,c,d]

References/Source	Concentration
Dudka and Miller (1999)	μg/g, ppm, ≥0.03 ppm
Barley	0.21
Lettuce	0.19
Potatoes	0.21
Spinach	0.53
Carrots	0.05
Onions	0.04
Soybeans	0.04
Tomatoes	0.03
FDA TDS	μg/serving, ≥2.0 μg/serving
Applesauce, canned	8.5
Fruit cocktail, canned	7.1
Spinach, fresh	2.4
Peaches, canned	6.0
Peas, canned	4.9
Apple juice, bottled	2.6
Wine	7.7

[a]*Adapted from data in Dudka and Miller (1999) and FDA TDS, Juberg et al. (1997).*
[b]*For uncontaminated soils.*
[c]*Detection limits in original sources.*
[d]*Reported as either unit of mass (ppm) or serving (FDA TDS).*

containers for canned foods, and the reduction of lead in typical food production and transport.

As noted earlier, using lead-seamed cans for canned foods was, in some cases, the major contributor to Pb contamination of foodstuffs in cans. Bolger et al. (1991), for example, reported Pb concentrations in canned foods with and without lead-seamed construction, as seen in Table 6.42. Ratios of Pb levels in lead-seamed can contents were manyfold higher than for containers without Pb. For example, canned vegetable soup Pb levels in lead-seamed cans were 18-fold higher, 0.18, versus 0.01 ppm Pb in cans not lead-seamed.

TABLE 6.41 Lead Levels in Selected Food Items for the U.S. FDA Total Diet Survey (2004)[a,b,c]

Food Number	Food Name	Average Concentration/ppm
1	Milk, whole fluid	TR 0.002
3	Milk, chocolate, lowfat, fluid	TR 0.002
10	Cheese, American, processed	TR 0.005
12	Cheese, cheddar, natural (sharp/mild)	0.000
13	Beef, ground, regular, pan-cooked	TR 0.006
17	Ham, cured (not canned), baked	0.000
18	Pork chop, pan-cooked with oil	TR 0.003
20	Pork bacon, oven-cooked	0.000
21	Pork roast, loin, oven-roasted	0.000
26	Turkey breast, oven-roasted	0.000
29	Bologna (beef/pork)	TR 0.005
34	Fish sticks or patty, frozen, oven-cooked	TR 0.003
35	Eggs, scrambled with oil	0.000
39	Pork and beans, canned	0.000
42	Lima beans, immature, frozen, boiled	0.000
46	Peas, green, frozen, boiled	TR 0.002
47	Peanut butter, creamy	TR 0.008
50	Rice, white, enriched, cooked	0.000
54	Corn, fresh/frozen, boiled	0.000
55	Corn, canned	TR 0.002
58	Bread, white, enriched	0.000
69	Noodles, egg, enriched, boiled	TR 0.004
71	Corn flakes cereal	TR 0.005
73	Shredded wheat cereal	TR 0.006
78	Apple (red), raw (with peel)	0.000
80	Banana, raw	0.000

(Continued)

TABLE 6.41 Lead Levels in Selected Food Items for the U.S. FDA Total Diet Survey (2004)[a,b,c]—(cont.)

Food Number	Food Name	Average Concentration/ppm
83	Peach, raw/frozen	0.000
85	Pear, raw (with peel)	0.000
87	Fruit cocktail, canned in light syrup	TR 0.009
93	Pineapple, canned in juice	TR 0.010
98	Orange juice, frozen conc., reconstituted	0.000
99	Apple juice, bottled	TR 0.005
107	Spinach, fresh/frozen, boiled	TR 0.005
108	Collards, fresh/frozen, boiled	TR 0.005
109	Lettuce, iceberg, raw	0.000
110	Cabbage, fresh, boiled	0.000
113	Broccoli, fresh/frozen, boiled	0.000
114	Celery, raw	0.000
117	Tomato, raw	0.000
119	Tomato sauce, plain, bottled	TR 0.003
122	Green beans, canned	TR 0.003
125	Pepper, sweet, green, raw	TR 0.002
128	Onion, mature, raw	0.000
131	Beets, canned	TR 0.004
136	Potato, boiled (without peel)	0.000
137	Potato, baked (with peel)	TR 0.011
148	Meatloaf, beef, homemade	0.000
156	Soup, tomato, canned, condensed, prepared with water	0.000
164	Butter, regular (salted)	0.000
166	Mayonnaise, regular, bottled	0.000
169	Sugar, white, granulated	0.000
170	Syrup, pancake	0.000

(*Continued*)

TABLE 6.41 Lead Levels in Selected Food Items for the U.S. FDA Total Diet Survey (2004)[a,b,c]—(cont.)

Food Number	Food Name	Average Concentration/ppm
187	Candy bar, milk chocolate, plain	TR 0.015
191	Carbonated beverage, cola, regular	0.000
194	Carbonated beverage, cola, low-calorie	0.000
197	Tea, from tea bag	0.000
198	Beer	0.000
199	Wine, dry table, red/white	TR 0.008

[a]*Accessed from U.S. FDA database online, November 22, 2010 http://www.fda.gov/Food/FoodSafety/FoodContaminantsAdulteration/TotalDietStudy/ucm184293.htm.*
[b]*Selected to represent illustrative major U.S. dietary components.*
[c]*TR = trace concentration.*

TABLE 6.42 U.S. Total Diet Pb Survey for Canned Foods (μg/g): Lead Levels in Pb-Seamed versus Non-Pb-Seamed Cans[a,b,c]

Canned Food Item	Pb Level, Lead Cans[d]	Pb Level, Nonlead Cans[d]
Tuna	0.77	0.05
Orange juice	0.08	0.02
Applesauce	0.26	0.08
Apple juice	0.10	0.02
Fruit punch	0.08	0.03
String beans	0.26	0.03
Baked beans	0.27	0.02
Tomatoes	0.27	0.03
Chicken noodle soup	0.12	N.R.[e]
Vegetable soup	0.18	0.01

[a]*Adapted from Capar (1990) and Bolger et al. (1991).*
[b]*Adult canned foods also eaten by young children.*
[c]*For fiscal years 1982–1985.*
[d]*Rounding.*
[e]*Not reported; Bolger et al. (1991).*

TABLE 6.43 Changes in Production of Lead-Soldered Food Cans (Millions) 1979–1990[a,b,c]

Year	Number of Pb-Soldered Cans (Millions)	% Pb-Seamed Units of Total Production[d]
1979	27,576	90.29
1980	24,405	85.84
1981	20,516	74.23
1982	17,412	63.21
1983	13,891	51.56
1984	11,683	41.55
1985	8,769	31.47
1986	6,775	24.31
1987	4,282	15.09
1988	1,626	5.79
1989	1,117	3.71
1990	210	3.07

[a]*Tabulation data provided by CMI; Communication of August 27, 1990 by R.R. Budway, CMI General Counsel, to P. Mushak.*
[b]*Only three-piece cans fabricated with Pb.*
[c]*Soft drink cans not in total counting.*
[d]*% of all two- and three-piece cans ever shipped.*

The impact of lead-seamed can contributions to Pb contents of canned foods declined significantly with the phasing out of those containers. Adams (1991) showed a precipitous decline in Pb levels in canned foods from the 1982–1983 to the 1988–1989 period, the time of maximal reductions in lead-seamed can production. In FY 1983–1984, the leaded container food contaminant level was only 60% that of a year earlier. Two years later, the contaminant level was down to 35%, and in 1988–1989, the level was around 10% of that found 6 years earlier.

These data are consistent with figures for production of Pb-seamed cans as a fraction of the total, tabulated in Table 6.43. These data were provided to this author by the Can Manufacturers Institute (CMI) as part of a communication from CMI General Counsel R.R. Budlay to P. Mushak (August 27, 1990). In 1979, Pb-seamed food cans represented 90.3%, or 27.576 billion units of total container production. By 1984, such containers represented only 41.55%, or 11.683 billion units, of total production. The 1989 fraction was only 1.117 billion containers or 3.71% of total can production.

The corresponding count for 1990, a year later, was only one-fifth of that, 210 million containers.

A number of surveys of dietary Pb have been done for other countries. The 1982 Nutrition Foundation report used for U.S. Pb content data also reported Pb levels in the Canadian food supply. Table 6.44 has illustrative Pb levels in the Canadian food supply. Twelve food groups were sampled and individual food representatives within the groups were analyzed with respect to both uncanned and, where available, canned food levels. Canadian data showed a similar elevation in Pb content for canned versus uncanned foods as seen in U.S. analyses. Among selected dairy items, cheese had the highest Pb level, 0.24 ppm. In the meat, poultry, fish category, uncanned values were significantly elevated and levels ranged from 0.05 to 0.17 ppm. Canning increased levels up to 0.26 ppm. Uncanned cereals, potatoes, and leafy vegetables as groups were the next highest in Pb concentration, with further Pb elevation seen for canned examples. Uncanned garden fruits and vegetables were moderately elevated in Pb, but significantly elevated when canned. Beverages, overall, had the lowest Pb content across the 12 groups.

The later Canadian survey of Dabeka et al. (1987) reported lead contents of about 10 diet categories including tap water and beverages. Illustrative median Pb values of items within these categories are in Table 6.45. Five dietary categories had medians of 0.030 ppm or more: cheeses, meat/fish/poultry, starch vegetables, cereals, and miscellaneous snacks and desserts. Cheese showed the highest Pb content, ranging up to 6.78 ppm.

The most wide-ranging dietary Pb survey was the food component of the Global Environmental Monitoring System (GEMS-Food, 1991), done through the UNEP. Data were gathered from 39 nations including Canada and the United States (Galal-Gorchev, 1991). Table 6.46 sets forth 11 food categories and the mean Pb content of each. Canned foods and beverages, as expected, have the highest Pb concentrations at 0.20 ppm in each case. Fish ranked next at 0.10 ppm.

The percentage contribution of illustrative food categories in each of five nations—Australia, Canada, Finland, the Netherlands, United Kingdom—are depicted in Table 6.47. Among Canadian dietary items, vegetables and the meat/fish/poultry categories each contributed 17% of total daily Pb intake, followed by beverages and cereals/cereal products at 15% each. The Finnish diet had the highest percentages of Pb intake via cereals/cereal products (24%) and fruits (22%). Beverages comprised 20% of those diets followed closely by milk products at 17%. Dutch dietary item Pb content was highest for cereals/cereal products at 17% followed by the vegetable category at 12%. For the United Kingdom, cereals (including breads) were highest at 15%, followed by potatoes at 14%.

Chen and Gao (1993) reported results of the Chinese TDS carried out in 1990. The food consumption pattern in the form of market-basket surveys and associated contaminant levels and their intakes were determined in four

TABLE 6.44 Lead Levels in Selected Canadian Dietary Items from the 1982 Nutrition Foundation Report on Safety of Lead and Lead Salts in Food[a,b,c]

		Lead Level (μg/g, ppm)	
Food Group	Food Item	Uncanned	Canned
Dairy products	Evaporated milk	–	0.04
	Whole milk	0.02	
	Cheese	0.24	
	Butter	0.05	
Meat, poultry, fish	Beef	0.09	
	Pork	0.10	
	Veal	0.05	
	Lamb	0.05	
	Poultry	0.08	
	Eggs	0.05	
	Fish, fresh	0.05	
	Tuna		0.26
	Salmon		0.16
	Shellfish	0.05	
	Organ meats	0.10–0.17	
	Prepared meat, poultry		0.25
Cereals	Breads and rolls	0.05	
	Flour	0.08	
	Breakfast cereals, all types	0.05	
	Rice	0.05	
	Pasta	0.05	0.10
	Corn	0.03	0.36
Potatoes	Potatoes, baked and boiled	0.06	0.12
	Sweet potatoes, yams	0.08	0.25

(Continued)

TABLE 6.44 Lead Levels in Selected Canadian Dietary Items from the 1982 Nutrition Foundation Report on Safety of Lead and Lead Salts in Food[a,b,c]—(cont.)

		Lead Level (μg/g, ppm)	
Food Group	**Food Item**	**Uncanned**	**Canned**
Leafy vegetables	Cabbage	0.05	
	Celery	0.03	
	Lettuce, other greens	0.07	0.13
	Asparagus	0.03	0.13
	Mushrooms	0.04	0.18
Legumes	Beans	0.03	0.16
	Peas	0.03	0.13
Root vegetables	Carrots	0.07	0.13
	Onions	0.05	
	Beets	0.05	0.17
	Turnips, parsnips	0.03	
Garden fruits	Tomatoes	0.04	0.25
	Cucumbers	0.03	
	Squash	0.03	
	Eggplant	0.03	
	Tomato juice		0.20
	Tomato paste		0.57
Fruits	Citrus	0.03	0.14
	Apples	0.14	0.17
	Apple juice		0.08
	Grapes	0.02	0.14
	Peaches	0.04	0.27
	Pears	0.04	0.22
	Pineapple	0.05	0.19
	Cherries	0.02	0.14
	Berries	0.05	0.14

(*Continued*)

TABLE 6.44 Lead Levels in Selected Canadian Dietary Items from the 1982 Nutrition Foundation Report on Safety of Lead and Lead Salts in Food[a,b,c]—(cont.)

		Lead Level (μg/g, ppm)	
Food Group	Food Item	Uncanned	Canned
	Nuts	0.05	
Oils and fats	Cooking oils	0.02	
	Salad dressings	0.05	
	Sugar	0.05	
	Syrup		0.17
Sugar and Adjuncts	Jams and jellies	0.05	
	Puddings	0.05	
	Candy	0.06	
	Coffee, tea	0.03	
Drinks	Soft drinks		0.05
	Alcoholic beverages	0.01	

[a]*Adapted from Canadian Food Surveys in Nutrition Foundation (1982) report.*
[b]*Twelve food groups, includes beverages.*
[c]*Uncanned levels plus canned levels where available.*

major geographic groupings in China with included provinces: North 1, Heilongjiang, Liaoning, Heibei; North 2, Henan, Shaanxi, Ningxia; South 1, Shanghai, Jiangxi, Fujian; South 2, Hubei, Sichuan, Guangxi. Each regional market basket consisted of 12 food composites. For Pb, the contributions of various food groups as a national percentage of the total national Pb intake were: cereals, 39.6; vegetables, 27.5; potatoes, 7.8; meats, 5.6; fruits, 5.4; eggs, 5.1; legumes and nuts, 4.4; aquatic foods, 2.8. Milk, water, prepared beverages, and milk and sugar added relatively minor amounts of Pb to daily intake.

The various national and international surveys included food crop data for food crop components produced in uncontaminated soils. Soils in urban areas or areas impacted by lead smelters not only have elevated Pb content, as is to be expected, but crops grown there also show Pb concentrations well above samples produced in uncontaminated soils. Table 6.48 depicts food crops with high Pb content under these conditions. Finster et al. (2004) reported that leafy and root vegetables grown in urban gardens in Chicago, IL, had Pb levels well above 10 ppm. Swiss chard was found at 22–24 ppm

TABLE 6.45 Illustrative Pb Levels in Various Food Categories of the Canadian Diet[a,b,c]

Food Category	Median (ppm)	Ranges (ppm)
Cheese (excluding cottage cheese)	0.034	0.028–6.78
Meat, fish, poultry, meat soups	0.031	0.011–0.121
Dairy products, eggs	0.003	0.001–0.082
Fruits, fruit juices, canned and fresh	0.008	0.002–0.109
Potatoes, rice, other starches	0.07	0.006–0.084
Nonstarch vegetables	0.032	0.001–0.254
Miscellaneous snacks, desserts	0.033	0.014–1.38
Cereals, bread	0.032	0.012–0.078
Beverages	0.009	<0.00005–0.029

[a] *Adapted from Dabeka et al. (1987) and World Health Organization (1995).*
[b] *Units of ppm.*
[c] *Canned and uncanned items combined.*

TABLE 6.46 Typical Pb Concentrations of Illustrative Foods in the GEMS-Food International Survey[a,b,c]

Food	Pb Level (μg/g, ppm)
Cereals	0.06
Meats	0.05
Organ foods	0.02
Fish	0.10
Shellfish	0.02
Vegetables	0.05
Fruits	0.05
Roots/tubers	0.05
Eggs	0.02
Canned foods	0.20
Canned beverages	0.20

[a] *Adapted from Galal-Gorchev (1991).*
[b] *GEMS-Food: Food component of the Global Environmental Monitoring System, UNEP.*
[c] *Data for 39 countries, including the United States.*

TABLE 6.47 Relative Contributions of Dietary Pb to Intake as a Function of Food Group Among Indicated Countries[a,b,c]

Country	Food Group	% Total Pb Intake
Canada	Vegetables	17
	Meat/fish/poultry	17
	Beverages	15
	Cereals/products	15
	Fruits, juices	10
Finland	Cereals/products	24
	Fruits	22
	Beverages, etc.	20
	Milk/products	17
The Netherlands	Cereals/products	17
	Vegetables	12
United Kingdom	Breads/cereals	15
	Beverages	14
	Potatoes	10
Australia	Tea	20

[a]*Adapted from Galal-Gorchev (1991), from the GEMS-Food International Survey.*
[b]*Reported for adults.*
[c]*Food groups contributing ≥10% total Pb intake.*

Pb, while onion samples had 21 ppm. Moseholm et al. (1992) showed lettuce grown close to a Pb smelter had a Pb level in the range of 0.7–1.3 ppm, while kale grown close to the smelter had Pb levels 0.6–2.4 ppm.

6.3.6 Lead in Drinking Water

Drinking water Pb has had a long history of producing both lead exposures and associated lead poisonings. References appeared in Greco-Roman times to the health hazards of using waters flowing in lead piping versus waters in wood pipes (see Chapter 2). This history extended into more recent decades as cities and towns appeared and grew in industrializing countries. Growth in urbanization was accompanied by growth in setting up and maintaining public drinking water systems, typically established as public utilities operated by public agencies and drawing water from both surface water and

TABLE 6.48 Selected Lead Levels of Garden Crops Produced in Contaminated Soils[a,b,c]

Locale	Food Items	Pb Level (μg/g, ppm)[d]	References
Chicago, IL (urban gardens)	Cilantro	49	Finster et al. (2004)
	Collard greens	12	
	Coriander	39	
	Ipasote	14	
	Lemon balm	20	
	Mint	<10–60	
	Rhubarb	<10–36	
	Swiss chard	22–24	
	Carrot	10	
	Onion	21	
	Radish	12–18	
Copenhagen, Denmark, close to Pb smelter	Lettuce	0.7–1.3	Moseholm et al. (1992)
	Carrots	0.07–0.28	
	Potatoes	0.6–2.4	
	Kale	1.4–9.3	

[a] *Adapted from original articles.*
[b] *Contamination from mobile or point sources.*
[c] *Different crops tested in indicated locales.*
[d] *Levels ≥10 ppm.*

groundwater. Areas in suburban or remote locales, by contrast, often rely on private water sources such as wells.

The topic of Pb in drinking water, technologically, is a complex one in both the number of discrete steps needed to produce potable drinking water for human populations and the myriad of physical, physicochemical, and chemical factors that work to produce low, moderate, or high concentrations of Pb in drinking water sources for these populations. The environmental physical factors typically include the nature of the water chemistry, the nature of the components of water transport from sources to the residences and/or public areas such as schools for human populations, the characteristics of residential plumbing systems, and finally, patterns of water use in different risk groups in the population.

Natural levels of lead in both surface water and groundwater free of lead contamination from anthropogenic activities or arising from contamination in the course of storage, transport, and distribution to consuming populations are extremely low. A range of 0.005–10 μg/l was estimated by NAS/NRC (1980) but the upper end of the range likely reflects artifactual contamination by Pb during laboratory testing. U.S. EPA (1986b) settled on an upper limit of 0.02 μg/l.

Untreated surface waters averaged 3.9 μg Pb/l for 749 surface water samples, with a range of 1–55 μg/l in the report by Durum et al. (1971). NAS/NRC (1980) reviewed surface water levels in the literature, a range from undetected to 890 μg/l, with the upper end of the range likely showing considerable contamination. U.S. EPA (1986b) selected the Patterson concentration of 0.006–0.05 μg/l for remote streams. Proximity to sewage treatment, urban runoff, and industrial waste pollution resulted in much higher values, around 100 μg/l. Groundwater lead content has been reported in the range of 1–100 μg/l (NAS/NRC (1980).

A relatively high fraction of U.S. households, 84%, receive treated drinking water supplies from public or private surface water or groundwater systems. Water wells are the second largest source (U.S. EPA, 1986b). Waters leaving treatment plants in U.S. and Canadian systems are low, 2–3 μg/l Pb (Dabeka et al., 1987; U.S. EPA, 1991).

Water Pb levels at the tap are typically much higher than concentrations of Pb in piped water leaving treatment plants (Isaac et al., 1997; Mushak and Crocetti, 1990; NAS/NRC, 1993; U.S. EPA, 1986b, 2006; WHO, 1995). Isaac et al. (1997) reported that ratios of water Pb levels to Pb concentrations measured at the tap with variable standing times ranged from 0.17 to 0.69 in water flowing into Massachusetts homes. Lead enters household and public water plumbing at a number of sites before exiting the tap. First, there are lead connectors, "pigtails," joining water lines to household plumbing; in the past, lengths of lead pipe were used in areas with highly corrosive soils for other forms of metal plumbing (U.S. EPA, 1985). Lead piping has largely been abandoned in favor of copper or nonmetal plumbing materials and one only finds lead plumbing in the oldest housing in older areas of countries such as the United States. Overall, the principal contributor of leachable lead in household plumbing is corrosible 50:50 lead–tin soldered plumbing joints usually found in copper plumbing. Pb-alloyed brass household water fixtures, e.g., faucets in kitchens and bathrooms, are a second significant source.

Pb leaching into household and public building plumbing from lead–tin (Pb–Sn) 50:50 solder in copper water lines has been the subject of a number of studies. When used with copper plumbing lines, Pb–Sn solder is affected by electrochemical, i.e., galvanic, reactions which produce divalent Pb ion. In general, the most significant factors affecting pipe joint solder Pb release are aggressive water, i.e., low pH, and low "hardness" or carbonate level

(U.S. EPA, 1986b), standing time in unit plumbing lines, temperature, and relative efficiency of the soldering. Other factors include age of a given type of plumbing and interior surface coating (deMora and Harrison, 1984). Subramanian et al. (1994) were able to show that any of a number of non-lead fitting solders was sufficiently resistant to galvanic leaching to be safe to use. In addition to the most widely used alternative, a 95:5 tin–antimony solder, tested materials included 94:6 tin–silver, 96:4 tin–silver, and 95.5:4.0:0.5 tin–copper–silver solders.

Quantification of Pb leach rates from simulated plumbing systems and also household plumbing systems with lead solder, copper piping, and treated or well waters have been reported. Treweek et al. (1985) used pilot-plant simulations of corrosion in domestic pipe materials to evaluate the impact of different water quality variables on releases of lead and copper from soft, low-alkalinity water consumed in Portland, OR, over a test time of 18 months. The lead–tin solder-coated copper coiling tested with chloraminated water, with ample standing time of 8 hours, showed lead leaching at the outset exceeding the then MCL of 50 μg/l, while chlorinated water only showed exceedences over the MCL for copper. Over time, the Pb leach rates declined. These findings led to a health advisory for residents to run the tap water before using to flush standing water Pb accumulations.

Thompson and Sosnin (1985) evaluated the corrosion characteristics of 50:50 Pb–Sn solder in household plumbing, using either short-term static tests in lead-soldered copper test cells (15 days) or mock-up plumbing assemblies (40 days) and three water sources differing in pH and corrosiveness. Water Pb buildup in acidic, most aggressive test, water at room temperature was initially 14 μg/l/hour with a final rate of 11 μg/l/hour rate. Hot water testing (60°C) for this same set of parameters gave an initial leaching rate of 48 μg/l/hour and a final value of 55 μg/l/hour. Birden et al. (1985) studied Pb leaching from lead-soldered joints using simulated household plumbing assemblies with multiple soldered joints and several public well water sources in New England differing in relative corrosivity. One well water sample had a pH of 6.7 and low total hardness of 47 mg/l versus the second water with pH of 7.8 and hardness of 110 mg/l. The lower pH and hardness water had a Pb leaching from 50:50 tin–lead solder of 364 μg/l compared to 80 μg/l for the second system.

A second source of galvanic Pb leaching in domestic plumbing systems, in addition to that from lead–tin soldered joints, is household plumbing fixtures. Lead sources in fixtures include both soldered connections and the brass alloy composition with a permissible significant Pb content (3–8% Pb). Samuels and Méranger (1984) tested lead leaching from commercially available kitchen faucets with varying details of construction in presence or absence of leaded solder in the connecting tube assemblies. Waters were tested as available from the Ottawa, Canada, water supply, in addition to a well water source and aqueous fulvic acid solution. Filtered water Pb leaching from those kitchen faucets with lead-soldered joints was extensive, ranging from 4 to 55 mg/l (4,000–55,000 μg/l), for an initial 24-hour stagnation (contact) period.

Several surveys of Pb in drinking water, mainly in the United States and Canada, have been carried out. These surveys, differing in their geographic and environmental scope, were done from the early 1980s to 2006. The reported water Pb levels in these surveys were illustrative of tap water statistics such as mean water Pb across cities, states, and countries and prevalence rates of water Pb exceedences for standards and guidelines as they existed at the time. For example, the U.S. EPA had a Pb in drinking water supply MCL of 50 μg/l until 1991. In 1991, this regulatory approach changed to a two-part control protocol. First, U.S. EPA promulgated a tap water action level of 15 μg/l, not to be exceeded at the 90th percentile, for the sampled community in order to forestall some form of advisory or treatment technology change and, second, an enforceable MCL standard of 5 μg/l (5 ppb) for water exiting public water treatment plants.

Table 6.49 presents tap water Pb levels in selected U.S., Dutch, and English water systems reported by various authors. Sampling methodology used standing, flushed, and composited samples and various types of water piping. It is clear that stagnant water had much larger amounts of Pb than flushed samples. Composite samples from random collections were intermediate in Pb concentration. Using plastic pipe resulted in relatively lower Pb content in household water, but levels were still measurable. Running, flushed water line samples depicted water system background Pb content as treated waters left the plant. The highest values in Table 6.49, not unexpectedly, were the mean of 1,075 μg/l and the maximum figure of 2,826 μg/l for standing water samples collected in English households having lead water lines.

More recent selected household water surveys in U.S. and Canadian households showed marked reductions in both standing and running water samples. In Table 6.50, geographic locales with soft waters and a history of aggressive plumbing corrosion showed the highest concentrations, e.g., the selected Massachusetts communities. One still saw a distinction between standing and flushed water samples, with U.S. Midwest standing levels being about fivefold higher than flushed samples.

Distribution of water Pb elevations in water systems across geography and time have been reported. Table 6.51 shows the water Pb content of U.S. urban water systems at the community-based water Pb action level of 15 μg/l at three times: 1992, 1993, and more recently, from 1999 to 2006, with most for the period 2000–2003. Overall, the 90th percentile water Pb levels are lower in the early 2000s than in the early 1990s when the 1991 U.S. EPA Pb–Cu rule was implemented.

TABLE 6.49 Selected Drinking Water Pb Concentrations in U.S., English, and Dutch Plumbing Systems[a,b,c,d]

Locale	Plumbing/Sample Conditions	Concentration (μg/l) Mean	Concentration (μg/l) Maximum	References
Seattle, WA	Standing, overnight	4.3	11.5	Herrera et al. (1982)
University of Washington	Standing, overnight	3.9	170	Dangel (1975), as cited in Ohanian (1986)
Tolt River	Standing, overnight, 30 second flush	5	17	
Boston, MA	Composite sample	30	1,510	Karalekas et al. (1976)
Boston, MA	Standing, overnight	96	1,108	Worth et al. (1981)
	Running, 5-minute flush	17	208	
South Carolina	Metal pipes	20.7	163	Sandhu et al. (1977)
	Plastic pipes	18.7	73	
Morris County, NJ	Composite	44	260	Benson and Klein (1983)
England, various water lines for estate plumbing	Replacement Pb piping			Thomas et al. (1979)
	a. Standing, overnight	1,075	2,826	
	b. Running, 5-minute flush	139	354	
	Copper piping			
	a. Standing, overnight	4		
	b. Running, 5-minute flush	3		
The Netherlands	Composite samples	81	180	Zoetman and Haring (1978), as cited in Ohanian (1986)

[a]*Adapted from Ohanian (1986), covering various reports.*
[b]*Various water Pb leaching conditions.*
[c]*U.S. tap water Pb action level (1991) = 15 μg/l at 90th percentile community prevalence.*
[d]*WHO (1993) water Pb guideline = 10 μg/l.*

TABLE 6.50 Illustrative Drinking Water Pb Levels Reported for U.S. and Canadian Communities

Locale	Occupancy Type	Water Status/Other	Pb–W (μg/l)	References
Midwestern United States	Residences	Standing	3.9	Clayton et al. (1999)
		Flushed	0.8	
Massachusetts				
Gardner	Residences	Standing	25.0	Isaac et al. (1997)
Fall River			15.3	
New Bedford			11.6	
Clinton			7.7	
Arizona	Residences	–	0.7	Sofuoglu et al. (2003)
Halifax, Canada	Homes	Standing	16	Moir et al. (1996)
		Running	8	
	Apartments	Standing	3	
		Running	2	
	School	Standing	6	
		Running	5	
Vancouver, Canada	Apartments	Copper or plastic pipe	20	Singh and Mavinic (1991)
	Houses		13	

TABLE 6.51 Illustrative U.S. City Drinking Water Pb Levels at the 90th Percentile over Three Monitoring Periods[a,b,c]

Locale	1992 90th %ile (μg/l)	1993 90th %ile (μg/l)	1999–2006 Monitoring Interval (μg/l)
Chicago, IL	20	10	7
Philadelphia, PA	15	322	13
Washington, DC	39	18	63
Detroit, MI	15	21	12
Minneapolis, MN	32	19	6
St. Paul, MN	28	54	11
Portland, OR	53	41	8
Phoenix, AZ	11	19	1
Yonkers, NY	110	68	18
Syracuse, NY	40	50	25
Galveston, TX	6	18	2
Miami Beach, FL	4	27	8
Richmond, VA	25	16	4
Tacoma, WA	17	32	12

[a]*Adapted from U.S. EPA (2006).*
[b]*Cities are those exceeding the EPA 1991 action level in 1992 or 1993.*
[c]*Recent monitoring mainly in years 2000, 2001, 2002, 2003, except for one city in 1999–2001, one city in 2003–2006.*

Percentages of water Pb collected through the day and lying within successively higher water Pb ranges have been reported in Great Britain. The data are in Table 6.52. The distribution statistics totaled for all three countries—England, Scotland, Wales—showed 25.3% of households in the 10–50 μg/l range. The highest values, 301 μg/l and above, were found for about 1% of samples in the three countries. The data for Scotland, with its historically very corrosive water supply and lead-containing water collection systems, were the main contributor to the high exceedences of Pb in drinking water. For example, 16% of Scottish water samples were in the range of 101–300 μg/l, and 5% were 301 μg/l and above.

TABLE 6.52 Late Twentieth Century Distribution of Lead Concentrations in Tap Water Samples in Great Britain[a]

	Percent Households			
Pb Level (μg/l)	England	Scotland	Wales	All
0–9	66.0	46.4	70.5	64.4
10–50	26.2	19.2	20.7	25.3
51–100	5.2	13.4	6.5	6.0
101–300	2.2	16.0	1.5	3.4
≥301	0.4	5.0	0.8	0.9
% Total	100.0	100.0	100.0	100.0

Source: *WRC Water Research Centre (1983, Table 1).*
[a]*Daytime water draws randomly consumed.*

REFERENCES

Adams, M.A., 1991. FDA total diet study: dietary intakes of lead and other chemicals. Chem. Speciation Bioavailability 3, 37–41.

Adgate, J.L., Willis, R.D., Buckley, T.J., Chow, J.C., Watson, J.G., Rhoads, G.G., et al., 1998. Chemical mass balance source apportionment of lead in house dust. Environ. Sci. Technol. 32, 108–114.

Al-Chalabi, A.S., Hawker, D., 1997. Response of vehicular lead to the presence of street dust in the atmospheric environment of major roads. Sci. Total Environ. 206, 195–202.

Angle, C.R., McIntire, M.S., 1979. Environmental lead and children: the Omaha study. J. Toxicol. Environ. Health 5, 855–870.

Angle, C.R., McIntire, M.S., 1982. Children, the barometer of environmental lead. Adv. Pediatr. 27, 3–31.

Badawy, S.H., Helal, M.I.D., Chaudri, A.M., Lawlor, K., McGrath, S.P., 2002. Soil solid-phase controls lead activity in soil solution. J. Environ. Qual. 31, 162–167.

Barltrop, D., Strehlow, C.D., Thornton, I., Webb, J.S., 1975. Absorption of lead from dust and soil. Postgrad. Med. J. 51, 801–804.

Bates, M., Malcolm, M., Wyatt, R., Garrett, N., Galloway, Y., Speir, T., et al., 1995. Lead in children from older housing areas in the Wellington region. N. Z. Med. J. 108, 400–404.

Beloian, A., 1985. Model system for use of dietary survey data to determine lead exposure from food. In: Mahaffey, K.R. (Ed.), Dietary and Environmental Lead: Human Health Effects. Elsevier, Amsterdam, the Netherlands, pp. 109–155.

Benson, J.A., Klein, H., 1983. Lead in drinking water. Investigation of a corrosive water supply. J. Environ. Health 45, 179–181.

Birden Jr., H.H., Calabrese, E.J., Stoddard, A., 1985. Lead dissolution from soldered joints. J. Am. Water Works Assn. 77, 66–70.

Bolger, P.M., Carrington, C.D., Capar, S.G., Adams, M.A., 1991. Reductions in dietary lead exposure in the United States. Chem. Speciation Bioavailability 3, 31–36.

Bornschein, R., Clark, S., Pan, W., Succop, P., 1991. Midvale community lead study. Chem. Speciation Bioavailability 3, 149–162.

Brewer, R., Belzer, W., 2001. Assessment of metal concentrations in atmospheric particles from Barnaby Lake, British Columbia, Canada. Atmos. Environ. 35, 5223–5233.

Brunekreef, B., Noy, D., Biersteker, K., Boleij, J., 1983. Blood lead levels of Dutch city children and their relationship to lead in the environment. J. Air Pollut. Control Assoc. 33, 872–876.

Capar, S., 1990. Survey of lead and cadmium in adult canned foods eaten by young children. J. Assoc. Off. Anal. Chem. 73, 357–364.

Capar S., 1991. Analytical method aspects of assessing dietary intake of trace elements. In: Subramanian, K., Iyengar, G., Okamoto, K. (Eds.), Biological Trace Element Research. ACS Symp. Ser. 445, 181–195.

Casteel, S., Weis, C.P., Henningsen, G.M., Brattin, W.J., 2006. Estimation of relative bioavailability of lead in soil and soil-like materials using young swine. Environ. Health Perspect. 114, 1162–1171.

Chan, L.Y., Kwok, W.S., Chan, C.Y., 2000. Human exposure to respirable suspended particulate and airborne lead in different roadside microenvironments. Chemosphere 41, 93–99.

Chaney, R., Mielke, H., 1986. Standards for soil lead limitations in the United States. Trace Subst. Environ. Health 20, 357–377.

Chartsias, B., Colombo, A., Hatzichritidis, D., Leyendecker, W., 1986. The impact of gasoline lead and blood lead in man. Sci. Total Environ. 55, 275–282.

Chen, J., Gao, J., 1993. The Chinese Total Diet Study in 1990. Part I. Chemical contaminants. J. AOAC Int. 76, 1193–1201.

Chiaradia, M., Cuppelin, F., 2000. Behavior of airborne lead and temporal variation of its source effects in Geneva (Switzerland): comparison of anthropogenic versus natural processes. Atmos. Environ. 34, 959–971.

Chirinje, T., Ma, L.Q., Reeves, M., Szulczewski, M., 2004. Lead distribution in near-surface soils of two Florida cities: Gainesville and Miami. Geoderma 119, 113–120.

Chow, J.C., Watson, J.G., Ashbaugh, L.L., Magliano, K.L., 2003. Similarities and differences in PM_{10} chemical source profiles for geological dust from the San Joaquin Valley, California. Atmos. Environ. 1317–1340.

Clark, S., Bornschein, R.L., Succop, P., Que Hee, S.S., Hammond, P.B., Peace, B., 1985. Condition and type of housing as an indicator of potential environmental lead exposure and pediatric blood lead levels. Environ. Res. 38, 46–53.

Clark, S., Grote, J., Wilson, J., Succop, P., Chen, M., Galke, W., et al., 2004. Occurrence and determinants of increases in blood lead levels in children shortly after lead hazard control activities. Environ. Res. 96, 196–205.

Clayton, C.A., Pellizzari, E.D., Whitmore, R.W., Perritt, R.L., Quackenboss, J.J., 1999. National human exposure assessment survey (NHEXAS): distributions and associations of lead, arsenic, and volatile organic compounds in EPA Region 5. J. Exposure Anal. Environ. Epidemiol. 9, 381–392.

Colorado Department of Health, 1990. Leadville Metals Exposure Study. Division of Disease Control and Environmental Epidemiology. With University of Colorado at Denver and the U.S. Agency for Toxic Substances and Disease Registry, April.

Culbard, E.B., Thornton, L., Watt, J., Wheatley, M., Moorcroft, S., Thompson, M., 1988. Metal contamination in British suburban dusts and soils. J. Environ. Qual. 17, 226–234.

Dabeka, R.W., McKensie, A.D., Lacroix, G.M.A., 1987. Dietary intakes of lead, cadmium, arsenic and fluoride by Canadian adults: a 24-hour duplicate diet study. Food Addit. Contam. 4, 89–102.

Davidson, C.I., Chu, L., Grimm, T.C., Nasta, M.A., Qamoos, M.P., 1981. Wet and dry deposition of trace elements onto the Greenland ice sheet. Atmos. Environ. 15, 1429–1437.

Davies, B.E., 1983. A graphical estimation of the normal lead content of some British soils. Geoderma 29, 67–75.

Davies, D.J.A., Watt, J.M., Thornton, I., 1987. Air lead concentrations in Birmingham, England—a comparison between levels inside and outside inner-city homes. Environ. Geochem. Health 9, 3–7.

Day, J.P., Hart, M., Robinson, M.S., 1975. Lead in urban street dust. Nature 253, 343–345.

Del Delumyea, R., Kalivretenos, A., 1987. Elemental carbon and lead content of fine particles from American and French cities of comparable size and industry, 1985. Atmos. Environ. 21, 1643–1647.

DeMiguel, E., Llamas, J.F., Chacon, E., Berg, T., Larssen, S., Royset, O., et al., 1997. Origin and patterns of distribution of trace elements in street dust: unleaded petrol and urban lead. Atmos. Environ. 31, 2733–2740.

DeMora, S.J., Harrison, R., 1984. Lead in tap water: contamination and chemistry. Chem. Br. 900–906.

Deroanne-Bauvin, J., Delcarte, E., Impins, R., 1987. Monitoring of lead composition near highways—a ten year study. Sci. Total Environ. 59, 257–266.

Diemel, J.A.L., Brunekreef, B., Boleij, J.S.M., Biersteker, K., Veenstra, S.J., 1981. The Arnhem lead study: II. Indoor pollution and indoor/outdoor relationships. Environ. Res. 25, 449–456.

Ducoffre, G., Claeys, F., Bruaux, P., 1990. Lowering time trend of blood lead levels in Belgium since 1978. Environ. Res. 51, 25–34.

Dudka, S., Miller, W.P., 1999. Accumulation of potentially toxic elements in plants and their transfer to human food chains. J. Environ. Sci. Health, Part B 34, 681–708.

Durum, W.H., Hem, J.D., Heidl, S.G., 1971. Reconnaissance of selected minor elements in surface waters of the United States. USGS Circular No. 643. United States Geological Survey, Washington, DC.

Dzubay, T.G., Hines, L.E., Stevens, R.K., 1976. Particle bounce errors in cascade impactors. Atmos. Environ. 10, 229–234.

Elinder, C.-G., Friberg, L., Lind, B., 1986. Decreased blood levels in residents of Stockholm for the period 1970–1984. Scand. J. Work Environ. Health. 12, 114–120.

El-Shobokshy, M.S., 1985. The dependence of airborne particulate deposition on atmospheric stability and surface conditions. Atmos. Environ. 19, 1191–1197.

Erel, Y., Veron, A., Halicz, L., 1997. Tracing the transport of anthropogenic lead in the atmosphere and in soils using isotopic ratios. Geochim. Cosmochim. Acta 58, 4495–4505.

Erel, Y., Axelrod, T., Veron, A., Mahrer, Y., Katsafados, P., Dayan, U., 2002. Transboundary atmospheric lead pollution. Environ. Sci. Technol. 36, 3230–3233.

Ewers, L., Clark, S., Menrath, W., Succop, P., Bornschein, R., 1994. Clean-up of lead in household carpet and floor dust. Am. Ind. Hyg. Assoc. J. 55, 650–657.

Facchetti, S., 1989. Lead in petrol. The isotopic lead experiment. Acc. Chem. Res. 22, 370–374.

Facchetti S., Geiss F., 1982. Isotopic lead experiment: status report. Commission of the European Communities, Publication No. Eur 8352 EN. CEC, Luxembourg.

Farfel, M.R., Bannon, D., Chisolm Jr., J.J., Lees, P.S.J., Lim, B.S., Rohde, C.A., 1994. Comparison of a wipe and a vacuum collection method for the determination of lead in residential dusts. Environ. Res. 65, 291–301.

Fergusson, J.E., Ryan, D.E., 1984. The elemental composition of street dust from large and small urban areas related to city type, source and particle size. Sci. Total Environ. 34, 101–116.

Fergusson, J.E., Schroeder, R.J., 1985. Lead in house dust of Christchurch, New Zealand. Sci. Total Environ. 46, 61–72.

Finster, M.E., Gray, K., Binns, H.J., 2004. Lead levels of edibles grown in contaminated residential soils: a field survey. Sci. Total Environ. 320, 245–257.

Freitas, H., Prasad, M.N.V., Pratas, J., 2004. Plant community tolerant to trace elements growing on the degraded soils of Sao Domingos mine in the south east of Portugal: environmental implications. Environ. Int. 30, 65–72.

Fugas, M., Wilder, B., Paukovic, R., Hrsak, J., Steiner-Skreb, D., 1973. Concentration levels and particle size distribution of lead in the air of an urban and an industrial area as a basis for the calculation of population exposure. In: Barth, D., Berlin, A., Engel, R., Recht, P., Smeets, J. (Eds.), The Environmental Health Aspects of Lead: Proceedings of an International Symposium, October, 1972, Amsterdam, the Netherlands. Commission of the European Communities, Luxembourg, pp. 961–968.

Galal-Gorchev, H., 1991. Global overview of dietary lead exposure. Chem. Speciation Bioavailability 3, 5–11.

Galke, W.A., Hammer, D.I., Keil, J.E., Lawrence, S.W., 1975. Environmental determinants of lead burdens in children. In: Hutchinson, T.C., Epstein, S., Page, A.L., Van Loon, J. (Eds.), International Conference on Heavy Metals in the Environment. Proceedings, vol. 3. Toronto, Canada. Institute for Environmental Studies, pp. 53–74.

GEMS-Food, 1991. Summary of 1986–1988 monitoring data. FAO/WHO Collaborating Centres for Food Contamination Monitoring. Geneva, Switzerland.

Gillies, J.A., O'Connor, C.M., Mamane, Y., Gertler, A.W., 1999. Chemical profiles for characterizing dust sources in an urban area, western Nevada, USA. In: Livingstone, I., (Ed.), Aeolian geomorphology. Papers from the 4th International Conference on Aeolian Research, 1998. Oxford, United Kingdom. Z. Geomorph. 116 (Suppl.), 19–44.

Godin, P.M., Feinberg, M.H., Ducauze, C.J., 1985. Modeling of soil contamination by airborne lead and cadmium around several emission sources. Environ. Pollut. Ser. B 10, 97–114.

Gulson, B.L., Mahaffey, K.R., Mizon, K.J., Korsch, M.J., Cameron, M.A., Vimpani, G., 1995. Contribution of tissue lead to blood lead in adult female subjects based on stable lead isotope methods. J. Lab. Clin. Med. 125, 703–712.

Gulson, B.L., Jameson, C.W., Mahaffey, K.R., Mizon, K.J., Korsch, M.J., Vimpani, G., 1997. Pregnancy increases mobilization of lead from maternal skeleton. J. Lab. Clin. Med. 130, 51–62.

Harrison, R.M., 1979. Toxic metals in street and household dusts. Sci. Total Environ. 11, 89–97.

Harrison, R.M., Tilling, R., Callen Romero, M.S., Harrad, S., Jarvis, K., 2003. A study of trace metals and polycyclic aromatic hydrocarbons in the roadside environment. Atmos. Environ. 37, 2391–2402.

Hartwell, T.D., Handy, R.W., Harris, B.S., Williams, S.R., Gehlbach, S.H., 1983. Heavy metal exposure in populations living around zinc and copper smelters. Arch. Environ. Health 38, 284–295.

Herrera, C.E., Ferguson, J.F., Benjamin, M.M., 1982. Evaluating the potential for contaminating drinking water from the corrosion of tin-antimony solder. J. Am. Water Works Assn. 74, 368–375.

Hertzman, C., Ward, H., Ames, N., Kelly, S., Yates, C., 1991. Childhood lead exposure in Trail revisited. Can. J. Public Health 82, 385–391.

Hilts, S.R., 2003. Effect of smelter emission reductions on children's blood lead levels. Sci. Total Environ. 303, 51–58.

Ho, K.F., Lee, S.C., Chow, J.C., Watson, J.G., 2003. Characterization of PM_{10} and $PM_{2.5}$ source profiles for fugitive dust in Hong Kong. Atmos. Environ. 37, 1023–1032.

Hui, C.A., 2002. Concentrations of chromium, manganese, and lead in air and in avian eggs. Environ. Pollut. 120, 201–206.

Hunt, W.F., Pinkerton, C., McNulty, O., Creason, J., 1971. A study in trace element pollution of air in 77 midwestern cities. Trace Subst. Environ. Health 4, 56–58.

Isaac, R.A., Gil, L., Cooperman, A.N., Hulme, K.I., Eddy, B., Ruiz, M., et al., 1997. Corrosion in drinking water distribution systems: a major contributor to copper and lead in wastewater and effluents. Environ. Sci. Technol 31, 3198–3203.

Jacobs, D.E., Clickner, R., Zhou, J., Viet, S.M., Marker, D.A., Rogers, J.W., et al., 2002. The prevalence of lead-based paint hazards in U.S. housing. Environ. Health Perspect. 110, A599–A606.

Jensen, R.A., Laxen, D.P.H., 1985. Sources of lead in urban dust: identification of a contribution from newspaper printworks. Sci. Total Environ. 46, 19–27.

Juberg, D.R., Kleinman, C.F., Kwon, S.C., 1997. Position paper of the American Council on Science and Health: lead and human health. Ecotoxicol. Environ. Saf. 38, 162–180.

Kapaki, E.N., Varelas, P.N., Syrigou, A.I., Spanaki, M.V., Andreadou, E., Kakami, A.E., et al., 1998. Blood lead levels of traffic- and gasoline exposed professionals in the city of Athens. Arch. Environ. Health 53, 286–291.

Karalekas, P.C., Craun, G.F., Hammonds, A.F., 1976. Lead and other trace metals in drinking water in the Boston metropolitan area. J. Am. Water Works Assn. 90, 150–172.

Kimbrough, D.E., Suffet, I.H., 1995. Off-site forensic determination of airborne elemental emissions by multi-media analysis: a case study of two secondary lead smelters. Environ. Sci. Toxicol. 29, 2217–2221.

Kurkjian, R., Dunlap, C., Flegal, A.R., 2002. Lead isotope tracking of atmospheric response to post-industrial conditions in Yerevan, Armenia. Atmos. Environ. 36, 1421–1429.

Landrigan, P.J., Gehlbach, S.H., Rosenblum, B.F., Shoults, J.M., Candelaria, R.M., Barthel, W.F., et al., 1975. Epidemic lead absorption near an ore smelter: the role of particulate lead. N. Engl. J. Med. 292, 123–129.

Landrigan, P.J., Baker Jr., E.L., Himmelstein, J.S., Stein, G.F., Wedding, J.P., Straub, W.E., 1982. Exposure to lead from the Mystic River bridge: the dilemma of deleading 306, 673–676N. Engl. J. Med. 306, 673–676.

Lanphear, B.P., Edmond, M., Jacobs, D.E., Weitzman, M., Tanner, M., Winter, N.L., et al., 1995. A side-by-side comparison of dust collection methods for sampling lead-contaminated house dust. Environ. Res. 68, 114–123.

Lanphear, B.P., Weitzman, M., Winter, N.L., Eberly, S., Yakir, B., Tanner, M., et al., 1996. Lead-contaminated house dust and urban children's blood lead levels. Am. J. Public Health 86, 1416–1421.

Lau, W.M., Wong, H.M., 1982. An ecological survey of lead contents in roadside dusts and soils in Hong Kong. Environ. Res. 28, 39–54.

Laxen, D.P.H., Raab, G.M., Fulton, M., 1987. Children's blood lead and exposure to lead in household dust and water—a basis for an environmental standard for lead in dust. Sci. Total Environ. 66, 235–244.

Leggett, R.W., 1993. An age-specific kinetic model of lead metabolism in humans. Environ. Health Perspect. 101, 598–616.

Linzon, S.N., Chai, B.L., Temple, P.J., Pearson, R.G., Smith, M.L., 1976. Lead contamination in urban soils and vegetation by emissions from secondary lead industries. J. Air. Pollut. Control Assoc. 26, 650–654.

Liu, Z.P., 2003. Lead poisoning combined with cadmium in sheep and horses in the vicinity of non-ferrous metal smelters. Sci. Total Environ. 309, 117–126.

Loo, B.W., Adachi, R.S., Cork, C.P., Goulding, S.S., Jaklevic, J.N., Landis, D.A., et al., 1979. A second generation dichotomous sampler for large-scale monitoring of airborne particulate matter. Lawrence Berkeley Laboratory. Report No. 8725. As cited in U.S. EPA 1986a, Ch. 4.

Maenhaut, W., Zoller, W.H., Duce, R.A., Hoffman, G.L., 1979. Concentration and size distribution of particulate trace elements in the south polar atmosphere. J. Geophys. Res. 84, 2421–2431.

Malcoe, L.H., Lynch, R.A., Kegler, M.C., Skaggs, V.J., 2002. Lead sources, behaviors, and socioeconomic factors in relation to blood lead of Native American and white children: a community-based assessment of a former mining area. Environ. Health Perspect. 110 (Suppl. 2), 221–231.

Markowitz, M.E., Weinberger, H.L., 1990. Immobilization-related lead toxicity in previously lead-poisoned children. Pediatrics 86, 455–457.

McGrath, S.P., 1986. The range of metal concentrations in topsoils of England and Wales in relation to soil protection guidelines. In: Hemphill, D.D. (Ed.), Trace Substances in Environmental Health, 20. pp. 242–252.

McIntire, M.S., Angle, C.R., 1973. Air lead/blood lead in G-6-PD deficient black school children. In: Barth, D., Berlin, A., Engel, R., Recht, P., Smeets, J. (Eds.), Environmental Health Aspects of Lead. Proceedings of an International Symposium, October, 1972, Amsterdam, the Netherlands. Commission of the European Communities, Luxembourg, pp. 749–761.

McKeague, J.A., Wolynetz, M.S., 1980. Background levels of minor elements in some Canadian soils. Geoderma 24, 299–307.

Milar, C.R., Mushak, P., 1982. Lead-contaminated house dust: hazard, management, and decontamination. In: Chisolm, J.J. Jr., O'Hara, D.M. (Eds.), Lead Absorption in Children: Management, Clinical, and Environmental Aspects. Proceedings of a Conference, November, 1979. Urban and Schwartzenberg, Baltimore, MD, pp. 143–152.

Mitchell, D.G., Aldous, K.M., 1974. Lead content of foodstuffs. Environ. Health Perspect. 7, 59–64.

Moir, C.M., Freedman, B., McCurdy, R., 1996. Metal mobilization from water-distribution systems of buildings serviced by lead-pipe mains. Can. Water Resour. J. 21, 45–52.

Moseholm, L., Larsen, E.H., Andersen, B., Nielsen, M.M., 1992. Atmospheric deposition of trace elements around point sources and human health risk assessment. I. Impact zones near a source of lead emissions. Sci. Total Environ. 126, 243–262.

Murgueytio, A.M., Evans, R.G., Daryl, R., 1998. Relationship between soil and dust lead in a lead mining area and blood lead levels. J. Exposure Anal. Environ. Epidemiol. 8, 173–186.

Mushak, P., 1991. Gastrointestinal absorption of lead in children and in adults: overview of biological and biophysico-chemical aspects. Chem. Speciation Bioavailability 3, 87–104.

Mushak, P., 1993. New directions in the toxicokinetics of human lead exposure. Neurotoxicology 14, 29–42.

Mushak, P., 1998. Uses and limits of empirical data in measuring and modeling human lead exposure. Environ. Health Perspect. 106 (Suppl. 6), 1467–1484.

Mushak, P., Crocetti, A.F., 1990. Methods for reducing lead exposure in young children and other risk groups: an integrated summary of a report to the U.S. Congress on childhood lead poisoning. Environ. Health Perspect. 89, 125–135.

Mushak, P., Mushak, E.W., 2000. Lessons learned: a comparative analysis of the evolution of lead and mercury as public health hazards. In: Nriagu, J.O. (Ed.), 11th International Conference on Heavy Metals in the Environment. CD-ROM: Manuscript No. 1445. University of Michigan, Ann Arbor, MI: August 6–10, 2000.

Mushak, P., Schroeder, C., 1980. Multiple media pollutant exposures and their regulation. Report to the National Commission on Air Quality. Contract No. 23a-AQ-6981. Discussed and cited in the 1981 National Commission on Air Quality report: To Breathe Clean Air, pp. 2.1.9, 2.2–4, 3.1-30, 3.1-34.

National Academy of Sciences: National Research Council, 1980. Lead in the Human Environment. National Academy Press, Washington, DC.

National Academy of Sciences: National Research Council, 1993. Measuring Lead Exposure in Infants, Children, and Other Sensitive Populations. National Academy Press, Washington, DC.

National Air Pollution Surveillance Reports (NAPS), 1971–1976. Report Series No. EPS-5-AP. Air Pollution Control Directorate, Ottawa, Canada.

Needleman, H.L., Davidson, I., Sewell, E.M., Shapiro, I.M., 1974. Subclinical lead exposure in Philadelphia school children: identification by dentine lead analysis. N. Engl. J. Med. 290, 245–258.

Neri, L.C., Johansen, H.L., Schmitt, N., Pagan, R.T., Hewitt, D., 1978. Blood lead levels in children in two British Columbia communities. Trace Subst. Environ. Health 12, 403–410.

Nriagu, J.O., 1978. Lead in the atmosphere. In: Nriagu, J.O. (Ed.), The Biogeochemistry of Lead in the Environment: Part A, Ecological Cycles. Elsevier/North Holland Biomedical Press, Amsterdam, the Netherlands, pp. 137–184.

Nriagu, J.O., 1979. Global inventory of natural and anthropogenic emissions of trace metals to the atmosphere. Nature 279, 409–411.

Nriagu, J.O., 1983a. Lead and Lead Poisoning in Antiquity. John Wiley & Sons, New York.

Nriagu, J.O., 1983b. Saturnine gout among Roman aristocrats: did lead poisoning contribute to the fall of the Empire? N. Engl. J. Med. 308, 660–663.

Nriagu, J.O., 1985. Historical perspective on the contamination of food and beverages with lead. In: Mahaffey, K.R. (Ed.), Dietary and Environmental Lead: Human Health Effects. Elsevier, New York, pp. 1–41.

Nutrition Foundation, 1982. Assessment of the safety of lead and lead salts in food: a Report of the Nutrition Foundation's Expert Advisory Committee, Washington, DC.

O'Flaherty, E.J., 1998. A physiologically based kinetic model for lead in children and adults. Environ. Health Perspect. 106 (Suppl. 6), 1495–1503.

Ohanian, E.V., 1986. Health effects of corrosion products in drinking water. Trace Subst. Environ. Health 20, 122–138.

O'Heany, J., Kusiak, R., Duncan, C.E., Smith, L.F., Spielberg, L., 1988. Blood lead and associated risk factors in Ontario children. Sci. Total Environ. 71, 477–483.

Oliver, D.P., McLaughlin, M.J., Naidu, R., Smith, L.H., Maynard, E.J., Calder, I.C., 1999. Measuring lead bioavailability from household dusts using an in vitro model. Environ. Sci. 33, 4434–4439.

Page, R.A., Cawse, P.A., Baker, S.J., 1988. The effect of reducing petrol lead on airborne lead in Wales, U.K.. Sci. Total Environ. 68, 71–77.

Pao, E.M., 1989. Obtaining data on food intake. In: Livingston, G.E. (Ed.), Nutritional Status Assessment of the Individual. Food and Nutrition Press, Trumbull, CT.

Patterson, C.C., 1982. Natural Levels of Lead in Humans. Institute for Environmental Studies, University of North Carolina, Chapel Hill, NC, Carolina, Environmental Essay Series, vol.3.

Patterson, C.C., 1983. Criticism of "flow of metals into the global atmosphere." Geochim. Cosmochim. Acta 47, 1163–1168.

Pennington, J.A.T., 1983. Revision of the Total Diet Study food list and diet. J. Am. Diet. Assoc. 82, 166–173.

Pennington, J.A.T., Gunderson, E., 1987. History of the Food and Drug Administration's Total Diet Study—1961 to 1987. J. Assoc. Off. Anal. Chem. 70, 772–782.

Pinkerton, C., Creason, J.P., Hammer, D.I., Colucci, A.V., 1973. Multi-media indices of environmental trace-metal exposure in humans. In: Hoekstra, W.G., Suttie, J.W., Ganther, H.E., Mertz, W. (Eds.), Trace Element Metabolism in Animals—2: Proceedings of the 2nd International Symposium on Trace Element Metabolism in Animals, Madison, WI. University Park Press, Baltimore, MD, pp. 465–469.

Ponka, A., Salminen, E., Ahonen, S., 1991. Lead in the ambient air and blood specimens of children in Helsinki. Sci. Total Environ. 138, 301–308.

Rabinowitz, M.B., Bellinger, D.C., 1988. Soil lead-blood lead relationship among Boston children. Bull. Environ. Contam. Toxicol. 41, 791–797.

Rabinowitz, M.B., Wetherill, G.W., Kopple, J.D., 1976. Kinetic analysis of lead metabolism in healthy humans. J. Clin. Invest. 58, 260–270.

Rameau, J.T.L.B., 1973. Lead as an environmental pollutant. In: Barth, D., Berlin, A., Engel, R., Recht, P., Smeets, J. (Eds.), Environmental Health Aspects of Lead. Proceedings of an International Symposium, October, 1972, Amsterdam, the Netherlands. Commission of the European Communities, Luxembourg, pp. 189–200.

Rieuwerts, J.S., Farago, M., 1996. Heavy metal pollution in the vicinity of a secondary lead smelter in Czech Republic. Appl. Geochem. 11, 17–23.

Roels, H.A., Buchet, J.P., Lauwerys, R.R., Bruaux, P., Claeys-Thoreau, F., Lafontaine, A., et al., 1980. Exposure to lead by the oral and pulmonary routes of children living in the vicinity of a primary lead smelter. Environ. Res. 22, 81–94.

Samuels, E.R., Méranger, J.C., 1984. Preliminary studies on the leaching of some metals from kitchen faucets. Water Res. 18, 75–80.

Sandhu, S.S., Warren, W.J., Nelson, P., 1977. Inorganic contaminants in rural drinking water. J. Am. Water Works Assoc. 69, 219–222.

Schroeder, H.A., Balassa, J.J., 1961. Abnormal trace metals in man: Lead. J. Chronic Dis. 14, 408–425.

Settle, D.M., Patterson, C.C., 1980. Lead in albacore: guide to lead pollution in the Americas. Science 207, 1167–1176.

Settle, D.M., Patterson, C.C., 1982. Magnitude and sources of precipitation and dry deposition fluxes of industrial and natural leads to the North Pacific at Eniwetok. J. Geophys. Res. 87, 8857–8869.

Shacklette, H.T., Hamilton, J.C., Boerngen, J.G., Bowles, J.M., 1971. Elemental composition of surficial materials in the conterminous United States: an account of the amounts of certain elements in samples of soils and other regoliths. Geological Survey Professional Paper No. 574-D. U.S. Geological Survey, Washington, DC.

Schilling, R.J., Bain, R.P., 1988. Prediction of children's blood lead levels on the basis of household-specific soil lead levels. Am. J. Epidemiol. 128, 197–205.

Silbergeld, E.K., Schwartz, J., Mahaffey, K.R., 1988. Lead and osteoporosis: mobilization of lead from bone to blood in postmenopausal women. Environ. Res. 47, 79–94.

Singh, I., Mavinic, D.S., 1991. Significance of buildings and plumbing specifics on trace metal concentrations in drinking water. Can. J. Civ. Eng. 18, 893–903.

Sofuoglu, S.C., Lebowitz, M.D., O'Rourke, M.K., Robertson, G.L., Dellarco, M., Moschandreas, D.J., 2003. Exposure and risk estimates for Arizona drinking water. J. Am. Water Works Assoc. 95, 67–79.

Stark, A.D., Quah, R.F., Meigs, J.W., DeLouise, E.R., 1982. The relationship of environmental lead to blood-lead levels in children. Environ. Res. 27, 372–383.

Sternbeck, J., Sjodin, A., Andreasson, K., 2002. Metal emissions from road traffic and the influence of resuspension—results from two tunnel studies. Atmos. Environ. 36, 4735–4744.

Stewart, D.D., 1887. Notes on some obscure cases of poisoning by lead chromate; manifested chiefly by encephalopathy. Medical News 18 (June), 676–681.

Stewart, D.D., 1895. Lead convulsions. A study of sixteen cases. Am. J. Med. Sci. CLX, 286–306.

Subramanian, K.S., Sastri, V.S., Connor, J.W., 1994. Drinking water quality: impact of non-lead based plumbing solders. Toxicol. Environ. Chem. 44, 11–20.

Succop, P., Bornschein, R., Brown, K., Tseng, C.-Y., 1998. An empirical comparison of lead exposure pathway models. Environ. Health Perspect. 106 (Suppl. 6), 1577–1583.

Sutherland, R.A., Day, J.P., Bussen, J.O., 2003. Lead concentrations, isotope ratios, and source apportionment in road deposited elements, Honolulu, Oahu, Hawaii. Water Air Soil Pollut. 142, 165–186.

Sutton, P.M., Athanasoulis, M., Flessel, P., Guirguis, G., Haan, M., Schlag, R., et al., 1995. Lead levels in the household environment of children in three high-risk communities in California. Environ. Res. 68, 45–57.

Tepper, L.B., Levin, L.S., 1975. A survey of air and population lead levels in selected American communities. Environ. Qual. Saf. Suppl. II: Lead, 152–195.

Ter Haar, G., Aranow, R., 1974. New information on lead in dirt and dust as related to the childhood lead problem. Environ. Health Perspect. 7, 83–89.

TerraGraphics Environmental Engineering, 2000. 1999 Populated Area Five Year Study Review, vol. II. 1999 Five Year Review Report: Bunker Hill Superfund Site, Table 2.3.

Thomas, H.F., Elwood, P.C., Welsby St., E., Leger, A.S., 1979. Relationship of blood lead in women and children to domestic water lead. Nature 282, 712–713.

Thompson, N.G., Sosnin, H.A., 1985. Corrosion of 50–50 tin–lead solder in household plumbing. Welding J. April, 20–24.

Thornton, I., 1988. Metal content of soils and dusts. Sci. Total Environ. 75, 21–39.

Torfs, K., Van Grieken, R., 1997. Chemical relations between atmospheric aerosols deposition and stone decay layers on historic buildings at the Mediterranean coast. Atmos. Environ. 31, 2179–2192.

Treweek, G.P., Glicker, J., Chow, B., Sprinkler, M., 1985. Pilot-plant simulation of corrosion in domestic pipe materials. J. Am. Water Works Assoc. 77, 74–82.

Turer, D., Maynard, J.B., Sansalone, J.J., 2001. Heavy metal contamination in soils of urban highways: comparison between runoff and soil concentrations at Cincinnati, Ohio. Water Air Soil Pollut. 132, 293–314.

U.S. Centers for Disease Control, 1985. Preventing Lead Poisoning in Young Children. A Statement by the Centers for Disease Control. U.S. Department of Health and Human Services, Atlanta, GA.

U.S. Centers for Disease Control, 1991. Preventing Lead Poisoning in Children. A Statement by the Centers for Disease Control. U.S. Department of Health and Human Services, Atlanta, GA.

U.S. Centers for Disease Control, 2005. Preventing Lead Poisoning in Young Children. A Statement by the Centers for Disease Control. Department of Health and Human Services, Atlanta, GA.

U.S. Department of Housing and Urban Development, 1995. Guidelines for the Evaluation and Control of Lead-Based Paint Hazards in Housing. Office of Lead Hazard Control, Washington, DC.

U.S. Department of Housing and Urban Development, 1999. Requirements for Notification, Evaluation and Reduction of Lead-Based Paint Hazards in Federally-Owned Residential Property and Housing Receiving Federal Assistance. Final Rule.

U.S. Department of Housing and Urban Development, 2001. National Survey of Lead and Allergens in Housing. Final Report.

U.S. Environmental Protection Agency, 1971. Guidelines: Air Quality Surveillance Networks. Publication No. AP-98. Research Triangle Park, NC: Office of Air Programs. As cited in: U.S. EPA, 1977, Ch. 4.

U.S. Environmental Protection Agency, 1977. Air quality criteria for lead. Report No. EPA-600/8-77-017. Office of Research and Development, Washington, DC.

U.S. Environmental Protection Agency, 1985. Occurrence of lead in drinking water, food, and air. Report by JRB Associates. Office of Drinking Water, Washington, DC. As cited in U.S. EPA, 1986b.

U.S. Environmental Protection Agency. 1986a. Air Quality Criteria for Lead. 4 Vols. Report No. EPA/600/8-83/028bF. Environmental Criteria and Assessment Office, Washington, DC.

U.S. Environmental Protection Agency, 1986b. Reducing lead in drinking water: a benefit analysis. Report No. EPA-230-09-86-019. Office of Policy Planning and Assessment, Washington, DC.

U.S. Environmental Protection Agency, 1989. Risk assessment guidance for Superfund, vol. 1—Human Health Evaluation Manual (Part A). Report No. EPA/540/1-89/002. Office of Emergency and Remedial Response, Washington, DC.

U.S. Environmental Protection Agency, 1994. Guidance Manual for the Integrated Exposure-Uptake Biokinetic Model for Lead in Children. EPA/540-R93/081. Office of Research and Development, Washington, DC.

U.S. Environmental Protection Agency, 1995. National Air Quality and Emissions Trends Report, 1994. Report No. EPA 454/R-95-014. Office of Air and Radiation, Washington, DC.

U.S. Environmental Protection Agency, 2001. Lead: Identification of Dangerous Levels of Lead: Final Rule. 40 CFR Part 745. FR 66: 1206–1240, January 5, 2001.

U.S. Environmental Protection Agency, 2006. Air Quality Criteria for Lead, vol. 1. Report No. EPA/600/R-05/144aF. National Center for Environmental Assessment, Washington, DC.

U.S. Environmental Protection Agency, 2007a. Air Toxics Data Archive. STAPPA/ALAPCO/U.S. EPA Air Toxics Monitoring Subcommittee. <http://vista.cira.colostate.edu/atda/> (accessed October 30, 2010).

U.S. Environmental Protection Agency, 2007b. Lead: Human Exposure and Health Risk Assessments for Selected Case Studies, vol.1. Human Exposure and Health Risk Assessments—Full Scale. Report No. EPA-452/R-07-014a. Office of Air Quality Planning and Standards, Research Triangle Park, NC.

U.S. Environmental Protection Agency, 2008a. Child-Specific Exposure Factors Handbook. Report No. EPA/600/R-06/096F. National Center for Environmental Assessment, Washington, DC.

U.S. Environmental Protection Agency, 2008b. Office of the Administrator. National Ambient Air Quality Standards for Lead. Final Rule. Federal Register 73 FR 66964-67062, November 12, 2008.

U.S. Food and Drug Administration, 1985. Market-basket survey: preliminary results for lead analysis. As cited in U.S. EPA, 1986a, Ch. 7.

U.S. Food and Drug Administration, 2004. U.S. FDA Total Diet Study—Analytical Results. <http://www.fda.gov/Food/FoodSafety/FoodContaminantsAdulteration/TotalDietStudy/ucm184293.htm> (accessed November 3, 2010).

von Lindern, I.H., Spalinger, S.M., Bero, B.N., Petrosyan, V., von Braun, M.C., 2003a. The influence of soil remediation on lead in house dust. Sci. Total Environ. 303, 59–78.

von Lindern, I., Spalinger, S., Petrosyan, V., von Braun, M.C., 2003b. Assessing remedial effectiveness through the blood lead soil/dust relationship at the Bunker Hill site Superfund site in the Silver Valley of Idaho. Sci. Total Environ. 303, 139–170.

Walter, S.D., Yankel, A.J., von Lindern, I.H., 1980. Age-specific risk factors for lead absorption in children. Arch. Environ. Health 35, 53–58.

Water Research Centre, 1983. Lead in Drinking Water. Burke, T. (Ed.), Technical Report No. TR187. WRC Environmental Protection, Marlow, Bucks, United Kingdom.

Welter, E., Calmano, W., Mangold, S., Tröger, L., 1999. Chemical speciation of heavy metals in soils by use of XAFS spectrometry and electron microscopic techniques. Fresenius J. Anal. Chem. 364, 238–244.

World Health Organization, 1993. Guidelines for Drinking Water Quality. second ed. WHO, Geneva, Switzerland, vol. 1: Recommendations.

World Health Organization, 1995. Environmental Health Criteria 165: Inorganic Lead. International Programme on Chemical Safety, Geneva, Switzerland.

Worth, D., Matranga, A., Lieberman, M., Devas, E., Karalekas, P., Ryan, C., et al., 1981. Lead in drinking water: the contribution of household tap water to blood lead levels. In: Lynam, D.R., Piantinida, L.E., Cole, J.F. (Eds.), Environmental Lead. Proceedings of the Second International Symposium on Environmental Lead Research. Academic Press, New York, pp. 199–225.

Yankel, A.J., von Lindern, I.H., Walter, S.D., 1977. The Silver Valley lead study: the relationship between childhood blood lead levels and environmental exposure. J. Air Pollut. Control Assoc. 27, 763–767.

Yocum, J.E., 1982. Indoor–outdoor air quality relationships: a critical review. J. Air Pollut. Control Assoc. 32, 500–520.

Young, T.M., Heeraman, D.A., Sirin, G., Ashbaugh, L.L., 2002. Resuspension of soil as a source of airborne lead near industrial facilities and highways. Environ. Sci. Technol. 36, 2484–2490.

Part 2

Lead Exposure in Human Populations

Chapter 7

Lead Exposure in Human Populations: Lead Intakes

7.1 INTRODUCTION

Chapters in this part of the book present and discuss lead exposure in human populations. Exposure here refers to actual, not potential, contact with lead in one or more environmental media by human populations. Furthermore, the term here also refers to both intake of media lead into various receiving body compartments and subsequent uptake into the human body. Other definitions of the term exist in the lead literature. For example, exposure is confined to lead intake into receiving body compartments and lead contact by an absorbing surface.

Four chapters comprise Part 2 of this monograph: Chapter 7 describes lead intakes in U.S. and other populations; Chapter 8 addresses Pb toxicokinetics and biological markers of human lead exposure; Chapter 9 presents modeling data reported for human lead exposures; and Chapter 10 describes the environmental epidemiology of human lead exposures.

Chapter 7 is the toxicological interface between lead in the external environment and lead in the human body; that is, it connects Chapters 6 and 8. Chapter 6 addressed lead levels in environmental media relevant to human contact. Chapter 8, the internal component of the interface, deals with lead disposition in the body. The latter describes the various kinetic and metabolic processes that govern Pb entry into, distribution within, and subsequent retention or excretion of internal lead burdens in exposed human populations. The older literature regarding lead did not always make clear distinctions between lead intake and lead uptake, which are two distinct processes in human lead exposures. Lead uptake deals with lead absorption into the bloodstream from receiving body compartments. In most cases, the distinction between the processes of intake and uptake are physiologically and toxicologically clear. Ingestion or inhalation of media-specific lead into those respective receiving body compartments occurs first, followed by absorption of some fraction of lead in those intakes into the bloodstream. In the case of

Trace Metals and other Contaminants in the Environment, Volume 10
ISSN: 1875-1121 DOI: 10.1016/B978-0-444-51554-4.00007-9

dermal contact, the distinction between the two steps is blurred since intake through skin is not fully distinct from uptake. With parenteral, e.g., intravenous, administration, the two merge into a single step.

Lead intake by inhalation occurs in two stages. First, some fraction of inhaled lead is retained and deposited in various compartments of the respiratory tract. The balance is exhaled. With inhalation, depositions in different parts of the respiratory tract yield different overall exposure results. Larger particles containing lead are deposited in the upper tract and undergo ciliary removal and are swallowed, passing into the gastrointestinal tract. The smallest particles containing lead are deposited in the pulmonary compartment. Sites for actual lead uptake into the bloodstream and the uptake characteristics for these different sites influence uptake rates.

TABLE 7.1 **Illustrative Combinations of Pb Concentrations and Daily Environmental Media Intake Amounts Demonstrating Equivalent Intake Parameters**[a–c]

Environmental Medium	Media Pb Concentration (ppm or μg/m^3)	Daily Intake Mass (g) or Volume (m^3, l)	Daily Pb Intake (μg/day)
Lead paint dust	5,000 (0.5%)	0.001 g	5
Lead paint dust	10,000 (1.0%)	0.001 g	10
Lead paint dust	25,000 (2.5%)	0.001 g	25
Lead in diet	0.0025	2,000 g	5
Lead in diet	0.005	2,000 g	10
Lead in diet	0.010	2,000 g	20
Lead in soil	500	0.010 g	5
Lead in soil	1,000	0.010 g	10
Lead in soil	2,000	0.010 g	20
Lead in tap water	0.005	1.0 l	5
Lead in tap water	0.010	1.0 l	10
Lead in tap water	0.025	1.0 l	25
Lead in air	1.0	5 m^3	5
Lead in air	2.0	5 m^3	10
Lead in air	5.0	5 m^3	25

[a]*Media Pb levels for paint, soil, and tap water at or above regulatory levels.*
[b]*Ingested amounts typical for young children.*
[c]*Air ventilation rate = 5 m^3 for young children.*

Intake of lead or other substances in humans is typically indexed as daily intake. However, intakes scaled for different time frames have been employed in different settings over the years, e.g., weekly intakes (FAO/WHO, 1993) of Pb and other contaminants per body weight. This specifically applies to ingestion, inhalation, and, in some cases, dermal application. Daily lead intake into body compartments is the product of lead concentration in some medium and the mass (diet Pb) or volume (air Pb, water Pb) of lead-containing medium taken in daily. High levels of lead in an environmental medium can be quite toxic when ingested in relatively modest quantities daily.

On the other hand, low concentrations of lead can be of concern when large amounts of the lead-containing medium are ingested. The relationships are depicted in Table 7.1 where a high-lead medium such as lead paint dust ingested in small amounts can pose the same risk as low lead in the daily diet consumed at typical food intake levels. Table 7.1 shows that, because of either high Pb concentrations or high amounts of daily media mass intake, equivalent amounts of Pb can hypothetically enter receiving compartments daily regardless of the nature of the contact medium.

7.2 DAILY LEAD INTAKES BY U.S. AND OTHER POPULATIONS

Chapter 6 had numerous tabulations of lead concentrations in environmental media encountered by human populations in the United States and around the world. This chapter presents estimates for daily Pb intakes using Chapter 6 information combined with information on media-specific intake amounts. Intakes are provided for the major media-specific routes of exposure as they are available for various segments of human populations. Relevance of lead-containing environmental media differs across segments of populations, with given concentrations of lead in some contact medium translating into quite different degrees of lead exposure, depending on population and individual characteristics. For example, infants and toddlers ingest food at much higher rates on some body metric basis than older children or adults. Infants and toddlers ingest more of some particular Pb-contaminated environmental medium than do older individuals. A notable example is dust or soil by way of normal hand–mouth activity in this developmental age band. Dietary lead intakes differ with age as a function of type of dietary item, caloric requirements, etc. Groups of individuals with the same demographic characteristic but affected by different external factors such as climate and temperature will ingest different amounts of Pb-contaminated media, e.g., volumes of water.

A major determinant of lead intakes among human populations has been the period for which data were gathered. In past decades when lead levels were much higher than today in canned foods, in ambient air, in crops, and in household painted surfaces, lead intake estimates were much higher. To the extent possible, available data for various periods are included in this chapter's tabulated intake estimates in order to show temporal changes in exposures. The reason is the same as for using older data in Chapter 6. Bone Pb in older subjects will reflect lead intakes from prior decades.

Lead-containing environmental media ingestion or inhalation rates are presented here for ambient air, diverse human diets, drinking water, and dust and/or soil. Dermal contact and potential dermal uptake into the bloodstream for the inorganic forms of environmental Pb encountered by nonoccupational populations are very low compared to the other routes of exposure, and this route is not addressed.

Various primary sources for estimating the exposure parameters associated with daily intake quantities of (lead-containing) media were used and are noted. A child exposure-specific source of parameter selection for media of interest was the U.S. EPA "Child-Specific Exposure Factors Handbook" (2008).

7.2.1 Daily Human Intakes of Ambient Air Lead

Daily intakes of lead through inhalation by human or experimental populations are the product of ventilation rate and air Pb concentration, further adjusted for lead deposition rates of the inhaled volumes in the respiratory tract. The ventilation rate, i.e., total volumes of inhaled air per unit time, typically 24 hours, can be determined by either direct measurement or via food-energy intakes. The latter may also be employed to arrive at the ratios of total energy expenditure to basal metabolic rate (BMR).

Layton (1993) carried out a series of studies with U.S. subjects of different ages for long-term and short-term inhalation rates estimated via various methodologies. Choice of methodology depended on the activity being evaluated. Daily long-term inhalation rates, which are the estimates most relevant to this chapter, were calculated from food-energy intakes. Table 7.2 presents the daily inhalation (ventilation) rates for children up to 18 years of age. Children under 9 years of age showed no gender-specific inhalation rates, but those 9 years and older showed a measurable difference favoring a higher rate for male children. Table 7.2 shows that inhalation rates for older male children were highest between 15 and 18 years, and for females, between 9 and 11 years. For those youngest children not identified by gender, the highest inhalation rates were calculated between the ages of 6 and 8.

Table 7.3 depicts estimates of long-term daily deposition rates of Pb in young children who varied in age and gender using available empirical data. Pb deposition rates increased with child's age in the range 1.4–3.2 μg Pb/day. Arcus-Arth and Blaisdell (2007) reported the statistical distributions of daily breathing rates for narrow age groups of infants and children.

TABLE 7.2 Estimated Daily Long-Term Inhalation Rates for U.S. Children of Differing Ages[a–c]

Group/Age (years)	Duration[d]	Mean Daily Inhalation Rate (m^3/day)
Children		
<1	1	4.5
1–2	2	6.8
3–5	3	8.3
6–8	3	19
Children—Males		
9–11	3	14
12–14	3	15
15–18	4	17
Children—Females		
9–11	3	13
12–14	3	12
15–18	4	12

[a]*Adapted from Layton (1993) and U.S. EPA (2008).*
[b]*Based on food-energy intake rates from Layton (1993).*
[c]*Gender-specific data for children ≥9 years of age and adults.*
[d]*Number of years/cohort.*

Tables 7.4 and 7.5 present the calculated daily inhalation rates and deposition rates for Pb at a time of maximum air Pb in the 1970s and then in 1994, when air Pb concentrations had essentially declined to their lowest levels. The pronounced distinctions in inhaled and deposited Pb in children at these two time points are striking, showing a 20-fold or more decline in deposition rates for urban areas. Again, the air Pb declines and associated inhaled and deposited Pb reflected the leaded gasoline ban in the United States and other industrialized nations. Daily inhaled Pb is deposited in various parts of the respiratory tract at a rate of 30% for children and adults (U.S. EPA, 1986, Ch. 10). Male children show a somewhat higher daily deposition rate than females. While the tables are for U.S. estimates, the values for air Pb outside the United States and those that are tabulated in Chapter 6 would show similar results.

Data given in Tables 7.2–7.4 are not expressed normalized to body weight or some other somatic metric. However, the nature of the

TABLE 7.3 Estimated Long-Term Daily Inhaled and Deposited Pb in U.S. Children of Various Ages for 1970–1974[a]

Group/Age (years)	Daily Inhalation (m^3/day)[b]	Daily Inhaled Pb (μg/day)[c]	Daily Deposited Pb (μg/day)[d]
Children			
<1	4.5	4.8	1.4
1–2	6.8	7.3	2.2
3–5	8.3	8.9	2.7
6–8	10	10.7	3.2
Males			
9–11	14	15.0	4.5
12–14	15	16.1	4.8
15–18	17	18.2	5.5
Females			
9–11	13	13.9	4.2
12–14	12	12.8	3.8
15–18	12	12.8	3.8

[a] *Average of five urban U.S. arithmetic air Pb mean, 1970–1974, is given in Table 6.3: 1.07 μg/m^3.*
[b] *Adapted from Layton (1993) and U.S. EPA (2008).*
[c] *Air Pb mean (μg/m^3) × Daily Inhalation Rate (m^3/day).*
[d] *Daily inhaled Pb (μg/day) × 0.3 (deposition rate).*

methodology employed by Layton (1993), involving the dependence of air inhalation rates on food-energy intake rates, scaled total inhalation values in a manner that reflected body mass; consequently, when one adjusts these data for body weight, there is limited change across children's ages.

Adult inhalation rates for males and females generally cluster around 20 m^3/day for individuals with typical occupational activity patterns in temperate climates. One can estimate the daily Pb inhalation and deposition rates for nonoccupational inhalation Pb exposures by using this value to arrive at values corresponding to those in Tables 7.3 and 7.4 for children.

7.2.2 Daily Human Intakes of Dietary Lead

Dietary lead concentrations relevant to human populations are presented in Chapter 6 and include periodic surveys from various agencies such as the U.S. FDA. This chapter has estimates of ingested daily amounts of dietary

TABLE 7.4 Estimated Daily Long-Term Inhaled and Deposited Pb in U.S. Children of Various Ages for 1994[a]

Group/Age (years)	Daily Inhalation Rate (m^3/day)[b]	Daily Inhaled Pb (μg/day)[c]	Daily Deposited Pb (μg/day)[d]
Children			
<1	4.5	0.23	0.07
1–2	6.8	0.34	0.10
3–5	8.3	0.42	0.13
6–8	10	0.50	0.15
Males			
9–11	14	0.70	0.21
12–14	15	0.75	0.23
15–18	17	0.85	0.26
Females			
9–11	13	0.65	0.20
12–14	12	0.60	0.18
15–18	12	0.60	0.18

[a]*Urban mean air Pb for U.S. sites, U.S. EPA (1995), Table A-10.*
[b]*Adapted from Layton (1993) and U.S. EPA (2008).*
[c]*Air Pb mean (μg/m^3) × Daily Inhalation Rate (m^3/day).*
[d]*Daily Inhaled Pb (μg/day) × 0.3 (deposition rate).*

components and, based on these amounts and food group Pb levels given in the previous chapter, estimates of intakes of daily dietary Pb.

Human exposures to lead in the diet historically comprised a major fraction of overall lead exposures in the United States and elsewhere, especially for adults. However, in a number of instances, such as infant consumption of Pb in evaporated milk from lead-seamed cans, many children also sustained significant dietary Pb exposures. Dietary lead intakes, for purposes of this chapter, are separated from drinking beverages prepared from drinking water, but include foods cooked in tap water containing Pb.

Dietary components showed major reductions in Pb content over the last several decades, and Pb intakes through the diet obviously track these declines. Dietary Pb intake estimates in earlier decades were poorly recorded. Available data, however, clearly indicate that dietary Pb intakes during the era of leaded gasoline and lead-seamed cans for foods combined

TABLE 7.5 U.S. Daily Intakes of Food and Beverages During the Early 1980s[a–c]

Food/Beverage Group	Child, 2 Years Old	14–16 Years Old		25–30 Years Old		60–65 Years Old	
		Male	Female	Male	Female	Male	Female
Meats/meat products	133	269	182	319	194	252	172
Food crops	282	528	386	518	390	532	437
Dairy	390	645	405	351	245	279	208
Canned foods	72	104	77	103	73	119	99
Canned juices	54	30	28	27	28	12	17
Frozen juices	65	75	53	73	66	61	72
Soda	65	274	232	315	228	85	78
Canned beer	0	17	0	318	51	116	18
Water	441	743	596	1,061	903	1,244	1,166
Total	1,502	2,685	1,959	3,086	2,178	2,700	2,267

[a]*Adapted from U.S. EPA (1986), and using summary data of Pennington (1983).*
[b]*Units of g/day.*
[c]*"Water" category includes beverages such as tea, coffee, and powdered drinks.*

to produce huge daily Pb intakes in diet because dietary components were contaminated by Pb.

Analogous to the approach for ambient air, this section first addresses estimates of the total amounts of diet ingested daily and components within the total dietary profiles for various groups within the U.S. and other populations. The increased uncertainty and variability in more generalized approaches for quantification of Pb in foodstuffs is amplified when approaching dietary intake amounts on a total or group-specific mass basis. For example, despite more public interest in "healthy" foods, a high per capita consumption rate of "fast" foods further complicates comparisons of historical dietary Pb intakes with those of recent years.

Food supplies of developed, industrialized countries in North America and Europe have centralized distribution systems and one would expect that the level of heterogeneity for lead distribution in dietary components would vary less across smaller subdivisions of nations than, say, ambient air Pb, dust or soil Pb, and tap water Pb. Variability and uncertainty remain nonetheless. First, there are socioeconomic components to diets within populations that translate to different amounts of Pb intakes and exposures.

Low-income subsets are more likely to produce more homegrown diet components in gardens. Groups of the population who rely on subsistence fishing and harvest catches where aquatic lead contamination is likely or of concern will ingest amounts of the element significantly above mean population intakes. Ethnic food preferences can also raise the likelihood of higher Pb content in certain imported canned foods.

The principal sources of systemic assessments of dietary intakes of food groups in the United States are the USDA's NFCS and the USDA Continuing Survey of Food Intakes by Individuals (CSFII) (U.S. Department of Agriculture, 1990). The NFCS databases include figures for 1987/1988. The CSFII values are for two more recent periods, 1989–1991 and 1994–1996, and were used for a number of estimated intakes. The U.S. EPA exposure factors handbook for children (U.S. EPA, 2008) was also used. Tables 7.5–7.8 present estimates of daily amounts of food groups ingested in various U.S. population age groups. Values are in two forms, as daily weights per person and/or dietary amounts on a body weight basis. Table 7.5 shows that a typical 2-year-old child ingests around 1 kg of various food groups and about 440 ml of water. With age, both food and water amounts consumed daily increase, peaking in the 25- to 30-year-old U.S. male at around 3.1 kg and in the U.S. female at 60–65 years of age.

Table 7.6 shows daily amounts of food ingested by U.S. children in the overall age range of <1 year old to 19 years old. The teenaged band comprises the highest total daily intake group, showing 200 g more on average than infants. However, when intakes are adjusted to body weight, the highest intake rate is in early infants and least in the 12 to 19-year-olds, a factor of about five higher in the infants. Distribution of ingested daily amounts of food into food group categories as a function of body weight is tabulated in Table 7.7. While absolute daily intakes differ across ages, all children consume the same food group in proportional amounts in terms of mean and 95th percentile estimates. Dairy products rank highest across all ages in terms of daily intake while the "fish" category ranks lowest. Table 7.8 presents the amounts of daily food group intake for the "homegrown" food items. The first part gives the fraction of children consuming homegrown foods while the remainder tabulates mean and 95th percentile intakes as a function of body weight.

Daily intakes of dietary Pb vary with age and gender. U.S. dietary Pb intakes are presented in Table 7.9 for the early 1980s. This period reflected some persisting impact of leaded gasoline use in the nation, based on data presented in Chapter 6 for food group lead concentrations over the decades. Teenagers and young adults consumed more Pb on a total daily intake basis than young children or the elderly, as shown in Table 7.9. Gender differences in daily U.S. dietary Pb intakes are seen in the teen, young adult, and older

TABLE 7.6 U.S. Per Capita Total Diet Intakes for Children at Different Ages[a–c]

Child Age Band (years)	Mean	95th Percentile
Diet intake, g/day, as consumed		
<1	1,000	1,800
1–2	1,100	1,800
3–5	1,000	1,700
6–11	1,100	1,900
12–19	1,200	2,300
Diet intake, g/kg-day, as consumed		
<1	140	240
1–2	84	150
3–5	55	100
6–11	36	69
12–19	26	40

[a]*Adapted from U.S. EPA (2008) based on 1994/1996 USDA CSFII.*
[b]*Intake amounts as consumed.*
[c]*Mean/95th percentile values.*

TABLE 7.7 U.S. Per Capita Total Intakes (Mean/95th Percentile) of Different Consumer Food Categories for Children of Various Ages[a–c]

	Children's Age Band (years)				
Food Group	<1	1–2	3–5	6–11	12–19
Meat	1.1/5.9	4.4/10.2	4.1/9.4	2.9/6.8	2.2/4.9
Fish	0.1/0.5	0.4/1.8	0.3/1.7	0.3/1.4	0.2/1.1
Vegetables	6.9/24.2	9.5/23.3	7.3/18.3	5.3/13.5	4.0/9.3
Grains	4.1/20.2	11.2/24.7	10.3/21.1	7.2/15.6	4.4/9.7
Dairy	111/235	37.5/90.2	20.9/48.8	13.9/33.5	6.1/17.8
Fruits	13.2/41.2	19.3/53.9	11.0/32.7	5.4/18.0	2.8/11.0

[a]*Adapted from U.S. EPA (2008) and using USDA CSFII 1994–1996 data.*
[b]*g/kg-day food group.*
[c]*General population members, <1 to 19 years of age.*

TABLE 7.8 Mean U.S. Consumption of Homegrown Foods from Different Food Groups by Children of Various Ages[a–c]

	% Children Consuming by Age (years)				Mean and 95th Percentile by Age (years)			
Food Group	1–2	3–5	6–11	12–19	1–2	3–5	6–11	12–19
Homegrown fruits	6.3	6.8	6.3	5.8	8.7/61	4.1/8.9	3.6/16	1.9/8.3
Homegrown vegetables	16.7	15.2	18.1	16.1	5.2/19.6	2.5/7.7	2.0/6.2	1.5/6.0
Home-produced meats	4.8	4.9	6.4	6.2	3.7/10	3.6/9.1	3.7/14	1.7/4.3
Home-caught fish	1.4	1.8	2.3	1.7	–	–	2.8/7.1	1.5/4.7

[a]*Adapted from U.S. EPA (2008), using USDA NCFS for 1987–1988.*
[b]*g/kg-day.*
[c]*Insufficient data for groups not indicated.*

TABLE 7.9 Total Pb Intake (μg/day), by Age and Sex, of Food and Beverages in the Early 1980s[a–c]

		14–16 Years Old		25–30 Years Old		60–65 Years Old	
Food Category	Child, 2 Years Old	Male	Female	Male	Female	Male	Female
Meat/meat products	3.4	7.4	4.8	7.4	5.0	5.4	4.0
Food crops	5.5	11.7	8.1	11.3	7.9	9.6	7.8
Dairy	2.8	5.4	3.5	3.4	2.5	3.1	2.3
Canned foods	7.3	11.8	8.1	12.0	8.8	14.4	11.6
Canned juices	2.7	1.5	1.4	1.4	4.4	0.6	0.9
Frozen juices	0.5	0.7	0.5	0.7	0.6	0.5	0.7
Soda	0.7	3.0	2.3	2.9	2.1	0.9	0.9
Canned beer	0.0	0.1	0.0	2.5	0.7	1.0	0.3
Water	2.1	3.2	2.5	3.6	3.0	4.2	3.9
Total	25.0	44.8	31.2	45.2	32.0	39.7	42.4

[a]*Adapted from U.S. EPA (1986) and using data of Pennington (1983).*
[b]*Units of μg/day.*
[c]*"Water" category includes coffee, tea, and powdered drinks.*

adult groups, with males ingesting more Pb. No gender difference was seen with infants or toddlers.

U.S. population daily diet Pb intake declines with the phase out of leaded gasoline and lead seams in food or beverage cans are evident from the national data. Chapter 6 noted Pb concentration declines in various food groups from the 1970s to more recent years. In terms of total daily Pb intakes, a marked decline has been observed for all age and gender groups. Adams (1991) reported that for the relatively short period from 1982 to 1988, a period associated with reductions in leaded gasoline and lead-seamed can use, toddlers showed about a fivefold reduction in daily intakes from 25 to 5 μg/day. A similar relative reduction was seen with adult women from 36 to 8 μg/day.

Egan et al. (2002) reported lower daily U.S. diet Pb intakes into the 1990s. The 1994–1996 period showed daily dietary Pb intake ranges for (1) infants up to 11 months, (2) 2 year olds, and (3) older children or adults as 0.8–5.7, 2.4–10.1, and 4–19 μg/day, respectively. For 1982–1984, corresponding figures for the above age groups were (mean): 17, 23, and 29–41 μg/day. The 2004 U.S. FDA survey noted in Chapter 6 recorded virtually all categories of food groups for U.S. population segments as being below the measurement detection limit. Consequently, any current estimates of daily diet Pb would reflect statistically projected figures with considerable uncertainty. Suffice it to say that today's figures would likely be lower than those noted by Egan et al. (2002) for the mid-1990s period.

The trend of declining Pb in dietary daily intakes is readily apparent in U.S. EPA figures selected as exposure input values for dietary Pb to the agency's Integrated Exposure-Uptake Biokinetic Model for estimating PbB values in children (U.S. EPA, 2009). Table 7.10 summarizes Pb intakes for 12-month age bands in children up to 84 months of age. In infants of 12–23 months old, Pb intake declined about 65% between pre-1991 data and 1995–2003 data. Similar declines were seen in other age bands.

International daily Pb intake estimates for diet track a similar array of older values as in the United States and were summarized for the 1980s by Galal-Gorchev (1991). Table 7.11 summarizes the international picture for daily diet Pb intakes by children on the basis of body weight for nine countries representing various locales and demographic profiles. The U.S. weekly and daily diet Pb intakes adjusted to body weight were the lowest, 3.1 μg/week and 0.4 μg/day, respectively. Poland ranked highest, at 28 μg/week (4 μg/day). Corresponding figures for adults are presented in Table 7.12. India, Italy, and Cuba showed the highest diet Pb intakes on the basis of body weight, with the United States showing mean data (rounding) of <0.02.

Some subsets of human populations have relatively limited dietary Pb sources. Nursing infants, for example, will ingest any Pb passing into breast milk from the nursing mother while bottle-fed infants can consume any tap

TABLE 7.10 Total Pb Intake (μg/day) by Child Age Band Based on Various U.S. EPA Updatings of Dietary Pb[a–c]

Age Band (months)	Diet Pb Intake		
	Pre-1991	1991–1999 TDS Data	1995–2003 Data
0–11	5.5	3.2	2.3
12–23	5.8	2.6	2.0
24–35	6.5	2.9	2.1
36–47	6.2	2.7	2.0
48–59	6.0	2.6	2.0
60–71	6.3	2.7	2.1
72–84	7.0	3.0	2.2

[a]*Based on U.S. EPA (1994) model manual diet Pb estimates for 1980s.*
[b]*Based on U.S. FDA Total Diet Study estimates: 1991–1999 and 1995–2003.*
[c]*Total age band, 0–84 months.*

TABLE 7.11 Daily Pb Intakes in Child Population for Various Countries Surveyed in the UNEP/GEMS Program[a–c]

Country	Total Weekly Pb Intake (μg/kg body weight)	Total Daily Pb Intake (μg/kg body weight)
Poland	27.7	4.0
Germany[d]	26.9	3.8
Hungary	25.8	3.7
Canada	17.6	2.5
Sweden	15.1	2.2
Philippines	12.6	1.8
United Kingdom	11.8	1.7
Cuba	7.3	1.0
United States	3.1	0.4

[a]*Adapted and estimated from values in Galal-Gorchev (1991) to daily intakes.*
[b]*Infants and older children up to age 12, 1980–1988.*
[c]*μg/kg body weight.*
[d]*Pre-reunification figures for Federal Republic of Germany.*

TABLE 7.12 Daily Pb Intakes in Adult Population for Various Countries Surveyed in the UNEP/GEMS Program[a–c]

Country	Total Weekly Pb Intake (μg/kg body weight)	Total Daily Pb Intake (μg/kg body weight)
India	60.0	8.6
Italy	59.3	8.5
Cuba	63.3	9.0
Germany[d]	26.7	3.8
France	19.6	2.8
Poland	18.3	2.6
Japan	9.8	1.4
United Kingdom	7.2	1.0
Canada	5.7	0.8
Sweden	2.6	0.4
United States	0.02	0.0[e]

[a]*Adapted and estimated from values in Galal-Gorchev (1991) for weekly intakes.*
[b]*Adults for the period 1980–1988.*
[c]*μg/kg body weight.*
[d]*Pre-reunification figures for the Federal Republic of Germany.*
[e]*Rounding.*

water Pb in formula mixtures. It is not now possible to provide an accurate estimate of daily Pb intakes by nursing. Earlier literature indicated such Pb exposures for infants might be of some significance, but analytical method sensitivity and persisting sample lead contamination have been problematic. More recent and comparatively more reliable measurement data reported by Gulson et al. (2003) suggest a daily intake figure for this pathway of about 0.8 μg/day. The corresponding formula-based Pb intake is on the order of 1–2 μg/day.

The marked decline in dietary Pb daily intakes from the 1970s to the present has produced a marked shift downward in the relative fractional contribution of diet Pb to total daily Pb intakes. For example, Bolger et al. (1991) estimated that the percentage contribution to total Pb intake in 2 year olds attributable to their diet declined from 47% in 1986 to 16% in 1990. This threefold decline occurred with major declines in leaded gasoline and lead-seamed food can use but relatively little decline in other major Pb pathways for children, e.g., soil Pb.

TABLE 7.13 Estimated Direct Daily and Indirect Total Water Intakes by Water Source for the U.S. Population[a–d]

	Daily Water Intake (ml/day)		
Water Source	Mean	90th Percentile	95th Percentile
Community supply	927	2,016	2,544
Bottled water	161	591	1,036
Other sources	128	343	1,007
Missing sources	16	–	–
All sources	1,232	2,341	2,908

[a]Sources: *Data from USDA CSFII (1998) and U.S. EPA, 2008.*
[b]*Direct intakes = water consumed directly; indirect intakes = water used in food preparation.*
[c]*Based on two nonconsecutive days of reporting,* N = *15,303.*
[d]*For all ages in the population.*

7.2.3 Daily Human Intakes of Drinking Water Lead

Various U.S. and other surveys of drinking water consumption rates differ considerably depending on what subdivisions of total fluid volume intakes are included in the various totals being reported. Some deal solely with daily direct water intake volumes while others include water quantities consumed directly and volumes used in daily food preparation. Still other surveys include bottled water or water volumes used for beverages as well as tap water volumes or include just community water sources in the intake tallies. The various tabulations presented in this section note that these qualifications were feasible.

Consumption volumes differ across age and gender in human populations, reflecting factors such as climate, physical exertion, and body mass. It should be noted, nonetheless, that it is the variability of Pb content that is the larger parameter contributing to both variability and uncertainty compared to quantities of water drunk daily. For example, levels of Pb in drinking water, especially in homes with lead-soldered plumbing or lead service lines, can vary by an order of magnitude or even more, while the range of imbibed water volumes will be considerably less.

Table 7.13 tabulates estimates of total (direct and indirect) water intakes as a function of water source for all ages in the U.S. population as provided by surveys done by the U.S. EPA (2008) and the USDA CSFII (1998) survey programs. The daily intake mean volumes for community, bottled water, other sources, and unknown sources are 927, 161, 128, and 16 ml, respectively, producing a total of 1,232 ml. The corresponding 90th and 95th per-

TABLE 7.14 Estimated Daily Direct and Indirect Community Water Intakes by Children's Age Category[a–e]

Age Band (years)	Sample *N*	Mean	90th Percentile	95th Percentile
Intake (ml/day)				
0.5–0.9	160	412	884	1,101
1–3	1,834	313	691	942
4–6	1,203	420	917	1,165
7–10	943	453	978	1,219
15–19	825	760	1,610	2,062
Intake (ml/kg-day)				
0.5–0.9	153	45	103	122
1–3	1,752	23	51	67
4–6	1,103	21	44	64
7–10	879	15	32	39
15–19	816	12	25	32

[a]*Adapted from U.S. EPA (2008) and data from USDA-CSFII (1998).*
[b]*Direct intakes = water consumed directly from the tap; indirect intakes = water used in food preparation.*
[c]*Based on two nonconsecutive days of sampling.*
[d]*Sample size varies with age band.*
[e]*Reported as total daily intake/person or daily intake per kg body weight.*

centile values are 2,016 and 2,544 ml for community water and 591 and 1,036 ml for bottled water. The table shows that community water supplies, provided in large measure by public water treatment facilities, constitute the main source of the typical U.S. resident having public supply access. In terms of mean allocation of water volume intakes as a function of water sources, community supplies comprise about 75% of total water daily intake. For the 95th percentile of source-based water intakes, the corresponding figure is 87%.

Table 7.14 presents daily water intake estimates as a function of age within the childhood group, calculated both as total volumes per day and as a function of body weight. The source of water in these estimates is the community water system in the survey communities. Five age bands for children, along with their respective daily water intake volumes, are given in terms of total daily intakes or daily intakes on the basis of body weight. Total water intakes are highest in children of 15–19 years of age, 760 ml, and least for

TABLE 7.15 Tap Water Daily Intake Rates for Women of Childbearing Age (15–49 years)[a,b]

Reproductive Status	Mean	90th Percentile	95th Percentile
Daily intake, person (ml/day)			
Control	1,157	1,983	2,310
Pregnant	1,189	2,191	2,424
Lactating	1,310	1,945	2,191
Daily intake, person (ml/kg-day)			
Control	19.1	33.1	39.1
Pregnant	18.3	34.5	39.6
Lactating	21.4	35.1	37.4

[a]*Adapted from U.S. EPA (2008).*
[b]*Tap water fraction of daily fluid intake: control, 57.2%; pregnant, 54.1%; lactating, 57.0%.*

those 1–3 years of age. However, on the basis of body weight, the values are largely reversed. Similar relative rankings were found at the 90th and 95th percentiles.

Table 7.15 presents water volume intakes among women of childbearing age, pregnant women, and those who are lactating. Such intake estimates provide a measure of potential lead exposure risks to the fetus and the newborn when the concentrations of Pb in these water sources are known. In terms of total water daily consumption, the volumes are similar in all three categories for mean (1,157–1,310 ml/day) or higher percentile consumption rates.

The global figure for daily drinking water Pb intake reported from the GEMS/Food/UNEP program by Galal-Gorchev (1991) is 40 μg/day. This estimate is based on an international consumption volume across age groups, gender, and nationalities of 2,000 ml and a global average water concentration of 0.02 μg/ml or 20 μg Pb/l water. Daily tap water Pb consumption rates have varied with time in U.S. populations of varying age. For the early 1980s, prior to controls on lead content of plumbing materials and plumbing practices (banning use of 50–50 lead–tin solder), total water Pb intakes for various ages were 2.1 μg/day, 2-year-olds; 3.2 and 2.5 μg/day, 14- to 16-year-old males and females, respectively; 3.6 and 3.0 μg/day, 25- to 30-year-old males and females, respectively; 4.2 and 3.9 μg/day, 60- to 65-year-old males and females, respectively.

Estimates of daily water Pb intakes can be readily calculated for more recent times with data for water Pb levels for typical communities (U.S.

EPA, 2006) in combination with daily consumption volumes recorded in tables presented in this chapter (Table 7.14). Table 7.16 tabulates estimated daily water Pb intakes for indicated communities and residents of different ages in those communities. The cities indicated in the table were recorded with exceedances of the current drinking water lead action level at the tap, a level of 15 μg/l at the 90th percentile of distributions in survey results for these communities. These cities, therefore, represent more of a worst case scenario for U.S. urban community water supplies.

Table 7.16 presents data for two tap water surveys, in 1993 and some more recent survey results over the years 2001–2003 depending on the individual city. Older children 15–19 years old show the highest tap water lead intakes for both the 1993 and the more recent surveys, since they consume the highest volume of tap water among the given childhood age bands (0.76 l). Of the city tap water Pb levels recorded for 1993, Philadelphia, PA, had the highest intakes across the board, and was the highest among the 18 cities surveyed with action level exceedances. In Philadelphia, children of 15–19 years old experienced significant potential lead exposures from tap water, depending on household practices such as flushing or not flushing standing water from household plumbing. By contrast, during 2001–2003 this city's tap water Pb values were significantly lower and, consequently, children residing in this community had relatively lower intakes.

The tap water intakes for those 18 cities with action level exceedances for 2001–2003 were, overall, considerably below what they were in 1993. These reductions reflected a variety of water treatment and other reduction strategies implemented post-1993. Nonetheless, some communities, such as Washington, DC, still have the potential for tap water Pb exposures.

7.2.4 Daily Human Intakes of Soil Lead

Soil Pb intakes by segments of various populations can pose significant potential risks of exposures. However, the nature of soil Pb exposures are such that they present higher risks to sensitive groups such as young children than they do for adults. The reason for this is simply that the young child orally explores his/her physical environment and typically engages in various hand-to-mouth activities that include ingestion of varying amounts of soil. Older literature labeled all such normal hand-to-mouth activity as a "pica" behavior but that label is now reserved for "excessive" ingestion of nonfood materials. However, no clear clinical definition of, or criteria for, this behavior exists. Some literature has defined the behavior in statistical terms, e.g., some upper percentile of a distribution of observed daily ingestion rates within some study population. For discussing the broad topic of soil ingestion through normal child activity, it is probably best to dispense with the term and confine it to extreme intakes, e.g., gram quantities.

TABLE 7.16 Estimated Mean Daily Tap Water Pb Intakes for Children in the Indicated U.S. Communities for the Indicated Time Periods and Child Ages[a–d]

	Period									
	1993					2001–2003				
Community	0.5–0.9 Years	1–3 Years	4–6 Years	7–10 Years	15–19 Years	0.5–0.9 Years	1–3 Years	4–6 Years	7–10 Years	15–19 Years
Bayonne, NJ	7	6	8	8	14	7	6	8	8	14
Cedar Rapids, IA	33	25	34	36	61	2	2	3	3	5
Chicago, IL	4	3	4	5	8	3	2	3	3	5
Columbia, SC	16	12	17	18	30	2	2	3	3	5
Columbus, OH	6	5	6	7	11	0	0	0	0	1
Detroit, MI	9	7	9	9	16	5	4	5	5	9
Galveston, TX	7	6	8	8	14	1	1	1	1	2
Miami Beach, FL	11	8	11	12	21	3	2	3	4	6
Minneapolis, MN	8	6	8	9	14	2	2	3	3	5
Philadelphia, PA	132	100	135	145	245	5	4	5	6	10
Phoenix, AZ	8	6	8	9	14	0	0	0	0	1
Portland, OR	17	13	17	18	31	3	2	3	4	6

(*Continued*)

TABLE 7.16 Estimated Mean Daily Tap Water Pb Intakes for Children in the Indicated U.S. Communities for the Indicated Time Periods and Child Ages[a–d]—(cont.)

	Period									
	1993					2001–2003				
Community	0.5–0.9 Years	1–3 Years	4–6 Years	7–10 Years	15–19 Years	0.5–0.9 Years	1–3 Years	4–6 Years	7–10 Years	15–19 Years
Richmond, VA	7	5	7	7	12	2	1	2	2	3
St. Paul, MN	22	17	23	24	41	0	0	0	0	1
Syracuse, NY	21	16	21	23	38	10	8	11	11	19
Tacoma, WA	13	10	13	14	24	5	4	5	5	9
Yonkers, NY	28	21	29	31	52	7	6	8	8	14
Washington, DC	7	6	8	8	14	26	20	26	28	48

[a] *Communities noted in U.S. EPA (2006: Table 3.11) as exceeding EPA Pb action levels.*
[b] *90th percentile Pb levels (μg/l) for 1993 and most recent monitoring periods: 2001–2003.*
[c] *Volumes of water consumed at different ages by children as given in Table 7.13. Excludes bottled water and minor categories.*
[d] *Methods for tap water testing referenced in U.S. EPA (2006).*

Two parameters determine daily soil Pb intake, much the same way they do with other environmental media. They are the soil Pb concentration and the daily amounts of soil ingested. Daily ingested amounts of soils vary in children, but the range of variation is dwarfed by the huge range of soil Pb concentrations the children may encounter. The latter reflect all of the contamination histories of specific soils encountered by children, including pathways for contamination such as atmospheric Pb fallout onto yard soils, weathering or chalking of exterior lead paints onto adjacent soils, and Pb-bearing waters runoff onto adjacent soils. While ingested amounts across individuals from early childhood through adulthood will typically lie within an order of magnitude, 20–200 mg, Pb concentrations can range from background, uncontaminated levels in the range of 20–50 ppm or so to seriously contaminated soils having lead content at percentage levels.

The role of soil Pb exposures in human Pb toxicity is of more concern for young children and has prompted various studies of daily ingestion rates for soils containing Pb or other contaminants. Ethical considerations preclude feeding Pb-laced soils to children so typical methods in the relevant literature involve using adult healthy volunteers or doing excretion studies for soil Pb ingested in normal play activities. Both these approaches require elemental "tracers" which are ingested by children along with various soils but are assumed to have virtually no absorption themselves. Typical studies among these approaches are by Binder et al. (1986), Calabrese and Stanek (1995), Calabrese et al. (1989), Calabrese et al. (1991), Calabrese et al. (1997), Clausing et al. (1987), Davis et al. (1990), Stanek and Calabrese (1995a,b), and Van Wijnen et al. (1990).

Table 7.17 provides summary information derived from the above reports. Note that the range of intake amounts of soil not only varies across investigators' reports but also within studies for different tracers used to monitor intake-excretion balances. Mean daily ingestion values across studies for soil alone ranged from 66 to 271 mg/day, with the upper percentiles well over 1,000 mg/day. Stanek and Calabrese (1995a,b) saw variation with tracer of daily ingestion rates by children in the studies showing a range of 85–218 mg/day based on four different geochemical tracers. Some of the studies, e.g., those of Calabrese et al. (1989), Calabrese et al. (1997), and Davis et al. (1990), reported intake data for dust and soil combined.

The studies cited above typically reported intakes over limited single points in time, leaving open the question of stability of the results. Stanek and Calabrese (1995a) extended the utility of these studies by carrying out estimated distributions of individual mean daily soil intake rates projected over 365 days. The median value for this estimating exercise was 75 mg/day, while the 90th and 95th percentiles of daily ingestion were 1,190 and 1,751 mg/day, respectively. The range was 1–2,268 mg/day.

Few studies of daily soil ingestion actually precisely differentiated soil from either interior or exterior dusts. Children ingest dusts in sufficient

TABLE 7.17 Soil Lead Intake Rates for Children Reported in Indicated Tracer Studies[a–c]

Mean					Upper Percentile (mg/day)				
Al	Si	AIR[d]	Ti	Y	Al	Si	Ti	Y	References
181	184				584	578			Binder et al. (1986)
230		129							Clausing et al. (1987)
39	82		246						Davis et al. (1990)
65[e]	160[e]		268						Davis et al. (1990)
153	154		218	85	223	276	1,432	106	Calabrese et al. (1989)
154[e]	483[e]		170[e]	65[e]	478[e]	653[e]	1,059[e]	159[e]	Calabrese et al. (1989)
122	139		271	165	254	224	279	114	Stanek and Calabrese (1995a)
133[f]					217[f]				Stanek and Calabrese (1995a)
69–120									Van Wijnen et al. (1990)
66[f]					280[f]				Calabrese et al. (1997)
196[e]					994[e]				Calabrese et al. (1997)

[a]*Adapted from U.S. EPA (2008) and data contained in the indicated studies.*
[b]*"Upper percentile" as indicated in the included studies.*
[c]*Means and upper percentiles = soil or soil and dust combined.*
[d]*Acid-insoluble residue.*
[e]*Dust+soil combined.*
[f]*Best tracer method.*

TABLE 7.18 Estimated Daily Intakes of Dust (mg/day) by Young Children[a–c]

Methods	Median (50th Percentile)	90th Percentile	95th Percentile	Mean	Maximum
Median of best 4	−6	209	353	17	684
Best tracer	27	559	614	127	1,499
2nd best tracer	8	356	410	83	1,685

[a]*Data from Calabrese et al. (1997).*
[b]*Using "Best Trace Elements" method for 64 children in Anaconda, MT.*
[c]*Median of best 4, best tracer, and 2nd best tracer.*

quantities, especially interior dusts, that this pathway is considered a major one for child Pb exposure (Lanphear et al., 1998). Calabrese et al. (1997) estimated daily dust ingestion rates in young children using a combination of tracers they noted as "Best Tracer Method." Results are presented in Table 7.18. The median daily dust ingestion rate using "best tracer" was 27 mg/day, with the corresponding mean, maximum, 90th and 95th percentile values being 127, 1,499, 559, and 644 mg/day, respectively. Overall, it appears that amounts of dust ingested daily by children rival quantities of soil ingested.

REFERENCES

Adams, M.A., 1991. FDA total diet study: dietary intakes of lead and other chemicals. Chem. Speciation Bioavailability 3, 37–41.

Arcus-Arth, A., Blaisdell, R.J., 2007. Statistical distributions of daily breathing rates for narrow age groups of infants and children. Risk Anal. 27, 97–110.

Binder, S., Sokal, D., Maughan, D., 1986. Estimating soil ingestion: the use of tracer elements in estimating the amount of soil ingested by young children. Arch. Environ. Health 41, 341–345.

Bolger, P.M., Carrington, C.D., Capar, S.G., Adams, M.A., 1991. Reductions in dietary lead exposure in the United States. Chem. Speciation Bioavailability 3, 31–36.

Calabrese, E.J., Stanek III, E.J., 1995. Resolving intertracer inconsistencies in soil ingestion estimation. Environ. Health Perspect. 103, 454–457.

Calabrese, E.J., Barnes, R., Stanek III, E.J., Pastides, H., Gilbert, C.E., Veneman, P., et al., 1989. How much soil do children ingest: an epidemiological study. Regul. Toxicol. Pharmacol. 10, 123–137.

Calabrese, E.J., Stanek, E.J., Gilbert, C.E., 1991. Evidence of soil–pica behavior and quantification of soil ingested. Hum. Exp. Toxicol. 10, 245–249.

Calabrese, E.J., Stanek III, E.J., Pekow, P., Barnes, R., 1997. Soil ingestion estimates for children residing on a Superfund site. Ecotoxicol. Environ. Saf. 36, 258–268.

Clausing, P., Brunekreef, B., van Wijnin, J.H., 1987. A method for estimating soil ingestion by children. Int. Arch. Occup. Environ. Health 59, 73–82.

Davis, S., Waller, P., Buschbom, R., Ballou, J., White, P., 1990. Quantitative estimates of soil ingestion in normal children between the ages of 2 and 7 years: population-based estimates using aluminum, silicon, and titanium as soil tracer elements. Arch. Environ. Health 45, 112–122.

Egan, S.K., Tao, S.S.-H., Pennington, J.A.T., Bolger, P.M., 2002. U.S. Food and Drug Administration's Total Diet Study: intake of nutritional and toxic elements, 1991–1996. Food Addit. Contam. 19, 103–125.

Food and Agriculture Organization: World Health Organization, 1993. Evaluation of certain food additives and contaminants. Forty-First Report of the Joint FAO Expert Committee on Food Additives. Technical Report Series 837. Geneva, Switzerland.

Galal-Gorchev, H., 1991. Global overview of dietary lead exposure. Chem. Speciation Bioavailability 3, 5–11.

Gulson, B.L., Mizon, K.J., Korsch, M.J., Palmer, J.M., Donnelly, J.B., 2003. Mobilization of lead from human bone tissue during pregnancy and lactation—a summary of long-term research. Sci. Total Environ. 303, 79–104.

Lanphear, B.P., Matte, T.D., Rogers, J., Clickner, R.P., Dietz, B., Bornschein, R.L., et al., 1998. The contribution of lead-contaminated house dust and residential soil to children's blood lead levels: a pooled analysis of 12 epidemiological studies. Environ. Res. 79, 51–68.

Layton, D.W., 1993. Metabolically consistent breathing rates for use in dose assessments. Health Phys. 64, 23–36.

Pennington, J.A.T., 1983. Revision of the total diet study food list and diets. J. Am. Diet. Assoc. 82, 166–173.

Stanek III, E.J., Calabrese, E.J., 1995a. Daily estimates of soil ingestion in children. Environ. Health Perspect. 103, 276–285.

Stanek III, E.J., Calabrese, E.J., 1995b. Soil ingestion estimates for use in site evaluations based on the best tracer method. Hum. Ecol. Risk Assess. 1, 133–156.

U.S. Department of Agriculture: Agricultural Research Service, 1990. Continuing Survey of Food Intakes by Individuals 1987–1988. Food Surveys Research Group, Beltsville, MD.

U.S. Department of Agriculture: Agricultural Research Service, 1998. Continuing Survey of Food Intakes by Individuals 1994–96, 1998. Food Surveys Research Group, Beltsville, MD.

U.S. Environmental Protection Agency, 1986. Air Quality Criteria for Lead. Report No. EPA/600/8-83/028bF. Environmental Criteria and Assessment Office, Washington, DC (Chapter 10).

U.S. Environmental Protection Agency, 1994. Guidance Manual for the Integrated Exposure-Uptake Biokinetic Model for Lead in Children. Report No. EPA/540-R93/081. Office of Research and Development, Washington, DC.

U.S. Environmental Protection Agency, 1995. National Air Quality and Emissions Trends Report, 1994. Report No. EPA 454/R-95-014. Office of Air and Radiation, Washington, DC.

U.S. Environmental Protection Agency, 2006. Air Quality Criteria for Lead. Vol. 1. Report No. EPA/600/R-05/144aF. National Center for Environmental Assessment, Washington, DC.

U.S. Environmental Protection Agency, 2008. Child-Specific Exposure Factors Handbook. Report No. EPA/600/R-06/096F. National Center for Environmental Assessment, Washington, DC.

U.S. Environmental Protection Agency, 2009. IEUBK win32 Lead Model Version 1.1. Build 11. <http://www.epa.gov/superfund/health/contaminants/lead/index.htm> (accessed 09.03.11.).

U.S. Food and Drug Administration, 2004. U.S. FDA Total Diet Study—analytical results. <http://www.fda.gov/Food/FoodSafety/FoodContaminantsAdulteration/TotalDietStudy/ucm184293.htm> (accessed 03.11.10.).

Van Wijnen, J.H., Clausing, P., Brunekreef, B., 1990. Estimated soil ingestion by children. Environ. Res. 51, 147–162.

Chapter 8

Lead Exposure in Human Populations: Lead Toxicokinetics and Biomarkers of Lead Exposure

8.1 INTRODUCTION

Toxicokinetics describes the biokinetics of toxic substances. It includes the kinetic processes for toxic substances which govern the movement into, within, and from the bodies of human populations. The overall lead toxicokinetic process includes: (1) the uptake, i.e., absorption rate, of lead into the bloodstream from various body compartments such as the lung or GI tract; (2) movement within the bloodstream followed by transport internally to target tissues and their cellular components; (3) retention within one or more tissues; and finally (4) excretion from the body by various systemic pathways. Older literature made incorrect reference to lead toxicokinetics as lead "metabolism," but the latter term is more correctly employed with toxic substances undergoing actual chemical transformation within such processes as addition or removal of chemical groups and oxidative or reductive changes.

Lead toxicokinetics serves two critical roles in the delineation of lead toxicity. First, it provides the kinetic underpinnings for expressions of lead intoxication in humans and other species. The rate of Pb entry into, and deposition within, tissues and cellular organelles is a prerequisite for toxic expressions with differing Pb exposures.

Second, lead's kinetic behavior *in vivo* provides the means by which one can identify and exploit biomarkers of toxic lead exposures as well as determine the dose portion of critical dose–toxic response relationships for lead poisoning. Measurement of lead in whole blood and its relatively reliable use in determining both systemic lead exposure and the extent of toxic injury (dose–response relationships) is mainly feasible because we understand how Pb's toxicokinetic behavior in blood relates to the temporal and toxicological

Trace Metals and other Contaminants in the Environment, Volume 10
ISSN: 1875-1121 DOI: 10.1016/B978-0-444-51554-4.00008-0

characterization of lead toxicity. Bone Pb measurements increasingly provide a useful Pb-exposure biomarker for assessment of body lead burdens and lead accumulation over the long term. Finally, the toxicokinetic aspects of lead exposure measurements provide the wherewithal for developing and using various mechanistic/biokinetic models that simulate lead exposures.

8.2 ABSORPTION OF LEAD IN HUMANS AND ANIMALS

The previous chapter discussed the quantitative aspects of Pb intakes in human populations. Some fraction of this inhaled or ingested environmental lead in humans, in various chemical and physical forms, is absorbed or taken up into the bloodstream, followed by *in vivo* distribution, redistribution, retention, and excretion. Overall, fractional uptake of an ingested or inhaled quantity of Pb per unit time is termed the absorption rate. The propensity for some chemical species of Pb to be absorbed by the human body is termed bioavailability, and the percent bioavailability describes the extent of such uptake.

Bioavailability of lead in some environmental matrix is one of the parameters which define the extent to which lead in these media has the potential to pose toxicological harm by producing internal lead exposures (Mushak, 1991, 1998). The second parameter, often overlooked in assessing overall Pb uptake, is total Pb intake indexed on a daily or other time basis. Lead intake, as noted in Chapter 7, is determined by the lead concentration in an environmental medium and the mass of lead-containing medium ingested (Mushak, 1991, 1998). Across a group of subjects where daily intake masses are relatively stable, the Pb concentration becomes the chief variable, along with bioavailability, in determining total Pb uptake rates. These interrelationships can be illustrated with a simple calculation. Ingestion of 100 mg daily of lead in soil at a level of 10,000 ppm (1% by weight) and with a low bioavailability of 5% produces the same amount of absorbed Pb, 50 μg, as results from ingesting 100 mg of soil having Pb at 1,000 ppm and a bioavailability of 50%. For equivalent concentrations of Pb in two soils ingested to the same extent weight-wise, the exposure hazard posed by the sample having 10-fold higher bioavailability is also 10-fold higher.

8.2.1 GI Absorption of Lead in Humans and Animals

GI bioavailability of lead differs with exposed human host factors such as age and stage of physiological development. Infants and toddlers are widely assumed to have higher Pb uptake rates for ingested Pb than older children or adults, but the question remains about when in chronological development uptake rates for lead decline or begin to decline to those of older children or adults.

While infants and toddlers at some quite early age likely absorb more Pb due to higher bioavailability than older children or adults, what is the maximum bioavailability in this age band determinable in the peer-reviewed Pb literature? Maximum bioavailability, or uptake, is not 100%. This appears to be so even in the case of fully soluble forms of lead ingested by humans or experimental animals. The maximum bioavailability for soluble Pb species is often taken to be 50% of the total soluble amounts of Pb in solution under normal, i.e., nonfasting, conditions. Pb uptake with fasting can range above 50%.

Table 8.1 presents illustrative reports of Pb oral absorption figures for young children. Older literature indicates uptake rates on the order of 40–50% or so for infants and perhaps somewhat older children (Alexander et al., 1974; Ziegler et al., 1978). The Ziegler et al. estimate is more clearly confined to infants on the basis of study design, within the age range of 2 weeks to 2 years old, while the earlier Alexander et al. report had fewer numbers and a wider range of child ages, up to 8 years. The Barltrop and Strehlow (1978) findings provided inconclusive results. This can be

TABLE 8.1 Dietary Lead Absorption Rates in Young Children

Study Group	Percent Absorption	Comments	References
Metabolic balance study, 8 children 3 months–8 years old, Pb intake: 5–17 μg/kg/day	53	Variable diet versus age, older children in group	Alexander et al. (1974)
Infants, 2 weeks–2 years old, *N* = 12, 89 balance studies total	42	Infant diets	Ziegler et al. (1978)
Hospitalized children, *N* = 29, 3 weeks–14 years old	Variable, 15/29 in negative balance	Hospital diets, bone Pb releases	Barltrop and Strehlow (1978)
Children, 6–11 years old, variable diets	10–15%, 12–66% of total PbB is endogenous	Stable Pb isotope ratios; variable diets	Gulson et al. (1997)
Two groups of children: (1) *N* = 22, birth to 1–2 years; (2) *N* = 20, 2–4 years old	Maximum 12%, no bone Pb assessment, 1–5% bone Pb releases	Stable Pb isotopic ratios; infant diets	Manton et al. (2000)

attributed to confounding of uptake estimates by children's medical histories, including hospitalization for medical reasons with abrupt changes in lead exposures. Children were up to 14 years old, well beyond the upper age bound of interest for the very young exposure subjects.

More recent studies of Pb GI uptake in young children record lower uptake rates. Gulson et al. (1997) used stable Pb isotopic ratio methods that contrast current intake Pb with Pb released from children's bones to determine that children between 6 and 11 years old absorbed 10–15% of Pb from oral intakes. Of total PbB, furthermore, 12–66% is of endogenous origin. Children in this study were older than infants or toddlers, with the youngest being 6 years of age.

Manton et al. (2000) studied two groups of young children, one aged 0 months to 1–2 years (N = 22) and a second group, N = 20, 2–4 years of age. These workers estimated a maximum uptake of 12%, using the stable Pb isotopic ratio method and infant diets. If endogenous Pb releases from mineral tissue were taken into account, a low range of 1–5% would result. These estimates run counter to earlier data using metabolic balance studies. Specifically, these data of Manton et al. for the infants run counter to Ziegler et al. findings and require confirmation.

Overall, it does appear that infants and toddlers consuming diets common to this age band absorb more Pb by ingestion than do older children, and by 5–6 years of age, such children have GI Pb uptake rates comparable to that for adults, about 10% or somewhat less.

Adult GI uptake data for Pb have been gathered using relatively healthy adult volunteers under conditions of either normal meal consumption or fasting. Various isotopic methodologies have been used, either radioisotopic Pb or stable isotopic measurements. Illustrative study results are depicted in Table 8.2. Overall, Pb uptake in adults with regular access to diets and fluids ranges from ~3% to 10%. Under fasting conditions, however, various investigators found that uptakes increase to over 60%, values not materially different from uptakes noted for infants. The similarity of fasting values for Pb uptake with uptakes in the very young presents the possibility that it is diet composition and timing of ingestion in variably aged human populations that accounts for differences in uptake, and not intrinsic physiological, biochemical, or anatomical differences.

Drinking water Pb, under conditions of variable fasting after breakfast and using ^{203}Pb labeling, is highly absorbed with an average uptake rate of 70% (Blake et al., 1983). Shorter time between morning meal and ingestion of water Pb reduced uptake to 15%, while increasing the "hardness" by addition of inorganic hardness elements reduced uptake to 1–2% (Blake et al., 1983).

Table 8.3 presents experimental studies of Pb uptake under various Pb dosing regimens. Collectively, uptake data in this table support the findings in humans of higher uptakes of Pb in very young animals versus juvenile or

TABLE 8.2 Dietary Lead Absorption in Human Adults

Study Group	Percent Absorption	Comments	References
Fed subjects			
Adult volunteers, meals enriched with ^{204}Pb	8	Mass spectrometry using stable (^{204}Pb) isotope	Rabinowitz et al. (1976)
Adult volunteers, $N = 5$, meals enriched with ^{204}Pb	10	Mass spectrometry using stable (^{204}Pb) isotope	Rabinowitz et al. (1980)
Adult volunteers, $N = 6$	6	^{203}Pb label, includes fecal endogenous excretion calculation	Chamberlain et al. (1978)
Adult volunteers, $N = 23$	4	^{203}Pb label plus diet	James et al. (1985)
Adult volunteers, $N = 6$, male/female	25	Stable isotopic ratio 206/207 changes	Maddaloni et al. (1998)
Fasted subjects			
Adult volunteers, $N = 5$	35	Intake enriched with stable ^{204}Pb isotope	Rabinowitz et al. (1980)
Adult volunteers, $N = 8$	63	^{203}Pb isotopic tracer ingested	Heard and Chamberlain (1982)
Adult volunteers, $N = 6$	45	^{203}Pb isotopic tracer ingested	Chamberlain et al. (1978)
Adult volunteers, $N = 11$	21	Ingested ^{203}Pb isotopic tracer 2 hours postbreakfast	Blake (1976)
Adult volunteers, $N = 6$	26	Stable Pb isotopic ratios 206/207 measured	Maddaloni et al. (1998)
Adult male volunteers. ^{203}Pb in drinking water, various times postbreakfast	70 (67–76)	^{203}Pb radiolabel + water; shorter fasting showed 15% uptake; water "hardness" reduced uptake to 1–2%	Blake (1976), Blake et al. (1983)

adult animals. This is the case for young rodents, rabbits, or monkeys. For example, rat pups absorb up to 50% of Pb label (^{203}Pb) versus 1% in adults (Kostial et al., 1978). Infant monkeys absorb up to 85% Pb orally versus only 4% in adults (Munro et al., 1975).

Uptakes of oral Pb presented in different chemical forms to humans or animals are in Table 8.4. Human adult volunteers, ingesting different

TABLE 8.3 Experimental Models of Lead Absorption from the GI Tract

Test System	Dosing Conditions	Absorption Rate (%)	Comments	References
Rat pups or adults; pups preweaning	Laboratory chow plus radiolabel	1, adults; up to 50, pups	Used ^{203}Pb label	Kostial et al. (1978), Forbes and Reina (1972)
Infant monkeys	Diet with Pb	65–85 for infants versus 4 for adults		Munro et al. (1975)
Juvenile rhesus monkeys	Diet plus radiolabel	Young monkeys absorbed 50% more than adults	Used ^{210}Pb label	Pounds et al. (1978)
Rabbits, 5 weeks old	Fed Pb in five roadside soils	Up to 45% absolute uptake		Kierski (1992)
Adult rats, 4 months old	Radiolabel + small amounts of food	2	Used ^{203}Pb radiolabeled tracer	Quarterman and Morrison (1978)

chemical forms of tracer Pb, show little dependency of uptake rate on starting chemical form. Rabinowitz et al. (1980), for example, showed there's little difference in uptake for the sulfide, chloride, or amino acid complex forms of the element in adult volunteers. Barltrop and Meek (1975), however, found that rats fed various chemical forms of Pb absorbed more lead as the carbonate or thallate than as the sulfide, chromate, naphthenate or octoate relative to lead acetate as a reference ingestion form. Elevation of PbB was the exposure indicator. The latter chemical species showed uptake of 44–67% relative to the acetate. Ku et al. (1978) reported that rat pups fed Pb as the acetate or phospholipid complex showed no uptake differences from similarly dosed adults, using tissue Pb changes and other criteria.

The Pb-containing environmental matrix consumed by humans or experimental animals can affect the oral absorption rate. Most of the available information on this particular variable was gathered from experimental animal testings and is illustrated in Table 8.5. Dosing animals with Pb in a dietary matrix resembling human diet food groups showed higher uptakes versus conventional laboratory diets, up to 20% with the former versus around 1% with the latter (Kostial et al., 1978).

Lead in aged lead paint has been shown to be extensively absorbed when particles of this medium occur in soils from weathering. Casteel et al. (2006) showed that Pb paint powder in the form of an SRM (NIST) of certified Pb content had 86% relative bioavailability from that paint material added to

TABLE 8.4 Oral Lead Absorption Rate of Different Chemical Forms in Humans and Animals

Study Group	Absorption Rate	Comments	References
A. Humans			
Adult volunteers, $N = 5$, fasted or fed conditions, ingestion of sulfide, chloride, or cysteine complex	Fasted rate: 35% absorption of all three forms	Fed subjects also show no differences in uptake with species, ^{204}Pb stable tracer	Rabinowitz et al. (1980)
Adult volunteers, $N = 5$, fasted or fed conditions, sulfide or chloride forms	6–7% for both forms under fed conditions	Fasting showed different uptakes; ^{203}Pb radiolabel used as tracer	Chamberlain et al. (1978)
Adult volunteers, $N = 8$, lead given with or without radiolabel	Same uptake of Pb when meat plus label or meat incorporating label	^{203}Pb radiolabel used as tracer	Heard and Chamberlain (1982)
Adult volunteers, $N = 3$, lead given as chloride or alginate	14% uptake for either form	^{203}Pb radioisotopic tracer	Harrison et al. (1969)
B. Experimental animals			
Calves fed Pb as oxide, phosphate, basic carbonate, wet and dry paint films, finely ground galena ore	Comparable toxicity for all forms except galena ore, which was less toxic	Toxicity criteria: PbB, kidney Pb, survival rates	Allcroft (1950)
Rats fed lead as various chemical forms and compared to the acetate	Sulfide, chromate, naphthenate, and octoate 44–67% of acetate; higher uptake for carbonate, thallate	Uptake criteria: Pb content of blood and other tissues	Barltrop and Meek (1975)
Adult rats or pups fed Pb as acetate or phospholipid complex at 300 ppm Pb	No differences in uptake rates in either age group	Uptake criteria: weight change, tissue Pb, and ALA-U levels	Ku et al. (1978)

soils when compared to soluble lead acetate (43% absolute bioavailability). Dacre and Ter Haar (1977) reported that lead from Pb paint in soils had similar uptake to lead acetate when fed to rats. Paint Pb, studied as particles of varying sizes, showed an inverse relationship of uptake rate to particle size for the chromate or octoate forms in paint films (Barltrop and Meek, 1979).

TABLE 8.5 Experimental Studies of Oral Lead Absorption from Different Environmental Media

Study Group	Absorption Rate	Comments	References
A. Different diet types			
Rats fed laboratory chow versus feeding with "people" foods: fruits, baby food, liver, bread, milk	<1% laboratory chow; up to 20% with "people" foods	Lower uptakes in adult rodents more due to diet, not test species	Kostial and Kello (1979)
B. Lead paint forms			
Young swine, capsules of lead paint plus soil	86% = relative bioavailability versus soluble acetate; 43% = absolute uptake using PbB	Criteria: Pb in blood, liver, kidney, femur; powdered SRM/NIST material certified paint	Casteel et al. (2006)
Rats fed lead paint soils versus acetate	Uptake of Pb similar across groups	Uptake indexed as weight change, tissue Pb, ALAD activity	Dacre and Ter Haar (1977)
Rats fed lead paint as variably sized particles of Pb chromate or Pb octoate in paint, 36 animals: 6 each in 6 groups	Uptakes of Pb in paint particles higher for smaller size particles, <50 μm	Uptake indexed by PbB and kidney Pb (total)	Barltrop and Meek (1979)
C. Street dusts			
In vitro solubility testings of street dusts in London, England, using gastric acid pH simulations	60% solubility of dust Pb	pH ~ 1.0	Duggan and Williams (1977)
In vitro solubility testing of street dusts, Lancaster, England	77% solubility	pH ~ 1.0	Harrison (1979)
In vitro solubility testing of street dusts from England and New Zealand	90% solubility	pH range of 0–5	Day et al. (1979)
D. Lead in extractive industry-impacted media			

(Continued)

TABLE 8.5 Experimental Studies of Oral Lead Absorption from Different Environmental Media—(cont.)

Study Group	Absorption Rate	Comments	References
A series of studies using juvenile swine administered soils, slags, etc. from 17 U.S. mining, milling, smelting sites	Relative bioavailability (versus soluble acetate) ranges from 19% to 100% using PbB	Uptake indexed by Pb in blood, liver, kidney, femur	Casteel et al. (2006)
Review of animal model choices for evaluation of Pb uptakes from extractive industry-affected media	Juvenile swine are better choices than are rodents	Uptake discussed on physiological, anatomical, and behavioral grounds	Weis and Lavelle (1991)
Rats exposed to Pb in mining wastes	Relative bioavailability = 20% using PbB values		Freeman et al. (1992)

Dusts containing lead have been shown to be a significant pathway for childhood Pb exposures, notably dusts with which children readily and frequently come in contact. Dusts are composed of much smaller particles in more mobilizable form than soils and are also more readily transportable by various means within children's Pb environments. In the case of street dusts, *in vitro* testings show that simulated gastric acidity conditions in children as a solubilization surrogate for ingestion of leaded dusts by children readily mobilize street dusts. Across various studies, up to 90% Pb in street dusts can be dissolved (Table 8.5).

Lead in dusts and soils generated through various contamination processes in areas with mining, milling, and smelting emissions have been studied with respect to Pb bioavailability. The most comprehensive set of studies of this type were those supported by the U.S. EPA Region VIII and carried out by various collaborators, notably Casteel et al. In the summary report of multiple-site results, 17 U.S. extractive-industry waste sites with various types of Pb-laced geochemical media were investigated using juvenile swine fed Pb in different soil media. As noted in Table 8.5, Casteel et al. (2006) observed relative bioavailabilities ranging from 19% to 100% using PbB measurements as the exposure indicator. This corresponds to a range of 10–50% absolute bioavailability, the latter being equivalent to the absolute uptake for fully soluble Pb from the GI tract.

Studies of Pb bioavailability in extractive industry-impacted media have employed various experimental animal test species. Of the various species examined, the juvenile swine is described as the most appropriate one based on physiological, anatomical, and behavioral grounds (Mushak, 1991, 1998; Weis and LaVelle, 1991). The rodent animal model differs sufficiently from characteristics relevant to the young child orally exploring his/her environment that rodent uptake data, on balance, significantly understate uptake rates. The juvenile swine model involves intermittent feeding periods, much like infants and toddlers, but a feeding behavior distinct from rodents and lagomorphs (rabbits). Freeman et al. (1992) reported relative bioavailability of Pb in mining—waste media of 20% in rats; Casteel et al. (2006) typically reported higher values with the same Pb media in juvenile swine.

One parameter of interest for GI uptake of Pb in humans and experimental test animals is the relative uptake rate of Pb as a function of Pb levels in intake media and total daily Pb oral intakes. Such data are limited, but what is available would indicate that over a range of daily Pb intakes germane to intake media encountered by children and adults the uptake rates are stable. Flanagan et al. (1982), in a study with adult volunteers studied for Pb uptake at different Pb intakes, showed stable uptake up to total daily Pb intakes of 400 μg Pb. Experimental animals fed Pb (intubation) over a range greatly higher proportionately (1–10 mg/kg Pb) than the Flanagan et al. human intakes showed declining absorption rates of Pb with increasing oral doses.

8.2.2 Pulmonary Pb Absorption from the Respiratory Tract of Humans

Differential uptake of Pb from the lungs of very young children versus older children or adults is less clear. One complication, noted in Chapter 7, is that ventilation rates with associated deposition rates of Pb-containing particulates in air also differ considerably with age and respiratory tract anatomical and physiological characteristics. A second complication lies in where in the respiratory tracts of exposed individuals the Pb-containing particles are deposited. Deposition in the pulmonary tract leads to Pb uptake from the lung. Depositions high in the respiratory tract can be followed by mobilization of the particles via retrociliary action and by swallowing. This process results in some fractional uptake of inhaled lead via the gut. Localization in different parts of the tract means that not all originally inhaled Pb particles undergo the same Pb uptake rates. After deposition, deep alveolar uptake is apparently quantitative, while swallowed particles would be much less so, 10–50%.

Quantitative studies of human respiratory uptake, for obvious reasons, have been confined to chamber and other studies of relatively healthy adult volunteers. Assessments of children's respiratory Pb uptakes have been confined to modeling approaches using parameters for anatomical, physiological, and behavioral characteristics as the basis for distinctions. Overall, all Pb deposited

TABLE 8.6 Respiratory Lead Deposition/Absorption Rates[a] in Humans

Study Group	Study Details	Deposition/Absorption Rate	References
Chamber studies, adult subjects	$^{203}Pb_2O_3$ aerosol chamber with mouthpiece	30–80%; particles of 0.09, 0.04, 0.02 μm, respectively	Chamberlain et al. (1978)
Adult subjects breathing ambient air	Ambient air in U.K. urban areas: 2–20 μg/m^3, mainly 0.1 μm	60% fresh exhaust, 50% other air Pb conditions	Chamberlain et al. (1978)
Children particle dosimetry models	(1) Particle tracheobronchial or pulmonary deposition, children versus adults; (2) Total particle deposition	Pulmonary deposition lower in children versus adults, total deposition greatest in infants; decreases with age	Phalen and Oldham (2001), Asgharian et al. (2004)

[a]*Deposited Pb is totally absorbed; deposition rate = overall absorption rate.*

in the pulmonary tract of humans, both adults and children, is absorbed. Deposition rates, consequently, are taken to be equivalent to uptake rates.

Studies of respiratory Pb uptake in human adults have employed various methodologies, with chamber aerosol systems and their variations most widely used. For example, Chamberlain et al. (1978) studied air Pb depositions and uptakes under experimental chamber and ambient air conditions. Adults inhaling Pb aerosols (chamber method, using mouthpiece) containing ^{203}Pb-labeled Pb aerosols showed a range of 30–80% deposition depending inversely on particle size of Pb, sizes ranging from 0.09 (30%) to 0.02 μm (80%). Chamberlain et al. (1978) reported that ambient air Pb in urban U.K. locales underwent 60% deposition in the lung from fresh exhaust and 50% for air Pb particles suspended in the atmosphere longer. These deposition rates were for particles mainly 0.1 μm in diameter, at an air Pb level of 2–20 μg/m^3 (Table 8.6).

Pulmonary dosimetry modeling for children's respiratory Pb deposition and uptake shows that (1) pulmonary deposition is lower in children than adults, while (2) total deposition is greatest in infants and declines with age (Asgharian et al., 2004; Phalen and Oldham, 2001).

8.2.3 Dermal Absorption of Inorganic Lead in Humans

The inorganic salt forms of divalent lead, the species of direct interest for toxic exposures of nonoccupational human populations, appear to be very poorly absorbed through the skin in humans. Absorption in this context refers to transdermal passage and entry into the bloodstream. Intradermal binding of inorganic lead followed by exfoliation or extended retention would not be uptake in the systemic toxicological sense. Moore et al. (1980) examined ^{203}Pb-labeled lead acetate uptake in eight adult volunteers when contained in cosmetic preparations and applied in dry or wet forms. Absorption rates ranged from undetected to 0.3% based on label applied. These authors calculated that conventional use of such Pb-containing cosmetics would be associated with 0.06% uptake.

Older literature indicated that using long-chained organic salts of divalent lead for various purposes, such as lubricants, would result in higher percutaneous uptakes than for simple divalent Pb salts. For example, Rastogi and Clausen (1976) reported that higher uptakes of Pb occurred when applied to rat skin as the naphthenate salt rather than in the acetate form.

8.3 *IN VIVO* DISTRIBUTION OF LEAD IN HUMANS AND EXPERIMENTAL ANIMALS

Absorbed lead is distributed first to blood and then to diverse body compartments in humans and experimental animals. Entry into and movement out of subcompartments of blood are closely linked with the toxicokinetic correlates of toxic lead exposures and associated lead poisoning.

8.3.1 Lead Movement into and out of Whole Blood in Humans and Animals

Lead that enters the bloodstream of humans and experimental animals after absorption is first distributed to plasma/serum, followed by movement to erythrocyte and soft tissue binding sites with subsequent re-equilibration to longer-term binding sites in soft and mineralizing tissues. Lead entry into and removal from plasma/serum is rapid. Chamberlain et al. (1978) reported that injected radiolabeled Pb in adult volunteers entered and was removed from plasma/serum in less than 1 hour. Table 8.7 summarizes data for Pb distribution and clearance in human blood.

Almost all of the lead in whole blood under conditions of relatively stable exposure, i.e., near-steady-state conditions, is bound within erythrocytes of humans and test animals. The biomolecular site of Pb binding in the erythrocyte was long held to be a form of hemoglobin, Hb-A_2 (Bruenger et al., 1973; Moore, 1988). More recent evidence, however, identifies the erythrocyte enzyme, delta-aminolevulinic acid dehydratase (δ-ALAD), as the actual site (Bergdahl et al., 1997a, 1998). Other sites of binding in blood appear to play a role in both transport and a protective function in high lead exposures as occur among lead workers. Lolin and O'Gorman (1988) and

TABLE 8.7 Lead Distribution and Clearance in Blood of Human Populations

Study Group	Study Design	Results	References
Adult human volunteers or lead workers	Pb partitioning within whole blood	>99% of Pb in erythrocytes	DeSilva (1981), Everson and Patterson (1980), Manton and Cook (1979)
Adult human volunteers	Pb partitioning within whole blood	Older literature indicates main binding to hemoglobin (Hb-Az); recent evidence identifies ALAD as the principal binding site	Bruenger et al. (1973), Moore (1988), Bergdahl et al. (1997a, 1998)
Lead workers	Pb partitioning at high exposures within whole blood	Pb-binding protein probably Pb–metallothionein, inversely linked to level of Pb toxicity	Lolin and O'Gorman (1988), Raghavan et al. (1980)
Adult human volunteers	Injected ^{203}Pb tracer exposures	Pb first entering blood is removed from plasma with a half-life <1 hour	Chamberlain et al. (1978)
Adult Pb workers	Erythrocyte–plasma equilibria	Equilibrium at near steady state stable up to 40–50 μg Pb/dl whole blood; increased Pb-plasma at higher total Pb levels	DeSilva (1981), Manton and Malloy (1983), Manton and Cook (1984)
Adult human volunteers	Pb uptake via inhalation, ingestion, or injection	PbB half-life of ~30 days with modest exposures; Pb workers removed from exposure may show longer half-lives	Chamberlain et al. (1978), Griffin et al. (1975), Rabinowitz et al. (1976), O'Flaherty et al. (1982)
Pb-exposed children	Pb declines with, for example, age or changed residence	PbB half-life variable with exposure history, ranging up to months or several years	Succop et al. (1987), Manton et al. (2000), NAS/NRC (1993), Mushak (1989, 1992)

(Continued)

TABLE 8.7 Lead Distribution and Clearance in Blood of Human Populations—(cont.)

Study Group	Study Design	Results	References
Fetal Pb transfer and uptake	Fetal tissue and cord blood Pb analyses	Pb readily transferred transplacentally by end of first trimester through birth; Pb-cord blood = 80% of maternal PbB	Barltrop (1969), Ryu et al. (1978)
New lead workers	PbB increases to plateau at work site	PbB plateau with increased Pb exposure ~60 days	Griffin et al. (1975), Tola et al. (1973)

Raghavan et al. (1980) described a Pb-binding protein of molecular weight <10 kDa, likely a Pb–metallothionein, in the erythrocytes of lead workers. The Pb-binding protein level was inversely associated with the severity of lead's toxic effects.

The very small fraction of whole blood Pb contained in the plasma component is about 1% or so, based on a number of reports (Bergdahl et al., 1997b, 1999; DeSilva, 1981; Everson and Patterson, 1980; Smith et al., 2002). This fractional distribution of 99:1 persists up to a whole blood Pb concentration of 40–50 μg/dl, after which the relative fraction in plasma increases with increased PbB, reflecting increasing saturation of binding sites in the erythrocyte (DeSilva, 1981; Manton and Cook, 1984).

Plasma is the component of whole blood which is the Pb delivery mechanism for absorbed Pb or bone-resorbed Pb to target tissues. Theoretically, it would be a better dose metric than whole blood. However, the extremely small levels of Pb in plasma make it very difficult to use for routine measurement owing to Pb contamination and requirements for high-measurement sensitivity and specificity (Mushak, 1998; NAS/NRC, 1993; Smith et al., 2002). In addition, the risk of erythrocyte hemolysis with fragile cells produces the potential for artifactually elevated levels of Pb in plasma from Pb released from the erythrocyte. This Pb transferred to plasma will, of course, markedly affect estimates of Pb toxicokinetics. For example, losses of ~1% of erythrocyte Pb content to plasma through hemolysis will approximately double the measurable plasma Pb concentration.

Important aspects of the behavior of Pb in blood are how fast the level declines with reduced exposures and how fast it increases with newly increased exposures. This decline is linked to reduced input to blood with continuing removal from blood and excretion. The decline rate is toxicokinetically depicted as the "half-life" of PbB levels in exposed children or adults. The latter is typically evaluated for PbB plateaus seen in new workers producing or handling lead products.

The half-life of PbB decline with reduced Pb exposures in children and even adults is 30 or more days for light or moderate histories of exposure and for the "fast" kinetic component of PbB multicompartment kinetics (Mushak, 1989, 1991, 1998; NAS/NRC, 1993; U.S. CDC, 1991, 2005; U.S. EPA, 1986, 2006). The half-life is the time required for an initial PbB level to decline to half that value. Half-life determinations are mathematically complex since changes in PbB are a complex function of more rapid movements of Pb within and across soft tissues and slow Pb kinetics for various subcompartments of bone Pb.

Children with extended excessive Pb exposures who are then placed in lower exposure settings can show PbB decline on the order of months (Manton et al., 2000; Roberts et al., 2001; Succop et al., 1987). Frequently, serial testings are required to separate various kinetic compartments and their associated half-lives. However, the statistical design of such studies is typically one where kinetic components would not be fully identified, particularly the fast component. Slow kinetic components will reflect the higher relative input of bone Pb releases to PbB through resorption of prior accumulations of bone Pb versus ongoing exogenous Pb intakes and uptakes.

To illustrate the role of variable Pb kinetic compartments, lead workers who encounter abrupt reductions in Pb exposure through retirement, change in workplace Pb exposures, or job actions continue to show the first, rapid kinetic component. Subsequently, workers will show slow decline rates reflecting bone Pb releases. The latter would reflect employment histories and severity of Pb exposures. Nilsson et al. (1991) reported the half-life of the first, fast kinetic component in these workers to be as short as 7 days (Gerhardsson et al., 1995; Nilsson et al., 1991).

Several investigators have evaluated how fast subjects' PbB levels respond to increases in their Pb exposures through studies of adult volunteers and new lead workers. Tola et al. (1973) reported that new Pb workers showed elevated PbB responses up to a plateau value over an average of 60 days. Benson et al. (1976) found that PbB levels in a group of 20 new lead workers began to rise by 1 week of exposure and peaked on average by 3 weeks.

Schlegel and Kufner (1979) studied two adult volunteers who consumed daily oral doses of 5 mg/day. PbB values began to rise significantly in days, achieving a peak level in about 6 weeks with both subjects. Cools et al. (1976) reported that 11 volunteers ingesting Pb as the acetate achieved a PbB of 35 μg/dl in 15 days. Omokhodion and Crockford (1991) showed that adult volunteers ingesting acute doses of Pb chloride reached peak PbB values very rapidly and, similarly, showed declines to baseline in days. A decline in half-life can be estimated here for 6–7 days.

8.3.2 Distribution of Pb to Soft Tissues in Humans and Experimental Animals

Lead transfer to soft and mineral tissues occurs as a complex function of absorbed dose, nutritional status, hormonal activity, and relative stability of environmental exposures and resulting Pb intake/uptake rates. Details are presented later. Lead is transferred within exposed individuals' bodies and, in pregnancy, to the fetus. Lead enters the fetus through transplacental transfer during pregnancy, and the process is under way by the 12th week of gestation (Barltrop, 1969; Mayer-Popkin et al., 1986) or even earlier (Borella et al., 1986; Chaube et al., 1972). Fetal Pb uptake continues through the remainder of the pregnancy with Pb being measured in fetal bone, kidney, and liver, and lower levels in blood, brain, and heart. Many studies have reported that Pb is present in cord blood with cord PbB levels being somewhat less than ($\sim$0.90), and correlated with, maternal PbB values across various exposure groups (Carbone et al., 1998; Graziano et al., 1990). Table 8.8 depicts illustrative study results showing Pb distribution in human soft tissues.

Lead levels in various human soft tissues have mainly been gathered from autopsy samples with measurements dating to the 1960s. Some data represented tissue biopsy collection but that sampling was confined to adults, largely for assessing health risks associated with high toxic Pb exposures. Two factors affecting such measurements need to be kept in mind. First, histories of autopsy sampling may or may not have relevance to typical human exposures or steady-state toxicokinetics. Autopsies subsequent to fatal accidents or death from terminal, chronic diseases raise the question of any effect of circumstances of death on Pb levels *in vivo* prior to death. Second, older data gathered under problematic sampling and laboratory measurement protocols raise the question of sample contamination, measurement sensitivity, and method specificity. Given the limits on autopsy material analysis, variably exposed experimental animals were often used.

Acute Pb exposures produce *in vivo* distributions that would apply only to predistribution equilibrium (presteady-state) conditions and will potentially differ considerably from the toxicokinetic picture for chronic, long-term exposures. Acute, subacute, and subchronic exposures to Pb remain of interest to pediatric specialists concerned with individual child Pb poisoning cases, but the bulk of the public health-oriented interest is in chronic, low level Pb toxicity to human populations, particularly the very young. Consequently, *in vivo* movement of Pb after uptake under the latter circumstances is of principal interest.

The relative amounts of total body Pb sequestered in soft tissues under ambient (nonoccupational) exposure conditions in humans differ with age. Proportionately more Pb is lodged in soft tissues of young children, i.e.,

TABLE 8.8 Pb Distribution in Human Soft Tissues[a,b]

Study Subjects	Study Design	Results	References
Children, adults	Autopsy sample analysis, differing ages	94% of adult body Pb in bone; 73% Pb in children in bones; bone Pb increases with age up to middle age	Barry (1975, 1981); Drasch et al. (1987)
Adults >20 years old	Soft tissue Pb content	(1) Soft tissue Pb stabilizes by early adulthood (2) Most soft tissue Pb is in liver (33%), followed by skeletal muscle (18%), skin (16%), brain 2% (3) Highest Pb levels in liver and kidney cortex	Schroeder and Tipton (1968), Barry (1975, 1981), Gerhardsson et al. (1995)
Swedish adults	Kidney biopsy samples	Kidney cortex Pb = 0.2 μg/g (max = 0.6 μg/g)	Barregard et al. (1999)
Nonoccupational adult males	Pb relative distribution in brain regions	Hippocampus = amygdala > medulla oblongata > half brain > optic tract	Grandjean (1978)
Children with fatal Pb poisoning	Autopsy brain samples	Hippocampus > frontal cortex ≫ occipital white matter, pons	Okazaki et al. (1963)

[a]*Chronic exposures to Pb of mainly nonoccupational subjects.*
[b]*Measurements at different years; values reflect environmental Pb exposures.*

~30% versus ~70% in bone, than in adults, where bone contains ~94% and soft tissues and organs the balance (Barry, 1975, 1981; Drasch et al., 1987). Levels of Pb in all tissues in children and adults have declined markedly over the years, reflecting much reduced Pb intakes from much reduced environmental levels (Drasch and Ott, 1988; Drasch et al., 1987; NAS/NRC, 1993).

Early measurements of tissue Pb, when Pb exposures were much higher than currently observed, indicated that most soft tissues have <0.5 ppm Pb levels, with the possible exception of the kidney and aorta where some Pb binding and deposition might persist. Brain similarly is low with ambient Pb exposures, <0.2 ppm Pb, though this can double with higher exposures. While soft tissue Pb is higher in older children than in infants, little added Pb content occurs in adults versus older children (Barry, 1975, 1981). With severe Pb poisonings, the Pb content of children's brain tissue can reach 1 ppm or somewhat more. For relative distribution across soft tissues, 33% is lodged in liver, 18% in skeletal muscle, 16% in skin, and 2% in brain (Barry, 1975; Gerhardsson et al., 1995; Schroeder and Tipton, 1968). In terms of total Pb content, liver and kidney cortex have the highest levels (Barry, 1975; Gerhardsson et al., 1995), with kidney cortex biopsy levels averaging 0.2 ppm with a maximum of 0.6 ppm (Barregard et al., 1999). Soft tissue Pb in humans is relatively stable with age, beginning in older children and continuing into late adulthood. This is in contrast to bone Pb, where accumulation persists into late adulthood with males and into menopause with females.

The brain, especially the developing brain in children, is the principal target organ of Pb toxicity in humans and animals. Therefore, Pb distribution within brain is of interest. Autopsy brain samples from children with fatal Pb poisoning show, in order of declining Pb content: hippocampus > frontal cortex ≫ occipital white matter, pons (Okazaki et al., 1963). Grandjean (1978) reported for nonoccupationally exposed adult males: hippocampus = amygdala > medulla oblongata > half brain > optic tract.

Subcellular distribution of Pb in human soft tissues appears to mainly involve the mitochondria and nuclei, two organelles known to either be affected toxicologically by lead or be involved in Pb sequestration and toxicokinetics. Intranuclear inclusion bodies, for example, have long been known to form as a transitory protective mechanism for averting or delaying lead toxicity. Lead in relatively large amounts is sequestered in nuclear inclusions and the biochemical and structural characteristics of these bodies have been described (Carroll et al., 1970; Moore et al., 1973). Cramér et al. (1974) showed the formation of intranuclear inclusions as an early response to Pb in kidney proximal tubule cells in new lead workers.

8.3.3 Lead in Mineralizing Tissues of Humans and Experimental Animals

Mineralizing tissue in humans consists of bone and teeth. These biominerals differ in a number of ways with respect to lead deposition and lead toxicokinetics. By and large, bone is the larger repository of lead in humans and is the more complex mineralizing tissue in terms of deposited Pb. Bone Pb can be readily resorbed and serve as a source of endogenous Pb exposure long after initial transport to and deposition in the various bone subcompartments. Table 8.9 presents illustrative summaries of Pb distributions and accumulations in human mineralizing tissues.

Lead in Teeth of Human Populations

Lead deposition in deciduous teeth varies as a complex function of anatomical development, age, and level of Pb exposure postnatally (Gulson and Wilson, 1994; Rabinowitz et al., 1993; Steenhout and Pourtois, 1981), the

TABLE 8.9 Pb Distribution in Human Mineral Tissues[a,b]

Study Subjects	Study Design	Results	References
Fetal tissues	Mineralizing tissue Pb levels	Pb accumulates in fetal skeletal tissue	Barltrop (1969), Horiuchi et al. (1959)
Variably aged humans	Autopsy bone Pb levels at varying ages	~95% of body Pb in bone of adults and 73% in children; range up to 200 mg in elderly men; Pb accumulates in both cortical and trabecular bone	Barry (1975), Schroeder and Tipton (1968), Rabinowitz et al. (1976)
Variably aged humans	Autopsy bone Pb levels for midfemur and pelvic bone	Bone Pb levels off in middle age and then decreases	Drasch et al. (1987)
Variably aged humans	Age-related Pb accumulation in cortical and trabecular bone	Main Pb accumulation in children is in trabecular bone; in adults, in both trabecular and cortical bone	Aufderheide and Wittmers (1992)
Adults and children >10 years old	*In vivo* XRF measurement of bone Pb	Gradual increase in bone Pb after age 20	Kosnett et al. (1994)
Elderly adults	*In vivo* XRF measurements of bone Pb in multiyear study	Subjects showed changes in bone Pb with varying Pb exposures and other factors	Hu et al. (1995, 1996, 1998), Cheng et al. (1998)
Children	Pb in whole teeth or tooth sections/type	Pb accumulates in teeth with age and Pb exposure and is correlated with PbB; secondary dentine Pb is highest	Steenhout and Pourtois (1981), Needleman and Shapiro (1974)

[a]*Chronic Pb exposures of mainly nonoccupational subjects.*
[b]*Measurements at different years; values reflect differing Pb exposures.*

differing dentition anatomy within a tooth (Gulson and Wilson, 1994; Needleman and Shapiro, 1974; Rabinowitz, 1995), and the type of teeth and position within the jaw (Delves et al., 1982; Grandjean et al., 1986; Mackie et al., 1977). Dentine Pb levels exceed the levels in enamel, and circumpulpal dentine has the highest relative levels (Gulson and Wilson, 1994; Needleman and Shapiro, 1974). Enamel captures the Pb deposition *in utero* while dentine records accumulation from birth to shedding (Gulson and Wilson, 1994).

Teeth are generally considered to be a relatively inert repository for Pb compared to bone, although some slow and relatively modest inter-tissue exchange may occur (Gulson et al., 1997; Rabinowitz et al., 1993). However, Gulson et al. (1997) estimated, using stable Pb isotopic ratios for historic versus current Pb stores, that there was no exchange of Pb in enamel while dentine Pb had an exchange rate of about 1% per year. This overall slow exchange mobility for teeth Pb still allows using tooth dentine to track past lead exposures and to help quantify previous Pb accumulation in early childhood. Furthermore, the total Pb content of either exfoliated or permanent teeth in children and adults is quite small compared to the total lead content of the human skeletal system (encompassing both cortical and trabecular bone).

The toxicokinetic relationships between tooth Pb and PbB have been evaluated. Levels of lead in various tooth components as measured and reported were related to available data on PbB by Rabinowitz (1995). Dentine, whole-shed teeth or crowns, but not circumpulpal dentine, were included. Mean tooth Pb values varied from 2.8 to 12.7 ppm and PbB levels from 6.5 to 17 μg/dl. A good fit was found in plots of tooth versus PbB ($R^2 = 0.97$, $p < 0.0001$).

Lead in Bones of Human Populations

The human skeletal system accumulates lead and begins that accumulation *in utero*, continuing into at least the fifth or sixth decade in males and to the time of menopause in females. Some of the general aspects of lead's lifetime behavior in bone are presented in Table 8.9, while the complex toxicokinetic relationships of bone Pb deposition and resorption with other kinetic compartments such as whole blood and plasma are set forth in Table 8.10.

Lead accumulates in fetal tissue and reports of that accumulation date to the 1950s, beginning with fetal autopsy tissue measurements (Barltrop, 1969; Horiuchi et al., 1959). Postnatal autopsy sample Pb measurements with high variability as to Pb-exposure status, health, and nature of death produced the earliest results. Barry (1975, 1981) reported that nonoccupational Pb exposures through much of adulthood can result in total skeletal Pb burdens of up to 200 mg. Occupational exposures, particularly for those in the lead industry, can result in several times this value. These calculations of skeletal Pb burdens were for the period of highest lead exposures. More recent data, such as those of Drasch et al. (1987), show much lower skeletal total Pb loadings, consistent with known declines in environmental Pb emissions and associated exposures beginning in the 1970s. The Drasch et al. (1987)

TABLE 8.10 Illustrative Studies: Biokinetic Behavior of Pb in Bone and Interactions with Other Tissues

Study Subjects	Study Design	Results	References
Published studies on Pb in human bone	PB-PK modeling of human bone physiology versus Pb deposition–resorption	Bone Pb loss occurs via resorption and some diffusion	O'Flaherty (1993, 1995, 1998)
Various studies summarized by this author: cortical and trabecular bone	Bone types anatomically and metabolically; factors affecting metabolism	Cortical bone: 80% of total, dense, compacted, mineralized; trabecular bone: spongy, 20% of total, with greater turnover and metabolic activity. Hormonal control is via 1,25-$(OH)_2$-vitamin D, parathyroid hormone, calcitonin, estrogen, etc.	Hernandez-Avila et al. (2002), Illich and Kerstetter (2000), Hu et al. (1998)
Variably aged subjects and gender	Lead content of bone types and determinants	Male bone Pb increased greatly >40 years of age; female levels declined slightly	Kosnett et al. (1994)
Subjects varying by age	Incremental annual bone Pb changes	Range of 0.31 μg/g mineral/year to 0.46 μg/g mineral/year across studies	Gordon et al. (1993), Kosnett et al. (1994), Hu et al. (1990)
In vivo and cadaver bone Pb measurements	Measurement of homogeneity of bone Pb distribution	Overall, Pb distribution in tibia is uniform	Gordon et al. (1994), Wittmers et al. (1988), Todd et al. (2000)
Summary of various studies	Bone types and turnover rates	Adult tibia (95% cortical), 2% turnover/year; adult trabecular bone (calcaneus patella) 8% turnover/year	Rabinowitz (1991), Hu et al. (1998)
Variably Pb-exposed populations	Bone types and associated Pb half-lives	Trabecular bone Pb ~12–19 years; tibial bone Pb ~13–27 years; newer data show tibial shorter half-lives for younger than older workers: 5 years versus 14 years; also, half-lives much shorter with lower exposures	Brito et al. (2002), Bergdahl et al. (1998), Gerhardsson et al. (1993), Rabinowitz (1991)

(Continued)

TABLE 8.10 Illustrative Studies: Biokinetic Behavior of Pb in Bone and Interactions with Other Tissues—(cont.)

Study Subjects	Study Design	Results	References
Variably Pb-exposed populations	Persistence of unremodeled bone and bone Pb, childhood to adulthood	Approximately 35–40% of older adult skeleton mass has unremodeled bone from earlier years	Kim et al. (1995), McNeill et al. (2000)
Variably Pb-exposed human subjects	Bone Pb fractions of total PbB using various methods	In adults, bone Pb adds 40–90% of total Pb; in children, 40–90%	Manton (1985), Gulson et al. (1995, 1997, 1999), Gwiazda et al. (2005), Manton et al. (2000)
Variably Pb-exposed female subjects	PbB contributions versus bone Pb content	In females 46–74 years, bone Pb increase of 19 μg/g mineral = PbB increase of 1.7, ~1/10 ratio post-versus premenopausal women exposed in childhood: postmenopausal—0.13 μg/dl PbB/μg Pb/g mineral; premenopausal—0.07 μg/dl PbB/μg Pb/g mineral	Korrick et al. (2002), Popovic et al. (2005)
Variably exposed workers	EPA-estimated box plots for workers' exposure in various reports	Mean cortical Pb/current Pb = 1.2 (range 0.4–26, active workers), mean for retirees = 3.2 (range 2.0–5.3)	U.S. EPA (2006, Ch. 4)
Variably exposed subjects	Correlation of trabecular Pb to PbB versus tibial Pb to PbB	Relationship of trabecular (patella) Pb to PbB more robust than tibial Pb to PbB	Hernandez-Avila et al. (1996), Hu et al. (1998)
Variably exposed subjects	Mobilization of bone Pb back to blood	Mobilization can be significant for marked changes to lower exposures, pregnancy, menopause, etc.	Gulson et al. (1995), Nilsson et al. (1991), Markowitz and Weinberger (1990)

findings estimated total skeletal Pb of 41 mg for adult males and 24 mg for females. Furthermore, a decline over a 10-year period of up to 50% was observed by these investigators.

Proportional to total Pb content in other tissues, bone Pb clearly dwarfs any other. Collectively, the data of Schroeder and Tipton (1968), Barry (1975), and Rabinowitz et al. (1976) were the first to indicate that 95% of total body Pb in adults resides in bone, while a lower percentage, 73%, is found for children.

Aufderheide and Wittmers (1992) determined that during childhood, the principal type of bone accumulating Pb is trabecular, while in adults both cortical and trabecular bone accumulate Pb. The precise growth period for onset of bone accumulation of Pb is difficult to identify, but Kosnett et al. (1994) showed that overall accumulation is evident by young adulthood, i.e., after age 20. Behavior of Pb in bone in the late years of life is also of interest. Hu et al. (1995, 1996, 1998) and Cheng et al. (1998) used *in vivo* XRF measurement of bone Pb in aging subjects to establish the relationship of bone Pb to aging and other parameters such as nutrition, and to various outcome measures linked to Pb exposures.

Biokinetic Behavior of Pb in Bone and its Interactions with Other Tissues

The older scientific literature advanced the view that lead lodged in bone represented an inert fraction of body lead and, in fact, such deposition served as a means to attenuate Pb toxicity from what would otherwise be circulating Pb, the latter threatening toxicological targets among tissues and organs. Supporting this view were earlier findings that humans accumulated most of their body Pb in the skeleton and accumulation continued with aging. That is, ongoing Pb exposures were linked to the highest Pb fraction in bone at all ages and bone Pb increased across study population ages because Pb exposures continued across ages, particularly with occupational exposures to Pb.

This view first began to change with the findings from occupational epidemiology that long after lead workers retired, their PbB values declined only slowly from active work levels, remained elevated above concentrations seen in nonlead workers, or above values explicable by environmental Pb intakes (Alessio et al., 1976). This evidence of a statistical association was supported by Christoffersson et al. (1984), who found that bone Pb was strongly correlated with PbB in retired Pb workers, while PbB was little correlated with bone Pb in active workers with elevated job Pb exposures. Furthermore, several studies in the 1980s indicated that, for workers retired or idled by job actions, the half-life of declines in workers' PbB levels was directly proportional to how long they had workplace exposures (Hryhorczuk et al., 1985; O'Flaherty et al., 1982). These studies concluded that the source of the continued elevation in PbB was bone Pb released to blood via resorption. Significant in toxicological terms to any interchange of bone Pb with blood Pb was the finding of Alessio et al. (1976) that erythrocyte protoporphyrin (EP) elevation, a toxicity indicator for impaired heme biosynthesis, was much more correlated with chelatable lead (which probes Pb stores in parts of bone) than with PbB.

Firmer evidence of bone Pb mobility evolved with use of stable Pb isotope ratio analyses, where historical (bone) Pb in the body isotopically distinct from current, external Pb intakes can be employed to estimate endogenous, bone Pb contributions to current PbB under both (near) steady-state conditions and metabolic disturbances of steady state. Manton (1985), in an early use of the approach, noted that for two subjects, bone Pb contributed 70% of measured PbB over a 9-year period. Gulson et al. (1995, 1997, 1998, 1999) carried out a series of studies with Australian immigrants having bone Pb isotopic profiles different from the Pb absorbed from their new sources. Female subjects were studied prepregnancy and during pregnancy and lactation.

Table 8.10 presents studies which illustrate the direct relationship of bone Pb and PbB. Both tibial and trabecular bone Pb are correlated with PbB, and the correlation is more robust for trabecular bone Pb. Pb sequestered in trabecular bone is known to be more mobile and has a much lower half-life of release than does Pb in tibial or other cortical bone (Hernandez-Avila et al., 1996; Hu et al., 1998). As noted in Table 8.10, bone Pb via resorptive processes contributes 40–90% of measured (total) PbB. In essence, equilibria exist between bone subcompartments, other tissues, and blood/plasma.

Various studies have attempted to address the interrelationship of bone Pb to both PbB and plasma Pb (PbP), and also PbB to PbP. PbP is now considered to be the circulating, central biokinetic compartment providing Pb to affected tissues and providing the dose metric most relevant for dose–toxic response relationships (NAS/NRC, 1993; U.S. EPA, 2006). Hernandez-Avila et al. (1998) showed that PbB was highly correlated with PbP, PbB explaining 95% of PbP variability. However, in multivariate regression modeling that included PbB, trabecular (patella) bone Pb concentration was independently associated with PbP.

Chuang et al. (2001) extended this line of studies to examine the variables that contribute to the relationship of Pb-exposure biomarkers to fetal Pb exposure, indexed as cord blood. Maternal PbB, tibial and patella bone Pb, and measures of exogenous Pb intakes via lead-glazed ceramics (LGCs) and ambient air Pb were then modeled for impact on maternal PbP, maternal PbP being the Pb delivery pathway to the fetus. Lead in both bone types, in LGCs, and in ambient air were all significant contributors to PbB. Their analyses showed that maternal PbP varied independently of PbB. Comparative contributions of endogenous (bone) Pb and exogenous Pb (air and ceramics) were about equal. These investigators also showed that Pb moved from plasma to erythrocytes.

While it is clear that bone Pb resorption occurs and elevates both PbP and PbB, particularly during metabolic stresses such as pregnancy, lactation, and onset of menopause in women, some have probed the question of whether the partitioning of resorbed Pb into plasma provides more direct movement of released Pb to tissues than like amounts absorbed from external media. If so, does transport selectivity carry added toxicological risk from resorbed Pb? That is to say, does endogenous Pb of like concentration pose a greater toxicity risk than absorbed exogenous Pb?

Several reports have attempted to address this question. Cake et al. (1996) asserted that plasma lead originating from bone Pb resorption partitioned differently between erythrocytes and plasma. Bergdahl and Skerfving (1997) refuted these results, arguing that there is no biokinetic difference or, implicitly, no toxicological difference. Gulson et al. (2000) reported absence of any difference using urinary Pb isotopic ratio analyses during pregnancy and postpartum. Urinary isotopic ratios were used as a surrogate for plasma values. The data of Chuang et al., described earlier, indicate equilibrium for resorbed Pb in plasma that favors endogenous Pb movement to erythrocytes. The Pb mobilization test in lead workers administered Ca $(Na)_2$ EDTA by Araki et al. (1984) and Ishihara et al. (1984) resulted in Pb level spiking in plasma while PbB remained unchanged or showed some decline. Given the relatively much smaller amounts of Pb in plasma versus erythrocytes, responses of these two Pb compartments are difficult to interpret other than to show chelant-induced plumburesis was occurring via plasma filtration. The known direct relationship of bone Pb to either PbP or PbB has led some investigators to attempt to describe the potential of bone Pb to produce belated toxicity risk through resorption, specifically by examining the extent to which PbB values are maintained at elevated concentrations using simple ratios of PbB to Pb-bone.

Such efforts are, however, fraught with uncertainty. In most cases outside of occupational Pb exposures, little is known about Pb-exposure histories of human subjects or populations being studied. That is, PbB is affected in adults, particularly older adults, by both exogenous Pb ingested with ongoing exposures and bone Pb releases to PbB by resorption. Second, any large range of PbB in a population will have other toxicokinetic changes occurring, such as blood Pb values being impacted by changes in urinary clearance (Chamberlain, 1983).

One way around the problem is analysis of data sets where relatively high trabecular and cortical bone Pb levels are compared to relatively low exogenous, current Pb uptakes and contributions to PbB for purposes of estimating predominantly endogenous (resorptive) Pb contributions to PbB. Some illustrative ratios (slope values) of PbB to Pb-bone in the literature are provided in Table 8.11 for various Pb toxicokinetic scenarios: pre- versus postmenopausal changes in bone Pb releases to PbB of women, lead worker exposures during work versus job actions, men and women with heavy Pb exposures as children, and ratios of cortical or trabecular bone Pb to PbB.

Popovic et al. (2005) reported that in menopause, a metabolic change associated with osteoporosis, bone mineral loss, and resorptive Pb releases in

TABLE 8.11 Illustrative PbB/Pb-Bone Ratios (Slopes) in Adult Human Populations with Elevated Pb-Exposure Histories

Study Group	Study Design	Results	Study
Premenopausal retired female Pb workers	Measured PbB and cortical bone Pb	0.067	Popovic et al. (2005)
Postmenopausal retired Pb workers	Measured PbB and cortical bone Pb	0.132	Popovic et al. (2005)
Women with high childhood Pb exposures	Measured PbB and cortical bone Pb	0.052	McNeill et al. (2000)
Men with high childhood Pb exposures	Measured PbB and cortical bone Pb	0.126	McNeill et al. (2000)
Lead workers	Measured PbB and either cortical or trabecular bone Pb	0.161 (cortical bone); 0.142 (trabecular)	Cake et al. (1996)
Older men, 45–75 years old	Measured PbB and trabecular bone Pb	0.083	Hu et al. (1996)
Lead workers	Measured PbB and either cortical or trabecular bone Pb	0.136 (cortical bone); 0.078 (trabecular)	Fleming et al. (1999)

women, and the blood to bone Pb ratio (0.132) is twice that in premenopausal subjects (Table 8.11). This difference reflects a higher PbB in postmenopausal women. The gender variable in bone Pb accumulation favoring males is readily apparent in the studies of McNeil et al. (2000). Young adult men with childhood lead exposures had an associated ratio ~2.5-fold higher than for young women with a similar Pb history. Among lead workers, both Cake et al. (1996) and Fleming et al. (1999) noted that cortical bone in both worker study groups produced higher ratios of PbB to bone Pb than did trabecular bone, particularly in the Fleming data set. Impact of bone Pb was twice that for premenopausal women.

8.3.4 Chelatable Lead in Human Populations

One toxicologically and biokinetically distinct fraction of Pb in humans is chelatable Pb. This fraction of lead in the bodies of exposed humans derives from Pb mobilization in response to administration of lead-binding chelating agents, within defined medical and diagnostic regimens, to acutely or chronically lead-exposed children and adults. It is quantifiable on a case-specific basis by measuring plumburesis after giving chelant, the amount of chelant depending on whether diagnostic, i.e., challenge chelation or chelation therapy as a medical intervention in serious Pb poisoning was carried out. Chelatable Pb does not define metabolic and anatomically distinct sites in the manner that Pb in soft and mineralizing sites does. The term is chiefly characterized biokinetically within regimens using different chelants and different chelants mobilizing Pb in a complex mix of mineralizing and soft tissue. Chelatable Pb has long offered a more toxicologically relevant picture of that fraction of total body Pb which is available to produce and be more closely correlated with toxic responses than even blood lead. In that sense, chelatable Pb straddles two fractions of body Pb: the spontaneously mobile and circulating Pb fraction and the total lead body burden. The latter likely involves chelation-mobilizable Pb from more mobile bone subcompartments, e.g., the periosteum and trabecular bone.

Two levels of chelation-mobilizable Pb fractions can be discerned and discussed, differing in the amounts of a particular chelant administered to produce chemical decorporation of Pb from the body via movement into urine. Table 8.12 summarizes some illustrative cases highlighting the chelatable lead fraction of body Pb. A major argument used for challenge chelation is that amounts of Pb mobilized into urine by chelant administration in asymptomatic subjects, particularly Pb-exposed children with elevated PbB levels, will often rival those determined in cases of Pb poisoning. This happenstance is toxicokinetically rooted in part in the ability of chelants to mobilize lead in more active skeletal subcompartments, likely the periosteal and trabecular fractions, so that there is a logarithmic relationship of chelatable Pb to PbB (Chisolm et al., 1976; Saenger et al., 1982).

A bone subcompartment accessible to chelants is evident in the adult data of Araki and Ushio (1982), where the extent of plumburesis in human subjects rises with age, an observation consistent with bone Pb increases with age but not the relatively stable soft tissue Pb levels over various age intervals of adulthood. In addition, the likelihood of chelants mobilizing Pb from high stores of Pb can also be seen in the "rebound" phenomenon using these agents, where an initial significant decline in PbB and increased plumburesis is followed by a rise in PbB owing to re-equilibrating Pb compartments (Chisolm and Barltrop, 1979; Chisolm et al., 1985).

Piomelli et al. (1984) reported that the frequency of exceeding some medical intervention level, e.g., 0.6 μg Pb/mg Ca $(Na)_2$ EDTA/8 hours, rises markedly along with PbB. Asymptomatic children having PbB 40–49 μg/dl showed 38% exceeding the indicated action level, exceedences foretelling elevated poisoning risk. As in children, provocative (challenge) chelation in lead workers produced significant urinary Pb excretion exceeding an action level even at moderate PbB levels.

The findings of Cory-Slechta et al. (1987) opened to question the continued diagnostic use of Ca $(Na)_2$ EDTA, and to some extent its therapeutic use

TABLE 8.12 Illustrative Studies of Chelatable Lead Toxicokinetics in Human Populations

Study Subjects	Study Design	Results	References
Lead-exposed children	Provocative chelation testing using CaN_2 EDTA at lower, subtherapeutic doses	Logarithmic relationship, chelatable to PbB; nonsymptomatic children show chelation plumburesis similar to that in lead-poisoned ones	Chisolm et al. (1976), Saenger et al. (1982)
Lead-exposed children	Time course of PbB versus chelant treatment (CaN_2 EDTA or $CaNa_2$ EDTA + BAL)	A rebound in PbB after initial significant decline	Chisolm and Barltrop (1979), Chisolm et al. (1985)
Lead-exposed children, $N = 41$	Time course of PbB versus chelant treatment (oral D-penicillamine)	Decline from original mean 53 μg/dl PbB but elevation still present	Marcus (1982)
Lead-exposed children, $N = 210$, PbB 30–39 μg/dl and 40–49 μg/dl	Frequency of exceeding reference ratio, μg Pb/mg EDTA/8 hours	PbBs 40–49 μg/dl, 38% > action ratio of 0.6; for 30–39 μg/dl, 12%	Piomelli et al. (1984)
Nonexposed adults	Extent of chelatable urinary Pb versus age	Amounts of urinary Pb increase with age	Araki and Ushio (1982)
Lead workers	Chelatable Pb amounts over 24 hours versus action level of 35 μg/dl	Action level exceeded at significant frequency despite only moderate PbBs	Hansen et al. (1981)

with pediatric inpatients and others, based on the use of this chelant for Pb mobilization studies in rats. In particular, they observed two troublesome features. First, initial treatment with the chelant remobilized and transferred Pb from bone and kidney to brain and liver. Second, there was no net removal of Pb from brain sites over an entire course of the 5× treatment protocol. Additional chelants have been introduced since the findings of Cory-Slechta et al. (1987), but their use has largely been therapeutic rather than diagnostic. Specifically, oral D-penicillamine and dimercaptosuccinic acid (DMSA, Succimer, Chemet), and particularly the latter, are being used as outpatient drugs subsequent to medical intervention for chronic or acute childhood lead poisoning (O'Connor and Rich, 1999; Rogan et al., 2001; U.S. CDC, 1991). One concern with use of these oral outpatient agents is whether such agents do or do not enhance ingested Pb absorption when administered in a contaminated environment and with problematic or minimal supervision (Kapoor et al., 1989; U.S. CDC, 1991).

A remaining issue with use of the traditional as well as the newer, outpatient chelants is the "rebound" phenomenon. Some insight about rebound biokinetics may be found in the Cory-Slechta et al. (1987) report discussed above and other citations there. They noted that alternate administrations across a serial dosing protocol, i.e., injections 1 and 3, produced bone Pb reductions but injections 2, 4, and 5 did not. They postulated return of mobilized Pb back to bone stores. If such redeposited Pb is more mobile than initially unmobilized Pb, this fraction's biokinetics of Pb movement to blood may differ from overall bone Pb mobility.

8.4 RETENTION AND EXCRETION OF LEAD IN HUMAN POPULATIONS

Unabsorbed ingested Pb is simply excreted with feces. Absorbed Pb is retained in the short term and long term. Short-term retention studies in humans have principally involved adult volunteers ingesting isotopic Pb labels. The limited information for young children is derived from metabolic balance studies. Estimates of retention in the near term depend on the methodology employed in the estimates. Long-term retention of absorbed Pb in humans and experimental animal models entails deposition in bone and teeth, topics presented earlier in this chapter.

Illustrative studies of Pb retention and excretion in human populations are presented in Table 8.13. Booker et al. (1969) found that 4.4% of injected ^{212}Pb appeared in urine within 24 hours, followed later by equal amounts appearing in urine and feces. Rabinowitz et al. (1973) reported that when the stable Pb isotope ^{204}Pb was ingested with meals, 76% of the label appeared in urine, 16% in feces (endogenous excretion of absorbed Pb), and 8% in minor excretory routes: hair, nails, sweat. Endogenous fecal excretion of Pb in humans as studied by Chamberlain et al. (1978) is approximately 50% that of Pb excretion in urine.

Failure to account for the endogenous fecal Pb fraction will result in erroneous estimates of overall uptake. Chamberlain et al. (1978) noted that short-term Pb excretion in adult humans is 50–60% of absorbed fractions, with long-term retention amounting to about 25%. Ziegler et al. (1978) reported 34% of absorbed Pb was retained in infants. With regard to infant Pb excretion rates, U.S. EPA (1986, Table 10.3) estimated a daily infant Pb excretion rate of 1.50 μg/kg body weight per day. Chamberlain (1983), using various data sets for Pb exposures in adults, reported that the clearance rate (kg/day) increased with Pb exposure and PbB, up to a PbB of 80 μg/dl, the highest level evaluated.

TABLE 8.13 Illustrative Studies of Lead Retention and Excretion in Human Populations

Study Subjects	Study Design	Results	References
Adult volunteers, $N = 2$	Injected ^{212}Pb label in urine and feces	In 24 hours, 4.4% of dose, then equal amounts in urine, feces	Booker et al. (1969)
Adult volunteers	Ingestion of ^{204}Pb label and excretion	Urinary Pb = 38 μg/day, fecal Pb = 8 μg/day (76% and 16% recovery, respectively); hair, sweat, nails = 8%	Rabinowitz et al. (1973)
Adult volunteers	^{203}Pb-label injected, inhaled, ingested	With oral or injected intake, urine to feces Pb ~ 2:1	Chamberlain et al. (1978)
Various adult studies	Metabolic or isotopic label studies in adult volunteers	Long-term retention = 25% of absorbed Pb; short-term Pb excretion ~ 50–60% of absorbed fraction; excretion half-life ~ 19 days	Chamberlain et al. (1978), U.S. EPA (1986, Ch. 10)
Infants	Metabolic studies, estimates calculated in U.S. EPA (1986)	Infant Pb excretion rate = 1.50 μg Pb/kg body weight; ~34% Pb intake retained	U.S. EPA (1986, Table 10.3), Ziegler et al. (1978)
Children up to 8 years old	Metabolic balance studies	Approximately 18% retained	Alexander et al. (1974)
Various adult Pb-exposure groups	Urinary Pb excretion rate versus PbB levels	Clearance rate (μg/day) increases with Pb exposure and PbB, up to 80 μg/dl, highest level studied	Chamberlain (1983)

Table 8.14 depicts reported estimates of urinary Pb (PbU) and PbU excretion rates. The most reliable measure of urinary Pb excretion is its indexing to grams of daily creatinine, rather than daily total Pb or Pb/liter urine. Such indexing minimizes confounding such as body mass/area and urine volume differences. (Araki et al., 1990; Diamond, 1988). German children exposed in the 1980s to Pb emissions from a smelter showed a urinary Pb excretion

TABLE 8.14 PbU Toxicokinetics in Human Populations

Subject Group	Study Design	Results	References
German children impacted by Pb smelter	PbU/g creatinine	Geometric mean PbU (μg Pb/g creatinine) = 7–10	Brockhaus et al. (1988)
Lead workers, various reports	PbU, μg Pb/day	>200 μg/day	Wedeen et al. (1975), Lin et al. (2001), Biagini et al. (1977)
Various adult groups, different studies	Daily urinary Pb clearance, from plasma, l/day	Mean = 18 l/day (13–22 range)	Chamberlain et al. (1978), Manton and Cook (1984)
Pb workers	Utility of PbU excretion rate versus PbU concentration	PbU excretion rate is more reliable than PbU level owing to factors affecting urinary volume outputs	Araki et al. (1990), Diamond (1988)
U.S. children and adults	Geometric mean urine Pb (μg Pb/g creatinine), NHANES IV: 1999–2002	Children 6–11 years: 1.17 (1999–2000) 0.92 (2001–2002) Children 12–19 years: 0.50 (1999–2000) 0.40 (2001–2002) Adults 20 years and above: 0.72 (1999–2000) 0.66 (2001–2002)	U.S. CDC (2005), U.S. EPA (2006)

rate of 7–10 μg/g creatinine (geometric means) (Brockhaus et al., 1988). By contrast, the U.S. National Health and Nutrition Examination Survey (NHANES) IV of Pb exposures in residents of various ages showed markedly lower rates. Children 6–11 years of age had excretion rates of 1.17 and 0.92 μg Pb/g creatinine in the 1999–2000 and 2001–2002 periods, respectively. Corresponding figures for subjects 12–19 years of age in those same periods are 0.50 and 0.40 μg Pb/g creatinine, respectively. Adults in the same periods showed 0.72 and 0.66 μg/g creatinine, respectively.

Among lead workers, daily urinary Pb outputs exceed 200 μg/day based on various reports spanning a number of years (Biagini et al., 1977; Lin et al., 2001; Wedeen et al., 1975). Urinary Pb clearance in adult subjects from various reports was a mean of 18 l/day, with a range of 13–22 l/day (Chamberlain, 1983; Chamberlain et al., 1978; Manton and Cook, 1984).

8.5 LEAD–NUTRIENT INTERACTIONS IN HUMAN POPULATIONS

Lead's biokinetic behavior does not occur in a metabolic vacuum but is affected by extrinsic host factors such as the nutritional status of human populations and various physiological changes and stresses. The interactions of lead with typical nutrients in humans occur at several levels. These interactions include impacts on Pb uptake, distribution or retention, and redistribution of absorbed Pb from such body stores as bone Pb. Such interactions also vary with varying physiological status of human subjects. For example, pregnancy will induce some level of Ca resorption from bone, depending on dietary Ca intakes. Such remobilization also releases co-deposited Pb, as described earlier. Finally, the complexity of the interaction will vary, ranging from those for single nutrients to those for complete meals and diets. While a number of nutritional factors are known to affect Pb toxicokinetics, those involving calcium, iron, phosphorus, and vitamin D are best known for human subjects. Other interactions have principally been described for experimental animal models. The extensive literature on this topic is illustrated by selected studies covered in Table 8.15.

The interactions of Pb with nutrients in human populations, notably interactions with inorganic nutrients, generally appear to be in the direction of nutrient status affecting Pb toxicokinetics and toxicity rather than being fully reciprocal. That is, changes in the amounts of Pb available for intake and uptake are not generally assumed to affect uptake and utilization of calcium, iron, phosphorus, etc., compared to the reverse. While Fe deficiency is linked to increased PbB, high intakes and uptakes of Pb do not induce Fe deficiency. We would not expect fully reciprocal interactive relationships, given that the nutrients of interest are essential substances and are under homeostatic control physiologically.

Children as a group are of particular concern with regard to lead–nutrient interactions for various reasons. First, their nutrient requirements for development are critical to optimal health. At the same time, they tend in many nutritional scenarios to be Ca deficient. Second, there is a strong interactive, antagonistic relationship between Ca or Fe nutrition and Pb toxicokinetics, with deficiencies in these elements promoting Pb uptake and retention, while repletion tends to reduce Pb uptake and/or retention.

The second risk population for interactive relationships and toxicokinetics of Pb is the fetus, with respect to gestational nutritional requirements *vis-a-vis* Pb uptake into fetal tissues. Closely linked to fetal Pb exposures are increased Pb exposures of the mothers during pregnancy. Fetal skeletal

TABLE 8.15 Illustrative Dietary Interactions of Ingested Pb with Nutrient and Other Dietary Components in Human Subjects[a,b,c]

Study Group	Study Design	Results	References
Children 1–4 years old, low-income, inner-city families	PbB levels versus Ca, P intakes	PbBs > 40 μg/dl had lower Ca, P in diet ($p < 0.01$)	Mahaffey et al. (1976)
Children 1–4 years old, low-income, inner-city families	PbB levels versus Ca, vitamin D	PbB inversely related to Ca intakes; PbBs > 60 μg/dl had lower Ca and vitamin D ($p < 0.001$)	Sorrell et al. (1977)
43 children 1–6 years old, variable exposures	PbB versus Ca intake	Inverse association, $p < 0.05$, $r = -0.327$; high Pb had lower Zn intake	Johnson and Tenuta (1979)
Infants up to 24 months old	Metabolic studies, Pb retention versus diet	Pb retention inversely related to Ca intake, $r = 0.279$, $p < 0.01$	Ziegler et al. (1978)
Adult volunteers, $N = 8$	^{203}Pb label, GI Pb uptake, fasting conditions ± Ca, P supplements	60% Pb label uptake without supplement intake; 10% uptake at 140 mg P, 200 mg Ca, respectively	Heard and Chamberlain (1982)
Pregnant women followed in pregnancy	Stable isotope labeling of PbB versus Ca intakes	Pregnant women ingesting high Ca delayed PbB elevation to 6–8 months gestation	Gulson et al. (2004)
1,155 Chinese adults in various residential districts	PbB versus tofu-based Ca intakes (PbB = 8–15 μg/dl)	≥750 g/week of tofu intake reduced PbB by 1.4 μg/dl versus 250 g/week	Chen et al. (2001)
U.S. children in NHANES II, White and Black, ages 1–11 years	PbB versus Ca intake	PbB inversely correlated to Ca ($p = 0.028$); White children had higher Ca intake ($p < 0.001$)	Mahaffey et al. (1986)
Children ages 1–8 years, African-American ($N = 91$) and Hispanic ($N = 51$)	PbB versus vitamin D nutrition (serum 25-OH-D) in summer and winter	Seasonal increases in PbB and serum 25-OH-D in older children 4–8 years of age were significantly associated	Kemp et al. (2007)
Elderly Caucasian males, mean age = 67 years (49–93 years)	Bone Pb accumulation versus vitamin D intake	Bone Pb decreased with higher Ca intakes	Cheng et al. (1998)

(Continued)

TABLE 8.15 Illustrative Dietary Interactions of Ingested Pb with Nutrient and Other Dietary Components in Human Subjects[a,b,c]—(cont.)

Study Group	Study Design	Results	References
Adult males, evaluated in winter and summer	Bone Pb levels versus decreased vitamin D biosynthesis	PbB levels higher in winter, attributed to higher bone Pb resorption due to lower vitamin D formation	Oliveira et al. (2002)
Lead workers	Dietary Ca intake versus PbB	No statistically significant effect	Kristal-Boneh et al. (1998)
43 Children PbB > 30 μg/dl, EP > 35 μg/dl	Fe deficiency frequency versus elevation in PbB	Fe deficiency frequency increased as PbB increased	Yip et al. (1981)
66 Children, chelatable Pb and Fe	Indexed by urinary ALA levels	Inverse relationship between chelatable Fe and chelatable Pb	Chisolm (1981)
California children aged 1–6	Association of Fe deficiency (≥12 ng/ml ferritin) with PbB	Higher PbB levels (GM = 5.6 μg/dl) in Fe-deficient children	Bradman et al. (2001)
Albany, NY: low socioeconomic status (SES) infants/toddlers, at 6 and 12 months	PbB versus Fe intake	Low Fe intake linked to elevated PbB (GM = 5.1 μg/dl, 12 months)	Schell et al. (2004)
Elderly adults in the Veterans Administration Normative Aging Study	Fe intake versus PbB level	Lower Fe intakes (11 mg/day) linked to increased PbB versus higher Fe intakes (23.5 mg/day)	Cheng et al. (1998)
U.S. children	NHANES II: FE status versus Pb dose–response	Low Fe increases Pb toxicity curve	Marcus and Schwartz (1987)
U.S. children	NHANES II: toxicity of Pb alone versus Pb+low Fe	Fe deficiency+Pb ≫ hematotoxic versus Pb alone	Mahaffey and Annest (1986)
U.S. children and adults	NHANES II: Fe status versus frequency of elevated PbB	Low Fe children had higher prevalence of elevated PbB; adult women showed no association, men showed a weak association	Yip and Dallman (1984)

[a]*Reports using complete diet or dietary components.*
[b]*Various statistical analyses employed.*
[c]*Various study protocols employed: isotopic labeling, metabolic studies, etc.*

growth requirements, especially in the third trimester, unleash a high demand for body Ca, either through the diet or from bone Ca (and Pb) resorption. Pregnancies in the face of Ca deficiencies increase the amounts of Pb presented to *both* fetus and mother for added toxic risk.

8.5.1 Lead–Calcium Interactions

Considerably more is known about the Pb–Ca interaction than about other nutritional interactive pairings (see Table 8.15). Such knowledge dates back to the early part of the twentieth century with the observation that Pb follows calcium. That is, the two substances are closely related physiologically and toxicologically.

A number of studies have described the Pb–Ca interaction in infants, toddlers, and older children. Illustrative examples are given in Table 8.15. Many of the children studied have been high-exposure risk children by virtue of risk factors such as being in a low-income family and having significant Ca deficiency. Children 1–4 years of age and in low-income, inner-city families showed statistically significant inverse relationships of PbB elevations and dietary Ca (and P or vitamin D) intakes in a number of the older studies (Johnson and Tenuta, 1979; Mahaffey et al., 1976; Sorrell et al., 1977; Ziegler et al., 1978). In some cases, other nutrients such as phosphorus, vitamin D, and zinc were examined. White and African-American children, ages 1–8 years, examined as to their Ca intakes and PbB levels in the U.S. NHANES II (1976–1980), were shown by Mahaffey et al. (1986) to present a statistically significant inverse relationship.

Gulson et al. (2004) examined the relationship of maternal PbB (and thereby fetal PbB and fetal toxicity risk) and use of Ca supplements during pregnancy. Using the stable Pb isotopic ratio approach, they found that the higher Ca intake group did not show elevated PbBs until 6–8 months into pregnancy, while the low-Ca group showed elevations much earlier. The high fetal demand for Ca associated with the high pulse of skeletal growth in the final trimester benefited from the supplementation versus those without these Ca intakes. Hernandez-Avila et al. (1996) studied use of Ca supplementation in postpartum women (1 month forward) and reported there was a statistically significant reduction in patellar Pb in high-Ca women. There were reductions in PbB and tibia, but these did not reach statistical significance. The Gulson et al. data can be compared to findings of Hertz-Picciotto et al. (2000) that intake of Ca at the highest level (>2,000 mg Ca) in a pregnancy cohort (N = 195) at two age bands had a protective effect in the latter stage of pregnancy, and did so by shifting the onset of PbB increase in older women (age 38) while more significantly attenuating PbB increases in pregnancies in the younger group (age 18).

The efficacy of adequate Ca intake for reducing Pb uptake and retention in adults other than in pregnant women has not been widely studied. Korrick et al. (2002) studied the impact of Ca nutritional status among Boston women 30–50 years of age and those over 50 over five ranges of daily intake, ranging from a low of 321 mg Ca/day to 2,583 mg Ca/day in the highest intake group. They reported no statistically significant alterations in PbB, tibial Pb, or trabecular Pb under the study conditions. Chen et al. (2001) reported that 1,155 Chinese adults in different districts showed reduced PbB in the amount of 1.4 μg/dl (PbB range, 8–15 μg/dl) for high Ca intakes (in tofu, 750 g/week) versus low Ca intakes (in tofu, 250 g/week). Heard and Chamberlain (1982) found that the uptake of radiolabeled (^{203}Pb) Pb in adult volunteers ingesting the Pb label with and without Ca/P supplements (fasting) was significant at 60% without supplement and decreased to 10% when the subjects took 200 mg Ca and 140 mg P. There was no statistically significant effect of dietary Ca intake on PbB in lead workers (Kristal-Boneh et al., 1998).

A number of experimental animal studies confirm and clarify the inverse relationship between Pb uptake rate and Ca intake under controlled experimental conditions (Mushak and Crocetti, 1996a,b). Experimental animal studies of the inverse relationship between Pb uptake and Ca status date to results of Sobel et al. (1940) who demonstrated Pb uptake and both PbB and Pb-bone levels varied inversely with Ca intake.

Since these early findings, the inverse PbB to Ca intake relationship has been amply documented in rats (Barton et al., 1978a; Mahaffey et al., 1973), the pig (Hsu et al., 1975), the horse (Willoughby et al., 1972), and in lambs (Morrison et al., 1977).

Test animal data are particularly useful in elucidating the mechanisms of the nutrient–Pb interactions. Part of the effect is at the level of Pb uptake across the gut wall and part appears to operate with absorbed Pb retention rates. Barton et al. (1978a,b) showed a role for divalent metal transporter (DMT) substances at the level of gut wall transport while Smith et al. (1978) reported that Ca and Pb are taken up at different sites, Pb uptake occurring at the P site, suggesting a complex ternary interaction of Pb, Ca, and P. Precipitation of Pb with Ca–P complexes may be operative. Ca intake levels also affect Pb retention, with both deficiency and excess in intake increasing retention, i.e., suppressing Pb excretion rate. In the case of high Ca intakes, a role for bone Pb and Ca resorption suppression may exist (Goyer, 1978). Fullmer and Rosen (1990) reviewed the experimental data on intestinal interactions of Pb and Ca, with an emphasis on data from Pb-exposed chicks. This review further related the Pb–Ca interaction to vitamin D biosynthesis, *in vivo* interactions occurring in the face of already high body Pb burdens.

8.5.2 Lead–Iron Interactions

The Pb–Fe interaction, like that for Ca, is largely antagonistic in nature; amounts of each that are present inversely affect the other. From the available evidence, iron deficiency apparently leads to increased Pb uptake, while dietary Fe sufficiency reduces those uptakes. Iron status, like Ca status, is principally seen affecting Pb toxicokinetics and toxicity in young children rather than in older children or adults.

Yip et al. (1981) were among the first to report that in young children with elevations in both PbB and EP—elevations in EP being a biomarker for both Fe deficiency and Pb exposure/toxicity—the frequency of Fe deficiency was associated with increased frequency of elevated PbB. Another earlier paper on the topic, that of Chisolm (1981), found that in 66 young children with elevated PbB and hematotoxicity indexed by elevated urinary ALA, there was an inverse relationship between chelatable Fe and chelatable Pb. Bradman et al. (2001) found that a study population of Fe-deficient California children aged 1–6 showed an inverse relationship between Fe deficiency (ferritin ≤12 ng/ml) and PbB, with the geometric mean PbB in the deficient children being 5.6 μg/dl.

Several studies of U.S. children based on results of NHANES II reported associations of Fe status with PbB or Pb toxicity. Mahaffey and Annest (1986) noted that the combination of Pb (PbB elevation) and Fe deficiency was much more hematotoxic than Pb alone. Similar results involving Fe status of NHANES II children and Pb dose–hematotoxic response or frequency of elevated PbB have appeared (Marcus and Schwartz, 1987; Yip and Dallman, 1984).

The inverse Pb–Fe status relationship can be seen in early infancy. Low socioeconomic infants in an urban area (Albany, NY) at both 6 and 12 months showed that the level of Fe intake was inversely associated with elevated PbB (Schell et al., 2004). At 12 months, the geometric mean PbB was 5.1 μg/dl.

The Pb–Fe interaction in adults has not been well studied, and available data are mixed. Yip and Dallman (1984) reported very weak (males) or nonexistent (females) relationships between iron status and PbB values. Elderly adults in the "Normative Aging Study" of the U.S. Veterans Administration showed that PbB means at Fe intake of 24 mg/day were higher than at lower intakes, 11 mg Fe/day (Cheng et al., 1998).

The findings of Pb–Fe interactions in children have been supported by results of experimental animal studies. Mahaffey-Six and Goyer (1972) noted that Fe deficiency in rats increased tissue levels and relative toxicity of Pb. Rats studied by Barton et al. (1978b) showed, via everted duodenal sac, that at the level of the gut, reduced Fe increased Pb uptake and vice versa. These data support the operation across the gut wall of a DMT system, a mechanism also described by Morrison and Quarterman (1987) for intestinal loops in rats. Cerklewski (1980) noted that Fe deficiency increases Pb transfer to suckling rats while, in mice, Hamilton (1978) showed Fe deficiency had no effect on Pb retention. The latter finding suggested that a principal point of interaction is at the uptake step rather than at retention. Relative to this,

Flanagan et al. (1979) concluded from their studies that duodenal Pb is absorbed by two processes: one is controlled by Pb content in the lumen and the second involves a carrier-mediated transport showing saturation kinetics. The latter mechanism is activated by low-Fe diet and is inhibited by Fe. See, however, the more recent findings described later.

Mechanism of the Iron–Lead Interaction

A number of recent studies established the basis of the toxicokinetic interaction of Fe with Pb. This interaction appears to occur through operation of the divalent metal transporter 1 (DMT1), also known as the natural resistance-associated macrophage protein 2 (NRAMP2), the divalent cation transporter 1 (DCT1) (Bressler et al., 2004, 2007; Gu et al., 2009; Kayaalti et al., 2010; Mims and Prchal, 2005).

DMT1 functions in the transport of ferrous iron (Fe II) and certain nonessential divalent ions, e.g., lead and cadmium, across the plasma membrane and out of the endosomal compartment in a pH-dependent fashion. In the mammalian GI tract, DMT1 mediates apical Fe uptake into duodenal enterocytes lining the small intestine and also serves a general cellular role in tandem with the transferrin receptor (TFR) by using Fe-loaded diferric transferrin at the plasma membrane. It transports the iron via the endosome into cytosol.

In Fe-deficient diets or cells, Fe depletion upregulates expression of DMT1 at the mRNA level. Increased DMT1 with reduced Fe intake permits opportunistic binding and enteric transport of other divalent ions, including Pb. This behavior provides a molecular basis for the Pb–Fe antagonistic interaction long observed experimentally and epidemiologically.

8.5.3 Pb Interactions with Other Nutrient Factors

Among other nutritional factors potentially interacting with Pb, those of the vitamin D endocrine system are probably the more significant. Unlike Ca and Fe, however, the vitamin D interaction has been much less studied in human populations and is even less understood. Vitamin D in the form of a dietary nutrient or as an endogenous substance biosynthesized *in situ* in amounts proportional to the intensity of solar stimulation has been reported to be variably associated with Pb uptake, retention, and toxicity.

Earlier data include those of Sorrell et al. (1977), indicating that at very high elevations in PbB (>60 μg/dl) there is inverse relationship ($p < 0.001$) with intakes of both Ca and vitamin D. Rosen et al. (1980) reported that children having elevated PbB in the range of 33–120 μg/dl had significantly lower serum levels of the vitamin D metabolite, 1,25-dihydroxyvitamin D compared with age-matched controls. There was also an inverse (negative) statistical relationship between PbB in the indicated range and serum levels.

More recent investigations support an endocrinological connection while extending the complexity of the interaction to include bone Pb stores.

Kemp et al. (2007) reported that in African-American and Hispanic children 1–8 years of age evaluated for PbB and serum vitamin D, children 1–3 years of age had a 32% increase in PbB in summer. Older children had a smaller change (13%). However, it was only in the older children that a positive statistical association was detected, even though it was younger children showing the higher relative increase in PbB. The study indicates a seasonal association related to vitamin D metabolism in the older children, but does not rule one out for the younger set, since dietary vitamin D intakes may be relatively invariant in the younger versus older children, thus attenuating vitamin D biosynthesis rates. Oliveira et al. (2002) reported that in adult males, PbB levels measured higher in winter than in summer, the opposite of young children. The researchers ascribed higher PbB levels in winter to reduced vitamin D biosynthesis, forcing higher bone resorption of Ca and thereby Pb.

The Pb–vitamin D interactive relationship has long been confounded by the role of seasonality in its biosynthesis via differing solar intensity with season. That is, more vitamin D biosynthesis would be expected to occur in summer. The role of relative differences in vitamin D biosynthesis in lead poisoning, long termed the "summer disease," remains problematic to some extent. There has been a school of thought arguing that PbB spiking in and around the summer season is not endocrinological, but rather due to the combined Pb-exposure mechanisms of more outdoor or like increases in Pb contact and children's vacation schedules increasing home Pb exposures (Yin et al., 2000). It is likely, from the evidence, that both factors play an etiological role in producing increased Pb exposures in children in warmer months.

Interactions of P with Pb appear to be ternary rather than binary in nature, requiring Ca to be present in order to enhance the full interactive effect in adult human volunteers (Heard and Chamberlain, 1982). In these adult subjects, the interactive impact on Pb uptake of increased Ca and P together was a sixfold reduction, whereas either element alone had only about one-fifth the combined impact.

Interactions of Pb toxicokinetics with nutrients such as proteins, fats, zinc and lead have been reported but an actual, consequential role in humans for interactions that have been seen mainly in experimental animals remains to be established.

8.6 THE ROLE OF GENETIC VARIABILITY IN Pb TOXICOKINETICS

Nutrient–Pb interactions as discussed earlier illustrate the role of extrinsic influences on Pb toxicokinetics. Extrinsic factors are those which are not intrinsic to the biology or physiology of Pb-exposed human populations and are therefore mutable. The nature and extent of nutrient–Pb interactions change with nutritional status and particular physiological stresses or aging milestones. At the same time, such interactions can be offset by interventions that alter nutritional status, among other things.

By contrast, certain intrinsic characteristics of human populations affect different subsets of those populations in varying ways. Two forms of intrinsic host or receptor human populations exist with respect to Pb exposure: (1) those characteristics shared by all and homogeneously distributed, e.g., all members of a living population at some time were, are, or will be children 5 years of age; and (2) intrinsic characteristics heterogeneously distributed in a human population. An important characteristic in this latter category versus Pb toxicokinetics is genetic variability in such biomarkers as PbB. The extent to which such variability may enhance or attenuate toxic harm in respective genotypes and phenotypes will in turn help characterize genetic susceptibility. Table 8.16 provides examples of studies of the impact of certain genetically variable metabolic indices with Pb exposure indexed as PbB.

Hematology in human populations has a number of genetic determinants, and the genetic polymorphism for various components of human blood compartments has the potential to affect the binding of lead in blood, the subsequent dose–toxic response relationships for lead from such alterations, and to enhance the variability between subjects with the same overall external lead contact.

Various genetically polymorphic components of blood claimed in various publications to affect lead binding, its toxicokinetics, and/or its toxicity are described later. Of considerable concern is the extent to which populations having potential genetic polymorphism for lead binding in blood are known to also have elevated levels of lead in their immediate environment. Links from these population-variant substances to lead toxicokinetics or toxicology differ greatly in terms of the strength of the connection. In a number of cases, there is a clear relationship while in others a toxicokinetic or toxic consequence of such genetic–lead interactions is unclear.

8.6.1 Genetic Polymorphism in δ-ALAD and Relationships to PbB

The most studied of the genetic polymorphisms for blood components in potentially lead-exposed populations and relevant to Pb toxicokinetics is that for the erythrocyte heme pathway enzyme δ-ALAD (EC 4.2.1.24). This enzyme, also known as porphobilinogen synthase (PBG-S), participates in the heme biosynthesis pathway, catalyzing the cyclodehydration of two units of δ-ALA to PBG. Inhibition by lead leads to the accumulation of δ-ALA in plasma and urine. Excess substrate is neurotoxic in animals and may play a role in manifestations of lead poisoning (Audeskirk, 1985) and, presumably, in the genetic disorder acute intermittent porphyria (Bonkowsky, 1982). It is

TABLE 8.16 Studies of ALAD Genetic Variability in Pb Toxicokinetics

Study Subjects	Study Design	Results	References
New York City children, $N = 1{,}000$	Pb screening program testing blood for Pb and ALAD genotype	Children with the ALAD2 allele had a fourfold higher frequency of PbB ≥ 30 μg/dl than the ALAD 1-1 children	Astrin et al. (1987)
New York City children, $N = 1{,}278$	Pb screening testing PbB and isozyme genotype	Children with the 1-2 or 2-2 isozymes have PbB about 10 μg/dl higher than with the 1-1 genotype	Wetmur et al. (1991b)
Individuals homozygotic for the ALAD2 allele	Reverse transcription and sequencing of the ALAD2 allele	Difference with ALAD1 is a single substitution of asparagine for lysine	Wetmur et al. (1991a)
German Pb workers, $N = 202$	Studied for PbB versus ALAD isozyme genotype	Workers with ALAD2 allele had significantly higher PbB, ~10 μg/dl	Ziemsen et al. (1986)
Canadian lead–zinc smelter workers, $N = 134$	Measurement of PbB, heme toxicity indices, and ALAD alleles	No significant link between isozyme variant and PbB ($p = 0.08$)	Alexander et al. (1998)
Construction workers, $N = 691$	Measurement of PbB, ALAD allele, indices of kidney function	No significant link between isozyme variant and PbB ($p = 0.73$), borderline association with bone Pb ($p = 0.06$)	Smith et al. (1995b)
Canadian lead smelter workers, $N = 381$	Measurement of PbB, serum Pb, PbBO, ALAD allele	ALAD1 workers showed higher slope of PbBO to integrated PbB/time versus ALAD2 workers; ALAD 1-2 or ALAD 2-2 had higher PbB	Fleming et al. (1999)
South Korean lead-acid battery workers, $N = 57$	Measurement of urine Pb, PbB, ALAD allele forms after DMSA dosing, 4 hours	Workers with the ALAD 1-2 isozyme excreted less Pb at 4-hour urine collection ($p = 0.05$)	Schwartz et al. (1997)
South Korean Pb workers ($N = 798$) plus controls ($N = 135$)	Measurement of tibia Pb, PbB, DMSA-induced urine Pb, ALAD isozyme variant	Marginally elevated PbBs with ALAD 1-2 variant; no difference in tibia Pb	Schwartz et al. (2000)
Korean Pb workers, $N = 652$	Measurement of ALAD2 alleles, other polymorphisms, and patellar Pb	No association of patellar Pb with ALAD isozyme variants	Theppyang et al. (2004)

a zinc-requiring sulfhydryl enzyme, is homooctameric with eight zinc atoms, of which four are required for activity, and is extremely sensitive to lead, showing a Pb Ki of 0.07 pM (Simons, 1995). The enzyme is vestigial in the erythrocyte cytosol, and is retained during the complex sequence of cell formation and maturation in erythroid tissue.

The enzyme shows genetic polymorphism in certain human populations (Astrin et al., 1987; Wetmur, 1994; Wetmur et al., 1991a). The gene encoding δ-ALAD is localized on chromosome 9q34 and is expressed as two codominant gene alleles, ALAD1 and ALAD2. Expression of these two alleles produces three ALAD isozyme genotypes: ALAD 1-1 (normal), ALAD 1-2 (heterozygote variant), and the relatively uncommon ALAD 2-2 homozygote variant. The gene frequencies are 0.9 and 0.1 for ALAD1 and ALAD2 alleles, respectively. Distribution of the isozyme ALAD variants differs with race. Caucasians show about 80% of ALAD 1-1, 18–19% ALAD 1-2, and 1–2% of the homozygotic isozyme 2-2. This accords with the two gene frequencies of 0.9 and 0.1. Afrocentric populations show little of the variant allele expression and associated isozyme variability, while those of Asian descent appear intermediate in the fractions of isozyme variants. Wetmur et al. (1991a) characterized the molecular genetic basis of the genotypic distinctions, showing through reverse transcription and sequencing of the ALAD2 allele from a homozygotic 2-2 individual that this isozyme presents a single nucleotide difference from the common allele, a G-to-C transversion of coding nucleotide 177, consistent with substitution of an asparagine unit for a lysine in residue 59 (K59N) of the subunit. This substitution is not proximate to the enzyme's active site, suggesting that a conformational change leads to increased lead binding in the variant.

Genetic polymorphism expressed as ALAD isozyme variants has been linked to differential Pb binding in the erythrocyte. Such binding differences could be significant to lead–blood interaction, lead body burden toxicokinetics, and expressions of lead toxicity. Bergdahl et al. (1998) showed that the principal binding site of lead, ~80%, is δ-ALAD in human erythrocytes, rather than the long-assumed two hemoglobin isoforms. Second, epidemiological data suggest that children (Astrin et al., 1987; Wetmur, 1994) and adults (Bergdahl et al., 1997a; Schwartz et al., 1997; Smith et al., 1995a; Wetmur et al., 1991a,b; Ziemsen et al., 1986) with the ALAD2 variant have higher levels of blood lead, and lead workers with the variant accumulate higher bone Pb in their bones with age (Smith et al., 1995a). Thirdly, differences in lead–ALAD isozyme variant interactions appear to influence the severity of toxic effects, but in inconsistent ways (Bergdahl et al., 1997a; Schwartz et al., 1995, 1997; Smith et al., 1995b).

Astrin et al. (1987) tested 1,000 children for ALAD genotype and PbB concentrations in a lead screening program in New York City. They found that, among these children, individuals with the ALAD2 allele had a fourfold higher frequency (30% versus 8%) of a PbB of 30 μg/dl or higher. Among

African-American children the difference was more dramatic, with 30% of ALAD2 children having PbB ≥ 30 μg/dl, versus 3% for the normal form. As conceded by the authors, this group of children reflected a selection bias, in that entry to PbB testing was contingent on an initial elevated free erythrocyte protoporphyrin (FEP) measurement, an elevation that is the result of lead exposure and/or iron deficiency. There is no indication at present that FEP elevation is genetically related to ALAD isozyme genotype, but at the least it is clear that at lead exposures sufficient to be statistically associated with elevated FEP, lead binding by ALAD within the erythrocyte will be robust enough to reveal and amplify binding differences among variant forms.

The related study by Wetmur et al. (1991b) of 1,278 screened children showed that children with the ALAD 1-2 isozyme variant had, on average, 10 μg/dl higher PbB than those with the 1-1 form.

A number of studies have attempted to characterize the impact of ALAD1 and ALAD2 allele forms on lead binding in adult blood, and summary data are included in Table 8.16. These studies have generally involved a look at lead workers or other occupational groups where the overall lead exposure would be quite elevated compared to the general adult population. Lower lead exposures in groups of adult workers show less ALAD polymorphic linkage, at least with regard to PbB. There is the same intrinsic bias to these worker studies as exists for lead screening of children, i.e., ALAD genetic polymorphism is being examined at the high end of the external lead/blood lead curve. However, if the binding differences are operative at high lead exposures, that is where the public and occupational health attention should be.

A group of 202 German lead workers were studied with regard to both PbB and ALAD isozyme genotype by Ziemsen et al. (1986). They showed that employees with the ALAD2 allele had a statistically significantly higher PbB level than those with the ALAD1 form, on average 10 μg/dl higher PbB, similar to the value noted above for children. Later studies did not report a consistent simple statistical association between genotype and PbB, but more complex associations were noted.

Bergdahl et al. (1997a) found no association between ALAD isozyme genotype and either PbB or bone lead in 89 lead workers, while 134 lead smelter workers tested for both genotypes, PbB, and heme toxicity markers by Alexander et al. (1998) showed no statistically significant association between the two measures ($p = 0.08$). With low level lead exposure in a group of 691 building trades workers tested for ALAD genotype, PbB, and bone lead, Smith et al. (1995b) showed no significant differences in blood lead but a borderline significance for bone lead as a function of ALAD allele. Schwartz et al. (1995) reported that in a group of Korean lead battery workers, no overall association of ALAD genotype with PbB was found but for those blood levels at or above 40 μg/dl, the association existed with an odds ratio of 2.6. The authors also noted the significant finding that such variables as work duration, factory, and reduced zinc protoporphyrin (ZPP) with increased PbB in ALAD2 subjects indicated that there was a selection factor operative *vis-a-vis* ALAD2 subjects that meant those subjects enjoyed a net protective effect of the allele variant, not an increased risk.

Several studies probing more complex lead biokinetic questions with reference to ALAD genotype have appeared. For example, Schwartz et al. (1997) concluded that, in a group of 57 lead workers with either the ALAD 1-1 or 1-2 isozymes, those individuals with the variant 1-2 isozyme had lower bioavailable lead stores than the 1-1 isozyme subjects when challenged with DMSA as lead mobilizing agent.

Do ALAD genotypes affect the degree of toxic expression of lead in exposed subjects? The results of various studies present a mixed picture. Smith et al. (1995b) found no association between genotype, PbB, or bone lead and hematocrit or hemoglobin level. This group of subjects had moderately elevated workplace exposure, but exposures much lower than those of lead workers. In this same group of workers, the same researchers showed borderline significance between biomarkers of chronic kidney toxicity, blood urea nitrogen (BUN) and uric acid, and ALAD genotype. The renal biochemical endpoints, urinary calcium and creatinine, were reported to be significantly lower in lead workers with the ALAD2 genotype by Bergdahl et al. (1997a). On the other hand, Schwartz et al. (1995) found that in high lead exposures of Korean lead battery workers, the workers with the ALAD2 genotype had the lower ZPP levels when controlling for blood lead. That is, the variant genotype seemed to protect this group of workers from the hematotoxic effects of lead as expressed in the heme biosynthesis pathway.

More recent studies of effect modification of ALAD genotypes continue a generally mixed picture (Table 8.17). Lee et al. (2001) compared hematopoietic outcomes in Korean Pb workers (N = 798, versus 135 controls) tested as to various dose Pb markers and ALAD genotype. Tibia Pb was linked to all five hematotoxicity endpoints: ZPP, Hb, Hct, plasma, and urinary ALA. PbB and chelatable Pb were only linked to three: ZPP and the two ALA measures. There was no effect of ALAD isozyme variant on these associations. For this same worker cohort, Lee et al. (2001) reported no effect modification by ALAD variant on the PbB to blood pressure (BP) relationship. Weaver et al. (2003) reported that, for the higher PbB subjects in this cohort of Korean workers, the ALAD 1-2 variant was linked to lower BUN and serum creatinine, and higher creatinine clearance. Weaver et al. (2005), in a related study, showed an effect of ALAD genotype on the Pb dose–uric acid relationship. ALAD 1-1 subjects had higher uric acid with elevated PbB, while ALAD 1-2 subjects showed the reverse. Weaver et al. (2006) extended the above efforts to 647 Korean Pb workers through measuring PbB and patellar Pb, genotyping ALAD variants, and five measures of renal function: BUN, creatinine, creatinine clearance, *N*-acetyl glucosaminadase, and retinol-binding protein (RBP).

TABLE 8.17 Studies of ALAD Genetic Variability as a Factor in Lead Dose–Toxicity Relationships

Study Subjects	Study Design	Results	References
Canadian lead–zinc smelter workers, $N = 134$	Measured PbB, coproporphyrin, and ZPP levels	PbB levels ≥40 μg/dl showed higher ZPP ($p < 0.03$) with ALAD1 isovariant subtypes and CP ($p < 0.01$) versus ALAD 1-2 individuals	Alexander et al. (1998)
Unionized construction workers, $N = 691$	Measured PbB, bone Pb; ALAD genotyping	No association between ALAD genotype and either hemoglobin or hematocrit	Smith et al. (1995b)
Unionized construction workers, $N = 691$	Measured PbB, bone Pb, indices of kidney function, ALAD genotypes	Borderline association ($p = 0.06$) of ALAD2 isozyme variant with BUN	Smith et al. (1995a)
Korean Pb workers, $N = 691$	Measured PbB, urinary and plasma ALA, ALAD genotypes	Significantly higher plasma ALA ($p < 0.03$) in ALAD 1-1 subjects	Sithisarankul et al. (1997)
Korean Pb workers, $N = 798$	Measured PbB, bone Pb, chelatable Pb; genotyping of ALAD variants and other biochemical factors	ALAD 1-2 variant subjects showed higher Pb linked to lower BUN and serum creatinine, higher creatinine clearance	Weaver et al. (2003)
Korean Pb workers, $N = 798$ plus 135 controls	Measured Pb biomarkers, ALAD variants, hematopoietic outcomes	Tibia Pb associated with five hematopoietic outcomes: ZPP, hemoglobin, hematocrit, urinary and plasma ALA; PbB and chelatable Pb linked to ALA and ZPP measures; no clear effect of ALAD genotype on associations	Lee et al. (2001)
Korean Pb workers, $N = 798$ plus 135 controls	Measured BP and hypertension, dose measures, and ALAD genotype	ALAD genotype was not associated with BP measures and did not modify the lead-outcome relationships	Lee et al. (2001)

(Continued)

TABLE 8.17 Studies of ALAD Genetic Variability as a Factor in Lead Dose–Toxicity Relationships—(cont.)

Study Subjects	Study Design	Results	References
Korean Pb workers, $N = 798$ current and former workers	Measured Pb dose, ALAD genotype, uric acid	Older workers (Mean = 40.6 years old) showed an effect of ALAD genotype on the Pb dose–uric acid relationship; ALAD 1-1 genotype subjects had higher uric acid with higher dose while ALAD 1-2 individuals showed the reverse	Weaver et al. (2003)
Korean Pb workers, $N = 647$	Measured PbB, patellar Pb, genotyped ALAD variants, and renal function: BUN, creatinine, creatinine clearance, and RBP	Limited effect of observed ALAD 1-2 variant on calculated creatinine clearance	Weaver et al. (2006)

8.6.2 Genetic Hemopathic Disorders and Lead–Blood Interactions

The role of genetic variability in other hematological factors affecting lead biokinetics or expression of lead toxicity is less well known or in any event appears less well accepted. Sickle cell anemia among individuals of African ancestry has been claimed to be a risk factor that may affect lead's toxic effects in African-American children based on reports dating mainly to the 1970s. Sickle cell anemia is one of several genetically based hemoglobinopathies, along with the thalassemic (alpha, beta) syndromes and glucose-6-phosphate dehydrogenase (G6PD) deficiency. As with sickle cell disease, sporadic reports have appeared on the topic of lead and these latter two disorders. These disorders and their potential relationships to lead are discussed together.

Sickle cell disease describes several genetically determined disorders of which the principal one is sickle cell anemia, where individuals are homozygotic (HbSS or HbCS) for defective hemoglobin production in the beta chain (HbS). In the United States, about 50,000 African-Americans in the late

1980s were reported to have sickle cell disease (Galloway and Harwood-Nuss, 1988). Those with sickle cell trait, having inherited only one gene for the defective hemoglobin variant (HbAS), number about 8% of African-Americans (Galloway and Harwood-Nuss, 1988). The G6PD gene has about 300 variants in its base sequence and some of the variants are linked to lack of enzymatic activity. The disorder occurs in certain population groups, including residents of Sardinia, Italy, or those of Sardinian ancestry, and in African-Americans. In a more common form of the thalassemias, beta-thalassemia, a single nucleotide substitution or deletion on the beta-globin gene produces a net excess of alpha-globin chains for which no pairing beta chains exist. Homozygotic individuals, such as those with Cooley's anemia, have hemoglobin A reduced to very low levels.

A potential relationship of lead to these genetic disorders springs in theory from several clinical or epidemiological realities that were recognized early. Clinically, adverse interactions between the indicated hemoglobinopathies and the widely recognized hematotoxicant, lead, would create the possibility for one potentiating the toxicity of the other in the blood-forming system as well as in other organs and systems. For example, lead reduces both heme and globin production, produces erythrocyte hemolysis, and produces a long-recognized anemia in children and lead workers. Sickle cell anemia is a chronic hemolytic anemia. The epidemiological concern traces to the fact that significant fractions of individuals within various populations having the genetic hemoglobinopathies also typically had and continue to have concurrent elevated lead exposures. This is especially true for low-income, inner-city, African-American children. Alterations in lead biokinetics in the blood compartment can arise, given that a number of these hemoglobinopathies entail disorders such as chronic erythrocyte hemolysis. Reduced cell stability would then affect erythrocyte to plasma lead distribution and subsequent lead removal rates from the circulating lead compartment.

Does the coexistence of elevated lead body burden and these hemoglobinopathies alter lead biokinetics with reference to PbB changes? The available information is mixed and inconclusive. The largest study of children in whom this relationship would likely be present, urban African-Americans in older American cities, was that of Adebonojo (1974). A group of 1,559 Black children residing in Philadelphia in 1972–1973 was tested with regard to blood lead concentrations in children 6 months to 18 years of age, along with the frequency of sickle cell trait or the actual disease, and the presence or absence of G6PD deficiency. The author's data included tabulated blood lead mean levels in various childhood age bands (Adebonojo, 1974, Table 5), as a function of sickle cell trait or the disease and the presence or absence of G6PD deficiency. He noted that the presence or absence of disease (HbSS) or sickle cell trait (HbAS) did not appear to affect mean blood lead levels, nor did the presence or absence of G6PD deficiency. However, the author's use of mean values for all children within an age band, with its associated high standard deviation, would have obscured any relationship at more elevated PbB values. Stratification of PbB by deciles and statistical testing would have been more desirable and more likely to reveal inherent differences.

The findings of Adebonojo showing no relationship between PbB and hematological disorders are at odds with the inverse correlation findings of Adams et al. (1974) in HbA children and those of McIntire and Angle (1972) for G6PD deficiency. However, these latter studies were quite small compared to the study population of Adebonojo.

8.6.3 Genetic Polymorphism in the Vitamin D Receptor and Pb Toxicokinetics

Reports by Schwartz et al. confined to a South Korean lead workers' cohort have appeared on the genetic variability in the vitamin D receptor (VDR), its link to Pb-exposure biomarkers, and its modification of Pb–toxic response relationships. The VDR is the receptor controlling calcium absorption and bone mineralization (Onalaja and Claudio, 2000). Its common allele is VDR-bb and its genotypic variant is bB and much less commonly, VDR BB. Given the multiple forms of Pb–Ca interactions in humans, the potential role of VDR genetic polymorphism is of considerable interest.

Schwartz et al. (2000) noted that the VDR allele in these Korean Pb workers was linked to higher PbB, tibial Pb, and chelatable Pb. Lee et al. (2001) showed that there was no clear effect modification of the VDR genotype on dose–responses for Pb and heme disturbances, while they did demonstrate that workers with the Bb allele (Bb or BB) showed higher systolic and diastolic blood pressure (DBP) and frequency of hypertension than did workers with the common variant, VDR-bb. Patellar Pb showed a positive association with the VDR variant, VDR-B, at 25% higher ($p < 0.05$), and the VDR-B allele modified the association of patellar Pb with age in these workers. Weaver et al. (2003, 2005, 2006) evaluated the relationship of biomarkers of Pb exposures and measures of kidney function with the VDR allelic variant in these Korean Pb workers. They noted no effect of VDR genotype (Weaver et al., 2003), while higher tibia Pb, but not PbB or chelatable Pb, was associated with higher uric acid in those with the VDR-B genotype (Weaver et al., 2005). Higher Pb dose via PbB and chelatable Pb, but not patellar Pb, was linked to worsened kidney function in subjects with the VDR-B allelic variant (Weaver et al., 2006).

8.7 TOXICOKINETICS AND BIOMARKERS OF HUMAN Pb EXPOSURE

Lead toxicokinetics provides the underpinning for using biological markers of human Pb exposures (Pb-exposure biomarkers). Biomarkers of Pb exposure are the dose/exposure metrics for quantifying the amounts of Pb in biological compartments that are linked to one or more toxicological endpoints. Different Pb-exposure biomarkers are available for different Pb-exposure characterizations. In particular, we are concerned with biomarkers that capture different time intervals: for exposures which are remote in time or that provide endogenous Pb exposures.

Three criteria are applied to choices of Pb-exposure biomarkers in human populations: (1) ease and reliability of measurement of Pb in complex biological matrices; (2) relative reliability and stability of the measurement as the "dose" (exposure metric) portion of dose–toxic response relationships for Pb in humans (including the quantitative indexing of the Pb-exposure range and temporal reach of Pb exposures); and (3) the relative acceptance of Pb-exposure biomarkers in medical diagnostic, environmental, epidemiological, and forensic scientific settings. Biomarkers discussed in this section include: (1) Pb in whole blood, PbB; (2) Pb in plasma, PbP; (3) Pb in bone, PbBO; and (4) Pb in urine, PbU. Pb in hair is not discussed in any detail owing to the many methodological problems associated with hair sample collection and analysis. Related external/internal mixed biological media such as nails and sweat are similarly not given much attention.

These criteria are obviously interrelated. An excellent choice of a Pb-exposure biomarker on diagnostic or predictive grounds is only acceptable if it can be broadly applied and quantified in the biological medium of interest. Methodology for sampling and measurement must be sensitive, specific, and have steps that minimize random or systematic errors, such as random or systematic Pb contamination of samples. Exposure measurements which can be quantified must also be linked to dose–toxic response relationships and have a wide level of acceptance in terms of diagnostic, epidemiological, or forensic applicability.

The biomarker PbB meets two criteria completely and one criterion mostly. Certified laboratories, with reference to external proficiency and overall performance evaluations for both pediatric and occupational Pb-exposure determinations, are presently able to quantify Pb in validly collected samples. Blood samples using appropriate protocols can be reliably collected, preserved, and provided for laboratory testing. Values of PbB are widely accepted in clinical, epidemiological, and forensic settings as evidence of lead exposures. A PbB measurement is also widely assumed to be a useful if not always total surrogate for Pb toxicity to target organs and tissues in exposed human and experimental animal populations. Other biomarkers of Pb exposure, listed above, can be similarly ranked using such criteria.

8.7.1 Lead in Whole Blood

The measurement and/or diagnostic utility of PbB has been discussed at length in both expert consensus documents (NAS/NRC, 1993; U.S. CDC,

1985, 1991, 2005; U.S. EPA, 1986, 2006; WHO, 1995) and individual reviews (Baloh, 1974; Barbosa et al., 2005; Mushak, 1989, 1992, 1993, 1998, 2003).

Measurement of PbB in Human Populations

Blood sampling and laboratory measurement methods for PbB were first given systematic, expert consensus evaluation in Chapter 9 of the U.S. EPA's (1986) Pb criteria document. Methodologies first described there were both expanded upon and further assessed in the NAS/NRC's (1993) expert committee report on Pb exposures in sensitive human populations. The most recent update on the topic is in the U.S. EPA's 2006 update of the 1986 four-volume compendium (U.S. EPA, 2006).

Whole blood is collected either through venous blood samples (more reliable in some cases) from decubital vein puncture with rigorous skin decontamination of any surface Pb through several methods, or by capillary blood (finger stick) collection. Samples go into heparinized, low-lead blood tubes with care taken to assure adequate mixing and avoid microclotting. The latter complication will confound the accuracy of the laboratory measurement. Collections of capillary blood samples are often done in screening exposure populations such as young urban children. Confining capillary blood collection using microtubes and small collection volumes to experienced phlebotomists or otherwise competent hands will minimize the likelihood of either contamination by Pb or underestimation of PbB level by collection of too much tissue serum from fingertip pressure. A common practice, and one recommended by the U.S. CDC (1991), is to confirm an initial capillary positive result with a venous puncture sample. Schlenker et al. (1994), subsequent to the 1991 CDC recommendations, showed in a large pediatric comparison study funded by CDC that in competent hands, capillary sample results are sufficiently accurate to be used interchangeably with venous puncture sample results.

Various laboratory techniques can be used to quantify Pb content in whole blood, typically reported in units of μg Pb/dl blood, μmol/l, or μg/l. Quantitative methods for analysis of Pb biomarkers divide themselves into definitive and reference methods. These two method classes were discussed earlier for environmental measurements (Chapter 6). In expert consensus evaluations of PbB methods, the basic differences of these two methods were first discussed in the 1986 EPA Pb criteria document and were given further scrutiny in the 1993 NAS/NRC report.

A definitive method is one which establishes the accuracy and precision for all other methodologies. It is also operationally defined as one where all principal variables are connected by hard evidence to the absolute mass of the substance (Pb in this case) with an extremely high level of confidence. Given the cost, extreme demands on laboratory performance in terms of steps and freedom from Pb contamination, definitive procedures are confined to such uses as certification of reference materials of the U.S. NIST, formerly the National Bureau of Standards, or other international technical bodies. Reference methods are those validated by a definitive method providing accuracy through consensus evaluation of performance sufficient for use in certified or research laboratories.

The definitive method for Pb quantification in biological media, and notably whole blood, is IDMS. IDMS' accuracy traces to the fact all analytical manipulations are on a weight basis and entail simple procedures (NAS/NRC, 1993; U.S. EPA, 1986). There are essentially two reference methods, both validated with IDMS and in widespread use for routine measurements in environmental and occupational epidemiology and clinical applications (NAS/NRC, 1993; U.S. EPA, 2006). These are a spectroscopic method, graphite-furnace atomic absorption spectrometry (GF-AAS), and an electrochemical approach, ASV. Both ASV and GF-AAS demonstrate the requisite "track record" in terms of accuracy, precision, time requirements, and cost-effectiveness for routine but reliable methodology (Flegal and Smith, 1995; NAS/NRC, 1993). ICP-MS can also be viewed as a reference method for those analytical settings where costs for operator expertise, instrumentation, and procedures are not critical, such as academic research laboratories.

Method detection limit (MDL) and practical quantitation limit (PQL) for current reference methods are adequate for typical screening or survey (epidemiological) and medical diagnostic purposes.

Interpretive Characteristics of PbB Values for Dose–Response Relationships

An analytically valid PbB measurement is typically taken to be a reflection of active, ongoing Pb exposure, e.g., circulating Pb being delivered to target tissues and posing toxicological risk in real time, rather than depicting total Pb body burden, i.e., the overall mass of Pb within the bodies of exposed individuals, regardless of whether that body burden is toxicologically active in real time or not. However, the source of the "circulating" Pb in blood will vary among exposed human populations in ways requiring knowledge of Pb-exposure histories for accurate and useful interpretations. Lead workers whose current elevated Pb exposures are high enough to mainly reflect and closely correlate with actual ongoing exposures will have PbB values that are simultaneously poorly correlated with cortical and trabecular bone Pb stores, even though such bone Pb burdens are the principal measure of Pb "body burden." The situation, furthermore, reverses and becomes more complex with retirement or other causes for removal of the workers from active, ongoing external Pb contact. A number of studies show that both PbB and early toxicity biomarkers, e.g., EP elevation, remain elevated and above referent (control) levels and are principally derived from historical Pb accumulation in bone (Alessio et al., 1976; Christoffersson et al., 1984; Nilsson et al., 1991; Skerfving et al., 1988). In these cases, bone Pb serves to provide both circulating active Pb exposure (persistent elevated PbB) and the main source of Pb body burden.

The utility of PbB as an exposure biomarker requires an understanding of its constraints to wide use.

A single PbB measurement taken in isolation in terms of relative interpretive stability is problematic. First, a number of factors can affect the extent to which a single measurement is a surrogate for serial measurements or measurements for more than one individual: time of year, age of tested subjects, relationship of extent of external Pb contact to PbB measurement, etc. The hazards of such single measurements have been discussed (Mushak, 1998; U.S. EPA, 2006). Refer to Figure 4.6, U.S. EPA (2006), for a graphic illustration of this. Serial PbB measurements have been employed in a number of epidemiological prospective studies of developmental Pb toxicity from around the world and associated dose–response relationships are detailed in a later chapter. An important utility of serial PbB measurements on a group basis is determining the most statistically robust measure of the PbB dose metric: peak value, lifetime average value, concurrent value, etc. For example, it has been shown that in grouped PbB data for children followed over time, PbB values over time show rank ordering, such that some initial PbB ranking is preserved (Otto et al., 1985). There are several reasons for this. First, one can argue that children in a relatively invariant Pb contamination setting continue to be exposed externally in relation to original exposures. Second, bone Pb values acquired in proportion to exposures at an earlier age may produce resorbed Pb releases to blood in proportion to their original sequestration. Or, a combination of the two may operate.

Lanphear et al. (2005) evaluated various expressions of Pb exposure and reported a relatively higher association of concurrent PbB (at 5–7 years of age) with children's IQ decrements compared to other measures, e.g., lifetime average and peak values.

Recently, Hornung et al. (2009) noted that this closer link of concurrent PbB with IQ decrement is not attributable to correlated serial PbB measurements. This conclusion was based on a statistical approach where ratios of children's PbB at 6 years versus 2 years were factored into the regression models.

Such epidemiological studies entail cohorts with moderately elevated PbB with no history of chelation treatment or undefined earlier exposures. Children sustaining excessive exposures above medical guidelines during such serial studies would be removed. By contrast, high-risk children presenting with PbB levels requiring that same level of intervention appear to show that maximum or peak Pb is the exposure metric with the closest association with toxic effects (Chisolm, 1982; Chisolm and Harrison, 1956).

The above discussions make it clear that the relative robustness of PbB metrics versus toxic outcomes depends on the exposure setting and the exposure populations.

PbB has the diagnostic virtue of responding relatively rapidly to increases in Pb exposures, particularly in the case of childhood exposure where such rapid biokinetic response is crucial to medical detection and intervention. Depending to some extent on amounts of increased exposures, responses can occur in hours or several days (see earlier discussion).

8.7.2 Lead in Plasma—PbP

The measurement and/or diagnostic utility of PbP has been discussed at length in both expert consensus documents (NAS/NRC, 1993; U.S. EPA, 1986, 2006; WHO, 1995) and individual reviews (Barbosa et al., 2005; Mushak, 1989, 1992, 1993, 1998, 2003).

PbP is the pathway by which Pb that is absorbed from receiving compartments or is resorbed from bone stores is transferred to target tissues or is excreted. Compared to the history of PbB as a widely accepted biomarker of Pb exposure, consensus acceptance of PbP as a potentially useful biomarker is relatively recent (NAS/NRC, 1993; U.S. EPA, 1986). There are two reasons for this. First, analytical methodological problems for measuring PbP on a routine basis are daunting. Second, while PbP is qualitatively assumed to be the more direct biomarker of dose in dose–toxic response relationships than PbB, quantitative expressions of these relationships are largely lacking. That is, what level of PbP is linked to what toxic expression(s) in dose–toxic response relationships?

Measurement of PbP in Human Populations

Whether PbP as a Pb-exposure biomarker will ever enjoy routine use in conventional laboratories, even those certified in external proficiency programs, is unclear. It is entirely likely that such measurements will only be done in special or research laboratories. Plasma Pb content is ~0.5–1.0% of that measured in whole blood, the remainder sequestered in the erythrocyte. In fact, this fractional figure is relatively recent, having first been reported by Everson and Patterson (1980) using ultraclean room techniques and high accuracy and precision measurement methodology. Earlier literature had indicated higher PbP values. A PbB value of 20 μg/dl in children, the risk guideline employed by the U.S. CDC for defining a threshold for medical concern, translates to a PbP of 0.2 or 0.3 μg/dl plasma content. Such values are at the edge of reliable quantitation in many laboratories, using flameless AAS or ASV. At a minimum, the more complex and expensive methodology of ICP-MS would be required (Barbosa et al., 2005).

Noted earlier, contamination of plasma by either erythrocyte or external Pb would be very problematic with many plasma measurements, especially at lower exposures now encountered by human risk populations. Smith et al. (2002) included a measurement of hemoglobin content of plasma as one means of adjusting for this artifact in their PbP measurement protocol. External contamination is also very troublesome at the sample collection, laboratory manipulation, and measurement steps. Conventional field collection protocols and laboratory procedures are of limited use for PbP measurement. Precautions include use of extremely low Pb content reagents and sampling with special low-Pb blood tubes. NAS/NRC (1993) and Mushak (1998) have illustrated the huge impacts on accuracy of PbB measurement due to these artifacts.

Interpretive Characteristics of PbP Values for Dose–Response Relationships

PbP is a relatively rapid reflection of Pb uptake and distribution toxicokinetics in human populations (NAS/NRC, 1993; U.S. EPA, 2006) and is the *in vivo* medium by which Pb is excreted to urine through glomerular filtration in humans. This behavior in terms of rapid exchange of Pb with target tissues and PbP makes the latter a more temporally sensitive biomarker for toxicokinetics and toxicodynamics. Little has evolved in the more current toxicological literature on Pb to quantify dose–response relationships using PbP as the dose metric beyond attempts at elucidating the exposure marker trio of PbB, PbP, and Pb in bone.

Two factors will likely continue to make PbP problematic as the dose metric in toxicity relationships. One, discussed already, is methodological. The second, in cases where PbP measurements are actually doable with acceptable accuracy and precision, has to do with the level of variability and uncertainty attending a given PbP measurement. A single PbP measurement in isolation is fraught with problems relative to serial PbP testings for many of the same reasons presented earlier for PbB single-shot results. Variability and uncertainty, in turn, would arise from the fact that PbP, like PbB, reflects both exogenous and endogenous contributions, but quantitation of the endogenous, i.e., bone Pb resorption to plasma, component would be difficult.

PbP appears to be variable across individuals who otherwise have quite similar PbB measurements. Smith et al. (2002) noted that for a given PbB, PbP can vary among subjects by two- to fourfold. However, this variability is greatly outweighed by the very robust curvilinear relationship between PbP and PbB data pairs, a strong relationship that persists across a broad range of PbB values rather than only within segments of the total data set (Bergdahl et al., 1999; DeSilva, 1981; Manton and Cook, 1984; Smith et al., 2002).

Smith et al. (2002) made the observation that over the low end of PbB values in their Mexico City study subjects (total $N = 63$), plotting percentage PbP versus PbB showed a significant negative relationship ($p < 0.001$), such that % PbP decreased in the range 2–10 μg/dl PbB. It is of interest to consider that the different relationships of the two variables, PbB ≤ 10 μg/dl and PbB ≥ 10 μg/dl, may relate to dose–toxic response relationships. That is, relative availability of PbP at sub-10 μg/dl levels of PbB may relate to the findings in the international pooled analysis of Lanphear et al. (2005) showing that the slope of the IQ loss versus PbB relationship is much steeper <10 μg/dl than at higher values.

Hernandez-Avila et al. (1998) reported that plasma Pb, expressed as the ratio of PbP/PbB is more strongly correlated with bone Pb levels than is PbB. It is not clear that this indicates preferential transfer of bone Pb to plasma, since Gulson et al. (2000) found no preferential release of bone Pb to putative PbP, using urinary Pb as a surrogate for PbP. In any event, given the critical role of bone Pb as a huge reservoir of endogenous Pb for resorptive release during developmental milestones or stress, it is imperative to have the best biomarker of resorbed Pb for dose–toxic effect relationships.

8.7.3 Lead in Bone—PbBO

The measurement and/or diagnostic utility of Pb in bone and bone subcompartments has been discussed at length in both expert consensus documents (NAS/NRC, 1993; U.S. ATSDR, 2007; U.S. CDC, 1991; U.S. EPA, 1986, 2006; WHO, 1995) and individual reviews (Barbosa et al., 2005; Hu et al., 1998, 2007; Mushak, 1989, 1992, 1993, 1998, 2003; Todd and Chettle, 1994).

Measurement of PbBO in Human Populations

Measuring Pb in bone compartments shares some characteristics with measuring PbP. For one thing, it is arguable whether PbBO as a Pb-exposure biomarker will enjoy routine measurement and use in conventional laboratories, even those certified in various proficiency programs. Like PbP, it is entirely likely that such measurements will continue to be done in special or research laboratories. It is more likely, however, that Pb measurements in bone would enjoy more use in population screenings than in any individual case management or patient diagnostic scenario. The same would apply in forensic settings.

The nature of bone as a functional biological matrix in individuals and the division of human bone types with their accumulating Pb contents into cortical/dense and trabecular/spongy anatomical categories obviously affect both the methodology of bone Pb measurement and interpretation of those results. The earliest measurements of bone Pb were postmortem, autopsy samples gathered for specific study protocols, and for purely toxicokinetic elucidation and clarification purposes. As noted already, occupational Pb exposures and lead accumulation in young children and Pb release at later ages were of principal concern.

In vivo measurement of bone Pb was obviously compelled by the nature of the methods for measurement and its interpretive purposes. Biopsy samples analyzed by accepted reference methods, e.g., AAS, while of some use in validating other and much less invasive approaches, were considered limited in widespread use early on and were/are confined to validating purposes and only in adult populations.

Evolution of the current methodology for *in vivo* testings, i.e., K X-ray fluorescence (K-XRF) spectrometry, is of relatively recent vintage, with much of the developmental and standardization work done over the past several decades. Early in method development, both the K-line and L-line XRF techniques enjoyed considerable interest and their relative technical and instrumentation aspects have been compared (NAS/NRC, 1993). The L-line XRF variation has largely been supplanted by K-line excitation methodologies. In practice, the required K-line excitation of Pb in bone matrix for subsequent fluorescence has been achieved by use of 88 keV gamma rays from a cadmium radioisotope, ^{109}Cd, with appropriate detectors of given configurations and geometries (Chettle et al., 2003; Somervaille et al., 1989; Todd and Chettle, 2003; Todd et al., 2000). As irradiation sites, the preference has been mid-tibia for the cortical bone Pb content and, for spongy (trabecular) bone, the patella. Patella versus calcaneus measurements (representing spongy bone) favor the patellar site for various reasons.

There have been two major areas of evaluation and interpretive interest for bone Pb quantitation in various human populations. The first has to do with calibration, sensitivity, and other method-specific matters. The second has to do with parameters rooted in the presence of Pb in bone and characteristics of its distribution.

K-XRF methods are typically calibrated using phantoms of known Pb level comprised of plaster of Paris or similar mineral material (Chettle et al., 2003; Todd and Chettle, 2003). Todd et al. (2002) observed that such simulations of bone matrix provide potentially erroneous calibrations since they are not true bone matrix. The former is mainly calcium sulfate hemihydrate while bone mineralogy is a heterogeneous matrix of hydroxyapatite (basic calcium orthophosphate, among other formulations).

A second matter that was of some concern was the question of what expression of detection limits to use. This was central to applying K-XRF methodology to bone Pb burdens in human populations having lower, environmentally derived Pb levels, versus those in workplace settings. As originally developed in occupational health settings, lead worker XRF bone Pb measurements expressed MDLs in the conventional manner, as multiples of the standard deviation of background, twice the median error, etc. This approach has given way to one in which detection limits are probabilistic and not point estimate in statistical nature; that is, all measurements—negative and positive—about a central estimate are employed. This approach had the effect of reducing detection limits below those estimated conventionally (Kim et al., 1995).

How reliable are K-XRF measurements for bone Pb with respect to intra-subject uncertainty and variability, intersite-within-subject variability, etc.? This has been addressed in the aggregate by various investigators (Chettle et al., 2003; Todd, 2000; Todd and Chettle, 2003; Todd et al., 2003). Uppermost were parameters such as subject placement during site irradiation, the influence of overlying skin/tissue, irradiation time, stability of measurement in the near term, etc.

Interpretive Characteristics of PbBO Values for Dose–Response Relationships

The available evidence for PbBO as a dose metric in Pb toxicology and epidemiology documents that it: (1) is a cumulative but not inert dose/exposure metric which both serves to quantify stored Pb and, in metabolically diverse settings, released Pb as an endogenous source of systemic Pb exposure; (2) may be a better Pb exposure correlate with various toxic effects than PbB in settings where resorptive releases are significant or long term, e.g., in retired Pb workers or in children, older children, or adults whose Pb-exposure histories foretell sizeable stores of bone Pb; and (3) is likely a measure of the source of much of the Pb that is released on provocative chelation and that better reflects potential toxicity risk than PbB. However, the relative robustness of PbBO in dose–response relationships versus other measures of toxic dose, e.g., PbB or chelatable Pb, will vary case to case.

It is arguable whether PbBO, in contrast to PbB measurements, will be routinely used as a Pb-exposure biomarker in even certified, competent laboratories. It is, rather, likely that such measurements will be done in special or research laboratories. We are particularly referring to limited use in medical or clinical settings where bone Pb would help inform a diagnosis. It should certainly be the case that one could survey or screen groups of individuals with a known or suspected history of relatively high or even moderate Pb exposures earlier in life (Stokes et al. 1998).

As noted earlier, the skeletal system in human populations contains the main body store of Pb. Also discussed were the quantitative characteristics of this repository and its toxicological implications. The accumulation of Pb in bone with age would predict that with aging, the role of PbBO in cortical or trabecular bone would increase in importance, particularly in the onset of menopause and osteoporosis in women. Also noted were the studies indicating that even in childhood some sizeable fraction of total PbB appears to derive from releases of PbBO. Also of concern are findings that PbB values at steady state, in later childhood and in adulthood, become increasingly insensitive statistically to the potential for bone Pb accumulations to produce endogenous toxic Pb exposures.

Bone Pb shares one feature with PbP that currently limits its routine use as a Pb-exposure biomarker. That is, we presently have no widely accepted schemata for stable dose–toxic response relationships using PbBO measurement as dose and significant toxicity endpoints as response. Recent studies have attempted to close that gap. These studies are discussed in later chapters dealing with Pb toxicity and environmental epidemiology. While we can recognize the overall utility of PbBO versus health endpoints that exist in populations such as the elderly with earlier higher environmental Pb exposures (Hu et al., 1998, 2007) and retired Pb workers with persisting elevated PbB levels from bone Pb resorption (Alessio et al., 1976), it is still difficult to say that some tibial or patellar Pb concentration is quantitatively associated with some toxic endpoint in a significant fraction of an affected human population. Bone Pb is not always the better dose metric in certain cases. Schwartz et al. (2001) reported that, among a large cohort of South Korean Pb workers ($N = 803$), PbB was a better predictor of effects on neurobehavioral performance, as in executive abilities, manual dexterity, and peripheral motor strength, than either tibial or chelatable Pb. In a cohort subset, Weaver et al. (2003) found PbB but not patellar Pb associated with systolic blood pressure (SBP), with neither measure linked to diastolic pressure or hypertension status.

The Schwartz et al. (2001) and Weaver et al. (2003) results may simply confirm earlier observations. That is, with active workplace Pb exposure, PbB is better than other exposure measures (see earlier sections). Such a complication would not apply to nonoccupational exposures of general populations.

Shih et al. (2007) reviewed studies that employed both bone Pb and PbB versus cognitive function in adult subjects. Such studies were consistent with the relationship of Pb-exposure biomarkers to other effects, such as those on the hematopoietic system, in that retired Pb workers showed cognitive effects more highly linked to PbBO than PbB while current workers with ongoing exposures showed the closer association with the nearer-term measure, PbB.

8.7.4 Lead in Urine—PbU

The measurement and/or diagnostic utility of Pb in urine generated either spontaneously or as plumburesis in response to chelation have been discussed at length in both expert consensus documents (NAS/NRC, 1993; U.S. ATSDR, 1999; U.S. CDC, 1991; U.S. EPA, 1986, 2006) and individual reviews (Barbosa et al., 2005; Diamond, 1988; Mushak, 1989, 1992, 1993, 1998, 2003).

PbU as a measure of spontaneous, i.e., unstimulated, Pb excretion has not typically been a Pb-exposure biomarker with wide laboratory or diagnostic popularity for various reasons. First, early efforts at analytical measurement were plagued by various technical problems of matrix complexity, sensitivity, and specificity along with questions about reliable sample collection protocols and the most reliable way to index a PbU measurement: 24-hour collections in outpatients, 24-hour collections adjusted for specific gravity, PbU level/liter urine, and PbU excretion rate per unit mass of creatinine. PbU as an adjunct measure in chelation therapy or provocative Pb chelation for evaluating the toxicologically active fraction has been less of a problem, inasmuch as the amounts of PbU were much higher than encountered in spontaneous excretions.

With regard to diagnostic utility, early studies of PbU in childhood Pb poisoning populations screened for toxicity frequencies or for medical diagnostic purposes provided mixed results in terms of false-positive and false-negative rates (U.S. EPA, 1977). Testing for plumburesis in chelation protocols for both Pb poisoned/exposed children and lead workers had wider acceptability (Piomelli et al., 1984; U.S. CDC, 1985, 1991).

Measurement of PbU in Human Populations

A long-standing question with PbU analyses was what type of sample collection best reflected ongoing Pb exposures. Hospitalized children or even adults being treated via chelation therapy require collection of total plumburesis samples over various time frames to determine excretion rates as a function of administered chelant (U.S. CDC, 1985, 1991). To measure provocative chelation in children, i.e., the Pb mobilization test, the practice was to determine the "lead excretion ratio." Piomelli et al. (1984) calculated this by measuring total Pb excretion over 8 hours (Pb level × volume) divided by the amount of chelant (mg). Values >0.60 for tested children were taken to mean excessive Pb exposure. For current applications of spontaneous PbU measurement, the common sampling and normalization method is to determine excretion rate per unit mass creatinine given the route of Pb excretion via glomerular filtration. The use of Pb excretion "rate" adjusted to creatinine output versus just concentration may make the use of random (spot) versus timed collections more reliable in childhood lead exposure screening (Berger et al., 1990). A similar utility for creatinine adjustment appears to be the case with long-term workplace exposures (Hirata et al., 1995; Tsaih et al., 1999).

Instrumentation for PbU analysis includes the conventional reference methods GF-AAS and ASV, with some preference given to the former. While reference material for calibrating and validating measurements can be made available, the urinary matrix over time can change.

Interpretive Characteristics of PbU Values for Dose–Response Relationships

PbU has some interpretive features shared with other biomarkers of Pb exposure, especially PbB and PbP. Notably, single PbU measurements done in isolation do not offer helpful data on the nature of Pb exposures or their sources. Serial measurements are considerably more useful. Such an application would be more of an advantage in occupational exposure monitoring for lead workers than in monitoring short-term or even chronic exposures in the very young.

PbU has the significant virtue of closely reflecting glomerular filtration of plasma Pb loadings over various time frames. The review of Diamond (1988) can be consulted for quantitative linkages. The use of PbU as a surrogate for PbP was actually employed in the study of Gulson et al. (2000), who used stable Pb isotopic ratio methods to evaluate any selective partitioning of resorbed bone Pb into plasma.

8.7.5 Pb in Miscellaneous Biological Materials from Human Populations

Among the lesser-used biological media studied as Pb-exposure markers are those with characteristics such as noninvasive accessibility that would appear to offer a simpler, inexpensive alternative. Mushak (1992) described the relative advantages and limitations of hair sampling as a Pb-exposure biomarker. Hair collection is noninvasive and offers an indefinitely stable medium. But hair is also a very problematic matrix for external surface contamination. Of most concern is the extreme difficulty of chemically debriding hair surfaces of any external Pb without simultaneously removing Pb from internal deposits in hair. Despite claims of effective contamination removal protocols, it remains to be established that the current state of the measurement art encourages hair Pb analyses (Barbosa et al., 2005; U.S. EPA, 2006).

There are intrinsic problems with hair Pb in terms of understanding Pb toxicokinetics in hair or other keratinizing tissue. Hair Pb loadings appear to vary with hair color, highly variable deposition rates, etc. (Barbosa et al., 2005; Mushak, 1992; U.S. ATSDR, 2007; U.S. EPA, 2006). Of particular concern, Pb-hair is poorly correlated with PbB, especially at low Pb levels currently encountered by nonoccupational human populations (Campbell and Toribara, 2001). The latter is to be expected, given what each biomarker is reflecting in time scales of exposures.

Overall, problems of surface contamination, poorly understood hair Pb toxicokinetics, and interrelationships of hair Pb with other biomarkers would argue against widespread use of this indicator in the environmental epidemiology of Pb.

REFERENCES

Adams III, J.G., DeSimone, J., Orbach, H., Blanksma, L., Westclitz, M., Heller, P., 1974. The effect of lead on the proportion of hemoglobin S in individuals with sickle cell trait. Blood 44, 944.

Adebonojo, F.O., 1974. Hematological status of urban black children in Philadelphia. Clin. Pediatr. 13, 874–888.

Alessio, L., Bertazzi, P.A., Monelli, O., Toffoletto, F., 1976. Free erythrocyte protoporphyrin as an indicator of the biological effect of lead in adult males. III. Behavior of free erythrocyte protoporphyrin in workers with past lead exposure. Int. Arch. Occup. Environ. Health 38, 77–86.

Alexander, F.W., Clayton, B.E., Delves, H.T., 1974. Mineral and trace-metal balances in children receiving normal and synthetic diets. QJM 43, 89–111.

Alexander, B.H., Checkoway, H., Costa-Mallen, P., Faustman, E.M., Woods, J.S., Kelsey, K.T., et al., 1998. Interaction of blood lead and δ-aminolevulinic acid dehydratase genotype on markers of heme synthesis and sperm production in lead smelter workers. Environ. Health Perspect. 106, 213–216.

Allcroft, R., 1950. Lead as a nutritional hazard to farm livestock. IV. Distribution of lead in the tissues of bovines after ingestion of various lead compounds. J. Comp. Pathol. 60, 190–208.

Araki, S., Ushio, K., 1982. Assessment of the body burden of chelatable lead: a model and its application to lead workers. Br. J. Ind. Med. 39, 157–160.

Araki, S., Aono, H., Fukahori, M., Tabuki, K., 1984. Behavior of lead and zinc in plasma, erythrocytes, and urine and ALAD in erythrocytes following intravenous infusion of CaEDTA in lead workers. Arch. Environ. Health 39, 363–367.

Araki, S., Sata, F., Murata, K., 1990. Adjustment for urinary flow rate: an improved approach to biological monitoring. Int. Arch. Occup. Environ. Health 62, 471–477.

Asgharian, B., Ménache, M.G., Miller, F.J., 2004. Modeling age-related particle deposition in humans. J. Aerosol. Med. 17, 213–224.

Astrin, K.H., Bishop, D.F., Wetmur, J.G., Kaul, B., Davidow, B., Desnick, R.J., 1987. δ-aminolevulinic acid dehydratase isozymes and lead toxicity. Ann. N.Y. Acad. Sci. 514, 23–29.

Audeskirk, G., 1985. Effects of lead exposure on the physiology of neurons. Prog. Neurobiol. 24, 199–231.

Aufderheide, A.C., Wittmers Jr., L.E., 1992. Selected aspects of the spatial distribution of lead in bone. Neurotoxicology 13, 809–819.

Baloh, R.W., 1974. Laboratory diagnosis of increased lead absorption. Arch. Environ. Health 28, 198–208.

Barbosa Jr., F., Tanus-Santos, J.E., Gerlach, R.F., Parsons, P.J., 2005. A critical review of biomarkers used for monitoring human exposure to lead: advantages, limitations, and future needs. Environ. Health Perspect. 113, 1669–1674.

Barltrop, D., 1969. Transfer of lead to the human foetus. In: Barltrop, D., Burland, W.L. (Eds.), Mineral Metabolism in Pediatrics. Davis, Philadelphia, PA, pp. 135–151.

Barltrop, D., Meek, F., 1975. Absorption of different lead compounds. Postgrad. Med. J. 51, 805–809.

Barltrop, D., Meek, F., 1979. Effect of particle size on lead absorption from the gut. Arch. Environ. Health 34, 280–285.

Barltrop, D., Strehlow, C.D., 1978. The absorption of lead by children. In: Kirchgessner, M. (Ed.), Trace Element Metabolism in Man and Animals—3: Proceedings of the Third International Symposium, July, 1977. Freising-Weihenstephan, West Germany, Technische Universität, München, pp. 2–334.

Barregard, L., Svalander, C., Schütz, A., Westberg, G., Sällsten, G., Blohmé, I., et al., 1999. Cadmium, mercury, and lead in kidney cortex of the general Swedish population: a study of biopsies from living kidney donors. Environ. Health Perspect. 107, 867–871.

Barry, P.S.I., 1975. A comparison of concentrations of lead in human tissues. Br. J. Ind. Med. 32, 119–139.

Barry, P.S.I., 1981. Concentrations of lead in the tissues of children. Br. J. Ind. Med. 38, 61–71.

Barton, J.C., Conrad, M.E., Harrison, L., Nuby, S., 1978a. Effects of calcium on the absorption and retention of lead. J. Lab. Clin. Med. 91, 366–376.

Barton, J.C., Conrad, M.E., Nuby, S., Harrison, L., 1978b. Effects of iron on the absorption and retention of lead. J. Lab. Clin. Med. 92, 536–547.

Benson, G.I., George, W.H.S., Litchfield, M.H., Seaborn, D.J., 1976. Biochemical changes during the initial stages of industrial lead exposure. Br. J. Ind. Med. 33, 29–35.

Bergdahl, I.A., Skerfving, S., 1997. Partition of circulating lead between plasma and red cells does not seem to be different for internal and external sources of lead. Am. J. Ind. Med. 32, 317–318.

Bergdahl, I.A., Gerhardsson, L., Schütz, A., Desnick, R.J., Wetmur, J.G., Skerfving, S., 1997a. Delta-aminolevulinic acid dehydratase polymorphism: influence on lead levels and kidney function in humans. Arch. Environ. Health 52, 91–96.

Bergdahl, I.A., Schütz, A., Gerhardsson, L., Jensen, A., Skerfving, S., 1997b. Lead concentrations in human plasma, urine and whole blood. Scand. J. Work Environ. Health 23, 359–363.

Bergdahl, I.A., Sheveleva, M., Schütz, A., Artamonova, V.G., Skerfving, S., 1998. Plasma and blood lead in humans: capacity-limited binding to δ-aminolevulinic acid dehydratase and other lead-binding components. Toxicol. Sci. 46, 247–253.

Bergdahl, I.A., Vahter, M., Counter, S.A., Schütz, A., Buchanan, L.H., Ortega, F., et al., 1999. Lead in plasma and whole blood from lead-exposed children. Environ. Res. 80, 25–33.

Berger, O.G., Gregg, D.J., Succop, P.A., 1990. Using unstimulated urinary lead excretion to assess the need for chelation in the treatment of lead poisoning. J. Pediatr. 116, 46–51.

Biagini, G., Caudarela, R., Vangelista, A., 1977. Renal morphological and functional modification in chronic lead poisoning. In: Brown, S.S. (Ed.), Clinical Chemistry and Chemical Toxicology of Metals. Elsevier/North Holland Biomedical Press, Amsterdam, The Netherlands, pp. 123–126.

Blake, K.C.H., 1976. Absorption of ^{203}Pb from gastrointestinal tract of man. Environ. Res. 11, 1–4.

Blake, K.C.H., Barbezat, G.O., Mann, M., 1983. Effect of dietary constituents on the gastrointestinal absorption of ^{203}Pb in man. Environ. Res. 30, 182–187.

Bonkowsky, H.L., 1982. Porphyrin and heme metabolism and the porphyrias. In: Zakim, D., Boyer, T.D. (Eds.), Hepatology: A Textbook of Liver Disease. Saunders, Philadelphia, PA, pp. 351–393.

Booker, D.V., Chamberlain, A.C., Newton, D., Stott, A.N.B., 1969. Uptake of radioactive lead following inhalation and injection. Br. J. Radiol. 42, 457–466.

Borella, P., Picco, P., Masellis, G., 1986. Lead content in abortion material from urban women in early pregnancy. Int. Arch. Occup. Environ. Health 57, 93–99.

Bradman, A., Eskenazi, B., Sutton, P., Athanasoulis, M., Goldman, L.R., 2001. Iron deficiency associated with higher blood lead in children living in contaminated environments. Environ. Health Perspect. 109, 1079–1084.

Bressler, J.P., Olivi, L., Cheong, J.H., Kim, Y., Bannon, D., 2004. Divalent metal transporter 1 in lead and cadmium transport. Ann. N.Y. Acad. Sci. 1012, 142–152.

Bressler, J.P., Olivi, L., Cheong, J.H., Kim, Y., Maerten, A., Bannon, D., 2007. Metal transporters in intestine and brain: their involvement in metal-associated neurotoxicities. Hum. Exp. Toxicol. 26, 221–229.

Brito, J., McNeill, F., Webber, C., Wells, S., Richard, N., Carvalhoa, M., et al., 2002. Evaluation of a novel structural model to describe the endogenous release of lead from bone. J. Environ. Monit. 4, 184–192.

Brockhaus, A., Collet, W., Dolgner, R., Engleke, R., Ewers, U., Freier, I., et al., 1988. Exposure to lead and cadmium of children living in different areas of north-west Germany: results of biological monitoring studies 1982–1986. Int. Arch. Occup. Environ. Health 60, 211–222.

Bruenger, F.W., Stevens, W., Stover, B.J., 1973. The association of ^{210}Pb with constituents of erythrocytes. Health Phys. 25, 37–42.

Cake, K.M., Bowins, R.J., Vaillancourt, C., Gordon, C.L., McNutt, R.H., Laporte, R., et al., 1996. Partitioning of circulating lead between serum and red cells is different for internal and external sources of lead. Am. J. Ind. Med. 29, 440–445.

Campbell, J.R., Toribara, T.Y., 2001. Hair-root lead to screen for lead toxicity. J. Trace Elem. Exp. Med. 14, 69–72.

Carbone, R., Laforgia, N., Crollo, E., Mautone, A., Iolascon, A., 1998. Maternal and neonatal lead exposure in southern Italy. Biol. Neonate 72, 362–366.

Carroll, K.G., Spinelli, F.R., Goyer, R.A., 1970. Electron probe microanalyzer localization of lead in kidney tissue of poisoned rats. Nature 227, 1056.

Casteel, S.W., Weis, C.P., Henningsen, G.M., Brattin, W.L., 2006. Estimation of relative bioavailability of lead in soil and soil-like materials using young swine. Environ. Health Perspect. 114, 1162–1171.

Cerklewski, F.L., 1980. Reduction in neonatal lead exposure by supplementary dietary iron during gestation and lactation in the rat. J. Nutr. 110, 1453–1457.

Chamberlain, A.C., 1983. Effect of airborne lead on blood lead. Atmos. Environ. 17, 677–692.

Chamberlain, A.C., Heard, M.J., Little, P., Newton, D., Wells, A.C., Wiffen, R.D., 1978. Investigations into lead from motor vehicles. Report No. AERE-R9198. United Kingdom Atomic Energy Authority, Harwell, UK.

Chaube, S., Swinyard, C.A., Nishimura, H., 1972. A quantitative study of human embryonic and fetal lead with considerations of maternal fetal lead gradients and the effects of lead on human reproduction. Teratology 5, 253.

Chen, C., Wang, X., Chen, D., Li, G., Ronnenberg, A., Watanabe, H., et al., 2001. Tofu consumption and blood lead levels in young Chinese adults. Am. J. Epidemiol. 153, 1206–1212.

Cheng, Y., Willett, W.C., Schwartz, J., Sparrow, W., Weiss, S., Hu, H., 1998. Relation of nutrition to bone lead and blood lead levels in middle-aged to elderly men. The Normative Age Study. Am. J. Epidemiol. 147, 1162–1174.

Chettle, D.R., Arnold, M.L., Aro, A.C.A., Fleming, D.E.B., Kondrashov, V.S., McNeill, F.E., et al., 2003. An agreed statement on calculating lead concentration and uncertainty in XRF in vivo bone lead analysis. Appl. Radiat. Isot. 58, 603–605.

Chisolm, Jr., J.J. 1981. Dose–effect relationship for lead in young children: evidence in children for interactions among lead, zinc, and iron. In: Lynam, D.T.R., Piantinada, L.G., Cole, J.F. (Eds.), Environmental Lead: Proceedings of the Second International Symposium on Environmental Lead Research, December 1978, Cincinnati, OH. Academic Press, New York, pp. 1–7.

Chisolm Jr., J.J., 1982. Management of increased lead absorption—illustrative cases. In: Chisolm Jr., J.J., O'Hara, D.M. (Eds.), Lead Absorption in Children: Management, Clinical, and Environmental Aspects. Urban and Schwarzenberg, Baltimore, MD, pp. 171–188.

Chisolm Jr., J.J., Barltrop, D., 1979. Recognition and management of children with increased lead absorption. Arch. Dis. Child. 54, 249–262.

Chisolm Jr., J.J., Harrison, H.E., 1956. Quantitative urinary coproporphyrin excretion and its relation to edathemil calcium disodium administration in children with acute lead intoxication. J. Clin. Invest. 35, 1131–1138.

Chisolm, Jr., J.J. Mellits, E.D., Barrett, M.B., 1976. Interrelationships among blood lead concentration, quantitative daily ALA-U and urinary lead output following calcium EDTA. In: Nordberg, G.F. (Ed.), Proceedings of the Third Meeting of the Subcommittee on the Toxicology of Metals, November 1974, Tokyo, Japan. Elsevier Publishing Co., Amsterdam, The Netherlands, pp. 416–433.

Chisolm Jr., J.J., Mellits, E.D., Quaskey, S.A., 1985. The relationship between the level of lead absorption in children and the age, type, and condition of housing. Environ. Res. 38, 31–45.

Christoffersson, J.O., Schütz, A., Ahlgren, L., Haeger-Aronsen, B., Mattsson, S., Skerfving, S., 1984. Lead in finger-bone analyzed in vivo in active and retired lead workers. Am. J. Ind. Med. 6, 447–457.

Chuang, H.-Y., Schwartz, J., Gonzales-Cossio, T., Lugo, M.C., Palazuelos, E., et al., 2001. Interrelations of lead levels in bone, venous blood, and umbilical cord blood with exogenous lead exposure through maternal plasma lead in peripartum women. Environ. Health Perspect. 109, 527–532.

Cools, A., Salle, H.J.A., Verberk, M.M., Zielhuis, R.L., 1976. Biochemical response of male volunteers ingesting inorganic lead for 49 days. Int. Arch. Occup. Environ. Health 38, 129–139.

Cory-Slechta, D.A., Weiss, B., Cox, C., 1987. Mobilization and redistribution of lead over the course of calcium disodium ethylene diamine tetraacetate chelation therapy. J. Pharmacol. Exp. Ther. 243, 804–813.

Cramér, K., Goyer, R.A., Jagenburg, R., Wilson, M.H., 1974. Renal ultrastructure, renal function, and parameters of lead toxicity in workers with different periods of lead exposure. Br. J. Ind. Med. 31, 113–127.

Dacre, J.C., Ter Haar, G.L., 1977. Lead levels in tissues from rats fed soils containing lead. Arch. Environ. Contam. Toxicol. 6, 111–119.

Day, J.P., Fergusson, J.E., Chee, T.M., 1979. Solubility and potential toxicity of lead in urban street dust. Bull. Environ. Contam. Toxicol. 23, 497–502.

Delves, H.T., Clayton, B.E., Carmichael, A., Bubear, M., Smith, M., 1982. An appraisal of the analytical significance of tooth-lead measurements as possible indices of environmental exposure of children to lead. Ann. Clin. Biochem. 19, 329–337.

DeSilva, P.E., 1981. Determination of lead in plasma and its relationship to lead in erythrocytes. Br. J. Ind. Med. 38, 209–217.

Diamond, G.L., 1988. Biological monitoring of urine for exposure to toxic metals. In: Clarkson, T.W., Nordberg, G., Sager, P. (Eds.), Scientific Basis and Practical Applications of Biological Monitoring of Toxic Metals. Plenum Press, New York, pp. 515–529.

Drasch, G.A., Ott, J., 1988. Lead in human bones. Investigations of an occupationally non-exposed population in southern Bavaria (F.R.G.): II. Children. Sci. Total Environ. 68, 61–69.

Drasch, G.A., Bohm, J., Bauer, C., 1987. Lead in human bones. Investigations of an occupationally non-exposed population in southern Bavaria (F.R.G.). I. Adults. Sci. Total Environ. 64, 303–315.

Duggan, M.J., Williams, S., 1977. Lead in dust in city streets. Sci. Total Environ. 7, 91–97.

Everson, J., Patterson, C.C., 1980. "Ultra-clean" isotope dilution/mass spectrometric analyses for lead in human blood plasma indicate that most reported values are artificially high. Clin. Chem. 26, 1603–1607.

Flanagan, P.R., Hamilton, D.L., Haist, J., Valberg, L.S., 1979. Inter-relationships between iron and lead absorption in iron-deficient mice. Gastroenterology 77, 1074–1081.

Flanagan, P.R., Chamberlain, M.J., Valberg, L.S., 1982. The relationship between iron and lead absorption in humans. Am. J. Clin. Nutr. 36, 823–829.

Flegal, A.R., Smith, D.R., 1995. Measurements of environmental lead contamination and human exposure. Rev. Environ. Contam. Toxicol. 143, 1–45.

Fleming, D.E., Chettle, D.R., Webber, C.E., O'Flaherty, E.J., 1999. The O'Flaherty model of lead kinetics: an evaluation using data from a lead smelter population. Toxicol. Appl. Pharmacol. 161, 100–109.

Forbes, G.B., Reina, J.C., 1972. Effect of age on gastrointestinal absorption (Fe, Sr, Pb) in the rat. J. Nutr. 102, 647–652.

Freeman, G.B., Johnson, J.D., Killinger, J.M., Liao, S.C., Feder, P.I., Davis, A.O., et al., 1992. Relative bioavailability of lead from mining waste soil in rats. Fundam. Appl. Toxicol. 19, 388–398.

Fullmer, C.S., Rosen, J.F., 1990. Effect of dietary calcium and lead status on intestinal calcium absorption. Environ. Res. 51, 91–99.

Galloway, S.J., Harwood-Nuss, A.L., 1988. Sickle-cell anemia—a review. J. Emerg. Med. 6, 213–226.

Gerhardsson, L., Attewell, R., Chettle, D.R., Englyst, V., Lundstrom, N.-G., Nordberg, G.F., et al., 1993. In vivo measurements of lead in bone in long-term exposed lead smelter workers. Arch. Environ. Health 48, 141–156.

Gerhardsson, L., Englyst, V., Lundström, N.G., Nordberg, G., Sandberg, S., Steinvall, F., 1995. Lead in tissues of deceased lead smelter workers. J. Trace Elem. Med. Biol. 9, 136–143.

Gordon, C.L., Chettle, D.R., Webber, C.T., 1993. An improved instrument for the in vivo detection of bone lead. Br. J. Ind. Med. 50, 637–641.

Gordon, C.L., Webber, C.T., Chettle, D.R., 1994. The reproducibility of ^{109}Cd-based X-ray fluorescence measurements of bone lead. Environ. Health Perspect. 102, 690–694.

Goyer, R.A., 1978. Calcium and lead interactions: some new insights. J. Lab. Clin. Med. 91, 363–365.

Grandjean, P., 1978. Regional distribution of lead in human brains. Toxicol. Lett. 2, 65–69.

Grandjean, P., Hansen, O.N., Lyngbye, K., 1986. Lead concentration in deciduous teeth: variation related to tooth type and analytical technique. J. Toxicol. Environ. Health 19, 437–445.

Graziano, J.H., Popovac, D., Factor-Litvak, P., Shrout, P., Kline, J., Murphy, M.J., et al., 1990. Determinants of elevated blood lead during pregnancy in a population surrounding a lead smelter in Kosovo, Yugoslavia. Environ. Health Perspect. 89, 95–100.

Griffin, T.B., Coulston, F., Goldberg, L., Wills, H., Russell, J.C., Knelson, J.H., 1975. Clinical studies of men continuously exposed to airborne particulate lead. In: Griffin, T.B., Knelson, J.H. (Eds.), Lead—Environmental Quality and Safety. Georg Thieme Publishers, Stuttgart, Germany, pp. 221–240.

Gu, C., Chen, S., Xu, X., Zheng, L., Li, Y., Wu, K., et al., 2009. Lead and cadmium synergistically enhance the expression of divalent metal transporter 1 protein in central nervous system of developing rats. Neurochem. Res. 34, 1150–1156.

Gulson, B.L., Wilson, D., 1994. History of lead exposure in children revealed from isotopic analyses of teeth. Arch. Environ. Health 49, 279–283.

Gulson, B.L., Mahaffey, K.R., Mizon, K.J., Korsch, M.J., Cameron, M.A., Vimpani, G., 1995. Contribution of tissue lead to blood lead in adult female subjects based on stable lead isotope methods. J. Lab. Clin. Med. 125, 703–712.

Gulson, B.L., Jameson, C.W., Mahaffey, K.R., Mizon, K.J., Korsch, M.J., Vimpani, G., 1997. Pregnancy increases mobilization of lead from maternal skeleton. J. Lab. Clin. Med. 130, 51–62.

Gulson, B.L., Mahaffey, K.R., Jameson, C.W., Mizon, K.J., Korsch, M.J., Cameron, M.A., et al., 1998. Mobilization of lead from the skeleton during the post-natal period is larger than during pregnancy. J. Lab. Clin. Med. 131, 324–339.

Gulson, B.L., Mahaffey, K.R., Jameson, C.W., Patison, N., Law, A.J., Mizon, K.J., et al., 1999. Impact of diet on lead in blood and urine in female adults and relevance to mobilization of lead from bone stores. Environ. Health Perspect. 107, 257–263.

Gulson, B.L., Mizon, K.J., Palmer, J.M., Korsch, M.J., Donnelly, J.B., 2000. Urinary excretion of lead during pregnancy and postpartum. Sci. Total Environ. 262, 49–55.

Gulson, B.L., Mizon, K.J., Korsch, M.J., Palmer, J.M., Donnelly, J.B., 2004. Blood lead changes during pregnancy and postpartum with calcium supplementation. Environ. Health Perspect. 112, 1499–1507.

Gwiazda, R., Campbell, C., Smith, D., 2005. A noninvasive isotopic approach to estimate the bone lead contribution to blood in children: implications for assessing the efficacy of lead abatement. Environ. Health Perspect. 113, 104–110.

Hansen, J.P.B., Dossing, M., Paulev, P.-E., 1981. Chelatable lead body burden (by calcium-disodium EDTA) and blood lead concentration in man. J. Occup. Med. 23, 39–43.

Hamilton, D.L., 1978. Interrelationship of lead and iron retention in iron-deficient mice. Toxicol. Appl. Pharmacol. 46, 651–661.

Harrison, R.M., 1979. Toxic metals in street and household dusts. Sci. Total Environ. 11, 89–97.

Harrison, G.E., Carr, T.E.F., Sutton, A., Humphreys, E.R., 1969. Effect of alginate on the absorption of lead in man. Nature 224, 1115–1116.

Heard, M.J., Chamberlain, A.C., 1982. Effect of minerals and food on uptake of lead from the gastrointestinal tract in humans. Hum. Toxicol. 1, 411–415.

Hernandez-Avila, M., Gonzalez-Cossio, T., Palazuelos, E., Romieu, I., Aro, A., Fishbein, E., et al., 1996. Dietary and environmental determinants of blood and bone levels in lactating postpartum women living in Mexico City. Environ. Health Perspect. 104, 1076–1082.

Hernandez-Avila, M., Smith, D., Meneses, F., Sanin, L.H., Hu, H., 1998. The influence of bone and blood lead on plasma lead levels in environmentally exposed adults. Environ. Health Perspect. 106, 473–477.

Hernandez-Avila, M., Peterson, K.E., Gonzalez-Cossio, T., Sanin, L.H., Aro, A., Schnaas, L., et al., 2002. Effect of maternal bone length on length and circumference of newborns and 1-month-old infants. Arch. Environ. Health 57, 482–488.

Hertz-Picciotto, I., Schramm, M., Watt-Morse, M., Chantala, K., Anderson, J., Osterloh, J., 2000. Patterns and determinants of blood lead during pregnancy. Am. J. Epidemiol. 152, 829–837.

Hirata, M., Yoshida, T., Miyajima, K., Kosada, H., Tabuchi, T., 1995. Correlation between lead in plasma and other indicators of lead exposure among lead exposed workers. Int. Arch. Occup. Environ. Health 68, 58–63.

Horiuchi, K., Horiguchi, S., Suekane, M., 1959. Studies on the industrial lead poisoning. 1: Absorption, transportation, deposition and excretion of lead. 6: The lead contents in organ-tissues of the normal Japanese. Osaka City Med. J. 5, 41–70.

Hornung, R.W., Lanphear, B.P., Dietrich, K.N., 2009. Age of greatest susceptibility to childhood lead exposure: A new statistical approach. Environ. Health Perspect. 117, 1309–1312.

Hryhorczuk, D.O., Rabinowitz, M.B., Hessl, S.M., Hoffman, D., Hogan, M.M., Mallin, K., et al., 1985. Elimination kinetics of blood lead in workers with chronic lead intoxication. Am. J. Ind. Med. 8, 33–42.

Hsu, F.S., Krook, L., Pond, W.G., Duncan, J.R., 1975. Interactions of dietary calcium with toxic levels of lead and zinc in pigs. J. Nutr. 105, 112–118.

Hu, H., Milder, F.L., Burger, D.E., 1990. X-ray fluorescence measurements of lead burden in subjects with low-level community lead exposure. Arch. Environ. Health 45, 335–341.

Hu, H., Aro, A., Rotnitzky, A., 1995. Bone lead measured by X-ray fluorescence: epidemiological methods. Environ. Health Perspect. 103 (Suppl. 1), 105–110.

Hu, H., Aro, A., Payton, M., Korrick, S., Sparrow, D., Weiss, S.T., et al., 1996. The relationship of bone and blood lead to hypertension. The Normative Aging Study. JAMA 275, 1171–1176.

Hu, H., Rabinowitz, M., Smith, D., 1998. Bone lead as a biological marker in epidemiological studies of chronic toxicity: conceptual paradigms. Environ. Health Perspect. 106, 1–8.

Hu, H., Shih, R., Rothenberg, S., Schwartz, B.S., 2007. The epidemiology of lead toxicity in adults: measuring dose and consideration of other methodologic issues. Environ. Health Perspect. 115, 455–462.

Illich, J.Z., Kerstetter, J.E., 2000. Nutrition in bone health revisited. A story beyond calcium. J. Am. Coll. Nutr. 19, 715–737.

Ishihara, N., Shiojima, S., Hasegawa, K., 1984. Lead and zinc concentrations in plasma, erythrocytes and urine in relation to ALA-D activity after intravenous infusion of Ca-EDTA. Br. J. Ind. Med. 41, 235–240.

James, H.M., Hilburn, M.E., Blair, J.A., 1985. Effects of meals and meal times on uptake of lead from the gastrointestinal tract of humans. Hum. Exp. Toxicol. 4, 401–407.

Johnson, N.E., Tenuta, K., 1979. Diets and lead blood levels of children who practice pica. Environ. Res. 18, 369–376.

Kapoor, S.C., Wielopowski, L., Graziano, J.H., LoIacono, N.J., 1989. Influence of 2,3-dimercaptosuccinic acid on gastrointestinal lead absorption and whole-body lead retention. Toxicol. Appl. Pharmacol. 97, 525–529.

Kayaalti, Z., Odabaşi, M., Söylemezoğlu, T., 2011. Genotype and allele frequencies of divalent metal transporter 1 polymorphism in Turkish population. Mol. Biol. Rep. 38 (4), 2679–2684.

Kemp, F.W., Neti, P.V.S.V., Howell, R.W., Wenger, P., Louria, D.B., Bogden, J.D., 2007. Elevated blood lead concentrations and vitamin D deficiency in winter and summer in young urban children. Environ. Health Perspect. 115, 630–635.

Kierski, M.W. 1992. The oral bioavailability of soil-lead in the weanling rabbit. Ph.D. Thesis. University of Minnesota, Minneapolis, MN, Diss. Abs. Int. B 53, pp. 2819–2820.

Kim, R., Aro, A., Rotnitsky, A., Amarasiriwardena, C., Hu, H., 1995. K X-ray fluorescence measurements of bone lead concentration: the analysis of low-level data. Phys. Med. Biol. 40, 1475–1485.

Korrick, S.A., Schwartz, J., Tsaih, S.-W., Hunter, D.J., Aro, A., Rosner, B., et al., 2002. Correlates of bone and blood lead levels among middle-aged and elderly women. Am. J. Epidemiol. 156, 335–343.

Kosnett, M.J., Becker, C.E., Osterloh, J.D., Kelly, T.J., Pasta, D.J., 1994. Factors influencing bone lead concentration in a suburban community assessed by noninvasive K X-ray fluorescence. JAMA 271, 197–203.

Kostial, K., Kello, D., 1979. Bioavailability of lead in rats fed "human" diets. Bull. Environ. Contam. Toxicol. 21, 312–314.

Kostial, K., Kello, D., Jugo, S., Rabar, I., Maljkovic, T., 1978. Influence of age on metal metabolism and toxicity. Environ. Health Perspect. 25, 81–86.

Kristal-Boneh, E., Froom, P., Yerushalmi, N., Harari, G., Ribak, J., 1998. Calcitropic hormones and occupational lead exposure. Am. J. Epidemiol. 147, 458–463.

Ku, Y., Alvarez, G.H., Mahaffey, K.R., 1978. Comparative effects of feeding lead acetate and phospholipid-bound lead on blood and tissue lead concentrations in young and adult rats. Bull. Environ. Contam. Toxicol. 20, 561–567.

Lanphear, B.P., Hornung, R., Khoury, J., Yolton, K., Baghurst, P., Bellinger, D.C., et al., 2005. Low-level environmental lead exposure and children's intellectual function: an international pooled analysis. Environ. Health Perspect. 113, 894–899.

Lee, B.-K., Lee, G.-S., Stewart, F.S., Ahn, K.-D., Simon, D., Kelsey, K.T., et al., 2001. Associations of blood pressure and hypertension with lead dose measures and polymorphisms in the vitamin D receptor and δ-aminolevulinic acid dehydratase genes. Environ. Health Perspect. 109, 383–389.

Lin, J.-L., Kim, R., Tsaih, S.-W., Sparrow, D., Hu, H., 2001. Lead chelation therapy and urate excretion in patients with chronic kidney diseases and gout. Kidney Int. 60, 266–271.

Lolin, Y., O'Gorman, P., 1988. An intra-erythrocytic low molecular weight lead-binding protein in acute and chronic lead exposure and its protective role in lead toxicity. Ann. Clin. Biochem. 25, 688–697.

Mackie, A.C., Stephens, R., Townsend, A., Waldron, H.A., 1977. Tooth lead levels in Birmingham children. Arch. Environ. Health 32, 178–185.

Maddaloni, M., Loiacono, M., Manton, W., Blum, C., Drexler, J., Graziano, J., 1998. Bioavailability of soil borne lead in adults by stable isotope dilution. Environ. Health Perspect. 106 (Suppl. 6), 1589–1594.

Mahaffey-Six, K., Goyer, R.A., 1972. The influence of iron deficiency on tissue content and toxicity of ingested lead in the rat. J. Lab. Clin. Med. 79, 128–136.

Mahaffey, K.R., Annest, J.L., 1986. Association of erythrocyte protoporphyrin with blood lead level and iron status in the second national health and nutrition examination survey, 1976–1980. Environ. Res. 41, 327–338.

Mahaffey, K.R., Goyer, R., Haseman, J.K., 1973. Dose-response to lead ingestion in rats fed low dietary calcium. J. Lab. Clin. Med. 82, 92–100.

Mahaffey, K.R., Treloar, S., Banks, T.A., Peacock, B.J., Parekh, L.E., 1976. Differences in dietary intake of calcium, phosphorus and iron of children having normal and elevated blood lead concentrations. J. Nutr. 106, 7.

Mahaffey, K.R., Gartside, P.S., Glueck, C.J., 1986. Blood lead levels and dietary calcium intake in 1 to 11-year old children: the Second National Health and Nutrition Examination Survey, 1976–1980. Pediatrics 78, 257–262.

Manton, W.I., 1985. Total contribution of air lead to blood lead. Br. J. Ind. Med. 42, 168–172.

Manton, W.I., Cook, J.D., 1979. Lead content of cerebrospinal fluid and other tissue in amyotrophic lateral sclerosis (ALS). Neurology 29, 611–612.

Manton, W.I., Cook, J.D., 1984. High accuracy (stable isotope dilution) measurements of lead in serum and cerebrospinal fluid. Br. J. Ind. Med. 41, 313–319.

Manton, W.I., Malloy, C.R., 1983. Distribution of lead in body fluids after ingestion of soft solder. Br. J. Ind. Med. 40, 51–57.

Manton, W.I., Angle, C.R., Stanek, K.L., Reese, Y.R., Kuehnemann, T.J., 2000. Acquisition and retention of lead by young children. Environ. Res. 82, 60–80.

Marcus, A.H., Schwartz, J., 1987. Dose-response curves for erythrocyte protoporphyrin vs. blood lead: effects of iron status. Environ. Res. 44, 221–227.

Marcus, S.M., 1982. Experience with D-penicillamine in treating lead poisoning. Vet. Hum. Toxicol. 24, 18–20.

Markowitz, M.E., Weinberger, H.L., 1990. Immobilization-related lead toxicity in previously lead-poisoned children. Pediatrics 86, 455–457.

Mayer-Popkin, O., Denkhaus, W., Konietzko, H., 1986. Lead content of fetal tissues after maternal intoxication. Arch. Toxicol. 58, 203–204.

McIntire, M.S., Angle, C.R., 1972. Air lead: relation to lead in blood of black school children deficient in glucose-6-phosphate dehydrogenase. Science 177, 520–522.

McNeill, F.E., Stokes, L., Chettle Brito, J.A., Chettle, D.R., Kaye, W.E., 2000. ^{109}Cd K X-ray fluorescence measurements of tibial lead content in young adults exposed to lead in early childhood. Occup. Environ. Med. 57, 465–471.

Mims, M.P., Prchal, J.T., 2005. Divalent metal transporter 1. Hematology 10, 339–345.

Moore, M.R., 1988. Haematological effects of lead. Sci. Total Environ. 71, 419–431.

Moore, J.F., Goyer, R.A., Wilson, M., 1973. Lead-induced inclusion bodies: solubility, amino-acid content and relationship to residual acidic nuclear proteins. Lab. Invest. 29, 484–494.

Moore, M.R., Meredith, P.A., Watson, W.S., Summer, D.J., Taylor, M.K., Goldberg, A., 1980. The percutaneous absorption of lead-203 in humans from cosmetic preparations containing lead acetate, as assessed by whole-body counting and other techniques. Food Cosmet. Toxicol. 18, 399–405.

Morrison, J.N., Quarterman, J., 1987. The relationship between iron status and lead absorption in rats. Biol. Trace Elem. Res. 14, 115–126.

Morrison, J.N., Quarterman, J., Humphries, W.R., 1977. The effect of dietary calcium and phosphate on lead poisoning in lambs. J. Comp. Pathol. 87, 417–429.

Munro, I.C., Willes, R.F., Truelove, J.F., 1975. Absorption and tissue distribution of inorganic lead in the developing infant monkey (*Macaca irus*). Toxicol. Appl. Pharmacol. 33, 128–129.

Mushak, P., 1989. Biological monitoring of lead exposure in children: overview of selected biokinetic and toxicological issues. In: Smith, M., Grant, L.D., Sors, A. (Eds.), Lead Exposure and Child Development: An International Assessment. Kluwer Academic Press, Lancaster, U.K., pp. 129–145.

Mushak, P., 1991. Gastrointestinal absorption of lead in children and adults: overview of biological and biophysico-chemical aspects. Chem. Speciation Bioavailability 3, 87–104.

Mushak, P., 1992. The monitoring of human lead exposure. In: Needleman, H.L. (Ed.), Human Lead Exposure. CRC Press, Boca Raton, FL, pp. 45–64.

Mushak, P., 1993. New directions in the toxicokinetics of human lead exposure. Neurotoxicology 14, 29–42.

Mushak, P., 1998. Uses and limits of empirical data in measuring and modeling human lead exposure. Environ. Health Perspect. 106 (Suppl. 6), 1467–1484.

Mushak, P., 2003. Lead remediation and changes in human lead exposure: some physiological and biokinetic dimensions. Sci. Total Environ. 303, 35–50.

Mushak, P., Crocetti, A.F., 1996a. Lead and nutrition. Part I. Biological interactions of lead with nutrients. Nutr. Today 31, 12–17.

Mushak, P., Crocetti, A.F., 1996b. Lead and nutrition. Part II. Some potential impacts of lead-nutrient interactions in US populations at risk. Nutr. Today 31, 115–122.

National Academy of Sciences: National Research Council, 1993. Measuring Lead Exposure in Infants, Children, and Other Sensitive Populations. National Academy Press, Washington, DC.

Needleman, H.L., Shapiro, I.M., 1974. Dentine lead levels in asymptomatic Philadelphia school children: subclinical exposure in high and low risk groups. Environ. Health Perspect. 7, 27–31.

Nilsson, U., Attewell, R., Christoffersson, J.-O., Schütz, A., Ahlgren, L., Skerfving, S., et al., 1991. Kinetics of lead in bone and blood after end of occupational exposure. Pharmacol. Toxicol. 69, 477–484.

O'Connor, M.E., Rich, D., 1999. Children with moderately elevated lead levels: is chelation with DMSA helpful? Clin. Pediatr. 38, 325–331.

O'Flaherty, E.J., 1993. Physiologically-based models for bone-seeking elements. IV. Kinetics of lead disposition in humans. Toxicol. Appl. Pharmacol. 118, 16–29.

O'Flaherty, E.J., 1995. Physiologically-based models for bone-seeking elements. V. Lead absorption and disposition in childhood. Toxicol. Appl. Pharmacol. 131, 297–308.

O'Flaherty, E.J., 1998. A physiologically based kinetic model for lead in children and adults. Environ. Health Perspect. 106 (Suppl. 6), 1495–1503.

O'Flaherty, E.J., Hammond, P.B., Lerner, S.I., 1982. Dependence of apparent blood lead half-life on the length of previous lead exposure in humans. Fundam. Appl. Toxicol. 2, 49–54.

Okazaki, H., Aronson, S.M., DiMaio, D.J., Olivera, J.E., 1963. Acute lead encephalopathy of childhood: histological and chemical studies with particular reference to angiopathic aspects. In: Yahr, M.D. (Ed.), Transactions of the American Neurological Association, vol. 88. Springer Publishing Co., New York, pp. 248–250.

Oliveira, S., Aro, A., Sparrow, D., Hu, H., 2002. Season modifies the relationship between bone and blood lead levels: the Normative Aging Study. Arch. Environ. Health 57, 466–472.

Omokhodion, F.O., Crockford, G.W., 1991. Sweat lead levels in persons with high blood lead levels: experimental elevation of blood lead by ingestion of lead chloride. Sci. Total Environ. 108, 235–242.

Onalaja, A.O., Claudio, L., 2000. Genetic susceptibility to lead poisoning. Environ. Health Perspect. 108 (Suppl.), 23–28.

Otto, D.A., Robinson, S., Baumann, S., Schroeder, S., Mushak, P., Kleinbaum, D., Barton, C., Boone, L., 1985. Five-year follow-up study of children with low-to-moderate lead absorption: electrophysiological evaluation. Environ. Res. 63, 105–113.

Phalen, R.F., Oldham, M.J., 2001. Methods for modeling particle deposition as a function of age. Respir. Physiol. 128, 119–130.

Piomelli, S., Rosen, J.F., Chisolm Jr., J.J., Graef, J.W., 1984. Management of childhood lead poisoning. J. Pediatr. 105, 523–532.

Popovic, M., McNeill, F.E., Chettle, D.R., Webber, C.E., Lee, C.V., Kaye, W.E., 2005. Impact of occupational exposure on lead levels in women. Environ. Health Perspect. 113, 478–484.

Pounds, J.G., Marlar, R.J., Allen, J.R., 1978. Metabolism of lead-210 in juvenile and adult rhesus monkeys (*Macaca mulata*). Bull. Environ. Contam. Toxicol. 19, 684–691.

Quarterman, J., Morrison, E., 1978. The effect of age on the absorption and excretion of lead. Environ. Res. 17, 78–83.

Rabinowitz, M.B., 1991. Toxicokinetics of bone lead. Environ. Health Perspect. 91, 33–37.

Rabinowitz, M.B., 1995. Relating tooth and blood lead levels in children. Bull. Environ. Contam. Toxicol. 55, 853–857.

Rabinowitz, M.B., Wetherill, G.W., Kopple, J.D., 1973. Lead metabolism in the normal human: stable isotope studies. Science 182, 725–727.

Rabinowitz, M.B., Wetherill, G., Kopple, J.D., 1976. Kinetic analysis of lead metabolism in healthy humans. J. Clin. Invest. 58, 260–270.

Rabinowitz, M.B., Kopple, J.D., Wetherill, G., 1980. Effect of food intake and fasting on gastrointestinal lead absorption in humans. Am. J. Clin. Nutr. 33, 1784–1788.

Rabinowitz, M.B., Leviton, A., Bellinger, D., 1993. Relationships between several blood lead levels and exfoliated tooth dentin lead levels: models of tooth lead kinetics. Calcif. Tissue Int. 53, 338–341.

Raghavan, S.R.V., Culver, B.D., Gonick, H.C., 1980. Erythrocyte lead-binding protein after occupational exposure. Environ. Res. 22, 264–270.

Rastogi, S.C., Clausen, J., 1976. Absorption of lead through the skin. Toxicology 6, 371–376.

Roberts, J.R., Reigert, J.R., Ebeling, M., Hulsey, T.C., 2001. Time required for blood lead levels to decline in non-chelated children. Clin. Toxicol. 39, 153–160.

Rogan, W.J., Dietrich, K.N., Ware, J.H., Dockery, D.W., Salganik, M., Radcliffe, J., et al., 2001. The effect of chelation therapy with Succimer on neuropsychological development of children exposed to lead. N. Engl. J. Med. 344, 1421–1426.

Rosen, J.F., Chesney, R.W., Hamstra, A.J., DeLuca, H.F., Mahaffey, K.R., 1980. Reduction in 1,25-dihydroxyvitamin D in children with increased lead absorption. N. Engl. J. Med. 302, 1128–1131.

Ryu, J.E., Ziegler, E.E., Fomon, S.J., 1978. Maternal lead exposure and blood lead concentration in infancy. J. Pediatr. 93, 476–478.

Saenger, P., Rosen, J.F., Markowitz, M.E., 1982. Diagnostic significance of edetate disodium calcium testing in children with increased lead absorption. Am. J. Dis. Child. 136, 312–315.

Schell, L.M., Denham, M., Stark, A.D., Ravencraft, J., Parsons, P., Schulte, E., 2004. Relationship between blood lead concentration and dietary intakes in infants from 3 to 12 months of age. Environ. Res. 96, 264–273.

Schlegel, H., Kufner, G., 1979. Long-term observation of biochemical effects of lead in human experiments. J. Clin. Chem. Clin. Biochem. 17, 225–233.

Schlenker, T.L., Fritz, C.J., Mark, D., Layde, M., Linke, G., Murphy, A., et al., 1994. Screening for pediatric lead poisoning. Comparability of simultaneously drawn capillary and venous blood samples. JAMA 271, 1346–1348.

Schroeder, H.A., Tipton, I.H., 1968. the human body burden of lead. Arch. Environ. Health 7, 965–978.

Schwartz, B.S., Lee, B.-K., Stewart, W., Ahn, K.-D., Springer, K., Kelsey, K., 1995. Associations of δ-aminolevulinic acid dehydratase genotype with plant, exposure duration, and blood lead and zinc protoporphyrin levels in Korean lead workers. Am. J. Epidemiol. 142, 738–745.

Schwartz, B.S., Lee, B.-K., Stewart, W., Sithisarankul, P., Strickland, P.T., Ahn, K.-D., et al., 1997. δ-aminolevulinic acid dehydratase genotype modifies four hour urinary lead excretion after oral administration of dimercaptosuccinic acid. Occup. Environ. Med. 54, 241–246.

Schwartz, B.S., Lee, B.-K., Lee, G.-S., Stewart, W.F., Simon, D., Kelsey, K., et al., 2000. Associations of blood lead, dimercaptosuccinic acid-chelatable lead, and tibia lead with polymorphisms in the vitamin D receptor and δ-aminolevulinic acid dehydratase genes. Environ. Health Perspect. 108, 949–954.

Schwartz, B.S., Lee, B.-K., Lee, G.-S., Stewart, W.F., Lee, S.S., Hwang, K.Y., et al., 2001. Associations of blood lead, dimercaptosuccinic acid-chelatable lead, and tibia lead with neurobehavioral test scores in South Korean lead workers. Am. J. Epidemiol. 153, 453–464.

Shih, R.A., Hu, H., Weisskopf, M.G., Schwartz, B.S., 2007. Cumulative lead dose and cognitive function in adults: a review of studies that measured both blood lead and bone lead. Environ. Health Perspect. 115, 483–492.

Simons, T.J.B., 1995. The affinity of human erythrocyte porphobilinogen synthase for Zn^{2+} and Pb^{2+}. Eur. J. Biochem. 234, 178–183.

Sithisarankul, P., Schwartz, B.S., Lee, B.-K., Kelsy, K.T., Strickland, P.T., 1997. Aminolevulinic acid dehydratase genotype mediates plasma levels of the neurotoxin, 5-aminolevulinic acid, in lead-exposed workers. Am. J. Ind. Med. 32, 15–20.

Skerfving, S., 1988. Biological monitoring of exposure to lead. In: Clarkson, T.W., Nordberg, G. F., Sager, P.R. (Eds.), Biological Monitoring of Toxic Metals: Rochester Series on Environmental Toxicity. Plenum Press, New York, pp. 169–197.

Smith, C.M., DeLuca, H.F., Tanaka, Y., Mahaffey, K.R., 1978. Stimulation of lead absorption by vitamin D administration. J. Nutr. 108, 843–847.

Smith, C.M., Wang, X., Hu, H., Kelsey, K.T., 1995a. A polymorphism in the δ-aminolevulinic acid dehydratase gene may modify the pharmacokinetics and toxicity of lead. Environ. Health Perspect. 103, 248–253.

Smith, C.M., Hu, H., Wang, X., Kelsey, K.T., 1995b. ALA-D genotype is not associated with HT or HB levels among workers exposed to low levels of lead. Med. Lav. 86, 229–235.

Smith, D., Hernandez-Avila, M., Telléz-Rojo, M.M., Mercado, A., Hu, H., 2002. The relationship between lead in plasma and whole blood in women. Environ. Health Perspect. 110, 263–268.

Sobel, A.E., Yuska, H., Peters, D.D., Kramer, B., 1940. The biochemical behavior of lead. I. Influence of calcium, phosphorus, and vitamin D on lead in blood and bone. J. Biol. Chem. 132, 239–265.

Somervaille, L.J., Nilsson, U., Chettle, D.R., Tell, I., Scott, M.C., Schütz, A., et al., 1989. *In vivo* measurements of bone lead—a comparison of two X-ray fluorescence techniques used at three different bone sites. Phys. Med. Biol. 34, 1833–1845.

Sorrell, M., Rosen, J.F., Roginsky, M., 1977. Interactions of lead, calcium, vitamin D, and nutrition in lead-burdened children. Arch. Environ. Health 32, 160–164.

Steenhout, A., Pourtois, M., 1981. Lead accumulation in teeth as a function of age with different exposures. Br. J. Ind. Med. 38, 297–303.

Stokes, L., Letz, R., Gerr, F., Kolczak, M., McNeill, F.E., Chettle, D.R., et al., 1998. Neurotoxicity in young adults 20 years after childhood exposure to lead: the Bunker Hill experience. Occup. Environ. Med. 55, 507–516.

Succop, P.E., O'Flaherty, E.J., Bornschein, R.L., Clark, C.S., Krafft, K., Hammond, P.B., et al., 1987. A kinetic model for estimating changes in the concentration of lead in the blood of young children. In: Lindberg, S.E., Hutchinson, T.C. (Eds.), International Conference on Heavy Metals in the Environment, vol. 2. CEP Consultants, Edinburgh, U.K., pp. 289–291.

Theppyang, K., Schwartz, B.S., Lee, B.-K., 2004. Associations of patella lead with polymorphisms in the Vitamin D receptor, δ-aminolevulinic acid dehydratase and endothelial nitric oxide synthase genes. J. Occup. Med. 46, 528–537.

Todd, A.C., 2000. Coherent scattering and matrix correction in bone-lead measurements. Phys. Med. Biol. 45, 1953–1963.

Todd, A.C., Chettle, D.R., 1994. In vivo X-ray fluorescence of lead in bone: review and current issues. Environ. Health Perspect. 102, 172–177.

Todd, A.C., Chettle, D.R., 2003. Calculating the uncertainty in lead concentrations for *in-vivo* bone lead X-ray fluorescence. Phys. Med. Biol. 48, 2033–2039.

Todd, A.C., Carroll, S., Godbold, J.H., Moshier, E.L., Khan, F.A., 2000. Variability in XRF-measured tibia lead levels. Phys. Med. Biol. 45, 3737–3748.

Todd, A.C., Carroll, S., Geraghty, C., Khan, F.A., Moshier, E.L., Tang, S., et al., 2002. L-shell X-ray fluorescence measurements of lead in bone: accuracy and precision. Phys. Med. Biol. 47, 1399–1419.

Todd, A.C., Arnold, M., Aro, A., Chettle, D.R., Fleming, D.E.B., McNeill, F.E., et al., 2003. Corrections to "How to calculate lead concentration and concentration uncertainty in XRF in vivo bone lead analysis" by Kondrashov and Rothenberg. Appl. Radiat. Isot. 51, 41–50.

Tola, S., Hernberg, S., Asp, S., Nikkanen, J., 1973. Parameters indicative of absorption and biological effect in new lead exposure: a prospective study. Br. J. Ind. Med. 30, 134–141.

Tsaih, S.-W., Schwartz, J., Lee, M.-L.T., Amarasiriwardena, C., Aro, A., Sparrow, D., et al., 1999. The independent contribution of bone and erythrocyte lead to urinary lead among middle-aged and elderly men: the Normative Aging Study. Environ. Health Perspect. 107, 391–396.

U.S. Agency for Toxic Substances and Disease Registry, 1999. Toxicological Profile for lead. Department of Health and Human Services, Atlanta, GA.

U.S. Agency for Toxic Substances and Disease Registry, 2007. Toxicologic Profile for Lead. U.S. Centers for Disease Control, Atlanta, GA.

U.S. Centers for Disease Control, 1985. Preventing Lead Poisoning in Young Children. A Statement by the Centers for Disease Control. U.S. Department of Health and Human Services, Atlanta, GA.

U.S. Centers for Disease Control, 1991. Preventing Lead Poisoning in Young Children. A Statement by the Centers for Disease Control. U.S. Department of Health and Human Services, Atlanta, GA.

U.S. Centers for Disease Control, 2005. Preventing Lead Poisoning in Young Children. A Statement by the Centers for Disease Control. U.S. Department of Health and Human Services, Atlanta, GA.

U.S. Environmental Protection Agency, 1977. Air quality criteria for lead. Report No. EPA-600/8-77-017. Office of Research and Development, Washington, DC.

U.S. Environmental Protection Agency, 1986. Air quality criteria for lead, 4 vols. EPA 600/8-83/028bF. Environmental Criteria and Assessment Office, Washington, DC.

U.S. Environmental Protection Agency, 2006. Air quality criteria for lead, vol. 1. Report No. EPA/600/R-05/144aF. National Center for Environmental Assessment, Washington, DC.

Weis, C., Lavelle, J.M., 1991. Characteristics to consider when choosing an animal model for the study of lead bioavailability. Chem. Speciation Bioavailability 3, 113–119.

Weaver, V.M., Schwartz, B.S., Ahn, K.-D., Stewart, W.F., Kelsey, K.T., Todd, A.C., et al., 2003. Associations of renal function with polymorphisms in the δ-aminolevulinic acid dehydratase, vitamin D receptor, and nitric oxide synthase genes in Korean lead workers. Environ. Health Perspect. 111, 1613–1619.

Weaver, V.M., Schwartz, B.S., Jaar, B.J., Ahn, K.-D., Todd, A.C., Lee, S.S., et al., 2005. Associations of uric acid with polymorphisms in the δ-aminolevulinic acid dehydratase, vitamin D receptor, and nitric oxide synthase genes in Korean lead workers. Environ. Health Perspect. 113, 1509–1515.

Weaver, V.M., Lee, B.-K., Todd, A.C., Ahn, K.-D., Shi, W., Jaar, B.J., et al., 2006. Effect modification by δ-aminolevulinic acid dehydratase, vitamin D receptor and nitric oxide synthase gene polymorphisms on associations between patella lead and renal function in lead workers. Environ. Res. 102, 61–69.

Wedeen, R.P., Maesaka, J.K., Weiner, B., Lipat, G.A., Lyons, M.M., Vitale, L.F., et al., 1975. Occupational lead nephropathy. Am. J. Med. 59, 630–641.

Wetmur, J.G., 1994. Influence of the common human δ-aminolevulinate dehydratase polymorphism on lead body burden. Environ. Health Perspect. 102 (Suppl. 3), 215–219.

Wetmur, J.G., Kaya, A.H., Plewinska, M., Desnick, R.J., 1991a. Molecular characterization of the human delta-aminolevulinate dehydratase 2 (ALAD2) allele: implications for molecular screening of individuals for genetic susceptibility to lead poisoning. Am. J. Hum. Genet. 49, 757–783.

Wetmur, J.G., Lehnert, G., Desnick, R.J., 1991b. The δ-aminolevulinate dehydratase polymorphism: higher blood lead levels in lead workers and environmentally exposed children with the 1-2 and 2-2 isozymes. Environ. Res. 56, 109–119.

Willoughby, R.A., Thirapatsakun, T., McSherry, B.J., 1972. Influence of rations low in calcium and phosphorus on blood and tissue lead concentrations in the horse. Am. J. Vet. Res. 33, 1165–1173.

Wittmers, L.E., Wallgren, J., Alich, A., Aufderheide, A.C., Rapp, G., 1988. Lead in bone. IV. Distribution of lead in the human skeleton. Arch. Environ. Health 43, 381–391.

World Health Organization, 1995. Inorganic Lead. Environmental Health Criteria 165. International Programme on Chemical Safety. WHO, Geneva, Switzerland.

Yin, L.-M., Rhoads, G., Lioy, P., 2000. Seasonal influences on childhood lead exposure. Environ. Health Perspect. 108, 177–182.

Yip, R., Dallman, P.R., 1984. Developmental changes in erythrocyte protoporphyrin. Roles of iron deficiency and lead toxicity. J. Pediatr. 104, 710–713.

Yip, R., Norris, T.N., Anderson, A.S., 1981. Iron status of children with elevated blood lead concentrations. J. Pediatr. 98, 922–925.

Ziegler, E.E., Edwards, B.B., Jensen, R.L., Mahaffey, K.R., Fomon, S.J., 1978. Absorption and retention of lead by infants. Pediatr. Res. 12, 29–34.

Ziemsen, B., Angerer, J., Lehnert, G., Benkmann, H.-G., Goedde, H.W., 1986. Polymorphism of delta-aminolevulinic acid dehydratase in lead-exposed workers. Int. Arch. Occup. Environ. Health 58, 245–247.

Chapter 9

Predictive Modeling Approaches for Assessing Human Lead Exposure

9.1 INTRODUCTION

This chapter describes the various types of models and model methodologies that develop the exposure portion of various dose–response relationships. The exposure/dose portion of dose–responses can derive from data sets generated by measuring all variables or those generated through simulation of the various Pb exposure relationships with environmental, socioeconomic, or demographic input parameters. This chapter confines itself to using models for quantifying lead in some biological medium as the outcome variable, principally lead in whole blood. Early effect outcomes in humans from contact with Pb in various sources via different pathways, for example, early hematotoxic effect indicators, are not presented here but are discussed in some of the later chapters.

What are models, and specifically exposure models? Mushak (1989) offered a broad definition for models as being:

"... abstract constructions and depictions of complex systems that permit easier comprehension for study and application."

Confined to lead exposure model conceptualization, such models are constructions and depictions specifically intended to set forth toxicokinetic, biological, and toxicological grounds for the multiple sequenced relationships that link a potential lead source to a quantitative statement of lead's presence in some biological compartment.

Predictive Pb exposure models serve various purposes in lead environmental epidemiology and human lead risk assessments. For example, well-validated mechanistic, i.e., biokinetic, models can generate outputs that substitute for measurement data where gathering measurements may not be feasible or even possible. These constraints arise from realities of resource limits, poor subject accessibility, or inappropriate timing for gathering the

Trace Metals and other Contaminants in the Environment, Volume 10
ISSN: 1875-1121 DOI: 10.1016/B978-0-444-51554-4.00009-2

most relevant exposure data for determining maximal health risk. With respect to temporal limits, measuring desired exposure outcomes such as blood lead may not be temporally possible. This limit would be evident in cases where a regulatory risk assessor wants to know the consequences, i.e., future Pb exposures, for affected children if existing levels of Pb exposure were significantly reduced. We likewise cannot measure through impacted community screenings or surveys any historic prevalences of Pb exposures that existed for groups of interest earlier in lead contamination history or prior to the onset of any Pb contamination.

One can, however, attempt to sidestep some of the above constraints to acquiring measurement data in epidemiological studies through exposure simulations, if Pb input measurements are available. This would be the case with biokinetic or metabolic models. This would also be the case in some situations where *ad hoc* or statistical empirical models derived from a modeled relationship for a particular site and set of environmental Pb site parameters were applied to other sites very similarly situated. The relative flexibility of the *ad hoc* or setting-specific empirical models may or may not be less widely applicable, i.e., more problematic, than various metabolic models. The relative merits of these model forms emerge through comparisons contrasting measured to simulated or predicted data outputs.

Predictive Pb exposure modeling has been addressed to some extent in both public agency, i.e., expert consensus, and individual treatises and reviews. The former includes sections within U.S. EPA (1986, 1994a,b,c, 2006), and the latter reviews such as those of Mushak (1993, 1998), O'Flaherty (1998), White et al. (1998), and Hogan et al. (1998).

The simplest first-generation predictive Pb exposure models consisted of statistical or *ad hoc*, mathematical models. These came into wide use in the 1960s and 1970s with the development of reliable measurement methods for blood lead levels and for Pb in diverse environmental media: air, dust, diet, soil, sediment, and water. The models employed regression analyses of the relationship of lead in diverse environmental media to lead in blood of children or adults.

9.2 *AD HOC*/STATISTICAL MODELS

Statistical models of human Pb exposures can be subdivided into several mathematical categories. The earliest statistical Pb exposure models consisted of simple regression analyses of changes in the measured biomarker (PbB) outcome data and showed the relationship of PbB changes to changes in environmental Pb intakes and uptakes. Such models are the basis for some of the illustrative studies depicted in Chapter 10. As seen in Chapter 10, environmental variables show diverse mathematical relationships for Pb-bearing media and the PbB outcome measure.

There are several major limits of the simple regression-type statistical analysis model. One is that PbB outcomes for a specific location depend on the site-specific conditions governing the eventual statistical relationship. A second limit arises from typical absence of any means for evaluating Pb exposure pathways from source(s) to exposed individuals, their relative importance to the outcome measure, and evaluation of any role for host factors affecting the resulting PbB values, e.g., child age or family socioeconomic status.

For example, soil Pb affects children's PbB levels directly through Pb-containing soil ingestion and indirectly through the pathway of dust Pb, the latter as exterior and interior dust Pb. Ambient air Pb can be inhaled directly or can be deposited onto soils or as exterior or interior leaded dusts, followed by children ingesting these media. Paint Pb can be ingested as chips or flakes or can deteriorate into interior dust Pb and be ingested. Inability to identify the presence and extent of such pathways in simple, single-medium Pb regression analyses can misstate blood lead–environmental Pb relationships.

9.2.1 Structural Equation Modeling

More refined forms of regression analyses have emerged to better quantify source–exposure and source–pathway–exposure relationships for lead. A particularly useful form is a variation on the multiple regression technique of structural equation modeling (SEM). Figure 9.1 shows interior dust, the principal Pb exposure pathway for infants and toddlers, arising from exterior Pb dusts, interior Pb paint, or secondary occupational ("take-home") dusts.

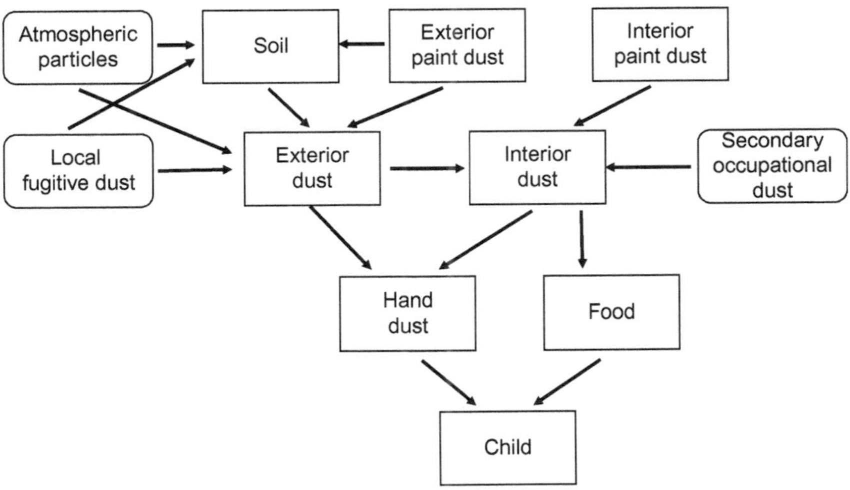

FIGURE 9.1 Typical pathways of childhood exposure to lead in dust and other Pb media inputs in structural equation modeling. *U.S. EPA (1996b), Figure 2.3.*

Exterior Pb dusts, in turn, arise from proximate fugitive (reentrained) Pb dusts, atmospheric Pb fallout, and soil Pb dusts. Figure 9.1 shows atmospheric Pb mainly entering interior dusts through exterior dusts, which is a valid depiction for the typical case where air Pb is currently relatively low. However, atmospheric Pb can also be incorporated in interior Pb dusts directly by entering residential interiors. Air Pb emissions from operating smelters and battery plants can be direct, significant contributors to interior dusts.

Where atmospheric Pb levels are high, this pathway is significant in terms of PbB contribution. Note that in this modeled depiction, a causality chain and pathway flow within that chain are implicitly required and explicitly depicted. For example, environmental Pb input to the human receptor goes from remote sources to proximate pathway. Hand Pb is proximate to the intake and uptake of Pb by children and lead workers with dusty Pb conditions in the workplace. This approach provides considerable mechanistic information on the specifics of how Pb exposures occur. Cross-sectional SEM models, the more typical form of this approach, probe the relationships among several environmental pathways simultaneously, as would be the case above linking soil and dust Pb to Pb via soil to dust, dust to blood, and the direct path of soil Pb to blood Pb.

SEM statistical frameworks for such studies were examined and first applied to these questions by researchers at the University of Cincinnati (Bornschein et al., 1985; Succop et al., 1998). SEM permits allocation of the variance in a set of interconnected outcomes to those directly affecting the dependent variable and those following indirect pathways. Put differently, the modeling approach stratifies the explanatory variables relevant to lead exposures into exogenous and endogenous variables. Endogenous intermediate variables include dust, soil, and paint Pb, and the principal outcome endogenous variable is PbB.

The composite structural statistical model developed by the Cincinnati group (Succop et al., 1998) to deal with 11 environmental epidemiological studies of different Pb contamination settings beginning in the 1980s is presented in summary form as Figure 2 and Table 3 in Succop et al. (1998). It provides much pathway and exposure mechanistic data. Across the 11 studies, significant tested direct pathways included interior and exterior paint Pb loading, soil Pb, entry dust Pb, floor dust Pb, and hand wipe Pb. As seen in Succop et al. (1998), the significant tested contributors to PbB variance for the 11 studies were floor dust Pb levels in 7 of 11 studies (64%); child hand wipe Pb in 4 of 6 studies (67%); soil lead level in 3 of 8 studies (38%); exterior paint Pb loading in 2 of 9 studies (22%); interior paint Pb loading in 1 of 10 studies (10%); and entry dust Pb level in 1 of 10 studies (10%).

Other researchers have employed the SEM approach to elucidate pathways and Pb contact mechanisms. For example, Lanphear and Roghmann (1997) evaluated the pathways for lead exposures in urban children with a particular focus on inputs to interior dust lead. Demographic variables were used only as exogenous factors while the intermediate endogenous variables were dust Pb and hand Pb, with PbB as the outcome endogenous variable. Other endogenous variables were painted surface status, soil Pb, paint Pb, and water Pb. While soil and paint Pb contributed to interior dust Pb levels, paint Pb was the more important contributor ($p < 0.001$). Dust Pb, in turn, was significantly linked to PbB both directly and indirectly via hand Pb. Mouthing behavior in this study was observed as an important variable as was race.

A second variation on single-site regression analyses is the use of pooled analyses of multisite Pb exposures, which serves to elucidate the broader applicability of *ad hoc*, statistical models. Lanphear et al. (1998) carried out a pooled analysis of 12 studies which met a set of criteria: well-defined sampling protocols, measurements of key environmental media such as dust Pb loadings or levels, soil Pb levels, paint Pb loadings, wipe samples for dust Pb included, contemporaneous collections of dust and blood Pb, and a child age pool 6–36 months old. This large study confirmed that house dust Pb, indexed as a Pb loading metric, is the principal pathway for Pb to PbB in very young children. Interior dust Pb loading was the index of environmental Pb and showed more robust association with children's PbB values. This pooled analysis with its high sample size ($N = 1{,}861$ children) further linked a child's age, race, and mouthing behaviors with PbB and showed that floor dust Pb loading was associated with increased frequency of elevated PbB at interior dust values of 5–10 $\mu g/m^2$.

Ad hoc, site-specific statistical exposure modeling approaches have often been applied in the form of slope-factor relationships, whereby Pb in some environmental medium is linked to PbB via a numerical "slope factor." In the case of air Pb, the slope factor is the ratio of community or larger area blood Pb level to an air Pb concentration. In their application, the product of a selected slope factor and air Pb value provides an estimated blood Pb figure for a cohort of Pb-impacted individuals. The currently reported range of slope factors using an air Pb is 5–10 and in some cases higher (Brunekreef, 1984; Brunekreef et al., 1983; Hayes et al., 1994; Hilts, 2003; U.S. EPA, 1986, Ch. 11, Vol. 3; WHO, 2000).

A second category of slope factor models employs Pb levels in some medium in combination with estimates of intakes, uptakes, and a simple "biokinetic" slope factor to estimate PbB levels. In a conceptual sense, the simpler slope factor approach, relating a medium Pb level like air lead to blood lead, is a variant of this second category. That is, an air Pb translating into intake and uptake quantities provides the same type of results as the earlier approach.

9.2.2 Slope Factor Models with a Biokinetic Factor

Slope factor biokinetic models are, as the label implies, hybrid models for prediction of Pb exposures. That is, they employ both empirical and biokinetic slope values derived from the full mechanistic biokinetic models. Typically, the extant slope factor models occur in algebraic equation form, and differ in the complexity of the equations with regard to the number of terms. Two published model approaches of the above type are the Stern (1994) and the Bowers (Bowers and Cohen, 1998; Bowers et al., 1994) models. Of these, the Bowers et al. model is better known and more widely used. The U.S. EPA (1996a) has adapted the slope factor mixed model approach to use in risk assessment for groups such as pregnant workers at Pb-contaminated regulatory sites.

A description and use of the Bowers model is set forth in Bowers and Cohen (1998). In their article, a median PbB can be calculated and algebraic terms include a PbB constant (baseline) derived for adults, terms for intake and uptake, terms for a biokinetic slope factor relating PbB to daily Pb uptake, terms for soil-related Pb intakes, exposure frequency, etc. A related equation can also be used to estimate target PbBs, such as PbB values at some percentile in the distribution, e.g., the 95th percentile, that could not be exceeded for protecting fetal health given such empirical data as the geometric standard deviation (GSD) for PbB among women in the childbearing age band.

The Bowers model, alone among biokinetic slope factor models, has the advantage of having been compared with measured PbB batched data at Superfund sites and with a full biokinetic model, the O'Flaherty simulation model. The four Superfund sites included were Midvale, UT; Butte, MT; Leadville, CO; and Granite City, IL. The use of EPA-selected parameters for its risk assessment gave a higher estimate of PbBs using the Bowers equation than did the model when using Bowers' recommended values, with the latter being closer to observed PbB values (Tables 7 and 8 in Bowers and Cohen, 1998). With regard to comparisons with the O'Flaherty model using the same inputs, the latter gave values that were similar to Bowers at age 20 but produced lower values at older ages (Table 2 in Bowers and Cohen, 1998).

How broad is the applicability of slope factor approaches? The multisite pooled analysis of dust Pb loadings reported by Lanphear et al. (1998) yielded results which appear to be consistent with more recent NHANES.

9.3 MECHANISTIC/BIOKINETIC MODELS

Much of the interest in using predictive modeling of human Pb exposures concerns mechanistic, i.e., biokinetic, models. Biokinetic models differ from site-specific or multisite *ad hoc*/slope factor models in a number of important ways. First, they are constructed (by definition) within a computational and biological/kinetic framework. That framework specifies inclusion of all identified parameters representing mechanisms by which environmental Pb is deposited in and removed from human organs and tissues. Mechanistic models are structured to be much more complex than the regression models described earlier. This complexity, however, permits the flexibility to project simulated Pb exposures across human populations, especially in the dimensions of time and space. Such flexibility is also valuable in combining epidemiological information with environmental data to construct dose–toxic response relationships and expressions of overall health risk from Pb exposure. Biokinetic models are especially flexible for regulatory settings, as noted earlier.

The evolution of biokinetic models of Pb exposures in human populations allows classification into models that were useful in their time but are now mainly of historical interest and those currently enjoying wide use. This evolution of contemporary exposure models has tracked scientific developments in the field of Pb toxicokinetics. Currently employed biokinetic models can be classified into those having different mathematical construction and predictive conceptualization underlying the model. One of the widely used models is the classical compartmental type. Others are PB-PK models.

Application of biokinetic models to actual exposure populations first requires steps to determine the relative reliability of the particular model. Two critical steps are sensitivity analysis and model evaluation. The first judges the assumptions within the model and magnitude of the predicted response. The latter assesses the degree of concordance between simulation and observation.

Generally, biokinetic model evaluation has both a validation step and a calibration stage. The former refers to overall computational soundness and plausibility of the biological simulations operating within the model. Model validation includes documenting the model's scientific basis and source code accuracy verification. The latter addresses adjusting or fine-tuning a generally valid and applicable model using reference data, typically from on-site measurements. The general assessment of the IEUBK model, a widely used computerized simulation model in human health risk assessments, has been carried out by EPA's Technical Review Workgroup for Lead (U.S. EPA, 1994b).

Efforts have been made to compare the predictive accuracy of the IEUBK model at extractive industry sites with single, unstructured PbB surveys (Bornschein et al., 1991, 1992; Hogan et al., 1998). These PbB surveys are themselves potentially problematic and must be shown to be reliable before use in calibrating or evaluating modeling results. What should the level of calibration be? Any model which must always be fine-tuned with single sets of currently gathered measurements ascribes little intrinsic value to the utility of that model across the whole spectrum of its likely uses. A major use of the EPA IEUBK model for risk assessors and risk managers is to predict PbB as output when detailed site characterizations by other means are not possible. As noted earlier in general comments on mechanistic models, these include future changes in contaminated land uses and future demographic or socioeconomic changes in affected populations.

Any disagreement between biokinetically predicted PbB results and accurate blood lead measurements indicates discordances in input parameters (assuming basic computational soundness). The EPA IEUBK model has a number of input steps where uncertainty and variability can arise. Two are significant: (1) the bioavailability or uptake rate of lead from media directly relevant to infants, toddlers, and older children and (2) the amount of lead-contaminated medium ingested.

There is also the question of range of applications for EPA's IEUBK and similar models. Few predictive exposure models for any substance are expected to be equally applicable for all environmental contamination settings; for that matter, neither is any model likely to be universally useful or adaptable. One obvious reason for any model's limits is the restricted scope of the empirical data typically used to evaluate models. The usual process is to evaluate models with data most directly relevant to the intended application(s). One then has the narrower task of determining whether a model such as the IEUBK model is appropriate to its actual uses.

As a simple illustration, EPA regional and state regulators use the IEUBK model to ascertain the baseline or "no action" blood lead levels in a community impacted by lead-containing extractive industry wastes. That same community may also be the site of a "single-shot," i.e., *ad hoc*, blood lead screening or survey done concurrently, paid for by affected interests who may be potentially liable for abatement of exposures associated with the site. A local health department may provide a site for the blood samplings and related work with environmental samplings. Modeled PbB data using environmental data from the site might be found on superficial examination to be different from the measured levels in the single-shot survey. This lack of agreement would raise the question of whether measured or simulated PbB data is the problem. At this point, one has to examine both sets of information critically.

How Good Is Good Enough?

One basic question associated with Pb exposure modeling is: How accurate does a model simulating lead exposure have to be for acceptance in some application? The question is logically subsumed under a broader one: What is the purpose to which the model is being put? It appears that Pb exposure models in current use each have some advantages and limits, depending on likely cases to which they may be applied.

If the model user seeks predictive estimates within a rather broad range and/or is using predictive data as but one element in a large cluster of criteria for risk assessment or risk management of a site, such as a hazardous waste site, the required level of agreement between empirical data and model outputs may not dictate very close concordance. If predictive data are being sought for some finite threshold value in permissible lead exposure or some cutoff level in lead exposure distributions, then more rigorous model performance may be required. As a specific example, is EPA's IEUBK model reliable for evaluating child lead exposures at hazardous waste sites?

A related qualifier is the extent to which some particular model for some site- or scenario-specific use is flexible for multiple uses by the regulator or risk manager. For example, biokinetic models such as EPA's IEUBK model for children (U.S. EPA, 1994a,b,c) or the O'Flaherty PB-PK model for children and adults (O'Flaherty, 1995, 1998) theoretically permit assessment of the results of altered land uses, altered population demographics, or impact of lead remedial actions when projecting into future years for body lead burdens indexed by PbB. Such applications should not require recalibration with empirical data for every use.

The EPA IEUBK model in particular is not hobbled by a discrete set of circumstances in the way that a single PbB survey would be. Even without the intervention of an abrupt artifact like parental awareness temporarily affecting children's PbB concentrations, PbB integrates mainly those exposure factors currently operating and those shown by past detailed inferential statistical analyses to explain statistically significant fractions of changes. These include the numbers of children currently residing in the impact zone, socioeconomic status of children in proximate areas to—e.g., Superfund sites, the quality of yards in terms of vegetative cover, and landscaping,—the distance of residences to the site, and relative location of the homes within air lead fallout zones as discernible in wind patterns. Factors operating for the site include the level of control of Pb releases offsite affecting the probability of runoff, contamination of groundwater and the frequency of fugitive dust Pb releases from unsecured on-site waste piles.

Gradual or abrupt changes in the character of the affected areas have the potential to change child lead exposures and PbB concentrations. When that occurs, the prior PbB survey becomes an inaccurate or even irrelevant marker of what is happening now. If, as examples, the community deteriorates, if home ownership declines with increase in poorly maintained rental housing, if transient families occupy the dwellings, and if more infants and toddlers occupy the dwellings, then lead exposures will increase and not be reflected in the earlier PbB survey. By contrast, the IEUBK model has the flexibility, with additional empirical data related to changed environmental circumstances, to estimate these increased exposures.

Current biokinetic models differ in their flexibility for use in other than stable, (near) steady-state Pb exposures of human risk populations. The IEUBK is only operative in cases where Pb exposures are stable and long term. By contrast, the O'Flaherty PB-PK model is constructed to reflect acute or episodic Pb exposures. Current biokinetic models also differ in the age ranges of exposed populations to be modeled. The IEUBK model is for children up to 84 months old. The PB-PK models are basically lifetime in their application.

One can also use biokinetic models for reconstructing lead exposures at remote times. Mushak (1993) used the IEUBK model to estimate the typical geometric mean blood lead of preindustrial children, based on archaeological and remote, uncontaminated environmental lead data collected over the years by Patterson and his colleagues and summarized in the NAS report on lead exposure (NAS/NRC, 1993).

9.3.1 Early Biokinetic Models

There are three early biokinetic forms of exposure modeling in the literature. These served to bridge the gap between statistical, regression modeling and the current biokinetic models enjoying relatively routine use. These are as follows: (1) the Rabinowitz et al. (1976) model; (2) the Marcus approach (Marcus, 1985a,b,c); and (3) the Bert model (Bert et al., 1989).

The Rabinowitz et al. (1976) Model

The Rabinowitz et al. model was perhaps the first widely recognized mechanistic model in the sense that it was constructed of kinetic components. It consists of a classical compartmental construction representing internal Pb movement under steady-state or near-steady-state conditions. Its principal features are set forth in Figure 9.2. Mathematically, the model employs coupled differential equations with linear exponential solutions and assumes connected, well-mixed pools for Pb deposition and interorgan movement. Three compartments, one of which is taken to be the central or blood compartment, were resolved via data gathered from ingestion of a stable Pb

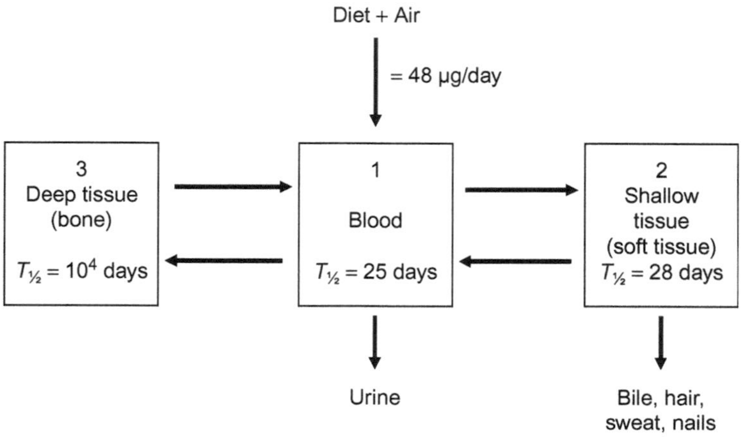

FIGURE 9.2 Structure of the Rabinowitz et al. (1976) compartmental Pb model with the indicated Pb half-times. *Adapted from U.S. EPA (2006), Figure 4-18.*

isotope by human volunteers. Besides blood, there were distinct soft-tissue and bone compartments, all three having different biological half-lives or mean lifetimes. In order of increasing intracompartmental retention, the respective half-lives were blood Pb (25 days), soft tissue Pb (28 days), and a bone compartment of ~10,000 days (~27 years). The corresponding kinetic Pb amounts in these three compartments were, respectively, 1.9, 0.6, and 200 mg Pb. This model was the first attempt to depict Pb dose internally in biokinetic and predictive terms, presenting a role for bone Pb accumulation and accounting for intake and excretion rates. It was limited by the small number of subjects, data that applied only to adults, no provision for accommodating Pb kinetics in children, and no biokinetic insight into what was occurring within the three major compartments.

The Marcus et al. Modeling Expansions on Rabinowitz et al.

The reports of Marcus (1985a,b,c) started with the Rabinowitz et al. model and expanded its conceptual and mathematical complexities in various ways. It remained, however, a classical compartment biokinetic model. Further compartmental analysis and expansions were used to produce multiple subcompartments of bone and blood for example. The bone compartment was stratified into two subcompartments: a fast, trabecular (spongy) bone Pb one with half-life of 100–700 days and a slow, cortical (dense) one with a much longer half-life, ca. 12,000–35,000 days. These kinetically distinct bone subcompartments are rationalized in terms of Pb diffusion rates across an extended cylindrical array of bone canaliculae, providing a link between canalicular fluid and the bone mineral volume proper. The Marcus refinements included refined modeling of known nonlinear kinetics for erythrocyte–plasma Pb exchange into diffusible plus complexed plasma Pb and intraerythrocyte deep and shallow subcompartments (Marcus, 1985c).

The Bert et al. Model

The Bert et al. (1989) approach largely recapitulates principal features of Rabinowitz et al. and Marcus, makes some added assumptions, and refines some clearance and excretory pathways for Pb via the GI tract. Data selected for the bone Pb loadings were those age-stratified autopsy figures from Barry (1975).

9.3.2 Contemporary Biokinetic Models of Human Pb Exposures

Current Pb exposure modeling through biokinetic approaches uses either classical compartment constructions with some PB-PK features or frameworks which are largely PB-PK in form. Evolution of these models from earlier precursors was greatly assisted by growing parallel knowledge and a growing literature about Pb biokinetics and other quantitative features of lead's behavior in various subsets of human exposure populations. Three models are employed currently, all computerized and each using different software. Each differs greatly in the number and detail for each of the necessary components in a useful exposure simulation model, extent and ease of application, extent of evaluation and verification of their respective source codes, extent of analyses accorded each in terms of agreement with measured data, and flexibility in assessing Pb exposures in different subgroups of exposed human populations.

The three models now of interest are: (1) the U.S. EPA IEUBK model (Hogan et al., 1998; U.S. EPA, 1994a,b; White et al., 1998); (2) the Leggett model (Leggett, 1993; Pounds and Leggett, 1998); and (3) the O'Flaherty model (O'Flaherty, 1993, 1995, 1998, 2000). The Pb exposure modeling literature characterizes the three contemporary models mathematically in varying ways.

The IEUBK model, specifically developed for and limited to simulations of childhood lead exposures up to 84 months of age, has largely compartmental features and is complex in structure and numbers of parameters. The Leggett and O'Flaherty models—simulating Pb exposures from birth through adulthood—more closely draw upon such physiological correlates as volumes, flows, composition, and metabolic activity and largely accord with a PB-PK model. The O'Flaherty and Leggett simulation systems are all-ages models.

The U.S. EPA IEUBK Model

The IEUBK model is age-limited, specifically to children's Pb exposures up to 84 months of age. The three conceptual and mathematical structural components of the IEUBK model are depicted in Figure 9.3: a front-end multimedia Pb intake module, followed by an uptake module, followed by a biokinetic endpoint module. While the name indicates three discrete simulation components, the model also has a statistical module. This module encompasses: (1) batch processing and analysis of simulation data, (2) some measure of interindividual variability, and (3) estimations of probabilities of some indicated PbB occurring in a group with comparable Pb exposures. At the time of this writing, the IEUBK model is the most complete in terms of ease of use and adaptability for human health risk assessment purposes if simulations of childhood Pb exposure are desired. Its inclusion of both exposure and statistical modules distinguishes it from the other two widely used contemporary exposure simulation models, which lack these components. The uses and limits of the IEUBK model in various Pb risk assessment scenarios have been described at length in the model's operating manual (U.S. EPA, 1994a) and independent articles (Hogan et al., 1998; White et al., 1998) and the reader is referred to those sources for detailed discussions on structure and use of this model.

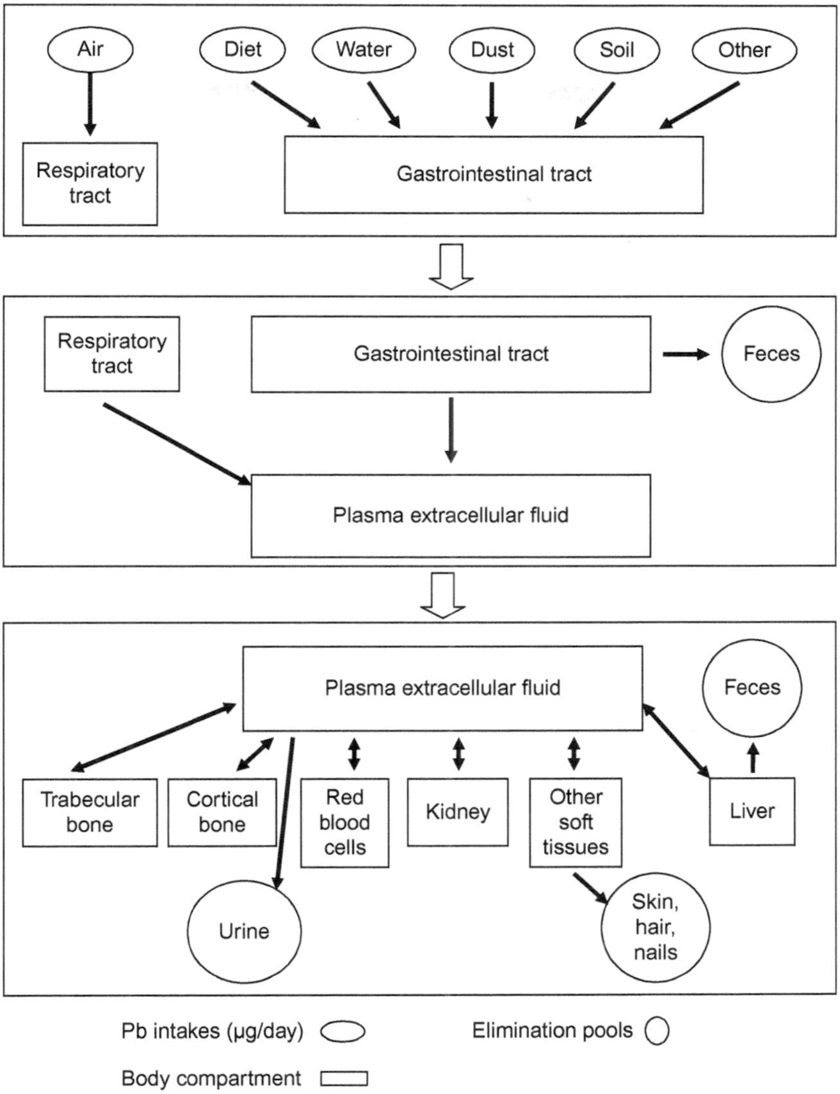

FIGURE 9.3 Structure of the integrated exposure uptake biokinetic model for lead in children. *Adapted from U.S. EPA (1994a).*

The IEUBK model, in common with the other two contemporary simulation methodologies, processes total daily amounts of Pb entering body compartments from various environmental media. It does this through a discrete component, the exposure module. Daily Pb intake amounts are calculated from two parameters, Pb concentration in the intake medium and the amounts of medium assimilated. Intake amounts are in units of volume for air and water, and mass for diet, soils, dusts, and miscellaneous sources. The exposure module also permits Pb inputs from additional sources such as homegrown food and interior dust Pb arising from take-home or secondary exposures. The model in its current incarnation stratifies intake amounts in terms of a full age-year for any of seven age bands indexed as years or months, e.g., 0.5–1 and 1–2 years. Tabular computational output data are rendered as years of age, while probabilistic distributions appear in graphic depictions as age in months.

As is the case with the model's biokinetic component, the IEUBK model was constructed to reflect steady-state Pb toxicokinetics and therefore relatively stable Pb exposures, preferably at least a year or so. Consequently, the model does not permit simulations of PbB levels where Pb exposures are either intermittent or acute.

The cumulative intake/exposure module provides either default selections or user-selected parameter values from available measurement data. Further intake parameter selections, such as water volumes consumed, amounts of dust ingested, and air volumes inhaled, vary across child ages and are either specified as default values or left to the user for selection. The IEUBK model's use for typical risk assessment purposes was designed and intended to be applied to typical children's residential settings as the spatial unit of exposure. The term typical, or hypothetical, child can include an individual typifying any of a group of children residing at a typical residence. EPA recommends (Hogan et al., 1998; U.S. EPA, 1994a; White et al., 1998) against use of the model for spatial reaches greater than a detached residence or other unit such as a caregiver's residence in the case of a typical child or a group of children. This caveat has to do with the significant increase in uncertainty that would accompany increased spatial scope for Pb exposures. The less sensitive the spatial contact parameter, the less the model is able to accommodate such anomalous exposure events as small but high Pb areas, i.e., Pb "hot spots." This caveat is also linked to the statistical assumption that daily intakes of multimedia Pb in the IEUBK model reach the exposure module as some measure of central tendency, specifically, median values.

The Pb uptake module, where Pb uptake is indexed as μg Pb absorbed per day, is the interface between the exposure module and the biokinetic processing of absorbed Pb quantities to provide an output as PbB, in the form of a probabilistic expression. The daily amounts of absorbed Pb are calculated for each environmental medium through computations of Pb levels in the medium, amount of the medium ingested, and the bioavailability or fraction of daily Pb intake that is absorbed. Bioavailability can be the default selection: diet and water Pb uptake of 50%; soils and dusts, 30%. Alternatively, users can select bioavailabilities other than default if additional data exist for other percentage uptake rates. The specific topic of bioavailability of lead as a function of environmental medium containing Pb was described in the previous chapter.

As noted earlier, the most significant determinants for Pb bioavailability modeling include chemical species of Pb, the overall particle size distributions for such media, and the micromineralogy of lead-containing particles. Bioavailabilities are technically not the same as solubilities or "bioaccessibilities." Bioaccessibilities are simplified surrogates for bioavailability. These test systems derive solubility relationships and measure the impact of amino acids, buffer mixtures at physiological pH, etc. (Drexler and Brattin, 2007; Drexler and Mushak, 1994, 1995; Ruby et al., 1993).

Enteric uptake of Pb by children is simulated in the IEUBK model by some known physiological characteristics of the Pb uptake step. Pb uptake occurs at different points in the element's movement through the GI tract, with the rates of uptake depending on factors such as gut pH and the binding site distributions for lead. At least two processes operate, one energy-dependent and one not. Pb diffuses across the gut wall via a concentration gradient from the luminal surface. This would presumably involve soluble divalent Pb, although other forms such as encasement in micelles can potentially participate (Mushak, 1991). These diffusion states could occur intracellularly or between cells at paracellular junctions (Mushak, 1991, 1998). Energy-dependent facilitative transmucosal transport entailing the divalent metal transporter system for nutrient ions calcium or iron (U.S. EPA, 1994a) comprises the second transport system. See the previous chapter for a discussion of current information on Pb uptake from the gut. In the design of the IEUBK model, the sum of both energy-requiring (active) and passive uptake mechanisms produce total uptakes and account for the nonlinear relationship of Pb intake amounts and PbB concentrations (White et al., 1998). That is, nonlinearity is consistent with saturability of the active transport system.

As noted in the previous chapter, respiratory uptake of Pb entails both a deposition step for Pb-bearing inhaled aerosols and an uptake step in the pulmonary tract, the uptake rate being a function of particle size and chemical composition. The IEUBK model employs a 30% deposition of Pb in the lung, and assumes there is 100% uptake of that fraction. Given the very low uptake of divalent inorganic lead from skin surfaces, a dermal uptake component is not included in this model.

The biokinetic component of the IEUBK model simulates Pb distribution to blood and other tissues and also the excretion of the toxicant, the latter mainly via urine and feces, and also by such minor routes as skin, nails, and hair. Conceptually and structurally, the model's biokinetic underpinning is a compartmental structure in which all compartments are taken to be kinetically homogeneous. Figure 9.3 graphically identifies seven compartments among which Pb transfers occur, including a central blood plasma–extracellular fluid (ECF) compartment, an erythrocyte compartment, five peripheral body compartments, and three excretion pools (White et al., 1998). Two of the compartments important to Pb deposition and release are subcompartments of bone: trabecular and cortical bone. Refer to Chapter 8 for a fuller description of Pb's behavior in these two bone subcompartments.

The compartmental nature of the IEUBK model springs from its construction using Pb levels in human tissues and organs to determine the size of intertissue and intercompartment transfer coefficients. Tissue Pb data were either determined directly from literature autopsy data for children or from allomerically scaled adult Pb findings. In that respect it differs from how PB-PK constructs for lead models function. In the latter, physiologically determined blood flow rates in tandem with various tissue membrane diffusion parameters control Pb movement. A characteristic of the IEUBK model in its current incarnation is the use of kinetic equations to ascertain masses, transfer and elimination rates at both the beginning and end of selected time intervals. From this a typical child's PbB at some time, *t*, is estimated. Critical to the choice of biokinetic parameters were the estimating processes in the sequence: (1) tissue/blood Pb level ratios were established; (2) compartmental concentration ratio calculations were then converted into ratios of masses of Pb using compartmental masses and/or sizes; (3) growth equations related body fluid volumes and organ weights as a function of child age; and (4) ratios of Pb in plasma and erythrocytes and transfer times between the two were calculated.

A central biokinetic parameter governing the operation of this compartmental model is the compartmental transfer time. These times cover Pb movement among the various compartments set forth in Figure 9.3. Times are based on plasma Pb. At steady state, the ratios of Pb masses in tissue compartments to plasma Pb masses are equivalent to the ratios of transfer times from tissues to plasma and ECF, and from plasma/ECF to tissues. Transfers are also assumed to be from the central to tissue compartments by a first-order kinetic process (White et al., 1998).

While the IEUBK model does not treat bone Pb kinetics in any detail in terms of transfer coefficients across all age bands, as required for a full PB-PK model, it attempts a limited approximation of bone Pb kinetics *vis-a-vis* blood Pb in young children. This age band in human populations is comparatively less impacted in terms of bone to blood Pb movement as a quantitative part of total Pb as measured by a PbB value. In the model, the cortical and trabecular bone subcompartments are assigned the same transfer coefficients (White et al., 1998).

The IEUBK model in its intake, uptake, and biokinetic components attempts to take some cognizance of interindividual variability of Pb in children's blood. Overall, this variability derives from both Pb exposure variation across children in a population as well as differences internally (biokinetically) in the form of differing Pb retention and excretion rates. Comparatively speaking, it is exposure variability that ranges more widely. Pb intake rates and subsequent uptake rates, for example, can differ by orders of magnitude across Pb-contaminated environmental media encountered by a typical group of children. At the same time, it is unlikely that biokinetic or other systemic factors will differ by more than one order of magnitude and, very likely, much less (Mushak, 1998; NAS/NRC, 1993).

The IEUBK model employs a log-normal Pb input distribution and computational approach using calculations of geometric means from an assumption of intakes and biokinetics based on median values, and employs a GSD. The model presents a default GSD for a typical array of exposed children and also offers user-selected GSD choices. The choice of GSD is critical in such log-normal relationships as Pb exposures in human populations, since in applications for risk assessment the higher the GSD for a given typical child or typical children in an exposed population, the higher the fraction of individuals in the upper tail of the distribution and vice versa.

The model in its current incarnation does not include a full probabilistic Pb exposure module with Monte Carlo simulations although attempts at such have been reported, apparently without wide acceptance (Bowers and Mattuck, 2001). The difficulty here, as with probabilistic methodologies in general, is the requirement for more empirical data for media Pb input and probabilistic distributions than is typically available in settings where this model is used.

The IEUBK model has been more exhaustively validated and calibrated than any like model being used today as a site-specific risk assessment tool by EPA or others. This evaluation ranged from verifying the basic computational soundness of the model to the range of its applications, independently evaluated by EPA's Science Advisory Board, Clean Air Scientific Advisory Committee (SAB/CASAC). SAB approved its use in stationary source air Pb exposure assessment and at hazardous waste sites containing lead (SAB/CASAC, 1989, 1992).

It is noteworthy how well the various input parameters used in default mode for the IEUBK model have held up to intense scrutiny. For example, the many studies so far using the only animal surrogate to pass scientific muster, the weanling swine, showed a marked agreement between absolute bioavailability of lead in diverse geochemical matrices and the IEUBK model default choice of 30% (Casteel et al., 2006).

A number of studies have appeared in which two types of comparisons were made: (1) modeled versus measured children's PbB values in batch simulations using data sets of environmental and blood Pb measurements and (2) comparisons across models using children's PbB simulations. For the first type, Hogan et al. (1998) compared simulated exposures as PbB GMs, with measured PbB values in three smelter communities (sample sizes in parentheses): Madison County, IL (333); Palmerton, PA (34); and the smelting/mining area of Galena, KS/Jasper County, MO (111), within the Tri-State lead-zinc belt. The measured geometric PbB means for the tested children in the KS/MO, Madison County, IL, and Palmerton, PA, groups were 5.2, 5.9, and 6.8 μg/dl, respectively. The modeled values for these same groups using the available environmental Pb and selected uptake/biokinetic parameters were in very good agreement, at 4.6, 5.9, and 7.5 μg/dl, respectively. Across the groups, deviation was <1 μg/dl. When the comparisons were made in terms of frequency of children's PbB values exceeding 10 μg/dl, measured values were 20%, 19%, and 29% for the KS/MO, Madison, IL, and Palmerton, PA, groups, respectively. The corresponding simulated percentages were, again, in good agreement, at 18%, 23%, and 31%.

These comparisons were for the entire age band, from <12 months of age to 84 months of age. The important stratification of this broad age band into subsets of infants and toddlers shows, for both simulation results and measured values, that those in the 1–2 years of age group had higher PbB results than did older children.

The IEUBK model has also been compared in terms of outputs to the two other biokinetic modeling approaches: the Leggett (Pounds and Leggett, 1998) and the O'Flaherty (1998) models. The O'Flaherty model provides estimates of children's PbB values that are quite close to the IEUBK while the Leggett biokinetic output estimates are about twofold greater than the others using the same set of intake and uptake conditions (O'Flaherty, 1998; Pounds and Leggett, 1998).

The O'Flaherty PB-PK Model

The O'Flaherty PB-PK model was constructed to simulate levels of Pb in the Pb exposure biomarkers blood and bone for all ages in Pb-exposed human populations from birth through adulthood. The model in its operational features was described in various articles and systematic reviews (O'Flaherty, 1993, 1995, 1998, 2000). Figure 9.4 depicts the operational structure of the model, which includes intakes, multitissue distributions, and routes of elimination.

The O'Flaherty model assumes two routes of Pb intake and exposure: GI and respiratory routes. As in the case of the EPA model, dermal Pb exposures are assumed to be comparatively negligible and are not considered. This model, when operating on the model's original time resolution of 1 year, processes Pb intakes as totals from media-specific sources in units of μg Pb/year. In operation, time resolutions as short as 1 day can be handled.

The uptake rates for Pb from the pulmonary and GI tracts differ in some major ways from values used in the IEUBK or the Leggett models. Like the other models, however, it accommodates the age-dependent GI uptake differences in exposed humans. Respiratory uptake of inhaled air Pb is taken to be independent of age. This model assumes 50% uptake of the amounts of inhaled Pb versus the IEUBK's 100% uptake of the initial 30% deposition in the pulmonary tract under ambient conditions.

The O'Flaherty model scales Pb uptakes from the GI tract on a sliding scale from 58% at birth to a minimum of 8% by 8 years of age. However, in

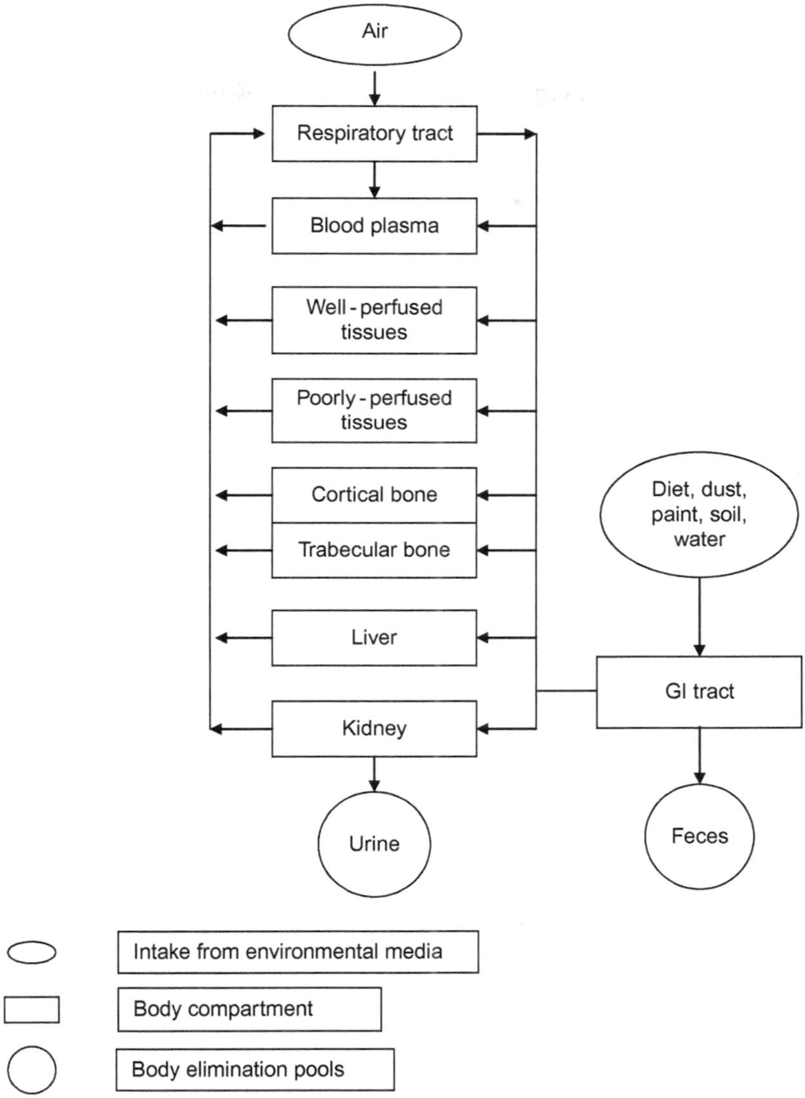

FIGURE 9.4 Structure of the O'Flaherty lead exposure biokinetic model. *Conceptualized by U.S. EPA (2006) from O'Flaherty (1993, 1995).*

the range of most interest for lead risk assessment, i.e., infants and toddlers 1–3 years of age, the O'Flaherty Pb uptake scaling allows uptake estimates of 55%, 45%, and 40% for 1-, 2-, and 3-year-old children, respectively. These values in this narrow age band are not materially different from the IEUBK rate of 50% for young children. For older children, the uptake parameters vary in the two models. At 5 years old, the O'Flaherty Pb uptake rate declines to about 15%, while 50% applies across all age ranges in the IEUBK model. User selection of uptake and biokinetic parameters beyond the default settings is not possible in the case of the desktop C-Basic version of this model. More freedom of choice is allowed with the Advanced Continuous Simulation Language (ACSL) computational version. This limit is especially problematic in the case of Pb uptake rates. At 8 years old and older, the fixed Pb GI uptake rate is 8% for all Pb exposure scenarios.

The O'Flaherty model includes simulations of somatic growth, bone growth and resorption processes, and various growth curves constructed to relate body weight to age via tissue growth and volume standardized growth relationships in either males or females. The central compartment differs from that of the IEUBK system. The central Pb exchange compartment in this PB-PK model, given its conceptual structure, is diffusible plasma Pb, where Pb flows from plasma to liver, kidney, highly perfused tissues, poorly perfused tissues, and cortical and trabecular bone compartments.

A principal driver of this PB-PK model is age-related bone formation and resorption, given that bone is the principal repository of Pb deposition and tardive releases relative to the central compartment or soft tissues. The simulations assume concurrent Pb movement in both cortical and trabecular bone, and account for the higher rates of bone turnover in immature bone (juvenile bone), the transition to mature bone with its lower turnover rates, and the activity in mature bone through late adulthood (O'Flaherty, 1998). Bone Pb uptake from and release to plasma is taken to occur via a heteroionic exchange process with other elements, notably calcium. Kinetically, this exchange of Pb is modeled as a radial diffusion in the volume of the osteon. Body weights and weight changes with age are included. For this model, formation and resorption rates at different ages for human bone Pb were first determined from available human data where stable calcium isotopic kinetics were evaluated.

The conceptual unit in the O'Flaherty approach is at the level of whole-body Pb kinetics. The principal parameter for the overall functioning of the O'Flaherty model is body weight. However, there is additional accounting for gender-based physiological differences apart from body weight. In the O'Flaherty model's construction, body weight is characterized by five constants: two for the growth phase in early childhood and three for both the adolescent growth spurt and stabilized mature adult weights. Once the growth curve has been specified, the anatomic and physiological components of it are defined as well.

Exchanges of Pb between plasma and tissues such as kidney and liver, given its PB-PK basis, are under the control of flow-limited processes. The O'Flaherty model, like the IEUBK model, takes account of the saturability of binding sites (see earlier) and depicts an overall curvilinearity to the erythrocyte–plasma relationship over a broad range of PbB and Pb intakes.

Age-dependent total Pb excretion rates for humans are simulated as combinations of kidney Pb clearance to urine and liver Pb clearance to bile. Proportionately, urinary and biliary clearance rates in humans are taken to be 70% and 30%, respectively, of total clearance.

The O'Flaherty model is not as complete operationally as the IEUBK system. Batch processing of multiple environmental data sets and multiple outputs for multiple exposure subjects, available in the current IEUBK version, is not present. The O'Flaherty model generates point estimate PbB values from point estimate inputs. Beck et al. (2001) described an exposure module add-on to accommodate variability (probabilistic) simulations for exposure, absorption and the nonlinearity due to erythrocyte binding, using a Monte Carlo approach. This adjustment has not been widely adopted. The O'Flaherty model is not equipped with a statistical module for doing descriptive and inferential statistics and graphics. The O'Flaherty approach also makes little provision for accommodating inter- and intraindividual variabilities.

Validation and calibration comparisons of the O'Flaherty model with measured exposure data have been reported, but fewer than were done for the IEUBK system. While the validation and calibration exercises carried out by O'Flaherty have included test animal dosing data and experimental PbB comparisons for rats (O'Flaherty, 1991) and nonhuman primates (Cynomolgus monkeys, Franklin et al., 1997), such efforts also included children and adults (O'Flaherty, 1993, 1995, 1998).

A useful evaluation of this system was the reported comparison of the O'Flaherty modeling results with measured PbB values from the University of Cincinnati prospective lead study (O'Flaherty, 1998). Simulated exposures of young children in this study, using measured environmental Pb levels, showed good agreement of PbB levels for the upper quartile of measured PbB concentrations. Of particular value in this comparison was the fact that the prospective study's pediatric subjects were followed through all early ages and the close agreement with modeled outputs over the age band indicated reliable age- and weight-variable parameter choices in construction of the model.

The model's Pb biomarker outputs accessible to the user and most useful for risk assessments include both PbB and bone Pb. An obviously valuable feature of this model is its applicability to simulating lead exposures in individuals older than 84 months, i.e., beyond the age band of the IEUBK model. Simulations by the O'Flaherty model of Pb exposures in adults, given major changes downward in Pb exposures in recent years in the United States and other industrialized countries, require availability of concentration-variable media Pb over the lifetime of individuals whose exposures are being simulated. Available data depicting earlier Pb exposures in individuals are typically limited so the model's use in reconstructing individuals' Pb exposures beginning when they were infants and toddlers may be limited. On the other hand, this model still has considerable value for simulating likely Pb exposures in the out years with better and more complete empirical data available for Pb inputs in various Pb media.

How does the O'Flaherty model perform when compared with the IEUBK system (for children) or the Leggett model? For children within the age band for which all three models are relevant and applicable, especially for infants and toddlers, the O'Flaherty and IEUBK models produce similar simulations (see, for example, O'Flaherty, 1998). By contrast, as discussed below, the Leggett model predicts higher PbB levels than does the O'Flaherty approach (O'Flaherty, 1998; Pounds and Leggett, 1998; U.S. EPA, 2006).

The Leggett Biokinetic Model

The Leggett model is constructed as a biokinetic system with mainly PB-PK features, with some compartmental kinetic characteristics. The model's structure and performance were originally developed for the purpose of calculating radiation doses from bone-seeking radionuclides and for use by the International Commission on Radiological Protection (ICRP) (Leggett, 1992, 1993). The earlier version of the model was modified in some of its features into the Pounds–Leggett model (Pounds and Leggett, 1998). The various components of the Leggett model are depicted graphically in Figure 9.5. Figure 9.5 has a central compartment, soft tissues depicted as kidney, liver and other soft tissues, and bone. Brain is singled out in the figure but without further biokinetic subdivision.

Overall, the model is constructed with 15 peripheral body compartments, and elimination of Pb is structured within three pools. Diffusible plasma Pb comprises the central compartment, with bound and free fractions, and most of the Pb is in bound form, in an approximate 5:1 ratio. Movement of Pb from plasma to tissues follows first-order kinetics, and transfer constants are age- and PbB-dependent. Dependency on PbB is built into this model as nonlinearity above a threshold level, analogous to the other systems, and accounts for the increased relative fraction of Pb in plasma versus erythrocytes noted in published data (Bergdahl et al., 1997; Manton and Cook, 1984; Manton et al., 2001). Concentrations of Pb in the blood compartment are simulated on the basis of age-dependent blood volume. With other tissues, only Pb masses are calculated. The temporal resolution of the Leggett model can be as short as 1 day, similar to the O'Flaherty model. Age dependency in the model is reflected in six age groups: infants; children; adolescents; and three subgroups of adults—young, middle-aged, and older adults.

Like the IEUBK and O'Flaherty models, Leggett considered the lung and the GI tract as the principal sites of Pb deposition for purposes of calculating intake. Dermal deposition and uptake were taken to be minor and not modeled. The respiratory tract is modeled as four compartments where Pb

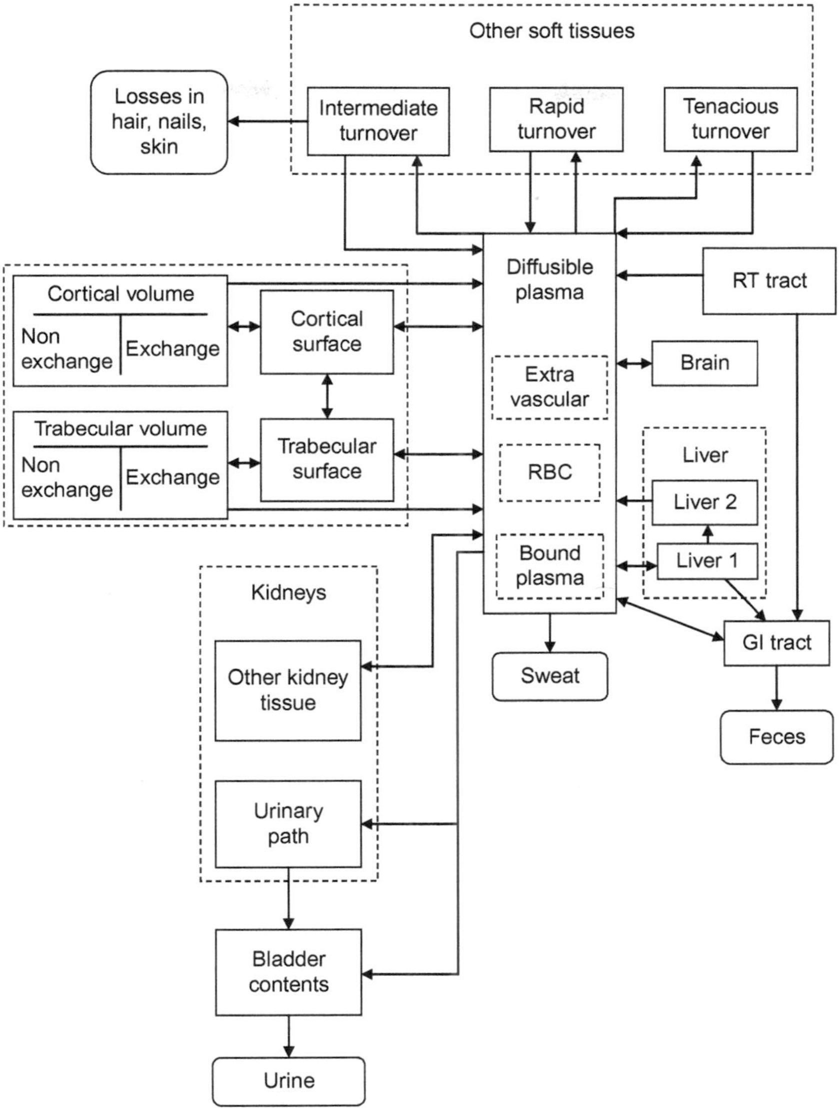

FIGURE 9.5 Structure of the Leggett lead biokinetic model (Leggett, 1993).

deposition and subsequent uptake occur with half-times of 1, 3, 10, and 48 hours, respectively. Only a small amount of inhaled Pb is assumed to enter the GI tract via ciliary clearance (~4%). Pb uptake from the GI tract is age-dependent, with 45% uptake to 1 year of age, falling to 30% by 15 years of age, and 15% afterward. Daily intakes are simulated as total daily amounts so that the quantities of media inhaled or ingested require knowledge of both Pb levels in assimilated media and volumes or masses of intake media. The Pb uptake rate is age-dependent but not media-dependent. This media Pb independency differs from the IEUBK model, where both default and user-selected uptakes for various media are selectable, but follows the media Pb independency in the O'Flaherty system.

The original construction of the model was based on various empirical data sets (Leggett, 1993; Pounds and Leggett, 1998): (1) adult human volunteers given labeled Pb via ingestion, inhalation, injection; (2) autopsy tissue measurements in variably aged subjects having environmental Pb exposures; (3) Pb balance studies; (4) autopsy tissue data in occupationally exposed subjects; (5) data from laboratory test animals; and (6) biokinetic testing results for elements similar to Pb in their bone-seeking behavior.

Bone is a major driver in the Leggett approach, as it is for the O'Flaherty model, with biokinetic subdivisions of Pb exchange rates into bone surface (rapid) and bone volume (much slower). Trabecular and cortical bone subcompartments are modeled separately. A total of six bone subcompartments were developed for the Leggett approach. The two main bone types—trabecular and cortical—are each subdivided into surface, exchangeable, and nonexchangeable subcompartments. The nonexchangeable subcompartment actually refers to that portion of bone-lodged Pb that is only released with resorption. The most rapid plasma Pb-bone Pb exchange occurs with Pb on the bone surface. The bone volume encases both the exchangeable and nonexchangeable main matrix Pb fractions. Half-lives for various Pb exchanges among bone subcompartments and with plasma have been determined for adults: plasma to bone surface, ~4 days; return to plasma from bone surface, 1.4 days; and bone surface to exchangeable subcompartment, 1.4 days. The half-life for return of exchangeable bone Pb in adults is, as expected, slower, ~30 days. The higher Pb masses in trabecular and cortical bone are taken to reside in the very kinetically slow nonexchangeable pools, with adult half-lives of 3.8 and 23 years for trabecular and cortical bone Pb, respectively.

Bone Pb transfer parameters are age dependent because bone turnover rates are age dependent, and typically are much faster in children. Compared to adults, for whom PbBO half-lives were given above, infants and children have much shorter half-lives for Pb mobility in the nonexchangeable bone volume compartment. For trabecular bone, half-lives at birth and 15 years of age are 68 and 725 days, respectively. For cortical bone, the half-lives at birth and 15 years of age are 68 and 1,354 days, respectively.

The liver is assigned two biokinetic compartments: a rapid compartment with a half-life of several days for movement to plasma and movement to the small intestine via biliary secretion. Similarly, there are two kinetic compartments for kidney Pb, the rapid one accounting for transfer of plasma Pb via the kidney to urine (adult half-life of ca. 5 days) and one representing Pb accumulation of 365 days. Three compartments are employed for all other soft tissues, generally depicted as involving rapid, intermediate, and slow turnover rates.

The outputs for the Leggett model most useful for the user are PbB and bone Pb. The Leggett model, like the O'Flaherty system, is not only applicable to infants and children but also to individuals older than children. This model is more difficult to use since it has a limited number of the accompanying modules in the other simulation systems. Intakes are modeled without an exposure module, and the user inputs age-specific point estimates expressed as units of μg/day. The lead intake time step is 1 day, meaning one can simulate abrupt, acute intermittent exposures (Khoury and Diamond, 2003), as one can also do with the O'Flaherty model. Dose reconstruction is straightforward, given data for Pb intakes. With regard to model construction, Pounds and Leggett (1998) noted that various software operating platforms can be used, including Windows, while Lorenzana et al. (2005) have reported a version built using ACSL.

Performance of the Leggett model has been compared with other simulation systems (Khoury and Diamond, 2003; Maddaloni et al., 2005; O'Flaherty, 1998; Pounds and Leggett, 1998). According to Pounds and Leggett (1998), the Leggett model, when adjusted for uptake differences with the IEUBK model, showed that PbB outputs for children 1–7 years of age were over twofold (2.2) higher than those of the IEUBK simulations. Since uptake differences were not present owing to adjustment, the differences are biokinetic in nature. A more recent comparison (U.S. EPA, 2007) showed the ratio of simulations for children's PbB levels when adjusting for uptake rates to be even higher, i.e., 3.3.

REFERENCES

Barry, P.S.I., 1975. A comparison of concentrations of lead in human tissues. Br. J. Ind. Med. 32, 119–139.

Beck, B.D., Mattuck, R.L., Bowers, T.S., Muir, B., 2001. The development of a stochastic physiologically-based pharmacokinetic model for lead. Sci. Total Environ. 274, 15–19.

Bergdahl, I.A., Schütz, A., Gerhardsson, L., Jensen, A., Skerfving, S., 1997. Lead concentrations in human plasma, urine and whole blood. Scand. J. Work Environ. Health 23, 359–363.

Bert, J.L., van Dusen, L.J., Grace, J.R., 1989. A generalized model for the prediction of lead body burdens. Environ. Res. 48, 117–127.

Bornschein, R.L., Succop, P., Dietrich, K.N., Clark, C.S., Que Hee, S., Hammond, P.B., 1985. The influence of social and environmental factors on dust lead, hand lead, and blood lead levels in young children. Environ. Res. 38, 108–118.

Bornschein, R.L., Clark, S., Pan, W., Succop, P., 1991. Midvale community lead study. Chem. Speciation Bioavailability 3, 149–162.

Bornschein R.L., Clark C.S., Pan W., 1992. The Butte-Silver Bow County Environmental Health Lead Study. Final Report. Butte-Silver Bow County Health Department and University of Cincinnati, Cincinnati, OH.

Bowers, T.S., Cohen, J.T., 1998. Blood lead slope factor models for adults: comparisons of observations and predictions. Environ. Health Perspect. 106 (Suppl. 6), 1569–1576.

Bowers, T.S., Mattuck, R.L., 2001. Further comparisons of empirical and epidemiological data with predictions of the Integrated Exposure Uptake Biokinetic Model for lead in children. Hum. Ecol. Risk Assess. 7, 1699–1713.

Bowers, T.S., Beck, B.D., Karam, H.S., 1994. Assessing the relationship between environmental lead concentrations and adult blood lead levels. Risk Anal. 14, 183–189.

Brunekreef, B., 1984. The relationship between air lead and blood lead in children: a critical review. Sci. Total Environ. 38, 79–123.

Brunekreef, B., Noy, D., Biersteker, K., Boleij, J., 1983. Blood lead levels of Dutch city children and their relationship to lead in the environment. J. Air Pollut. Control Assoc. 33, 872–876.

Casteel, S.W., Weis, C.P., Henningsen, G.M., Brattin, W.L., 2006. Estimation of relative bioavailability of lead in soil and soil-like materials using young swine. Environ. Health Perspect. 114, 1162–1171.

Drexler, J., Brattin, W.J., 2007. An *in vitro* procedure for estimation of lead relative bioavailability: with validation. Hum. Ecol. Risk Assess. 13, 383–401.

Drexler J., Mushak P., 1994. Health risks from extractive industry wastes: an approach to bioavailability of toxic metal and metalloidal contaminants. Annual Meeting of the Geological Society of America, October 24–27, 1994, Seattle, WA.

Drexler J., Mushak P., 1995. Health risks from extractive industry wastes: characterization of heavy metal contaminants and quantification of their bioavailability and bioaccessibility. Third International Conference on the Biogeochemistry of Trace Elements, May 15–19, 1995, Paris, France.

Franklin, C.A., Inskip, M.J., Baccanale, C.L., Edwards, C.M., Manton, W.I., Edwards, E., et al., 1997. Use of sequentially administered stable lead isotopes to investigate changes in blood lead during pregnancy in a nonhuman primate (*Macaca fascicularis*). Fundam. Appl. Toxicol. 39, 109–119.

Hayes, E.B., McElvaine, M.D., Orbach, H.G., Fernandez, A.M., Lyne, S., Matte, T.D., 1994. Long-term trends in blood lead levels among children in Chicago: relationship to air lead levels. Pediatrics 93, 195–200.

Hilts, S.R., 2003. Effect of smelter emission reductions on children's blood lead levels. Sci. Total Environ. 303, 51–58.

Hogan, K., Marcus, A., Smith, R., White, P., 1998. Integrated exposure uptake biokinetic model for lead in children: empirical comparisons with epidemiological data. Environ. Health Perspect. 106 (Suppl. 6), 1557–1567.

Khoury, G.A., Diamond, G.L., 2003. Risks to children from exposure to lead in air during remedial or removal activities at Superfund sites: a case study of the RSR lead smelter Superfund site. J. Expo. Anal. Environ. Epidemiol. 13, 51–65.

Lanphear, B.P., Roghmann, K.J., 1997. Pathways of lead exposure in urban children. Environ. Res. 74, 67–73.

Lanphear, B.P., Matte, T.D., Rogers, J., Clickner, R.P., Dietz, B., Bornschein, R.L., et al., 1998. The contribution of lead-contaminated house dust and residential soil to children's blood lead levels: a pooled analysis of 12 epidemiological studies. Environ. Res. 79, 51–68.

Leggett, R.W., 1992. A generic age-specific biokinetic model for calcium-like elements. Radiat. Prot. Dosim. 41, 183–198.

Leggett, R.W., 1993. An age-specific kinetic model of lead metabolism in humans. Environ. Health Perspect. 101, 598–616.

Lorenzana, R.M., Trost, R., Klotzbach, J.M., Follansbee, M.H., Diamond, G.L., 2005. Issues related to time averaging of exposure in modeling risks associated with intermittent exposures to lead. Risk Anal. 25, 169–178.

Maddaloni, M., Ballew, M., Diamond, G., Follansbee, M., Gefell, D., Goodrum, P., et al., 2005. Assessing non-residential lead risks at hazardous waste sites. Hum. Ecol. Risk Assess. 11, 967–1003.

Manton, W.I., Cook, J.D., 1984. High accuracy (stable isotope dilution) measurements of lead in serum and cerebrospinal fluid. Br. J. Ind. Med. 41, 313–319.

Manton, W.I., Rothenberg, S.J., Manalo, M., 2001. The lead content of blood serum. Environ. Res. 86, 263–273.

Marcus, A.H., 1985a. Multicompartment kinetic models for lead. I. Bone diffusion models for long-term retention. Environ. Res. 36, 441–458.

Marcus, A.H., 1985b. Multicompartment kinetic models for lead. II. Linear kinetics and variable absorption in humans without excessive lead exposures. Environ. Res. 36, 459–472.

Marcus, A.H., 1985c. Multicompartment kinetic models for lead. III. Lead in blood plasma and erythrocytes. Environ. Res. 36, 473–489.

Mushak, P., 1989. Biological monitoring of lead exposure in children: overview of selected biokinetic and toxicological issues. In: Smith, M., Grant, L.D., Sors, A. (Eds.), Lead Exposure and Child Development: An International Assessment. Kluwer Academic Press, Lancaster, UK, pp. 129–145.

Mushak, P., 1991. Gastrointestinal absorption of lead in children and adults: overview of biological and biophysico-chemical aspects. Chem. Speciation Bioavailability 3, 87–104.

Mushak, P., 1993. New directions in the toxicokinetics of human lead exposure. Neurotoxicology 14, 29–42.

Mushak, P., 1998. Uses and limits of empirical data in measuring and modeling human lead exposure. Environ. Health Perspect. 106 (Suppl. 6), 1467–1484.

National Academy of Sciences: National Research Council., 1993. Measuring Lead Exposure in Infants, Children, and Other Sensitive Populations. National Academy Press, Washington, DC.

O'Flaherty, E.J., 1991. Physiologically based models for bone-seeking elements. II. Kinetics of lead disposition in rats. Toxicol. Appl. Pharmacol. 111, 313–331.

O'Flaherty, E.J., 1993. Physiologically based models for bone-seeking elements. IV. Kinetics of lead disposition in humans. Toxicol. Appl. Pharmacol. 118, 16–29.

O'Flaherty, E.J., 1995. Physiologically based models for bone-seeking elements. V. Lead absorption and disposition in childhood. Toxicol. Appl. Pharmacol. 131, 297–308.

O'Flaherty, E.J., 1998. A physiologically based kinetic model for lead in children and adults. Environ. Health Perspect. 106 (Suppl. 6), 1495–1503.

O'Flaherty, E.J., 2000. Modeling normal aging bone loss, with consideration of bone loss in osteoporosis. Toxicol. Sci. 55, 171–188.

Pounds, J.G., Leggett, R.W., 1998. The ICRP age-specific biokinetic model for lead: validations, empirical comparisons, and explorations. Environ. Health Perspect. 106 (Suppl. 6), 1505–1511.

Rabinowitz, M.B., Wetherill, G., Kopple, J.D., 1976. Kinetic analysis of lead metabolism in healthy humans. J. Clin. Invest. 58, 260–270.

Ruby, M.V., Davis, A., Link, T.E., Schoof, R., Chaney, R.L., Freeman, G.B., et al., 1993. Development of an *in vitro* screening test to evaluate the *in vivo* bioaccessibility of ingested mine-waste lead. Environ. Sci. Technol. 27, 2870–2877.

Stern, A.H., 1994. Derivation of a target level of lead in soil at residential sites corresponding to a *de minimus* contribution to blood lead concentration. Risk Anal. 14, 1049–1056.

Succop, P., Bornschein, R., Brown, K., Tseng, C.-Y., 1998. An empirical comparison of lead exposure pathway models. Environ. Health Perspect. 106 (Suppl. 6), 1577–1583.

U.S. Environmental Protection Agency, 1986. Air Quality Criteria for Lead, 4 vols. Report No. 600/8-83/028bF. Environmental Criteria and Assessment Office, Washington, DC.

U.S. Environmental Protection Agency, 1994a. Guidance Manual for the Integrated Exposure-Uptake Biokinetic Model for Lead in Children, Report No. EPA/540-R93/081. Office of Research and Development, Washington, DC.

U.S. Environmental Protection Agency, 1994b. Technical Support Document: Parameters and Equations Used in the Integrated Exposure-Uptake Biokinetic Model (v 0.99d), Report No. EPA/540/R-94/040. Office of Research and Development, Washington, DC.

U.S. Environmental Protection Agency, 1994c. Validation Strategy for the Integrated Exposure-Uptake Biokinetic Model for Lead in Children, Report No. EPA 540-R94039. Office of Research and Development, Washington, DC.

U.S. Environmental Protection Agency, 1996a. Recommendations of the Technical Review Workgroup for Lead for an Interim Approach to Assessing Risks Associated with Adult Exposures to Lead. Draft Report, December.

U.S. Environmental Protection Agency, 1996b. Urban Soil Lead Abatement Demonstration Project. Vol. I: Integrated Report, Report No. EPA/600/P-93/001aF. Office of Research and Development, Washington, DC.

U.S. Environmental Protection Agency, 2006. Air Quality Criteria for Lead, Vol. 1, Report No. EPA/600/R-05/144aF. National Center for Environmental Assessment, Washington, DC.

U.S. Environmental Protection Agency, 2007. Lead: Human Exposure and Health Risk Assessments for Selected Case Studies, Report No. EPA-452/R-07-014a. Office of Air Quality Planning and Standards, Washington, DC.

U.S. Science Advisory Board, 1989. Clean Air Scientific Advisory Committee (CASAC) Report: Review of the Lead NAAQS Exposure Analysis Methodology and Evaluation, Report No. EPA SAB-CASAC-89-018. U.S. Environmental Protection Agency, Washington, DC.

U.S. Science Advisory Board, 1992. Indoor Air Quality and Total Human Exposure Committee. Review of the Uptake Biokinetic (UBK) Model for Lead, Report No. EPA SAB-IAQC-92.

White, P.D., Van Leeuwen, P., Davis, B.D., Maddaloni, M., Hogan, K.A., Marcus, A.H., et al., 1998. The conceptual structure of the Integrated Exposure Uptake Biokinetic Model for lead in children. Environ. Health Perspect. 106 (Suppl. 6), 1513–1530.

World Health Organization, 2000. Air Quality Guidelines for Europe, second ed. Inorganic Pollutants. Lead. Ch. 6.7. WHO Regional Office for Europe: European Series No. 91, Copenhagen, Denmark.

Chapter 10

The Environmental Epidemiology of Human Lead Exposure

10.1 INTRODUCTION

This chapter deals with epidemiological studies of human Pb exposure using biomarkers of such exposure. These studies deal with various parameters and correlates of human Pb exposure from demographic and environmental measurements. While much of the available data has been generated in the United States, illustrative international studies are also presented where relevant. Later chapters deal with epidemiological studies of dose–response relationships for Pb and an array of toxicity outcome measures addressing organ and tissue toxicities.

Several types of epidemiological data for human exposures are presented here. First are descriptive depictions of the prevalence of distributions of measured biomarkers of Pb exposure. Prevalence and distribution data include their statistical design for the various reported surveys and screenings. Surveys are typically structured to be both maximally representative of Pb exposures and representative in large geographic, socioeconomic, and demographic dimensions. Using repeat or serial surveys over time can further allow one to ascertain trends and the health or regulatory policy implications of these trends.

A second form of epidemiological study uses inferential statistical analyses where the epidemiology is of two types: exposure biomarkers such as blood Pb studied for links to environmental sources and pathways, and studies to ascertain Pb-containing media that may be significant contributors to toxic exposures and toxicity in human populations.

Survey and screening epidemiological data include the various population and subpopulation surveys and screenings of socioeconomic and demographic strata carried out in the United States and elsewhere over the past several decades. Additional cumulative aspects of these studies addressed here are the epidemiological and public health messages for Pb exposure

Trace Metals and other Contaminants in the Environment, Volume 10
ISSN: 1875-1121 DOI: 10.1016/B978-0-444-51554-4.00010-9

trends in terms of geography, economic status, and time. Illustrative studies include meta-analyses.

"Surveys" of lead exposures are data gathering and statistical devices and they differ in design from "screenings" of Pb exposures, particularly of young children, in a number of ways. Distinctions between the two may account for differences seen in Pb exposure statistics with serial National Health and Nutrition Examination Survey (NHANES) efforts over the years compared with results from community-specific lead screening programs for identification, for example, of childhood exposures. The U.S. ATSDR, in its 1988 report to the U.S. Congress (Ch. 5 of ATSDR), noted that the NHANES results produced higher exposure frequencies for several reasons. The NHANES II in the 1980s and early 1990s used directly measured PbB levels rather than preliminary EP screening measurement, a use that led to false negatives. Second, surveys are more intensive than screenings, entailing more blanket canvassing versus screening. Screenings are typically conducted in clinics.

However, past childhood Pb exposure screenings in urban areas have tended to show higher frequencies than surveys inasmuch as screening programs target more of the high-risk children in terms of demographics and more risk intensive sources of lead exposures. Inner-city screening programs targeting children in older, deteriorating housing with lead paint hazards are an illustration.

National exposure surveys discussed here include demographic and socioeconomic snapshots of the prevalences and incidences of biomarkers of Pb exposures with reference to distributions of measured biomarkers of Pb exposure and their variation, gathered by intensive measurement efforts stratified along multiple socioeconomic and demographic lines.

10.2 EARLY EPIDEMIOLOGICAL STUDIES OF HUMAN LEAD EXPOSURE

This section recapitulates information on United States' and other populations' Pb exposures as they were noted in earlier decades, beginning in the 1960s and 1970s. There are a number of important reasons for including this historical exposure aspect, notably the biological fact that older segments of the population acquired relatively higher body lead burdens in those decades versus current exposures, and these burdens are being returned from bone lead accumulations back to blood. The evidence for this is considerable (Miranda et al., 2010; Silbergeld et al., 1988).

Lead exposure surveys and other wide-scale population exposure assessments in the United States and elsewhere have largely employed PbB sampling and measurement approaches or any of several early hematotoxic effect biomarkers, such as EP and ZPP. The combination of both came into use by the 1970s. Pb testing in whole blood seemed to offer the best tradeoff between cost-effectiveness and use as an index of internal Pb exposure subsequent to a positive EP value for diverse health scenarios.

Few measurements of PbB in any systematic fashion for survey or screening purposes had been recorded before the 1960s. There are several reasons for this. First—as was the case for environmental Pb measurements relevant for human health stated in Chapter 6—few reliable quantitative measurement methods of the requisite sensitivity, specificity, reliability, and logistics existed before this period. PbB measurements up to this time typically required large volumes of whole blood to use in multistep, marginally sensitive and specific procedures (see, for example, NAS/NRC, 1993; U.S. EPA, 1977, 1986). While adequate for clinical application in suspected serious lead poisonings, such methods were deficient for wide-scale screenings or surveys of child and other human populations. These limits eventually led to development and use of microscale test procedures, particularly micro testings of Pb exposures using small blood volumes to measure PbB or EP elevation.

A second factor was the absence, until recent decades, of an established need for systematic evaluations of population-wide lead exposures in the United States and elsewhere. Lead toxicity and the biomarkers of such toxicity had traditionally lain in the realm of clinical toxicology and pediatrics with its focus on symptomatic lead poisoning. Isolated cases of lead poisoning usually produced case reports when poisonings were identified. The notion of having to quantify risk of lead poisoning within large risk groups when clinical childhood lead poisoning was approaching epidemic proportions in the post-WWII United States was a newer arrival to the nation's child and public health matrix. A third factor in a number of countries was the establishment and rapid growth of Federal and other governmental granting agencies that collectively provided resources for studying the nature and extent of lead poisoning in children and other risk groups. Arm's-length research funding supported studies that increasingly provided evidence that symptomatic and asymptomatic lead poisonings, especially in industrialized nations, were a much larger health issue than earlier realized.

Results from such studies, furthermore, triggered various legislative and executive regulatory and health policy actions to support more research, actions then producing more resources and scientific results. These new results were followed by additional demands. For example, the October 1970 Statement of the U.S. Surgeon General, "Medical Aspects of Childhood Lead Poisoning," identified a PbB of 40 μg/dl in children as the upper limit of safe (U.S. CDC, 1975). Shortly after the Surgeon General's Statement, Congress passed the 1971 Lead-Based Paint Poisoning Prevention Act (LBPPPA), whose provisions included mandating that Federal agencies take steps to identify the scope of the lead poisoning problem and implement steps for prevention. The 1971 LBPPPA and its later amendments and reauthorizations were instrumental in catalyzing more Federal research and more Federal activities in U.S. communities, particularly in establishing and increasing high-risk lead exposure screenings in American urban areas.

Early efforts at examining human lead exposure in other than conventional poisoning settings were directed at some relatively narrow questions. One question arising early in the history of Pb exposure studies was linking the extent of individual body lead burdens in industrialized nations to the extent of industrial or other anthropogenic activities involving lead production and consumption. A geographic element of this interest consisted of a cluster of studies comparing lead levels in the bodies of the United States and other Western subjects exposed to Pb under industrial, urbanized conditions with lead levels in individuals assumed to be minimally exposed. The rationale for this interest was obvious for the then-current state of scientific evidence. If levels of Pb in industrially exposed individuals did not vary from those body burdens identified in remote, unacculturated populations, then anthropogenic activity was not linked to toxic Pb exposures and associated toxicity. That is, Pb was present universally at "normal" levels in all subsets of human populations, regardless of place or circumstance.

Several early studies of this type purported to show no significant differences between remote, undeveloped area populations and populations in developed, industrialized countries. These findings, however, are rendered largely meaningless by deep flaws in their design and execution. For example, Needleman (2000) cites some early data for remote villagers in Mexico who seemingly had very little in the way of industrial exposures but whose body Pb levels were not materially different from those of industrialized exposure residents. As Needleman noted, a lethal flaw in this study was the fact that these villagers sustained high lead exposures from the extensive use of poorly fired pottery with lead glazes for food preparation and storage. Such glazes readily leach Pb into any contents.

Later studies of various remote populations, done in the 1960s by Goldwater and Hoover (1967) and Stopps (1968), included results in aboriginal populations such as those in Papua, New Guinea. They reported that blood levels in these individuals were not significantly lower than those in Western urban and suburban locales. The New Guinea subjects presented with a mean PbB of 13 μg/dl. Such results, in retrospect, were made highly unreliable by faults in the collection and measurement protocols; such studies had little appreciation of, or steps taken to prevent, sample and laboratory contamination problems.

Numerous later studies confirmed those shortcomings and conclusively document the pervasive effects of ubiquitous global Pb pollution in increasing human body Pb levels orders of magnitude above prehistoric populations. For example, subsequent studies of remote populations with better attention to contamination and sample integrity problems showed significantly lower body Pb levels in remote versus industrialized populations (Piomelli et al., 1980; Poole et al., 1980). Patterson et al. (1991) estimated that modern

TABLE 10.1 Mean PbB Levels in Urban and Rural U.S. Areas in 1960[a,b]

Survey Locale[c]	Mean PbB (μg/dl)	Excess Urban Versus Rural (μg/dl)
Chicago, IL (97)	20	6
Cincinnati, OH (137)	20	6
Dallas, TX (128)	18	4
Denver, CO (131)	19	5
New Orleans, LA (130)	22	8
New York, NY (112)	20	6
Rural Central Ohio (162)	14	–

[a]*Adapted from U.S. EPA (1977).*
[b]*Various ages from data of Hofreuter et al. (1961).*
[c]*Sample size in parentheses.*

North Americans carried 300-fold higher body Pb burdens than prehistoric humans. Flegal and Smith (1992) reported that prehistoric adults would have typically had a PbB value of 0.016 μg/dl, using extrapolation estimates from bone Pb contents of preindustrial and contemporary samples and animal data. Mushak (1993), using a computerized biokinetic Pb exposure simulation model and published estimates of prehistoric Pb levels in human diet, water, ambient air, etc., calculated a PbB for prehistoric children in the narrow range of 0.06–0.12 μg/dl, in good accord with the estimate of Flegal and Smith (1992).

Other studies involving descriptive or inferential PbB statistics in the United States and other populations in the 1960s and early 1970s were relatively scattered and largely confined to local areas that addressed specific Pb exposure concerns. For example, some studies contrasted PbB distributions in the United States or other urban dwellers versus suburban or rural residents. Others compared PbB levels associated with specific exposure variables, such as air Pb content, traffic density, Pb emissions from stationary sources such as smelters, Pb levels in soils, dusts, and interior or exterior Pb paints. Tables 10.1–10.3 present some of these earlier studies for illustration. It must be kept in mind that data described in these tables reflect limits prevailing at the time in the protocols for environmental epidemiology studies of Pb exposure. Sample contamination, measurement sensitivity and specificity, sampling statistical validity, toxicokinetic characteristics of lead disposition in humans and especially in young children were only poorly understood.

TABLE 10.2 Mean PbB Levels in Children in Smaller Illinois Cities, 1971[a–c]

Illinois City	Percent Screened	Mean PbB (μg/dl)[d]
Aurora (449)	5.1	28
Carbondale (264)	17.4	29
Chicago Heights (537)	10.4	25
Decatur (793)	5.8	22
East Moline (298)	12.3	24
East St. Louis (376)	4.1	29
Joliet (383)	4.5	28
Peoria (387)	3.0	33
Rockford (1208)	7.3	20
Springfield (670)	7.3	32

[a]*Adapted from U.S. EPA (1977), based on data of Fine et al. (1972).*
[b]*Children aged 1–6 years old.*
[c]*Sample size in parentheses.*
[d]*Rounding.*

TABLE 10.3 Mean PbB Levels (μg/dl) in Black Inner-City Children in Philadelphia, 1972–1973[a,b]

Age of Children (*N*)	PbB (*N*)[c]	PbB (*N*)[d]
<1 (61)	19 (61)	18 (9)
1–3 (289)	29 (289)	33 (40)
4–8 (404)	25 (404)	26 (55)
9–13 (394)	21 (394)	23 (44)
14–16 (242)	19 (242)	19 (21)

[a]*Adapted from data of Adebonojo (1974).*
[b]*Children aged 6 months to 18 years, males plus females.*
[c]*Children's PbBs with normal G6PD status.*
[d]*PbBs for G6PD-deficient children.*

Table 10.1 presents mean PbB values for selected U.S. major cities and a rural area of central Ohio. Overall, the urban mean values for the six cities ranged from 18 to 22 μg/dl, significantly higher than the rural Ohio figure of 14 μg/dl. The maximum values in these data sets ranged from 38 to 60 μg/dl.

These data typified other localized evaluations of the relationship of PbB in individuals of varying age in the United States and elsewhere. Hasselblad and Nelson (1975) compared urban versus suburban PbB values for three cities: Chicago, Philadelphia, and New York City. The urban mean PbB values, whether compared individually or as a group, were significantly higher statistically than suburban subjects. The mean urban PbB level was 18 versus 16 μg/dl for suburban residents. Male adult residents' PbB values were studied in 19 intermediate-size U.S. cities (Kubota et al., 1968). Mean PbB concentrations varied from 7 μg/dl in Lafayette, LA, to 20 μg/dl in Jacksonville, FL. Only three cities were below 8 μg/dl: Lubbock, TX; Geneva, NY; and Lafayette, LA. Some European data on PbB concentrations appeared in these earlier years. Haegar-Aronsen et al. (1971), for example, reported that female adults in Sweden and Finland had an average PbB of 10 μg/dl. Northern Italian adults from the Milan area, by contrast, had very high PbB averages: 30 μg/dl for males and 23 μg/dl for females.

Some limited epidemiological studies appeared in the 1970s dealing with PbB levels in children. Fine et al. (1972) studied PbB mean values of children 1–6 years of age living in 14 intermediate-size Illinois communities. Ten of these communities with the largest sampling populations are tabulated in Table 10.2. For the 10 communities, mean levels ranged from 20 to 33 μg Pb/dl. The fraction screened varied considerably, potentially affecting the statistical reliability of the results, ranging from 4.1% to 17.5%. Adebonojo (1974) reported mean PbB concentrations for inner-city Philadelphia children aged 6 months to 16 years. Results are tabulated in Table 10.3.

Children aged 1–3 years had higher PbB mean values than those 6 months old, and higher means than older children, notably over 8 years of age. Similar results were seen with New York City childhood lead screenings between 1970 and 1976, as reported in Table 10.4 (U.S. EPA, 1977). Children in the two annual-segment age bands encompassing 13–36 months of age had higher PbB levels than the youngest subset and the older age bands. Across races, the highest PbB levels were seen for Black children. It should be noted, however, that regardless of age band and race, New York City young children all had significant elevations in PbB. The 61- to 72-month age band of all three groups in 1970 had PbB means in the 20s.

Young children outside of densely populated older urban areas in the 1970s were also seen to have elevated PbB levels if they lived close to metal smelters. Baker et al. (1977) examined PbB levels in children 1–5 years of age living near lead, copper, and zinc smelters and compared results with a control community. Results are summarized in Table 10.5.

Three control towns were examined, while 3 lead smelter towns, 11 copper smelter communities, and 5 zinc smelter locales were tested. As seen in the table, the zinc smelter towns were much higher in arithmetic and geometric mean values compared to copper and lead smelters. Particularly in Table 10.5, the frequencies >35 μg/dl for the various categories are shown. Compared to

TABLE 10.4 Selected Early PbB Screening Results in New York City Children, 1970–1976, for Blacks, Hispanics, and Whites[a–d]

Group/Year	PbB (μg/dl) by Age Bands (months)					
Blacks	1–12	13–24	25–36	37–48	49–60	61–72
1970	27	31	31	28	31	23
1971	25	30	30	30	29	28
1972	22	26	27	26	25	24
1973	23	27	26	26	25	24
1974	22	26	26	24	24	22
1975	21	23	23	22	22	22
1976	18	21	21	21	20	19
Hispanics	1–12	13–24	25–36	37–48	49–60	61–72
1970	22	25	26	24	24	25
1971	20	23	25	25	24	24
1972	19	21	22	23	22	22
1973	20	22	23	23	22	22
1974	20	21	23	23	22	20
1975	17	20	21	21	21	20
1976	18	19	19	19	19	18
Whites	1–12	13–24	25–36	37–48	49–60	61–72
1970	21	24	25	25	24	25
1971	20	23	23	23	24	22
1972	17	20	22	21	21	21
1973	20	22	22	22	21	22
1974	19	21	20	21	21	21
1975	19	20	19	18	20	18
1976	21	17	19	19	18	18

[a]*Children from birth to 72 months.*
[b]*Adapted from U.S. EPA (1977) for Billick et al. report data.*
[c]*Age bands selected are at greatest risk.*
[d]*PbB values are geometric means, rounding.*

TABLE 10.5 PbB Arithmetic and Geometric Mean Levels in Young Children in U.S. Smelter and Comparison Communities in 1975[a–c]

Smelter Group (Child, *N*)	Arithmetic Mean (μg/dl)	Geometric Mean (μg/dl)	% >35 μg/dl
Control towns (258)	17	15	1.2
Lead towns (158)	16	15	4.4
Copper towns (909)	16	15	2.0
Zinc towns (347)	21	19	9.5

[a]*Adapted from data of Baker et al. (1977).*
[b]*Total of 1,672 children, 1–5 years of age.*
[c]*Pb smelter towns: Bixby, MO; Glover, MO; Herculaneum, MO.*
Cu smelter towns: Azo, AZ; Anaconda, MT; Copper Hill, TN; Douglas, AZ; Hayden, AZ; Hurley, MN; McGill, NV; Morenci, AZ; San Manual, AZ; White Pine, MI. Zn smelter towns: Amarillo, TX; Bartlesville, OK; Corpus Christi, TX; Monaco, PA; Palmerton, PA. Control towns: Albuquerque, NM; Perryville, MO; Sofford, AZ.

the exceedence rate of 1.2% for control subjects, the copper, lead, and zinc towns were significantly higher at 2.0%, 4.4%, and 9.5%, respectively.

The El Paso, TX, primary lead smelting facility was the subject of a number of epidemiological studies including PbB surveys in the 1970s. A 1972 U.S. CDC evaluation (Landrigan et al., 1975) examined air, dust, and soil Pb levels along with children's PbB values. Data were stratified as a function of distance from the facility. Air Pb around this operation ranged from 0.5 to 75 $\mu g/m^3$ with a mean value of 6.6 $\mu g/m^3$. Geometric mean dust and soil Pb values were 4,022 and 1,791 ppm, respectively. The majority of children (69%) 1–4 years of age and living near the smelter had PbB levels >40 μg/dl. The fraction exceeding 60 μg/dl was 14%. As expected, the children in the high PbB group lived closest to the smelter and had the highest dust and soil Pb levels. Age of children was also a determinant of PbB, a relationship commonly seen in such communities.

A large lead smelter in Kellogg, ID, was another subject of studies in the 1970s, particularly after a major baghouse fire in 1973 which led to an abrupt increase in air Pb emissions and children being hospitalized with Pb poisoning. The U.S. CDC responded in 1974 with a joint study with the State. Children in the entire catchment area showed 42% with PbB ≥40 μg/dl, while 99% of children living within 1 mile (1.6 km) of the facility had PbB levels ≥40 μg/dl. Yankel et al. (1977) stratified geometric mean PbB levels for these children, finding that those living within 1.6 km had a mean PbB of 66 μg/dl and those living 1.6–4.0 km had a mean of 48 μg/dl. Associated estimates of air Pb were quite elevated at 18 and 14 $\mu g/m^3$,

TABLE 10.6 Early Studies of Association of PbB Levels with Mobile Pb Sources

Study Site/Subjects (*N*)	Pb Source/Design	Mean PbB (μg/dl)	References
Houston, TX			
Policemen (141)	Subjects exposed to auto exhaust Pb: matched by age, education, race; exposed worker PbBs significantly higher; female residents not different	23[a]	U.S. EPA (1977), Table 12.14
Controls (150)		18	
Parking Garage Attendants (119)		28[a]	
Controls (95)		21	
Female residents near freeway (120)		13	
Controls (117)		12	
Los Angeles, CA			
Males 1–16 years	PbB means for subjects exposed to high traffic (Los Angeles) versus low traffic (Lancaster) areas	24	Johnson et al. (1975)
Males 17–34 years		17	
Females 1–16 years		17	
Females 17–34 years		13	
Lancaster, CA			
Males 1–16 years	PbB means for subjects exposed to high traffic (Los Angeles) versus low traffic (Lancaster) areas	11	Johnson et al. (1975)
Males 17–34 years		12	
Females 1–16 years		10	
Females 17–34 years		9	

[a]$p = 0.05$.

respectively. Other smelter and related facilities studied in this interval were those in East Helena, MT (Hammer et al., 1973) and Omaha, NE (Angle and McIntire, 1979). In summary, Pb smelters were significant sources of elevated Pb exposures in children, including toxic exposures, and this source affected children's exposures as a function of distance from stack emissions.

A number of earlier studies examined the relationship of PbB levels in various individuals to types of lead sources. Most of these were epidemiological in nature, with a few being more clinical in design and execution. Table 10.6 typifies studies of mobile Pb source exposures of adults. It shows that Houston, TX, policemen and parking garage attendants had significantly higher PbB levels than did control subjects, while women living closer to traffic than controls did not show significant difference.

The means for the policemen and attendants were 23 and 28 μg/dl, respectively. The higher PbB values for workers exposed to leaded auto exhaust in garages relate to the higher local emissions persisting in enclosed parking structures (see Emissions and Fate and Transport, Chapters 4 and 5). Two California communities differing as to automobile traffic density and air Pb levels were examined by Johnson et al. (1975). Blood lead arithmetic and geometric mean levels in Los Angeles, a high density locale with mean freeway air Pb of 6.3 $\mu g/m^3$, were compared for gender and ages with corresponding levels of Lancaster, CA, a smaller community with much lower air Pb, 0.6 $\mu g/m^3$. Male Los Angeles children, aged 1–16 years, had arithmetic and geometric means of 24 and 21 μg/dl, respectively. For female children, the corresponding values were 17 and 15 μg/dl. In Lancaster, male children aged 1–16 years had much lower arithmetic and geometric mean PbB values, 11 and 10 μg/dl. Similarly, female children's PbB values were 10 and 10 μg/dl, respectively. PbB differences were highly statistically significant, $p < 0.001$.

Caprio et al. (1974) described a large 1971 study of 5,226 Newark, NJ, children aged 1–5 years of age where PbB mean values were compared to proximity of these children to various area roadways. Results were computed as percentages of children exceeding various PbB ceilings. The highest fraction of children, 57%, having PbB levels 40 μg/dl or higher corresponded to distances of <100 ft from roadways. For children living >200 ft from roads, the corresponding fraction with PbB >40 μg/dl declined to 32%. Waldron (1975) reported analogous results for variably aged men and women living within 800 m of a new major vehicular interchange in Birmingham, England. Mean male and female PbB values before the interchange opening were 14 and 11 μg/dl, respectively. After about 1 year of operation, the corresponding values were 24 and 19 μg/dl, respectively. Overall, these illustrative data provide early evidence of the significant role of air Pb and auto exhaust Pb emissions producing significant human Pb exposures.

10.3 CONTEMPORARY Pb EXPOSURE SURVEYS AND THEIR VARIOUS U.S. AND INTERNATIONAL CORRELATES

This is the first of two sections that deal with contemporary epidemiological studies of Pb exposure in the United States and elsewhere, with a principal but not exclusive focus on Pb exposures of young children. It describes the serial national and international surveys of Pb exposures that began in earnest in the late 1970s and that continue to the present. The multiple U.S. NHANES, begun in the 1970s, are the main focus. These efforts included survey data for PbB concentrations. In the main, these surveys for PbB and other measurements avail themselves of rigorous statistical designs, multiple national socioeconomic and demographic strata, a rigorous field protocol for survey and sample data collection, a rigorous laboratory protocol for optimized measurements of PbB and related other biomarkers (hematological indicators), and stringent statistical methodologies for epidemiological assessments.

10.3.1 The Serial U.S. NHANES

The U.S. NHANES efforts were designed and carried out by the National Center for Health Statistics, now a part of the U.S. CDC. Pb measurement is but one of many national population variables measured in these surveys.

The NHANES II and later statistical designs are a stratified, multistage aggregated probability cluster sample of U.S. households representing, for NHANES II, 72 distinct socioeconomic and demographic strata among the noninstitutionalized U.S. population. NHANES II efforts encompassed the years 1976–1980. A total of 16,563 persons ranging in age from 0.5 to 74 years were included, of whom 9,933 individuals provided PbB measurements for estimating national statistical data.

Table 10.7 presents selected overall U.S. population survey data from the NHANES II effort calculated at the midpoint of this survey, 1978. The original data set includes several more age bands, but this table conveys the salient information with respect to variables impacting U.S. PbB levels. The arithmetic means for White residents, 0.5–64 years of age, ranged from 13.0 to 15.0 μg/dl. For Black residents, the arithmetic mean levels ranged from 13.8 to 20.9 μg/dl.

Several significant differences are apparent in this table. First, mean PbB values differed with age regardless of race: highest in the youngest children, 0.5–2 years of age, declining to a minimum at 18–24 years, and then rising to the oldest group. Second, regardless of age, Blacks had higher PbB mean values at any particular age. The racial distinction is pronounced for children of 0.5–2 and 3–5 years. Across percentiles, the age and racial distinctions were also pronounced, especially at the 95th percentile. Black children aged 3–5 years at the 95th percentile had a PbB value of fully 10 points higher than White children in the comparable age group at the 95th percentile, 35 versus 25 μg/dl.

Table 10.8 presents selected NHANES II PbB U.S. statistics for residents varying as to gender, race, age, and percentages of subjects exceeding the indicated range of PbB values. Averages are expressed as both arithmetic and geometric means. Male adults, 18 years and older, had significantly higher arithmetic and geometric mean PbB concentrations than females, regardless of race. Blacks had higher arithmetic and geometric mean PbB levels than Whites, particularly in the childhood age band. Of particular interest are the percentage distributions of PbB within PbB ranges ≥10 μg/dl. Black children had 2.0% ≥40 μg/dl, and 40% in the 20–29 μg/dl range. This compares to White children's corresponding figures of 0.2% and 16.1%. Few Black children had PbB values <10 μg/dl, 2.5%, while White

TABLE 10.7 The NHANES II Blood Pb Data Set: Selected Arithmetic Mean PbB Levels and Selected Percentile Distributions for U.S. Residents, 1976–1980, by Age and Race[a–d]

Age (years)	Arithmetic Mean (μg/dl)	Percentile 10th	25th	50th	95th
White					
0.5–2	15.0	8.0	11.0	14.0	26.0
3–5	14.9	9.0	11.0	14.0	25.0
6–8	13.3	8.0	10.0	12.0	22.0
18–24	13.0	8.0	9.0	12.0	23.0
35–44	14.4	7.0	10.0	13.0	27.0
55–64	14.4	8.0	10.0	13.0	26.0
Black					
0.5–2	20.9	11.0	15.0	19.0	38.0
3–5	20.8	12.0	16.0	20.0	35.0
6–8	17.7	11.0	13.0	17.0	–
18–24	13.8	8.0	10.0	13.0	23.0
35–44	15.6	9.0	11.0	14.0	–
55–64	17.2	10.0	12.0	16.0	28.0

[a]*Adapted from Annest and Mahaffey (1984), Table 1.*
[b]*PbB values at midpoint of survey, 1978.*
[c]*PbB sampling by venipuncture.*
[d]*Original survey age band = 0.5–74 years of age.*

children had 14.5% <10 μg/dl. Sizeable percentages of children of both races had PbB values in the 10–19 μg/dl range.

Variation in geometric mean PbB levels for children, aged 1–5 years, by race and frequency ≥30 μg/dl are set forth in Table 10.9. Among Black children, geometric mean PbB is similar for ages 1–4, declining at age 5. Less pronounced change at the oldest age is seen for White children. The percentage of Black children having PbB ≥30 μg/dl at all ages is about sixfold higher than for White children, 4.6–18.2% in Blacks versus 0.8–2.8% in Whites. In the years of the NHANES II, the figure of 30 μg/dl was the U.S. CDC guideline for onset of concern for early adverse effects.

The NHANES II included many other demographic and socioeconomic strata, where the PbB levels and distributions were important to identify and

TABLE 10.8 NHANES II Blood Pb Data: Blood Pb Means and PbB Distributions for U.S. Residents, 0.5–74 Years of Age as a Function of Age and Race[a–c]

Age (years)	Arithmetic Mean (μg/dl)	Geometric Mean (μg/dl)	% at PbB of <10	10–19	20–29	40
White						
0.5–5	14.9	14.0	14.5	67.4	16.1	0.2
6–17	12.1	11.3	30.4	63.4	5.8	–
18–74						
Men	16.6	15.6	8.1	64.8	23.3	0.5
Women	11.7	10.9	34.6	59.9	5.0	–
Black						
0.5–5	20.9	19.6	2.5	45.3	40.0	2.0
6–17	14.8	14.0	12.8	70.9	15.6	–
18–74						
Men	19.1	18.1	2.3	56.4	34.9	1.9
Women	12.7	12.0	24.7	68.1	7.2	–

[a]*Adapted from Annest and Mahaffey (1984), Table 4.*
[b]*For midpoint, March 1978.*
[c]*Values based on venipuncture samples.*

consider, particularly where they addressed childhood lead exposures. Table 10.10 presents family income and the degree of urbanization as significant variables for childhood lead exposures, addressing Pb exposures in those 6 months to 5 years of age. Data are presented in the form of percentages of children in the survey who exceeded indicated PbB concentrations.

Black children from low-income families, <\$6,000, showed 61.0% ≥20 μg/dl, 34.2% ≥25 μg/dl, and 18.5% ≥30 μg/dl. White children at this same income stratum, by contrast, showed much lower percentages over the same PbB thresholds: 33.1% ≥20 μg/dl; 14.0% ≥25 μg/dl; 5.9% ≥30 μg/dl. Black children from the middle income stratum, i.e., \$6,000–14,999, still showed significant percentages above various exceedence thresholds in PbB: 51.1% ≥20 μg/dl; 23.6% ≥25 μg/dl; 12.1% ≥30 μg/dl. The highest family income stratum of ≥\$15,000 showed for Black children: 36.5% ≥20 μg/dl; 8.0% ≥25 μg/dl; and 2.8% ≥30 μg/dl.

The level of urbanization (Table 10.10) is another significant variable in the levels of childhood Pb exposures for both White and Black children.

TABLE 10.9 NHANES II Blood Pb Data: Selected Geometric Mean PbB Levels and PbB Exceedence Percentages for White and Black U.S. Young Children, 1–5 Years Old[a–c]

Age (years)	Geometric Mean (μg/dl)	Percent >30 μg/dl
White		
1	14.1	2.8
2	14.3	2.8
3	14.3	2.9
4	13.8	1.8
5	13.9	0.8
Black		
1	20.4	18.2
2	20.5	16.8
3	21.3	18.1
4	20.0	11.1
5	18.2	4.6

[a]*Adapted from Annest and Mahaffey (1984), Table A.*
[b]*Based on NHANES II midpoint of survey.*
[c]*PbB collected by venipuncture.*

Furthermore, there is some tracking of income and urbanization to the extent that densely urbanized communities are also communities with high numbers of low-income families, especially in the older parts of the nation. Black children living in heavily urbanized cities, with a population of 1 million or more, especially bear this out. The percentages of Black children having PbB ≥20, ≥25, and ≥30 μg/dl in the most densely populated cities are seen in Table 10.10 to be 59.1%, 28.5%, and 15.2%, respectively. Corresponding percentages for White children in these same levels of urbanization are much lower: 28.3, 9.8, and 4.0 μg/dl, respectively. For urban areas having <1 million residents, the percentages of Black children having elevated PbB concentrations are less, the corresponding figures for threshold PbB exceedences of 20, 25, and 30 μg/dl being, respectively, 51.4%, 14.1%, and 10.2%. As for the higher-density case, White children exceedence percentages are half or less than those of Black children, the figures being 20.6%, 5.7%, and 1.6% for the PbB thresholds of ≥20, ≥25, and ≥30 μg/dl, respectively.

Significant percentage exceedences above the indicated PbB thresholds for Black children persist into the "Rural" demographic category, reflecting

TABLE 10.10 NHANES II Blood Pb Data: Percentage of Black and White Children Exceeding PbB Ceilings (μg/dl) as a Function of Family Income and Urbanization[a–c]

Demographic Variable	Percentage of White Children			Percentage of Black Children		
	≥20	≥25	≥30	≥20	≥25	≥30
Family income						
<$6,000	33.1	14.0	5.9	61.0	34.2	18.5
$6,000–14,999	21.1	5.6	2.2	51.1	23.6	12.1
≥$15,000	10.5	3.3	0.7	36.5	8.0	2.8
Urbanization						
Urban, ≥1 million	28.3	9.8	4.0	59.1	28.5	15.2
Urban, <1 million	20.6	5.7	1.6	51.4	14.1	10.2
Rural	9.5	2.7	1.2	29.9	19.4	10.3

[a]*Adapted from Annest and Mahaffey (1984), Table B.*
[b]*Children 0.5–5 years of age.*
[c]*NHANES II, 1976–1980.*

in part the higher prevalence of deteriorated housing with lead paint problems in rural areas as a function of race (U.S. ATSDR, 1988). The percentage of rural Black children having PbBs at the indicated blood lead thresholds are 29.9%, 19.4%, and 10.3%, respectively. White children, by contrast, were very much lower proportionately. Percentages for these children at the indicated PbB ceiling values are 9.5%, 2.7%, and 1.2%, respectively.

The Hispanic HANES

A follow-up survey carried out in 1982–1984 was the Hispanic HANES (Carter-Pokras et al., 1990). This effort came about because of concerns that the previous national survey may have underrepresented the full Hispanic population in the overall U.S. population mix. This survey canvassed the age band of 4–11 years for three groups of U.S. Hispanics: Mexican-Americans, Puerto-Ricans, and Cubans (Table 10.11). PbB levels for children 0.5–3 years were also measured, but only capillary blood samplings could be done, owing to the potential for high refusal rates by caregivers. Capillary levels at the time of the survey were deemed to provide higher PbB levels than did venipuncture. Children 4–11 years of age were tested through venipuncture sample collection, and it is this portion of Hispanic HANES that provided the bulk of summary information on Pb exposures.

TABLE 10.11 Hispanic HANES PbB Summary Data by Age and Ancestry or National Origin, 1982–1984[a–c]

Category/Age	Geometric Mean	% ≥25 μg/dl[d]
Mexican-American		
4–5 years	12.2	4.9
6–8 years	10.4	0.2
9–11 years	9.5	0.5
Puerto-Rican		
4–5 years	14.5	10.6
6–8 years	10.9	1.1
9–11 years	10.8	4.0
Cuban		
4–5 years[e]	–	–
6–8 years	8.5	–
9–11 years	8.3	–

[a] *Adapted from Hispanic HANES Tables, Carter-Pokras et al. (1990).*
[b] *Indicated Hispanic ancestry groups are the three largest in the United States.*
[c] *Samples by venipuncture.*
[d] *CDC's 1990 definition of elevated PbB.*
[e] *No estimate given;* N < *25 children.*

Table 10.12 presents PbB levels as geometric means for selected childhood and adult age bands as a function of gender and three race/ethnicity categories: Non-Hispanic Blacks, Non-Hispanic Whites, and Mexican-Americans as representative of Hispanic U.S. residents. Five age bands are depicted in this table, versus seven bands in the original data set since the selected age ranges best capture the age dependency of Pb exposures in the U.S. population. Black males 1–2 and 3–5 years of age showed the highest geometric means at 6.2 and 5.0 μg/dl, respectively, followed by a decline until the 50–60 years of age group, when the geometric mean increases to 6.6 μg/dl. Blacks presented with higher PbB means at all age bands, male and female, followed by Mexican-Americans, and then non-Hispanic Whites. Mexican-American male children, aged 1–2 and 3–5 years, had corresponding geometric means of 4.3 and 4.6 μg/dl, while White male children in these age brackets had geometric means of 3.5 and 2.9 μg/dl. Female children up to 6–11 years of age showed only small differences with their male age and race/ethnicity groupings. This is consistent with data from other studies,

TABLE 10.12 NHANES III—Phase 1 Data: Selected Geometric Mean (GM) PbB Levels (μg/dl) for U.S. Residents, 1988–1991, by Gender, Age, and Race/Ethnicity[a–d]

Age (years)	Non-Hispanic White (GM)	Non-Hispanic Black (GM)	Mexican-American (GM)
Males			
1–2	3.5	6.2	4.1
3–5	2.9	5.0	3.9
6–11	2.3	4.3	3.1
20–49	3.7	4.3	4.3
50–69	4.6	6.6	4.6
Females			
1–2	3.7	5.8	4.8
3–5	2.9	5.0	3.5
6–11	1.9	3.7	2.7
20–49	1.7	2.3	2.1
50–69	2.1	4.1	3.3

[a] *Adapted from Brody et al. (1994), Table 2.*
[b] *Full age range in NHANES III, Phase 1: 1–70 years.*
[c] *PbB sampling by venipuncture, original values in μmol/l.*
[d] *1 μmol/l = 20.7 μg/dl.*

indicating that the gender difference seen in adults is largely absent in childhood.

The PbB values in these groups, as observed in other surveys and screenings, were generally higher in the youngest age subset (Table 10.11). There were too few Cuban children in the youngest age subset to provide necessary data. Puerto-Rican children had higher PbB levels, regardless of age, compared to Mexican-American or Cuban children. The basis of this difference appears to be linked to poverty level and prevalence of poverty among the groups. Table 10.11 shows that Puerto-Rican children 4–5 years of age had over twice the prevalence of PbB levels ≥25 μg/dl than did Mexican-American children of this age range, 10.6 versus 4.9 μg/dl, respectively.

NHANES III—Phases 1 and 2

NHANES II was followed by a third national U.S. survey, divided into two phases covering two time bands of 3 years each. Phase 1 was for the years

1988–1991, while Phase 2 was directed to the next 3-year block, 1991–1994. Both phases retained the statistical national PbB survey design, i.e., a stratified multistage probability cluster sample of households for noninstitutionalized U.S. residential samples consisting of largely the same number of socioeconomic and demographic strata as did the NHANES II effort. Summary data, i.e., summaries drawn from data tapes, are depicted in such articles as Pirkle et al. (1994) and Brody et al. (1994) for Phase I, while Phase 2 data are summarized in Pirkle et al. (1998).

NHANES III—Phase 1

Table 10.12 presents summary data for U.S. children for the years 1988–1991, arrayed by age group, gender, and race/ethnicity. The original data sets from this survey included several more age bands and those shown here show the salient results. As discussed later, mean PbB levels for all strata in 1988–1991 indicate significant declines from 1976 to 1980 in NHANES II. Within the 1988–1991 results, however, the younger two age bands show much higher PbB means than older children. This is followed by declines at older ages until the 50–69 (and 70-plus) age bands, at which point PbB increases to equal or even exceed the levels of childhood. A part of this phenomenon in older adults relates to onset of bone demineralization and associated release of Pb via resorption into the blood stream. These elevations are temporally latent in their correspondence with earlier exposures and are endogenous exposures in terms of dose–response relationships expected.

There is little change by gender in the youngest two age groups compared to the clearly higher PbB levels for adult males compared to females. Race/ethnicity continues to be a significant variable and mimics the NHANES II findings for U.S. Whites compared with Blacks, with Mexican-American Hispanics intermediate in mean levels. Non-Hispanic Black male children aged 1–2 and 3–5 years show PbB means of 6.2 and 5.0 μg/dl, respectively, while White children of these ages show values of 3.5 and 2.9 μg/dl (Table 10.12). Mexican-American children in this table have means intermediate in value, 4.1 and 3.9 μg/dl. Adults 50–69 years of age in all of the race/ethnicity categories still show that Non-Hispanic Black males have higher PbB mean levels than either of the other two categories.

Percentages of U.S. individuals in the NHANES III—Phase 1 survey exceeding measured PbB levels at various ceiling values are summarized in Table 10.13 for the two age bands of young U.S. children in the survey, 1–2 and 3–5 years of age, and as a function of race/ethnicity. Non-Hispanic Black children in both age groups greatly exceed the frequency of elevated PbB measurements at all five ceiling PbB levels for the other two indicated categories in Table 10.13, the 1- to 2-year-old children, for example, showing exceedence rates of 63.9%, 21.6%, 10.2%, 5.4%, and 1.4% for ≥5, ≥10,

TABLE 10.13 NHANES III—Phase 1 Data: Percentage of U.S. Children Above Indicated PbB Levels (μg/dl) by Age and Race/Ethnicity, 1988–1991[a–d]

Ethnicity/Race	Age (years)	% ≥PbB Level				
		≥5 μg/dl	≥10 μg/dl	≥15 μg/dl	≥20 μg/dl	≥25 μg/dl
Non-Hispanic White	1–2	34.2	8.5	2.1	0.8	0.4
	3–5	21.3	3.7	0.7	0.4	0.4
Non-Hispanic Black	1–2	63.9	21.6	10.2	5.4	1.4
	3–5	54.5	20.0	6.0	2.9	0.8
Mexican-American	1–2	41.4	10.1	2.9	1.9	1.0
	3–5	34.5	6.8	1.4	0.7	0.7

[a]*Adapted from Brody et al. (1994), Table 4.*
[b]*Two age ranges: 1–2 and 3–5 years.*
[c]*Derived from NHANES III—Phase 1 Survey.*
[d]*PbB sampling by venipuncture.*

≥15, ≥20, and ≥25 μg/dl, respectively. Non-Hispanic White children in this age band had the lowest prevalences of elevated PbB levels for the various ceilings in PbB, with exceedence rates, in this table, of 34.2%, 8.5%, 2.1%, 0.8%, and 0.4% for the ceiling values of ≥5, ≥10, ≥15, ≥20, and ≥25 μg/dl, respectively. Children 3–5 years of age had elevated prevalence rates that closely tracked those of the younger age group across race/ethnicity but were lower in absolute value (Table 10.13).

NHANES III—Phase 2

The second phase of NHANES III encompassed a field survey of the U.S. national picture for lead exposures in the 3 years, 1991–1994, using the same statistical and logistical designs as Phase 1 and the NHANES II effort.

Phase 2, like the earlier Phase 1, reported the mean PbB for the U.S. noninstitutionalized population aged 1 year and older. The entire population geometric mean level for this time interval was 2.3 μg/dl, with 2.2% having PbB ≥10 μg/dl. For the entire early child age band of 1–5 years, 2.7 μg/dl was the mean concentration, with 4.4% of U.S. children 1–5 years of age having or exceeding the 1991 U.S. CDC guidance PbB toxicity level of 10 μg/dl. PbB means and exceedences ranked by socioeconomic and demographic categories clearly show continued declines from Phase 1 and especially marked declines from the 1976 to 1980 period covered by NHANES II. These changes are discussed below.

TABLE 10.14 NHANES III—Phase 2 Data: Selected Geometric Mean and Exceedence PbB Levels of U.S. Residents, 1991–1994, by Age, Race/Ethnicity, and Gender[a–d]

Age (years)	Geometric Mean (μg/dl)	% ≥10 μg/dl
1–2	3.1	5.9
3–5	2.5	3.5
6–11	1.9	2.0
20–49	2.1	1.5
50–69	3.1	2.9
Race/ethnicity		
Black, Non-Hispanic	2.8	5.2
Mexican-American	2.4	2.9
White, Non-Hispanic	2.2	1.5
Other	2.3	3.0
Gender		
Female	1.9	0.9
Male	2.8	3.5

[a]*Adapted from Pirkle et al. (1998), Table 4.*
[b]*Full age range in original NHANES III—Phase 2 data set, 1991–1994: 1 to ≥70 years.*
[c]*PbB ceiling level, 10 μg/dl, based on CDC 1991 guidance value for onset of childhood toxicity.*
[d]*PbB samples by venipuncture.*

Geometric mean and PbB exceedence rates ≥10 μg/dl for NHANES III—Phase 2 are presented with reference to U.S. population age band, race/ethnicity, and gender in Table 10.14. The age bands in Table 10.14 were selected from all of the age groupings in the original data (Pirkle et al., 1998) to show the maximum effect of age. Children 1–2 years of age show the highest mean among children (3.1 μg/dl) with declines for older children and young adults, followed by a marked increase in older adults 50–69 years of age (3.1 μg/dl). The 1–2 year age band in children also shows the highest percentage, among all age groups, attaining or exceeding the CDC guideline of 10 μg/dl: 5.9%. By contrast, older adults show an exceedence of half that, 2.9%, even though these two groups have similar mean PbB values. U.S. Blacks, in the race/ethnicity category, show the highest mean PbB, 2.8 μg/dl, across ages and the highest percentage, 5.2%, attaining or exceeding the U.S. CDC guideline of 10 μg/dl. Across all age groups, males had higher mean

TABLE 10.15 NHANES III—Phase 2 Data: Selected Geometric Mean and Exceedence[a] PbB Levels[b] of U.S. Residents, 1991–1994, by Housing Age[c], Income, and Urban Status[d]

Housing Age (Year Built)	Geometric Mean (μg/dl)	% ≥10 μg/dl
Before 1946	2.6	3.3
1946–1973	2.3	2.2
After 1973	1.9	1.2
Income[e]		
Low	2.6	4.5
Middle	2.2	1.8
High	2.1	0.7
Urban status		
Population ≥1 million	2.4	2.3
Population <1 million	2.2	2.0

[a]*Exceedence value defined in Table 10.13.*
[b]*PbB samples by venipuncture.*
[c]*Housing age = year built.*
[d]*Adapted from Pirkle et al. (1998), Table 4.*
[e]*Income scaled to poverty level, poverty income ratio (PIR): low=PIR <1.300; middle = 1.301–3.500; high =>3.501.*

PbB, 2.8 versus 1.9 μg/dl, with a much higher percentage attaining or exceeding 10 μg/dl, 3.5% versus 0.9%.

Other demographic and socioeconomic categories—housing age (year built), family income scaled to poverty and urban status—also remain as important determinants of Pb exposure profiles in the U.S. population. Table 10.15 relates these variables to population mean and exceedence values. The oldest U.S. housing stock, i.e., housing built before 1946, shows the highest geometric mean PbB, 2.6 μg/dl, while housing built after 1973 is linked to a mean of 1.9 μg/dl. Corresponding percentages of levels ≥10 μg/dl are quite different as seen in this table, 3.3% versus 1.2%. Low-income individuals, in terms of severity of poverty, show higher mean PbB than those of highest relative incomes, 2.6 versus 2.1 μg/dl. When looking at the entire population, individuals living in large cities with populations of 1 million or more show moderate differences for PbB elevation and a moderate increase in percentages of elevated PbB when contrasted with areas having <1 million.

These three variables are much more influential when used for young children aged 1–5 years. As noted in Pirkle et al. (1998), Table 1, children in housing built before 1946, i.e., the housing most likely to have lead-based paint and to be in questionable states of repair, had a mean PbB almost twice that for child residents in units built after 1973, 3.8 versus 2.0 μg/dl, respectively. Furthermore, those in the oldest homes had more than five times the percentage of PbB levels ≥10 μg/dl than those in the newest, 8.6% versus 1.6%, respectively. Similarly striking differences were seen for children from low-income versus high-income families, i.e., high versus low relative poverty level, 3.8 versus 1.9 μg/dl, a twofold difference. The corresponding percentages with elevated PbB ≥10 μg/dl are 8.0 versus 1.0 μg/dl, an eightfold difference. Level of urbanization showed similar mean PbBs for young children, 2.7 and 2.8 μg/dl in areas with <1 million and ≥1 million, respectively. However, percentages with elevated PbB are significantly higher in the more densely populated stratum, 5.4% versus 3.3%.

NHANES IV was carried out in the period 1999–2002 in two phases: for the years 1999–2000 and 2001–2002. Table 10.16 shows geometric mean PbB values and percentile PbB values by age and year of survey testing. For children 1–5 years of age, the geometric mean PbBs for the 1999–2000 and 2001–2002 time frames are 2.2 and 1.7 μg/dl, respectively. PbB values at selected percentiles for this group of children for the two periods were as follows: 50th, 2.2 and 1.5 μg/dl; 75th, 3.3 and 2.5 μg/dl; 90th, 4.8 and 4.1 μg/dl; 95th, 7.0 and 5.8 μg/dl. Older children, 6–11 years of age, showed the following PbB levels at selected percentiles for the two periods: 50th, 1.3 and 1.1 μg/dl; 75th, 2.0 and 1.6 μg/dl; 90th, 3.3 and 2.7 μg/dl; 95th, 4.5 and 3.7 μg/dl.

Gender and race/ethnicity were also examined in terms of geometric means and percentile PbB values. Males across the surveyed ages, consistent with earlier national surveys and many other studies, showed higher mean PbB values than females. Table 10.16 indicates that males of all ages for 1999–2000 and 2001–2002 had respective PbB means of 2.0 and 1.8 μg/dl versus corresponding levels of 1.4 and 1.2 μg/dl for females.

Table 10.17 presents the U.S. NHANES IV data for percentages of segments of the U.S. population having PbB ≥10 μg/dl when classified by age, gender, and race/ethnicity. With children 1–5 years old, Mexican-American male children had a higher percentage of elevated PbB than either White or Black male children and a higher rate than female Mexican-American children. Black female children had the highest overall percentage of early childhood PbB elevations, 3.7%. The highest overall rate of elevation was 7.5%, attaching to older Black adults ≥60 years old. By contrast, the lowest percentage elevation of all groups was 0.1% for both older Mexican-American children and White women ≥60 years old.

TABLE 10.16 NHANES IV Blood Pb Data: Selected Geometric Mean PbB and Percentiles for U.S. Residents by Age, Gender, and Race/Ethnicity, 1999–2002[a,b]

Age (years)	Survey Years	Geometric Mean (μg/dl)[c]	Selected Percentiles 50th	75th	90th	95th
1–5	1999–2000	2.2	2.2	3.3	4.8	7.0
	2001–2002	1.7	1.5	2.5	4.1	5.8
6–11	1999–2000	1.5	1.3	2.0	3.3	4.5
	2001–2002	1.3	1.1	1.6	2.7	3.7
12–19	1999–2000	1.1	1.0	1.4	2.3	2.8
	2001–2002	0.9	0.8	1.2	1.9	2.7
≥20	1999–2000	1.8	1.7	2.5	3.9	5.2
	2001–2002	1.6	1.6	2.2	3.6	4.6
Gender						
Males	1999–2000	2.0	1.8	2.9	4.4	6.0
	2001–2002	1.8	1.7	2.7	3.9	5.3
Females	1999–2000	1.4	1.3	1.9	3.0	4.0
	2001–2002	1.2	1.1	1.8	2.6	3.6
Race/ethnicity						
Non-Hispanic Blacks	1999–2000	1.9	1.7	2.8	4.2	5.7
	2001–2002	1.7	1.6	2.5	4.2	5.7
Non-Hispanic Whites	1999–2000	1.6	1.6	2.4	3.6	5.0
	2001–2002	1.4	1.4	2.1	3.1	4.1
Mexican-Americans	1999–2000	1.8	1.8	2.7	4.2	5.8
	2001–2002	1.5	1.5	2.2	3.6	5.4

[a]*Adapted from CDC Third National Report on Human Exposure to Environmental Chemicals (2005), Table 16.*
[b]*Based on NHANES IV, 1999–2002.*
[c]*Rounding.*

Comparatively, the findings for this most recent U.S. survey period 1999–2002 showed continued declines in both mean PbB concentrations and percentages of elevated PbB ≥10 μg/dl. This temporal pattern is discussed later.

More recent NHANES data gathering has extended to 2008 (U.S. CDC, 2010) and is included in trend analyses of U.S. national PbB values by Jones et al. (2009). The percentage of children having PbB values above 20 μg/dl decreased to 1.3% while as of 2004, 1.4% had PbB levels >10 μg/dl. For the

TABLE 10.17 NHANES IV Blood Pb Data: Selected Percentages of U.S. Residents with PbB Levels ≥10 μg/dl by Gender, Age, and Race/Ethnicity[a–c]

	Race/Ethnicity (% Exceedence)		
Gender/Age (years)	Non-Hispanic Black	Non-Hispanic White	Mexican-American
Males			
1–5	2.5	1.4	3.2
6–19	0.9	0.2	0.1
20–59	2.3	1.0	3.0
≥60	7.5	0.7	2.2
Females			
1–5	3.7	1.3	0.7
6–19	0.9	0.2	0.5
20–59	0.5	0.3	0.9
≥60	0.9	0.1	1.6

[a]*Adapted from the U.S. CDC NHANES IV (1999–2002), Table 1.*
[b]*U.S. Centers for Disease Control (2005b).*
[c]*PbB samples collected by venipuncture.*

2005–2008 survey period, 95% of children up to 60 months old had a PbB <5 μg/dl.

10.3.2 International Epidemiological Studies of Human Lead Exposure

The serial U.S. NHANES and their results within time bands and over time were accompanied by similar international studies. A selected group of results from such studies are presented in Table 10.18 for mainly Europe, with data for Mexico and Canada included.

The UNEP/WHO conducted an early international study over two periods, 1981 and 1983. Four countries were selected and the population segment comprising the survey test group were nonsmoking male teachers, 25–50 years of age. Details and rationales for survey design are provided in Ducoffre et al. (1990). A summary of results is provided in Table 10.18. Geometric mean PbB concentrations are highest in the two time periods for Malta, 28.8 and 23.2 μg/dl, followed by Mexico, 25.9 and 18.9 μg/dl, and

TABLE 10.18 Selected International Studies of Population PbB Levels for Indicated Time Periods[a–c]

Country	Year(s) of Testing	Group	PbB (μg/dl)	Reported Percentiles	References
UNEP/WHO survey[d]					Claeys-Thoreau et al. (1987)
Belgium	1981	Nonsmoking teachers	16.2	21.7 (90th)	
	1983		13.5	16.5 (90th)	
Malta	1981		29.3	45.5 (90th)	
	1983		24.7	34.3 (90th)	
Mexico	1981		25.5	20.1 (90th)	
	1983		18.8	26.3 (90th)	
Sweden	1981		7.7	14.7 (90th)	
	1983		5.3	11.4 (90th)	
Ontario, Canada[e]	1984	Children by urbanization:			O'Heany et al. (1988)
		Urban	12.0	–	
		Suburban	10.0	–	
		Rural	8.9	–	
		Total	10.4	–	
Germany[e]					Schulz et al. (2007b)
Ger ESs					
West Germany, Ger ES I Adults	1985–1986	Total	6.2	6.2 (50th)	
				11.4 (95th)	
		Males	8.0	7.9 (50th)	
				14.6 (95th)	
		Females	5.9	6.0 (50th)	
				10.7 (95th)	
All of Germany, Ger ES II Adults	1990–1992	Total	4.6	6.0 (50th)	Schulz et al. (2007b)
				10.7 (95th)	
		Males	5.5	4.5 (50th)	
				10.5 (95th)	
		Females	3.8	5.4 (50th)	
				11.6 (95th)	

(*Continued*)

TABLE 10.18 Selected International Studies of Population PbB Levels for Indicated Time Periods[a–c]—(cont.)

Country	Year(s) of Testing	Group	PbB (μg/dl)	Reported Percentiles	References
Ger ES III Adults	1998	Total	3.2	3.3 (50th) 7.2 (95th)	
		Males	3.7	3.8 (50th) 8.0 (95th)	
		Females	2.7	2.8 (50th) 6.3 (95th)	
Ger ES II	1990–1992	Children 6–14 years old, all of Germany	3.2	3.3 (50th) 6.2 (95th)	Schulz et al. (2007b)
Ger ES IV	2003–2006	Children 6–14 years old, all of Germany	1.6	1.6 (50th) 3.2 (95th)	
Other international surveys					
Italy, multiple cities[f]	2000	Residents, various ages	Males 4.5	Males 10 (95th)	Apostoli et al. (2002)
			Females 3.6	Females 6 (95th)	
France, northern communities[e]	1996–1997	Children, 8–11 years old	Nonindustrial areas 3.1; Industrial areas 4.0	–	Leroyer et al. (2000)
England, eight regions[e]	1995	Adults, children	Adult males 3.5	10.1 (95th)	Delves et al. (1996)
			Adult females 2.7	6.8 (95th)	
			Male children 2.3	5.6 (95th)	
			Female children 1.7	3.1 (95th)	

(*Continued*)

TABLE 10.18 Selected International Studies of Population PbB Levels for Indicated Time Periods[a–c]—(cont.)

Country	Year(s) of Testing	Group	PbB (μg/dl)	Reported Percentiles	References
England, various regions: controls and exposed[e]	1984–1986	Adults; children 6–7 years of age	1984: Adult males 10.0 Adult females 8.0 Children 9.9 1985: Adult males 10.5 Adult females 7.7 Children 9.2 1986: Adult males 10.0 Adult females 7.7 Children 7.7	–	Quinn and Delves (1987, 1988, 1989)
Czech Republic, nationwide[d]	2001–2006	Adults and children	Adult males 3.7 Adult females 2.5 Children 3.1	–	Černá et al. (2007)

[a]*Adapted from data in cited references.*
[b]*Original data expressed in various units converted to μg/dl where necessary.*
[c]*Methods of collection in original reports.*
[d]*Median PbB.*
[e]*Geometric mean PbB.*
[f]*Arithmetic mean PbB.*

Belgium, 16.0 and 13.6 μg/dl. Lowest PbB values by a significant amount were measured for Sweden, 7.7 and 5.3 μg/dl.

The UNEP/WHO survey results showed mean PbB levels for adults more closely linked to level of overall development than to just industrialization *per se*. For example, more industrialized Belgium had lower mean and 90th percentile blood lead results than did Malta or Mexico. Mean PbB levels for the adult segment of these countries who also had no apparent occupational exposures show PbB values that are either well above the U.S. NHANES II data (Malta, Mexico) or below (Sweden). It is clear from the UNEP/WHO data that the PbB levels were declining significantly in the early 1980s, even over the 2 years of this survey. Mean and especially 90th percentile PbB results are lower across the board in 1983 than in 1981.

Canadian children residing in Ontario in communities differing as to urbanization were shown by O'Heany et al. (1988) with different mean PbB results over short periods. As seen in Table 10.18, all areas combined indicated a mean of 10.4 μg/dl stratified into urban, 12.0 μg/dl; suburban, 10.0 μg/dl; and rural, 8.9 μg/dl.

A series of four lead exposure (PbB) surveys, the German Environmental Surveys (Ger ESs I–IV) were carried out in Germany over the years 1985–2006. Schulz et al. (2007a,b) have described the design and methodology of the effort and summarized the serial survey results. The earliest effort, 1985–1986, was only done for West Germany prior to reunification, but the three later testings—1990–1992, 1998, and 2003–2006—applied to all of Germany. Adults were principally surveyed in 1985–1986, 1990–1992, and 1998. Children 6 years or older were included in part of the 1990–1992 survey groups and comprised all of the 2003–2006 subjects. Geometric mean, median, and 95th percentile results were recorded (Table 10.18).

Combined gender, male and female adult results as geometric mean PbBs for the 1985–1986, the 1990–1992, and 1998 surveys were as follows: 1985–1986, 6.2, 8.0, and 5.9 μg/dl; 1990–1992, 4.6, 5.5, and 3.8 μg/dl; 1998, 3.2, 3.7, 2.7 μg/dl. Children 6–14 years old in both Ger ES II and IV had the following mean PbBs, with no disaggregating for gender: 3.2 and 1.6 μg/dl. Notably, the mean levels in the Ger ES I survey, 1985–1986, were significantly lower than those of other countries in this time frame. However, German PbB survey data for more recent years began to resemble those for both the U.S. NHANES and surveys in other countries seen in this table. Despite lower means overall, one still sees a marked decline in PbB summary data going from 1985–1986 to 1998 for adults and from 1990–1992 to 2003–2006 for children 6–14 years old.

Two general surveys gathering geometric mean PbB data for the years 1984–1986 and 1995 were carried out in England. For the 1984–1986 surveys, both adults and children were surveyed, and adult results were determined for either males or females. For 1995, both adults and children were stratified as to PbB means by gender. The later English survey (Delves et al., 1996) reported both means and 95th percentiles, while the earlier efforts presented only mean results (Quinn and Delves, 1987, 1988, 1989). In 1984, 1985, and 1986, children had mean PbBs of 9.9, 9.2, and 7.7 μg/dl, respectively. Adult male mean levels for these 3 years were 10.9, 10.5, and 10.0 μg/dl, respectively, compared to female results in these years of 8.0, 7.7, and 7.2 μg/dl. The 1995 survey showed as geometric mean PbBs: adult males, 3.5 μg/dl; adult females, 2.7 μg/dl; male children, 2.3 μg/dl; and female children, 1.7 μg/dl. The corresponding 95th percentile values for these groups were 10.1, 6.8, 5.6, and 3.1 μg/dl.

Multiple cities in Italy were surveyed in 2000 and included individuals of various ages (Apostoli et al., 2002). Male and female arithmetic mean PbB were 4.5 and 3.6 μg/dl, respectively. Northern communities in France were surveyed in the 1996–1997 period, and the effort was focused on school children 8–11 years of age (Leroyer et al., 2000). The geometric mean PbB for children of both genders in industrial areas was 4.0 μg/dl and in nonindustrial locales, 3.1 μg/dl. A nationwide survey was done in the Czech Republic, 2003–2006 (Černá et al., 2007). Both adults and children were included, with the corresponding median PbBs for adult men, adult women, and children being 3.7, 2.5, and 3.1 μg/dl.

10.4 CHILDHOOD Pb SCREENING PROGRAMS AND THEIR VARIOUS CORRELATES

The second body of data under the category of environmental epidemiology of lead exposure consists of the results derived from lead exposure screening programs. By and large, such screening programs were established and still operate to monitor the extent of U.S. childhood lead exposures and to chart progress in the control of childhood lead toxicity prevalence and incidence in high-risk young children. The great bulk of screening activity is testing for elevated PbB in young U.S. children within the CDC's current high-risk age band of those <72 months of age. Consequently, U.S. data are the focus of this section of the chapter.

Screening for elevated PbB in high-risk U.S. children in any organized fashion is of relatively recent vintage, begun in the 1970s, compared to the much longer history for overall awareness of childhood lead poisoning that began by the 1920s and 1930s. The initial impetus for structured U.S. childhood lead screening activities was statutory, mandated, and funded by provisions of the 1971 Lead-Based Paint Poisoning Prevention Act [LBPPPA, 42 USC 4801 et seq.]. A detailed presentation of this and other statutory approaches for lead exposure management are presented in sections dealing with lead regulation in the United States and internationally.

Several factors impeded organized efforts to deal with U.S. childhood lead poisoning even after its recognition as a pediatric health issue. First, the level of Pb exposure considered hazardous to human health and the socioeconomic/demographic characteristics of lead-exposed individuals have often been changing variables in effectively addressing the childhood lead problem (Mushak, 1992; NAS/NRC, 1972, 1993). Second, in the first half of the twentieth century, lead poisoning in the United States was considered a disease rooted in poverty and the overall behaviors of certain minorities who lived in substandard housing, and their general environment. That is, lead poisoning was assumed not to be pervasive in the wider child population but a problem confined to urban slum areas of the nation's older Northeastern and Midwestern cities. This uninformed societal view has largely but not totally yielded to the health science informing our current awareness of lead exposures and lead poisoning. That evidence shows lead is hazardous to children across socioeconomic and demographic strata through multiple environmental media exposures.

Lead screenings in the United States have typically used guidelines from Federal health agencies, specifically the U.S. Public Health Service (U.S. PHS)/U.S. CDC in both identifying who are high-risk individuals and what level of lead exposure (in the form of some measurement) shows an elevated exposure and associated toxicity risk. Following enactment of the LBPPPA, the U.S. CDC published a series of Statements on childhood lead poisoning beginning in 1975, with a simplified and quite elevated set of PbB toxicity risk levels, and the Statements continued through the present (U.S. CDC, 1975, 1978, 1985, 1991, 2005a). Table 10.19 summarizes guidance in a poisoning risk classification number based on co-measured EP for several time periods: 1978–1985 and 1985–1991. The CDC guidance for childhood lead screening for the period 1975–1991 included a combination of values for preliminary EP and follow-up PbB measurements to confirm a risk for children with lead toxicity. Classification numbers rise with increased toxicity risk, with Class IV defining a medical emergency requiring immediate hospitalization in the earlier schemes, and Class V doing so in the 1991 Statement. Interested readers can consult the Statements for the details of intermediate classifications.

The 1991 Statement rescinded this two-part screening protocol and recommended sole use of PbB measurements. The principal reason for this was realization that at PbB levels ≤25 μg/dl, measured EP values produce an unacceptably high rate of false negatives, and a normal EP measurement would often be associated with an elevated but unidentified PbB. Table 10.19 notes that CDC's definition of a PbB elevation continued to decline across the period 1975/1978 to 1991, from 30 to 10 μg/dl in the 1991 Statement, along with downward revision in the associated screening level of EP. Consequently, the numbers provided from screening programs are specifically within the threshold values that existed at the time of those screenings.

Table 10.20 presents summary screening data for the various U.S. childhood PbB screening activities in 1985–1986 and as provided to the U.S. ATSDR for its 1988 Report to Congress on U.S. childhood lead poisoning. Data in the table represent results for different categories of screening operations: statewide, principal counties, city-specific. A simple rendering of the prevalence of PbB elevations in some programs was complicated by the fact that CDC's 1985 Statement revising its definition of elevated PbB downward from 30 to 25 units occurred during information gathering and changed the definition of elevated PbB and thresholds for lead toxicity risks. Where necessary, the pre-1985 and

TABLE 10.19 USCDC Childhood Lead Screening Classification Schemes: 1978–Present[a–c]

Blood Lead (μg/dl)	Erythrocyte Protoporphyrin (EP) (μg/dl)			
CDC 1978–1985	50–109	110–249	≥250	
30–49	II	III	IV	
50–69	III	III	IV	
≥70	Rarely seen	IV	IV	
	(a) By chemical analysis			
CDC 1985–1991	**<35**	**35–109**	**110–249**	**≥250**
Not done	I	–	–	–
≤24	I	I	I	EPP[d]
25–49	I	II	III	III
50–69	–[e]	III	III	IV
≥70	–[e]	–[e]	IV	IV
	(b) By hematofluorometer			
	<35	**35–74**	**75–174**	**≥175**
Not done	I	–[f]	–[f]	–[f]
≤24	I	I	I	EPP[d]
25–49	I	II	III	III
50–69	–[e]	III	III	IV
≥70	–[e]	–[e]	IV	IV
CDC 1991–Present	Categories: use PbB measurement only			
≤9	I			
10–14	IIA			
15–19	IIB			
20–44	III			
45–69	IV			
≥70	V			

[a]*Classification numbers: poisoning risk increases with number; IV or V corresponds to the highest poisoning risk.*
[b]*During 1978–19991 period, EP testing done first; +value led to PbB test.*
[c]*For 1985–1991, EP measured by laboratory analysis ("free" EP) or by hematofluorometer, zinc form or ZPP.*
[d]*The genetic disorder erythropoietic protoporphyria.*
[e]*Rarely observed.*
[f]*Requires PbB test.*

TABLE 10.20 U.S. Childhood Lead Screening Activities for PbB, 1985–1986, Reported to U.S. ATSDR by State or Local Programs[a–c]

Agency Program	Number of Children Screened (years)	Confirmed Number	% Pb Toxicity
Delaware	5,818 (FY 85/July–June)	130	2.2
Washington, DC	17,000 (FY 85/ October–September)	595	3.5
Augusta, GA	2,960 (FY 85/ October–September)	8081/ 8644	9.0 (Combined %, both cities)
Savannah, GA	5,684 (FY 85/ October–September)		
Chicago, IL	37,409 (CY 85)	693	1.8
Indiana	3,770 (FY 85)	17	0.5
12 Counties, IA	2,143 (CY 85)	28	1.3
Scott County, IA	897 (January–November 86)	9	1.0
Wyandotte, KS	5,098 (CY 85)	16	0.3
Worcester, KS	8,161 (FY 85)	71	0.9
	9,658 (FY 86)	72	0.8
Detroit, MI	20,248 (CY 85) 13,132 (January–August 86)	371 392	1.8 3.0
Hennepin County, MN (Minneapolis)	3563 (January 85–January 86)	26	0.7
St. Paul, MN	8555 (CY 85) 8553 (January–November 86)	64 41	0.8 0.5
Mississippi	3628 (FY 86, October–September)	29	0.8
St. Louis, MO	12,308 (CY 85) 9,758 (January–October 86)	1,356 1,653	11.0 16.0
Douglas County, NE	3,167 (FY 86/ October–September)	29	0.8
Baltimore, MD	30,583 (CY 85)	504	1.7
Remainder of Maryland	18,132 (CY 85)	46	0.3
Massachusetts Statewide	142,000 (FY 85) 166,900 (FY 86)	1,531 1,011	1.0 0.6

(Continued)

TABLE 10.20 U.S. Childhood Lead Screening Activities for PbB, 1985–1986, Reported to U.S. ATSDR by State or Local Programs[a–c]—(cont.)

Agency Program	Number of Children Screened (years)	Confirmed Number	% Pb Toxicity
Boston, MA	29,925 (FY 85) 29,356 (FY 86)	507 333	1.7 1.2
Merrimac Valley, MA	5,050 (FY 85) 3,619 (FY 86)	177 42	3.5 1.2
Springfield, MA	1,735 (FY 85) 352 (FY 86)	34 3	2.0 0.9
New Hampshire	5,021 (FY 85/July–June) 6,483 (FY 86/July–June)	24 46	0.5 0.7
New Jersey	58,080 (CY 85)	1,690	2.9
New York City, NY	206,467 (FY 85)	1,337	0.7
Bronx, NY	44,501	288	0.7
Brooklyn, NY	72,314	720	1.0
Manhattan, NY	47,456	154	0.3
Queens, NY	38,604	154	0.4
Richmond, NY	3,256	24	0.7
North Carolina	15, 567 (FY 85 October–September)	66	0.4
Pennsylvania	22,894	631	0.3
Philadelphia, PA	15,133	357	2.3
NE Philadelphia	983	8	0.8
Allegheny County	2,092	32	1.6
Harrisburg	2,026	101	4.9
Erie County	1,080	9	0.8
Rhode Island	14,640	280	2.0
South Carolina	64,993	920	1.2
Dallas, TX	35,000	350–700	1–2%
Vermont	402	1	0.3
Total	785,285	11,739	1.5

[a]*Adapted from U.S. ATSDR (1988), Table V.14.*
[b]*FY 85 programs used CDC 1978 Statement classification scheme; CY 85 programs used CDC 1985 Statement classification scheme.*
[c]*Risk Classification Scheme for 1978: PbB ≥30 μg/dl+various levels of EP; Risk Classification Scheme for 1985: PbB ≥25 μg/dl+various levels of EP.*

1985 enumerations are presented in Table 10.20. The U.S. screening total for 1985–1986 was 785,285 children, of whom 11,739 or 1.5% met the relevant criteria for elevated Pb exposure risk. Prevalences of PbB elevations ranged significantly across programs. In general, older communities with older housing stock and denser urban development had the higher rates of elevations. The highest prevalence, using the newer, 1985 CDC Statement classifications, was recorded for St. Louis, MO, at 16%.

The more recent U.S. childhood PbB screening data have been gathered by the U.S. CDC, including for the years 1997–2007 (U.S. CDC, 2009a). Table 10.21 presents U.S. figures for: (1) the tally of U.S. children <72 months old, (2) the number of these children tested, and (3) percentage showing a confirmed (venipuncture) PbB level of ≥10 μg/dl. Across these years, the total child population in the high-risk age band hovered around 23.0–24.8 million children (rounding). Over these same 11 screening years, the number of children screened almost doubled (Table 10.21), while the percentage of confirmed positive PbB values declined by fivefold, providing an approximate tally of 31,000 children, or 1.0% (rounding) of all screened children having elevated PbB.

TABLE 10.21 U.S. CDC Childhood Lead Exposure Screening Data, 1997–2007[a–c]

Year	Population <72 Months Old	Number Tested	% Positive PbB ≥10 μg/dl
1997	23,345,397	1,611,568	7.61
1998	23,143,133	1,761,674	6.50
1999	23,023,683	1,875,500	5.03
2000	23,304,631	2,216,700	3.96
2001	23,380,551	2,538,008	3.03
2002	23,380,855	2,652,964	2.56
2003	23,612,242	3,092,229	2.27
2004	23,903,723	3,250,848	1.76
2005	24,204,520	3,136,843	1.53
2006	24,507,506	3,578,103	1.31
2007	24,761,587	3,136,843	1.00

[a]*Adapted from CDC Surveillance Data (1997–2007): U.S. CDC, National Center for Environmental Health, Lead Poisoning Prevention Program, June 1, 2009.*
[b]*Percentage of elevated PbBs.*
[c]*Confirmed by venous puncture or second capillary sample within 12 weeks.*

10.5 EPIDEMIOLOGICAL STUDIES OF PbB/Pb SOURCE RELATIONSHIPS

Epidemiological studies of PbB levels *vis-a-vis* various sources of Pb contributing to those PbB levels have comprised a very large literature and selected examples are presented in this section. These studies differ greatly in scope. Some describe specific groups studied in specific locales, while others present broader relationships. The examples here largely deal with major sources of Pb exposures of human populations nationwide and/or worldwide, such as ambient air lead from vehicular exhaust emissions from burning leaded gasoline. Such emissions impact global populations in both direct and indirect ways. Direct ways include inhalation of air Pb by both children and adults. The indirect component of air Pb exposures describes the ingestion of air Pb fallout in dusts and soils. For children, postdeposition Pb exposure pathways typically produce more of a fractional contribution to PbB levels than direct inhalation. One index of this relative impact is the so-called PbB/air Pb ratio. The direct inhalation component accounted for a ratio on the order of 2, while the aggregated impact of both direct inhalation and postair Pb deposition contributions produces ratios of 5 and higher (see discussion of slope factor modeling of PbB in Chapter 9).

Another major Pb source, for which nationwide data for young children in older U.S. communities exist, is lead paint in varying stages of deterioration producing exposures by a number of pathways. No specific study has fully detailed the entire quantitative relationships of U.S. housing stock lead paint to prevalence rates of elevated PbBs, but data do exist relating housing age and housing lead paint hazards to the estimation and the probability of elevated PbB in children. Earlier chapters, particularly that on sources of environmental lead, provided details from the 2001 National Survey of Lead and Allergens in Housing (U.S. Department of Housing and Urban Development, 2001) that showed the multiple relationships of old housing in urban and other U.S. areas to the prevalence of hazards for children from lead paint and relationships of lead paint hazard characteristics to Pb in dusts and soils.

Table 10.22 sets forth illustrative internationally published studies of PbB levels versus ambient air Pb. The role of gasoline Pb is depicted in the table in two ways: by direct comparative relationships of PbB with ambient air concentrations or comparison with gasoline Pb (g Pb/l fuel) consumption. The best studied of the PbB/gasoline–Pb/ambient air relationships was the very close correlation between declines of PbB seen across all subsets of the U.S. population and the associated decline in marketed gasoline Pb concentrations and ambient air Pb seen during both the NHANES II and III (Annest and Mahaffey, 1984; Pirkle et al., 1994, 1998; U.S. EPA, 2006). In the years 1976–1980 of NHANES II, the mean PbB across all strata of the U.S. population was 15.9 μg/dl in 1976, declining to 9.5 μg/dl by 1980, and 2.8 μg/dl

TABLE 10.22 Selected International Studies of the Relationship of Changes in Population PbB Levels to Changes in Air Pb and Gasoline Pb Concentrations

Area Reported	Population	Design	Years	PbB[a] (μg/dl)	Air Pb[b] (μg/m^3)	Gasoline Pb[c] (g/l fuel)	References
United States	Residents, 1–74 years old	NHANES II, III, National Surveys	1976	15.9	1.0	0.4	Annest et al. (1983), Pirkle et al. (1994)
			1979	12.1	0.7	0.31	
			1980	9.5	0.5	0.30	
			1988–1991	2.8	0.1	0.00	
United Kingdom	Adults, children	Various regions of England	1984	8.0–10.9	–	0.38	Delves et al. (1996), Quinn and Delves (1987, 1988, 1989)
			1985	7.7–10.5	0.48	0.34[d]	
			1986	7.2–10.0	0.24	0.15	
			1995	1.7–3.5	–	0.06	
Greece	Adults	City of Athens, multiyear screening	1982	16.0	1.8	0.40	Kapaki et al. (1998)
			1984	11.8	0.9	0.22	
			1988	8.0	0.7	0.15	
			1993	5.5	0.4	0.14	
Canada	Children, 3–6 years old	Ontario Province, multiyear sampling	1984	11.9	–	0.30	Langlois et al. (1996), Loranger and Zayed (1994)
			1988	5.1	–	0.09	
			1990	3.6	–	0.04	
			1992	3.5	–	0.00	
Mexico	Children, 6–36 months old	Mexico City prospective data	1989	14.6	–	0.20	Driscoll et al. (1992), Rothenberg et al. (1998)
			1990	9.8	–	0.18	
			1991	8.6	–	0.08	
			1992	9.0	–	0.07	
			1993	7.0	–	0.06	

(Continued)

TABLE 10.22 Selected International Studies of the Relationship of Changes in Population PbB Levels to Changes in Air Pb and Gasoline Pb Concentrations—(cont.)

Area Reported	Population	Design	Years	PbB[a] (μg/dl)	Air Pb[b] (μg/m³)	Gasoline Pb[c] (g/l fuel)	References
Italy	Adults, 18 years old	City of Turin, Pb isotopic study and later screenings	1980 1985 1993	21.0 15.1 6.4	3.0 2.0 0.5	0.6 0.4 0.11	Bono et al. (1995), Facchetti (1989)
Finland	Children	City of Helsinki, screenings	1983 1988 1996	4.8 3.0 2.6	0.33 0.10 0.01	0.35 0.14 0.00	Pönkä (1998), Pönkä et al. (1993)
Belgium	Adults, 20 years old	National	1979 1983 1987	17.0 14.7 9.0	1.05 0.66 0.49	0.45 0.40 0.15	Ducoffre et al. (1990)

[a] *PbB central tendency metrics given in original papers, all converted to units of μg/dl, if necessary.*
[b] *Air Pb details in original reports.*
[c] *Gasoline Pb values as reported by Thomas et al. (1999) and Quinn and Delves (1989).*
[d] *Estimated as 90% of 1984 value (Quinn and Delves, 1989).*

in the 1988–1991 period. Concurrent with these exposure declines, air Pb declined from a national mean of 1.0 μg/m^3 in 1976 to 0.5 units by 1980 and to 0.1 μg/m^3 by 1991. Gasoline Pb levels, indexed as g Pb/l fuel, declined from 0.47 to 0.00 by the 1988–1991 time frame.

More recent NHANES data show continued significant declines in children's PbB levels. For the 1999–2004 period, the percentage of children, 12–60 months old, with PbB >10 μg/dl had declined to 1.4% (Jones et al., 2009) and for 2005–2008, to 1.3% (U.S. CDC, 2010).

Similar declines in PbB of residents of other countries were recorded, as noted in Table 10.22. Studies showing the PbB/air Pb relationship as well as linked temporal trends appeared for residents of Belgium, Canada, Finland, Greece, Italy, and the United Kingdom. Studies in other countries differed in their demographic and geographic reach, but all collectively showed what has been observed in U.S. surveys. British data, covering the years from 1984 to 1995 (Delves et al., 1996; Quinn and Delves, 1987, 1988, 1989) showed the PbB means declined from a range of 8.0–10.9 μg/dl for different strata in 1984 to 1.7–3.5 μg/dl in 1995. The Pb content of UK gasoline over this period declined from 0.38 to 0.06 g Pb/l, while air Pb declined 50% in just the 1-year span, 1985–1986, from 0.48 to 0.24 μg/m^3. Ontario, Canada, children, tested for PbB levels in the period 1984–1992 showed a decline in mean PbB from 11.9 to 3.5 μg/dl. Gasoline Pb content in this period declined from 0.30 to 0.00 g/l (Thomas et al., 1999).

The relationship of U.S. children's PbB values to lead paint, indexed via housing age as a surrogate for Pb paint in their environments, can be seen in the NHANES III, Phase 2 survey discussed earlier, using data summarized in Table 10.15. Geometric mean PbBs were lower for newer housing, indexed as year of construction. The residents in the newest housing were about one-third the number of occupants in the oldest housing, built before 1946. Table 2 of Pirkle et al. (1998) showed this dependency most pronounced for non-Hispanic Black children aged 1–5 years, the geometric mean ranging from 5.8 to 3.3 μg/dl in going from housing built before 1946 to residences built after 1973. Equally significant, the percentage of children with PbBs ≥10 μg/dl was over sixfold higher in the oldest housing, 21.9%, versus newest units, 3.4%.

The U.S. ATSDR in its Report to Congress in 1988 produced a series of estimates of national prevalences of elevated PbB values for U.S. children associated with Pb sources providing exposure through various pathways. These estimates were derived using statistical models and environmental Pb data sets. These PbB prevalences are summarized numerically in Table 10.23. Numbers in this table are presented in several ways. For peeling lead paint exposures, using approaches detailed in the ATSDR report, the numbers of U.S. children whose PbB values rise above indicated PbB ceiling values as a function of deteriorated housing age are shown. In the case of leaded gasoline, where Pb content is shown on a phase-out schedule from 1985 to 1990,

TABLE 10.23 Projected Estimates of Elevated PbB Level Prevalence from Exposures to Various Pb Sources and Pathways[a–c]

Source/Pathway	Total Children	Children with PbB (μg/dl) >15	>20	>25
A. Peeling lead paint				
Pre-1940	277,000	187,800	85,200	29,400
1940–1949	218,000	147,800	67,100	23,100
1960–1974	72,000	48,800	22,200	7,000
Total, Pre-1980	567,000	384,400	174,500	60,100
B. Stationary sources[d]				
Primary Smelters	21,000	–	–	3,800
Secondary Smelters	187,000	–	7,500	5,500
Total	208,000	–	–	–
C. Leaded gasoline: Pb content phase-out years[e]		Number of children with PbB declining below		
		15 μg/dl	20 μg/dl	25 μg/dl
1985	–	696,000	232,000	72,000
1986	–	1,726,000	5,663,000	172,000
1987	–	1,597,000	518,000	157,000
1988	–	1,476,000	476,000	144,000
1989	–	1,353,000	434,000	130,000
1990	–	1,252,000	400,000	119,000

[a]*Adapted from U.S. ATSDR (1988): Ch. VI; and Mushak and Crocetti (1989).*
[b]*Projected to 1984 based on NHANES II 1976–1980 data.*
[c]*Total children PbB and prevalences derived in U.S. ATSDR (1988).*
[d]*See U.S. ATSDR (1988), Ch. VI, for bases of prevalence estimates and total child numbers.*
[e]*Using U.S. EPA logistic regression analyses to estimate numbers of children falling below indicated PbB.*

the numbers in the estimates are those U.S. children whose PbB values will respond as Pb gasoline declines below the indicated cut points. These numbers, as detailed in the Federal report, were computed using logistic regression analysis. For metallurgical operations such as smelters, illustrative of major stationary lead sources, calculating prevalences of elevations in PbB for indicated values used major site-specific percentages affected combined with total child populations potentially impacted by these sources.

Table 10.23 depicts major numbers of U.S. children in the mid-1980s who had sustained elevations at different PbB ceilings as a function of living in deteriorating aged housing with peeling lead paint. Prevalence rates for PbB levels >15, >20, and >25 μg/dl were calculated as discrete percentages (rounding) for each PbB ceiling concentration in the category of peeling lead paint: 68% >15 μg/dl, 31% >20 μg/dl, and 11% >25 μg/dl. Children living in pre-1940 deteriorated housing with peeling lead paint numbered 277,000, of whom 187,800 had projected PbB levels >15 μg/dl, 85,200 had PbB values >20 μg/dl, and 29,400 had PbB concentrations >25 μg/dl. The corresponding totals and the counts at the three prevalence ceilings for 1940–1959 and 1960–1974 were as follows: 218,000 total; 147,800, 67,100, and 23,100; 72,000 total; 48,800, 22,200, and 7,000.

The estimated numbers of U.S. children living near metallurgical point sources—primary and secondary smelters—are also depicted in Table 10.23. For primary smelters in the 1980s, 21,000 children lived near those operations, and of these, estimated PbB tallies >25 μg/dl were in the range of 3,800–5,500. A total of 187,000 children resided around secondary smelters, and 7,500 of these individuals had PbB levels >20 μg/dl.

Table 10.23 also shows that, over the period for phasing down (and eventually phasing out) leaded gasoline in the United States, 1985–1990, U.S. children whose PbB values declined below 15 μg/dl over the period were 696,000 in 1985 and 1,252,000 in 1990. Corresponding figures are given for the other PbB ceilings. The cumulative tally for all children whose PbBs declined below 15 μg/dl with continued Pb gasoline phasing down numbered in the millions. The method for prevalence estimation used in the ATSDR report does not allow calculating how many children's PbBs declined into the lower categories, only that shifts below the indicated exceedence value occurred.

National or international surveys as a function of sources of Pb that are more localized in their variability in population impacts are not feasible nor have they been attempted. One example of environmental Pb media localized or stratified in human impacts is lead in dust and/or soil, and studies of these media have been either site-specific or subnational in scope, encompassing multiple sites. Some of these studies involve young children at more than one site and specific type of Pb exposure. Noted earlier in a different context were the multisite evaluations, in terms of statistical modeling, carried out by Succop et al. (1998). These investigators examined a total of 1,855 children <72 months old in a total of 11 discrete studies described in 16 published articles or reports, children who were geographically, demographically, and socioeconomically distributed across the U.S. Pb exposure sources largely consisted of lead paint in urban areas and mining or metallurgical site media in rural areas. Intermediate pathways for these sources included interior Pb dusts and children's hand lead.

These investigators reported, *inter alia*, that levels of Pb in floor dust were statistically significantly related to exterior lead paint in 2/9 (22%)

cases, interior Pb paint loadings in 4/10 (40%) cases, to soil Pb concentration in 6/8 (75%) studies, and entry dust Pb levels in 7/10 (70%) studies. Hand wipe Pb, the marker for the proximate pathway to child Pb ingestion, was predicted by interior Pb paint loading in 2/6 (33%) studies, by soil Pb in 2/3 (67%) of cases, by entry dust Pb levels in 1/6 (17%) instances and floor dust Pb concentration in 5/6 (83%) of cases.

Some of the significant structural equation-based regression meta-analyses results reported by Succop et al. (1998) for various Pb sources and pathways are given in Table 10.24. The analyses are for log-transformed PbB and are not adjusted for covariates. The most robust relationship for Pb source metrics to children's PbB in this series was estimated for interior floor dust Pb loading ($R^2 = 0.96$), followed by entry dust Pb level, $R^2 = 0.72$, perimeter (dripline) soil Pb level, $R^2 = 0.65$, interior dust Pb level, $R^2 = 0.58$, and interior dust Pb loading, $R^2 = 0.55$. These relationships were before adjustment for two statistical outliers, the adjustment leading to the interior floor dust Pb loading of $R^2 = 0.96$. The weakest associations were for interior and exterior paint Pb loadings, $R^2 = 0.07$ in both cases. As would be expected, the most significant proximate pathway contributor to PbB was hand wipe Pb, being the proximate path for Pb

TABLE 10.24 Meta-analyses of Relationships Between Children's Log-Transformed PbB and Various Environmental Pb Sources[a–d]

Independent Variable (Measurement Metric)	R^2	Slope Estimate	Number of Studies
ln interior dust Pb loading (μg/m^2)	0.96	0.474	8
ln hand wipe Pb (μg)	0.90	1.184	6
ln exterior entry dust Pb loading (μg/m^2)	0.72	0.435	10
ln perimeter soil Pb level (ppm)	0.65	0.233	6
ln interior dust Pb level (ppm)	0.58	0.529	10
ln interior dust Pb loading (μg/m^2)[e]	0.55	0.444	10
ln maximum interior paint Pb loading (mg/cm^2)	0.07	0.232	8
ln maximum exterior paint Pb loading (mg/cm^2)	0.07	0.152	9

[a]*Adapted from Succop et al. (1998).*
[b]*Not adjusted for covariates.*
[c]*Employing structural equation modeling across the studies.*
[d]*Listed in descending order of squared correlation.*
[e]*Excluding two project study results which were likely outliers.*

ingestion via normal hand–mouth activity in the very young, including infants and toddlers. The squared correlation was 0.90.

Lanphear et al. (1998) reported results of a pooled analysis of 12 epidemiological studies directed to the quantitative determination of contributions from dust and soil Pb to children's PbB concentrations. These authors calculated such relationships as: (1) the projected geometric mean PbBs of children 6–36 months of age as a function of interior dust Pb loading and soil Pb, (2) the likelihood of children having a PbB ≥10 μg/dl as a function of interior dust Pb loading and soil Pb concentration, and (3) the likelihood of children having a PbB ≥15 μg/dl as a function of such environmental Pb parameters. Table 10.25 presents some summary data from the pooled studies, giving geometric mean and exceedence probability estimates ≥10 μg/dl in terms of dust Pb loadings and concurrent soil Pb.

Table 10.25 shows that at even relatively low interior dust Pb loadings, e.g., 10–20 μg/ft^2, and lower soil Pb contents, there are significant fractions of children exceeding the U.S. CDC toxicity guideline of 10 μg/dl as of the time of this writing. It is notable that the dust loading at 40 units, the current dust Pb floor hazard promulgated by the U.S. EPA, shows significantly elevated PbBs and high risks of exceeding the U.S. CDC toxicity threshold. This clearly indicates a limited protective utility of this floor dust Pb standard. At the higher dust and soil Pb values, the PbB means approach the toxicity threshold and do so with a 50–50 chance of attaining that threshold.

TABLE 10.25 **Pooled Analysis Predictive Results for Geometric Mean PbB (μg/dl) and Probability Frequency ≥10 μg/dl in Children Exposed to Dust and Soil Pb[a–c]**

Dust Pb Loading (μg/ft^2)	Soil Pb (ppm)					
	100	500	1,000	1,500	2,000	4,000
	Geometric Mean PbB (Probability % ≥10 μg/dl)					
5	4.1 (5.0)	4.9 (9.3)	5.3 (12)	5.5 (14)	5.7 (15)	6.1 (18)
10	4.7 (8.3)	5.6 (14)	6.1 (18)	6.3 (20)	6.5 (22)	7.1 (26)
20	5.4 (13)	6.5 (21)	7.0 (26)	7.3 (28)	7.6 (30)	8.1 (35)
40	6.3 (20)	7.5 (30)	8.1 (35)	8.4 (38)	8.7 (40)	9.4 (45)
70	7.0 (26)	8.4 (37)	9.1 (43)	9.5 (46)	9.8 (48)	10.5 (54)
100	7.6 (31)	9.0 (43)	9.7 (48)	10.2 (51)	10.5 (54)	11.3 (59)

[a]*Adapted from data in Lanphear et al. (1998).*
[b]*Children 6–36 months of age.*
[c]*Pooled analyses of 12 epidemiological studies from urban and metallurgical industry sites.*

Multisite, subnational U.S. evaluations of the impacts of lead in paint and interior dusts included the remediation effects at multiple U.S. sites, where old dwellings with lead paint hazards were subjected to various Pb abatement strategies. Galke et al. (2001) reported that the combined interventions at 14 locales involving 2,600 units treated to control lead paint hazards, mainly in the form of window replacement and repair of painted surfaces, had reduced geometric mean PbB levels by 26%, from 11.0 to 8.2 μg/dl. Dust lead loadings had also been reduced and the reduction persisted at least 1 year.

Young children can be exposed to dusts from secondary exposures, i.e., "take home" exposures, where family members in the children's households bring home leaded dusts from workplaces in various forms, typically small-particle dusts adhering to shoes, clothing, skin, etc. Roscoe et al. (1999) carried out a meta-analysis of 10 reports of such secondary child Pb exposures in children 1−5 years of age in the period 1987−1994. Results indicated that the geometric mean children's PbB across the studies was 9.3 μg/dl versus a U.S. child population mean of 3.6 μg/dl, a significant difference of more than 2.5-fold. Of equal concern for exposures in these children, the percentages ≥10 and ≥15 μg/dl were 52% and 21%, respectively. These greatly exceeded U.S. children's prevalences of 8.9% and 1.1%, respectively.

A number of smaller, site-specific studies have described the inferential statistical relationship of PbB, particularly children's PbB, to various Pb sources and pathways. These have included dietary, tap water, soil, and dust Pb. Dietary lead has typically had a broader reach for human exposures in industrialized countries, where the food supply is more centralized than local or subsistence in nature. Details on the Pb content of dietary components and typical diets were provided earlier. Compendia of such localized efforts include the 1986 and 2006 U.S. EPA criteria documents (U.S. EPA, 1986, 2006).

Table 10.26 depicts some recent epidemiological studies relating children's PbB levels to Pb concentration in various external Pb sources and, in one case, the relationship of resorbing bone Pb levels to this biomarker.

The NHANES 2004 PbB survey was accompanied by household dust Pb sampling and the relationships between blood lead and household dust Pb analyzed through various statistical methods (Dixon et al., 2009; Gaitens et al., 2009). There was a statistical association of PbB to floor dust Pb. Child age and other child/family variables as well as year of house construction and window sill and floor Pb were significant predictors of PbB. These workers also calculated via logistic regression analyses that up to 5% of children in homes with floor dust Pb at 12 $\mu g/ft^2$ would have a PbB ≥10 μg/dl.

A large number of environmental epidemiological studies dating to the 1970s were carried out at a lead−zinc smelter complex in the Silver Valley of Idaho, continuing into more recent years at multiple towns within the Bunker Hill Superfund Site (Sheldrake and Stifelman, 2003; von Lindern

TABLE 10.26 Illustrative Recent Epidemiological Studies of Blood Lead—Lead Source Relationships

Pb-Containing Medium	PbB Metric	Study Details	Results	References
Home dust lead levels in the United States	Log PbB versus floor dust Pb ($N = 2{,}155$)	NHANES IV gathered Pb data in homes; home Pb source data versus NHANES PbB statistics; children, 1–5 years old	In linear models, PbB showed association with floor dust Pb to child age, year house built, economic status, race and ethnicity, window sill dust Pb; 5% of children Pb-exposed at ≥10 μg/dl PbB at a floor dust of 12 μg/ft^2	Dixon et al. (2009), Gaitens et al. (2009)
Air, soil dust Pb	Modeled children's PbB: IEUBK model, concurrent PbB form	USEPA case studies for PbB estimates and health risk assessments at three sites: primary smelter, secondary smelter, general urban case Pb sources and pathways– air, soil, dust Pb levels	Various Pb air quality scenarios developed for each case study Pb inputs to estimate PbB levels for three cases. Serial tabulations in U.S. EPA (2007): Chapters 3 and 5	U.S. EPA (2007)
Soil, dust Pb at residences, in community	Measured Superfund site PbB levels: Bunker Hill, ID	30 years of serial exposure assessments of children's PbBs; site soil, dust, garden Pb; multivariate Pb exposure regression models, e.g., slope factor models and structural equation	PbB to media Pb year-to-year determined; PbB declines with/without interventions were 7.5 μg/dl without and 2.5–15 μg/dl above 7.5 μg/dl with intervention for 2-year-olds; house dust Pb = 40–50% source of PbB; community soil = 30% source of PbB	von Lindern et al. (2003a), Spalinger et al., 2007

(*Continued*)

TABLE 10.26 Illustrative Recent Epidemiological Studies of Blood Lead—Lead Source Relationships—(cont.)

Pb-Containing Medium	PbB Metric	Study Details	Results	References
Soil, dust Pb at residences and in community	Measured Superfund site PbB levels: Bunker Hill, ID	Serial dust and soil Pb analyses; statistical relationship modeling of soil Pb to dust Pb	Soil Pb, household hygiene, number of adults, child playtimes outside contributed to variance in entry mat dust Pb loading rates	von Lindern et al. (2003b)
Soil Pb in metropolitan New Orleans, LA	Screening PbB level coupled with a soil Pb, $N = 5,467$	2000–2005 Census Tract stratification of data bases, permutation statistical methods	Curvilinear association of PbB with soil Pb, 2000–2005; 94% (rounding) of screening children had PbB ≥2 μg/dl	Mielke et al. (2007), Zahran et al., 2010
Tap water Pb	PbB screening database, $N = 7,270$	Screening PbB in a NC community versus GIS-based housing age, tap water sources, and change in water disinfection to using chloramines	PbB screening levels increased with introduction of chloramines	Miranda et al. (2007)

et al., 2003a,b). Serial statistical analyses showed significant declines in children's PbBs with soil and residential dust removal efforts. Both residential and community soils contributed to residential dust and to children's PbB levels. Without cleanup, PbB levels declined 7.5 μg/dl but up to an additional 15 μg/dl with interventions. The intervention–remediation experience also showed that contiguous, multiunit cleanup steps were more effective than the isolated intervention approach.

A series of modeled concurrent and average lifetime Pb exposures in children in three case studies were described as part of human health risk assessments carried out by the U.S. EPA's Office of Air Quality Planning and Standards (OAQPS: U.S. EPA, 2007). Inputs to the IEUBK model of environmental Pb levels came from data gathered in three case studies. Simulation exercises then determined geometric mean PbB estimates, which were tabulated in the OAQPS report. Modeled PbB levels were estimated for low and high environmental Pb inputs in the three cases: a primary smelter, a secondary smelter, and a "general urban case study." PbB levels were then estimated for the current air Pb standard and a series of alternative standard values and a series of alternative statistical forms of a given standard value were determined. A discussion of these case study results appear in Chapter 23.

Recent studies of environmental Pb impacts on children's blood Pb levels have included the relationship of potential changes in tap water lead with changes in community water treatments. Miranda et al. (2007) examined 7,270 PbB screening records in a North Carolina county and linked these to drinking water sources, age of residences, and other variables such as season of screening. Statistically significant associations with PbB included use of chloramines as disinfectants in water treatment.

10.6 TEMPORAL TRENDS

Population-wide and population-segment blood lead levels have been declining significantly for a number of years. This has been dramatically shown in the United States, Europe, and Japan. This temporal trend has been closely connected to companion reductions in various lead sources and pathways producing the PbB concentrations, as shown in previous tables. These close inferential statistical connections, of course, also provide further evidence for causality between Pb source and PbB. The major changes in Pb source levels, particularly those producing significant exposures of young children, have included worldwide reduction in ambient air Pb due to leaded gasoline phaseout internationally, controls on lead use in food production, and controls on industrial Pb emissions. Pockets of persisting elevated PbB rates in the United States, and likely elsewhere, can be identified within the larger picture of downward Pb exposure trends. Particularly notable are subsets of young children living in inner-city old housing with lead paint in problematic condition, whose PbB level decline rates trail those for other child segments of the national population (Jones et al., 2009; Pirkle et al., 1998).

The most comprehensive and valid statistical look at PbB declines in human populations in recent decades was made possible by tracking those declines within descriptive demographic and socioeconomic statistics for the various strata of the U.S. NHANES from 1976 to 2008. Given similarity among industrialized developed countries such as the United States, Canada, and much of Europe, we would expect NHANES trend data to be relevant for these other jurisdictions. In fact, data presented above in international efforts at evaluating relative changes in Pb sources for human PbB levels support this expectation.

Table 10.27 sets forth the geometric mean PbB values for U.S. children $\leq$5 years of age, presented in successive NHANES findings, 1976–2008. Over the period 1976–2008, GM PbB in children $\leq$5 years of age declined

TABLE 10.27 NHANES PbB Trend Data 1976–2008, Children, ≤5 Years Old

NHANES	Survey Years	Geometric Mean PbB, μg/dl (Age Range)	References
NHANES II	1976–1980	14.8 (0.5–5 years)	Annest and Mahaffey (1984)
NHANES III—Phase 1	1988–1991	3.6 (1–5 years)	Brody et al. (1994)
NHANES III—Phase 2	1991–1994	2.7 (1–5 years)	Pirkle et al. (1998)
NHANES IV	1999–2002	1.9 (1–5 years)	U.S. Centers for Disease Control (2005b)
Update NHANES	2003–2004	1.8 (1–5 years)	U.S. CDC (2009b), Fourth Exposure Report
Update NHANES	2005–2006	1.5 (1–5 years)	U.S. CDC, July 2010, Fourth Exposure Report, Updated Tables

from 14.9 to 1.9 μg/dl. This decline was most pronounced from the 1976 to 1980 period to the NHANES III, Phase 1 period of 1988–1991 at 76%, followed by 30% declines comparing Phase 2 with phase 1 of NHANES III, and NHANES IV with NHANES III—Phase 2. From 2002 forward, the slope of decline attenuates.

Additional important data from these serial surveys include the critical findings of declines in rates of PbB values at or above 10 μg/dl. For the periods of NHANES IV versus NHANES III—Phase 2, where CDC's definition of an elevation was ≥10 μg/dl, the percentage of exceedences declined from 4.4% to 1.6% for children 1–5 years of age. One can contrast either of these percentages with the very high percentages given from NHANES II reported values and seen in Table 10.9 for NHANES II. For Black children, 0.5–5 years of age, 97.5% had PbB ≥10 μg/dl, and for Whites, the corresponding figure was 85.5%.

REFERENCES

Adebonojo, F.O., 1974. Hematological status of urban black children in Philadelphia. Clin. Pediatr. 13, 874–888.

Angle, C.R., McIntire, M.S., 1979. Environmental lead and children: the Omaha study. J. Toxicol. Environ. Health 5, 855–870.

Annest, J.L., Mahaffey, K., 1984. Blood lead levels for persons ages 6 months–74 years, United States: 1976–1980. DHHS Publication No. (PHS): 84-1683. Vital and Health Statistics, Series 11, No. 233. National Center for Health Statistics, Washington, DC.

Annest, J.L., Pirkle, H.L., Makuc, D., Neese, J.W., Bayse, D.D., Kovar, M.L., 1983. Chronological trend in blood lead levels between 1976 and 1980. N. Engl. J. Med. 308, 1373–1377.

Apostoli, P., Baj, A., Bavazzano, P., Ganzi, A., Neri, A., Ronchi, L., et al., 2002. Blood lead reference values: the results of an Italian polycentric study. Sci. Total Environ. 287, 1–11.

Baker Jr., E.L., Hayes, C.G., Landrigan, P.J., Handke, J.L., Leger, R.T., Housworth, W.J., et al., 1977. A nationwide survey of heavy metal absorption in children living near primary copper, lead, and zinc smelters. Am. J. Epidemiol. 106, 261–273.

Bono, R., Pignata, C., Scursatore, E., Rovere, R., Natale, P., Gillies, G., 1995. Updating about reductions of air and blood lead concentrations in Turin, Italy, following reductions in the lead content of gasoline. Environ. Res. 70, 30–34.

Brody, D.J., Pirkle, J.L., Kramer, R.A., Flegal, K.M., Matte, T.D., Gunter, E.W., et al., 1994. Blood lead levels in the U.S. population: phase I of the third National Health and Nutrition Examination survey (NHANES III, 1988 to 1991). J. Am. Med. Assoc. 272, 277–283.

Caprio, R.J., Margulis, H.L., Joselow, M.M., 1974. Lead absorption in children and its relationship to urban traffic densities. Arch. Environ. Health 28, 195–197.

Carter-Pokras, O., Pirkle, J., Chavez, G., Gunter, E., 1990. Blood lead levels of 4–11 year old Mexican American, Puerto Rican and Cuban children. Publ. Health Rep. 105, 388–393.

Černá, M., Spěváčková, V., Batáriová, A., Šmid, J., Čejchanová, M., Očadliková, D., et al., 2007. Human biomonitoring system in the Czech Republic. Int. J. Hyg. Environ. Health 210, 495–499.

Claeys-Thoreau, F, Thiessen, L, Bruaux, P, Ducoffre, G, Verduyn, G., 1987. Assessment and comparison of human exposure to lead between Belgium, Malta, Mexico, and Sweden. Int. Arch. Occup. Environ. Health 59, 31–41.

Delves, H.T., Diaper, S.J., Oppert, S., Prescott-Clarke, P., Periam, J., Gompertz, D., 1996. Blood lead concentrations in United Kingdom have fallen substantially since 1984. BMJ 313, 883–884.

Dixon, S.L., Gaitens, J.M., Jacobs, D.E., Strauss, W., Nagaraja, J., Pivetz, T., et al., 2009. Exposure of U.S. children to residential dust lead, 1999–2004: II. The contribution of lead-contaminated dust to children's blood lead levels. Environ. Health Perspect. 117, 468–474.

Driscoll, W., Mushak, P., Garfias, J., Rothenberg, S.J., 1992. Reducing lead in gasoline: Mexico's experience. Environ. Sci. Technol. 26, 1702–1705.

Ducoffre, G., Claeys, F., Bruaux, P., 1990. Lowering time trend of blood lead levels in Belgium since 1978. Environ. Res. 51, 25–34.

Facchetti, S., 1989. Isotopic lead experiment. Sci. Total Environ. 93, 537–538.

Fine, P.R., Thomas, C.W., Suho, R.H., Cohnberg, R.E., Flashner, B.S., 1972. Pediatric blood lead levels. A study in 14 Illinois cities of intermediate population. J. Am. Med. Assoc. 221, 1475–1479.

Flegal, A.R., Smith, D.R., 1992. Lead levels in preindustrial humans. N. Engl. J. Med. 326, 1293–1294.

Gaitens, J.M., Dixon, S.L., Jacobs, D.E., Nagaraja, J., Strauss, W., Wilson, J.W., et al., 2009. Exposure of U.S. children to residential dust lead, 1999–2004: I. Housing and demographic factors. Environ. Health Perspect. 117, 461–467.

Galke, W., Clark, S., Wilson, J., Jacobs, D., Succop, P., Dixon, S., et al., 2001. Evaluation of the HUD Lead Hazard Control Grant Program: early overall findings. Environ. Res. 86, 149–156.

Goldwater, L.J., Hoover, A.W., 1967. An international study of normal levels of lead in blood and urine. Arch. Environ. Health 15, 60–63.

Haegar-Aronsen, B., Amdulla, M., Frisilett, B., 1971. Effect of lead on δ-aminolevulinic acid dehydrase activity in red blood cells. Arch. Environ. Health 23, 440–445.

Hammer, D.I., Finklea, J.F., Hendricks, R.H., Hinners, T.A., Riggan, W.B., Shy, C.M., 1973. Trace metals in human hair as a simple epidemiological monitor of environmental lead exposure. Trace Subst. Environ. Health X, 25–38.

Hasselblad, V., Nelson, W., 1975. Additional analyses of the Seven City Lead Study. Environ. Qual. Safety Suppl. II. Lead, 147–151.

Hofreuter, D.H., Cotcott, E.J., Keenan, R.G., Xinteras, C., 1961. The public health significance of atmospheric lead. Arch. Environ. Health 3, 568–574.

Johnson, D.E., Tillery, J.B., Prevost, R.J., 1975. Levels of platinum, palladium, and lead in populations of southern California. Environ. Health Perspect. 12, 27–33.

Jones, R.L., Homa, D.M., Meyer, P.A., Brody, D.J., Caldwell, K.L., Pirkle, J.L., et al., 2009. Trends in blood lead levels and blood lead testing among US children aged 1 to 5 years, 1988–2004. Pediatrics 123, e376–e385.

Kapaki, E., Varelas, P.M., Syrigou, A.I., Spanaki, M.V., Andreadou, E., Kakami, A.E., et al., 1998. Blood lead levels of traffic- and gasoline-exposed professionals in the city of Athens. Arch. Environ. Health 53, 287–291.

Kubota, J., Lazar, V.A., Losel, F., 1968. Copper, zinc, cadmium, and lead in human blood from 19 locations in the United States. Arch. Environ. Health 16, 788–793.

Landrigan, P.J., Gehlbach, S.H., Rosenblum, B.F., Shoults, J.M., Candelaria, R.M., Barthel, W.F., et al., 1975. Epidemic lead absorption near an ore smelter: the role of particulate lead. N. Engl. J. Med. 292, 123–129.

Langlois, P., Smith, L., Fleming, S., Gould, R., Goel, V., Gibson, B., 1996. Blood lead levels in Toronto children and abatement of lead-contaminated soil and house dust. Arch. Environ. Health 51, 59–67.

Lanphear, B.P., Matte, T.D., Rogers, J., Clickner, R.P., Dietz, B., Bornschein, R.L., et al., 1998. The contribution of lead-contaminated house dust and residential soil to children's blood lead levels: a pooled analysis of 12 epidemiological studies. Environ. Res. 79, 51–68.

Leroyer, A., Nisse, C., Hemon, D., Gruchociak, A., Salomez, J.L., Haguenoer, J.M., 2000. Environmental lead exposure in a population of children in northern France: factors affecting lead burden. Am. J. Ind. Med. 38, 281–289.

Loranger, S., Zayed, J., 1994. Manganese and lead concentrations in ambient air and emission rates from unleaded and leaded gasoline between 1981 and 1992 in Canada: a comparative study. Atmos. Environ. 28, 1645–1651.

Mielke, H.W., Gonzales, C.R., Powell, E., Jartun, M., Mielke Jr., P.W., 2007. Nonlinear association between soil lead and blood lead of children in metropolitan New Orleans, Louisiana: 2000–2005. Sci. Total Environ. 388, 43–53.

Miranda, M.L., Kim, D., Hull, A.P., Paul, C.J., Galeano, M.A.O., 2007. Changes in blood lead levels associated with use of chloramines in water treatment systems. Environ. Health Perspect. 115, 221–225.

Miranda, M.L., Edwards, S.E., Swamy, G.K., Paul, C.J., Neelon, B., 2010. Blood lead levels among pregnant women: historical versus contemporary exposures. Int. J. Environ. Res. Publ. Health 7, 1508–1519.

Mushak, P., 1992. Perspective. Defining lead as the premiere environmental health issue for children in America: criteria and their quantitative application. Environ. Res. 59, 281–309.

Mushak, P., 1993. New directions in the toxicokinetics of human lead exposure. NeuroToxicology 14, 29–42.

Mushak, P., Crocetti, A.F., 1989. Determination of numbers of lead-exposed children as a function of lead source: integrated summary of a report to the U.S. Congress on childhood lead poisoning. Environ. Res. 50, 210–229.

National Academy of Sciences, National Research Council, 1972. LEAD: Airborne Lead in Perspective. National Academy Press, Washington, DC.

National Academy of Sciences, National Research Council, 1993. Measuring Lead Exposure in Infants, Children, and Other Sensitive Populations. National Academy Press, Washington, DC.

Needleman, H.L., 2000. The removal of lead from gasoline: historical and personal reflections. Environ. Res. 84, 20–35.

O'Heany, J., Kusiak, R., Duncan, C.E., Smith, L.F., Spielberg, L., 1988. Blood lead and associated risk factors in Ontario children. Sci. Total Environ. 71, 477–483.

Patterson, C., Ericson, J., Manea-Krichten, M., Shirata, H., 1991. Natural skeletal levels of lead in *Homo sapiens sapiens* uncontaminated by technological lead. Sci. Total Environ. 107, 205–236.

Piomelli, S., Caresh, L., Corash, M.B., Seaman, C., Mushak, P., Glover, B., et al., 1980. Blood lead concentrations in a remote Himalayan population. Science 210, 1135–1137.

Pirkle, J.L., Brody, D.J., Gunter, E.W., Kramer, R.A., Paschal, D.C., Flegal, K.M., et al., 1994. The decline in blood lead levels in the United States: the National Health and Nutrition Examination Surveys. J. Am. Med. Assoc. 272, 284–291.

Pirkle, J.L., Kaufman, R.B., Brody, D.J., Hickman, T., Gunter, E.W., Paschal, D.C., 1998. Exposure of the U.S. population to lead, 1991–1994. Environ. Health Perspect. 106, 745–750.

Pönkä, A., 1998. Lead in the ambient air and blood of children in Helsinki. Sci. Total Environ. 219, 1–5.

Pönkä, A., Salminen, E., Ahonen, S., 1993. Lead in the ambient air and blood specimens of children in Helsinki. Sci. Total Environ. 138, 301–308.

Poole, C., Smythe, L.E., Alpers, M., 1980. Blood lead levels in Papua New Guinea children living in a remote area. Sci. Total Environ. 15, 17–24.

Quinn, M.J., Delves, H.T., 1987. The UK blood lead monitoring programme 1984–1987: protocol and results for 1984. Hum. Toxicol. 6, 459–474.

Quinn, M.J., Delves, H.T., 1988. The UK blood lead monitoring programme 1984–1987: results for 1985. Hum. Toxicol. 7, 105–123.

Quinn, M.J., Delves, H.T., 1989. The UK blood lead monitoring programme 1984–1987: results for 1986. Hum. Toxicol. 8, 205–220.

Roscoe, R.J., Gittleman, J.L., Deddens, J.A., Petersen, M.R., Helperin, W.E., 1999. Blood lead levels among children of lead exposed workers. A meta-analysis. Am. J. Ind. Med. 36, 475–481.

Rothenberg, S.J., Schnaas, L., Perroni, E., Hernandez, R.M., Karchmer, S., 1998. Arch. Environ. Health 53, 231–235.

Schulz, C., Angerer, J., Ewers, U., Kolossa-Ehring, M., 2007a. The German Human Biomonitoring Commission. Int. J. Hyg. Environ. Health 210, 373–382.

Schulz, C., Conrad, A., Becker, K., Kolossa-Ehring, M., Seifert, B., 2007b. Twenty years of the German Environmental Survey (GerES), Human biomonitoring-temporal and spatial (West Germany/East Germany) differences in population exposure. Int. J. Hyg. Environ. Health 210, 271–297.

Sheldrake, S., Stifelman, M., 2003. A case study of lead contamination cleanup effectiveness at Bunker Hill. Sci. Total Environ. 303, 105–123.

Silbergeld, E.K., Schwartz, J., Mahaffey, K., 1988. Lead and osteoporosis: mobilization of lead from bone in postmenopausal women. Environ. Res. 47, 79–94.

Spalinger, S.M., von Braun, M.C., Petrosyn, V., von Lindern, I.H., 2007. Northern Idaho house and soil lead levels compared to the Bunker Hill Superfund Site. Environ. Monit. Assessment 130, 57–72.

Stopps, G.J., 1968. Symposium for air quality criteria—lead. J. Occup. Med. 10, 550–564.

Succop, P., Bornschein, R., Brown, K., Tseng, C.-Y., 1998. An empirical comparison of lead exposure pathway models. Environ. Health Perspect. 106 (Suppl. 6), 1577–1583.

Thomas, V.M., Socolow, R.H., Fanelli, J.J., Spiro, T.G., 1999. Effects of reducing lead in gasoline: an analysis of the international experience. Environ. Sci. Technol. 33, 3942–3948.

U.S. Agency for Toxic Substances and Disease Registry, 1988. The Nature and Extent of Lead Poisoning in Children in the United States: A Report to Congress, July. U.S. Department of Health and Human Services, Atlanta, GA.

U.S. Centers for Disease Control, 1975. Increased Lead Absorption and Lead Poisoning in Young Children: A Statement by the Centers for Disease Control. U.S. Department of Health, Education, and Welfare, Atlanta, GA.

U.S. Centers for Disease Control, 1978. Preventing Lead Poisoning in Young Children. A Statement by the Centers for Disease Control, April. U.S. Department of Health and Human Services, Atlanta, GA.

U.S. Centers for Disease Control, 1985. Preventing Lead Poisoning in Young Children. A Statement by the Centers for Disease Control. U.S. Department of Health and Human Services, Atlanta, GA.

U.S. Centers for Disease Control, 1991. Preventing Lead Poisoning in Young Children. A Statement by the Centers for Disease Control. U.S. Department of Health and Human Services, Atlanta, GA.

U.S. Centers for Disease Control, 2005a. Preventing Lead Poisoning in Young Children. A Statement by the Centers for Disease Control. U.S. Department of Health and Human Services, Atlanta, GA.

U.S. Centers for Disease Control, 2005b. Blood Lead Levels—United States, 1999–2002. MMWR Morb. Mortal. Wkly. Rep. 54, 513–516.

U.S. Centers for Disease Control, 2009a. Lead. CDC's National Surveillance Data (1997–2007), June 1. Atlanta, GA: U.S. Department of Health and Human Services. <http://www.cdc.gov/nceh/lead/data/national.htm>, (accessed 11.12.10.).

U.S. Centers for Disease Control, 2009b. Fourth National Report on Human Exposure to Environmental Chemicals. U.S. Department of Health and Human Services, Atlanta, GA.

U.S. Centers for Disease Control, 2010. Fourth National Report on Human Exposure to Environmental Chemicals. Updated Tables, July. U.S. Department of Health and Human Services, Atlanta, GA. <http://www.cdc.gov/exposurereport/> (accessed 12.12.10.).

U.S. Department of Housing and Urban Development, 2001. National Survey of Lead and Allergens in Housing. Final Report. Volume I: Analysis of Lead Hazards. Office of Lead Hazard Control, Washington, DC.

U.S. Environmental Protection Agency, 1977. Air Quality Criteria for Lead. Report No. EPA-600/8-77-017. Office of Research and Development, Washington, DC.

U.S. Environmental Protection Agency, 1986. Air Quality Criteria for Lead, 4 vols. EPA 600/8-83/028bF. Environmental Criteria and Assessment Office, Washington, DC.

U.S. Environmental Protection Agency, 2006. Air Quality Criteria for Lead, Vol. 1. Report No. EPA/600/R-05/144aF. National Center for Environmental Assessment, Washington, DC.

U.S. Environmental Protection Agency, 2007. Lead: Human Exposure and Health Risk Assessments for Selected Case Studies, Vol. I. Human Exposure and Health Risk Assessments—Full Scale. Report No. EPA-452/R-07-014a, October. Office of Air Quality Planning and Standards, Research Triangle Park, NC.

von Lindern, I.H., Spalinger, S.M., Bero, B.N., Petrosyan, V., von Braun, M.C., 2003a. The influence of soil remediation on lead in house dust. Sci. Total Environ. 303, 59–78.

von Lindern, I., Spalinger, S., Petrosyan, V., von Braun, M.C., 2003b. Assessing remedial effectiveness through the blood lead soil/dust relationship at the Bunker Hill site Superfund site in the Silver Valley of Idaho. Sci. Total Environ. 303, 139–170.

Waldron, H.A., 1975. Lead levels in blood of residents near the M6-A38(M) interchange, Birmingham. Nature 253, 345–346.

Zahran, S., Mielke, H.W., Gonzales, C.R., Powell, E.T., Weiler, S., 2010. New Orleans before and after Hurricanes Katrina/Rita: a quasi-experiment of the association between soil lead and children's blood lead. Environ. Sci. Technol. 44, 4433–4440.

Part 3

Lead Toxicity in Humans

Chapter 11

Lead Toxicity in Humans: A Brief Historical Perspective and Public Health Context

11.1 INTRODUCTION

Chapter 2 provided a brief historical summary of lead as a mined and processed metallurgical and utilitarian commodity from ancient periods to the present. The chapter also identified the various ways by which production and uses of lead heralded its entry into the human environment as a contaminant and potential toxicant. That chapter, however, offered only passing comments about the available record for lead poisoning as a human health issue.

This chapter provides a companion historical reprise, presenting a brief summary of the historical record for lead toxicity in human populations. This health history of toxic impacts in humans focuses on lead's temporal reach and attempts a systematized look at the evolution of lead toxicology and epidemiology as a public health issue. This chapter also serves as the pathway to the following chapters detailing lead's human and experimental toxicology and epidemiology. Evolution of lead poisoning as a public health issue was guided by certain societal, scientific, and public health realities which served to determine the nature and extent to which communities and societies acquired information about lead exposures and lead poisoning and how they responded to such information through various reactive and proactive mechanisms.

Such realities included: (1) the comparative modern history of public and health scientists' awareness of lead's human health hazards versus earlier periods; (2) the contentious, often adversarial relationships among lead's various contending stakeholders *vis-a-vis* lead production, commerce, and diverse human health hazards illustrated in this chapter via several case studies; (3) the existence and nature of the limited sources of, and controls on,

Trace Metals and other Contaminants in the Environment, Volume 10
ISSN: 1875-1121 DOI: 10.1016/B978-0-444-51554-4.00011-0

information providing this history; and (4) the especially critical role of the current formal public health and regulatory structures to provide research support, to provide for dissemination of information to the public from such research, and to establish processes to inform and protect the public health from lead toxicity.

11.2 THE SCIENTIFIC AND PUBLIC HEALTH AWARENESS OF LEAD AS A HUMAN HEALTH HAZARD

One can stratify information about and the directions taken by the development of the human toxicological record for lead prior to the nineteenth century and that presented from the nineteenth century to the present.

11.2.1 Scientific and Public Health Awareness of Pb Poisoning Before the Nineteenth Century

The discussions in Chapter 2 made it clear that public and public health awareness of human lead exposures arising from increasing contamination was very limited and fragmented from ancient times to the eighteenth and early nineteenth centuries. Consequently, the full extent of awareness within early science and medicine of lead's potential for inducing diverse toxic signs and symptoms in even severe clinical poisonings with early lead use is difficult to quantify. Various early medical and other writers drew attention to the health hazards associated with particular uses, but by and large, these accounts say little that is either comprehensive or quantitative in modern toxicological or epidemiological terms. Major (1945), for example, summarized some of the classical but fragmentary accounts of what now would be viewed as clinical lead poisoning.

Table 11.1 tabulates some illustrative fragmented descriptions of Pb toxicity that appeared over the period from the Greco-Roman era to about the seventeenth century. Table 11.2 cites some scattered and illustrative poisoning reports in the seventeenth and eighteenth centuries.

What did the ancients know about lead poisoning and when did they know it? It is clear from writings of the time that early chroniclers, especially from the Greco-Roman era and including Rome's Imperial period, warned against the potential for lead poisoning associated with lead intakes and exposures via various pathways, especially through inhalation and ingestion. However, as noted in the NAS/NRC (1993) report, explicit conclusions of etiology linking specific instances of exposure to specific descriptions of classical acute or chronic lead poisonings are difficult to identify in period writings. Hippocrates (460–377 BCE) is claimed by some to have noted descriptions of what could be taken to be lead poisoning (Hunter, 1962; Sigerist, 1934; Waldron, 1973). The earliest description of adult human lead poisoning in this period has also been credited to the Greek poet and

TABLE 11.1 Illustrative Reports of Lead Poisonings in the Period from the Greco-Roman Era to ca. 1600

Time Period	Reported Signs and Symptoms	Clinical Characterization in Modern Terms	Primary Source(s)	Secondary Reference(s)
Greco-Roman era to ca. 200 CE	Acute abdominal distress, vomiting, muscle pain, paralysis, hallucination; death in extreme, untreated cases	Acute lead poisoning with acute encephalopathy and peripheral neuropathic features	Hippocrates Nikander Galen Vitruvius Dioscorides Pliny the Elder	Major (1945), Waldron (1973), Nriagu (1983a,b; 1985), Retief and Cilliers (2006)
Greco-Roman era to 600 CE and later	Abdominal colic, motor peripheral paralysis with foot and wrist drop, seizures, death	Chronic lead poisoning with chronic encephalopathic and peripheral neuropathic features; Pb chronic GI effects	Paul of Aegina Seventh century	Major (1945) Nriagu (1983a,b)
Greco-Roman era to Middle Ages	Claims of lowered fertility, increased miscarriages and abortions; gouty disease	Pb reproductive and developmental toxicity; saturnine gout secondary to Pb chronic nephropathy	Unclear for reproductive effects; Hippocrates referred to Pb-linked gout	Gilfillan (1965), Nriagu (1983b), Woolley (1984)
Middle Ages (ca. 1000 CE)	Pb colic	Pb GI toxicity: chronic constipation, severe gastric pain, nausea, vomiting	Avicienna 1000 CE	Aub et al. (1926)
ca. 1700 CE and before	Pb colic: colica Pictonum	Pb GI toxicity: chronic constipation, severe gastric pain, nausea, vomiting	Citois 1616: Poitiers, France physician	Aub et al. (1926)

TABLE 11.2 Illustrative Reports of Lead Poisonings in the Seventeenth and Eighteenth Centuries

Affected Population(s)	Reported Signs and Symptoms	Clinical Characterization in Modern Terms	Reference(s)
Individuals consuming adulterated beverages such as wine and cider	(a) Pb colic labeled with various geographic designations (b) Chronic gout	(a) Pb gastrointestinal toxicity: constipation, epigastric pain, nausea (b) Saturnine gout secondary to chronic Pb nephropathy	Waldron (1973), Major (1945), Nriagu (1983b,1985), Wedeen (1984), Eisinger (1982), McCord (1954b)
Individuals consuming drinking water in corrosive Pb plumbing	Paralysis and palsy; Pb colic	(a) Pb GI tract toxicity: constipation, epigastric pain, nausea, vomiting (b) Peripheral neuropathy, saturnine gout	McCord (1954a), Dana (1848), Nriagu (1985), Tronchin (1764)
Lead miners: Goslar, Germany	Pb colic	(a) Pb GI tract toxicity: constipation, epigastric pain, nausea, vomiting (b) Peripheral neuropathy, saturnine gout	Stockhausen, 1656, Edsall (1907: Chapter 5)

physician Nikander who described, in the second century BCE, signs and symptoms of what on balance appears to have been a case of advanced chronic lead poisoning with descriptions of enteric distress (lead colic), progressive motor peripheral neuropathy, myopathy, and added features of CNS involvement (Major, 1945; NAS/NRC, 1993). However, the level of Pb exposure or the nature of the exposure, e.g., occupational or iatrogenic, is not known. Retief and Cilliers (2006), however, view the descriptions of lead poisoning by Paul of Aegina in the seventh century CE as the first recording of chronic lead poisoning, with lead colic and peripheral/central neuropathy as core symptomatology.

Some writers (Gilfillan, 1965; Nriagu, 1983b; Woolley, 1984) have elicited heated debate by hypothesizing a potential etiological role for multisource toxic lead exposures among the Roman Imperial leadership in the deterioration and fall of the Roman Empire. One particular focus of the hypothesis has been the potential for reduced fertility rates, increased sterility, and other reproductive effects at projected lead exposures which in modern times have been described for comparably measured elevated lead exposures in human populations (see relevant chapters in Part 3). Within this constellation of potential toxicological harm, widely accepted neurocognitive and other neurobehavioral impairments beginning *in utero* and continuing postnatally at relatively low to moderate lead exposures can also be included.

This hypothesis has been challenged by others (Hodge, 1981; Needleman and Needleman, 1985; Retief and Cilliers, 2006; Scarborough, 1984) for reasons alleging that proponents overestimated the extent of actual lead exposures, underestimated alternative organizational and cultural explanations for the Imperial political decline and fall, and failed to reconcile significant prevalences and incidences of such toxic endpoints in the absence of archival data documenting their wide occurrence.

A critique of this issue is beyond the scope of this book but not irrelevant to its goals. It is valid and useful to ask whether class-stratified lead exposures in ancient cultures and societies defined macroscale dose–population responses for toxic endpoints and to define such endpoints at the higher end of the entire dose–response spectrum. Lead-associated impairments of function at executive and other levels would be consistent with, if not the sole basis of, the political and anthropological histories of the Imperial era. One can also argue that even a contributory role for lead reproductive and other systemic toxicity as a risk factor in the survival of lead-exposed populations merits consideration. At minimum, lead would be a serious risk factor for multiple reproductive, developmental neurotoxic, immunotoxic, and endocrinological effects.

The earliest records revealed, first, that lead was a human poison and, second, that it was a poison which affected some specific organs and systems and could eventually lead to fatal outcomes. Also present in these early observations was the conclusion that lead poisoning was a progressive disease with an increasingly dire prognosis attending continued (chronic) exposure.

Signs and symptoms noted by these ancient writers, when translated to current toxicological and epidemiological nomenclature, included central and peripheral nervous system (PNS) motor and sensory injury, manifest hematotoxicity as frank anemia, severe gastrointestinal effects (earliest references to "lead colic"), and in later centuries, saturnine gout attending lead nephrotoxicity.

As seen in Table 11.2, by the eighteenth century, a cluster of geographically diverse reports of episodic lead poisonings were accumulating and served to inform a medical scientific notion among some practitioners of toxic lead exposures consistently producing toxic outcomes. That is, causal connections were beginning to be drawn. These reports noted affected populations in Europe, in the American colonies, and in the Caribbean. The medical literature reference to the prevalence and incidence of lead's chronic GI toxicity (lead colic) from ingestion of lead not only through tainted water and other beverages but also from dietary contamination was typical. For example, it should be noted that the use of highly lead-contaminated food sweetener first described in Roman Imperial times continued well into more recent centuries (see Chapter 2).

11.2.2 Scientific and Public Health Awareness of Pb Poisoning in the Nineteenth and Early Twentieth Centuries

Table 11.3 presents some illustrations of the intersection in the nineteenth century of lead poisoning as a health issue and the industrial development of societies that produced greatly expanded exposures of diverse human populations to lead as a consequence. The nineteenth century also witnessed relatively rapid growth of medical and public health awareness and concern about lead's adverse effects in both environmental and occupational settings. This growth arose in part from the growth in the human toxicology and public health literature for lead exposures arising from the increased poisoning frequencies noted above. It also arose in part because of the larger developments occurring in scientific activity and a more credible and systematic practice of environmental medicine and clinical toxicology.

In the occupational sector, the classic occupational medical treatise of Tanquerel des Planches (1839) appeared in French in 1839 as *Traite des Maladies de Plumbe ou Saturnines*. This influential and detailed report on workers in lead-related occupations was a major benchmark for the nineteenth century and was translated into English almost a decade later by Dana (1848). Tanquerel des Planches studied over 1,200 workers, about two-thirds of whom were either lead paint pigment producers or painters using lead paint. His work provided a systematic lead poisoning case review from workplace reports through the various signs and symptoms medical practitioners needed to explore. This early occupational health researcher not only laid out the nature and extent of the lead paint problem as then applied to the workplace but also devoted part of his efforts to evaluating less hazardous substitutes such as various zinc pigments. For example, in the Dana translation (pp. 338–340) recommendations were made to use "zinc white" (zinc oxide) instead of "white lead," the latter invariably consisting of basic lead carbonate.

Tanquerel des Planches' urgings carried over into later organized regulatory efforts for controlling workplace paint lead exposures and uses of lead paint pigments that would also produce health threats to the general public using leaded paints. The overall thrust of the work predictably encountered resistance among lead paint producers, especially in the American paint industry, and this aspect of the findings are detailed elsewhere. The review of Stevenson (1949) provides more detail as to the overall history of lead paint hazards in the nineteenth century.

TABLE 11.3 Illustrative Reports of Lead Poisonings in the Nineteenth and Early Twentieth Centuries: Lead Toxicity in Industrializing Societies

Affected Population(s)	Poisoning Signs and Symptoms	Public Health Context	References
French lead workers: pigment production, painters	Pb colic ($N=1217$) Arthralgia ($N=755$) Paralysis ($N=127$) Encephalopathy ($N=72$)	First systematic attempt to categorize occupational Pb poisoning and exposure antecedents; reported distribution and tallies of toxic effects	Tanquerel des Planches (1839) [French], Dana (1848) [English trans.]
English women working in the lead trades	Worker effects typical of adult Pb poisoning; reproductive toxicity in the form of low fertility, miscarriages, stillbirths, failure to thrive, infant mortality	Confirmed the reproductive and developmental toxicity of Pb in pregnancy ascribed by modern writers to early toxicity in the Roman Empire; led to banning fertile women working in the Pb workplace	Paul (1860), Oliver (1914, 1911), Legge and Goadby (1912)
U.S. lead workers in specific categories to include smelter workers, battery workers, Pb pigment workers, ceramic workers, etc.	Worker effects typical of adult Pb poisoning, including fatal Pb poisoning	Hamilton's series put U.S. worker Pb mortality and morbidity in context with international worker poisoning statistics; series pointed to means to reduce workplace toxicity	Hamilton (1914a, b), Hamilton (1924, 1925)
Australian children exposed to chalking Pb paint	Acute and chronic encephalopathy, peripheral motor and sensory neuropathy, ocular neuritis	First modern documentation of young children as at high risk for Pb paint poisoning	Gibson et al. (1892), Gibson (1904)
Various exposure populations reviewed clinically	Multiple groups of poisoning subjects with range of signs and symptoms	Early examples of systemic assessment of occupational and environmental Pb exposure: diagnosis, treatment, hygiene	Aub et al. (1926), Edsall (1907)

(*Continued*)

TABLE 11.3 Illustrative Reports of Lead Poisonings in the Nineteenth and Early Twentieth Centuries: Lead Toxicity in Industrializing Societies—(cont.)

Affected Population(s)	Poisoning Signs and Symptoms	Public Health Context	References
American children with severe Pb poisoning, mainly due to Pb paint	Acute and chronic pediatric Pb encephalopathy	Early studies that established U.S. childhood Pb poisoning as a serious problem; confirmed the Australian child Pb poisoning findings; foretold the later epidemic of U.S. childhood Pb poisoning	Blackfan (1917), Thomas and Blackfan (1914), McKhann and Vogt (1926)
Low-level Pb poisoning from incremental intakes of small amounts of Pb	Early but demonstrable Pb poisoning	Showed there is a continuum of Pb intakes producing a continuum of poisoning, i.e., a dose–response relationship	Minot (1938)
American children surviving acute/chronic Pb poisoning	Acute/chronic Pb encephalopathy	Neurotoxic sequelae of initial Pb encephalopathic event: mental retardation, behavioral problems, inability to concentrate, etc.	Byers and Lord (1943)

Occupational lead poisoning, as a pervasive problem in other work settings, was also being recognized as the nineteenth century progressed. McCord (1953) cites Pb poisoning estimates contained in the notes of Rickard (1932), collected by the latter for his industrial text, *A History of American Mining*. These figures showed there were high rates of lead poisonings among hard rock silver miners in Utah and Colorado as well as in those working in the mines of Arizona, Idaho, Nevada, and New Mexico in the last quarter of the nineteenth and the early twentieth centuries. The earliest mining operations, directed at silver extraction from ore bodies, invariably entailed co-separation and processing of lead ore.

One reason for the high toll of lead poisoning among these early silver miners working in ore bodies of the Intermountain West was geological and geochemical. The principal lead ore being mined in the Western United States, lead carbonate or cerussite, was toxicologically distinct from and more bioavailable than the principal lead mineralogical form in Eastern and Midwestern ore bodies, i.e., galena or lead sulfide, which often occurred encased in a quartz silicate matrix. The silicate matrix of the galena assisted in producing a relatively more biologically inert form of lead at the particle size range typically generated and encountered during mining. This encased galena form of geochemical lead gave rise to the simplistic view that all mining of lead is less occupationally hazardous to health than its milling, smelting, and refining.

Poisoning records are principally known for early Utah silver mines in the period 1870–1900 in McCord's tallies, but we can assume that similar poisoning rates occurred in mining cerussite lead in Colorado and the other states. For example, cerussite was the main form of lead mined in the Smuggler Mountain area of Colorado. A group of eight mines near Salt Lake City in the period 1875–1898 produced about 9,000 severe lead poisoning cases. Similarly, The Park City district, east of Salt Lake City, was estimated to have produced 5,500 poisoning cases over a 20-year period among miners working in just three silver-rich mines: Silver King, Daly West, and Ontario. McCord provides a reasonable cumulative tally of about 30,000 lead poisoning cases just in Utah during 1875–1900, which would have included some repeat poisonings when miners returned to work. We can also assume, if only by inference, that the early, primitive milling and smelting methods during this time also would have produced extensive lead poisonings.

In the early twentieth century, occupational physician Alice Hamilton published multiple accounts of the occupational health hazards from production and use of numerous lead products in the United States and elsewhere, typified by morbidity statistics in employment categories such as lead smelter and refinery workers, lead paint pigment production workers, and structural painters using leaded paints and lead paint pigments (Hamilton, 1914a,b, 1919, 1924, 1925). Health hazards to workers in the painting trades were added catalysts for the organization of restrictions on uses of leaded paints occupationally and for the establishment of the White Lead Convention as a signed international code of regulations beginning in the 1920s (ILO, 1927). The history of occupational lead exposure leading up to the Convention and the chronology of ratification among some nations which became signatories is detailed in the 1927 report of the International Labor Conference in Geneva, Switzerland (ILO, 1927). Nation-specific ratifications of the Convention terms continued well past 1927.

English villages of the nineteenth and early twentieth centuries, where pottery and china works employed a significant fraction of the town's people, were known to have had high rates of lead poisonings among these workers (Legge, 1901; Oliver, 1911; Paul, 1860). These workers included men, women, and children. Exposures occurred from workplace practices such as applying and firing the lead glazes. The level of lead exposure and associated poisoning was sufficiently high that rather gross reproductive and developmental effects were noted in records of the time. Birth and infant survival rates plummeted in these villages compared to communities with little industrialization of this type.

Female workers in the lead trades generally were recognized in the early regulatory and medical communities as showing a variety of toxic reproductive effects. As summarized and discussed by Paul in 1860 and Oliver in 1911, these effects included premature delivery, abortion, stillbirths, and sterility. These findings were followed in time by formal restrictions on women working in occupations where they might be occupationally exposed to lead (see Chapter 25 for discussions of controls on female employment in various categories with occupational Pb exposure).

With regard to environmental lead exposures, Nriagu (1985) and McCord (1954a) described the increasing awareness in the nineteenth century of the health risk of drinking water distributed through lead plumbing. This was especially so for the New England area with its corrosive water sources, where Dana's studies of lead poisoning attributable to drinking water in Lowell, MA, made an impression on local health authorities.

Awareness that young children exposed to lead constitute a specific high-risk population developed around the turn of the twentieth century with the reports of Gibson et al. (1892) and Gibson (1904) from Queensland, Australia. These physicians showed that children ingesting weathered and chalking exterior lead paint on verandas become severely lead-poisoned. These findings were slow to be accepted elsewhere in Western medical and public health circles, as they were considered an idiosyncratic environmental health problem somehow peculiar to the climate and other conditions in Queensland. Nonetheless, these investigators provided the wherewithal that in the early twentieth century became the basis for assessing childhood lead poisoning from deteriorating lead paint in America's densely populated and older inner cities.

The first quarter of the twentieth century witnessed the growing acknowledgement in the medical and popular media that lead from more than one source could pose health threats, including to children's health. This awareness created various efforts to regulate these sources, as noted in a later chapter.

Moves were under way to respond to the lead paint problem in Europe even before the 1921 White Lead Convention. In 1915, the British government issued a report of a special committee appointed to evaluate dangers associated with using lead paint in buildings in Britain (see Chapter 26). Earlier, in 1906, Britain took the major step of including occupational lead poisoning among the injuries and diseases covered by workmen's compensation insurance. Austria, Belgium, France, Germany, Holland, and

Switzerland all implemented controls on lead use in the first decade of the twentieth century.

By the early twentieth century, there was a sizeable and credible body of scientific and medical information in the international professional literature. In 1922, Blänsdorf reported that the cumulative occupational medical and scientific literature on lead appearing over the recent centuries numbered 3,000 citations. The popular press was also paying more attention to lead, certainly the health hazards of lead paint to children in some of America's older cities, such as in Baltimore, MD (Fee, 1990).

The first quarter of the twentieth century also produced a set of key pioneering pediatric studies on paint lead poisoning of children in older American inner cities, beginning with studies of Thomas and Blackfan in 1914 and involving low-income children in Baltimore. While the multimedia nature of lead exposure and associated poisoning was vaguely being recognized by the early twentieth century, it is clear that the principal concern at the time was slowly being directed to lead paint poisoning. This source was pervasive in the contact environment of especially vulnerable risk groups, i.e., impoverished young children living in deteriorated housing that also had deteriorated lead paint. Exposures were also quite intense, producing in early decades multiple fatalities and severe, disabling sequelae among survivors. Poisoning victims in this case as noted by Fee (1990) were outside the mainstream of public and medical attention, so that preventive measures were not implemented until poisonings were occurring in epidemic numbers.

These early American results confirmed in their major features the findings of the Australian practitioners decades before. For reasons noted above, the complexity and perceived localization of the problem along demographic and socioeconomic lines did not create any immediate rush in medical and scientific circles to explore the extent of lead paint poisoning among American children. Additional difficulties in encouraging further assessments initially included the absence of a readily useful good measure of body lead burden for typical physicians and laboratories and the absence of pathognomonic (unique pathological features) signs and symptoms due only to lead.

11.3 EARLY EXPERIMENTAL ANIMAL STUDIES DOCUMENTING AND CHARACTERIZING LEAD TOXICITY

Whatever the level of knowledge internationally about human lead poisoning in this period, the first several decades of the twentieth century produced clear evidence of experimental lead poisoning. By the late 1930s, summaries of these reports appeared in the international lead toxicology literature and *Chemical Abstracts*, with numerous studies quantifying the nature and extent of toxic outcomes in a variety of experimental animals. Illustrative experimental animal studies are summarized in Table 11.4. These illustrative

TABLE 11.4 Illustrative Early Twentieth-Century Studies of Experimental Lead Poisoning

Animal Species	Pb Compound, Dosing	Results	References
Guinea pigs	Pb solution, 0.01 mg Pb/cc washed erythrocytes	Anemia and stippling of cells in dosed animals	Beresina and Engling (1912)
Guinea pigs	Chronic oral dosing ($N=28$) in food, 32 mg Pb carbonate, 3 × weekly	Severe anemia, nephritis, and cirrhosis; dosing ultimately fatal	Ophüls (1915)
Guinea pigs	Oral dosing, Pb carbonate	Pb-exposed males produced offspring with high mortality, retarded development	Weller (1916)
Guinea pigs	Oral dosing: Pb carbonate, sulfate $N=500$ animals	Carbonate and sulfate equally bioavailable, toxic	Fairhall et al. (1940)
Rabbits	Adult rabbits injected with Pb acetate 10×, 0.5 g/injection	Pb levels accumulated in various rabbit organs; marked anemia	Carcanague and Maurel (1913), Maurel and Carcanague (1913)
Rabbits	Oral dosing, Pb acetate	Pb-exposed males produced offspring with high mortality, smaller offspring	Cole and Bachhuber (1914)
Rabbits	Injections under skin ($N=2$) of Pb carbonate, Pb sulfate	Lethal outcome preceded by limb paralysis similar to that seen in cats	Straub (1913)
Rats	Pb carbonate: fed to either adult or young albino rats, up to 0.15 mg/kg in adults, up to 1.2 mg/kg in young animals, 8 weeks	Mortality high by 9–17 weeks older animals, no overt toxicity in younger rats	Sollmann (1922)
Rats	Pb carbonate or sulfate, fed chronically	Both forms equally bioavailable	Lehmann (1929)
Cats	Pb carbonate, Pb sulfate: injected under skin	By 2 months, paralysis of hind, front limbs, followed by death	Straub (1913)

(*Continued*)

TABLE 11.4 Illustrative Early Twentieth-Century Studies of Experimental Lead Poisoning—(cont.)

Animal Species	Pb Compound, Dosing	Results	References
Cats	Chronic oral dosing of adult cats ($N=6$) with Pb carbonate, 0.2 g in food	Pb encephalopathic and gastrointestinal effects (Pb colic)	Wassermann (1916)
Fowl	Oral dosing, Pb acetate	Pb-dosed males produced chicks with high mortality and retarded growth	Cole and Bachhuber (1914)
Frogs	Various injected doses ($N=32$), Pb acetate (saturated solution), followed by muscle stimulation tests	Pb did not target any muscle group; muscle fatigue is the factor in Pb palsy localization	Mellon (1913)

animal data in published studies by the first 30–40 years of the twentieth century also documented that lead paint pigments—and especially the white lead pigments basic lead carbonate and lead sulfate—were particularly potent toxicants for numerous species of test animals. The test species in the earliest studies included guinea pigs (Beresina and Engling, 1912; Ophüls, 1915; Weller, 1916), rabbits (Carcanague and Maurel, 1913; Maurel and Carcanague, 1913; Straub, 1913), dogs (Camus, 1913), frogs (Mellon, 1913), cats (Straub, 1913; Wassermann, 1916), fowl (Cole and Bachhuber, 1914), and rats (Lehmann, 1929; Sollmann, 1922).

These experiments were conducted with a range of goals, including experimental models of the toxic effects reported in humans, quantification of the range of doses required to induce some effect, and comparative species sensitivity to particular lead compound dosings. This interest in animal models clearly was evidence of a growing awareness and widespread concerns about human lead exposures from lead in commercial channels by the early twentieth century.

An excellent early comparative species analysis of chronic experimental lead poisoning is the 1922 paper of Sollmann on chronic intoxication induced by the white lead paint pigments basic white lead carbonate and lead sulfate. The data from this paper are presented below in the discussion dealing with experimental dose–toxic response relationships.

In the aggregate, by the 1920s there was a large and growing database showing that lead will produce toxic effects in organs and systems and that the nature of these effects buttress observations seen in human lead poisonings.

Of special relevance to this chapter and its focus on the toxicological hazards of lead for young children, a number of experimental test species clearly showed lead as salts, including the white lead pigments, to be a potent reproductive and developmental toxin. Lead toxicity covers and adversely impacts the whole spectrum of mammalian reproduction and development, beginning with gametotoxicity. For example, the gametotoxicity studies of Weller (1916) in guinea pigs indicated that progeny of males exposed to lead showed increased mortality rates and retarded development via such endpoints as permanent weight deficits. Similarly, Cole and Bachhuber (1914) showed that lead exposure of male parents of rabbits and fowl produced a reduced survival rate and reduced growth in surviving progeny. Early on, lead was also shown to affect young organisms post-reproduction. Dilling and Healey (1926), for example, demonstrated that lead salts were more potent than ionic salts of other metals in retarding the growth of tadpoles.

Adverse impacts of lead forms such as the white pigment basic lead carbonate and lead compounds used as glazes on the development of the human fetus were known to exist in the nineteenth century and the early decades of the twentieth century. The selective targeting of the CNS of young children, versus the (PNS) of adults, was noted early on (Strong, 1920). To the extent that the human CNS toxic vulnerability for various agents is most pronounced in the first several years of life, as stated by a number of pediatricians such as Rogers et al. (1934) early in the twentieth century, the message to lead paint pigment makers was clearly that the lead-exposed infants and toddlers were more neurologically vulnerable to lead, and was not equivalent to a small lead worker in terms of dose–neurotoxic response relationships. Rogers et al. (1934) stated clearly that the evidence in the experimental animal literature showed developing organisms to be more vulnerable on a dose basis and a toxic effect basis than adult organisms.

Experimental animal studies also served to characterize qualitatively and quantitatively severe toxic risks from both acute and chronic white lead paint pigment exposures. Experimental testing, where the durations and intensities (dosing levels) of controlled exposures were known, showed that both acute, high lead pigment intoxications and longer term, lower exposures to such substances induced specific forms of toxicity. For example, guinea pig studies cited above produced a constellation of toxic effects for both acute and chronic dosing conditions. The clear message for childhood lead poisonings from lead paint pigment ingestion was that both forms of lead contact, acute and chronic, posed health hazards for infants and toddlers.

Early experimental data also indicated interest in relative toxicity of different lead species in animals. By 1940, there were at least three adequately done studies that showed, using a variety of experimental animal test species, that white lead carbonate and lead sulfate were essentially equal in toxicological potency. Two studies were done in Germany in the 1920s (Koelsch et al., 1929; Lehmann, 1929) and one was done in the United States in 1940 (Fairhall et al., 1940). The Fairhall et al. study, for example, involved 500 guinea pigs dosed under various conditions.

Experimental animal studies indicated that there were clear dose–toxic response relationships that operated within a test animal species and across species when using white lead carbonate and lead sulfate as dosing substances. Injected lead carbonate or sulfate in the cat at a dose of 1–1.5 mg/kg body weight produced severe nervous system damage and death (Straub, 1913), while at the other end of the dosing route and dosing quantity spectrum, feeding lead carbonate to white rats at a level of 1–1.5 mg/kg body weight was without overt toxic effect in young rats (Sollmann, 1922).

11.4 SCIENTIFIC AND PUBLIC HEALTH AWARENESS OF PB POISONING FROM THE SECOND HALF OF THE TWENTIETH CENTURY TO THE PRESENT

Table 11.5 briefly depicts illustrative scientific and public health milestones in the second half of the twentieth century to the present that began to frame lead poisoning as a contemporary public health issue. The focus in this table is on Pb injury of the human central and PNS. An important factor in evaluating pediatric lead poisoning in the U.S. medical management system over the decades of the twentieth century was the increasing stringency of clinical and public health criteria for defining how bad lead poisoning had to be to be defined as such.

Originally, a child (and in the workplace, an adult) had lead poisoning if he/she died from lead exposure. This was followed by adding coma, convulsions, irritability, and colic to the list with mortality by the 1930s. This was then followed in the 1940s by the practice of searching for overt neurotoxic sequelae among children surviving the more acute encephalopathic episodes or such sequelae accompanying chronic pediatric encephalopathy (Byers and Lord, 1943).

Recognition of U.S. childhood lead poisonings as a chronically occurring disease, in contrast to a disease due to only episodic acute poisoning episodes, began in the 1950s and 1960s (Chisolm and Harrison, 1956). Criteria rooted in overt clinical Pb poisonings were succeeded by the findings in the latter part of the twentieth century that certain toxic effects were subclinical, i.e., not amenable to ready identification by overt signs and symptoms and through relatively simple medical diagnostic tools. Finally the findings were that certain toxic effects, e.g., developmental neurotoxicological effects, were both expressed with no discernible empirical threshold and persistent and likely irreversible in their health impact in children.

TABLE 11.5 Illustrative Reports and Summaries of Pb Poisoning from the Second Half of the Twentieth Century to the Present

Study Population	Pb Exposure(s)	Effect Endpoint(s)	Results	References
A. Clinical Pb poisoning				
Hospitalized or screened children	100–150+ μg/dl PbB	Developing brain injury (CNS)	Acute/chronic encephalopathy with high mortality before chelation use	Chisolm and Harrison (1956), NAS/NRC (1972, 1980, 1993), USCDC (1978)
Hospitalized or screened children	40 μg/dl-higher PbB	Kidney injury	Chronic and acute nephropathic effects, e.g., Fanconi syndrome, chronic aminoaciduria	Chisolm and Harrison (1956), Chisolm (1962), NAS/NRC (1972), USCDC (1978)
Hospitalized or screened children	40 μg/dl PbB	Hematological effects	Anemia, reduced hemoglobin, reduced erythrocyte survival via hemolysis	Chisolm and Barltrop (1979), Waldron (1966), NAS/NRC (1972, 1980), USCDC (1978)
Adults	Mainly occupational exposures: >120–150 μg/dl PbB	CNS affected at very high levels; chronic nervous system effects with more typical workplace levels	Adult chronic encephalopathy; significant sensory or motor nerve injury, e.g., wrist or foot drop	Cumings (1959), Feldman et al. (1977), NAS/NRC (1972, 1980)
Adults	≥60 μg/dl PbB	Chronic occupational Pb nephropathy	Mainly proximal tubular injury; hyperuricemia	Wedeen et al. (1986), Goyer (1989), NAS/NRC (1972)

(*Continued*)

TABLE 11.5 Illustrative Reports and Summaries of Pb Poisoning from the Second Half of the Twentieth Century to the Present—(cont.)

Study Population	Pb Exposure(s)	Effect Endpoint(s)	Results	References
Adults	50– to ≥60 μg/dl PbB	Chronic occupational Pb hematotoxicity	Impaired heme production, anemia hemolytic effects	Baker et al. (1979), NAS/NRC (1972)
B. Subclinical/low-level Pb poisoning				
1. Fetal study groups examined prospectively				
	Maternal PbB levels typically averaged <25 μg/dl	*In-utero* and early postnatal developmental indices	No association of low maternal PbB with spontaneous abortion or stillbirth; association with preterm delivery, birth weight in most studies; subsequent negative associations of fetal/ maternal PbB and early growth and cognitive development	Grant and Davis (1989), Ernhart et al. (1989), NAS/NRC (1993)
2. Postnatal study groups examined cross-sectionally				
Boston area children (*N*=158)	Shed teeth dentine Pb at two levels	Full scale and verbal IQ scores, WISC-R	Higher tooth Pb group had full scale IQ loss of 4.5 points	Needleman et al. (1979), Davis and Svendsgaard, 1987
German children	Shed teeth	German WISC	Inverse linkage of deficits in visual-spatial, visual-motor skills	Winneke et al. (1983, 1988)
Greek children (*N*=509)	PbB	WISC-R	Decrements in both verbal and performance versus PbB	Hatzakis et al. (1989)
Scottish children (*N*=501)	PbB	British Ability Scales (BAS)	Significant inverse association concurrent PbB versus BAS total and quantitative, reading scales scores	Fulton et al. (1987)

(Continued)

TABLE 11.5 Illustrative Reports and Summaries of Pb Poisoning from the Second Half of the Twentieth Century to the Present—(cont.)

Study Population	Pb Exposure(s)	Effect Endpoint(s)	Results	References
3. Postnatal study groups examined prospectively				
Boston area children (*N*=249)	Cord PbB and serial postnatal PbB: 6, 12, 18, 24, and 57 months, 10 years; postnatal mean levels: 6.2 (6 months) to 2.9 μg/dl (10 years)	BSID, 6–24 months; MSCA, 57 months; WISC-R, K-TEA, 10 years	At 6 and 24 months, MDI inversely linked to cord PbB; starting 10 μg/dl, McCarthy GCI scores inversely linked to 24 months PbB; subset of children with PbBs <10 μg/dl had higher IQ deficits versus those ≥10 μg/dl	Bellinger (1984, 1987, 1992)
Cincinnati, OH, children (*N*=305)	Maternal PbB+ cord, neonatal, 3 months, 1st year, 2nd year, 3rd year, 4th year, 5th year, 7th year, 15–17 years PbB levels: PbB level mean ranged up to 12, 5th year	BSID, 3–24 months; K-ABC, 48 and 60 months; WISC-R, 78 months	Inverse linkage prenatal PbB and 6 months BMDI; at 4 years, higher neonatal PbB inversely linked to deficits, all K-ABC scores; at 5 years, postnatal PbBs inversely linked to K-ABC results, closest with visual-spatial skills; 7 years, inverse link to WISC-R full and performance results; concurrent PbBs closest linkage to WISC-R results	Dietrich et al. (1987), Dietrich et al. (1992), Dietrich et al. (1993), Ris et al. (2004)
Port Pirie, Australia, children (*N*=723)	Maternal PbB+ cord, 6, 15, 24, 36, and 48 months PbB; integrated average to age 4; PbB 11–13 years; PbB levels: mean lifetime to age 7=9.1 μg/dl	24 months: BSID; 48 months: MSCA; 7 years: WISC-R; 11–13 years: WISC-R	Significant inverse linkages of WISC-R full scale and verbal scores with integrated PbB at 7–8 years; WISC-R full scale scores inversely linked with integrated PbBs at 11–13 years	McMichael et al. (1988), Baghurst et al. (1992), Tong et al. (1996)

(*Continued*)

TABLE 11.5 Illustrative Reports and Summaries of Pb Poisoning from the Second Half of the Twentieth Century to the Present—(cont.)

Study Population	Pb Exposure(s)	Effect Endpoint(s)	Results	References
International pooled analysis of children in seven prospective studies (*N*=1333): Boston, MA; Cincinnati, OH; Cleveland, OH; Rochester, NY; Mexico City; Port Pirie; Kosovo, Former Yugoslavia	Four PbB measures: concurrent, peak, average lifetime, and early (6–24 months) PbB	Full scale IQ at 7 years	Log-linear model showed significant inverse association of IQ decrement with concurrent PbB; decrement in IQ higher below 10 μg/dl	Lanphear et al. (2005)

Lead toxicity is presently understood to express its biological and toxicological characteristics across a broad spectrum of Pb exposures, specifically within a broad range of dose–response relationships applicable to subsets of human populations, such as young children, who are especially sensitive to toxic lead exposures. Clinical Pb poisoning, especially in sensitive subsets, is still a necessary topic for discussion regardless of current low frequency. First, clinical poisoning defines the nature of those toxic effects arising from the highest portion of the dose–response spectrum, including death as the ultimate outcome and one registered historically as occurring in quite a number of instances, especially in children in certain U.S. urban areas.

Secondly, the current low frequency of severe poisonings in certain human populations mainly traces to reduced human exposures because of reduced Pb amounts entering various exposure pathways. Such reductions in turn spring from current abatements in industrial activities using Pb, resulting in lower Pb emissions coupled with various efforts by health and regulatory agencies to control contact by risk groups with Pb. If Pb emissions were to increase significantly, as with emergence of new Pb-requiring technologies, or if controls on human contact with Pb were to decline or disappear, then the risk of all levels of Pb-poisoning prevalence and incidence would likely increase. Put differently, lead has not disappeared from the human environment; it is only being held in technological and regulatory check.

11.5 AWARENESS OF LEAD POISONING AMONG THE GENERAL PUBLIC AND ITS SOCIAL INSTITUTIONS

Public awareness about health hazards would have arguably been spotty, rarely shared, and largely ineffectual in the period from the Greco-Roman era up to about the seventeenth century. The means by which public awareness could occur after this period improved considerably, but it is of critical importance to point out that public awareness on any widespread or collective scale is of recent vintage. The reasons for this are complex and multifactorial and are provided in Mushak's perspectives and in summary form below.

It is not apparent from available records how extensively or effectively early medical and scientific writers' admonitions about lead's harmful health effects were disseminated to the political and economic leadership of peoples and nations from the Greco-Roman era up to the eighteenth century. We can plausibly assume that the social and political organizations of early cultures and empires were such that ordinary citizens and noncitizen classes (laborers, serfs, vassals, slaves, and prisoners of war) had little say in how lead was used in the broader society and equally little say in avoiding lead exposures from that use.

In many ways, this long-persisting relationship between general public comprehension of lead toxicity and the element's pervasive presence remained intact until very recent times. The powerlessness of ordinary citizens in Greco-Roman and later eras in averting lead uses and associated exposures and poisonings was aggravated by high illiteracy and technological ignorance among other than the very few. Even those knowledgeable about lead toxicity formed opinions from flawed observation and nonscientific methods rather than from scientific premises or systematically gathered data. Furthermore, the nature of lead exposures and poisonings differed greatly across socioeconomic and demographic strata. The Roman aristocracy has been held to have sustained toxic lead exposures through sources and routes different than those for plebeians and slaves.

Later, through the Middle Ages and extending into the Industrial Revolution and the nineteenth century, the scattered but primitive public health records are studded with isolated epidemics of lead poisonings, particularly among populations in Europe and in the early Americas. Again, as with the classical era toxicological chronicles, we have little idea how widespread or technically sophisticated public awareness of lead's hazards to health was in the Middle Ages and beyond. Here, also, effective communication about poisoning episodes and the level of public awareness would have been low for other than the communities most acutely affected. Lead poisoning epidemics were still being responded to by a monarchical bureaucratic political structure, especially in Europe. Illiteracy was still universal among townspeople and land-bound peasants. This is compellingly illustrated in Eisinger (1982), detailing German city–state edicts banning Pb-adulterated wines with harsh punishments by the end of the seventeenth century. Widespread illiteracy was coupled with the lack of written materials that could be widely disseminated warning about poisoning outbreaks.

The evolution and development of methods and mechanisms for communicating information about lead's toxic potential began in the seventeenth century, albeit haltingly and with marginal effectiveness. This information about lead and health was communicated via the early medical journal. However, Eisinger (1982) pointed out that many physicians were neither interested in nor did they pay close attention to early medical reports of lead poisonings in journals. Coupled with this mixed interest in medical journal reports on lead was the lack of interaction among physicians that may have impeded sharing lead poisoning case findings outside of the victim's family, friends, and neighbors.

In the eighteenth century, isolated medical reports from physicians and others established plausible pathways for the cause of poisoning epidemics through lead entering some contact medium. Findings were recorded in medical media, but data dissemination remained problematic.

By the latter part of the nineteenth century, a growing awareness and concern in the popular press was established for perhaps the first time about lead's health hazards among nonoccupational populations. A striking example of this appears in the form of an 1897 advertisement in the Diamond Jubilee issue of the Illustrated London News, touting the safety for the consumer of a particular brand of lead-free enamel widely sold in Europe and New York City. This advertising plate was reprinted as Figure I-6 in the 1993 National Academy of Science's report on lead exposure in sensitive populations (NAS/NRC, 1993). The advertising of non-lead paints and enamels on consumer safety grounds clearly suggests that this was a marketing strategy based on recognized health dangers of lead among the mass of the population of industrialized countries.

The extent to which considerable ignorance remained in some segments of the general public by the early twentieth century in the face of a growing written record on lead's health hazards is not quantifiable, but expected awareness limits in low-income and poorly educated pockets of risk populations (Fee, 1990; Rabin, 1989) reinforce the notion that effective dissemination of data is as important as the original generation of that data.

The second half of the twentieth century in the United States and Western Europe witnessed the development of a sustained interest in lead and other environmental pollutants first by the print and then by the electronic mass media. The comparative perspective of Mushak and Mushak

(2000) illustrated the interest of the print media from the 1950s through the 1970s in such topics as air pollution and airborne lead from high levels of leaded gasoline combustion and the continuing story of lead paint poisoning in America's inner cities.

Early efforts by local communities with an extensive lead poisoning problem, particularly from a deteriorating lead-painted housing inventory, to exert some controls on continued lead entry into consumer channels such as lead paint benefited considerably from dissemination of information via newspaper and radio, the media of the time. This was the case for Baltimore beginning in the 1930s and 1940s (Fee, 1990). The 1950s saw an expanded role for print and electronic media in assisting local governments in trying to control the entry of lead in paint into the retail pipeline as well as attempts to deal with deteriorating lead paint in slum housing. For example, by the 1950s both Baltimore and New York City required warning labels on new lead paint cans entering commerce.

Labor organizations comprised a mechanism for information to Pb workers. According to Mushak and Mushak (2000), labor unions would have been one means by which workers would have been alerted to and made aware of lead's health hazards. We can also assume that inroads made by labor unionization would have been a major factor in pushing measures for lead exposure reduction. The influence of unionization on expediting worker awareness of lead in the workplace and the publicity elsewhere about the problem of occupational lead poisoning could be discerned in the move to unionize hard rock miners in the American West. For example, Derickson (1987) described the development of miners' unions in certain Western mining states, noting especially the establishment of a hospital infrastructure for hard rock silver miners in the Coeur d'Alene valley of the Idaho Panhandle in the late nineteenth century. The hospitals dealt with mining accidents and occupational poisoning risks from lead and other toxic metal exposures. Two labor groups active in this regard were the Western Federation of Miners and, later, the International Union of Mine, Mill and Smelter Workers.

11.6 THE CONTENTIOUS RELATIONSHIPS BETWEEN LEAD'S ECONOMICS AND HUMAN HEALTH HAZARDS: LESSONS FROM CASE STUDIES

Struggles of an organized or concerted nature between lead's producers or users and those with a growing awareness of lead's potential health impacts largely trace to the latter part of the nineteenth century. They co-emerged and coevolved with the merging and consolidation of lead mining, production, use and marketing into large corporate entities (Knerr, 1992: sections on corporate emergence and consolidation of the U.S. lead industry in the late nineteenth to early twentieth centuries). Before the extractive industry corporation developed, the means employed by producers and users of lead to protect their activities were poorly documented. Limited information suggests these efforts were largely localized and variably effective. Production and use of lead were largely decentralized and entailed small levels of production and production highly dependent on local and regional markets from antiquity through the eighteenth century. On the other hand, the need for lead's producers and users to pursue economically defensive measures rarely arose in earlier times when public awareness was limited and any perceived need to regulate lead production and use was similarly limited.

That does not mean that earlier collisions between lead's economics and its health impacts did not occur. The eighteenth century had already witnessed the early development of controversies about which source or sources of lead might cause specific poisoning episodes. Fueling such controversies were the contentions of various affected interests that aimed to exculpate specific candidate sources, much as they continue to do to the present, as detailed in a later section. Typical of this early contentiousness were Baker's findings of the source of lead contamination producing the Devonshire Colic (see Chapter 2). Baker held to his findings that lead entered the cider pressings via lead fittings used to hold together the stone presses, a practice apparently unique to Devon cider producers.

Cider producers, in response, attacked Baker's conclusions about any lead source specific to the Devon cider industry and held that the source of lead was linked to the common practice of using lead shot as an abrasive for cleaning wine and cider bottles to debride residues. Attributing lead contamination to this cleaning practice had the self-serving virtue of exculpating any cider production practices unique to the Devon producers in favor of a more global candidate and one more beyond the control of local interests than Devon cider production itself. Contemporary medical accounts do not record the full resolution of these contending arguments about lead source, but it is noteworthy that when lead fittings in the stone presses were no longer used, the incidence of lead poisonings abated markedly.

The variably adversarial lead industry–health impact relationship, not surprisingly, was first expressed in the workplace, as described earlier. Occupational medicine in its early forms first had to deal with the adverse health impacts of lead among smelter and lead production workers and workers using lead products in many forms. Lead poisoning was an identified entity among lead workers in Europe by about 1840, but a label for occupational lead disease, *morbi metallica*, has existed since Greco-Roman times. The early part of the nineteenth century *vis-a-vis* occupational lead hazards could be characterized as poisonings endured out of economic necessity, in as much as the health risks of such work jousted with economic survival of the lead worker and his/her family.

Rapid industrialization of Europe and America in the second half of the nineteenth century, however, gave rise to the trade union movement as a countervailing force against the level of involuntary and indiscriminate worker exposures to lead at high, toxic levels.

In recent years, broader public concerns and control initiatives directed at lead's uses and its potential for environmental contamination have invariably jostled with various economic and industrial interests. These interests were those who either had a major stake in the preservation of lead's role in the mainstreams of commerce or those whose operations produced environmental lead emissions, regardless of the nature of their operations. Examples of the former would be lead mining, smelting, refining and recycling industries while other industrial emitters of lead would include operators of hazardous waste and municipal solid waste incinerators. This section therefore has to do with the means and methods employed by lead's economic interests in preserving those interests. General perspectives on the topic have been prepared by Graebner (1987) and Mushak (1992), while specific aspects of lead industry–public health conflicts over certain uses of the element have been described by Rosner and Markowitz (1985), Markowitz and Rosner (2002), Rabin (1989), Nriagu (1990), Fee (1990), and Needleman (2000).

One can best show the early interplay of public health, level of awareness and lead toxicity by several case studies: specifically, the lead paint problem as dealt with in one American city in past decades, and the various controversies and consequences attending the introduction and subsequent use for 60 years of lead additives such as TEL in gasoline in America and other industrialized nations. These examples differed in many ways, the fundamental one being that the paint lead problem was generated by previous near universal use in old residential units of lead paint that was now deteriorating.

11.6.1 Lead Paint as Case Study

Defense of continued American lead paint production and use in the face of growing evidence of lead poisoning from occupational exposures of production workers and painters and childhood lead poisoning from ingestion of paint lead from deteriorated lead-painted housing surfaces engaged the efforts of both the lead trade group, the Lead Industries Association (LIA) and the larger individual paint lead manufacturers such as the National Lead Co. (Markowitz and Rosner, 2002).

In the first half of the twentieth century, the lead paint industry was increasingly beleaguered by adverse developments. In the early 1920s, the White Lead Convention came into being under the principal auspices of European occupational health groups. This effort's aim was banning or controlling the use of the principal lead paint pigment, basic lead carbonate or "white lead," potentially affecting workers and the public. A number of European and other nations became signatories to the terms of the Convention either at the outset or over time. The United States, however, was not a signatory to the Convention. By the late 1920s, the first medical reports of childhood lead poisonings in the United States began to appear and their number increased greatly in the 1930s and 1940s.

The LIA became the principal vehicle representing the lead pigment and paint manufacturers' positions as these controversies and developments were occurring from the 1930s to the 1950s. According to Markowitz and Rosner (2002), defending lead in paints and some other uses was a principal task of the organization in those years. This LIA defense consisted of a vigorous public relations effort to promote the aesthetic and economic virtues of lead paint use and to challenge the validity of the early medical reports of childhood lead poisoning.

In 1933 and 1934, the LIA turned its attention to vigorous lobbying against any efforts in various state legislatures to limit uses of lead paint in new structures and to thwart workman's legislation having anything at all to do with treating lead poisoning on the job as a distinct disease compensable through workmen's compensation. Also noted was a putative effort by the Secretary of LIA at the time to head off paint lead restrictions being considered by the Massachusetts Department of Labor (Markowitz and Rosner, 2002).

Large lead paint producers also appeared to protect their markets against challenge on health hazard grounds by such means as aggressive advertising that showed the benefits of lead paint over any substitutes. The largest producer for many decades, the National Lead Co., employed that approach. The products of this company, notably Dutch Boy brand white lead paint, were heavily advertised in magazines of the 1920s. This company also produced promotional material directed at both painters and consumers. One example was a child's coloring book, touting the advantages of numerous lead products to be found in the middle-class home.

Fee (1990) described the public health and awareness history *vis-a-vis* lead paint poisoning for urban Baltimore in the period of the most explosive increase in the most serious forms of lead paint poisonings. Elements of the environmental health story for lead paint and pediatric health in Baltimore were multifaceted but clearly included the pivotal role of a strong health agency leadership in dealing with lead paint poisoning, the readiness of an affected population to respond to information about environmental health threats to their youngest family members, the role of the print and early electronic media in educating the public about the nature and magnitude of the problem, and the role of strong economic interests who had a financial stake in rental income-producing properties in Baltimore.

Baltimore's health system recognized early that the rash of poisonings being encountered in their inner-city clinics was due to lead and particularly to ingestion of lead paint. This recognition fed into a health infrastructure with a proactive health commissioner, Dr Huntington Williams, who also had the support of the city's political and civic establishment, including Mayor Howard Jackson. Despite this armoring for battle with the problem, health practitioners also recognized the complexity of any attempt to implement a comprehensive, very expensive lead paint abatement program. Such an approach would predictably elicit much hostility from influential political and economic interests if property owners were liable for lead paint hazard reduction. Conversely, any attempt at having lead paint abatement done at the city's expense would produce political and taxpayer hostility and resistance.

In lieu of lead abatement, Baltimore's health apparatus chose to deal with lead paint exposure and intoxication through local media involvement, community awareness, and resident family education. That approach was attacked by others on the obvious grounds that this approach meant that lead paint would remain *in situ*. On the other hand, it is likely that any attempt at direct paint lead removal at an abatement level that would have provided permanent freedom from future poisonings in Baltimore or any similar American city in the 1930s, 1940s, and 1950s would not have gone very far.

Public awareness and education to minimize contact with lead paint in those Baltimore residential units with young children was initiated and sustained in the city's high-risk neighborhoods by a weekly radio program, by health messages in the local press, and by a health department newsletter. Some messages with basic exposure prevention themes were distributed in large numbers. Implicit in the awareness and education approach was the rationale that one need only make residents aware of the problem. Afterwards, it would be the caregiver's responsibility to see that children were not exposed to paint lead. In the three decades that the city's public education and information apparatus were fully active, lead poisoning rates continued to grow, indicating that even when awareness and education about lead poisoning were intensively pursued, lead poisoning could not be prevented. Such a finding was replicated much later by Lanphear et al. (1996, 1999), who showed that interior leaded dust control efforts by families without systematic use of hazard controls on lead in these dusts failed so long as deteriorated lead painted surfaces remained.

Baltimore later chose statutory remedies to complement education programs. In 1941, the city enacted an ordinance, commonly ignored, that specified lead paint abatement if a child in a particular unit was diagnosed as lead-poisoned. The city also began to specify maximum lead content in paints entering retail pipelines in the city.

11.6.2 Leaded Gasoline as Case Study

The introduction of leaded gasoline in America in the mid-1920s differed as to context for awareness and concerns about lead poisoning from what existed for peeling lead paint in deteriorating Baltimore housing stock. Leaded gasoline emerged on the automotive engineering and sociocultural scene because it eliminated engine knocking and permitted the use of high-compression engines needed for heavier, faster vehicles. Its discovery for this use was neither particularly systematic scientifically nor did its promotion by various business interests at governmental health policy levels occur because there were simply no antiknock alternatives available (see below).

The potential health hazards of TEL and/or the resulting leaded fuel mixtures were largely unknown quantities to the general public at the time of their introduction. However, there were informed health scientists who were highly vocal in condemning TEL's introduction, predicting that among other hazards, lead emissions from combustion of leaded gasoline would become a pervasive environmental contamination problem (see the critiques of Needleman, 2000; Nriagu, 1990; Rosner and Markowitz, 1985).

Equally puzzling, besides those events attending the choice of an additive which proved lethal to workers in its early production and was condemned for potential environmental contamination, was the dismissal and *de facto* rejection of some already known, safer, and equally adequate alternatives at the time chemist Thomas Midgley first discovered TEL's antiknock property in 1921. Alternatives such as ethyl alcohol or benzene, an aromatic hydrocarbon, could be blended into gasoline and in fact were successful antiknock compound competitors in Europe and Latin America. The highly important technical and engineering issues of how this organolead antiknock choice prevailed in the United States over various alternatives, while the opposite occurred in other countries, have been addressed in some detail by Kovarick (1999, 2003), while Kitman (2000) published a general overview in the non-technical media.

For example, Kovarik (2003) noted that ethyl alcohol at 20–30% blending into gasoline was the European alternative to TEL gasoline in the 1920s and 1930s. Three years before Midgley's discovery of TEL's antiknock properties, the July 6, 1918 issue of *Scientific American* printed an article making it clear that ethyl alcohol blended into gasoline had been shown to be a satisfactory antiknock additive.

The initial production efforts for TEL were disastrous in terms of worker poisonings, a number of which were fatal. Many other workers sustained severe nervous system damage, including persisting hallucinations, delusions, and other psychotic behaviors. In one very short period, 5 fatalities and 35 other poisonings occurred at one TEL production facility (Needleman, 2000; Rosner and Markowitz, 1985). Publicity about such poisonings became widespread, and skepticism and alarm soon developed about both the safety of the additive itself and any forms of lead being emitted in vehicular exhaust.

TEL production was temporarily suspended while manufacturers dealt with growing general public and health practitioner alarm. TEL producers responded by bringing about one of the more bizarre government–industrial collaborations in twentieth-century American industrial history. The U.S. Bureau of Mines was commissioned to carry out a study where TEL makers would have heavy oversight over design, execution, and dissemination of results. This included a major say on what information would be released to the public. The limited study, involving an experimental exposure design that had many limits, soon was claimed to show no adverse effects of combusted leaded gasoline.

Numerous health and scientific critics remained unconvinced, and the struggle between these skeptics, and various industrial interests and those in the health community sympathetic to them, has been amply detailed in the history of this episode by Rosner and Markowitz (1985), Nriagu (1990), Needleman (2000), Kovarik (1999, 2003) and Kitman (2000). Eventually the TEL manufacturers, the oil and auto industries and others won the day, and TEL production and use in gasoline began again and would continue for decades. The interaction of lead economics with public health concerns in the specific case of leaded gasoline was largely rooted in various activities of the major producer of the TEL antiknock additive, the Ethyl Corporation. Ethyl's economic reason for being was the introduction and continued production of basically one group of organolead compounds, the lower alkyl lead antiknock additives, of which TEL was the dominant substance. Ethyl's formation and early corporate history are detailed in the 1983 corporate profile by Robert (1983), Rosner and Markowitz (1985), Graebner (1987) and Kovarik (1999).

Early on, Ethyl's industrial survival lay largely with its corporate parents, General Motors and DuPont. The latter could marshal a large array of links with opinion-molding strata of American scientists and public officials and effectively did so in pushing for the adoption and use nationwide of TEL as an additive whose properties would permit a social and economic revolution in those early decades of the automobile culture (Graebner, 1987; Kitman, 2000; Rosner and Markowitz, 1985).

Like LIA in its early involvement with lead paint controversies, Ethyl saw virtue in commissioned studies of TEL's potential toxicity under the publicly perceived mantle of scientific objectivity. It commissioned the first of several studies that purported to show that using TEL was without public health hazard, regardless of the earlier severe poisonings and deaths resulting from TEL's production and regardless of the many questions generated by the quality and circumstances of the early data generation.

Ethyl's role in research on leaded gasoline use and public health consequences in the coming four to six decades was essentially monopolistic, in which the only work directed at the adverse health consequences was left to a research apparatus at a single university and directed by a single medical researcher with close and long-lived professional ties to Ethyl.

Ethyl's reliance on its dominance of gasoline lead research was increasingly augmented from the 1960s onward due to threats from various regulatory and legislative developments, not to mention changes in public perceptions about lead in general and gasoline lead in particular. For example, Ethyl found it increasingly necessary to be involved in the political and policy end of the leaded gasoline struggle, beginning in 1965 with hearings first for California's pollution problems, and later for those of the nation as well.

Ethyl was a major participant in the struggles of the 1970s and 1980s against national regulation of leaded gasoline via legislative and regulatory mandates. It commissioned, with other affected parties, its own studies to counter those being used by the U.S. EPA to promulgate ambient and fuel additive standards. It joined in and actively attempted to guide workshops and meetings convened to consider regulatory options for leaded gasoline and their scientific underpinnings. In those same years, Ethyl's defensive actions included participating in litigation against regulation by the Federal government.

It is historically ironic that the last phase of Ethyl's defense of leaded gasoline use in America sprang not from Federal regulatory initiatives but from a far-reaching decision by one of its original corporate parents, General Motors (Robert, 1983). In January 1970, General Motors announced that it would meet newly promulgated exhaust standards for hydrocarbon emissions by use of catalytic converters. These converters would be poisoned by lead in vehicular exhaust and the increased production of converter-equipped cars in the nation's new car fleet would itself force removal of lead in gasoline.

11.7 SOURCES AND CONTROL OF HUMAN HEALTH HAZARD INFORMATION FOR LEAD

The question has already been posed (Graebner, 1987; Mushak, 1992; Needleman, 2000) as to why, in terms of societal process, lead's economic and sociocultural value dominated any concerns about public health for virtually all of lead's recorded history. This chapter permits the clear inference that one factor was simply the availability of information and the ability to act on that information. Other forces were certainly present. Two factors of some importance were the nature of the decision-making process and the genesis of information on lead's adverse impacts plus dissemination of this generated information.

Decisions having to do with lead production and use lay almost exclusively in the hands of the producers and users through the first half of the twentieth century and somewhat beyond. This was largely because there was not yet a role for outside, i.e., arm's-length, systematic evaluation of such decisions and their consequences, and no regulatory oversight existed for any stage of the production process. This absence of regulation ranged from the actual siting of a facility to the eventual assessment of any health and safety hazards associated with the products. Risk assessment as now understood and presented in this book was nonexistent with regard to informing any type of primary exposure prevention. Adverse impacts on public health therefore could only be identified *post hoc*.

The equally important matter of what information is produced about lead's potential for health impacts and who produces that data has been noted (Graebner, 1987; Mushak, 1992). For many years and until relatively recently, information on lead's environmental and toxicological behavior was usually generated through studies sponsored and supported by lead producers, industrial lead users, or their trade groups.

The lead industry's trade groups were of several types. The National Lead Institute, perhaps the earliest lead industry group, began to sponsor research on lead as a human health issue in the early 1920s. These efforts were prompted in part by questions about occupational and, somewhat later, childhood lead poisoning in the United States. Aub et al. (1924) were provided funds to study such topics as the behavior of lead in the human body. In 1928, the LIA was formed, with a number of the major lead producers and lead users as members. The LIA appears to have been the *de facto* organizational successor to the National Lead Institute. The LIA recently disbanded through a formal bankruptcy action.

The LIA during its existence appeared to have a number of roles and some of these were referenced in connection with the case study for Pb paint. From the 1920s through the 1950s, LIA mainly worked to promote the use of lead products as widely as possible and to defend such uses from attack by various outside interests. The latter included earlier medical researchers studying childhood lead poisoning and various public agencies attempting to impose some form of regulatory control. LIA's lobbying role appears to have grown over the decades as its promotional activities became less prominent. With the advent of organized environmental awareness, beginning in the late 1960s, the LIA worked with others to pursue a greater litigatory role in response to myriad regulatory efforts by the U.S. EPA. This included participating in litigation in the Federal courts against regulatory rulemaking for lead dispersal by Federal and other public regulatory bodies. A major role for LIA remained in the public relations aspect of lead and health, however. The LIA, for example, sponsored the Kettering Abstracts, an information base dealing with exposure, health effects, and related topics for lead. The information system was quartered at the Division of Bibliographic Research, Kettering Laboratory, University of Cincinnati and distributed to various academic medical and other libraries in the United States.

A later industry associated organization, the International Lead Zinc Research Organization (ILZRO), chartered in 1959, has functioned as a corporate entity since 1963 (Mushak, 1992). While ILZRO has shared some role with the LIA in promoting and defending lead production and use in various public and regulatory settings, its main functions appear to have been more technical than those of the LIA. It also disseminated production and consumption statistics on lead and zinc as part of its metal commodity promotional role. It has provided research funds to grantees for various lead-related topics and has sponsored technical meetings where grantee research results have been presented. Spokespersons for ILZRO have also participated in various regulatory hearings and scientific workshops on various topics having to do with lead.

In addition to the activities of lead trade groups, a number of the larger lead-producing or using companies have traditionally carried on their own activities directed at their own economic well-being.

The controls on information about lead remained intact until the 1960s and 1970s when public agencies, in their public health roles, began to fund and help disseminate arm's-length scientific data gathering on lead exposure and poisoning. This topic is addressed in the next section.

11.8 ESTABLISHMENT OF ACTIVE PUBLIC ROLES IN PB POISONING AWARENESS, IN ADVANCING PB-POISONING RESEARCH AND IN CREATING STABLE MECHANISMS FOR HUMAN HEALTH PROTECTION

Until the 1960s, any interplay of public awareness and concerns about lead's health hazards with attempts at control of lead exposures and associated lead poisonings was largely nonexistent; where present, these were only marginally effective. Individuals or communities might have been aware of the health risks posed by lead and its compounds, but there was little that could be done to turn these concerns into meaningful actions addressing Pb exposure and poisoning controls without viable lead health policy and regulatory structures to both initiate and sustain responses with critical scientific and public health information.

In order to understand how such concerns can be effectively used, it is useful to conceptually consider what has to happen before any effective or sustainable interactions among these topics can come about. What is required is effective operation of a feedback loop, as briefly depicted in Figure 1.1.

Mushak (1991) first presented one form of such a loop for lead that was later incorporated into the NAS/NRC (1993) report on lead exposures in sensitive populations. Typically, the loop for lead and health is initiated with the revelation of scientific research findings or a set of research results which then triggers public awareness or concern. When enough information has accumulated, the process yields a sense that action is needed. This reciprocally triggers the call for, and support of, new research and any additional information that flows from the research. Several elements sustain this loop. First is the accumulation of relevant information that confirms that there is a public health problem such as lead poisoning. However, the development and/or assimilation of information on lead or any other environmental contaminant by impacted individuals or populations typically does not occur from intermittent, unorganized, or indiscriminate attention to the contaminant. It rather requires the existence or implementation of a component of this feedback loop process operating to sustain attention to an environmental contaminant like lead.

Secondly, there is a scientific or public health research component which produces credible scientific information that is new and is added to existing data. Such research activity is sustained by public research support entities, entities which also have need of the information for various purposes. Research results then stimulate more interest in such research while concurrently helping quantify the magnitude of adverse risks to human health. The combined research and risk evaluation data are then provided to public entities which provide additional research support, maintaining the loop's operations.

The previous sections elucidating lead's health history clearly showed that elements for a lead-health loop did not exist until very recently. Once installed, its operation has generated an enormous amount of toxicological and epidemiological data with which the public health and regulatory communities could take actions to minimize health risks and exposures producing those health risks.

The continued operation of such a loop is strongly augmented by public sector actions that include legislative and regulatory initiatives. Several major developments in the late 1960s and early 1970s occurred to provide such a sustaining framework for this loop. These were the development of the environmental awareness movement in America and Western Europe, the enactment of comprehensive environmental protection legislation, the establishment of the U.S. Environmental Protection Agency in 1970 to regulate environmental contaminants such as lead, and the creation of the National Institute of Environmental Health Sciences which supports both intramural and extramural research on the topic.

REFERENCES

Aub, J.C., Minot, A.S., Fairhall, L.T., Reznikoff, P., 1924. Recent investigations of absorption and excretion of lead in the organism. JAMA 83, 588–593.

Aub, J.C., Fairhall, L.T., Minot, A.S., Reznikoff, P., 1926. Lead Poisoning. Medicine Monographs, vol. 7. The Williams and Wilkins Co, Baltimore, MD.

Baghurst, P.A., McMichael, A.J., Wigg, N.R., Vimpani, G.V., Robertson, E.F., Roberts, R.R., et al., 1992. Environmental exposure to lead and children's intelligence at the age of seven years: the Port Pirie study. N. Engl. J. Med. 327, 1279–1284.

Baker Jr., E.L., Landrigan, P.J., Barbour, A.G., Cox, D.H., Folland, D.S., Ligo, R.N., et al., 1979. Occupational lead poisoning in the United States: clinical and biochemical findings related to blood lead levels. Br. J. Ind. Med. 36, 314–322.

Bellinger, D., Leviton, A., Waternaux, C., Needleman, H., Rabinowitz, M., 1987. Longitudinal analysis of prenatal and postnatal lead exposure and early cognitive development. N. Engl. J. Med. 316, 1037–1043.

Bellinger, D.C., Needleman, H.L., Leviton, A., Waternaux, C., Rabinowitz, M.B., Nichols, M.L., 1984. Early sensory-motor development and prenatal exposure to lead. Neurobehav. Toxicol. Teratol. 6, 387–402.

Bellinger, D.C., Stiles, K.M., Needleman, H.L., 1992. Low-level lead exposure, intelligence, and academic achievement: a long-term follow-up study. Pediatrics 90, 855–861.

Beresina, E., Engling, M., 1912. Experimental studies on lead poisoning. Chem. Zeit. 35, 842.

Blackfan, K.D., 1917. Lead poisoning in children with especial reference to lead as a cause of convulsions. Am. J. Med. Sci. 153, 877–887.

Blänsdorf E. 1922. [Ger.]. The lead literature: cumulative publications in the area of occupational hygiene. VII: Berlin.

Byers, R.K., Lord, E.E., 1943. Late effects of lead poisoning on mental development. Am. J. Dis. Child. 66, 471–494.

Camus, J., 1913. Meningitis and lead intoxication. Compt. Rend. Soc. Biol. 72, 861–864, Chem. Abs. 8, 178.

Carcanague, M., 1913. Influence of saturnism on the weight of the rabbit. Compt. Rend. Soc. Biol. 74, 452–455.

Chisolm Jr, J.J., 1962. Aminoaciduria as a manifestation of renal tubular injury in lead intoxication and a comparison with patterns of aminoaciduria seen in other diseases. J. Pediatr. 60, 1–17.

Chisolm Jr., J.J., Barltrop, D., 1979. Recognition and management of children with increased lead absorption. Arch. Dis. Child. 54, 249–262.

Chisolm Jr., J.J., Harrison, H.E., 1956. The exposure of children to lead. Pediatrics 18, 943–958.

Cole, L.J., Bachhuber, L.J., 1914–1915. The effect of lead on the germ cells of the male rabbit and fowl as indicated by their progeny. Proc. Soc. Exp. Biol. Med. 12, 24–29.

Cumings, J.N., 1959. Part 3. Lead. Heavy Metals and the Brain. Blackwell Scientific Publications, Oxford, England, pp. 95–155.

Dana S.L. 1848. Lead Diseases (Editor and Translator) L. Tanquerel des Planches. 1839. Traite des Maladies de Plomb, ou Saturnines. A Treatise from the French. With Notes and Additions of the Use of Lead Pipe and its Substitutes. Daniel Bixby & Co., Lowell, MA.

Davis, J.M., Svendsgaard, D.J., 1987. Low level lead exposure and child development. Nature 329, 297–300.

Derickson, A., 1987. To be his own benefactor: the founding of the Coeur d'Alene Miners' Union Hospital, 1891. In: Rosner, D., Markowitz, G. (Eds.), Dying for Work. Workers' Safety and Health in Twentieth-Century America. Indiana University Press, Bloomington, IN.

Dietrich, K., Berger, O., Succop, P., 1993. Lead exposure and the motor developmental status of urban 6-year-old children in the Cincinnati Prospective Study. Pediatrics 91, 301–307.

Dietrich, K.N., Krafft, K.M., Bornschein, R.L., Hammond, P.B., Berger, O., Succop, P.A., et al., 1987. Low-level fetal lead exposure and effect on neurobehavioral development in early infancy. Pediatrics 80, 721–730.

Dietrich, K.N., Succop, P.A., Berger, O.G., Keith, R., 1992. Lead exposure and the central auditory processing abilities and cognitive development of urban children: the Cincinnati lead study cohort at age 5 years. Neurotoxicol. Teratol. 14, 51–56.

Dilling, W.J., Healey, C.W., 1926. Influence of lead and the metallic ions of copper, zinc, thorium, beryllium, and thallium on the germination of frog's spawn and on the growth of tadpoles. Ann. Appl. Biol. 13, 177–186, Chem. Abs. 20: 3749, 1926.

Edsall D.L. 1907. Chronic lead poisoning. In: Osler, W. (Ed.), Modern Medicine, Ch. V, vol. 1. Lea Brothers and Co., Philadelphia and New York, pp. 84–115.

Eisinger, J., 1982. Lead and wine: Eberhard Gockel and the colica Pictonum. Med. Hist. 26, 279–302.

Ernhart, C.B., Morrow-Tlucak, M., Wolf, A.W., Super, D., Drotar, D., 1989. Low level lead exposure in the prenatal and early preschool periods: intelligence prior to school entry. Neurotoxicol. Teratol. 11, 161–170.

Fairhall, L.T., Sayers, R.R., Miller, J.W., 1940. The relative toxicity of lead and some of its common compounds. Publ. Health Bull. 253, 1–40.

Fee, E., 1990. Public health in practice: an early confrontation with the 'silent epidemic' of childhood lead poisoning. J. Hist. Med. Allied Sci. 45, 570–606.

Feldman, R.G., Hayes, M.K., Younes, R., Aldrich, F.D., 1977. Lead neuropathy in adults and children. Arch. Neurol. 34, 481–488.

Fulton, M., Raab, G., Thomson, G., Laxen, D., Hunter, R., Hepburn, W., 1987. Influence of blood lead on the ability and attainment of children in Edinburgh. Lancet 1, 1221–1226.

Gibson, J.L., 1904. A plea for painted railings and painted walls of rooms as the source of lead poisoning among Queensland children. Aust. Med. Gazette 23, 149–153.

Gibson, J.L., Love, W., Hardine, D., Bancroft, P., Turner, A.J., 1892. In: Huxtable, L.R. (Ed.), Transactions of the Third Intercolonial Medical Congress of Australasia. Charles Potter, Sydney, pp. 76–83.

Gilfillan, S.C., 1965. Lead poisoning and the fall of Rome. J. Occup. Med. 7, 53–60.

Goyer, R.A., 1989. Mechanisms of lead and cadmium nephrotoxicity. Toxicol. Lett. 46, 153–162.

Graebner, W., 1987. Hegemony through science: information engineering and lead toxicology, 1925–1965. In: Rosner, D., Markowitz, G. (Eds.), Dying for Work: Workers' Safety and Health in Twentieth-Century America. Indiana University Press, Indianapolis, IN, pp. 140–159.

Grant, L.D., Davis, M.J., 1989. Effects of low-level lead exposure on paediatric neurobehavioral development: current findings and future directions. In: Smith, M.A., Grant, L.D., Sors, A.I. (Eds.), Lead Exposure and Child Development: An International Assessment. Kluwer Academic, Dordrecht, The Netherlands, pp. 49–115.

Hamilton, A., 1914a. Lead Poisoning in the Smelting and Refining of Lead. U.S. Department of Labor, Washington, DC, Bulletin of the U.S. Bureau of Labor Statistics, No. 141. February 17.

Hamilton, A., 1914b. Lead poisoning in the United States. Am. J. Publ. Health 4, 477–488.

Hamilton, A., 1919. Lead poisoning in American industry. J. Ind. Hyg. 1, 8–21.

Hamilton, A., 1924. The prevalence and distribution of industrial lead poisoning. JAMA 83, 583–588.

Hamilton, A., 1925. Industrial Poisons in the United States. The MacMillan Co., New York.

Hatzakis, A., Kokkevi, A., Maravelias, C., Katsouyanni, K., Salaminios, F., Kalandidi, A., et al., 1989. Psychometric intelligence deficits in lead-exposed children. In: Smith, M.A., Grant, L.D., Sors, A.I. (Eds.), Lead Exposure and Child Development: An International Assessment. Kluwer Academic, Dordrecht, The Netherlands, pp. 211–223.

Hodge, A.T., 1981. Vitruvius, lead pipes and lead poisoning. Am. J. Archaeol. 85, 486–491.

International Labour Office, 1927. White Lead. Data Collected by the International Labour Office in Regard to the Use of White Lead in the Painting Industry. Studies and Reports: Series F, Industrial Hygiene, No. 11. Geneva, Switzerland.

Hunter, D., 1962. The Diseases of Occupation. Livingstone, London.

Kitman J.L. 2000. The secret history of lead. The Nation 270, March 20.

Knerr, D., 1992. Eagle-Picher Industries. Strategies for Survival in the Industrial Marketplace, 1840–1980. Ohio State University Press, Columbus, OH.

Koelsch, F., Lederer, E., Koelsch, R., 1929. Comparative investigations on the toxicity of basic lead sulfate and basic lead carbonate. Arch. Hyg. 101, 234–256, Chem. Abs. 23: 5050, 1929.

Kovarik, B., 1999. Charles F. Kettering and the 1921 discovery of tetraethyl lead in the context of technological alternatives. Proceedings of the Conference, Society of Automotive Engineers, Fuels and Lubricants. Baltimore, MD, 1994, revised in 1999.

Kovarik, W., 2003. The 1920s environmental conflict over leaded gasoline and alternative fuels. Proceedings of the Annual Conference, American Society for Environmental History. Providence, RI, March 26–30.

Lanphear, B.P., Winter, N.L., Apetz, L., Eberly, S., Weitzman, M., 1996. A randomized trial of the effect of dust control on children's blood lead levels. Pediatrics 98, 35–40.

Lanphear, B.P., Howard, C., Eberly, S., Auinger, P., Kolassa, J., Weitzman, M., et al., 1999. Primary prevention of childhood lead exposure: a randomized trial of dust control. Pediatrics 103, 772–777.

Lanphear, B.P., Hornung, R., Khoury, J., Yolton, K., Baghurst, P., Bellinger, David C., et al., 2005. Low-level environmental lead exposure and children's intellectual function: an international pooled analysis. Environ. Health Perspect. 113, 894–899.

Legge, T.M., 1901. Industrial lead poisoning. J. Hyg. 1, 96–108.

Legge, T.M., Goadby, K.W., 1912. Lead Poisoning and Lead Absorption: The Symptoms, Pathology, and Prevention. Edward Arnold, London, U.K.

Lehmann, K.B., 1929. Comparative investigations on the toxicity of basic lead sulfate and basic lead carbonate. Arch. Hyg. 101, 197–204, Chem. Abs. 23, 5050, 1929.

Major, R.H., 1945. Classic Descriptions of Disease. III. Lead Poisoning. third ed. C.C. Thomas, Springfield, IL, pp. 312–325.

Markowitz, G.E., Rosner, D., 2002. Deceit and Denial: The Deadly Politics of Industrial Pollution. University of California Press, Los Angeles, The Milbank Foundation, New York.

Maurel, Carcanague, 1913. Relation between the distribution of lead in various organs and tissues after hypodermic injection and the order of sensitiveness of various anatomical elements. Comp. Rend. Soc. Biol. 73, 329–332, Chem. Abs. 8, 179, 1914.

McCord, C.P., 1953. Lead and lead poisoning in early America: lead mines and lead poisoning. Ind. Med. Surg. 22, 534–539.

McCord, C.P., 1954a. Lead and lead poisoning in early America: the lead pipe period. Ind. Med. Surg. 23, 27–31.

McCord, C.P., 1954b. Lead and lead poisoning in early America: clinical lead poisoning in the Colonies. Ind. Med. Surg. 23, 120–125.

McKhann, C.F., 1926. Lead poisoning in children, with notes on therapy. Am. J. Dis. Child. 32, 386–392.

McMichael, A., Baghurst, P., Wigg, N., Vimpani, G., Robertson, E., Roberts, R., 1988. Port Pirie cohort study: environmental exposure to lead and children's abilities at the age of four years. N. Engl. J. Med. 319, 468–475.

Mellon, R.R., 1913. The relation of fatigue to paralysis location in plumbism. Arch. Intern. Med. 12, 399–419.

Minot, A.S., 1938. The physiological effects of small amounts of lead: an evaluation of the lead hazard of the average individual. Physiol. Rev. 18, 554–574.

Mushak P. 1991. U.S. Agency for Toxic Substances and Disease Registry's Report to Congress on childhood lead poisoning in America: review and update. In: Proceedings of the First National Conference on Laboratory Issues in Childhood Lead Poisoning Prevention, Columbia, MD, pp. 79–104.

Mushak, P., 1992. Perspective: defining lead as the premiere environmental health issue for children in America: criteria and their quantitative application. Environ. Res. 59, 281–309.

Mushak P., Mushak E.W., 2000. A comparative analysis of the evolution of lead and mercury as public health hazards. In: 11th International Conference on Heavy Metals in the Environment (Nriagu JO, Ed.). CD-ROM: Manuscript No. 1445. University of Michigan, Ann Arbor, MI, August 6–10.

National Academy of Sciences: National Research Council, 1972. Lead. Airborne Lead in Perspective. National Academy Press, Washington, DC.

National Academy of Sciences: National Research Council, 1980. Lead in the Human Environment. National Academy Press, Washington, DC.

National Academy of Sciences: National Research Council, 1993. Measuring Lead Exposure in Infants, Children, and Other Sensitive Populations. National Academy Press, Washington, DC.

Needleman, H.L., 2000. The removal of lead from gasoline: historical and personal reflections. Environ. Res. 84, 20–35.

Needleman, L., Needleman, D., 1985. Lead poisoning and the decline of the Roman aristocracy. Classical Views 4, 63–94.

Needleman, H.L., Gunnoe, C., Leviton, A., Reed, R., Peresie, H., Maher, C., et al., 1979. Deficits in psychologic and classroom performance of children with elevated dentine lead levels. N. Engl. J. Med. 300, 689–695.

Nriagu, J.O., 1983a. Lead and Lead Poisoning in Antiquity. John Wiley & Sons, New York.

Nriagu, J.O., 1983b. Saturnine gout among Roman aristocrats: did lead poisoning contribute to the fall of the Empire? N. Engl. J. Med. 308, 660–663.

Nriagu, J.O., 1985. Historical perspective on the contamination of food and beverages with lead. In: Mahaffey, K.R. (Ed.), Dietary and Environmental Lead: Human Health Effects. Elsevier, New York, pp. 1–41.

Nriagu, J.O., 1990. The rise and fall of leaded gasoline. Sci. Total Environ. 92, 13–28.

Oliver, T., 1911. Lead poisoning and the race. BMJ 1, 1096–1098.

Oliver, T., 1914. Lead Poisoning: From the Industrial, Medical and Social Points of View: Lectures Delivered at the Royal Institute of Public Health. Paul B. Hoeber, New York.

Ophüls, W., 1915. Chronic lead poisoning in guinea-pigs: with special reference to nephritis, cirrhosis and polyserositis. Am. J. Med. Sci. 150, 518–541.

Paul, C., 1860. Study of the effect of slow lead intoxication on the product of conception [in French]. Arch. Gen. Med. 15, 513–533.

Rabin, R., 1989. Warning unheeded: a history of child lead poisoning. Am. J. Publ. Health 79, 1668–1674.

Retief, F.P., Cilliers, L., 2006. Lead poisoning in ancient Rome. Acta Theol. 26 (Suppl. 7), 147–164.

Rickard, T.A., 1932. A History of American Mining. McGraw-Hill, New York.

Ris, D., Dietrich, K.N., Succop, P.A., Berger, O.G., Bornschein, R.L., 2004. Early exposure to lead and neuropsychological outcomes in adolescence. J. Int. Neuropsychol. Soc. 10, 261–270.

Robert, J.C., 1983. Ethyl: A History of the Corporation and the People Who Made it. University Press of Virginia, Charlottesville, VA.

Rogers, T.S., Peck, J.R.S., Jupe, M.H., 1934. Lead poisoning in children with a case record. Lancet 2, 129–133.

Rosner, D., Markowitz, G., 1985. A 'gift of God'? The public health controversy over leaded gasoline during the 1920s. Am. J. Publ. Health 75, 344–352.

Scarborough, J., 1984. The myth of lead poisoning among the Romans: an essay review. J. Hist. Med. 39, 469–475.

Sigerist, H.E., 1934. The medical and scientific periodicals of the seventeenth and eighteenth centuries. Bull. Inst. Hist. Med. 2, 285–343.

Sollmann, T., 1922. Studies of chronic intoxications on albino rats. J. Pharmacol. Exp. Ther. 19, 375–391.

Stevenson L.G. 1949. A History of Lead Poisoning. Ph.D. Thesis. The Johns Hopkins University, Baltimore, MD.

Stockhausen S. 1656. [Lat.] Treatise on the noxious fumes of litharge, diseases caused by them and miner's asthma. Goslar, Germany. As cited in Eisinger 1982.

Straub, W., 1913. Poison and disease, according to observations on experimental lead poisoning. Münch. Med. Wochenschr. 61, 5–7, Chem. Abs. 8, 965–966.

Strong, R.A., 1920. Meningitis, caused by lead poisoning in a child of nineteen months. Arch. Pediatr. 37, 532–537.

Tanquerel des Planches L. 1839. Lead diseases (Translated, edited SL Dana, 1848). Daniel Bixby and Co., Lowell, MA.

Thomas, H., Blackfan, K., 1914. Recurrent meningitis due to lead, in a child of five years. Am. J. Dis. Child. 8, 377–380.

Tong, S., Baghurst, P., McMichael, A., Sawyer, M., Mudge, J., 1996. Lifetime exposure to environmental lead and children's intelligence at 11–13 years: the Port Pirie cohort study. BMJ 312, 1569–1575.

Tronchin T. 1764. A Treatise on the colica Pictonum; or the Dry-Belly-Ache (Schomberg R, Editor and Translator). W. Johnston, London, U.K.

U.S. Centers for Disease Control, 1978. Preventing Lead Poisoning in Young Children. A Statement by the Centers for Disease Control. U.S. Department of Health, Education, and Welfare, Atlanta, GA.

Waldron, H.A., 1966. The anaemia of lead poisoning: a review. Br. J. Ind. Med. 23, 83–100.

Waldron, H.A., 1973. Lead poisoning in the ancient world. Med. Hist. 17, 391–399.

Wassermann, F., 1916. Investigation of chronic lead poisoning in cats by the Roentgen rays. Arch. Exp. Pathol. Pharmacol. 79, 383–411, Chem. Abs. 10: 1680, 1916.

Wedeen, R.P., 1984. Poison in the Pot: The Legacy of Lead. Southern Illinois University Press, Carbondale, IL.

Wedeen, R.P., D'Haese, P., van de Vyver, F.L., Verpooten, G.A., DeBrae, M.E., 1986. Lead nephropathy. Am. J. Kidney Dis. 8, 380–383.

Weller, C.V., 1916. Histological study of the testes of guinea pigs showing lead blastophthoria. Proc. Soc. Exp. Biol. Med. 14, 14.

Winneke, G., Krämer, U., Brockhaus, A., Ewers, U., Kujanek, G., Lechner, H., et al., 1983. Neuropsychological studies in children with elevated tooth-lead concentrations: II. Extended studies. Int. Arch. Occup. Environ. Health 51, 231–252.

Winneke, G., Collet, W., Lilienthal, H., 1988. The effects of lead in laboratory animals and environmentally-exposed children. Toxicology 49, 291–298.

Woolley, D.E., 1984. A perspective of lead poisoning in antiquity and the present. Neurotoxicology 5, 353–362.

Chapter 12

Neurotoxicity of Lead in Human Populations

12.1 INTRODUCTION

This chapter presents the scientific and medical evidence documenting environmental lead as a particularly potent neurotoxin for injuring the developing brain in very young children and the fetal brain during elevated Pb exposures of the mother. It is now universally accepted that Pb produces neurotoxic harm across a wide range of Pb exposures, the nature of the harm ranging from fatal outcomes at the highest exposures to demonstrable reduction in normal neurocognitive and neurobehavioral function in very young children at very low exposures. However, no subset of the human population is free of neurotoxic harm if sufficient Pb exposure occurs.

Full understanding of this broad exposure–toxic outcome relationship did not occur overnight. Evolving comprehension is readily apparent in the historical record of lead and its various human health hazard relationships across decades and centuries (see previous chapter). This chapter, in common with the other chapters on toxicological harm from Pb exposures in humans, is not intended to be an encyclopedic recapitulation of the neurotoxicity record of Pb exposures in children, adults, and experimental test animals and wildlife. Rather, these chapters are distillations of the most reliable health science literature, i.e., literature that is most useful for addressing the principal end uses of the monograph: the science of lead as it elucidates and propels both human health risk assessment and regulatory initiatives.

This and other lead toxicology chapters use and merge two conceptual approaches: that of effect criticality and that embracing weight-of-evidence. The "criticality" concept is useful in elucidating lead's dose–toxic response relationships for an array of organ- and tissue-specific toxicological data on effects which are simultaneously highly significant and occur at or among the lowest doses. The criticality concept informs evidence of thresholds to one or more effects. In its broadest definition, weight-of-evidence quantifies multiple lines of evidence for lead toxicity.

The "criticality" concept, first articulated by Nordberg (1976), makes for ease of presentation within the rationale of what is most immediately

Trace Metals and other Contaminants in the Environment, Volume 10
ISSN: 1875-1121 DOI: 10.1016/B978-0-444-51554-4.00012-2

important as to sensitivity and importance of effects. Nordberg, summarizing the concept at an international meeting, defined such terms as "critical concentration," "critical organ," critical effect," etc. However, it has many quantitative drawbacks for full characterization of multisite, very potent toxicants like lead.

Current weight-of-evidence approaches for Pb show that we cannot assume that a single somatic target exists for Pb while all other target sites function with little or less sensitive responses at a given level of human exposure. Lead produces low-dose toxicity in multiple body sites and the respective threshold doses, i.e., "critical" doses, for onset of toxicity do not differ greatly.

For example, developmental neurotoxicological, cardiovascular, immunological and perhaps even nephrotoxic effects in high-risk populations appear to respond to Pb exposure within a relatively narrow dose–response range. Put differently, individuals exposed to Pb at low dose sustain neurotoxic insult in the face of other low-dose insults with the potential for at least additive toxicity and even synergistic interactions for toxicological dose–responses. The latter toxicological multiplier effect is still largely opaque to risk assessment methods, confined as they are to critical effects in critical organs/tissues at critical threshold doses.

There are a number of aspects of lead neurotoxicology presented in this chapter. The first is the clear distinction between Pb toxicity observed in young children qualitatively and quantitatively and what comprises the typical case for older children and adults. That is, the nature of Pb-associated neurotoxic injury is different and the onset of neurotoxic injury in children is known to occur at lower exposures than in adults and older children. This chapter is consequently divided into sections dealing separately with childhood versus adult Pb toxicity in humans.

The second is the marked shift to less deadly outcomes in the definitions of Pb poisoning over the decades. Here, the criteria are confined to developmental neurotoxicity, as noted in the previous chapter. In the late nineteenth and early twentieth centuries, the sole diagnostic feature for a child's Pb poisoning, typically clinically unmanaged encephalopathy, was death. If children survived encephalopathic episodes, they were assumed not to be Pb-poisoned. This draconian diagnostic marker with accompanying postmortem support gave way to a cluster of overt, fulminant signs and symptoms often seen in pediatric victims on the path to a likely fatal outcome, e.g., convulsions and coma. This still drastic set of frank characteristics eventually expanded to include grave and permanent sequelae, such as clinically defined mental retardation. Still later, overt toxicity was extended to signs and symptoms other than acute and/or chronic encephalopathy, particularly other neurotoxic effects and nephrotoxicity and hematotoxicity.

Most of the attention to neurotoxic effects of lead in recent decades has been to more subtle but still serious outcomes requiring more sophisticated tools for identifying and detecting more subtle effects in subsets of child and adult populations as well as in individuals. Mushak (1992) posed the argument that the relative severity of toxic harm in humans from lead should not be determined solely by the crudeness of the diagnostic and management tools at hand. While it is true that cruder diagnostic tools identify more severe effects, using logical extension of this argument to increasing crudeness of methods would define lethal outcomes as the most reliable measure.

Furthermore, the focus has increasingly been on an epidemiological, public health model of childhood Pb poisoning occurring in populations while not dismissing the classical medical, physician–patient model for lead poisoning intervention. Epidemiological approaches to, and evaluations of, childhood Pb poisonings complement and provide the underpinnings and rationales for clinical diagnostics. The practitioner, in a diagnostic role, then serves as the gatekeeper for integrating epidemiological evidence with other elements of evidence-based medicine in the course of poisoning management.

This chapter's characterization of lead as a neurotoxic hazard does not include detailed dose–response relationships with various levels of biomarkers such as PbB linked to various neurotoxic outcomes. The topics of dose/exposure metrics and defining full-spectrum dose–response relationships are presented in the next part, the section dealing with the elements of human health risk assessment for environmental lead. Here, for ease of discussion, only a broad yardstick is provided for toxic lead exposures. Specifically, general PbB ranges associated with the various categories of lead neurotoxicity, especially in children, are noted.

12.2 LEAD NEUROTOXICITY IN CHILDREN

Lead neurotoxicity in young children is described in this chapter in terms of four levels of Pb exposure, beginning with the most severe effects induced by very high Pb intakes, typically in disadvantaged, high-risk children from ingestion of lead paint containing high Pb loadings. The presentation follows both the historical record and the forms of neurotoxicity associated with the continued decline of blood lead levels deemed to be neurotoxicity thresholds for diverse endpoints in young children. These, along with a general PbB range, are: (1) clinical Pb poisoning induced with highly elevated blood lead levels, associated with PbB levels ranging upward from 80 to 100 μg/dl; (2) neurotoxic effects still measurable "clinically" but in children asymptomatic for classical encephalopathic toxic insults, associated with PbB values somewhat lower than for fulminant clinical effects, i.e., 40–80 μg/dl or somewhat lower; (3) more subtle neurotoxicity in largely asymptomatic children and effects elicited from cross-sectional epidemiological studies, associated with PbB levels ≤ 30–40 μg/dl; and (4) prospective/longitudinal epidemiological studies of subtle but still significant neurological harm in asymptomatic children, associated with serially measured PbB levels ≤ 10 and even ≤ 5 μg/dl.

Within the exposure categories, (3) and (4) are subcategories characterized by the types of low-level neurodevelopmental toxic effects, e.g., neurocognitive impairments, mobility, measures of academic achievement, impaired social conduct, physioanatomical measures of development and activity.

Included for discussion in both the child and adult neurotoxicity sections are some critical topics which inform continuing or new directions for the topic. The topics include effect persistence or irreversibility, late life and tardive effects from earlier developmental exposures, and the putative lack of a measured threshold to neurotoxicity versus PbB.

12.2.1 Clinical Pb Poisoning in Children

Although fulminant brain injury from very high Pb exposures has become less common in recent years, such episodes occasionally occur and would occur more frequently if Pb exposures rose significantly. Historically, such acute poisonings were the first described in the medical literature.

Childhood clinical Pb poisoning in its most fulminant forms has typically presented in the medical literature as a cluster of overt injuries in principally three systems: CNS, kidneys, and the blood-forming system. Other signs and symptoms arose from effects on the GI tract and the cardiovascular system. Some of the principal clinical characteristics of overt neurotoxicity are presented in Table 12.1. The primary sources date to the early to mid-twentieth century, the time when recognized clinical, severe lead poisonings were at their peak in the United States.

The most vulnerable target in severe childhood lead poisonings is the still-developing CNS. As shown in Table 12.1, CNS injury was manifested in the early twentieth century medical literature in two forms: acute and chronic encephalopathy (see, for example, Chisolm, 1968; Chisolm and Harrison, 1956). Acute encephalopathy which had progressed to intractable seizures and coma also had a high risk of cardiopulmonary arrest and death. Prior to chelation as therapeutic intervention, mortalities of up to 65% were reported in acute or mixed acute/chronic encephalopathy. Chelation-based therapeutic interventions, however, still had mortality rates ranging from 5% to 39% (Greengard et al., 1965; Niklowitz and Mandybur, 1975). Preceding this progression to catastrophic injury manifesting as acute encephalopathy, victims suffered vomiting, apathy, stupor, and hyperirritability.

Histopathological and morphological findings with fatal encephalopathy presented in the form of cerebral edema, endothelial hypertrophy and hyperplasia, and perivascular glial proliferation. Various ischemic disorders were also reported: cell necrosis and neuronal loss in isocortex and basal ganglia

TABLE 12.1 Clinical Neurological Signs and Symptoms in Childhood Lead Poisoning

Clinical Toxicological/ Diagnostic Endpoint	Signs, Symptoms, Histopathology	References
Acute encephalopathy	Intractable seizures, coma, cardiopulmonary arrest, death; precedents: vomiting, apathy, stupor, hyperirritability, vascular (cerebral edema) and particular neuronal injury in cerebellum	American Academy of Pediatrics (1969), Chisolm (1968)
Acute encephalopathic sequelae	Severe permanent neurological injury: cortical atrophy, convulsive disorder, mental retardation, blindness. Frequency: at least 25% of survivors	Chisolm and Harrison (1956), Smith (1964), Byers (1959)
Chronic encephalopathy (children >3 years old)	Loss of motor skills, speech; severe hyperkinetic, convulsive and behavioral disorders; a mixed picture with recurrent episodes of acute poisonings	NAS/NRC (1972), Byers and Lord (1943), Perlstein and Attala (1966)
Mixed acute/chronic encephalopathy	Mortality rates in young children: ~65% before chelation therapy; a range in mortality with chelation therapeutic intervention of 5–39%	Greengard et al. (1965), Wiklowitz and Mandybur (1975)
Fatal encephalopathy	Morphological findings: cerebral edema, endothelial hypertrophy and hyperplasia, perivascular glial proliferation; injury secondary to cerebral vasculopathy; ischemic disorders: cell necrosis, neuronal loss in isocortex and basal ganglia	Pentschew (1965), Popoff et al. (1963)

(Pentschew, 1965; Popoff et al., 1963). Overall, a number of these findings appear as secondary to a cerebral vasculopathy (Pentschew, 1965).

Acute encephalopathic sequelae took the form of severe and permanent neurological injury, cortical atrophy, convulsive disorder, mental retardation, and blindness in about one-quarter of the survivors (Byers, 1959; Chisolm and Harrison, 1956; Smith, 1964).

Chronic encephalopathy was characterized as affecting older children (>3 years of age) with loss of motor skills, speech, and the onset of severe hyperkinetic, convulsive, and behavioral disorders. An insidious interplay between acute and chronic encephalopathic forms of childhood lead poisonings often occurred. That is, recurring episodes of acute encephalopathy in seriously lead-contaminated surroundings would be triggered by an ongoing chronic encephalopathy reverting to the acute form with abrupt, high toxic lead exposures.

Some estimates attempted to link PbB values with fulminant neurological injuries such as acute and chronic encephalopathy, but those attempts were highly problematic on several counts. First, the earliest clinical records for childhood Pb poisoning were hobbled by lack of reliable methods to quantify the nature and extent of Pb exposure. While procedures such as colorimetry had been developed in the 1930s, their use was confined to specific facilities as largely a research or medical center tool. Second, time lags attending the exposure event and the symptomatology and/or intervention confounded Pb toxicokinetic relationships and the form of the dose–toxic response relationships.

Nonencephalopathic, clinically overt neurotoxic effects of lead in children were described in the pediatric plumbism report of Byers and Lord (1943). This was followed by a set of key reports during the 1950s to the early 1970s (Perlstein and Attala, 1966). These investigations can be collectively viewed as the next step in the evolution of lead neurotoxicity as a public health issue, specifically with regard to declining dose–neurotoxic response relationships accepted as adverse. Some of the significant nonencephalopathic overt neurotoxic effects published in that period are presented in Table 12.2. While these studies were accompanied by measures of exposure in most cases, those measures are presented in a more quantitative context in a later chapter on dose–response relationships.

Byers and Lord (1943) examined children surviving the acute, lethal phase of lead poisoning subsequent to ingestion of lead paint. These children ($N = 20$) presented with sensorimotor deficits, attention deficits, and marked behavioral disturbances. One or more sequelae in other studies such as sensory perception, motor coordination, reduced learning, and attention span collectively were observed in follow-up studies of children originally presenting with nonencephalopathic clinical lead poisoning (Chisolm, 1962, 1968; Cohen and Ahrens, 1959; Kline, 1960). Children with a history of clinical poisoning onset as ataxia, seizures, and GI symptoms displayed sequelae such as mental retardation, seizures, cerebral palsy, or optic atrophy (NAS/NRC, 1972, Table 4.6; Perlstein and Attala, 1966).

Children tested for clinical neurological injury from Pb in the form of overt peripheral neuropathy (Erenberg et al., 1974; Seto and Freeman, 1964) showed tremor, wrist and foot drop as determined by electromyography and nerve conduction velocity studies.

The focus in studies and medical management of neurological effects of Pb in children began slowly shifting from symptomatic to asymptomatic

TABLE 12.2 Nonencephalopathic Overt Neurotoxic Effects of Pb in Children

Study Group (*N*)	Pb Poisoning History	Neurotoxic Response(s)	References
Follow-up children (20)	Pb poisoning without encephalopathy, Pb paint ingestion	Sensorimotor deficits, Henton deficits, behavioral disturbances	Byers and Lord (1943)
Follow-up children (157)	Children with mode of onset = seizures, ataxia, GI symptoms	One or more sequelae as retardation, seizures, cerebral palsy, or optic atrophy	Adapted from Perlstein and Attala (1966), NAS/NRC (1972), Table 4.6
Follow-up children (variable)	Children with initial nonencephalopathic clinical Pb poisoning	One or more sequelae found as sensory perception and motor coordination impairments, reduced learning and attention span	Chisolm (1962, 1968), Kline (1960), Cohen and Ahrens (1959)
Children tested for frank Pb peripheral neuropathy	Pb-poisoned children presenting with overt peripheral neurological signs: tremor, wrist and foot drop	Peripheral neuropathy was determined via electromyography and nerve conduction velocity measures; sickle cell disease may be risk factor	Seto and Freeman (1964), Erenberg et al. (1974)

signs and symptoms in the early to mid-1970s. This evolving change in attention was the first major development in fleshing out the full dose–response spectrum of adverse lead effects at lower exposures since the Byers and Lord (1943) study. An interim period in the evolution of lead epidemiology occurred, which could still be characterized as being rooted in histories of earlier clinical lead poisonings and continued management in a clinical setting. In terms of exposures, these studies of general behavioral, psychomotor, and neurocognitive endpoints were done in children whose PbB histories revealed quite elevated levels. In some children, blood Pb values ranged from 30 to 80–100 μg/dl.

De la Burde and Choate (1972) reported that among 70 inner-city Richmond, VA, children with a Pb exposure history, subnormal fine motor function, disturbed behavior, and some subnormal IQ results were measured. In follow-up studies of this group, these same workers (De la Burde and Choate, 1975) reported abnormal psychometric measures and impaired auditory, vocal, tactile recognition, and behavioral measures.

Kotok and coworkers (Kotok, 1972; Kotok et al., 1977) evaluated inner-city children from two Northeastern cities. New Haven, CT, children, mean age 2.8 years with a high mean PbB (80 μg/dl), were reported to have significantly reduced fine motor and language performance on the Denver Developmental Test (Kotok, 1972). A follow-up study (Kotok et al., 1977) at 3.6 years, with mean PbB=80 μg/dl, showed reduced performance, but the reduction was not statistically significantly different. Perino and Ernhart (1974) examined the effects of relatively high Pb exposures (40–70 μg/dl) in New York City children, showing significantly lower performance in various McCarthy Scales.

12.2.2 Asymptomatic Pb Poisoning in Children

The "clinical" period of childhood lead poisoning as the principal clinical diagnostic and management focus gave way to neurotoxic injury whose manifestations could be collectively labeled asymptomatic or "subclinical" in nature (Table 12.3).

TABLE 12.3 Neurotoxic Effects of Pb in Exposed Asymptomatic Children at Moderate to High PbB Levels

Child Study Group (*N*)	Pb Exposure History	Neurotoxic Endpoints	References
Inner-city children from Richmond, VA (70)	PbB range 30–100 μg/dl	Significant subnormal fine motor function, disturbed behavior, subnormal Stanford–Binet IQ results	De la Burde and Choate (1972)
Same group	3–4 year follow-up, high tooth Pb = 202 ppm	Significantly abnormal psychometrics such as WISC, full-scale Bender-Gestalt. Also, significantly abnormal auditory, vocal, tactile recognition, and behavioral measures	De la Burde and Choate (1975)
Inner-city children from New Haven, CT (24)	Children aged 2.8 years, mean PbB 81 μg/dl	Significantly reduced fine motor performance and language performance on Denver Developmental Test	Kotok (1972)
Inner-city children from Rochester, NY (31)	Children aged 3.6 years; PbB = 80 μg/dl	Reduced performance but not statistically significant from controls	Kotak et al. 1977
Inner-city children from New York City (50)	Exposure PbB 40–70 μg/dl	Significantly lower performance in various McCarthy Scales	Perino and Ernhart (1974)

Cross-Sectional Epidemiological Studies of Asymptomatic Lead Neurotoxicity in Children: Neurocognitive Ability

The next battery of epidemiological assessments of Pb effects in terms of child health chronology consisted of various cross-sectional studies of asymptomatic children in the United States and internationally. These studies succeeded the clinic-based evaluations noted above and in most cases preceded the design and execution of a number of international prospective studies. Cross-sectional studies as presented here illustratively attempted: (1) much closer attention to Pb exposures representing typical child populations exposed to general or localized Pb emissions; (2) using more rigorous statistical measures to better control for confounding factors; (3) using relatively stable, specific indices of Pb exposure; and (4) using standardized effect measures. Illustrative U.S. studies are tabulated in Table 12.4, while results for those evaluations outside the United States are presented in Table 12.5.

Needleman et al. (1979) first reported cross-sectional evaluations of Pb-exposed children in the general population. This study was groundbreaking on a number of counts. It took account of and attempted to avoid epidemiological and statistical shortcomings in earlier studies. It also employed a stable, cumulative measure of Pb exposure and standard pediatric psychometric measures of general neurocognitive ability and classroom behavior (Table 12.4). Boston, MA, children with a Pb exposure history, indexed as the cumulative Pb biomarker dentine Pb, revealed significant lowering of test score results for WISC full scale and verbal IQ scores, Seashore Test, Token Test, sentence repetition, delayed reaction time and ratings from teachers.

Several studies were carried out with North Carolina rural and urban children. The first study was for children ($N = 104$) aged 10 months to 6.5 years and having a PbB range of 6–59 μg/dl. A fraction sustained secondary Pb exposure via "take-home" occupational workplace dusts. These children showed a significant inverse association of PbB with measures of cognition (Schroeder et al., 1985). A partial replication analysis using a more homogeneous study group ($N = 75$) showed a highly significant inverse association of concurrent PbB to IQ score (Stanford–Binet) (Schroeder and Hawk, 1987). The strength of the association was greater than for peak or mean PbB.

These children were also evaluated through electrophysiological means. Otto et al. (1981) reported a significant linear association between PbB and slow-wave voltage. In a follow-up of this group after 5 years (Otto et al., 1985), the relationship of slow-wave responses to PbB along with a significant association of pattern-reversal evoked potentials (PREPs) and brainstem auditory evoked potentials (BAEPs) with PbB persisted.

TABLE 12.4 Illustrative Cross-Sectional Epidemiological Studies of Low-Level Pb Exposures in Asymptomatic U.S. Children

Study Group (*N*)	Pb Exposure History	Pb Neurotoxic Effects	References
Boston, MA, children (58 exposed, 100 controls)	7-year-old children classified by tooth Pb level: <10 ppm, >20 ppm	Significant lowering of WISC (full scale, verbal) IQ, Seashore Test, Token Test, sentence repetition, delayed reaction time and ratings from teachers	Needleman et al. (1979)
Rural/urban North Carolina (NC) (104)	Age 10 months–6.5 years, PbB range: 6–59; Pb exposure in deteriorated housing	Significant linear relationship, Bailey MDI or Stanford–Binet IQ relationship versus PbB	Schroeder et al. (1985)
Rural/ urban children ($N = 75$) in two NC counties	PbB mean—22; Age 3–7 years	Highly significant association, current PbB versus IQ (Stanford–Binet)	Schroeder and Hawk (1987)
El Paso, TX, children (46 exposed, 78 control)	Pb exposure from a nearby Pb smelter; exposed PbB range = 40–68 μg/dl	IQ significantly lower on WISC + WPPSI subscale IQ tests	Landrigan et al. (1975)
Rural/urban NC children (variable *N*)	Children electrophysiologically tested at several ages; Pb exposures from mainly deteriorated environments. PbB range 6–59 μg/dl	Significant linear association between PbB and slow-wave voltage	Otto et al. (1981)
Follow-up to above, 2 years later (28)	Same exposure profile; PbB mean = 21.1	Persistence of slow-wave response with PbB	Otto et al. (1982)
Follow-up, 5 years later	Same exposure profile	Persistence of slow-wave response with PbB; association of PREP and BAEP with PbB	Otto et al. (1985)
Idaho children in smelter communities	Children exposed from Pb smelter	Significant negative correlation of peroneal NCVs with PbB	Landrigan et al. (1976)

(*Continued*)

TABLE 12.4 Illustrative Cross-Sectional Epidemiological Studies of Low-Level Pb Exposures in Asymptomatic U.S. Children—(cont.)

Study Group (*N*)	Pb Exposure History	Pb Neurotoxic Effects	References
Boston urban and Maine rural children aged 6–10 years enrolled in the New England Children's Amalgam Trial (534)	Urban or rural community Pb exposures: PbB 5–10 μg/dl versus PbB 1–2 μg/dl	WISC-III Full Scale IQ and Wechsler Individual Achievement scores were affected; significant negative association of IQ scores with higher versus lower PbBs; achievement test scores strongly negatively associated with PbBs 5–10 μg/dl	Surkan et al. (2007)
Detroit, MI, African-American children from the inner city, community exposures (237)	PbB mean = 5.4 μg/dl at 7.5 years old	WISC-III, 7.5 years old: Adjusted full-scale, verbal and performance IQ scores significantly associated with PbB levels, performance IQ still associated <7.5 μg/dl	Chiodo et al. (2004)

El Paso, TX, children (exposed=46, controls=78) exposed to Pb emissions from a nearby smelter and with PbB over the range 40–68 μg/dl showed significantly lower test scores for WISC and WPPSI Subscale Tests (Landrigan et al., 1975). Landrigan et al. (1976) investigated a second group of Pb smelter-impacted children in Idaho, showing a significant negative correlation of peroneal nerve conduction velocities to PbB.

Several more recent cross-sectional studies document persistence of significant negative associations of neurocognitive decrements with PbB < 10 μg/dl. Detroit, MI, African-American children ($N = 237$) having a mean PbB=5.4 μg/dl at 7.5 years of age showed statistically significant persistence of the association of WISC-III decrements in adjusted full-scale, verbal, and performance IQ scores to PbB levels <7.5 μg/dl (Chiodo et al., 2004). Boston urban and Maine rural children enrolled in the New England Children's Amalgam Trial ($N = 534$) and having general community Pb exposures were also tested for associations of IQ test scores with PbB at two ranges: 5–10 versus 1–2 μg/dl (Surkan et al., 2007). Significant negative associations of IQ scores with the higher versus the lower PbB ranges were

TABLE 12.5 Illustrative International Cross-Sectional Epidemiological Studies of Low-Level Pb Exposures in Asymptomatic Children

Study Group (*N*)	Pb Poisoning History	Neurotoxic Responses	References
England: London children, 4 PbB groups (Total *N* = 166)	Urban community exposures: low PbB, 8.8 μg/dl; 3 higher groups, mean = 11.6–19.6	Significant changes versus PbB seen for Conners Teacher Ratings and Rutter Teacher Ratings	Yule et al. (1984)
England: Urban children with recent PbB testing (78)	Urban community exposures, mean PbB 9.5 μg/dl (4.9–17.7 μg/dl). Tested in a six-part battery of automated tests	Children with PbB ≥ 11 μg/dl performed more poorly in the Paired Associate Learning and the Speed of Information tests	Yule et al. (1989)
England: Institute of Child Health/ Southampton Study: London 6 years old (402)	Urban community exposures; Pb indexed by three Pb ranges in shed teeth (crowns)	Significant negative Pb-teeth versus IQ relationship in boys but not girls	Pocock et al. (1989)
Scotland: Edinburgh children, age 6–9 years (501)	Urban community exposures; geometric mean PbB = 10.4 μg/dl	Significant negative relationships between log PbB and British Ability Scales: combined score number skills and reading	Raab et al. (1989)
New Zealand: Dunedin children, age 11 years (579)	Urban community exposures: mean PbB 11.1 μg/dl	Significant association of PbB with behavioral, inattention, and hyperactivity rating by parents and/or teachers	Silva et al. (1988)
Denmark: Aarhus children, first grade (156)	Urban community Pb exposures: circumpulpal dentine Pb mean 10.7 μg/g; high Pb ≥ 18.7 μg/g, low Pb ≤ 5 μg/g; PbB mean = 5.1 μg/dl	Significant association of Pb-dentine with Bender, Full-Scale WISC and verbal WISC	Hansen et al. (1989)
Germany: Nordenham children retested at age 9 (76)	Children exposed from a nearby lead–zinc smelter; PbB geometric mean = 7.8 μg/dl (from 1985). Pb-tooth geometric mean = 4.7 μg/g	Significant association of PbB and Pb-teeth with Vienna Reaction Test and Visual Evoked Potential responses	Winneke et al. (1989)

(*Continued*)

TABLE 12.5 Illustrative International Cross-Sectional Epidemiological Studies of Low-Level Pb Exposures in Asymptomatic Children—(cont.)

Study Group (*N*)	Pb Poisoning History	Neurotoxic Responses	References
Germany: Stolburg children tested at 9.4 years (89)	Children exposed from a nearby lead smelter. PbB mean = 14.3 μg/dl; Pb-teeth 6.2 ppm	Significant association of PbB and Pb-teeth with Vienna Reaction Test and visual evoked potential responses	Winneke et al. (1983)
Germany: Duisburg children tested at 8 years (control 26, exposed 26)	Children exposed from a nearby lead smelter. Pb-teeth, control: 2.4 ppm; Pb-teeth, exposed: 9.2 ppm	Significant association of Bender-Gestalt Test deficits with exposure	Winneke et al. (1982)
Greece: Lavrion children in elementary school (509)	Children exposed from a nearby lead–zinc smelter; PbB mean = 23.7 μg/dl	Highly significant association between full-scale WISC-R verbal and performance IQ with PbB	Hatzakis et al. (1989)
Italy: Sassuolo children, mean age = 7 years, 8 months (314)	Children exposed in an industrial zone, numerous ceramics operations. Pb analyzed in blood, hair, teeth. PbB = 11.5 μg/dl, Pb-teeth = 6.8 μg/g	Significant association between full-scale and verbal WISC IQ and Toulouse-Pieron Tests with tooth Pb	Vivali et al. (1989)
Germany: German children, three cities, age 6 years (384)	PbB mean, 6 years: 4.3 μg/dl (95th percentile, 8.9 μg/dl)	German WISC subtests, Vocabulary, Block Design and Summed Index. Adjusted vocabulary and summed index scores significantly inversely linked with PbB	Walkowiak et al. (1998)
Mexico: Torreón schoolchildren near lead facility (602)	PbB mean = 11.5 μg/dl	Peabody Picture Vocabulary test (Spanish) and WISC-R subtests significantly linked to concurrent PbB	Kordas et al. (2004)

found. The Wechsler Individual Achievement Test scores were strongly negatively associated with PbBs at the higher range.

The U.S. cross-sectional lead epidemiological findings were generally consistent with results of related efforts internationally (Table 12.5). Reports appeared from Denmark (Hansen et al., 1989), England (Yule et al., 1984, 1989; Pocock et al., 1989), Scotland (Raab et al., 1989), Germany (Winneke et al., 1982, 1983, 1988; Walkowiak et al., 1998), Greece (Hatzakis et al., 1989), Italy (Vivoli et al., 1989), Mexico (Kordas et al., 2004), and New Zealand (Silva et al., 1988). Urban community Pb exposures via ambient air and water Pb provided sources in Denmark, England, Scotland, and New Zealand. Pb exposures of child subjects in the studies from Germany, Greece, Italy, and Mexico arose from Pb emissions from smelter or mixed industrial point Pb sources.

Studies of children with urban community exposures are presented first. Yule et al. (1984) reported that there were significant changes versus PbB measurements for Conners and Rutter Teacher Ratings in those children whose earlier Pb exposures were elevated at 12–20 μg/dl (rounding) versus the low Pb group, mean=9 (rounding) (Yule et al., 1981). Psychometric outcomes in the children when younger included significant decrements in WISC-R Full Scale and Verbal IQ versus increasing PbBs as well as shortfalls in reading and spelling performance. By contrast, Lansdown et al. (1986) did not find significant differences when 82 higher PbB children (13–24 μg/dl) were compared with 60 lower exposure children (7–12 μg/dl).

Smith et al. (1983) used shed crowns to measure cumulative Pb exposure via teeth (Pb-T) levels and related the various Pb-T groupings to decrements in a large battery of psychometric tests of neurocognitive ability. One explanation of the finding of nonsignificance versus the opposite findings of Needleman et al. (1979) can be related to the fact that use of crowns by Smith et al. (versus dentine content by the Needleman group) made for a much less sensitive and more variable measure of Pb exposure than dentine. Crowns include low-lead enamel with variable remaining amounts of necrotic pulp containing Pb. However, a later evaluation of this data set, the Southampton Lead Study, showed child gender had a significant interaction with Pb-T, being stronger in boys than in girls. The association was negatively significant in boys at $p = 0.01$.

The Scottish cross-sectional study of Raab et al. (1989) involved 501 Edinburgh children, aged 6–9 years with community Pb exposures expressed as a geometric mean, PbB = 10.4 μg/dl. The children were tested with various parts of the British Ability Scales (BAS), showing that there were significant negative associations between log PbB and BAS combined score, number skills, and reading.

Danish schoolchildren (Aarhus, first grade, $N = 156$) having urban community Pb exposures were tested using circumpulpal dentine Pb levels as exposure marker and various psychometric measures (Hansen et al., 1989). The Bender, WISC-R Full Scale, and Verbal Test IQ decrements were significantly associated with the dentine Pb levels. Urban community Pb exposures in Dunedin, New Zealand, children (age 11, $N = 579$) were evaluated by

Silva et al. (1988) using PbB measurement (mean = 11.1 μg/dl; range 4–50), psychometric tests, and parental or teacher rating tests. WISC-R Full Scale, Verbal, and Performance IQ Tests were all nonsignificant, while parent or teacher rating scales of the Rutter Behavioral Rating Test versus PbB were statistically significant, and the Teacher portion of the Inattention Rating Test and both portions of the Hyperactivity Rating Test were all statistically significantly linked to PbB.

Children having Pb exposures from emissions from lead smelters and other industrial facilities were the subject of a number of international studies outside the United States. Winneke et al. evaluated children in several German lead smelter communities: Nordenheim (Winneke et al., 1989), Stolburg (Winneke et al., 1983), and Duisburg (Winneke et al., 1982). The smallest of the studies ($N = 52$, 26 exposed, 26 controls) involved Duisburg children tested at 8 years of age. Pb-T levels in control (2.4 ppm Pb) and exposed (9.2 ppm Pb) subjects served as the (cumulative) exposure biomarker and the Bender-Gestalt as the psychometric endpoint, with findings of a statistically significant association of the endpoint with the elevated Pb-T concentrations.

Stolburg children ($N = 89$) were tested at 9.4 years using both PbB (mean = 14 μg/dl) and Pb-T (mean = 6.2 ppm) as exposure biomarkers and the Bender-Gestalt Test as endpoint. Statistically significant associations in test deficits with both exposure biomarker elevations were recorded. Nordenheim children ($N = 76$) tested at 9 years of age using both PbB and Pb-T measurements (geometric mean PbB 7.8 μg/dl; geometric mean Pb-T 6.2 ppm) as exposure indices and the Vienna Reaction Test and visually evoked potential responses as endpoints had statistically significant associations of decrements in both endpoints with increases in exposure index.

The impact of a lead–zinc smelter in Lavrion, Greece, was evaluated in terms of impacts of its Pb emissions on the neurobehavioral health of children in the nearby community. Elementary schoolchildren ($N = 509$) with a mean PbB of 24 μg/dl (rounding) showed a highly significant inverse association between Full Scale, Verbal, and Performance WISC-R IQ Test results and increasing PbB. Italian children, age 7 years, 8 months, impacted by Pb releases from industrial facilities in Sassuolo, Italy, were studied (Vivoli et al., 1989). Exposure was indexed by both PbB (mean, 11.5 μg/dl) and Pb-T (mean, 6.8 ppm) measurements and several psychometric tests as endpoint. Decrements in Full Scale and Verbal WISC-R IQ and the Toulouse-Pieron Tests were all significantly associated with increasing Pb exposure indexed as Pb-T.

Kordas et al. (2004) studied Mexican schoolchildren ($N = 602$) impacted by a lead smelter in Torréon, Mexico, for associations of PbB elevation (mean = 11.5 μg/dl) and test outcomes on various psychometric tests. The Spanish version of the Peabody Picture Vocabulary Test and the subtests of the WISC-R IQ Test were significantly inversely associated with concurrent PbB measurements. Cho et al. (2010) reported that concurrent PbB in 667

Korean schoolchildren 8–11 years old showed a significant relationship with measures of Attention Deficit Hyperactivity Disorder (ADHD). The mean PbB was quite low, i.e., 1.9 μg/dl.

Prospective Epidemiological Studies of Asymptomatic Lead Neurotoxicity in Children: Neurocognitive Ability

The most current and also most reliably designed epidemiological studies of lead neurotoxicity in young children are a group of international evaluations that collectively offered a number of interpretive advantages over earlier cross-sectional efforts described above. First, this collection of studies benefited from using common epidemiological and statistical design approaches through investigator agreements both for quantifying exposures and using common psychometric and other neurobehavioral measures across studies. Lead exposures were determined through serial PbB measurements traversing the spectrum of neurodevelopment from gestation through the postnatal years, in some cases through the teen years and even beyond. Different commonly accepted instruments appropriate to the developmental milestones for the cohorts were employed.

In addition to providing advanced measures of both the lead dose and the toxic response portions of dose–neural response relationships, these prospective approaches helped to resolve interpretive questions not answered in earlier cross-sectional studies. For example, the inherent stability of the exposure measure is more discernible and quantifiable with serial measures of PbB or combined measures of blood and bone than isolated PbB measures done concurrently. One can also employ different statistical expressions of the exposure metric, e.g., average PbB, peak PbB, concurrent PbB, or lifetime integrated expression of exposure to estimate relative robustness of dose–response. Second, the potential confounding of dose–response relationships by any reverse causality is more reliably addressed. Reverse causality would hold that the dose term is elevated because of the higher likelihood of Pb exposures due to existing impairment of neurobehavioral function; in other words, PbB levels rise with a higher frequency of abnormal hand–mouth activity because of neurobehaviors unrelated to Pb exposures.

Summary tabulations of the main results of the individual prospective studies of lead effects on children's neurocognitive abilities are contained in Tables 12.6–12.11. Table 12.12 tabulates pooled analysis and meta-analysis of both prospective and within-study cross-sectional evaluations, while Table 12.13 summarizes epidemiological findings of prospectively examined lead-associated neurobehavioral endpoints other than cognitive abilities. The categories include social malbehavior, academic achievement, and lead's impacts on executive functions, attention, and other specific abilities.

The Boston Study is a prospective evaluation of a cohort representing children born in 1979–1980 who can be characterized as from relatively

TABLE 12.6 Prospective Epidemiological Assessment of Low-Level Lead Exposure on Neurocognitive Abilities in Young Children: The Boston Studies

Study Cohort (*N*)	Test Instrument/ Age	PbB	Outcome(s)	References
Boston, MA, children (249), almost all Caucasian, relatively affluent, born in 1979–1980	Bayley Scales of Infant Development (BSID); available PbB semiannually 6 months–2 years	Cord PbB > 10 μg/dl versus < 10 μg/dl. Postnatal PbBs at 2 years—Low PbB mean: 5.4 μg/dl, Median PbB mean: 7.2 μg/dl, High PbB mean: 7.7 μg/dl	Poorer BSID, > 10 μg/dl; no BSID association	Bellinger et al. (1987)
Boston, MA, children (170) in above cohort	McCarthy Scales at ~5 years of age; PbBs at 6, 12, 18, 24, and 57 months	Mean PbB = 6.4 μg/dl at 5 years	GCI scores significantly inversely associated with PbB at 2 years	Bellinger et al. (1991)
Boston, MA, children (148) in above cohort	WISC-R Scales, Battery Composite scores on Kaufman K-TEA	Mean PbB = 2.9 μg/dl at 10 years	WISC-R full scale and verbal IQ significantly inversely linked to PbB at 2 years	Bellinger et al. (1992)
Reanalysis of the 10-year group data (148)	Nonparametric smoothing techniques for IQ at 10 years, PbB at 2 years	–	Above inverse association persists at < 5 μg/dl	Schwartz (1994)
Reanalysis of data for a subset of cohort (48)	–	PbB ≤ 10 μg/dl	Significant association of IQ loss with 2-year PbB; slope at <10 μg/dl threefold steeper	Bellinger and Needleman (2003)

affluent families, of relatively high SES and 95% Caucasian (Bellinger and Needleman, 2003; Bellinger et al., 1987, 1991, 1992; Schwartz, 1994).

The cohort ($N = 249$) were tested using the Bayley Scales of Infant Development (BSID) at 2 years of age for associations with several measures

TABLE 12.7 Prospective Epidemiological Assessments of Low-Level Lead Exposures in Young Children: The Cincinnati, OH, Studies

Study Cohort (*N*)	Test Instrument/Age Timeline	PbB	Outcome(s)	References
Cincinnati, OH, children (300), mainly low-income African-American subjects	Bayley Scales of Infant Development (MDI): 3 and 6 months	Prenatal: 8.0 μg/dl; Cord: 6.3 μg/dl; 10-day: 4.6 μg/dl; 3-month concurrent: 5.9 μg/dl	Prenatal, cord PbB inversely associated with adjusted 3-month MDI	Dietrich et al. (1987)
Cincinnati child cohort (297)	Bayley MDI at 2 years	11 different prenatal or postnatal measures	No significant inverse linkage of 11 PbB measures versus Bayley MDI scales	Dietrich et al. (1990)
Cincinnati child cohort (258)	Kaufman Assessment Battery for Children (K-ABC) at 4 years	Various maternal and postnatal PbB levels, postnatal means or peak values	Significant inverse association between neonate (10-day) PbBs and all K-ABC subscales, but only for poorest families	Dietrich et al. (1991)
Cincinnati child cohort (259)	K-ABC scores at 5 years	Neonatal (10 days) and postnatal PbB levels	No significant inverse association between neonate postnatal PbBs and K-ABC results except one subtest score	Dietrich et al. (1992)
Cincinnati child cohort (253)	Revised Wechsler Intelligence Scale for Children (WISC-R) at 6.5 years of age	Arithmetic mean Pb 51–60 months: 11.8 μg/dl	Significant inverse relationship of performance IQ with postnatal PbBs at 3,4, 5,6 years or lifetime mean. For full-scale IQ, associations with PbBs at 5 and 6 years; mean lifetime PbB > 20 μg/dl produces 7 IQ point decrease versus 0–10 μg/dl group	Dietrich et al. (1993)
Cincinnati child cohort	Neurodevelopmental test battery at 15–17 years: Wide Range Achievement Test—Third Edition, Block Design Subtest and WISC-III, vocabulary subtest	PbB mean, 78 months	Significant inverse association between Learning/IQ Factor and fine-motor factor scores and 78-month PbBs; lower SES children more affected	Ris et al. (2004)

TABLE 12.8 Prospective Epidemiological Assessments of Low-Level Lead Exposures in Young Children: The Port Pirie, Australia, Studies

Study Cohort (*N*)	Test Instrument/ Age Timelines	Pb Level (μg/dl or ppm)	Outcomes	References
Port Pirie, Australia, children (600) exposed from a Pb smelter	BSID, 2 years old	PbB means: 14.3, 6 months 20.8, 15 months 21.2, 24 months	No significant inverse associations when adjustments done	Wigg et al. (1988)
Port Pirie, Australia, children (537) exposed from a Pb smelter	McCarthy Scales (MSCA), 4 years old	Integrated and year-specific PbB means by quartile	Integrated PbB most strongly inversely associated with GCI scores	McMichael et al. (1988)
Port Pirie, Australia, children (494) exposed from a Pb smelter	Revised Wechsler Intelligence Scale for Children (WISC-R), 7 years old	Integrated and year-specific PbB means	Integrated PbB most strongly inversely associated with verbal and full-scale IQ	Baghurst et al. (1992)
Port Pirie, Australia, children (375) exposed from a Pb smelter	WISC-R IQ testings, 11–13 years old	Concurrent and lifetime average PbB to age 11–13	Lifetime PbB to age 11–13 strongly inversely associated with WISC-R scores at 11–13 years	Tong et al. (1996)
Port Pirie, Australia, children (262)	WISC-R IQ Scales, 7 years	Pb-teeth levels, upper incisors 3–22 ppm, 90% of cohort	Full-scale WISC-R scores inversely linked to Pb-T values	McMichael et al. (1994)

of PbB: cord PbB ≤ 10 versus > 10 μg/dl, and three groupings of 2-year PbB levels, mean from 5.4 to 7.7 μg/dl (Bellinger et al., 1987). Poorer scores were determined for infants at the higher cord PbB, while postnatal measures showed no BSID association. A subset of the original cohort ($N = 170$) was tested psychometrically for general neurocognitive abilities using the McCarthy Scales at about 5 years of age, using as the exposure variable all of the serial measures at 6, 12, 18, 24, and 57 months of age (Bellinger et al., 1991). At testing, the mean PbB was 6.4 μg/dl. General Cognitive Index scores were statistically significantly associated inversely with children's PbB value at 24 months of age.

TABLE 12.9 Prospective Epidemiological Assessments of Low-Level Lead Exposure in Young Children: The Kosovo, Former Yugoslavia, Studies

Study Cohort (*N*)	Test Instrument/ Age at Testing	PbB (μg/dl)	Outcomes	References
Children in two Kosovo, former Yugoslavia, towns: a smelter, refinery, battery operation—Mitrovica—and the control site, Pristina (577)	Bayley MDI scores: 2 years old	Exposed group mean at 2 years: 35.5 μg/dl; control, 8.4 μg/dl	Statistically significant association of MDI and PbB, at 2 years	Wasserman et al. (1992)
Children in the two Kosovo communities (332)	McCarthy Scales, 4 years old	At 4 years, PbB mean, smelter town: 39.9 μg/dl; PbB mean, control town: 9.6 μg/dl	McCarthy Scales GCI scores inversely linked to all postnatal PbB measures and prenatal PbB > 20 μg/dl	Wasserman et al. (1994)
Children in the two Kosovo communities (301)	Wechsler Third Edition, WISC-OOO, IQ test, 7 years old	PbB Area-Under-the-Curve, AUC, 5–7 years	Adjusted PbB AUC significantly associated with IQ scores, most with performance IQ	Factor-Litvak et al. (1999)
Children in the two Kosovo communities (290 PbB; 167 bone Pb)	WISC-III IQ tests, 10–12 years old	Lifetime and concurrent PbB, tibial Pb	Lifetime and concurrent PbB both significantly inversely associated with IQ in Mitrovica, not Pristina; bone Pb more strongly linked to IQ scores	Wasserman et al. (2003)

A second cohort subset (*N* = 148) was tested at about 10 years of age, using as psychometric instruments the WISC-R Scales and Battery Composite scores on the Kaufman K-TEA test (Bellinger et al., 1992). By 10 years of age, the mean PbB had declined to 2.9 μg/dl, half that at 5 years and about one-third that at 24 months. WISC-R full scale and verbal IQ

TABLE 12.10 Prospective Epidemiological Assessments of Low-Level Lead Exposures in Young Children: The Mexico City, Mexico, Studies

Study Cohort (*N*)	Test Instrument/Age Timelines	PbB (μg/dl)	Outcomes	References
A. Prospective study				
Mexico City, Mexico, Prospective Pb Study (321): lower SES subjects, community Pb exposures: includes lead-glazed dinnerware	Brazelton Neonatal Behavior Assessment Scale (BNBAs): 48 hours, 15 and 30 days ($N = 42$)	Mean cord: 13.8 μg/dl	Cord PbB > maternal PbB associated with several adverse changes: Regulation of States, Autonomic Regulation, Gestation Age	Rothenberg et al. (1989)
Mexico City cohort (112)	McCarthy Scales GCI, 54 months	Geometric mean, 24–36 months, 9.7 μg/dl	GCI adjusted score at 48 months decreases 4 points for PbB doubling, 24–36 months	Schnaas et al. (2000)
Mexico City cohort (150)	Spanish language WISC-R, full scale, 6–10 years	Mean: pregnancy, 9.0 μg/dl; 1–5 years, 9.9 μg/dl; 6–10 years, 6.2 μg/dl	Adjusted IQ at 6–10 years significantly associated with third trimester maternal PbB	Schnaas et al. (2006)
B. Fetal neurodevelopment study				
Mexico City children, low to middle class (197)	Spanish version: Bayley Scales for both mental and psychomotor development	Cord: 6.7 μg/dl, 12 months: 7.2 μg/dl, 24 months: 8.4 μg/dl, bone (tibia): 17.8 μg/g, patella: 11.5 μg/g	Adjusted Bayley MDI scores inversely associated with cord PbB, with maternal patellar (trabecular) Pb	Gomaa et al. (2002)
C. The two-cohort study				
Two cohorts of Mexico City toddlers, those in B above plus an unnamed group (294), community Pb exposures	Spanish version: Bayley Scales, MDI and PDI, 12 and 24 months	Cord, 12- and 24-month levels: both postnatal PbBs < 5 μg/dl	Significant associations between MDI or PDI and 24-month PbB; steeper coefficients for PbB < 10 μg/dl	Tellez-Rojo et al. (2006)

TABLE 12.11 Other Prospective Epidemiological Assessments of Low-Level Lead Exposures in Young Children: Rochester, NY; Cleveland, OH; and Sydney, Australia

Study Cohort (*N*)	Test Instrument/Age Timelines	PbB/Other Pb Index	Outcome(s)	References
Rochester, NY: Low SES, mostly African-American children, enrolled at 5–7 months; community Pb exposures (172)	Stanford–Binet Intelligence Scale—Fourth Edition (SBIS-4), 3 and 5 years old	PbB peak, 9.7 μg/dl at 2 years. Lifetime PbB means: 7.7 μg/dl, 3 years; 7.4 μg/dl, 5 years. Peak PbB at 5 years: <10 μg/dl, 56% of children	Adjusted significant inverse associations of PbB and full-scale IQ, both years	Canfield et al. (2003)
Rochester, NY, cohort (174)	Wechsler Preschool and Primary Scales of Intelligence—Revised (WPPSI-R), 6 years old	Lifetime PbB mean = 7.7 μg/dl; Peak PbB = 11.4 μg/dl	Adjusted WPPSI-R full-scale and performance scores significantly inversely associated with lifetime average levels of PbB (5–9.9 versus <5 μg/dl). Inverse link of full-scale scores with peak PbB to 2.1 μg/dl	Jusko et al. (2008)
Cleveland, OH: Low SES children, 50% of mothers had alcohol abuse histories; community lead exposures (285)	Bayley Scales at 6 months, 2 and 3 years; KID scales at 6 months; Stanford–Binet IQ at 3 years	PbB measured versus maternal, cord Pb, postnatal to 3 years. Maternal Pb mean 6.5 μg/dl	No associations except maternal PbB effect on MDI, PDI, and KID score variance	Ernhart et al. (1987)
Cleveland, OH: subset of cohort (242)	WPPSI scales at 4.8 years	Maternal, cord, preschool postnatal PbB levels	No statistical associations	Ernhart et al. (1989)
Cleveland, OH: low SES children [providing shed incisor teeth] (164)	WPPSI test: full-scale, verbal, and performance IQs at 4 years, 10 months	Pb measured as circumpulpal dentine Pb	Significant inverse association of dentine Pb with verbal IQ; marginal association with full-scale IQ	Greene and Ernhart (1993)
Sydney, Australia: Sydney children (298)	McCarthy Scales, 4 years; WISC-R test at 7 years	PbB collected: cord PbB, 8.1 μg/dl; PbB peak, 18 months, 16.4 μg/dl; 48 months, 10.1 μg/dl; 7 years, 7.7 μg/dl	No significant associations of PbB index at any age with IQ measures	Cooney et al. (1989, 1991)

TABLE 12.12 Pooled- and Meta-Analyses of Effects of Lead on Children's Neurocognitive Abilities

Subjects/ Study Type	Endpoint	Exposure	Results	References
Full group, PbB: Children in seven prospective studies, total $N = 1,300+$	Full-scale IQ loss at school age, 6.9 years old	Four measures of PbB: concurrent, incremental lifetime, mean, peak blood, early childhood	IQ loss best correlated with concurrent PbB: 1.9 points, 10–20 μg/dl; 6.7 points, 1–10 μg/dl	Lanphear et al. (2005)
Children in seven prospective studies, pooled international analyses, $N = 244$	Neurocognitive abilities (IQ); IQ decrements at low PbB	PbB < 10 μg/dl; Interval: 4 years 10 months to 10 years	Incremental IQ loss greater at lower PbB; linear model: −0.8; Log-linear model: −0.4	Lanphear et al. (2005)
Children in 12 studies, meta-analyses	Neurocognitive abilities (IQ measures)	PbB, Pb-teeth	Highly significant associations between IQ and PbB or Pb-teeth; joint p (PbB) = 0.0001 by two methods	Needleman and Gatsonis (1990)
Children in six studies: three prospective, three cross-sectional, meta-analyses	Neurocognitive abilities (IQ measures); change in full-scale IQ for rise of 10–20 μg/dl PbB	PbB; prospective data, PbB = 24 months, average = first 3 years	An increase in PbB of 10–20 μg/dl reduced IQ 2.6 points	Schwartz (1994)
Children in five prospective, 14 cross-sectional studies, meta-analyses	Neurocognitive abilities (IQ); change for PbB (10 units) or Pb-teeth (5–10 ppm)	PbB or Pb-teeth; PbB at 2 years	Doubling of PbB or Pb-teeth reduced IQ 1–2 points	Pocock et al. (1994)

scores were significantly linked statistically and inversely with 24-month PbB measurements. Schwartz (1994) statistically reanalyzed the 10-year group using nonparametric smoothing techniques for the IQ–24-month PbB relationships. The statistically significant inverse association was reported to persist to exposures <5 μg/dl.

TABLE 12.13 Illustrative Epidemiological Studies of Neurobehavioral Indices Other Than Neurocognitive Abilities

Study Group	Endpoint	Pb Exposure Tests	Results	References
Academic achievement				
North Carolina selected county grade-schoolchildren (4th grade) 8,603 reading, 8,627 mathematics	Reading and mathematics scores	PbB measured via screening; community exposures	Early PbB levels are significantly inversely related to both reading and mathematics scores, with reading association as low as 2 μg/dl	Miranda et al. (2007)
North Carolina statewide grade schoolchildren	Reading and mathematics scores	PbB measured via screening	High PbB children's test scores distribution more impacted; Pb more toxic at lower tail of distribution	Miranda et al. (2009)
U.S. NHANES III (1988–1994) children 6–16 years old ($N=4{,}853$)	Reading and mathematics skills via Wide Range Achievement Test—Revised	PbB mean = 1.9 μg/dl; 63% < 2.5 μg/dl. Community exposures	Reading and mathematics scores inversely related to highest (> 3 μg/dl) versus lowest (< 1 μg/dl) quartiles	Lanphear et al. (2000)
Boston, MA, children born in 1979 and 1980 in one hospital ($N=2{,}000$)	Boston Teacher Questionnaire, given at 8 years	Pb-dentine levels	Significant association of reading and spelling deficits with elevated dentine Pb in girls	Leviton et al. (1993)
Christchurch, New Zealand, children ($N=1{,}200$)	Academic achievement and classroom performance measures, 12–13 years, e.g., Burt Reading Test, Progressive Achievement Test, and teacher ratings	Pb-dentine at 6–8 years; mean = 6.2 μg/g	Dentine Pb associated with deficits in all rating instruments and teacher ratings	Fergusson et al. (1993)
Social malbehavior/conduct				
Philadelphia African-American residents, birth–22 years old	Predictive measures of violent/delinquent behavior	"Higher" PbB levels	Elevated PbB one of the most significant predictors of criminality	Denno (1990)

Prospectively studied children in the Pittsburgh Youth Study at high risk of delinquency ($N = 503$)	Antisocial behavior and delinquency, tested at 7 and 11 years old	Pb in tibia, XRF analysis, 7 and 11 years	Parental and teacher reports of delinquency and aggressive behavior related to bone Pb	Needleman et al. (1996)
Prospectively studied children in the Cincinnati Lead Study	Parent and self-reported antisocial and delinquent behaviors, measured at 16–17 years old	Prenatal and postnatal PbB levels	Highest prenatal PbB category (>10 μg/dl) linked to more delinquent acts versus lower PbB; 78-month PbB also linked to more delinquent acts	Dietrich et al. (2001)
Group of adjudicated Pittsburgh delinquents versus controls (194 versus 146)	Delinquent status in the juvenile court system	Tibial bone Pb via XRF	Cases had significantly higher tibial Pb than controls. Bone Pb ≥25 μg/g 4 times more likely to be adjudicated for delinquency	Needleman et al. (2002)
Executive functions, attention, and other specific abilities				
Subset of original Boston cohort at 19–20 years old	Tests of spatial memory, memory span, cognitive flexibility, planning	Pb-teeth	Higher Pb-teeth significantly linked to poorer scores on certain battery parts	Bellinger et al. (1994)
Rochester Lead Study children	Executive function/learning at 48 and 54 months, Shape School Task	PbB mean = 6.5 μg/dl; 80% <10 μg/dl	PbB at 48 months inversely linked with focused attention, color naming, and responding inhibition	Canfield et al. (2003)
Torréon, Mexico, children affected by metal facility ($N = 594$)	Battery of tests for distractibility, memory sequencing, and others	PbB mean = 11.4 at testing	Performance on Sternberg Memory Test significantly associated with PbB	Kordas et al. (2006)

(*Continued*)

TABLE 12.13 Illustrative Epidemiological Studies of Neurobehavioral Indices Other Than Neurocognitive Abilities—(cont.)

Study Group	Endpoint	Pb Exposure Tests	Results	References
Older children, Cincinnati Lead Study, 15–17 years old	Executive functions: WISC Card Sorting Test, Rey–Osterruth Complex Figure; others such as attention, memory, fine-motor coordination	PbB, 30% ≥ 25 μg/dl	Attention component had most significant association with PbB	Ris et al. (2004)
Brain anatomical development				
Subset of young adults, Cincinnati Lead Study ($N = 48$)	Language functions	fMRI on 20–23 year-olds to test language function	Higher mean PbB levels associated with reduced activation, Broca's area, in left hemisphere and increased activation in temporal lobe	Yuan et al. (2006)
Subset of young adults, Cincinnati Lead Study ($N = 157$)	Reductions in adult gray matter volume	Relationships of neuropsychological outcomes to reduction in adult gray matter volume	Frontal gray matter volume reduction most pronounced in brain areas harboring executive functions, mood control, decision making	Cecil et al. (2008)
Subset of young adults, Cincinnati Lead Study	Relationship of childhood PbB (1–6 years) and gray matter volume loss	Study subjects' prior PbBs	Significant inverse associations between gray matter volume loss and mean PbB 3–6 years	Brubaker et al. (2010)
Subset of young adults, Cincinnati Lead Study ($N = 91$)	Diffusion tensor imaging to relate PbB to changes in white matter structure	Study subjects' prior PbBs	Decreased fractional anisotropy with increased PbB through white matter amount	Brubaker et al. (2009)

Bellinger and Needleman (2003) reanalyzed their earlier cohort data, specifically stratifying the outcomes of the psychometric tests and dose–response strengths for 24-month PbB levels into those < 10 μg/dl ($N = 48$) and those ≥ 10 μg/dl. These authors reported that the dose–response slope, IQ loss versus PbB increase, was threefold steeper at PbB < 10 μg/dl than for blood Pb values at 10 units or higher. This curvilinear relationship across the dose–response spectrum at lower versus higher dose has been documented in other individual and aggregated studies noted below.

A second U.S. study cohort evaluated prospectively for lead versus neurocognitive and neurobehavioral relationships was that in the Cincinnati Lead Study. Pregnant women and the later prospective child cohort recruited into this study had relatively high PbB levels and correspondingly lower SES characteristics than individuals in the Boston group. The principal prospectively monitored outcomes for general neurocognitive development in this study series are contained in multiple reports: Dietrich et al. (1987, 1991, 1992, 1993) and Ris et al. (2004).

The study cohort ($N = 300$) was first studied using the Bayley Scales at 3 and 6 months (Dietrich et al., 1987). Assessments of exposures to Pb included a prenatal, mean PbB of 8.0 μg/dl in cord blood, and both 10-day and 3-month PbB (concurrent with the first testing). The principal finding was that prenatal PbB (cord blood) was statistically inversely associated with the 3-month MDI portion after adjustment for confounders. This early association for Bayley Scales did not persist at 2 years ($N = 297$), using 11 different prenatal and postnatal measures (Dietrich et al., 1990). At 4 years of age, the cohort ($N = 258$) was tested using the Kaufman Assessment Battery for Children (K-ABC) and various maternal and postnatal PbB levels, either as mean or peak levels (Dietrich et al., 1991). The authors reported statistically significant inverse associations between 10-day PbBs and all K-ABC subscales, but only for the poorest subset of children.

Retesting using the K-ABC instrument at 5 years ($N = 259$) and the neonatal/10-day and postnatal PbB levels produced significant associations in only one subtest of the K-ABC Test. More robust responses are noted for this cohort ($N = 253$) at 6.5 years of age using the WISC-R Intelligence Scales and concurrent mean at 3, 4, 5, or 6 years or lifetime mean (Dietrich et al., 1993). The arithmetic mean PbB at 4–5 years was 1.8 μg/dl. Full-Scale IQ decrements were inversely related to PbB at 5 and 6 years of age while Performance IQ was impacted inversely with postnatal PbB at 3, 4, 5, or 6 years, or as lifetime mean. The mean lifetime Pb exposure index for values > 20 μg/dl produces a 7 IQ point decrease for a 10-unit change in PbB up to 10 μg/dl. Ris et al. (2004) examined the remaining subjects in the Cincinnati cohort at 15–17 years of age using a neurodevelopmental test battery that included the Wide Range Achievement Test—Third Edition, Revised, the Block Design Subtest and the WISC-III, Vocabulary Subtest. The reference PbB was the mean at 78 months. There were statistically significant inverse associations between Learning/IQ factor and fine motor factor scores and 78-month PbB levels. Lower SES children were the most affected.

The Port Pirie prospective evaluation of general neurocognitive abilities involved children exposed to Pb from a lead smelter ($N > 600$). Unlike the Boston and Cincinnati studies where Pb exposures were urban community based, this cohort sustained exposures from a major point source, the lead smelter. The major neurobehavioral findings with this Pb exposure cohort are contained in four reports: Wigg et al. (1988), McMichael et al. (1988), Baghurst et al. (1992), and Tong et al. (1996). As done for the other prospective evaluation cohorts, the investigators determined prenatal exposures among the mother–infant pairs. In this case, two measures were gathered: a maternal sample about mid-pregnancy and a cord blood level at delivery. PbB measures included age and serially gathered specific levels but also an aggregated measure of Pb exposure, the integrated or lifetime measurement.

Wigg et al. (1988) tested cohort children at 2 years of age using the Bayley Scales. Mean PbB values at ages 6, 15, and 24 months were used as the exposure metric. At 24 months, the PbB mean was 21.2 μg/dl. When adjusted for confounding factors, no PbB value was significantly inversely linked to Bayley scores. At 4 years of age, this cohort ($N = 537$) was tested using the McCarthy Scales and the PbB measures included year-specific PbB means and "integrated" PbB for the 4 years of life. The most robust inverse association with the McCarthy GCI was the integrated PbB value. Also, there were significant associations for PbB values at 6, 24, and 36 months.

These children ($N = 494$) were retested at 7 years of age using the WISC-R and, for the exposure measure, both year-specific and integrated PbB concentrations were included. As with the younger children, the most robust association was between the integrated values of PbB and both Verbal and Full-Scale scores for the WISC-R testings. This cohort at the same age provided shed teeth (central upper incisors) and the Pb-T values were also analyzed for associations. Full-Scale WISC-R scores were significantly associated with Pb-T values (McMichael et al., 1994).

Children living near a lead smelter in Kosovo Province in the former Yugoslavia were the subject of a series of prospective assessments of lead exposure following the protocols of the other international efforts (Factor-Litvak et al., 1999; Wasserman et al., 1992, 1994, 2003). Results are summarized in Table 12.9. Two communities were evaluated: Mitrovica, the site of a lead smelter, lead refinery, and battery operation, and a control community, Pristina. The beginning cohort was sizeable, $N = 577$, while the mean Pb exposures for infants in the lead operations community at 24 months were the highest of all the longitudinal studies, mean PbB = 35.5 μg/dl. This value was approximately fourfold higher than exposures in the control community (mean = 8.4 μg/dl).

Bayley Scales scores were found to be statistically inversely associated with 24-month PbB levels in the Mitrovica children (Wasserman et al., 1992).

A subset of the cohort ($N = 332$) was tested at 4 years of age using the McCarthy Scales. By this age, the smelter group mean PbB had risen to 40 μg/dl (rounding) versus controls' mean of 10 μg/dl (rounding). GCI scores were statistically inversely associated with all postnatal exposure markers and with prenatal values > 20 μg/dl (Wasserman et al., 1994). At 7 years of age, children ($N = 301$) were tested using the Third Edition, Wechsler (WISC-III) IQ scales and various exposure markers (Factor-Litvak et al., 1999). Adjusted for confounders, there was a statistically significant inverse association of Performance IQ on the WISC-III tests with an integrated exposure measure, the lifetime Area-Under-the-Curve measurements, at 5–7 years.

These children were tested at 10–12 years of age with reference to both a cumulative exposure indicator, tibial bone Pb concentration, and two PbB measures, lifetime and concurrent levels (Wasserman et al., 2003). Both lifetime and concurrent PbB values were statistically inversely linked to WISC-III IQ scores, while strongest associations were seen with the bone measure, tibial Pb.

Two longitudinal studies of low-level lead exposures in children were done in Mexico City, Mexico (Table 12.10). The first was the Mexico City Prospective Study (Rothenberg et al., 1989; Schnaas et al., 2000, 2006), followed by the Fetal Development Study (Gomaa et al., 2002; Tellez-Rojo et al., 2006). The first longitudinal effort entailed children with urban community Pb exposures, some common across the international studies but also including lead-glazed dinnerware. Children were lower SES subjects.

In a pilot report, Rothenberg et al. (1989) employed the Brazelton Neonatal Behavior Assessment Scales (BNBAS) as the endpoint measure and cord Pb as the exposure metric using 321 subjects. Cord PbB was more significant than maternal PbB in showing associations with several adverse BNBAS outcomes such as gestational age and autonomic regulation. A subset ($N = 112$) was tested at age 54 months using the McCarthy Scales (Schnaas et al., 2000). The measure of exposure was PbB, the group having a geometric mean PbB of 10 μg/dl (rounding). These investigators reported that the adjusted GCI scores at testing decreased four points for a doubling of PbB in the 24- to 36-month age band. At 6–10 years of age, a subset of the children ($N = 150$) tested with the Spanish version of WISC-R, after IQ adjustments for confounders, had inversely significant associations in Full Scale IQ scores with third-trimester maternal PbB levels. PbB means (rounding) were pregnancy, 8.0 μg/dl; 1–5 years, 10 μg/dl; and 6–10 years, 6.0 μg/dl.

The fetal neurodevelopmental study looked at lower- to middle-class Mexico City children ($N = 197$) with respect to the impact of both PbB and bone Pb (tibia, patella) on Bayley Scales (Gomaa et al., 2002). The Spanish version of the Bayley Scales was used at 24 months, while exposure measures were maternal tibial and patellar Pb level, cord Pb, 12- and 24-month PbB concentrations. Maternal patellar Pb and cord Pb were statistically significantly associated with adjusted MDI scores.

A subsequent investigation combined the data from the Gomaa et al. subjects with a second cohort ($N = 294$) of toddlers who were not characterized in the published material (Tellez-Rojo et al., 2006). This latter investigation permitted the overall tally of children to be stratified into those with PbB < 10 μg/dl and those at 10 and higher. The testings entailed the Spanish version of the Bayley Scales at 12 and 24 months, over which period cord and both postnatal PbB mean values were <5 μg/dl. There were statistical significant inverse associations of scores at 24 months with the 24-month PbBs. Slope coefficients showed a steeper dose–response relationship at PbB concentrations <10 μg/dl.

Table 12.11 summarizes results from other prospective studies in Rochester, NY; Cleveland, OH; and Sydney, Australia. These are discussed together. The Rochester, NY, study entailed a less extensive but still valid protocol, while the cohorts in the Sydney and Cleveland groups had limited evaluations overall or had problems with design or execution particulars.

Low SES, mainly African-American, children living in Rochester, NY ($N = 172$), were enrolled at 5–7 months of age in the Rochester study and followed to 6 years of age. Lead contact involved typical urban community exposure sources. Canfield et al. (2003) examined this cohort using as outcome marker the Stanford–Binet Intelligence Scale—Fourth Edition at both 3 and 5 years of age. The exposure biomarkers were multiple: PbB peaks at either 2 or 5 years; lifetime PbB means at 3 or 5 years; and individual serial measures at 6, 12, 18, 24, 36, 48, and 60 months of age. After adjustments for confounders, PbB values expressed as concurrent, lifetime average, average in infancy or peak PbB were all shown to be statistically significantly associated with the psychometric measure scores. Stratifying PbB levels into those never exceeding 10 μg/dl and those that did showed a larger effect size, such that the first decile of lifetime mean PbB up to 10 units was linked to an IQ decrement of 7.4 units.

Jusko et al. (2008) evaluated this cohort at 6 years of age, using the Revised WPPSI as the psychometric instrument and four forms of the Pb biomarker: lifetime average, infancy average, peak PbB, and concurrent concentrations. Adjusted IQ scores were lower by about 5 IQ points for those children with PbB 5.0–9.9 μg/dl than for those <5 μg/dl, using lifetime average PbB. A further statistical analysis showed the inverse exposure–IQ decrement relationship persisted to the lowest tested PbB value, i.e., 2.2 μg/dl. In terms of the overall dose–response relationship versus the lowest segment thereof, the ≤10 μg/dl subset showed an IQ decrement response much higher than those in the region of 20–30 μg/dl.

Ernhart et al. (1987) examined low SES inner-city Cleveland children ($N = 285$) having urban community Pb exposures. A complication here is that 50% of the mothers had alcohol abuse histories. Testing instruments included Bayley Scales at 6 months, 1 and 2 years, the Kent Infant

Development (KID) Scale at 3 years, and the Stanford–Binet IQ Test at 3 years. There were no statistical associations except for maternal PbB effects on the MDI, PDI, and KID variance.

A subset of this cohort ($N = 242$) was tested at 4.8 years of age using the Full Scale, Verbal, and Performance Scales of the WPPSI IQ Test as the outcome and maternal, cord, and early postnatal PbB values as exposure indicators (Ernhart et al., 1989). No statistically significant associations were reported. A further subset ($N = 164$) of this cohort was similarly tested at the same age using a cumulative Pb exposure measure, circumpulpal dentine Pb measured from shed incisors (Greene and Ernhart, 1993). They reported significant inverse statistical associations of Verbal IQ and marginal association of Full Scale IQ with the Pb-T measurements.

Cooney et al. (1989) reported their results from a prospective study of 298 Sydney, Australia, children at 4 years of age, using the McCarthy Scales as the psychometric outcome measure and PbB levels measured at various age points. No statistically significant associations were recorded for PbB levels with any of the following: cord, peak (18 months), or concurrent (48 months) PbB levels. Repeat testing using the WISC-R IQ tests at 7 years of age, where the mean PbB was 8 μg/dl (rounding), similarly yielded no significant associations (Cooney et al., 1991). This particular set of results was complicated by changes in the methods of PbB sampling in the course of the study.

A large U.S. study which is epidemiologically important in this category is the Treatment of Lead-exposed Children (TLC) effort, with 780 children followed in a multicenter, randomized placebo-controlled clinical trial and where PbB values were below those triggering medical management via chelation therapy, i.e., 20–44 μg/dl. An earlier report of study results noted that chelation treatment using Succimer® did not result in improved IQ test scores, although a transitory reduction (~10 months) in PbB occurred (Rogan et al., 2001; Rosen and Mushak, 2001).

These findings led to the use of the TLC cohort as a prospectively evaluated population where childhood exposures at various developmental periods could be related to IQ changes. Chen et al. (2005) reported that children in this cohort followed from 2 to 7 years of age with serial PbB and IQ score measures produced evidence that cross-sectional associations increased with age, rather than remaining stable versus earlier exposure measures. Specifically, any peaks in PbB around 24 months of age did not appear to be the principal dose marker in these associations; in actuality, concurrent PbB was the more robust exposure term. Hornung et al. (2009) employed a statistical approach to probe the concurrency robustness of the PbB metric versus cognitive deficits in a way minimizing rank ordering from earlier exposures. They found, using two prospective cohorts, concurrent PbB largely independently affecting these associations.

Pooled Analysis and Meta-Analysis Studies of Cross-Sectional and Prospective Child Cohorts

Several pooled analysis and meta-analysis studies for multiple cross-sectional (meta-analysis) and prospective (pooled analysis) cohort evaluations have been published. Lanphear et al. (2005) reported on a pooled analysis of seven of eight longitudinal studies in the literature that followed children until at least 5 years of age. Various criteria were employed for inclusion, such as complete analyses for covariates. The final sample size was more than 1,300 children drawn from the following cohorts: Boston, MA; Cincinnati, OH; Cleveland, OH; Rochester, NY; Mexico City, Mexico; Port Pirie, Australia; and Kosovo, former Yugoslavia. Various forms of the exposure measure were included: concurrent, maximum, lifetime average, and early childhood PbB. The psychometric outcome was full scale IQ. Within the combined data points for exposures, a subset of 103 children had maximum PbB $< 7.5\ \mu g/dl$.

A number of important findings sprang from the pooled analysis. First, concurrent PbB levels were the PbB measurement most strongly inversely associated with and that most explained IQ decrements in these children from full-scale testings. However, other forms of the dose function were also statistically significantly associated with IQ changes. A log-linear model was identified as the function most closely matching the shape of the dose–response curve. Performance and verbal IQ scores were also significantly linked with the exposure measures. A particularly valuable finding was the presence of a much steeper dose–response slope at sub-10 μg/dl values versus higher values.

Four meta-analyses of the more complete and reliable data contained in both cross-sectional and/or longitudinal evaluations have been published. Needleman and Gatsonis (1990) included seven PbB and five dentine Pb studies, which in turn were from studies generated over 17 years, 1972–1989. The levels of statistical significance for the exposure–neurobehavioral decrement relationship for PbB and dentine Pb as the dose measure were estimated as joint *p*-values. They found very low joint *p*-values.

The meta-analysis of Pocock et al. (1994) addressed cross-sectional studies using both blood and dentine Pb measures as the dose variable. A statistically significant association was noted for 2-year PbB values and IQ decrements, but no relationships were reported for prenatal or perinatal PbB values.

Schwartz (1994) did a meta-analysis on three longitudinal and four cross-sectional studies relating PbB to full scale IQ changes in schoolchildren. The methodology involved inverse variance weighting in a random effects model. Schwartz reported that an increase of PbB of 10–20 μg/dl produced a decline of 2.6 (rounding) IQ points for all seven original data sets. The effect of magnitude for the longitudinal efforts (3.0 IQ point loss, rounding) were moderately higher than for the cross-sectional reports (2.7 point loss, rounding). The report of Thacker et al. (1992) focused on five longitudinal studies expressed within 35 publications as reviewed by a peer review panel blinded to the data origin. In the judgment of the panel, the inconsistencies across studies make quantitative analyses problematic.

The Thacker et al. (1992) conclusions did not factor in the role of exposure profile and Pb toxicokinetic differences in affecting the form of the most significant results within and across studies. Mushak (1993) analyzed differences across the early results in phases of the prospective studies and demonstrated that the toxicokinetic and exposure differences do in fact account for dose–response differences.

Epidemiological Studies of Asymptomatic Lead Neurotoxicity in Children: Academic Achievement, Social Malbehavior, and Specific Ability Domains

The principal focus of epidemiological studies of lead developmental neurotoxicity entailed evaluations of general neurocognitive function. However, other endpoints have been evaluated with a number of significant results in areas such as lead's effects on academic achievement, social malbehavior, and specific domains of neural function in children. Illustrative studies are summarized in Table 12.13.

Academic achievement by school-age children and effects of Pb on achievement were generally probed in terms of impacts on reading and mathematical skills. The studies in Table 12.13, dating from the early 1990s forward, are notable for large sample sizes from large child population bases, for using various measures of Pb exposure, and for manifesting significant associations of these achievement endpoints at relatively low Pb exposure.

Miranda et al. (2007) reported on reading and mathematics scores for North Carolina fourth-grade children having community lead exposures of diverse types ($N = 8{,}603$, reading test; 8,627, mathematics). The Pb exposure biomarker was PbB measurement gathered in the course of screening. PbB levels measured in early childhood were statistically significantly inversely associated with deficits in reading and mathematics scores, with association persisting to exposure values as low as 2 μg/dl.

A follow-up study (Miranda et al., 2009) drew on North Carolina statewide screening data using quantitative regression. This approach predicts percentiles of an outcome variable, using a set of explanatory variables. Higher exposure children in terms of distributions were more spread out than low PbB children. The predictability with this technique for distribution tails gives a relative look at impacts of exposures at either end of the exposure range. More toxicity is apparent at the low-performance region.

A U.S. national evaluation of school-age children was reported by Lanphear et al. (2000) for subjects in the NHANES III (1988–1994; ages 6–16 years old; $N = 4{,}853$), associating PbB levels with reading and mathematics skills using the Wide Range Achievement Test—Revised. These researchers reported that both reading and mathematics scores were inversely related to the highest PbB group quartile, >3 μg/dl, versus the lowest quartile, <1 μg/dl.

Boston, MA, children born in 1979–1980 in one hospital ($N = 2{,}000$) were evaluated at 8 years of age for academic achievement using reading and spelling scores as endpoints via the Boston Teachers Questionnaire (Leviton et al., 1993). Exposure was determined through dentine Pb measurements. A statistically significant inverse association of reading and spelling deficits with increasing dentine Pb concentrations was reported.

Fergusson et al. (1993) evaluated 1,200 Christchurch, New Zealand, children using the Burt Reading Test, the Progressive Achievement Test, teacher ratings, and impacts of dentine Pb levels. Dentine Pb values were gathered at 6–8 years, with a mean value of 6.2 μg/g. Dentine Pb was statistically inversely linked to deficits in all formal rating instruments and teacher ratings.

Min et al. (2009) examined children's IQ and academic achievement at 4, 9, and 11 years in a group of African-American children ($N = 278$) with reference to PbB levels at 4 years. The mean PbB was 7 μg/dl. A nonsignificant difference in slope with higher PbB ≤ 7 μg/dl was observed versus higher levels. Lower reading scores were observed at 9 and 11 years and lower math scores were evident at 11 years.

Populations of inner-urban children from several large cities have been studied with reference to the role of early lead exposure in later social conduct and malbehavior problems, i.e., social dysfunction such as delinquency and presence in the criminal justice system. Of particular note are results from assessments in children in Pittsburgh and Philadelphia, PA, and Cincinnati, OH. Denno (1990) noted that African-American residents of Philadelphia, followed from birth through 22 years of age, presented with elevated PbB in childhood as one of the most significant predictors of criminal behavior in later years. Prospectively studied children in the Pittsburgh Youth Study ($N = 503$) at high risk of delinquency (Needleman et al., 1996) were evaluated for antisocial behavior and delinquency at 7 and 11 years of age, using bone Pb levels in tibia at both ages as the exposure biomarker. They found that parental and teacher reports of delinquency and aggressive behavior were directly related to bone Pb concentrations.

A later study of Pittsburgh children ($N = 194$ delinquents versus 46 controls) looked at the relationship of tibial bone Pb levels to delinquency status for a group of juveniles in the Pittsburgh juvenile court system (Needleman et al., 2002). Delinquency cases had significantly higher tibial Pb content than did control children. Children with bone Pb values ≥ 25 μg/g had a fourfold higher likelihood of being adjudicated as delinquent. Dietrich et al. (2001) noted that children in the Cincinnati prospective Pb study examined for delinquency status at older ages versus PbB levels measured pre- and postnatally had similar outcomes. Delinquency was determined through parent- and self-reported antisocial and delinquent behaviors recorded at 16–17 years of age. Both levels in the highest prenatal Pb group, PbB > 10 μg/dl, and the highest 78-month postnatal results were reported to be linked to more delinquent acts versus lower concentrations. The Cincinnati cohort was further examined as young adults by Wright et al. (2008) showing similar results, i.e., prenatal and postnatal PbB are linked to higher rates of arrests or violence.

Of particular interest is the meta-analysis of Marcus et al. (2010), which examined 19 studies having 8,561 children and adolescents (3.5–18.4 years old) with reference to PbB and conduct problems. There was a consistent direct link between elevated PbB and increased conduct difficulties. Braun et al. (2008) evaluated NHANES 2001–2005 data for conduct disorder–PbB relationships. Children in the fourth quartile had almost two-fold higher (1.73) rates of conduct disorders. Children with PbB > 1.5 μg/dl had an almost ninefold (8.64) increased odds of a conduct disorder in the past year versus those with PbBs 0.2–0.7 μg/dl.

Selected prospective developmental neurotoxicological epidemiology efforts focused on injury to specific ability domains such as executive functions, attention and neuroanatomical imaging evidence of physical neurological injury are noted in Table 12.13. Tests of these responses versus Pb exposure histories were done across a relatively large age range, the generally positive data showing impairments consistent with persistence and perhaps irreversibility of neurotoxic harm sustained much earlier, and with ongoing exposures.

Children in the Rochester, NY, study were tested at 48 and 54 months for executive function and learning, using the Shape School Task. This group had relatively low PbB levels, mean = 6.5 μg/dl, and 80% were <10 μg/dl. PbB levels at 48 months were statistically significantly inversely linked with decrements in focused attention, color naming, and inhibition to automatic responding. Torréon, Mexico, first-grade schoolchildren ($N = 594$) exposed to a metal facility's lead emissions were tested using a battery for distractibility, memory sequencing, and other components and evaluated versus PbB (mean = 11.4 μg/dl at testing; Kordas et al., 2006). There was a significant inverse link of the Sternberg Memory Test with PbB.

Teenage children in the Cincinnati Lead Study were evaluated at 15–17 years of age on executive function measures: WISC Card Sorting Test; Rey-Osterrieth Complex Figure; other tests for attention, memory, and fine-motor coordination (Ris et al., 2004). Within this exposure group, 30% of PbB values ≥ 25 μg/dl were most strongly inversely associated with the attention component of the testings.

Those individuals who were in several cohorts of the prospective evaluations were subsequently evaluated as young adults. A subset of the Boston cohort was tested at 19–20 years old using tests of spatial memory, memory span, cognitive flexibility, and planning as the neurobehavioral endpoint and levels of dentine teeth Pb as exposure marker. They found that higher tooth Pb was statistically significantly associated with poorer scores on Focus-Execute and Shift of the test battery, consistent with impaired executive functions. A group of young adults in the Cincinnati Lead Study ($N = 48$) were evaluated as 20- to 23-year-olds using functional MRI (fMRI) as to language function, specifically using a verb generation/finger tapping regimen (Yuan et al., 2006). Higher mean PbB values were associated with reduced activation within Broca's area, a region linked to speech production in the left hemisphere independent of IQ results, and increased activation in an area of the right temporal lobe.

Relationships Among Lead Exposure, Neurobehavioral and Other Deficits, and fMRI/MRS Correlates in Lead-Exposed Children

Several studies within the Cincinnati prospective Pb cohort evaluated a subset of subjects at ages 20–23 years. Yuan et al. (2006) reported that Pb-exposed children showed correlations of impaired language function with activation in portions of the brain (Table 12.13). Cecil et al. (2008) examined 157 cohort members for association of neurological assessment and region-specific adult brain gray matter volume. Frontal gray matter was the most affected, with subareas associated with executive functions, mood control, and decision making.

Brubaker et al. (2010) reported relationship of adult gray matter volume loss and PbB levels at 1–6 years of age. Significant inverse associations between gray matter volume loss and yearly mean PbB levels at 3–6 years of age were noted. Brubaker et al. (2009) also used diffusion tensor imaging (DTI) to assess relationships of mean childhood PbB levels to changes in brain white matter integrity in 91 cohort adults. Mean childhood PbB values in this group were associated with decreased functional anisotropy throughout the white matter and other changes in certain regions of white matter. Multiple insults appeared to have occurred.

12.3 LEAD NEUROTOXICITY IN ADULTS

The neurotoxicological and neuromechanistic evidence for adverse effects in adult human populations is considerably less complete and more fragmentary than for children. This comparative status of adults versus children exists despite adult exposures, mainly in occupational settings, having been the first area of interest for health science investigators. The reasons are many and understandable. These are the objective comparative toxicological and exposure sensitivity criteria that are now widely known and accepted as favoring exposure prevention in children, especially in infancy and toddlers, and in fetal development. The genesis of differential toxicological sensitivity is not cleanly established. However, children as an anthropometric group are Pb-exposed more than adults on a same concentration basis.

Are there intrinsic distinctions in injury sensitivity at the cellular, subcellular, or biomolecular level or are the accepted distinctions quantitative expressions of varying *in vivo* toxicokinetics? Taken to the next step, are differences in toxicokinetics intrinsic or merely extrinsic via behavioral differences? This is an open question to some degree. Even among adults exposed to Pb occupationally, the analysis is complicated by the "healthy worker" effect, i.e., workers showing any toxicological responses to Pb above the norm (hyperresponsive) are more likely to change employment, while those genetically or otherwise predisposed to be hyporesponsive remain in that setting. This topic is addressed in a subsequent chapter on risk assessment issues.

Are the differences in toxicological sensitivity between young children and adults sufficiently great that there is a scientific basis for different regulatory approaches for lead exposure control? One factor muddling the discussion is a societal judgment. That is, children are appropriately held to be involuntary subjects of their exposures with little or no compensatory cost–benefit return, while an adult lead worker is assumed to choose toxicological risk in exchange for some economic gain, e.g., a higher wage versus other workers in the facility or even the general workforce. Female workers who are in their childbearing years largely render this line of reasoning moot.

An additional complication is the conceptual nature of the neurometric approaches to evaluating adults having Pb exposure, especially older adults, versus the young child. As noted in the 2006 U.S. EPA lead criteria document (page 78 of Chapter 6), the evaluative process in children is geared to assessing the lack of attainment of neurocognitive and behavioral abilities. In adults, one addresses loss of abilities that were once present in the person. A second factor is the existence of a level of brain "reserve capacity" which itself must be significantly diminished before observing overt, clinical neurotoxicity (Bleecker et al., 2002, 2003, 2005). Unlike the high frequency of very high exposures in the early decades of lead's public health history, we would expect a lower frequency in more recent times.

The notion of reserve capacity is not confined to one toxicant or target organ. For example, the renal toxicant cadmium imparts nephrotoxicity both directly in the form of proximal tubule dysfunction and through reducing the kidney's significant reserve capacity to deal with myriad other factors affecting kidney function as one ages (WHO, 1992).

The peripheral nervous system of adults, particularly those having relatively high Pb exposures in the workplace, has been a major focus for neurological and neurophysiological assessments. The studies on nerve conduction velocities associated with "wrist-drop" and "foot-drop" as to threshold and frequency of those effects are tabulated in the 1986 U.S. EPA lead criteria document, Chapter 12: Vol. IV. Overall, there appears to be a threshold for such effects near 30 μg/dl in workers. Some variability across studies has been seen in terms of nerves affected, tracing at least in part to limited sample size and Pb history variables such as bone Pb burdens. More recent peripheral nerve testing instruments implicate sensory nerve neuropathy (median nerve) at a PbB threshold value of 28–30 μg/dl, considerably below those levels long linked to motor impairments in high exposures, but within the threshold for reported nerve conduction velocity slowing.

More recent studies of adult Pb neurotoxicity have probed the CNS in both occupational and environmental exposures. Interestingly, a number of these studies report associations with PbB levels which fall near or below regulatory triggers for worker removal from exposure in the United States. For example, worker performances on psychomotor, speed, and dexterity tests show a decline onset threshold of 18 μg/dl (Schwartz et al., 2001). An even lower threshold of 12 units among workers was apparent in the neurological studies of Lucchini et al. (2000), indexing outcomes such as difficulty in concentrating and irritability. These findings have not been universal, however. A complicating factor is indexing body lead burden. Where attempts were made to employ cumulative measures versus current concentrations, Bleecker et al. (2005) determined positive associations with outcome measures.

Much larger numbers of adult subjects were assessed for environmental exposures and in some cases, e.g., the VA Normative Aging Study, subjects were followed over extended times. A critical upshot of these environmental population studies was the replication to a degree of what was seen with workers over extended times. That is, measures of exposure which are cumulative and long term, e.g., bone Pb levels, showed multiple associations with neurocognitive tests (Rhodes et al., 2003) while current PbB values were negative in their associations with neurotoxic outcomes in the U.S. NHANES III (Krieg et al., 2005).

It is not surprising that the NHANES III negative results for adult associations with current PbB are at odds with the VA Normative Aging Study where both bone and PbB measures were used. There are several reasons for this. These two groups reflect an integration of high body Pb burdens over their earlier years, integration not captured effectively in current PbB values alone (NHANES III) but more effectively in bone Pb values (Rhodes et al., 2003). It is also not surprising that Lanphear et al. (2000) found significant positive associations of schoolchildren's PbB with achievement tests of reading and mathematics but negative outcomes for neurocognitive tests in NHANES III older adults (Krieg et al., 2005). Children's PbB reflects current exposures and is the sensitive dose measure. Adult PbBs reflect mainly variable bone Pb accumulations.

12.4 EXPERIMENTAL ANIMAL MODELS OF LEAD NEUROTOXICITY IN HUMAN POPULATIONS

A sizeable body of experimental animal data has been documented for neurotoxic effects of Pb in the developing and developed nervous systems. These studies ran the gamut in terms of experimental quality, i.e., adequate dosing and endpoint testing quality control and assurance, adequate numbers of test animals, adequate numbers of time points. Of particular importance, experimental data also differed as to direct relevance to Pb neurotoxicity in humans. Many experimental systems have been described using subcellular and suborganismal systems under a variety of relatively artificial conditions in terms of the realities of lead environmental epidemiology described earlier in this chapter.

This portion of the chapter confines itself to studies using intact animals to define toxicological phenomenology, limiting use of biochemical and subcellular data mainly to probe potential molecular and toxicological mechanisms. Only certain test species are of interest for assessing lead neurotoxicity—nonhuman primates and to some extent laboratory rats. Nonhuman primate data are of particular relevance. Those reports where dosing regimens provide administered or internal doses that approximate typically encountered exposures indexed as PbB values linked to lower-level effects in more toxicologically sensitive human populations are also examined. Illustrative animal studies are summarized in Table 12.14.

The most extensive and long-maintained lead exposure models of human neurobehavioral effects of Pb were those designed and evolved by Rice et al. using the cynomolgus monkey (*Macaca fascicularis*) exposed under various Pb dosing regimens over periods lasting as long as 9 years or more and tested using neurobehavioral instruments of diverse complexity. Variables tested, besides outcome regimens, included a range of Pb doses and different dosing periods over the total experimental lifetime. Three categories of exposure schedules typified the most relevant Rice studies: (1) several low dosing rates; (2) single-level dosing at different times; and (3) repeated testings with single exposure. The results of these investigations were distributed across six key reports published over 8–9 years.

A number of the Rice reports dealt with reinforcement regimens and discrimination tasks presented to the monkeys in varied, complex ways. Rice (1984) reported results of fixed-interval (FI) reinforcement schedule testing of the cynomolgus monkeys at 3 years of age using controls and two levels of dosing: 50 and 100 μg/kg/day. In these animals, the corresponding PbB peaks were 2, 15, and 25 μg/dl; at steady state, these peaks declined to 3, 11, and 13 units. Median interresponse times decreased and numbers of responses increased with increasing PbB. Differential reinforcement of low rate (DRL) assessment at 3 years of age under the same dosing regimen produced the finding that PbB increase impaired the ability of the animals to

TABLE 12.14 Selected Experimental Animal Model Data in Support of Lead Neuroepidemiological Findings

Test Group	Dosing/ Outcome Design	PbB (μg/dl)	Results	References
Nonhuman primates				
Cynomolgus monkeys (*Macaca fascicularis*)	0, 50, 100 μg/kg/ day Pb/fixed interval (FI) schedule of reinforcement, 3 years	PbB peak: 2, 15, 25 μg/dl; PbB steady state: 3, 11, 13	With increased PbB, median interresponse times of exposed monkeys decreased; number of responses increased	Rice (1984)
Cynomolgus monkeys (*Macaca fascicularis*)	Same dosings/ differential reinforcement of low rate (DRL), 3 years	PbB peak: 2, 15, 25 μg/dl; PbB steady state: 3, 11, 13	Increased PbB in Pb-exposed monkeys impaired ability to learn inhibition of response	Rice and Gilbert (1985)
Cynomolgus monkeys (*Macaca fascicularis*)	Same dosings/ several discrimination tasks at 3 years: nonspatial form, nonspatial color discrimination with irrelevant form cues, and nonspatial form discrimination with irrelevant color cues	PbB peak: 2, 15, 25 μg/dl; PbB steady state: 3, 11, 13	Dose-related impairment in performance, LOAEL for combined errors in reversals was the peak PbB 15 μg/dl and concurrent PbB of 11 μg/dl	Rice (1985)
Cynomolgus monkeys (*Macaca fascicularis*)	Same dosings/ spatial discrimination tasks at 9–10 years, with and without use of irrelevant cues	PbB peak: 2, 15, 25 μg/dl; PbB steady state: 3, 11, 13	Pb exposure produced impaired performance on tasks only in the presence of irrelevant cues	Gilbert and Rice (1987)

(*Continued*)

TABLE 12.14 Selected Experimental Animal Model Data in Support of Lead Neuroepidemiological Findings—(cont.)

Test Group	Dosing/ Outcome Design	PbB (μg/dl)	Results	References
Cynomolgus monkeys (*Macaca fascicularis*)	1.5 mg Pb/kg/ day, four groups (*N* = 13 each): exposed from birth, infancy exposure only up to 400 days; exposed after infancy (post 300 days). Tested using FI and fixed ratio (FR) schedules: 3 and 7–8 years	PbB, 32–36 μg/dl preweaning with Pb dosing; 19–26 μg/dl postweaning	At older age, Pb exposure produced increased response rates for all three dosing protocols	Rice (1992a)
Cynomolgus monkeys (*Macaca fascicularis*)	Same dosing/ nonspatial discrimination reversal tasks, 5–6 years	PbB, 32–36 μg/dl preweaning with Pb dosing; 19–26 μg/dl postweaning	Continuous and postinfancy exposures caused impairments in all tasks with continuous exposure producing most effect	Rice and Gilbert (1990c)
Cynomolgus monkeys (*Macaca fascicularis*)	Same dosing/ spatial discrimination reversal, 7–8 years with and without irrelevant cues	PbB, 32–36 μg/dl preweaning with Pb dosing; 19–26 μg/dl postweaning	Continuous exposure impaired responses in the absence of irrelevant cues; in the presence of cues, all three exposure groups similarly affected	Rice and Gilbert (1990b)
Cynomolgus monkeys (*Macaca fascicularis*)	Same dosing/ delayed spatial alteration tasks, 6–7 years	PbB, 32–36 μg/dl preweaning with Pb dosing; 19–26 μg/dl postweaning	All three variable exposure period groups similarly affected	Rice and Gilbert (1990a)

(*Continued*)

TABLE 12.14 Selected Experimental Animal Model Data in Support of Lead Neuroepidemiological Findings—(cont.)

Test Group	Dosing/ Outcome Design	PbB (μg/dl)	Results	References
Cynomolgus monkeys (*Macaca fascicularis*)	Single dose, 2 mg Pb/kg/day, from birth; testing at 60 days, 2.5–3 years, and 7–7.5 years; $N = 6$ each group. Various spatial and nonspatial discrimination tasks	PbB peak, 115 μg/dl, decline to steady state, 33 μg/dl	Juveniles (2.5–3 years) showed impairments in the presence of irrelevant form cues. Exposed adults were impaired on DRL and visual discrimination tasks	Rice (1992b)
Cynomolgus monkeys (*Macaca fascicularis*)	Same dosing/ concurrent discrimination performance tasks, 8–9 years	PbB, 32–36 μg/dl preweaning with Pb dosing; 19–26 μg/dl postweaning	All three variable exposure period groups showed decreased performance with continuous exposure producing most effect	Rice (1992c)
Female rhesus monkeys (*Macaca mulatta*) $N = 48$	Dosing to achieve a predetermined PbB	PbB: 35–40 μg/dl by 12 weeks	Pb exposure produced more fear, agitation, and cage side climbing than controls	Lasky and Laughlin (2001)
Rhesus monkeys, 4 years old	Chronic 0.7 mg/kg/day dosing + 2 pulses, year 1; exploration and activity behavior	PbB, 5 weeks: 55 μg/dl; 36 μg/dl for year 1	Exposed monkeys took longer to enter open area, had increased duration of exploring surroundings, and failed to habituate	Ferguson and Bowman (1990)

(*Continued*)

TABLE 12.14 Selected Experimental Animal Model Data in Support of Lead Neuroepidemiological Findings—(cont.)

Test Group	Dosing/ Outcome Design	PbB (μg/dl)	Results	References
Rhesus monkeys, 4 years old	Same dosing/ tested for muscle tone, agitation, other sensorimotor measures at 14 days, 6 weeks, 2 months	PbB, 5 weeks: 55 μg/dl; 36 μg/dl for year 1	At 6 weeks of exposure, lower muscle tone and greater agitation; at 2 months, decreased visual attentiveness in visual exploration tasks	Levin et al. (1988)
Laboratory rat strains				
Male weanling Long–Evans hooded rats, $N = 12$/group	Drinking water Pb at four doses: 25, 50, 100, 500 ppm/FI schedule tasks, food reinforcement, 1-minute intervals	25 ppm Pb, 15–20 μg/dl; 50 ppm Pb, 20–30 μg/dl	Dose-dependent increase in response rate, LOAEL 25 ppm Pb	Cory-Slechta et al. (1985)
Male weanling Sprague–Dawley rats	Drinking water Pb at three doses: 50, 300, 1,000 PbOAc; 30-second-FI schedule tasks	50 ppm Pb, <10 μg/dl	Significant increase in FI response rates for the 50 ppm group	Cory-Slechta and Thompson (1979)
Male weanling Long–Evans rats	Drinking water Pb at 150 ppm; fixed-ratio schedules	PbB, 26 μg/dl	Increased response rates and shorter waiting times with Pb exposure. Effects reversed by quinpirole	Brockel and Cory-Slechta (1998)
Weanling Long–Evans rats	Drinking water Pb, 50 ppm for 8–11 months; FI performance tests for 1-minute schedule	PbB, ~20 μg/dl	Exposure decreased the FI response rates	Cory-Slechta (1990)

(*Continued*)

TABLE 12.14 Selected Experimental Animal Model Data in Support of Lead Neuroepidemiological Findings—(cont.)

Test Group	Dosing/ Outcome Design	PbB (μg/dl)	Results	References
Variable age Fischer 344 rats	2 or 10 mg/kg/ day for different periods; FI and variable interval (VI) performance tested at 21 days, 8 months, 16 months	Steady-state PbB = 13–18 μg/dl	21-day and 16-month-old rats showed increased VI and FI response rates, adults decreased rates; FI seen at the 2 mg dosing	Cory-Slechta and Pokora (1991)
Other test animal species				
Golden hamsters	100 ppm $Pb(OAc)_2$ exposed day 8 gestation to 42 days of age; tests of aggressive behavior	Administered 10–15 μg Pb/ day	Exposed show slower growth rate and, by 5 days of age, increase in aggression	Delville (1999)
Male Dutch Belted rabbits	Paternal $Pb(OAc)_2$ exposure; exploratory behavior tests	20, 40, and 80 μg/dl	At 25 days postnatal, impaired exploratory behavior	Nelson et al. (1997)

learn inhibition of responses (Rice and Gilbert, 1985). Also at 3 years of age, several discrimination reversal tasks (nonspatial form discrimination, nonspatial color discrimination with irrelevant form cues, nonspatial form discrimination with irrelevant color cues) indicated dose-dependent impairments in performance, measured as combined errors, had a lowest observed adverse effect level (LOAEL) of 15 μg/dl at peak and the concurrent PbB of 11 μg/dl (Rice, 1985). Retesting the monkeys at 9–10 years of age under the same exposure and using spatial discrimination tasks with or without irrelevant cues produced impaired performance only in the presence of irrelevant cues (Gilbert and Rice, 1987).

A number of reports from the Rice group (Rice, 1990, 1992a,b; Rice and Gilbert, 1990a,b,c) described results from monkeys variably evaluated at different ages for a dosing level of 1.5 mg Pb/kg/day during one of three different intervals: dosing from birth; exposure only in infancy, up to 400 days; and only postinfancy, after 300 days. Two periods of PbB level measurement were reported: preweaning and postweaning. Use of FI and fixed-ratio (FR) reinforcement schedules at 3 and 7–8 years of age showed no Pb effect at 3 years for the FR tasks for any exposure time interval, while the FI schedule at 7–8 years was associated with increased response rates regardless of dosing schedule (Rice, 1992c). Nonspatial discrimination reversal tasks under these dosing conditions at 5–6 years of age indicated that the continuous and post-infancy times for Pb exposure caused impairments in all tasks, with the most robust effect linked to continuous exposure (Rice and Gilbert, 1990c).

Spatial discrimination reversal testing at 7–8 years with or without irrelevant cues showed impaired responses in the absence of cues at continuous exposure, and all three forms of exposure impaired responses in the presence of such cues (Rice and Gilbert, 1990b). Rice and Gilbert (1990a) found that all three exposure intervals impaired response at 6–7 years using the delayed spatial alteration tasks. Similarly, concurrent discrimination tasks at 8–9 years were affected with all three dosing regimens but continuous exposure produced the greatest impairment (Rice, 1992c).

The fixed dose-repeat testing regimen as reported by Rice (1992b) employed a dosing level of 2 mg/kg/day, administered from birth and using various spatial and nonspatial discrimination tasks with and without irrelevant cues, and tested at 60 days, 2.5–3 years, and 7–7.5 years. Steady-state PbB was 33 μg/dl, following a peak of 115 μg/dl. Juvenile animals (2.5–3 years of age) showed impairments in the presence of irrelevant form cues, while adults were impaired on DRL and visual discrimination tasks.

The Rhesus monkey (*Macaca mulatta*) has been employed as an experimental animal model in other investigations. In contrast to the Rice and coworker studies, there was less attention to preserving low PbB values during the task evaluations, and in those cases, PbB levels ranged from 35 to 40 μg/dl. Time frames for dosing regimens were significantly shorter.

Lasky and Laughlin (2001) reported that female monkeys orally dosed with boluses of Pb solution from 8 days forward, to a steady-state value of 35–40 μg/dl at a second stage of an ongoing study, showed disturbances in spatial exploration behavior during 24 weeks of exposure, showing more fear, agitation, and escape climbing behaviors. Exploratory and behavior testing results at higher PbB concentrations were also reported. Levin et al. (1988) tested Rhesus monkeys for muscle tone, agitation, and other sensorimotor measures at three times: 14 days, 6 weeks, and 2 months. PbB steady state for the entire year was 36 μg/dl, declining from 55 units at 5 weeks. By 6 weeks, there was lowered muscle tone and greater agitation. By 2 months, there was decreased visual attentiveness in visual exploration tasks. At 4 years of age, Rhesus monkeys exposed to earlier doses of Pb at 0.7 mg Pb/kg/day in two pulses over the first year (and producing the same PbB values stated above) showed more reluctance and took longer to enter open areas, took increased time to explore surroundings and failed to habituate (Ferguson and Bowman, 1990).

The second most frequently employed experimental animal model of Pb developmental neurotoxicity in humans is the rat, using three strains: Long–Evans, Sprague–Dawley, and Fischer 344 under different dosing and testing protocols. As for the nonhuman primate, studies noted here are for results associated with PbB levels approximating the low values described in the neuroepidemiology of childhood Pb exposure and established behavioral test measures, using administration relevant to human intakes, i.e., oral Pb dosings. The majority of studies within these criteria were reported by Cory-Slechta et al.

Male weanling Long–Evans rats orally dosed (water) at 25, 50, 100, or 500 ppm Pb as acetate were tested using FI schedule tasks and food reinforcement, 1-minute intervals (Cory-Slechta et al., 1985). There were dose-dependent increases in response rate, with a LOAEL of 25 ppm as acetate (14.3 ppm as elemental Pb). This administration rate provided PbB values approaching those low values seen in young children for neurotoxicological effects, 15–20 μg/dl. A similar protocol with water exposures of weanling Sprague–Dawley rats at 50, 300, and 1,000 ppm as the acetate and 30-second intervals showed significant increases in FI response rates at the lowest exposure, 50 ppm as acetate (Cory-Slechta and Thompson, 1979). This dosing level translated to < 10 μg/dl. Exposure of weanling Long–Evans rats at 150 ppm using an FR schedule produced increased response rates and shorter waiting times with exposure (Brockel and Cory-Slechta, 1998).

Animal age at time of testing after long-term Pb intakes from water was a significant factor in responses. This was seen with both the Long–Evans and Fischer 344 strains. When the former were exposed to 50 ppm Pb as acetate and tested after 8–11 months of exposure using FI/1-minute performance testing, exposure decreased the FI response rate (Cory-Slechta, 1990), compared to increase in young animals (Brockel and Cory-Slechta, 1998). Variably aged Fischer 344 rats exposed at 2 or 10 mg/kg/day Pb and tested at 21 days, 8 months, or 16 months showed age-dependent responses in FI and Variable-Interval (VI) performance tests. The PbB steady-state level was 13–18 μg/dl for the exposed groups. The two younger groups showed increased response rates, while the oldest group showed reduced rates.

12.5 GENERAL AND NEURAL-SPECIFIC MECHANISMS OF Pb TOXICITY IN THE HUMAN NERVOUS SYSTEM

The mechanisms of Pb developmental neurotoxicity in particular and multi-system toxicity in general are likely multiple. To date, no global, single mechanism explaining the element's toxicity, and particularly its neurotoxicity, has emerged. Rather, organ- and system-specific effects appear to be linked to specific mechanisms. In the case of neurotoxic effects, and all other effects in general, studies have typically focused on processes at the subcellular and biomolecular level to explain whole-animal results. The focus of this section is on those mechanistic models involving exposures of intact animals. Some studies employed *in vitro* approaches, but in those cases there is the relatively intractable question of relevance to *in vivo* toxic activity.

Earlier efforts at identifying toxicological mechanisms of Pb's effects focused on general roles of the element in interfering with essential or other processes by subcellular and molecular interactions. As noted in the 1986 U.S. EPA lead criteria compendium (Vol. IV), such interactions would include inhibiting Ca–Mg ion enzymes at membrane sites. Somewhat later, interest focused on calcium-specific interference at the level of classes of proteins (e.g., protein kinase C) (Markovac and Goldstein, 1988). More recent efforts homed in on neurochemical and related processes that are site-congruent with sites of toxic response, specifically effects on neurons, in the hippocampus, sensory-specific areas such as retinal rod cells and epigenetic changes affecting, for example, the aging nervous system.

Table 12.15 presents illustrative results of studies directed at multiple organ- and tissue-specific sites. Table 12.16 is a more focused presentation of data for neurotoxic mechanisms. Table 12.15 depicts six categories of toxic mechanisms generally determined to be operating for Pb toxicity across organs and systems. They are as follows: (1) altered ion homeostasis and function; (2) oxidative stress; (3) inflammatory responses; (4) genotoxicity and apoptosis; (5) endocrine disruption; and (6) induced Pb complex formation. Details of their expression are given in Tables 12.12–12.15. Not all categories are of comparable scope *in vivo* or comparable significance in impact.

Neural process categories evaluated within this approach included long-term potentiation of plasticity, neurotransmission effects, apoptosis, and evoked potentials as a correlate of sensory functions, and epigenetics, particularly neural integrity of the aging brain. A particular focus has been on long-term potential (LTP) activity linked with plasticity in neural tissue. Lasley and Gilbert (2002) investigated rat neurons from Pb-exposed animals for *in vivo* neurotransmission effects by using intracerebral dialysis in animals exposed to Pb sufficient to generate PbB levels of 20–40 μg/dl to assess Ca-independent neurotransmitter releases. Pb had a dual effect, increasing or reducing stimulated transmitter release of glutamate and GABA depending on exposure level. Further use of intracerebral dialysis by these authors revealed Ca–Pb interactions via inhibitory effects on influx through voltage Ca^{2+} and Ca^{2+}-mimicking effects of Pb.

Mechanisms that reside in the hippocampal region are of particular interest and have been studied by various authors *in vivo*, inasmuch as this brain area and its functions have long been linked with lead's neurobehavioral toxicity in humans and animals. A number of investigations addressed the matter of long-term potentiation of neural plasticity (LTP). Nihei et al. (2000) studied *in vivo* Pb effects on NMDA receptor expression and function in rats dosed to produce PbB values 25–32 μg/dl. Exposed rats showed decreased

TABLE 12.15 Illustrative Multisystem Toxicological Mechanisms for Lead Effects Relevant to CNS of Humans and Animal Models of Lead Toxicity: *In Vitro* and *In Vivo* Studies

Mechanistic Category	Mechanistic Pathway(s)	Results	Illustrative References
Altered ion homeostasis and functions	Ca^{2+} and other ions disrupted in various functions, e.g., essential metal ion transport, signal transduction	Pb induces disrupted homeostasis in various cells and cell preparations	Pentyala et al. (2010), Markovac and Goldstein (1988)
Oxidative stress	Excess formation of reactive oxygen species (ROS): superoxide; hydroxyl radicals; hydrogen peroxide	Reduced production and disturbances in NO signaling by ROS; upregulation of NO synthase; disturbance of enzymes mediating antioxidant processes; increased activation of NAD(P)H oxidase; membrane/lipid peroxidation	Dogru et al. (2008), Quintanar-Escorza et al. (2007)
Endocrine disruption	Disturbances through the hypothalamic-pituitary-gonadal (HPG) and hypothalamic-pituitary-thyroid (HPT) axes	Pb produces reductions in FSH, LH, testosterone, estradiol via HPG axis or steroidogenic protein, altered number of estrogen receptors; there is a negative association, PbB versus T_4 and free T_4, via the HPT axis	Srivastava et al. (2004), Biswas and Ghosh (2006), Dundar et al. (2006)
Inflammatory responses	Immunotoxic mechanisms: production and modulation of proinflammatory cytokines with ROS overproduction	Enhanced inflammation and tissue damage where immune cell mobilization occurs; Pb in rat brain produces neutrophil-derived inflammation and apoptosis	Chetty et al. (2005), Kibayashi et al. (2010)

(*Continued*)

TABLE 12.15 Illustrative Multisystem Toxicological Mechanisms for Lead Effects Relevant to CNS of Humans and Animal Models of Lead Toxicity: *In Vitro* and *In Vivo* Studies—(cont.)

Mechanistic Category	Mechanistic Pathway(s)	Results	Illustrative References
Genotoxicity and apoptosis	DNA damage, mutagenicity, clastogenicity, epigenetic effects	Pb workers at high exposures show DNA strand breaks, inconsistent reports; most studies show no Pb mutagenicity; observed clastogenic effects for workers confounded by multiexposures	Palus et al. (2003), Van Larabeke et al. (2004), Wang et al. (2008), Ustundag and Dudyu (2007)
Induced Pb complex formation	Pb binding to various sites on proteins: cytosol, erythrocyte, intranuclear	Net effect on cellular processes unclear; some effects adverse, some protective	Navarro-Moreno et al. (2009), Goyer and Rhyne (1973), Shelton et al. (1990)

hippocampal NR1 subunit mRNA and protein in animals having deficits in LTP and spatial learning. LTP was evaluated in rat hippocampal regions in terms of *in vivo* effects in animals dosed to produce 27−62 μg/dl and exposed from gestation day 16 to 120−180 days old (Gilbert et al., 1999) or exposed at a much lower level, 16 μg/dl from gestation day 0 through 90−130 days old (Gutowski et al., 1998). In the former regimen, Pb produced an elevated induction threshold and diminished magnitude in the effect in the dentate gyrus while causing diminished magnitude of effect in the latter in the CA1 subregion.

Low-frequency electrical stimulation in this rat mechanistic model in the CA1 and dentate gyrus subregions and where rat PbB values averaged 30 μg/dl showed exposed animals had decreased magnitude of LTP response (Zhao et al., 1999). Lasley and Gilbert (2002) reported that *in vivo* evoked glutamate releases versus LTP in rats having PbB 27−40 μg/dl produced a biphasic response, with animals having lower PbB showing decrease in stimulated release and little release response at values 62−117 μg/dl.

Fox et al. (2008) reported that in rat retinal rod cells, apoptosis accompanied high Pb exposures *in vivo* while low exposures induced cell proliferation. There were also decreases in Na^+ and K^+-ATPase activity. Wu et al. (2008) probed epigenetic changes in the brains of aging monkeys by using the epigenetic endpoints DNA methyltransferase and mRNA expression of β-amyloid precursor protein, transcription factor Sp1. Infancy exposures

TABLE 12.16 **Illustrative Potential *In Vivo* Neurotoxicologic Mechanisms Underlying the Developmental Neuroepidemiology of Lead in Humans**

Biological Site	Core Process	Exposure/Endpoint Design	Results	References
Rat neurons	*In vivo* neurotransmission effects via changes in transmitter release	PbB exposed animals = 20–40 μg/dl; Ca-independent neurotransmitter release using intracerebral dialysis	Biphasic: Pb exposure reduces or increases stimulated transmitter release of glutamate and GABA depending on PbB level	Lasley and Gilbert (2002)
Rat neurons	*In vivo* Ca^{2+}–Pb^{2+} interactions via effects on Ca^{2+} influx through voltage Ca^{2+} and Ca^{2+}-mimicking effects of Pb	PbB exposed animals = 20–40 μg/dl; Ca–Pb interactions via intracerebral dialysis	Pb exposures inhibited Ca^{2+} influx via Ca mimicking Pb and via voltage-sensitive channels	Lasley and Gilbert (2002)
Rat hippocampus	*In vivo* effects on NMDA receptor expression and function	PbB exposed animals = 25–32 μg/dl	Exposure decreased hippocampal NR1 subunit mRNA and protein in animals with LTP and spatial learning deficits	Nihei et al. (2000)
Rat hippocampal subregions	*In vivo* LTP/synaptic plasticity	27–62 μg/dl/Dentate gyrus LTP, gestation day 16 to 120–180 days old	Pb-induced elevated induction threshold and diminished magnitude	Gilbert et al. (1999)
Rat hippocampal subregions	*In vivo* LTP/synaptic plasticity	16 μg/dl/hippocampal CA1, gestation day 0 through 90–130 days of age	Pb diminished magnitude	Gutowski et al. (1998)

Rat hippocampal subregions	*In vivo* long-term depression (LTD) synaptic plasticity	Chronic exposure PbB = 30 μg/dl; low-frequency electrical stimulation in CAI and dentate gyrus subregions	Pb exposure decreased the magnitude of LTD form of plasticity	Zhao et al. (1999)
Rat hippocampal subregions	*In vivo* evoked glutamate releases versus LTP	Chronic exposure, PbB = 27–40 μg/dl dose versus stimulated glutamate release	Pb exposure produces biphasic changes in stimulated release; decrease, 27–40 μg/dl, no decrease at 62–117 μg/dl	Lasley and Gilbert (2002)
Rat retinal rod cells	Apoptosis	Variable Pb exposure/ evaluation of apoptosis in the Pb range and changes in retinal Na^+, K^+-ATPase activity	High Pb *in vivo* exposure produces apoptosis; low level causes proliferation. *In vivo* exposures decrease Na^+, K^+-ATPase activity	Fox et al. (1991, 2008)
Aged monkeys tested for DNA methyltransferase, mRNA expression of β-amyloid precursor protein, transcription factor Sp1	Epigenetic changes in aging brains	Infancy PbB levels reported elsewhere	Exposure in infancy increased mRNA expression, β-amyloid precursor protein, transcription factor Sp1, DNA methyltransferase	Wu et al. (2008)

(values noted earlier) increased mRNA expression, β-amyloid precursor protein transcription factor, and DNA methyl transferase.

12.6 SOME PERSISTING AND NEW DIRECTIONS FOR LEAD NEUROTOXICITY IN HUMAN POPULATIONS

A number of neurotoxicity and related issues for human Pb exposure remain, along with newly evolving questions. These are highlighted in Table 12.17.

Among the persisting matters are a series of framing questions: (1) Are there thresholds to Pb toxicity? (2) Are there peak periods to human toxic vulnerability to Pb? (3) Are there ideal measures of internal Pb exposure? (4) Are current regulatory definitions of and practices for workplace Pb exposures of adults adequate, given new data? (5) Are there neurotoxic effect modifiers as well as Pb exposure modifiers? (6) Are neurocognitive and neurobehavioral effects of Pb in humans persistent and even irreversible? And, (7) are there tardive, late life sequelae of developmental Pb exposures in humans?

No threshold to neurocognitive and other Pb-induced neurotoxic impairment in young children has been empirically demonstrated. A number of studies noted in this chapter show persisting dose–neurotoxic response relationships to well below the USCDC level of concern (10 μg/dl), to 2–3 μg/dl. If a threshold exists, it is within ~2–3 μg/dl of zero. This dose–response persistence raises a number of questions, not the least of which is how to effect adequate regulation. Nonthreshold toxics are typically regulated in a prescribed fashion while Pb is regulated as though it were a noncarcinogenic substance having a threshold for effects.

The matter of peak periods for neurotoxic vulnerability can be addressed by assuming much of early childhood is a high vulnerability/high toxic risk period, with one or more parts of this age band being somewhat more sensitive to Pb exposure. Large, pooled analyses indicate that no childhood Pb exposure period is free of elevated poisoning risks.

The question of an ideal dose measure has several dimensions. One is that of epidemiological adequacy. The other is the reliability of a readily available dose metric for translation to target-organ levels, e.g., brain Pb concentrations. The general view is that PbB is an acceptable Pb exposure surrogate for nervous system Pb though some information indicates relative lags in levels under defined conditions in experimental animals (Stangle et al., 2004).

The question of the validity of adherence to bifurcated dose–toxic response relationships remains: one for nonoccupational adult and childhood Pb exposures and one for workers exposed to Pb. Evidence continues to accumulate for persisting neurotoxic and other adverse outcomes in adults at workplace Pb exposure levels considerably below regulatory levels now in place for removal and return.

TABLE 12.17 Some Persisting and New Directions for Lead Neurotoxicity in Human Populations[a]

What Is the Question?	Current Evidence	Implications for Public Health
Does Pb developmental toxicity have a threshold to its effects?	Developmental neurotoxic effects persist well below Pb exposure threshold (as PbB) of 10 μg/dl in a number of studies	Pb exposure appears to produce neurotoxic risk in children without an empirically observed threshold; any threshold would be between 0 and 2–3 μg/dl PbB, approaching practical quantitation limits; environmental Pb is still regulated as a threshold toxicant rather than as a categorical nonthreshold toxic, for example, genotoxic carcinogens
Are there peak periods of neurotoxic vulnerability?	Current evidence is mixed for prenatal Pb effects, but fetal Pb effects can be potentially significant in certain population settings; evidence from larger studies and grouped analyses indicate, on balance, that neurotoxicity begins early in postnatal life and continues in childhood	Periods of peak sensitivity likely exist across childhood but detecting those depends on the nature of the child population and its statistical modeling. Concurrent PbB beyond 2–3 years is strongly linked to neurocognitive decrements and other neurotoxic endpoints, indicating no period in early childhood is refractory to neurotoxic insult. Where high exposures persist into older ages, for example, in smelter areas, continuing (concurrency) elevated PbB produces particular harm
Are there ideal measures of Pb exposure?	Evidence with Question 2 indicates concurrent PbB and PbB response remain useful for dose–neurotoxic relationships; evidence from studies with adult neurotoxicity outcomes indicate preferential use of serial PbB or bone Pb measures	Screening older children as well as infants and toddlers appears advisable since absence of screening for older children allows continuing toxic harm to go undetected

(Continued)

TABLE 12.17 Some Persisting and New Directions for Lead Neurotoxicity in Human Populations[a]—(cont.)

What Is the Question?	Current Evidence	Implications for Public Health
Is there still justification for higher permissible PbB levels as lead worker regulatory limits than for nonlead worker adults or even children?	Evidence in various chapters shows increasing toxic Pb effects among adults at lower and lower PbB levels, e.g., adult cardiovascular and nephrotoxic dose–responses for Pb	Retaining increasingly obsolete and decreasingly protective workplace Pb limits (PbB removal and return practices) is increasingly less defensible
Are there modifiers of neurotoxic effects in high-risk children analogous to modifiers of Pb exposure?	Current evidence identifies a number of effect modifiers in children: social stress and environmental enrichment; gender differences; race as a factor may operate through higher exposures in U.S. African-American children. Similarly, socioeconomic status affects both race/ethnicity outcomes and relative Pb exposures; studies adjusting for exposure differences report effect modification by SES	Subsets of human populations, particularly high-risk children, would appear to be at higher risk by virtue of factors that influence the neurobehavioral measure and the exposure level
Are developmental neurotoxic effects of Pb persistent and reversible?	Current evidence indicates persistence and likely irreversibility of Pb's early neurotoxic effects, including tardive, late life outcomes from developmental Pb exposures	Requires primary prevention of Pb exposure in young children as the only viable health and regulatory policy

[a] *This chapter and others.*

Current evidence noted in this chapter shows there are modifiers of neurotoxic effect as well as of Pb exposure *per se*. Such modifiers, collectively, define an added contribution to higher risk in children. See, for example, the data of Kim et al. (2009) for Korean school children exposed to both lead and manganese.

Current evidence also suggests that developmental neurotoxic effects of Pb exposure when expressed at peak periods of vulnerability, whether empirically discernible or not, persist over long periods, while the findings of neurodegenerative disease development in nonhuman primates traceable to early life developmental period Pb exposure suggest irreversibility (Rogan et al., 2001; Rosen and Mushak, 2001; Wu et al., 2008; Table 12.6). Wu et al. (2008) showed aged monkeys with persisting epigenetic changes in their aging brains traced to developmental Pb exposures.

The ability of various epidemiological methods to tease out the persistence of developmental neurotoxic effects in humans will arguably depend on the Pb exposure and demographic/socioeconomic profiles of study cohorts. Differences seen across studies in terms of identified persisting effects do not rule out effects being present (see Mushak, 1993, for differences in outcomes across studies).

The principal consequence of persisting and irreversible early neurotoxic effects in humans is that the outcome requires primary prevention of Pb exposure as the reigning health and regulatory policy.

REFERENCES

American Academy of Pediatrics, 1969. Prevention, diagnosis, and treatment of lead poisoning in childhood. Pediatrics 44, 291–298.

Baghurst, P.A., McMichael, A.J., Wigg, N.R., Vimpani, G.V., Robertson, E.F., Roberts, R.R., et al., 1992. Environmental exposure to lead and children's intelligence at the age of seven years: The Port Pirie Study. N. Engl. J. Med. 327, 1279–1284.

Bellinger, D., Leviton, A., Waternaux, C., Needleman, H., Rabinowitz, M., 1987. Longitudinal analysis of prenatal and postnatal lead exposure and early cognitive development. N. Engl. J. Med. 316, 1037–1043.

Bellinger, D., Sloman, J., Leviton, A., Rabinowitz, M., Needleman, H.L., Waternaux, C., 1991. Low-level lead exposure and children's cognitive function in the preschool years. Pediatrics 87, 219–227.

Bellinger, D., Hu, H., Titlebaum, L., Needleman, H.L., 1994. Attention correlates of dentin and bone lead levels in adolescents. Arch. Environ. Health 49, 98–105.

Bellinger, D.C., Needleman, H.L., 2003. Intellectual impairment and blood lead levels [letter]. N. Engl. J. Med. 349, 500.

Bellinger, D.C., Stiles, K.M., Needleman, H.L., 1992. Low-level lead exposure, intelligence, and academic achievement: a long-term follow-up study. Pediatrics 90, 855–861.

Biswas, N.M., Ghosh, P.K., 2006. Protection of adrenal and male gonadal functions by androgen in lead-treated rats. Kathmandu Univ. Med. J. 4, 218–221.

Bleecker, M.L., Lindgren, K.N., Ford, D.P., Tiburzi, M.J., 2002. The interaction of education and cumulative lead exposure on the mini-mental state examination. J. Occup. Environ. Med. 44, 574–578.

Bleecker, M.L., Lindgren, K.N., Scheetz, K., Tiburzi, M.J., 2003. Associations of chronic and current measures of lead exposure with different components of brainstem auditory evoked potentials. Neurotoxicology 24, 625–631.

Bleecker, M.L., Ford, D.P., Lindgren, K.N., Hoese, V.M., Walsh, K.S., Vaughan, C.G., 2005. Differential effects of lead exposure on components of verbal memory. Occup. Environ. Med. 62, 181–187.

Braun, J.M., Froelich, T.E., Daniels, J.L., Dietrich, K.N., Hornung, R., Auinger, P., et al., 2008. Association of environmental toxicants and conduct disorder in U.S. children: NHANES 2001–2004. Environ. Health Perspect. 116, 956–962.

Brubaker, C.J., Schmithorst, V.J., Haynes, E.N., Dietrich, K.N., Cecil, K.M., 2009. Altered myelination and axonal integrity in adults with childhood Pb exposure: a diffusion tensor imaging study. Neurotoxicology 30, 867–875.

Brubaker, C.J., Dietrich, K.N., Lanphear, B.P., Cecil, K.M., 2010. The influence of age of Pb exposure on adult gray matter volume. Neurotoxicology 31, 259–266.

Brockel, B.J., Cory-Slechta, D.A., 1998. Lead, attention, and impulsive behavior: changes in a fixed-ratio waiting-for-reward paradigm. Pharmacol. Biochem. Behav. 60, 545–552.

Byers, R.K., 1959. Lead poisoning. Review of the literature and report on 45 cases. Pediatrics 23, 585–603.

Byers, R.K., Lord, E.E., 1943. Late effects of lead poisoning on mental development. Am. J. Dis. Child. 66, 471–494.

Canfield, R.L., Henderson Jr., C.R., Cory-Slechta, D.A., Jusko, T.A., Lanphear, B.P., 2003. Intellectual impairment in children with blood lead concentrations below 10 μg per deciliter. N. Engl. J. Med. 348, 1517–1526.

Cecil, K.M., Brubaker, C.J., Adler, C.A., Dietrich, K.N., Altaye, M., Egelhoff, J.C., et al., 2008. Decreased brain volume in adults with childhood Pb exposure. PLoS Med. 5 (5), e212.

Chen, A., Dietrich, K.N., Ware, J.H., Radcliffe, J., Rogan, W.J., 2005. IQ and blood lead from 2 to 7 years of age: are the effects in older children the residual of high blood lead concentrations in 2-year-olds? Environ. Health Perspect. 113, 597–601.

Chetty, C.S., Vermuri, M.C., Campbell, K., Suresh, C., 2005. Lead-induced cell death of human neuroblastoma cells involves GSH deprivation. Cell Mol. Biol. Lett. 10, 413–423.

Chiodo, L.M., Jacobson, S.W., Jacobson, J.L., 2004. Neurodevelopmental effects of postnatal lead exposure at very low levels. Neurotoxicol. Teratol. 26, 359–371.

Chisolm Jr., J.J., 1962. Aminoaciduria as a manifestation of renal tubular injury in lead intoxication and a comparison with patterns of aminoaciduria seen in other diseases. J. Pediatr. 60, 1–17.

Chisolm Jr., J.J., 1968. The use of chelating agents in the treatment of acute and chronic lead intoxication in childhood. J. Pediatr. 73, 1–38.

Chisolm Jr., J.J., Harrison, H.E., 1956. The exposure of children to lead. Pediatrics 18, 943–958.

Cho, S.-C., Kim, B.-N., Hong, Y.-C., Shin, M.-S., Yoo, H.J., Kim, J.-W., et al., 2010. Effect of environmental exposure to Pb and tobacco smoke on inattentive and hyperactive symptoms and neurocognitive performance in children. J. Child. Psychol. Psychiatry 51, 1050–1057.

Cohen, G.J., Ahrens, W.E., 1959. Chronic lead poisoning: a review of seven years' experience at the Children's Hospital, District of Columbia. J. Pediatr. 54, 271–284.

Cooney, G., Bell, A., Stavrou, C., 1991. Low level lead exposures and neurobehavioral development: the Sydney study at seven years. In: Farmer, J.G. (Ed.), International Conference: Heavy Metals in the Environment, vol. 1. CEP Consultants, Ltd., Edinburgh, UK, pp. 16–19.

Cooney, G.H., Bell, A., McBride, W., Carter, C., 1989. Low-level exposures to lead: the Sydney lead study. Dev. Med. Child Neurol. 31, 640–649.

Cory-Slechta, D.A., 1990. Exposure duration modifies the effects of low-level lead on fixed-interval performance. Neurotoxicology 11, 427–441.

Cory-Slechta, D.A., Pokora, M.J., 1991. Behavioral manifestations of prolonged lead exposure initiated at different stages of the life cycle. 1. Scheduled-controlled responding. Neurotoxicology 12, 745–760.

Cory-Slechta, D.A., Thompson, T., 1979. Behavioral toxicity of chronic postweaning lead exposure in the rat. Toxicol. Appl. Pharmacol. 47, 151–159.

Cory-Slechta, D.A., Weiss, B., Cox, C., 1985. Performance and exposure indices of rats exposed to low concentrations of lead. Toxicol. Appl. Pharmacol. 78, 291–299.

De la Burde, B., Choate Jr., M.S., 1972. Does asymptomatic lead exposure in children have latent sequelae? J. Pediatr. 81, 1088–1091.

De la Burde, B., Choate Jr., M.S., 1975. Early asymptomatic lead exposure and development at school age. J. Pediatr. 87, 638–642.

Delville, Y., 1999. Exposure to lead during development alters aggressive behavior in golden hamsters. Neurotoxicol. Teratol. 21, 445–449.

Denno, D., 1990. Biology and Violence. From Birth to Adulthood. Cambridge University Press, New York.

Dietrich, K.N., Krafft, K.M., Bornschein, R.L., Hammond, P.B., Berger, O., Succop, P.A., et al., 1987. Low-level fetal lead exposure effect on neurobehavioral development in early infancy. Pediatrics 80, 721–730.

Dietrich, K.N., Succop, P.A., Bornschein, R.L., Krafft, K.M., Berger, O., Hammond, P.B., et al., 1990. Lead exposure and neurobehavioral development in later infancy. Environ. Health Perspect. 89, 13–19.

Dietrich, K.N., Succop, P.A., Berger, O.G., Hammond, P.B., Bornschein, R.L., 1991. Lead exposure and the cognitive development of urban preschool children: The Cincinnati lead study cohort at age 4 years. Neurotoxicol. Teratol. 13, 203–212.

Dietrich, K.N., Succop, P.A., Berger, O.G., Keith, R., 1992. Lead exposure and the central auditory processing abilities and cognitive development of urban children: The Cincinnati lead study cohort at age 5 years. Neurotoxicol. Teratol. 14, 51–56.

Dietrich, K., Berger, O., Succop, P., 1993. Lead exposure and the motor developmental status of urban 6-year-old children in the Cincinnati Prospective Study. Pediatrics 91, 301–307.

Dietrich, K.N., Ris, M.D., Succop, P.A., Berger, O.G., Bornschein, R.L., 2001. Early exposure to lead and juvenile delinquency. Neurotoxicol. Teratol. 23, 511–518.

Dogru, M.K., Dogru, A.K., Gul, M., Esrefoglu, M., Yurekli, M., Erdogan, S., et al., 2008. The effect of adrenomedullin on rats exposed to lead. J. Appl. Toxicol. 28, 140–146.

Dundar, B., Oktem, F., Arslan, M.K., Delibas, N., Baykal, B., Arslan, C., et al., 2006. The effect of long-term low-dose lead exposure on thyroid function in adolescents. Environ. Res. 101, 140–145.

Erenberg, G., Rinsler, S.S., Fish, B.G., 1974. Lead neuropathy and sickle cell disease. Pediatrics 54, 438–441.

Ernhart, C.B., Morrow-Tlucak, M., Marler, M.R., Wolf, A.W., 1987. Low level lead exposure in the prenatal and early preschool periods: early preschool development. Neurotoxicol. Teratol. 9, 259–270.

Ernhart, C.B., Morrow-Tlucak, M., Wolf, A.W., Super, D., Drotar, D., 1989. Low level lead exposure in the prenatal and early preschool periods: intelligence prior to school entry. Neurotoxicol. Teratol. 11, 161–170.

Factor-Litvak, P., Wasserman, G., Kline, J.K., Graziano, J., 1999. The Yugoslavia prospective study of environmental lead exposures. Environ. Health Perspect. 107, 9–15.

Ferguson, S.A., Bowman, R.E., 1990. Effects of postnatal lead exposure on open field behavior in monkeys. Neurotoxicol. Teratol. 12, 91–97.

Fergusson, D.M., Horwood, L.J., Lynsky, M.T., 1993. Early dentine lead levels and subsequent cognitive and behavioral development. J. Child. Psychol. Psychiatry 34, 215–227.

Fox, D.A., Kala, S.V., Hamilton, W.R., Johnson, J.E., O'Callaghan, J.P., 2008. Low-level human equivalent gestational lead exposure produces supernormal scotopic electroretinograms, increased retinal neurogenesis and decreased dopamine utilization in rats. Environ. Health Perspect. 116, 618–625.

Fox, D.A., Katz, L.M., Farber, D.B., 1991. Low-level developmental lead exposure decreases the sensitivity, amplitude, and temporal resolution of rods. Neurotoxicol. 12, 641–654.

Gilbert, S.G., Rice, D.C., 1987. Low-level lifetime lead exposure produces behavioral toxicity (spatial discrimination reversal) in adult monkeys. Toxicol. Appl. Pharmacol. 91, 484–490.

Gilbert, M.E., Mack, C.M., Lasley, S.M., 1999. Chronic developmental lead exposure and hippocampal long-term potentiation: biphasic dose–response relationship. Neurotoxicology 20, 71–82.

Gomaa, A., Hu, H., Bellinger, D., Schwartz, J., Tsaih, S.-W., Gonzalez-Cossio, T., et al., 2002. Maternal bone lead as an independent risk factor for fetal neurotoxicity: a prospective study. Pediatrics 110, 110–118.

Goyer, R.A., Rhyne, B.C., 1973. Pathological effects of lead. Int. Rev. Exp. Pathol. 12, 1–77.

Greene, T., Ernhart, C.B., 1993. Dentine lead and intelligence prior to school entry: a statistical sensitivity analysis. J. Clin. Epidemiol. 46, 323–339.

Greengard, J., Adams, B., Berman, E., 1965. Acute lead encephalopathy in young children: evaluation of therapy with a corticosteroid and moderate hypothermia. J. Pediatr. 66, 707–711.

Gutowski, M., Altmann, L., Svensson, K., Wiegand, H., 1998. Synaptic plasticity in the CA1 and CA3 hippocampal region of pre- and postnatally lead-exposed rats. Toxicol. Lett. 95, 195–203.

Hansen, O.N., Trillingsgaard, A., Beese, I., Lyngbye, T., Grandjean, P., 1989. Neuropsychological profile of children in relation to dentine lead level and socioeconomic group. In: Smith, M.A., Grant, L.D., Sors, A.I. (Eds.). Lead Exposure and Child Development: An International Assessment. Kluwer Academic, Dordrecht, The Netherlands, pp. 240–250.

Hatzakis, A., Kokkevi, A., Maravelias, C., Katsouyanni, K., Salaminios, F., Kalandidi, A., et al., 1989. Psychometric intelligence deficits in lead-exposed children. In: Smith, M.A., Grant, L.D., Sors, A.I. (Eds.), Lead Exposure and Child Development: An International Assessment. Kluwer Academic, Dordrecht, The Netherlands, pp. 211–223.

Hornung, R.W., Lanphear, B.P., Dietrich, K.N., 2009. Age of greatest susceptibility to childhood lead exposure: a new statistical approach. Environ. Health Perspect. 117, 1309–1312.

Jusko, T.A., Henderson Jr., C.R., Lanphear, B.P., Cory-Slechta, D.A., Parsons, P.J., Canfield, R.L., 2008. Blood lead concentrations <10 μg/dL and child intelligence at 6 years of age. Environ. Health Perspect. 116, 243–248.

Kibayashi, K., Nakao, K.-I., Taki, T., Koyama, H., 2010. Changes in the brain after intracerebral implantation of a lead pellet in the rat. J. Neurotrauma 27, 1925–1934.

Kim, Y., Kim, B.-N., Hong, Y.-C., Shin, M.-S., Yoo, H.-J., Kim, J.-W., et al., 2009. Co-exposure to environmental Pb and manganese affects the intelligence of school-aged children. Neurotoxicology 30, 564–571.

Kline, T.S., 1960. Myocardial changes in lead poisoning. Am. J. Dis. Child. 99, 48–54.

Kordas, K., Canfield, R.L., Rosado, J.L., Vargas, G.G., Cebrian, M.E., Rico, J.A., et al., 2006. Deficits in cognitive function and achievement in Mexican first-graders with low blood lead concentrations. Environ. Res. 100, 371–386.

Kordas, K., Lopez, P., Rosado, J.L., Vargas, G.G., Rico, J.A., Ronquillo, D., et al., 2004. Blood lead, anemia, and short stature are independently associated with cognitive performance in Mexican school children. J. Nutr. 134, 363–371.

Kotok, D., 1972. Development of children with elevated blood lead levels: a controlled study. J. Pediatr. 80, 57–61.

Kotok, D., Kotok, R., Heriot, T., 1977. Cognitive evaluation of children with elevated blood lead levels. Am. J. Dis. Child. 131, 791–793.

Krieg Jr., E.F., Chrislip, D.W., Crespo, C.J., Brightwell, W.S., Ehrenberg, R.L., Otto, D.A., 2005. The relationship between blood lead levels and neurobehavioral test performance in NHANES III and related occupational studies. Public Health Rep. 120, 240–251.

Landrigan, P.J., Gehlbach, S.H., Rosenblum, B.F., Shoults, J.M., Candelaria, R.M., Barthel, W.F., et al., 1975. Epidemic lead absorption near an ore smelter: the role of particulate lead. N. Engl. J. Med. 292, 123–129.

Landrigan, P.J., Baker Jr., E.L., Feldman, R.G., Cox, D.H., Eden, K.V., Orenstein, W., et al., 1976. Increased lead absorption with anemia and slowed nerve conduction velocity in children near a lead smelter. J. Pediatr. 89, 904–910.

Lanphear, B.P., Dietrich, K., Auinger, P., Cox, C., 2000. Cognitive deficits associated with blood lead concentrations <10 µg/dl in U.S. children and adolescents. Public Health Rep. 115, 521–529.

Lanphear, B.P., Hornung, R., Khoury, J., Yolton, K., Baghurst, P., Bellinger, D.C., et al., 2005. Low-level environmental lead exposure and children's intellectual function: an international pooled analysis. Environ. Health Perspect. 113, 894–899.

Lansdown, R., Yule, W., Urbanowicz, M.A., Hunter, J., 1986. The relationship between blood lead concentrations, intelligence, attainment and behavior in a school population: the second London study. Int. Arch. Occup. Environ. Health 57, 225–235.

Lasky, R.E., Laughlin, N.K., 2001. Exploring a partially enclosed space by lead-exposed female rhesus monkeys. Neurotoxicol. Teratol. 23, 177–183.

Lasley, S.M., Gilbert, M.E., 2002. Rat hippocampal glutamate and GABA release exhibit biphasic effects as a function of chronic lead exposure level. Toxicol. Sci. 66, 139–147.

Levin, E.D., Schneider, M.L., Ferguson, S.A., Schantz, S.L., Bowman, R.E., 1988. Behavioral effects of developmental lead exposure in rhesus monkeys. Dev. Psychobiol. 21, 371–382.

Leviton, A., Bellinger, D., Allred, E.N., Rabinowitz, M., Needleman, H., Schoenbaum, S., 1993. Pre- and postnatal low-level lead exposure and children's dysfunction in school. Environ. Res. 60, 30–43.

Lucchini, R., Albini, E., Cortesi, I., Placidi, D., Bergamaschi, E., Traversa, F., et al., 2000. Assessment of neurobehavioral performance as a function of current and cumulative occupational lead exposure. Neurotoxicology 21, 805–811.

Marcus, D.K., Fulton, J.J., Clark, E.J., 2010. Pb and conduct problems: a meta-analysis. J. Clin. Child Adolesc. Psychol. 39, 234–241.

Markovac, J., Goldstein, G.W., 1988. Picomolar concentrations of lead stimulate brain protein kinase C. Nature 334, 71–73.
McMichael, A.J., Baghurst, P.A., Wigg, N.R., Vimpani, G.V., Robertson, E.F., Roberts, R.J., 1988. Port Pirie cohort study: environmental exposure to lead and children's abilities at the age of four years. N. Engl. J. Med. 319, 468–475.
McMichael, A.J., Baghurst, P.A., Wigg, N.R., Vimpani, G.V., Robertson, E.F., Tong, S., 1994. Tooth lead levels and IQ in school-age children: the Port Pirie cohort study. Am. J. Epidemiol. 140, 489–499.
Min, M.O., Singer, L.T., Kirchner, H.L., Minnes, S., Short, E., Hussain, Z., et al., 2009. Cognitive development and low-level Pb exposure in poly-drug exposed children. Neurotoxicol. Teratol. 31, 225–231.
Miranda, M.L., Kim, D., Overstreet Galeano, M.A., Paul, C.J., Hull, A.P., Morgan, S.P., 2007. The relationship between early childhood blood lead levels and performance on end of grade tests. Environ. Health Perspect. 115, 1242–1247.
Miranda, M.L., Kim, D., Reiter, J., Overstreet Galeano, M.A., Maxson, P., 2009. Environmental contributors to the achievement gap. Neurotoxicology 30, 1019–1024.
Mushak, P., 1992. Perspective: Defining lead as the premiere environmental health issue for children in America: Criteria and their quantitative application. Environ. Res. 59, 281–309.
Mushak, P., 1993. New directions in the toxicokinetics of human lead exposure. Neurotoxicol. 14, 29–42.
National Academy of Sciences: National Research Council, 1972. Lead: Airborne Lead in Perspective. National Academy Press, Washington, DC.
Navarro-Moreno, L.G., Quintanar-Escorza, M.A., Gonzalez, S., Mondragon, R., Cerbon-Solorzano, J., Valdes, J., et al., 2009. Effects of lead intoxication on intercellular junctions and biochemical alterations of the renal proximal tubule cells. Toxicol. In Vitro 23, 1298–1304.
Needleman, H.L., Gatsonis, C.A., 1990. Low level lead exposure and the IQ of children: a meta-analysis of modern studies. JAMA 263, 673–678.
Needleman, H.L., Gunnoe, C., Leviton, A., Reed, R., Peresie, H., Maher, C., et al., 1979. Deficits in psychologic and classroom performance of children with elevated dentine lead levels. N. Engl. J. Med. 300, 689–695.
Needleman, H.L., Riess, J.A., Tobin, M.J., Biesecker, G.E., Greenhouse, J.B., 1996. Bone lead levels and delinquent behavior. JAMA 275, 363–369.
Needleman, H.L., McFarland, C., Ness, R.B., Fienberg, S.E., Tobin, M.J., 2002. Bone lead levels in adjudicated delinquents. A case control study. Neurotoxicol. Teratol. 24, 711–717.
Nelson, B.K., Moorman, W.J., Schrader, S.M., Shaw, P.B., Kreig Jr., E.F., 1997. Paternal exposure of rabbits to lead: behavioral deficits in offspring. Neurotoxicol. Teratol. 19, 191–198.
Nihei, M.K., Desmond, N.L., McGlothan, J.L., Kuhlmann, A.C., Guilarte, T.R., 2000. *N*-methyl-D-aspartate receptor subunit changes are associated with lead-induced deficits of long-term potentiation and spatial learning. Neuroscience 99, 233–242.
Niklowitz, W.J., Mandybur, T.I., 1975. Neurofibrillary changes following childhood lead encephalopathy: case report. J. Neuropathol. Exp. Neurol. 34, 445–455.
Nordberg, G.F., 1976. Effects and dose-response relationships of toxic metals. A report from an international meeting. Scand. J. Work Environ. Health 2, 37–43.
Otto, D., Benignus, V., Muller, K., Barton, C., Seiple, K., Prak, J., et al., 1982. Effects of low to moderate lead exposure on slow control potentials in young children: two year follow-up study. Neurobehav. Toxicol. Teratol. 4, 733–737.

Otto, D., Robinson, G., Baumann, S., Schroeder, S., Mushak, P., Kleinbaum, D., et al., 1985. Five-year follow-up study of children with low-to-moderate lead absorption: electrophysiological evaluation. Environ. Res. 38, 168–186.

Otto, D.A., Benignus, V.A., Muller, K.E., Barton, C.N., 1981. Effects of age and body lead burden on CNS function in young children. I. Slow cortical potentials. Electroencephalogr. Clin. Neurophysiol. 52, 229–239.

Palus, J., Rydzynski, K., Dziubaltowska, E., Wyszynska, K., Natarajan, A.T., Nilsson, R., 2003. Genotoxic effects of occupational exposure to lead and cadmium. Mutat. Res. 540, 19–28.

Pentschew, A., 1965. Morphology and morphogenesis of lead encephalopathy. Acta Neuropathol. 5, 133–160.

Pentyala, S., Ruggeri, J., Veerraju, A., Yu, Z.Z., Bhatia, A., Desaiah, D., et al., 2010. Microsomal Ca2+ flux modulation as an indicator of heavy metal toxicity. Indian J. Exp. Biol. 48, 737–743.

Perino, J., Ernhart, C.B., 1974. The relation of subclinical lead level to cognitive and sensorimotor impairment in black preschoolers. J. Learn. Disabil. 7, 616–620.

Perlstein, M.A., Attala, R., 1966. Neurologic sequelae of plumbism in children. Clin. Pediatr. 5, 292–298.

Pocock, S.J., Ashby, D., Smith, M.A., 1989. Lead exposure and children's intellectual performance: the Institute of Child Health/Southampton study. In: Smith, M.A., Grant, L.D., Sors, A.I. (Eds.), Lead Exposure and Child Development: An International Assessment. Kluwer Academic, Dordrecht, The Netherlands, pp. 149–165.

Pocock, S.J., Smith, M., Baghurst, P., 1994. Environmental lead and children's intelligence: a systematic review of the epidemiological evidence. BMJ 309, 1189–1197.

Popoff, N., Weinberg, S., Feigin, I., 1963. Pathologic observations in lead encephalopathy: with special reference to the vascular changes. Neurology 13, 101–112.

Quintanar-Escorza, M.A., González-Martinez, M.T., Navarro, L., Maldonado, M., Arévalo, B., Calderón-Salinas, J.V., 2007. Intracellular free calcium concentration and calcium transport in human erythrocytes of lead-exposed workers. Toxicol. Appl. Pharmacol. 220, 1–8.

Raab, G.M., Fulton, M., Thomson, G.O.B., Laxen, D.P.H., 1989. Blood lead and other influences on mental abilities—results from the Edinburgh lead study. In: Smith, M.A., Grant, L.D., Sors, A.I. (Eds.), Lead Exposure and Child Development: An International Assessment. Kluwer Academic, Dordrecht, The Netherlands, pp. 183–200.

Rhodes, D., Spiro III, A., Aro, A., Hu, H., 2003. Relationship of bone and blood lead levels to psychiatric symptoms: the Normative Aging Study. J. Occup. Environ. Med. 45, 1144–1151.

Rice, D.C., 1984. Behavioral deficit (delayed matching to sample) in monkeys exposed from birth to low levels of lead. Toxicol. Appl. Pharmacol. 75, 337–345.

Rice, D.C., 1985. Chronic low-lead exposure from birth produces deficits in discrimination reversal in monkeys. Toxicol. Appl. Pharmacol. 77, 201–210.

Rice, D.C., 1990. Lead-induced behavioral impairment on a spatial discrimination reversal task in monkeys exposed during different periods of development. Toxicol. Appl. Pharmacol. 106, 327–333.

Rice, D.C., 1992a. Lead exposure during different developmental periods produces different effects on FI performance in monkeys tested as juveniles and adults. Neurotoxicology 9, 75–88.

Rice, D.C., 1992b. Behavioral effects of lead in monkeys tested during infancy and adulthood. Neurotoxicol. Teratol. 14, 235–245.

Rice, D.C., 1992c. Effect of lead during different developmental periods in the monkey on concurrent discrimination performance. Neurotoxicology 13, 583–592.

Rice, D.C., Gilbert, S.G., 1985. Low lead exposure from birth produces behavioral toxicity (RLL) in monkeys. Toxicol. Appl. Pharmacol. 80, 421–426.

Rice, D.C., Gilbert, S.G., 1990a. Sensitive periods for lead-induced behavior impairment on a spatial delayed alternation task in monkeys. Toxicol. Appl. Pharmacol. 103, 364–373.

Rice, D.C., Gilbert, S.G., 1990b. Lead-induced behavioral impairment on a spatial discrimination reversal task in monkeys exposed during different periods of development. Toxicol. Appl. Pharmacol. 106, 327–333.

Rice, D.C., Gilbert, S.G., 1990c. Sensitive periods for lead-induced impairment (nonspatial discrimination reversal) in monkeys. Toxicol. Appl. Pharmacol. 102, 101–109.

Ris, M.D., Dietrich, K.N., Succop, P.A., Berger, O.G., Bornschein, R.L., 2004. Early exposure to lead and neuropsychological outcome in adolescence. J. Int. Neuropsychol. Soc. 10, 261–270.

Rogan, W.J., Dietrich, K.N., Ware, J.H., Dockery, D.W., Salganik, M., Radcliffe, J., et al., 2001. The effect of chelation therapy with Succimer® on neuropsychological development in children exposed to lead. N. Engl. J. Med. 344, 1421–1426.

Rosen, J.F., Mushak, P., 2001. Primary prevention of childhood lead poisoning—the only solution. N. Engl. J. Med. 344, 1470–1471.

Rothenberg, S.J., Schnaas, L., Cansino-Ortiz, S., Perroni-Hernández, E., de la Torre, P., Neri-Méndez, C., et al., 1989. Neurobehavioral deficits after low level lead exposure in neonates: the Mexico City pilot study. Neurotoxicol. Teratol. 11, 85–93.

Schnaas, L., Rothenberg, S.J., Perroni, E., Martinez, S., Hernandez, C., Hernandez, R.M., 2000. Temporal pattern in the effect of postnatal blood lead level on intellectual development of young children. Neurotoxicol. Teratol. 22, 805–810.

Schnaas, L., Rothenberg, S.J., Flores, M.-F., Martinez, S., Hernandez, C., Osorio, E., et al., 2006. Reduced intellectual development in children with prenatal lead exposure. Environ. Health Perspect. 114, 791–797.

Schroeder, S.R., Hawk, B., 1987. Psycho-social factors, lead exposure, and IQ. In: Schroeder, S. R. (Ed.), Toxic Substances and Mental Retardation: Neurobehavioral Toxicology and Teratology. American Association on Mental Deficiency, Washington, DC, pp. 97–137.

Schroeder, S.R., Hawk, B., Otto, D.A., Mushak, P., Hicks, R.E., 1985. Separating the effects of lead and social factors on IQ. Environ. Res. 38, 144–154.

Schwartz, J., 1994. Low-level lead exposure and children's IQ: a meta-analysis and search for a threshold. Environ. Res. 65, 42–55.

Schwartz, B.S., Stewart, W.F., Bolla, K.I., Simon, P.D., Bandeen-Roche, K., Gordon, P.B., et al., 2001. Past adult lead exposure is associated with longitudinal decline in cognitive function. Neurology 55, 1144–1150 [erratum at Neurology 56: 283]

Seto, D.S.Y., Freeman, J.M., 1964. Lead neuropathy in childhood. Am. J. Dis. Child. 107, 337–342.

Shelton, K.R., Cunningham, J.G., Klann, E., Merchant, R.E., Egle, P.M., Bigbee, J.W., 1990. Low-abundance 32-kilodalton nuclear protein specifically enriched in the central nervous system. J. Neurosci. Res. 25, 287–294.

Silva, P.A., Hughes, P., Williams, S., Faed, J.M., 1988. Blood lead, intelligence, reading attainment, and behavior in eleven year old children in Dunedin, New Zealand. J. Child Psychol. Psychiatry 29, 43–52.

Smith, H.D., 1964. Pediatric lead poisoning. Arch. Environ. Health 8, 256–261.

Smith, M., Delves, T., Lansdown, R., Clayton, B., Graham, P., 1983. The effects of lead exposure on urban children: the Institute of Child Health/Southampton study. Dev. Med. Child Neurol. 25 (Suppl. 47), 1–54.

Srivastava, D., Dearth, R.K., Hiney, J.K., Ramirez, L.M., Bratton, G.R., Dees, W., 2004. The effects of low-level Pb on steroidogenic acute regulatory protein (StAR) in the prepubertal rat ovary. Toxicol. Sci. 77, 35–40.

Stangle, D.E., Strawderman, M.S., Smith, D., Kuypers, M., Strupp, B.J., 2004. Reductions in blood lead overestimate reductions in brain lead after repeated Succimer® regimens in a rodent model of childhood lead exposure. Environ. Health Perspect. 112, 302–308.

Surkan, P.J., Zhang, A., Trachtenberg, F., Daniel, D.B., McKinle, S., Bellinger, D.C., 2007. Neuropsychological function in children with blood lead levels <10 μg/dL. Neurotoxicology 28, 1170–1177.

Tellez-Rojo, M.M., Bellinger, D.C., Arroyo-Quiroz, C., Lamadrid-Fiqueroa, H., Mercado-Garcia, A., Schnaas-Arrieta, L., et al., 2006. Longitudinal associations between blood lead concentrations <10 μg/dL and neurobehavioral development in environmentally-exposed children in Mexico City. Pediatrics 118, e323–e330.

Thacker, S.B., Hoffman, D.A., Smith, J., Steinberg, K., Zack, M., 1992. Effect of low-level body burdens of lead on the mental development of children: limitations of meta-analysis in a review of longitudinal data. Arch. Environ. Health 47, 336–346.

Tong, S., Baghurst, P., McMichael, A., Sawyer, M., Mudge, J., 1996. Lifetime exposure to environmental lead and children's intelligence at 11–13 years: the Port Pirie cohort study. BMJ 312, 1569–1575.

Ustundag, A., Dudyu, Y., 2007. The influence of melatonin and *N*-acetylcysteine in delta-aminolevulinic acid and lead-induced genotoxicity in lymphocytes *in vitro*. Biol. Trace Elem. Res. 117, 53–64.

Van Larabeke, N., Koppen, G., Nelen, V., Schoeters, G., Van Loon, H., Albering, H., et al., 2004. Differences in HPRT mutant frequency among middle-aged Flemish women in association with area of residence and blood lead levels. Biomarkers 9, 71–84.

Vivoli, G., Bergomi, M., Borella, P., Fantuzzi, G., Simoni, L., Catelli, D., et al., 1989. Evaluation of different biological indicators of lead exposure related to neuropsychological effects in children. In: Smith, M.A., Grant, L.D., Sors, A.I. (Eds.), Lead Exposure and Child Development: An International Assessment. Kluwer Academic, Dordrecht, The Netherlands, pp. 224–239.

Walkowiak, J., Altmann, L., Krämer, U., Sveinsson, K., Turfeld, M., Weishoff-Houben, M., et al., 1998. Cognitive and sensorimotor functions in 6-year-old children in relation to lead and mercury levels: adjustment for intelligence and contrast sensitivity in computerized testing. Neurotoxicol. Teratol. 20, 511–521.

Wang, C.Y., Lin, Y.W., Yang, J.L., 2008. Activation of protein kinase C alpha signaling prevents cytotoxicity and mutagenicity following lead acetate in CL3 human lung cancer cells. Toxicology 250, 55–61.

Wasserman, G., Graziano, J.H., Factor-Litvak, P., Popovac, D., Morina, N., Musabegovic, A., et al., 1992. Independent effects of lead exposure and iron deficiency anemia on developmental outcome at age 2 years. J. Pediatr. 121, 695–703.

Wasserman, G.A., Graziano, J.H., Factor-Litvak, P., Popavac, D., Morina, N., Musabegovic, A., et al., 1994. Consequences of lead exposure and iron supplementation on childhood development at age 4 years. Neurotoxicol. Teratol. 16, 233–240.

Wasserman, G.A., Factor-Litvak, P., Liu, X., Todd, A.C., Kline, J.K., Slavkovich, V., et al., 2003. The relationship between blood lead, bone lead and child intelligence. Child Neuropsychol. 9, 22–34.

Wigg, N.R., Vimpani, G.V., McMichael, A.J., Baghurst, P.A., Robertson, E.F., Roberts, R.J., 1988. Port Pirie cohort study: childhood blood lead and neuropsychological development at two years. J. Epidemiol. Community Health 42, 213–219.

Winneke, G., Hrdina, K.-G., Brockhaus, A., 1982. Neuropsychological studies in children with elevated tooth-lead concentrations: I. Pilot study. Int. Arch. Occup. Environ. Health 51, 169–183.

Winneke, G., Krämer, U., Brockhaus, A., Ewers, U., Kujanek, G., Lechner, H., et al., 1983. Neuropsychological studies in children with elevated tooth-lead concentrations: II. Extended studies. Int. Arch. Occup. Environ. Health 51, 231–252.

Winneke, G., Collet, W., Lilienthal, H., 1988. The effects of lead in laboratory animals and environmentally-exposed children. Toxicology 49, 291–298.

Winneke, G., Collet, W., Krämer, U., Brockhaus, A., Ewert, T., Krause, C., 1989. Follow-up studies in lead-exposed children. In: Smith, M.A., Grant, L.D., Sors, A.I. (Eds.), Lead Exposure and Child Development: An International Assessment. Kluwer Academic, Dordrecht, The Netherlands, pp. 260–270.

World Health Organization, 1992. Cadmium. Environmental Health Criteria 134. International Programme on Chemical Safety, Geneva, Switzerland.

Wright, J.P., Dietrich, K.N., Ris, M.D., Hornung, R.W., Wessel, S.D., Lanphear, B.P., et al., 2008. Association of prenatal and childhood blood Pb concentrations with criminal arrests in early adulthood. PLoS Med. 5 (5), e101.

Wu, J., Riyaz Masha, M., Brock, B., Cox, D.P., Cardozo-Pelaez, F., McPherson, C.A., et al., 2008. Alzheimer's disease—like pathology in aged monkeys after infantile exposure to environmental metal (Pb): evidence for a developmental origin and environmental link for AD. J. Neurosci. 28, 4405–4407.

Yuan, W., Holland, S.K., Cecil, K.M., Dietrich, K.N., Wessel, S.D., Altaye, M., et al., 2006. The impact of early childhood lead exposure on brain reorganization: a functional magnetic resonance imaging study of language function. Pediatrics 118, 971–977.

Yule, W., Lansdown, R., Millar, I.B., Urbanowicz, M.-A., 1981. The relationship between blood lead concentrations, intelligence and attainment in a school population: a pilot study. Dev. Med. Child Neurol. 23, 567–576.

Yule, W., Urbanowicz, M.-A., Lansdown, R., Millar, I.B., 1984. Teachers' ratings of children's behavior in relation to blood lead levels. Br. J. Dev. Psychol. 2, 295–305.

Yule, W., Lansdown, R., Urbanowicz, M., Muddiman, D., Hunter, J., 1989. Automated assessment of attention, vigilance and learning in relation to children's lead levels. In: Smith, M. A., Grant, L.D., Sors, A.I. (Eds.), Lead Exposure and Child Development: An International Assessment. Kluwer Academic, Dordrecht, The Netherlands, pp. 271–278.

Zhao, W.-F., Ruan, D.-Y., Xy, Y.-Z., Chen, J.-T., Wang, M., Ge, S.Y., 1999. The effects of chronic lead exposure on long-term depression in area CA1 and dentate gyrus of rat hippocampus in vitro. Brain Res. 818, 153–159.

Chapter 13

Cardiovascular Toxicity of Lead in Human Populations

13.1 INTRODUCTION

Lead has long been known to produce toxic effects on the heart and blood vessels in human populations and to show links to cardiovascular morbidity and mortality. A number of public agency expert consensus documents have discussed the topic, including IARC (2006), NAS/NRC (1993), U.S. EPA (1977, 1986, 2006), U.S. ATSDR (2007), and WHO (1995). In addition, individual critical reviews have appeared (Navas-Acien et al., 2008, 2007; Vaziri, 2008). This chapter presents some of the more salient toxicological, epidemiological, and mechanistic aspects of these effects and a contextual comparison with other key toxic endpoints of lead exposure.

Cardiovascular effects differ in a number of ways from other toxic endpoints associated with human Pb exposures, particularly with respect to Pb neurotoxicity. The focus of attention for neurotoxic effects epidemiologically, medically, and societally has invariably been the very young child, infants and toddlers, and on parallel concerns about prenatal exposures to lead. Lead neurotoxicity in adult humans, by contrast, has had a less voluminous and comprehensive literature for reasons explained in Chapter 12. By contrast, cardiovascular effects in humans have almost always been evaluated in adult human populations. Data are limited and significant risk analysis attention has not been given to cardiovascular effects in children.

This difference in attention to risk populations for lead neurotoxicity versus cardiovascular and cerebrovascular toxicity traces in part to what available data currently permit one to conclude comparatively about these serious adverse effects. Childhood Pb neurotoxicity is expressed through a myriad of developmental neurocognitive and neurobehavioral mechanisms centered on the developing brain for which adult neural analogs have not been as well conceptualized.

Cardiovascular toxicity, by contrast, most commonly employs blood pressure changes as the sentinel effect marker across adult groups. That endpoint

Trace Metals and other Contaminants in the Environment, Volume 10
ISSN: 1875-1121 DOI: 10.1016/B978-0-444-51554-4.00013-4

measure has been unremarkable in children but has repeatedly been shown to be extensively associated with Pb exposures in adult populations.

Comparative mechanistic rationales for cardiovascular effects and associated endpoints being present more in adults than in children have not been convincingly presented. A limited database exists linking associations of childhood lead exposures with cardiovascular effects versus the huge literature on the topic for adults sustaining occupational and/or environmental exposures. The toxicokinetics of Pb in young children differs significantly from that in adults in ways that may affect toxic endpoints that are highly labile, measures of SBPs and DBPs.

Dose–toxic response relationships for lead and cardiovascular toxicity versus neurotoxicity show some similarities and some differences across the Pb exposure spectrum. Low Pb exposures in young children in the sub-10 μg/dl region are linked to a more robust response in IQ decrements (steeper slope) relative to the higher portion of the dose curve (see Chapter 12). Dose–response relationships, however, continue in curvilinear form across the entire exposure (PbB) spectrum, with severe brain damage and a risk of or actual death as PbB levels exceed 80–100 μg/dl. Such cardiovascular effects as increased blood pressure, typically of SBP, are seen in adults to be mainly present at lower environmental Pb exposures. The increases in SBP seen at lower exposures are attenuated in lead workers, especially at PbB levels >40 μg/dl. This is not to argue that cardiovascular effects do not proceed with increasing exposures in occupational settings in adults, only that dose-dependent SBP changes at higher exposures are relatively less than at lower ones.

13.2 EPIDEMIOLOGY OF CARDIOVASCULAR TOXICITY IN ADULT OCCUPATIONAL POPULATIONS

Two discrete subsets of adults among human populations have been described in the literature with respect to Pb-associated cardiovascular and/or cerebrovascular morbidity and mortality. The earlier groups were those having chronic, relatively high lead exposure in the workplace. These worker cohorts are, however, complex in terms of cardiovascular risk owing to such factors as the "healthy worker" effect. This effect is an occupational health phenomenon arising from the fact that hyperresponsive or normoresponsive individuals who begin Pb-connected work are likely to leave that work relatively quickly compared to hyporesponders to adverse effects. The workplace fraction of hyporesponders would then increase over time and the hyperresponsive remainder would continue to decline. On the other hand, the more significant segment of the adult human population in terms of aggregate public health impacts are those having relatively low, nonoccupational Pb exposures.

13.2.1 Early Reports of Cardiovascular Toxicity in Adults Having High Workplace Pb Exposures

Some of the earliest epidemiological studies of Pb effects on cardiovascular function in adults evaluated evidence for Pb-associated hypertension and hypertension-linked cardiovascular and cerebrovascular mortality in lead workers. Summaries of these findings are in Table 13.1. There were several problems with these earlier efforts. The results were gathered for workers having very high exposures in the first quarter of the twentieth century, exposures much greater than those in the workplace in more recent decades. A second difficulty with results of those relatively crude approaches was reliable indexing of internal lead exposures via PbB, resulting in poor exposure classification and stratification of subcohorts.

Dingwall-Fordyce and Lane (1963) published findings that illustrate the impact of relatively higher exposures in earlier decades, reporting higher cerebrovascular mortality rates and higher rates of cerebral thrombosis and atherosclerosis for these earlier workers, while finding no differences in individuals exposed to workplace Pb in recent periods. Vigdartchik (1935) described Pb cardiovascular toxicity in earlier decades of the twentieth century, including the higher incidence of hypertension in Pb workers who had high exposures in the 1930s.

Positive findings of a cardiovascular effect with likely high Pb exposures in earlier decades were not often reported. As seen in Table 13.1, no associations between Pb exposure and cardiovascular disease were reported in the papers of Cramér and Dahlberg (1966), Dreessen et al. (1941), Kósmider and Petelenz (1962), Neal et al. (1941), and Belknap (1936). No increase in hypertension incidence with increased exposures was the most common negative finding. Difficulties in comparing these high-exposure study results included absence of good exposure measures and relatively small sample size for total or stratified cohorts.

These reports illustrate one difficulty arising with collective evaluation of cerebrovascular and cardiovascular effects in the global literature along a very elevated segment of Pb dose–responses, that of seeking consistency in core adverse response relationships, i.e., the strength and incidence of hypertension and blood pressure elevations with increases in exposure biomarkers such as PbB. That portion of the cardiovascular dose–response relationship for lead at moderate, environmental exposures generally displays a significant positive association. This relationship at quite elevated Pb exposures is less robust or declines in positive association with still further elevations in exposure, including the range of exposures encountered with moderate workplace exposures. With even higher exposures, like those in the data of Dingwall-Fordyce and Lane (1963) and Vigdartchik (1935), there is return of a positive association of hypertension and increased blood pressure with increasing PbB.

TABLE 13.1 Early Epidemiological Studies of Cardiovascular Toxicity with Occupational Lead Exposure

Study Group	Pb Exposure Level	Toxic Endpoint	Results	References
Lead workers, different times of employment, two groups of pensioner death rates: 1926–1961, and currently employed, 1946–1961	High exposures in first quarter of the twentieth century, lower exposures in more recent years	Cerebrovascular mortality rates: cerebral hemorrhage, cerebral thrombosis, cerebral arteriosclerosis	Earlier, higher exposures associated with higher mortality versus background; no differences in more recent years	Dingwall-Fordyce and Lane (1963)
Lead workers	High Pb exposures in the 1930s	Incidence of hypertension	Increased incidence of hypertension among Pb workers	Vigdartchik (1935)
Industrial workers ($N = 364$) with inclusion of Pb workers ($N = 273$)	Variable chronic Pb exposures in Pb workers ranked by coproporphyrin level	Incidence of hypertension	No increases in hypertension versus non-Pb workers	Cramér and Dahlberg (1966)
Pb workers with high exposures in earlier twentieth century ($N = 766$); 12% > 20 years exposure, 75% > 5 years in lead–acid battery production	High Pb exposures: $\geq 150\ \mu g/m^3$ versus $< 150\ \mu g/m^3$	Incidence of hypertension and arteriosclerotic heart disease	No increases in end point measures versus non-Pb workers	Dreessen et al. (1941)
Pb workers > 46 years old ($N = 38$)	History of Pb poisoning	EKG changes	66% had changes twofold that of nonworkers	Kósmider and Petelenz (1962)
Orchard workers using lead arsenate pesticide	Exposure to lead arsenate at three PbB and urinary Pb levels	Heart disease and hypertension	No endpoint changes linked to the pesticide exposure	NAS/NRC (1972), Neal et al. (1941)

(Continued)

TABLE 13.1 Early Epidemiological Studies of Cardiovascular Toxicity with Occupational Lead Exposure—(cont.)

Study Group	Pb Exposure Level	Toxic Endpoint	Results	References
Lead and battery workers versus other employees	High Pb exposures in early decades of the twentieth century	Hypertension	No increased hypertension rates over those with other work histories	Belknap (1936)

Cardiotoxicity in older studies in the form of electrocardiographic (EKG) abnormalities has been observed in Pb workers (Table 13.1). Two-thirds of a small group of employees older than 46 years of age showed EKG changes fourfold greater than in non-Pb employees.

13.2.2 More Recent Reports of Cardiovascular Toxicity in Adults Having Elevated Workplace Pb Exposures

More recent epidemiological studies of Pb exposures in the workplace shed considerable light on the various endogenous and exogenous factors that define the nature of the seemingly highly variable cause—effect relationship for lead and cardiovascular toxicity. Table 13.2 presents more recent cross-sectional data, while Table 13.3 sets out longitudinal study results. Studies summarized in these two occupational epidemiological data tables were selected with an eye to those efforts which minimized the chance of Type II statistical errors, i.e., failures to detect dose—response relationships where those were present. For example, sample sizes should be maximized, the exposure biomarker should be as toxicokinetically stable as possible, and the endpoint measure should be methodologically stable. As to endpoints, blood pressure measurements, in particular, are highly subject to the effects of measurement method differences across studies.

Few studies in the earlier literature used pulse pressure (PP) measures (SBP—DBP). As to the exposure biomarkers, those should be more toxicokinetically accurate in capturing both body lead burden and changes in that burden. While bone Pb measurements may have more scatter to their precision, they are more accurate in reflecting the most predictive exposure biomarker, the more important parameter.

Table 13.2 tabulates various cross-sectional studies of effects of workplace exposures to Pb at lower levels than prevailed in the early industrial workplace. Those levels were still significantly above ambient, environmental conditions that prevailed in most developed nations. Occupational

TABLE 13.2 Selected Cross-Sectional Epidemiological Studies of Lead-Associated Blood Pressure and Hypertension Effects in More Recent Occupational Exposures

Worker Cohort	Pb Exposure	Endpoint Measure	Results	References
White male police officers ($N = 431$). 24–55 years old	Exposure during work with elevated Pb-air	SBP, DBP measurements	Both BP measures increased with elevated PbB	Orssaud et al. (1985)
Pb–acid battery workers ($N = 270$) and unexposed employees ($N = 158$)	Pb worker mean PbB = 40 μg/dl; Non-Pb exposure mean PbB = 7 μg/dl	SBP, DBP measurements	Small, nonsignificant association, BP versus PbB, after adjustments	Parkinson et al. (1987)
Lead workers, long-term exposures	Pb smelter workers heavily exposed, 9–45 years	BP, other cardiovascular risk factors	Pb exposure associated with those EKG changes linked to ischemic heart disease	Kirby and Gyntelberg (1985)
Korean Pb workers, 26 South Korean facilities; mean age = 40.5 years ($N = 798$), controls: no Pb exposures	PbB levels, chelatable Pb (Succimer®)	SBP, DBP, and hypertension	Significant positive associations, chelatable Pb versus SBP; nonsignificant association, PbB and SBP	Lee et al. (2001)
Former U.S. organolead workers ($N = 543$)	PbB	SBP, DBP	SBP significantly associated with PbB	Schwartz et al. (2000)
Former U.S. organolead workers ($N = 543$), mean of 18 years since exposures	Tibial bone Pb, chelatable Pb levels	SBP, DBP, and hypertension	No significant association of BP with bone Pb measure	Schwartz et al. (2000)
U.S. members of a labor union	Tibial, trabecular bone Pb	SBP, DBP, hypertension	No significant associations for any endpoint with either bone Pb level	Hu et al. (1994)

(Continued)

TABLE 13.2 Selected Cross-Sectional Epidemiological Studies of Lead-Associated Blood Pressure and Hypertension Effects in More Recent Occupational Exposures—(cont.)

Worker Cohort	Pb Exposure	Endpoint Measure	Results	References
Birmingham, England male factory workers (N = 809)	PbB levels < 21 μg/dl – >50 μg/dl	SBP, DBP	No relationships; use of ZnPP in models biased to a null result	Maheswaran et al. (1993)
Japanese female crystal glass workers with controls (N = 193, aged 16–58 years)	PbB, 3.8–99.4 μg/dl	SBP, DBP	Significant associations: SBP, 1.26 mmHg/10 μg/dl; DBP, 1.05 mmHg/10 μg/dl	Nomiyama et al. (2002)

exposure study cohorts represented various work classifications, both genders, and assorted nationalities and races. They included South Korean battery workers, British factory workers, female Japanese leaded crystal makers, male French police officers, and former U.S. organolead production workers.

Biomarkers of Pb exposure in Table 13.2 include both PbB and bone Pb, the latter using cortical bone (tibia) and trabecular (e.g., patella, calcaneus) bone types. Blood pressure was the principal outcome measure, but the incidence of hypertension as defined by the various authors and described as a quantified endpoint in their reports is included. Table 13.2 depicts nine studies, five relying principally on PbB as the exposure biomarker, two using bone Pb measurements, one using both PbB and chelatable Pb (Succimer® as chelant), and one using environmental Pb measurements in the work environment.

Blood Pb, overall, is inconsistent in significant association with SBP and/or DBP or other cardiovascular endpoint such as hypertension or ischemic heart disease in occupational Pb exposures. Orssaud et al. (1985) noted that both BP measures were significantly increased with elevated air Pb in the work environment of French police officers, while PbB measured in the range of 3.8–99.4 μg/dl in Japanese glass workers (Nomiyama et al., 2002) was significantly associated with SBP and DBP. For each 1.0 μg/dl increase in PbB, the SBP increased by 1.26 millimeters of Hg and the DBP increased by 1.05 millimeters of Hg. Schwartz et al. (2000) reported that PbB in former organolead production workers was significantly linked to SBP but not DBP. Kirby and Gyntelberg (1985) showed that Pb exposure was associated with those EKG changes linked to ischemic heart disease.

TABLE 13.3 Selected Longitudinal Epidemiological Studies of Blood Pressure and Hypertension Effects of Occupational Lead Exposures

Worker Cohort	Pb Exposure	Endpoint Measure	Results	References
Korean lead workers at 26 South Korean lead facilities ($N = 575$), followed 3 years	PbB, tibial Pb levels	SBP	Significant positive longitudinal association of PbB with SBP	Glenn et al. (2006)
Korean lead workers ($N = 652$) of Glenn et al. (2006), followed cross sectionally third year	PbB, patellar Pb levels	SBP, DBP, hypertension, tests of genotype interactions	PbB significantly associated with SBP; no effects for either Pb marker on hypertension	Weaver et al. (2008)
Former U.S. organolead workers ($N = 496$); exposure ceased an average of = 18 years ago, followed 2 years	Baseline PbB = 4.6 μg/dl; PbB, tibia mean, tibia peak	SBP, DBP	Significant positive associations of PbB, peak tibia Pb with SBP, SBP increased 0.25 mmHg/SD PbB	Glenn et al. (2003)
Boston, MA, middle-aged police officers (47 years old), $N = 70$, followed 5 years	PbB at three exposures: < 20 μg/dl, 20–30 μg/dl, >30 μg/dl	SBP, DBP	Significant positive association of SBP with PbB >30 μg/dl	Weiss et al. (1986)
Canadian foundry workers ($N = 288$), followed 6 years	PbB	SBP, DBP	Significant positive association of PbB with DBP; nonsignificant positive association of PbB with SBP; coexposures to other vasoactive metals probable	Neri et al. (1988)

By contrast, Korean Pb workers at 26 different facilities only showed a nonsignificant link between PbB and SBP (Lee et al., 2001), but a significant positive association of chelatable Pb with that endpoint. Parkinson et al. (1987) found that Pb workers with a mean PbB of 40 μg/dl showed a small, nonsignificant association after adjustments of BP with PbB. No association of PbB was reported for male factory workers in Birmingham, England, with PbB levels ranging widely to above 50 μg/dl (Maheswaran et al., 1993).

Interestingly, no significant associations with SBP/DBP or hypertension were reported using bone Pb values in the studies of Schwartz et al. (2000) or Hu et al. (1994).

It is not surprising that a labile endpoint such as blood pressure and, to some extent, hypertension would show variable relationships with PbB and other exposure indicators when using cross-sectional approaches. Both PbB as exposure marker and BP as cardiovascular endpoint indicator are themselves quite temporally variable, particularly in workplace settings, so that single measures in each variable evaluated cross sectionally at one point in time would be expected to vary considerably. Some of the uncertainties in these studies relate to their overall design. For example, the Maheswaran et al. (1993) study of English factory workers employed an epidemiological model that controlled for zinc protoporphyrin levels in blood. This hematological substrate is itself directly associated with PbB level, such that over-control of Pb exposure is a result. These difficulties have promoted a higher interest in the use of longitudinal epidemiological studies to evaluate Pb–cardiovascular toxicological relationships under occupational conditions.

Table 13.3 summarizes results of selected longitudinal occupational assessments of the lead–cardiovascular toxicity relationship. Time spans in the longitudinal study designs differed from 2 (Glenn et al., 2003) to 6 years (Neri et al., 1988). Two studies employed PbB as the exposure biomarker (Neri et al., 1988; Weiss et al., 1986), while the others included both PbB and bone Pb (tibia Pb) measurements. The cardiovascular effect endpoints in all cases were SBP and DBP. The degree of Pb exposure differed across studies, with the baseline PbB for the former organolead workers as the lowest, 4.6 μg/dl. This is not surprising, given the large time gap since workplace exposures for the group, a mean period of 18 years. These particular individuals are toxicokinetically anomalous, having had more than just environmental Pb exposures in terms of body Pb burdens (bone Pb stores), but with those exposures being relatively remote in time.

All of the studies in Table 13.3 describe significant positive associations of the exposure biomarker with BP. Three entries show significant associations of lead with SBP (Glenn et al., 2003, 2006; Weiss et al., 1986), and one with only DBP, systolic pressure being nonsignificantly associated with PbB (Neri et al., 1988). The significant association of tibial Pb with BP was only for tibial peak Pb, not tibial mean level (Glenn et al., 2003), not unexpected given the many years of postemployment changes in bone Pb in these subjects. In this study, SBP increased 0.25 mmHg/1 SD PbB. The Boston, MA, middle-aged police officer study showed only the presence of a statistical association with the highest stratum of PbB, >30 μg/dl. No associations were seen with those in the PbB strata <20 and 20–30 μg/dl.

It is not surprising that longitudinal assessment of the blood Pb–BP relationship is consistently positive across studies. Multiple testings across time for both exposure and outcome measures sufficiently attenuated uncertainty and variability to permit detection of associations relative to a cross-sectional epidemiological design.

Occupational Pb exposure data would be expected to be of value in defining an upper bound to cardiovascular toxicity in terms of Pb exposure for human populations at large, depicting as it does much higher exposures than encountered by nonoccupational populations in developed countries. However, that particular virtue is compromised in terms of cardiovascular activity owing to apparent variation in relative strength of associations across a broad Pb exposure spectrum. Second, while more recent occupational data were gathered from larger sample sizes than in the older literature, worker study populations are still small compared to the total population. Constraints in the worker data have focused attention on nonworker general population studies. These have been of both the cross-sectional and the prospective, serial testing epidemiological type.

13.3 Pb CARDIOVASCULAR EFFECTS IN GENERAL ADULT POPULATIONS

Studies in the general adult population consist of approaches using either cross-sectional or prospective designs.

13.3.1 Cross-Sectional Epidemiological Studies of Lead's Cardiovascular Effects in General, Nonoccupational Populations

Tables 13.4 and 13.5 tabulate cross-sectional epidemiological data for nonoccupational populations sustaining cardiovascular effects from lead exposures using either PbB (Table 13.4) or bone Pb measurements (Table 13.5) as the exposure indicator and BP and/or hypertension as the toxic endpoints. The largest cohorts among these studies used the exposure marker PbB: the U.S. NHANES II, NHANES III, and more recent NHANES nationwide surveys (Den Hond et al., 2002; Muntner et al., 2005; Nash et al., 2003; Schwartz 1988; Scinicariello et al., 2010; Vupputuri et al., 2003) and the international surveys, Health Survey for England (HSE; Bost et al., 1999), the British Regional Heart Study (BRHS; Pocock et al., 1988), and the Belgian Cadmibel Study (BCS; Staessen et al., 1993).

Several conclusions can readily be drawn from the various sets of data in Table 13.4. First and most clearly, accessibility to and assessments of large cohorts of human subjects can demonstrate associations of blood pressure measures with PbB, with stratification into large population segments. Studies where sample sizes were more limited produced less robust associations. Vupputuri et al. (2003) analyzed four strata of NHANES III subjects—White males, females; Black males, females—using PbB with arithmetic means given in the table and SBP, DBP, and hypertension as the outcome

TABLE 13.4 Illustrative Cross-Sectional Epidemiological Studies of Cardiovascular Effects of Lead in Nonoccupational Populations: Blood Pb as Exposure Biomarker

Study Cohort	Pb Exposure Mean (μg/dl)	Endpoint Measures	Results	References
Four subgroups of adults in U.S. NHANES III, 1988–1994:		SBP, DBP, hypertension	Black males and females had significant effects associated with PbB levels for SBP, DBP. Linear significant association of hypertension with PbB in Black females	Vupputuri et al. (2003)
White males ($N = 5{,}360$)	4.4			
White females ($N = 5{,}188$)	3.0			
Black males ($N = 2{,}104$)	5.4			
Black females ($N = 2{,}300$)	3.4			
Four subgroups from U.S. NHANES III:		SBP, DBP	Black males and females had significant associations between PbB and SBP; Black females had significant inverse association between this PbB and DBP	Den Hond et al. (2002)
White males ($N = 4{,}685$)	3.6			
White females ($N = 5{,}138$)	2.1			
Black males ($N = 1{,}761$)	4.2			
Black females ($N = 2{,}197$)	2.3			
Women aged 40–69 years: U.S. NHANES III ($N = 1{,}786$ SBP and DBP, $N = 2{,}165$, hypertension)	2.9	SBP, DBP, hypertension	Linear PbB significantly linked to SBP for total sample, 1 μg/dl PbB = 0.25 mmHg	Nash et al. (2003)
Boston, MA, women, middle-aged ($N = 284$)	3.0	Hypertension	No significant association of PbB and hypertension; low statistical power	Korrick et al. (1999)
Males, females 19–29 years old ($N = 502$, 50% lived by smelter as children)	2.2	SBP, DBP	No significant association of PbB and BP, complicated by bone Pb inclusion in model	Gerr et al. (2002)

(Continued)

TABLE 13.4 **Illustrative Cross-Sectional Epidemiological Studies of Cardiovascular Effects of Lead in Nonoccupational Populations: Blood Pb as Exposure Biomarker—(cont.)**

Study Cohort	Pb Exposure Mean (μg/dl)	Endpoint Measures	Results	References
1995 Health survey for English subjects, males ($N = 2{,}563$) and females ($N = 2{,}763$), middle-aged (mean = 47.6 years)	Males: geometric mean 3.7 μg/dl; females: geometric mean 2.6 μg/dl	SBP, DBP, use of antihypertensives	For females, significant association of PbB with DBP, use of antihypertensive medicines; no links with SBP	Bost et al. (1999)
BCS, males ($N = 827$) and females ($N = 821$), ages 20–88 years; part lived near nonferrous smelters	PbB range, men: 2.7–84.9 μg/dl; women: 1.3–42.4 μg/dl	SBP, DBP	No association of PbB with SBP for women; no association of DBP for men or women; significant decrease of SBP with PbB in men	Staessen et al. (1993)
BRHS subjects ($N = 7{,}735$): males aged 40–59 years, 24 British towns. Analyses and reanalyses for adjusting for town, smoking, alcohol consumption	PbB median=14.5 μg/dl in 1978–1980	SBP, DBP	When adjusted for all confounders, PbB significantly linked to both SBP and DBP. PbB doubling elevated SBP, 1.5 units, DBP, 1.3 units	Pocock et al. (1988)
U.S. NHANES II subjects ($N = 5{,}803$ men and women) aged 12–74 years	PbB geometric mean = 15 μg/dl men, 11 μg/dl women	SBP, DBP	When adjusted for age, race, BMI, time trend, site, White males 40–59 years had significant association of PbB with BP	Schwartz (1988)

(*Continued*)

TABLE 13.4 Illustrative Cross-Sectional Epidemiological Studies of Cardiovascular Effects of Lead in Nonoccupational Populations: Blood Pb as Exposure Biomarker—(cont.)

Study Cohort	Pb Exposure Mean (μg/dl)	Endpoint Measures	Results	References
NHANES 1999–2002 U.S. adults, *N* (total for all studies) = 9,961	PbB-based quartiles (μg/dl) < 1.06, 1.06–1.63, 1.63–2.47, > 2.47, total geometric mean = 1.64 μg/dl	Hypertension (based on antihypertensive drug use or SBP >140 or DBP >90)	Mexican-American non-Hispanic Blacks: significant increase in odds ratio with PbB quartile	Muntner et al. (2005)
NHANES III subjects, *N* = 6,016 males, females, >17 years old	PbB quartiles: 0.7–1.4, 1.5–2.3, 2.4–3.7, 3.8–52.9 μg/dl	Hypertension (based on antihypertensive drug use or SBP >140 or DBP >90); ALAD genotype main effects	Prevalence odds ratio increased only in Blacks with PbB quartile; no ALAD genotype main effect seen	Scinicariello et al. (2010)

measures. Arithmetic means for the four strata of U.S. NHANES III showed Black males and females with higher mean values than the corresponding genders among Whites. Black males and females showed significant associations between SBP or DBP and PbB, with a significant linear association of hypertension and PbB in Black females. Three other data set analyses retrieved from the NHANES III are those of Scinicariello et al. (2010), Den Hond et al. (2002), and Nash et al. (2003). Den Hond et al. (2002) examined the same subcategories of NHANES III subjects as did Vupputuri et al. (2003) but used somewhat different sample sizes. These authors also found significant direct associations of SBP with PbB in Black men and women but an inverse linkage between PbB and DBP in Black females.

Women aged 40–59 years in the NHANES III, examined as to either BP and hypertension (*N* = 2,165) or elevated BP (*N* = 1,786), showed a significant linear association of PbB with SBP for the whole sample (Nash et al., 2003). The increase of 1 μg/dl PbB raised SBP by 0.25 mmHg. The earlier U.S. NHANES II, conducted in 1976–1980, was examined for similar associations in the age band 12–74 years of age, men and women, and a sample *N* of 5,803 subjects (Schwartz, 1988). The geometric mean for this earlier time period was 15 μg/dl in men and 11 μg/dl in women. Using SBP and DBP as the endpoints, with adjustments for age, race, BMI, time trend, and

TABLE 13.5 Illustrative Cross-Sectional Epidemiological Studies of Cardiovascular Effects of Lead in Nonoccupational Populations: Bone Pb as Exposure Biomarker

Study Group	Pb Exposure	Endpoint Measures	Results	References
Boston, MA, NAS subjects, mainly Caucasian males ($N = 590$), middle-aged, elderly	Trabecular, cortical bone Pb levels	Hypertension	No significant association of either bone Pb level versus hypertension	Hu et al. (1996)
Boston, MA, NAS subjects ($N = 519$) for cross-sectional analyses	Trabecular, cortical bone Pb levels	SBP, DBP as continuous endpoint	Significant positive association between tibial Pb and SBP, insignificant positive linkage for patellar Pb and SBP	Cheng et al. (2001)
Boston, MA, NAS subjects ($N = 593$) for cross-sectional analysis	Quintiles of PbB, trabecular, cortical bone Pb levels	Pulse pressure (PP) = SBP − DBP	Adjusted differences in PP as tibia Pb, PbB show significant trend in PP with tibial Pb ($p = 0.02$)	Perlstein et al. (2007)
Boston, MA, Nurses' Health Study subjects: middle-aged, elderly women (total $N = 284$, 89 cases, 195 controls)	Trabecular, cortical bone Pb levels	Hypertension	Significant adjusted positive association between patellar Pb and risk of hypertension	Korrick et al. (1999)
Idaho smelter-area Pb exposure group and age-matched control group, young adults 19–29 years, exposed as children, males and females	Tibial bone Pb	SBP, DBP	Significant positive associations between maximum tibial Pb group and both SBP, DBP	Gerr et al. (2002)

(*Continued*)

TABLE 13.5 Illustrative Cross-Sectional Epidemiological Studies of Cardiovascular Effects of Lead in Nonoccupational Populations: Bone Pb as Exposure Biomarker—(cont.)

Study Group	Pb Exposure	Endpoint Measures	Results	References
Pregnant Latina women ($N = 1{,}006$), Los Angeles, 3rd trimester and postpartum	cortical, trabecular bone Pb levels	SBP, DBP, hypertension	Significant positive association between trabecular Pb and 3rd trimester SBP, DBP, hypertension	Rothenberg et al. (2002)
Baltimore, MD, Memory Study participants ($N = 964$), cross-sectional portion, mixed gender, mixed ethnic older subjects	Tibial Pb levels	SBP, DBP, hypertension	Nonsignificant association between tibial Pb and hypertension, no association of tibial Pb with BP	Martin et al. (2006)

testing site, White males aged 40–59 years old had a significant association of PbB with BP.

Scinicariello et al. (2010) examined NHANES III data for 6,016 males and females >17 years old, using quartiles of PbB and prevalence odds ratio for hypertension versus ALAD genotype for main effect assessment. Prevalence odds ratio increased with the PbB quartile. No ALAD genotype main effect was noted.

Muntner et al. (2005) examined a U.S. cohort of men and women (total N, all studies = 9,961) in an NHANES data set covering 1999–2002. Quartiles of PbB comprised stratified Pb exposure and hypertension outcome measures. Significant increases in odds ratio for hypertension with increased PbB quartile was noted for Mexican-Americans and non-Hispanic Blacks.

Larger population studies outside of the United States, notably the 1995 HSE and the BRHS in the late 1970s–early 1980s showed positive associations with PbB. Pocock et al. (1988), in their analysis and reanalysis of the data from the BRHS, noted that PbB was significantly linked to both SBP and DBP, with a doubling of PbB producing 1.5 and 1.3 mmHg increases, respectively. BRHS subjects numbered 7,735 males aged 40–59 years residing in 24 British towns. The reanalyses adjusted for town, smoking, alcohol consumption and for median PbB of 14.5 μg/dl in 1978–1980. In the 1995 HSE study of middle-aged men (mean age 47.6 years, $N = 2{,}563$) and women ($N = 2{,}763$) tested for relationships of SBP, DPB, and use of antihypertensives to PbB (geometric mean 3.7 μg/dl men; 2.6 μg/dl women), significant associations were found only in women, in the form of PbB versus either DBP and/or use of antihypertensives (Bost et al., 1999).

In the BCS, men and women ranging in age from 20 to 80 years and living near nonferrous smelters were evaluated for relationship of SBP and DBP with PbB (Staessen et al., 1993). No link was observed for DBP in men or women, while an anomalous inverse relationship was seen in men for PbB and SBP and no relationship with SBP and PbB in women. Sample sizes in this study were 827 and 821 for men and women, respectively.

Table 13.5 summarizes results of six cross-sectional studies from U.S. locales, where the Pb exposure biomarker was bone Pb rather than PbB. Sample sizes varied as did the nature and extent of cardiovascular effect relationships to this more toxicokinetically stable biomarker. However, statistically significant associations of BP or hypertension with bone Pb were not consistently detected. Hu et al. (1996) analyzed for relationships of trabecular or tibial bone values with hypertension in mainly aging Boston, MA, Caucasian males enrolled in the VA NAS. No significant link of either bone Pb type with hypertension was identified. However, a second evaluation of male Boston enrollees in the NAS ($N = 519$, cross-sectional analyses) using the same bone Pb biomarkers but SBP and DBP as continuous endpoint variables (Cheng et al., 2001) produced a positive significant association of tibial Pb and SBP but only an insignificant positive association of SBP with patellar Pb.

Korrick et al. (1999) examined a relatively small group of middle-aged and elderly women enrolled in the Boston, MA, Nurses' Health Study (89 cases, 195 controls). Risk of hypertension was significantly associated with patellar Pb. By contrast, enrollees in the Baltimore, MD, Memory Study evaluated for blood pressure increases and hypertension using tibial Pb as exposure marker presented with a nonsignificant association between tibial Pb and hypertension (Martin et al., 2006), despite a larger sample size (N=964) and older subjects. However, this study differed from the NAS results of Cheng et al. (2001) in having both men and women and various ethnic representations among the group.

Two of the studies in Table 13.5 described data for populations in which the use of bone Pb as dose measure for studies of cardiovascular toxicity was done in the face of bone Pb burdens affected by exogenous or endogenous factors. One case represented a history of excessive early Pb exposure in a smelter community, while the other involved individuals, where the Pb-bone was undergoing toxicokinetic flux from the metabolic stresses of pregnancy and the postpartum period. Gerr et al. (2002) found significant positive associations between the maximum tibial Pb group and both SBP and DBP elevations in adults exposed to Pb as children in a smelter community. Latina Los Angeles, CA, residents followed during and after pregnancy in terms of SBP and DBP showed significant links between trabecular bone Pb and third trimester SBP, DBP, and hypertension (Rothenberg et al., 2002).

The NAS cohort in the hands of Perlstein et al. (2007) was analyzed for bone Pb relationship to changes in pulse pressure (SBP−DBP) as outcome measure. A significant trend ($p = 0.02$) in PP versus quintile of bone Pb was reported.

13.3.2 Longitudinal Epidemiological Studies of Lead Cardiovascular Effects in General, Nonoccupational Adult Populations

Three longitudinal epidemiological studies of the cardiovascular effects of Pb on humans, representing both European and U.S. subjects, are summarized in Table 13.6. Follow-up intervals extended to 5 years, sample sizes were moderate and the exposure marker was either PbB alone (Moller and Kristensen, 1992; Staessen et al.,1996) or blood and bone Pb (Cheng et al., 2001). The follow-up assessment of subjects in the Cadmibel Study, the PheeCad Study (Staessen et al., 1996), showed no significant association of PbB with hypertension or DBP and SBP increases. There were associations of PbB with both SBP and DBP in the Danish Glastrop Population Study (Moller and Kristensen, 1992). For a doubling of PbB in men and women, the corresponding SBP was 3.4 and 1.9 mmHg, respectively. For DBP, the changes for men and women were 3.0 and 1.1 mmHg, respectively. Part of this anomalous gender difference was due to the fact that this Danish study reported data for men gathered in 1981 and women in 1976, a 5-year interval when the PbB measure declined 40%. Finally, Cheng et al. (2001), reporting prospective results for the VA NAS effort, saw no link for PbB and risk of hypertension, while trabecular Pb levels were significantly positively linked to risk of hypertension.

The effects of Pb exposure on cardiovascular outcomes besides BP and hypertension have been reported and illustrative findings are shown in Table 13.7. These outcomes are ischemic heart disease, cardiovascular mortality, peripheral arterial disease (PAD), and other effects. Three reports concerned ischemic heart disease, and of particular note are the data retrieved from the U.S. NHANES III. Schwartz (1991) reported that EKG evidence of left ventricular hypertrophy (LVH) in 9,900 subjects aged 25−74 years indicate that for a 10-μg/dl PbB increase, the LVH odds ratio was 1.33. Gustavsson et al. (2001) reported the relative risks (RR) for myocardial infarctions among Swedish subjects in the Stockholm Heart Epidemiology Program with increases in Pb levels in ambient dusts to be little changed from unexposed: 0.88 and 1.03 for the two elevated dust Pb groups. Boston, MA, NAS male subjects aged <65 versus aged >65 years (mean = 68 years) were examined with respect to any associations of EKG changes and PbB or patellar and tibial bone Pb changes (Cheng et al., 1998). No associations

TABLE 13.6 Illustrative Longitudinal Epidemiological Studies of Cardiovascular Effects of Lead in Nonoccupational Populations

Study Group	Pb Exposure	Endpoint Measures	Results	References
Phee Cad, Belgium Study subjects, men and women ($N = 728$); randomly selected adults 20–82 years of age, median follow-up, 5.2 years	Pb mean: start, 8.7 μg/dl; end, 2.9 μg/dl	SBP, DBP, hypertension	No significant association of BP with hypertension; no association between BP change and PbB change	Staessen et al. (1996)
Boston, MA, NAS subjects, longitudinal part of study, males ($N = 474$)	Baseline PbB mean = 5.9 μg/dl	Risk of hypertension	Trabecular bone Pb significantly positively linked to risk of hypertension; no association of PbB with risk of hypertension	Cheng et al. (2001)
Danish Glostrup Population Study: Copenhagen, Denmark residents born in 1936 ($N = 1{,}009$), followed from 1976 to 1981 (women), to 1987 (men)	PbB range 2–62 μg/dl, declined 40% over period	SBP, DBP	Associations of PbB changes with BP changes—women, 1976: 2× PbB = SBP +3.4 mmHg, DBP +3.0 mmHg; men, 1981: 2× PbB = SBP +1.9 mmHg, DBP +1.1 mmHg	Moller and Kristensen (1992)

between PbB and any EKG measures were seen, while tibial and patellar Pb levels were significantly associated with increased QT and QRS intervals, but not arrhythmias, in the younger group.

Three studies provide a body of evidence for Pb-associated mortality outcomes for nonoccupational adult Pb exposures in the United States. Two employed NHANES III databases, and one evaluated mortality data from the ongoing NAS assessment of elderly Boston, MA, males.

Schober et al. (2006) examined cardiovascular mortality rates among U.S. NHANES III subjects aged 40 years followed 8.6 years ($N = 9{,}757$) and ranked by three levels of PbB: <5, 5–9, and ≥ 10 μg/dl. Increased cardiovascular mortality risk was associated with the two highest PbB categories, the trend test being significant. The previous survey, NHANES II, included data analyzed by Lustberg and Silbergeld (2002) for mortality rates as

TABLE 13.7 Illustrative Epidemiological Studies of Cardiovascular Effects of Lead Other than Blood Pressure and Hypertension

Study Group	Pb Exposure	Endpoint Measures	Results	References
A. Ischemic heart disease				
Stockholm, Sweden Heart Epidemiology Program subjects	Three Pb levels in ambient dusts: unexposed, mean Pb-dust 0.01 mg/m^3, mean Pb-dust 0.1 mg/m^3	RR for myocardial infarctions	RR versus the two dust Pb categories: 0.88 and 1.03 versus unexposed	Gustavsson et al. (2001)
U.S. NHANES II subjects (*N* ca. 9,900), aged 25–74 years	PbB levels, range unreported	LVH via EKGs	LVH odds ratio = 1.33 for 10 μg/dl PbB increase	Schwartz (1991)
NAS subjects, males <65 years old, >65years old (*N* = 775), mean age = 68 years	PbB and tibial, patellar bone Pb levels	EKG changes: QT, QRS intervals, atrioventricular conduction defects, presence of arrhythmias	Tibial, patellar Pb levels significantly linked to increased QT and QRS intervals in subjects <65 years old but not to arrhythmias; no association of linear PbB with any outcomes	Cheng et al. (1998)
B. Cardiovascular/circulatory mortality				
U.S. NHANES II subjects (*N* = 4,190) aged 30–74 years, followed to 1992	Three PbB (μg/dl) groups: <10, 10–19, 20–29	Circulatory mortality rates for the two higher PbB groups versus the lowest group	PbB 20–29 μg/dl showed significantly elevated mortality risk ratio of 1.39	Lustberg and Silbergeld (2002)
U.S. NHANES III subjects (*N* = 9,757) aged ≥40, followed 8.6 years	PbB (μg/dl) at three levels: <5, 5–9, ≥10	Three categories of mortality rates: all cause, cardiovascular, cancer	PbB in the higher two groups associated with increased cardiovascular mortality; trend test was significant	Schober et al. (2006)

(*Continued*)

TABLE 13.7 Illustrative Epidemiological Studies of Cardiovascular Effects of Lead Other than Blood Pressure and Hypertension—(cont.)

Study Group	Pb Exposure	Endpoint Measures	Results	References
U.S. NHANES III subjects ($N = 13{,}946$), age >17 years	PbB (μg/dl) at three levels: 1st tercile, <1.93 μg/dl, 1.93–3.63, and 3.63–10 μg/dl	Three categories of mortality rates: all cause, cardiovascular, myocardial infarction	Trends of increasing HR with increasing PbB significant ($p < 0.017$) for all models	Menke et al. (2006)
NAS longitudinal groups ($N = 860$)	Patellar Pb (μg/g) at three levels at baseline: <22, 22–36, >35 μg/g bone	Three mortality categories: all cause, all cardiovascular, cancer	Adjusted HR (p, trend): 0.02, all causes; 0.003 all cardiovascular; 0.14 cancer	Weisskopf et al. (2009)
Prospective cohort study, U.S. women ($N = 533$) aged 65–87 years, osteoporotic fractures subjects	PbB mean = 5.3 μg/dl	Cox proportional analyses: PbB ≥8 μg/dl versus PbB <8 μg/dl	Women with PbB ≥ 8 μg/dl had a HR for heart disease mortality of 3.08 ($p < 0.016$) versus <8 μg/dl	Khalil et al. (2009)
Copenhagen, Denmark residents ($N = 1{,}052$), followed 1976–1990	PbB, log-transformed	Fatal, nonfatal cardiovascular disease rates	No associations of PbB with risks of all outcomes	Moller and Kristensen (1992)
C. Peripheral arterial disease (PAD) and other outcomes				
U.S. NHANES IV subjects, 1999–2000 (total $N = 2{,}125$, 139 with PAD)	PbB (μg/dl) by quartiles: 1st, <1.4, 4th, >2.9	PAD	Fourth quartile PbB significantly associated with PAD: odds ratio 3.78	Navas-Acien et al. (2004)
NHANES 1999–2002, U.S. adults ($N = 9{,}961$ total studies)	PbB-based quartiles (μg/dl): <1.06, 1.06–1.63, 1.63–2.47, >2.47. Total GM = 1.64 μg/dl	Prevalence and adjusted odds ratios for PAD by quartile	Significant trends in prevalence and OR of PAD with PbB; prevalence 2.6–7.7% ($p < 0.001$ in both cases)	Muntner et al. (2005)

(Continued)

TABLE 13.7 Illustrative Epidemiological Studies of Cardiovascular Effects of Lead Other than Blood Pressure and Hypertension—(cont.)

Study Group	Pb Exposure	Endpoint Measures	Results	References
Boston, MA, Hospital for Women, women in labor ($N = 3{,}851$)	Cord blood Pb mean = 6.9 μg/dl	Pregnancy hypertension, delivery BP	Cord Pb levels significantly linked to SBP during labor and with pregnancy hypertension; 10 μg/dl increase = 3 mmHg SBP; at 15 μg/dl, risk is 2.2 versus 0.7 μg/dl	Rabinowitz et al. (1987)

followed up to 1992 in 4,190 individuals aged 30–74 years. The study group was stratified into one of three groups based on PbB: <10, 10–19, and 20–29 μg/dl. Individuals at the highest exposure level showed a significantly elevated risk ratio of 1.39.

Menke et al. (2006) examined an NHANES III cohort >17 years old ($N = 13{,}946$) using PbB stratified into terciles: <1.93, 1.93–3.63, >3.63–10 μg/dl, and analyzed for three categories of mortality rates versus PbB tercile: all cause, cardiovascular and myocardial infarction, and stroke. Third tercile PbB subjects (versus those in the lowest PbB group) had significant trends of increasing adjusted hazard ratios (HRs) for all cardiovascular mortality models evaluated ($p < 0.017$).

The NAS researchers longitudinally studied 860 subjects for Pb exposure versus patellar bone levels at three baseline terciles: <22, 22–35, >35 μg Pb/g bone (Weisskopf et al., 2009). Three mortality classifications were all cause, all cardiovascular, and cancer. The adjusted HR for all cardiovascular mortality versus highest tercile was 5.63, excluding baseline of existing cardiovascular disease or stroke subjects. Trend significance level for all cardiovascular mortality was $p < 0.003$.

Moller and Kristensen (1992) found no associations between log-transformed PbB and fatal or nonfatal cardiovascular disease rates, using a much smaller sample size ($N = 1{,}052$) and followed 1976–1990. In addition to a smaller sampling, this effort covered a 14-year period in which there was a pronounced overall decline in PbB associated with European controls on leaded gasoline. The NHANES II was carried out in a similar time frame with a concomitant fall in strata-variable mean PbB values, but the fourfold larger sample size and stratification of PbB permitted a more sensitive analysis in the NHANES II database.

A significant association with an odds ratio of 3.78 for PAD was reported with the fourth quartile of Pb exposure by Navas-Acien et al. (2004) for U.S. NHANES IV subjects ($N = 2{,}125$, 139 with PAD, 1999–2000) ranked by quartiles of PbB. The fourth quartile of subjects had PbB levels >2.9 μg/dl.

Pregnancy hypertension as a recognized risk factor for both maternal and fetal health was studied by Rabinowitz et al. (1987) using a cohort of pregnant women in a Boston, MA, hospital ($N = 3{,}851$) studied for pregnancy hypertension and delivery BP. The cord PbB mean was 6.9 μg/dl, and cord levels were significantly associated with SBP during labor and with pregnancy hypertension, with a 10-unit PbB increase associated with a 3-mmHg SBP rise. Compared to risk at 0.7 μg/dl, the risk ratio for 15 μg/dl was 2.2.

Given the complexity and to some extent the variability of the epidemiological relationships among cardiovascular effects and Pb exposure in human populations, researchers have resorted to meta-analyses. Four particularly significant meta-analyses and their summary data are presented in Table 13.8. Navas-Acien et al. (2008) analyzed eight studies. The principal exposure marker was tibial Pb (six studies), while BP and hypertension risk were endpoints. Overall, both SBP and hypertension risk increased with bone Pb increases. Nawrot et al. (2002) carried out a meta-analysis using 31 studies encompassing more than 58,000 adults. The 31 studies consisted of 19 general population studies, four studies of white-collar workers and eight of blue-collar workers. Association of PbB and BP was positive and statistically significant, the effect size being more robust for SBP.

Two earlier meta-analyses are those of Schwartz (1995) for adult men in 15 studies and Staessen et al. (1995) using adult men and women ($N = 33{,}141$) from general population (13 studies) and worker (10 studies) groups. Schwartz found highly significant and consistent associations between PbB and SBP for the entire group, and he also reported that decreasing PbB by half reduced SBP by 1.25 mmHg. Staessen et al. (1995) noted that for all studies, doubling PbB resulted in increases of 1.0 and 0.6 mmHg for SBP and DBP, respectively, the figure for SBP similar to that for the relationship in Schwartz (1995).

13.4 EPIDEMIOLOGICAL STUDIES OF LEAD'S CARDIOVASCULAR EFFECTS IN CHILDREN

Studies of cardiovascular effects of lead in children under typical environmental Pb exposure conditions are few and available data are noted in Table 13.9. Noted earlier were some intrinsic factors underlying both the relative absence of observed cardiotoxicity in children and a paucity of studies. Those few studies in the literature had small sample sizes and a wide range of Pb exposures studied at quite young ages, as young as 12 months. The largest of these studies had a cohort of 780 children evaluated as part of the

TABLE 13.8 Meta-analyses of Effects of Lead Exposure on Blood Pressure in Human Populations

Scope of Meta-Analysis	Nature of Studies Evaluated	Meta-Analytical Results	References
Total of eight studies used but 10 available; two not used for statistical reasons	Evaluated mainly tibia Pb versus blood pressure (six studies)	SBP and hypertension risk increased positively with bone Pb	Navas-Acien et al. (2008)
Total of 31 studies, $N > 58{,}000$ individuals, adults only	Evaluated PbB versus blood pressure; 19: general population; 4: white-collar workers; 8: blue-collar workers	Statistically significant positive association, PbB with BP; effect size greater for SBP versus DBP	Nawrot et al. (2002)
Total of 15 studies used initially; subsequent analyses with removal of main study and addition of eight negative reports	PbB versus SBP, DBP using data only on men	Highly significant and consistent association found between PbB and SBP; no effect of adding negative studies having PbB 10–15 μg/dl, lower SBP by 1.25 mmHg	Schwartz (1995)
Total of 23 studies ($N = 33{,}141$) from general population (135 studies) and worker groups (10 studies)	PbB versus SBP, DBP using data for men and women	For all studies, doubling PbB increased SBP 1 mmHg and DBP 0.6 mmHg	Staessen et al. (1994)

TLC Study described earlier, where children having PbB 20–44 μg/dl had their SBP and DBP measured versus PbB, with or without chelation therapy (Chen et al., 2006). No significant association of SBP or DBP with PbB or with chelation was noted.

Similarly, Factor-Litvak et al. (1999) showed a positive but nonsignificant association of PbB with BP in a total of 282 children in a smelter town versus a control community as part of a prospective study done in the former Yugoslavia. Here, the PbB range was quite wide, 1–77 μg/dl, and SBP and DBP were measured at 5 years of age. Gump et al. (2005) in their study of Oswego, NY, boys and girls ($N = 122$) measured SBP and DBP at 9.5 years of age using stress tasks to induce pressure increases and related these responses to cord PbB and 2.2-year-olds' PbB levels. They noted that cord PbB was significantly associated with resting BP but recorded no effect on BP with acute stress testing.

TABLE 13.9 Cardiovascular Effects of Lead in Young Children

Study Group	Pb Exposure	Endpoint	Results	References
TLC study group children (*N* = 280) with and without chelation followed for 5 years; entry at 12–33 months old; majority, African-American (77%)	20–44 μg/dl	SBP, DBP versus PbB, chelation therapy	No significant association of SBP, DBP with PbB or chelation	Chen et al. (2006)
Former Yugoslavia Prospective Pb Study (smelter town, control town) (*N* = 282)	PbB range: 1–70 μg/dl	SBP, DBP measured at 5 years old	Nonsignificant positive association of PbB with BP	Factor-Litvak et al. (1999)
Oswego Children's Study participants (*N* = 122, 46% male)	Cord Pb mean = 2.0 μg/dl; mean at 2.2 years = 4.6 μg/dl	SBP, DBP tested at 9.5 years old using stress tasks	Cord Pb significantly associated positively with resting BP; no effect on BP with acute stress	Gump et al. (2005)

13.5 EXPERIMENTAL ANIMAL STUDIES OF CARDIOVASCULAR EFFECTS OF LEAD

Experimental data are strongly supportive of a role for Pb as a vasoactive agent, and such cardiovascular effects have long been known in the literature. Table 13.10 presents illustrative summary information for effects of Pb at wide-ranging dosing levels on BP in laboratory rats. Experimental cardiovascular effects do not appear to be strain specific. Study results, overall, confirmed the apparent variable vascular response to Pb reported in epidemiological reports, in that low Pb exposures produced a statistically significant BP response while associations tended to attenuate with increasing doses.

Rats given Pb orally in water at low dosing (100 ppm) and BP evaluated as experimental endpoint showed statistically significant increases in SBP by 8 weeks, while PbB remained quite low, 2.4–3.2 μg/dl at 2–3 months (Attri et al., 2003; Ding et al., 1998). Data in Table 13.10 indicated that Pb oral dosing at 15 ppm in drinking water for as long as 18 months did not affect BP, while levels of 30 and 50 ppm did (Boscolo and Carmignani, 1988). Expression of an apparent dual effect may be evident in the study of Khalil-Manesh et al. (1994), where water Pb administration at 0.5% Pb

TABLE 13.10 Selected Recent Experimental Animal Models of Cardiovascular Effects of Lead in Humans

Testing Species	Experimental Design/ Pb Exposures	Results	References
Adult male Wistar–Kyoto rats, control and Pb-exposed groups	$N = 10$, control or dosed; Pb = 1 ml/100 ppm Pb acetate/d orally, up to 3 months; PbB = 2.4 μg/dl, 2 months; BP as endpoint	Dosed animals had statistically increased SBP	Attri et al. (2003)
Adult male Sprague–Dawley rats, control and dosed groups	$N = 7$, control or dosed; Pb = 100 ppm in drinking water, 12 weeks; PbB = 3.2 μg/dl at 14 weeks; BP as endpoint	Mean SBP in dosed animals statistically significantly higher than control by 8 weeks, rising to 14 weeks	Ding et al. (1998)
Adult male Sprague–Dawley rats, control and dosed groups	Control and dosed: Pb dosing (1) 100 ppm in drinking water, 6 months, water 6 months; or (2) 0.5% Pb water 6 months, Succimer® administered intermittently; BP as endpoint	BP increased with 0.01% Pb but not 0.5% Pb; Succimer® lowered BP in both dosing groups	Khalil-Manesh et al. (1994)
Adult male Sprague–Dawley rats, control and dosed groups	Dosing: 15, 30, 60 μg/ml Pb in drinking water, 18 months; BP as endpoint	BP increased with the 30 and 60 μg/ml Pb dosing groups	Boscolo and Carmignani (1988)

produced no increase in BP in male Sprague–Dawley rats, while a 50-fold lower dosing level did. In both cases, use of Succimer® chelating agent lowered BP.

13.6 SOME LIKELY TOXICOLOGICAL MECHANISMS OF Pb-ASSOCIATED CARDIOVASCULAR TOXICITY IN HUMANS AND TEST ANIMALS

Blood pressure (BP) and hypertension associated with Pb exposure appear to be related to toxic Pb effects on vascular reactivity, oxidative stress responses, and the renin–angiotensin–aldosterone system (RAAS).

Lead exposure affects BP through the renin–angiotensin system as studied in male adult rats (Boscolo and Carmignani, 1988; Rodriguez-Iturbe et al., 2005). Rodriguez-Iturbe et al. noted that the effects of Pb on the angiotensin II system included tubulointerstital inflammation, cells expressing angiotensin II, and activation of NF-kappa B. The RAA system is clearly involved in Pb-linked hypertension, especially in young animals (Bravo et al., 2007), the RAAS working via angiotensin II (Robles et al., 2007). Pb exposure in rats and humans has been linked to altered cardiac function and production of atherosclerosis via such pathways as internal medial thickening (Zeller et al., 2010). Pb exposure has also been studied with regard to vascular endothelial dysfunction, arising from Pb effects on inhibition of endothelium repair and impaired proliferation via endoplasmic reticulum stress (Shinkai et al., 2010).

Pb may induce increased reactivity to such catecholamines as norepinephrine, experimental data indicating a response at 50 but not 100 ppm Pb (Zhang et al., 2009). Furthermore, Pb particularly appears to affect α-adrenergic and β-adrenergic receptors, with antagonists to those receptors abolishing such effects as aortic contraction (Heydari et al., 2006). Male lead workers exposed for an average period of 6 years showed plasma norepinephrine and β-2-adrenoreceptor density highly correlated with PbB and BP, while PbB was also closely associated with BP (Chang et al., 1996).

There also appears to be some role for Pb on •NO in its role as endothelium-derived relaxing factor (Silveira et al., 2010). Pb exposure during gestation produced altered reactivity at least in part because of altered •NO release (Grizzo and Cordellini, 2008).

Pb also functions in experimental hypertensive and other vascular disease systems by inducing oxidative stress via ROS and nitric oxide, with increased superoxide, elevated malondialdehyde (MDA) and increased 3-nitrotyrosine. Vaziri et al. (1997) dosed adult male rats with low levels of Pb in drinking water (100 ppm, 12 weeks) followed by treatment with the antioxidant lazaroid. Pb treatment produced experimental hypertension in concert with increased ROS level and decreased NO_x excretion. The antioxidant abolished the Pb-induced hypertension and normalized both plasma MDA and urinary NO_x. Similarly, these workers (Vaziri et al., 1999) showed that vitamin E obliterated the hypertension while normalizing nitrotyrosine and urinary NO_x levels. Table 13.11 also notes the role of tempol, a superoxide dismutase (SOD)-mimetic drug, in modifying the oxidative stress induction of the hypertensive response in Pb-exposed rats (Vaziri et al., 2001). In particular, the drug reversed the upregulation of NO synthetase isoforms.

Stimulation of NAD(P)H oxidase with Pb exposure increases, and inhibition of the enzyme by antioxidants reduces this effect. Similarly, Pb exposure increases antioxidant enzymes, especially in female rats.

Table 13.11 lists, in summary form, some of the studies relevant to likely mechanisms of Pb cardiovascular toxicity. The studies are from the more recent literature and are not meant to be exhaustive. Most but not all of the data elaborating likely mechanisms arise from experimental animal models,

TABLE 13.11 **Illustrative Likely Mechanisms of Lead-Associated Hypertension and Other Cardiovascular Effects in Human and Experimental Populations**

Study System	Study Design	Pb Exposure	Results	References
A. *In vivo* studies				
Lead workers ($N = 27$, 26 males); mean exposure = 6 years	Evaluation of β-2-adrenoreceptor density and catecholamine responses	Workplace exposure	Plasma norepinephrine and β-2-adrenoreceptor density highly correlated with PbB and BP, PbB closely associated with BP	Chang et al. (1996)
Male Sprague–Dawley adult rats, controls and exposed ($N = 6$)	Low Pb exposure followed by treatment with antioxidant lazaroid (5 mg/kg, i.p., 2 × daily) for 2 weeks	100 ppm in drinking water, 12 weeks	Lazaroid abolished the hypertension, normalized plasma MDA and urinary NO_x. Pb exposure produced hypertension in concert with increased reactive oxygen species production and decreased NO_x excretion	Vaziri et al. (1997)
Male Sprague–Dawley adult rats, controls and exposed	Low Pb exposure, with or without vitamin E-fortified diet	100 ppm in drinking water, 12 weeks	Pb exposure without vitamin E produced hypertension increase in tissue nitrotyrosine, decrease in urinary NO_x. Vitamin E use halted hypertension and normalized tissue nitrotyrosine and urinary NO_x	Vaziri et al. (1999)
Male Sprague–Dawley adult rats, controls and exposed	Low Pb exposure, with or without tempol, a SOD-mimetic drug (15 mmol/kg/day, i.p., 2 weeks	100 ppm in drinking water, 12 weeks, then treatment	Pb exposure increased BP, reduced many NO metabolites, and upregulated NO synthase (NOS) isoforms; tempol halted hypertension, increased urinary NO metabolites and reversed the upregulation of NOS	Vaziri et al. (2001)

(*Continued*)

TABLE 13.11 Illustrative Likely Mechanisms of Lead-Associated Hypertension and Other Cardiovascular Effects in Human and Experimental Populations—(cont.)

Study System	Study Design	Pb Exposure	Results	References
Adult male Sprague–Dawley rats, control and dosed groups	Evaluation of the rennin–angiotensin system	Dosing: 15, 30, 60 μg/ml Pb in drinking water, 18 months	Lead exposure at 30 and 60 μg/ml increased BP by affecting the rennin–angiotensin system; increased responsiveness of α-2-adrenoreceptors; increased reactivity to stimulation and cardiac and vascular β-adrenergic and dopaminergic receptors	Boscolo and Carmignani (1988)
Adult male Sprague–Dawley rats, control and dosed groups	Evaluation of renal interstitial inflammation, nuclear factor-Kappa β (NF-Kappa β) activation, and expression of angiotensin II	Dosing: 100 ppm Pb acetate in drinking water	Exposed animals showed tubule interstitial inflammation, cells expressing angiotensin II and activation of NF-Kappa β	Rodriguez-Iturbe et al. (2005)
Adult male Sprague–Dawley rats, control and dosed groups	Evaluation of variable Pb dosing effects on NO_x and vascular responsiveness	100 ppm Pb, drinking water, levels short term: 4, 8 weeks; subchronic, 12 weeks	Short-term effects: NO_x redirection, elevated BP; subchronic effects: reversal of NO_x redirection, continued Pb increase; significant change in adrenergic and NO pathways	Heydari et al. (2006)
Adult male Wistar rats, control and dosed groups	Evaluation of Pb effects on angiotensin II-produced renal vasoconstriction	100 ppm Pb, drinking water, 12 weeks: presence and absence of L-NAME	Pb exposure increased BP endothelial NO synthase protein, oxidative stress and vascular reactivity to angiotensin II; L-NAME had no effect on the Pb group, nitrites were lower	Robles et al. (2007)
Adult male rats, control and dosed groups	Evaluation of Pb effects on BP with and without immunosuppression	100 ppm Pb drinking water; 14 weeks with/without mycophenolate mofetil	Mycophenolate mofetil suppressed the tubulo-interstitial accumulation of lymphocytes, macrophages; prevented oxidative stress and hypertension; NF-Kappa B activated	Bravo et al. (2007)

Wistar weaned rats (23 days old), control and dosed groups	*In utero* and perinatal Pb exposure: intact or denuded aortic rings precontracted, phenylephine then L-NAME, acetylcholine treatments	Dams, pregnancy and lactation: 1000 ppm Pb, drinking water	Prenatal Pb caused hypertension, changes in the production, release of basal, stimulated endothelium relaxing factors: Pb affects NO and cyclooxygenase pathways	Grizzo and Cordellini (2008)
Adult male Wistar rats, control and dosed groups	Pb-associated changes in tail BP, 5-HT, 5-HT 2B receptor, aortic HT response, pathological changes	100 ppm Pb, drinking water, 10 months	Pb chronic exposure increases BP, aortic contractility to 5-HT was decreased: aortic media and medial-lumen ratio of exposed group increased	Zhang et al. (2009)
B. *In vitro* studies				
Vascular bovine endothelial cells	Induced expression of endoplasmic reticulum chaperones GRP 78, GRP 94, and JIVK-AP-1 pathways	Various medium doses of Pb nitrate	Pb induces ER stress in a dose–response fashion, but stress is blocked by the chaperone protein upregulation	Shinkai et al. (2010)
Perfused adult male Wistar rat tails, with/without Pb	Isolated rat tail vasoconstriction with acute Pb treatment; role of endothelium	100 μM Pb acetate solution, 1 hour perfusion; relaxation agents, phenylephrine pre-contraction, L-NAME, other agents	Acute Pb treatment targets endothelium, releasing cyclooxygenase vasoconstrictors and generating ROS	Silveira et al. (2010)

such as male adult Sprague–Dawley rats, given low levels of Pb in drinking water (Boscolo and Carmignani, 1988; Rodriguez-Iturbe et al., 2005; Vaziri et al., 1997, 1999, 2001).

Human data are represented by results from lead workers.

REFERENCES

Attri, J., Dhawan, V., Mahmood, S., Pandhi, P., Parvana, H.K., Nath, R., 2003. Effect of vitamin C supplementation on oxidative DNA damage in an experimental model of lead-induced hypertension. Ann. Nutr. Metab. 47, 294–301.

Belknap, E.L., 1936. Clinical studies on lead absorption in the human: III. Blood pressure observations. J. Ind. Hyg. Toxicol. 18, 380–390.

Boscolo, P., Carmignani, M., 1988. Neurohumoral blood pressure regulation in lead exposure. Environ. Health Perspect. 78, 101–106.

Bost, L., Primatesta, P., Dong, W., Poulter, N., 1999. Blood lead and blood pressure: evidence from the Health Survey for England 1995. J. Hum. Hypertens. 13, 123–128.

Bravo, Y., Quiroz, Y., Ferrebuz, A., Vaziri, N.D., Rodriguez-Iturbe, B., 2007. Mycophenolate mofetil administration reduces renal inflammation, oxidative stress, and arterial pressure in rats with lead-induced hypertension. Am. J. Physiol. Renal Physiol. 293, F616–F623.

Chang, H.-R., Chen, S.-S., Chen, T.-J., Ho, C.-K., Chiang, C.-H., Yu, H.-S., 1996. Lymphocyte β2-adrenergic receptors and plasma catecholamine levels in lead-exposed workers. Toxicol. Appl. Pharmacol. 139, 1–5.

Chen, A., Rhoads, G.G., Cai, B., Salganik, M., Rogan, W.J., 2006. The effect of chelation on blood pressure in lead-exposed children: a randomized study. Environ. Health Perspect. 114, 579–583.

Cheng, Y., Schwartz, J., Vokonas, P.S., Weiss, S.T., Aro, A., Hu, H., 1998. Electrocardiographic conduction disturbances in association with low-level lead exposure (the Normative Aging Study). Am. J. Cardiol. 82, 594–599.

Cheng, Y., Schwartz, J., Sparrow, D., Aro, A., Weiss, S.T., Hu, H., 2001. Bone lead and blood lead levels in relation to baseline blood pressure and the prospective development of hypertension: the Normative Aging Study. Am. J. Epidemiol. 153, 164–171.

Cramér, K., Dahlberg, L., 1966. Incidence of hypertension among lead workers: a follow-up study based on regular control over 20 years. Br. J. Ind. Med. 23, 101–104.

Den Hond, E., Nawrot, T., Staessen, J.A., 2002. The relationship between blood pressure and blood lead in NHANES III. J. Hum. Hypertens. 16, 563–568.

Ding, Y., Vaziri, N.D., Gonick, H.C., 1998. Lead-induced hypertension. II. Response to sequential infusions of L-arginine, superoxide dismutase, and nitroprusside. Environ. Res. 76, 107–113.

Dingwall-Fordyce, I., Lane, R.E., 1963. A follow-up study of lead workers. Br. J. Ind. Med. 20 (20), 313–315.

Dreessen, W.C., Edwards, T.I., Reinhart, W.H., Rage, R.T., Webster, S.H., Armstrong, D.W., et al., The Control of the Lead Hazard in the Storage Battery Industry. U.S. Public Health Service, Washington, DC, Public Health Service Bull. 262.

Factor-Litvak, P., Wasserman, G., Kline, J.K., Graziano, J., 1999. The Yugoslavia prospective study of environmental lead exposures. Environ. Health Perspect. 107, 9–15.

Gerr, F., Letz, R., Stokes, L., Chettle, D., McNeil l, F., Kaye, W., 2002. Association between bone lead concentration and blood pressure among young adults. Am. J. Ind. Med. 42, 98–106.

Glenn, B.S., Bandeen-Roche, K., Lee, B.K., Weaver, V.M., Todd, A.C., Schwartz, B.S., 2006. Changes in systolic blood pressure associated with lead in blood and bone. Epidemiology 5, 538–544.

Glenn, B.S., Stewart, W.F., Links, J.M., Todd, A.C., Schwartz, B.S., 2003. The longitudinal association of lead with blood pressure. Epidemiology 14, 30–36.

Grizzo, L.T., Cordellini, S., 2008. Perinatal lead exposure affects nitric oxide and cyclooxygenase pathways in aorta of weaned rats. Toxicol. Lett. 103, 207–214.

Gump, B.B., Stewart, P., Reihman, J., Lonky, E., Darvill, T., Matthews, K.A., et al., 2005. Prenatal and early childhood blood lead levels and cardiovascular functioning in 9½ year old children. Neurotoxicol. Teratol. 27, 655–665.

Gustavsson, P., Plato, N., Hallqvist, J., Hogstedt, C., Lewne, M., Reuterwall, C., et al., 2001. A population-based case-referent study of myocardial infarction and occupational exposure to motor exhaust, other combustion products, organic solvents, lead, and dynamite. Stockholm Heart Epidemiology Program (SHEEP) study group. Epidemiology 12, 222–228.

Heydari, A., Norouzzadeh, A., Khoshbaten, A., Asgari, A., Ghasemi, A., Najaf, S., et al., 2006. Effects of short-term and subchronic lead poisoning on nitric oxide metabolites and vascular responsiveness in the rat. Toxicol. Lett. 166, 88–94.

Hu, H., Watanabe, H., Payton, M., Korrick, S., Rotnitsky, A., 1994. The relationship between bone lead and hemoglobin. J. Am. Med. Assoc. 272, 1512–1517.

Hu, H., Aro, A., Payton, M., Korrick, S., Sparrow, D., Weiss, S.T., et al., 1996. The relationship of bone and blood lead to hypertension. The Normative Aging Study. J. Am. Med. Assoc. 275, 1171–1176.

International Agency for Research on Cancer, 2006. IARC Monographs on the Evaluation of Carcinogenic Risks to Humans. Volume 87: Inorganic and Organic Lead Compounds. World Health Organization Press, Geneva, Switzerland.

Khalil, N., Wilson, J.W., Talbott, E.O., Morrow, L.A., Hochberg, M.C., Hillier, T.A., et al., 2009. Association of blood lead concentrations with mortality in older women: a prospective cohort study. Environ. Health Global Access Sci. Source 8, 15.

Khalil-Manesh, F., Gonick, H.C., Weiler, E.W.J., Prins, B., Weber, M.A., Purdy, R., et al., 1994. Effect of chelation treatment with dimercaptosuccinic acid (DMSA) on lead-related blood pressure changes. Environ. Res. 65, 86–99.

Kirby, H., Gyntelberg, F., 1985. Blood pressure and other cardiovascular risk factors of long-term exposure to lead. Scand. J. Work Environ. Health 11, 15–29.

Korrick, S.A., Hunter, D.J., Rotnitsky, A., Hu, H., Speizer, F.E., 1999. Lead and hypertension in a sample of middle-aged women. Am. J. Public Health 89, 330–335.

Kósmider, S., Petelenz, T., 1962. Electrocardiographic changes in elderly patients with chronic professional lead poisoning. Pol. Arch. Med. Wewn. 32, 437–442.

Lee, B.-K., Lee, G.-S., Stewart, W.F., Ahn, K.-D., Simon, D., Kelsey, K.T., et al., 2001. Associations of blood pressure and hypertension with lead dose measures and polymorphisms in the vitamin D receptor and δ-aminolevulinic acid dehydratase genes. Environ. Health Perspect. 109, 383–389.

Lustberg, M., Silbergeld, E.K., 2002. Blood lead levels and mortality. Arch. Intern. Med. 162, 2443–2449.

Maheswaran, R., Gill, J.S., Beevers, D.G., 1993. Blood pressure and industrial lead exposure. Am. J. Epidemiol. 137, 645–653.

Martin, D., Glass, T.A., Bandeen-Roche, K., Todd, A.C., Hi, W., Schwartz, B.S., 2006. Association of blood lead and tibia lead with blood pressure and hypertension in a community sample of older adults. Am. J. Epidemiol. 163, 467–478.

Menke, A., Muntner, P., Batuman, V., Silbergeld, E.K., Guallar, E., 2006. Blood lead below 0.48 micromol/L (10 microg/dL) and mortality among U.S. adults. Circulation 114, 1388–1394.

Moller, L., Kristensen, T.S., 1992. Blood lead as a cardiovascular risk factor. Am. J. Epidemiol. 136, 1091–1100.

Muntner, P., Menke, A., DeSalvo, K.B., Rabito, F.A., Batuman, V., 2005. Continued decline in blood lead levels among adults in the United States—the National Health and Nutrition Examination Surveys. Arch. Intern. Med. 165, 2155–2161.

Nash, D., Magder, L., Lustberg, M., Sherwin, R.W., Rubin, R.J., Kaufmann, R.B., et al., 2003. Blood lead, blood pressure, and hypertension in perimenopausal and postmenopausal women. J. Am. Med. Assoc. 289, 1523–1532.

National Academy of Sciences: National Research Council, 1972. Lead: Airborne Lead in Perspective. National Academy Press, Washington, DC.

National Academy of Sciences: National Research Council, 1993. Measuring Lead Exposure in Infants, Children, and Other Sensitive Populations. National Academy Press, Washington, DC.

Navas-Acien, A., Selvin, E., Sharrett, A.R., Calderon-Aranda, E., Silbergeld, E., Guallar, E., 2004. Lead, cadmium, smoking, and increased risk of peripheral arterial disease. Circulation 109, 3196–3201.

Navas-Acien, A., Guallar, E., Silbergeld, E.K., Rothenberg, S., 2007. Lead exposure and cardiovascular disease-a systematic review. Environ. Health Perspect. 115, 472–482.

Navas-Acien, A., Schwartz, B.S., Rothenberg, S.J., Hu, H., Silbergeld, E.K., Guallar, E., 2008. Bone lead levels and blood pressure endpoints: a meta-analysis. Epidemiology 19, 496–504.

Nawrot, T.S., Thijs, L., Den Hond, E.M., Roels, E.M., Roels, H.A., Staessen, J.A., 2002. An epidemiological re-appraisal of the association between blood pressure and blood lead: a meta-analysis. J. Hum. Hypertens. 16, 123–131.

Neal, P.A., Dreessen, W.C., Edwards, T.I., Reinhart, W.H., Webster, S.W., Castberg, H.T., et al., 1941. A Study of the Effect of Lead Arsenate Exposure on Orchardists and Consumers of Sprayed Fruit. U.S. Public Health Service, Washington, DC, Public Health Bulletin 267.

Neri, L., Hewitt, D., Orser, B., 1988. Blood lead and blood pressure: analysis of cross-sectional and longitudinal data from Canada. Environ. Health Perspect. 78, 123–126.

Nomiyama, K., Nomiyama, H., Liu, S.J., Tao, Y.X., Nomiyama, T., Omae, K., 2002. Lead induced increase of blood pressure in female lead workers. Occup. Environ Med. 59, 734–738.

Orssaud, G., Claude, J.R., Moreau, T., Lellouch, J., Juguet, B., Festy, B., 1985. Blood lead concentration and blood pressure [Letter]. BMJ 290, 244.

Parkinson, D.K., Hodgson, M.J., Bromet, E.J., Dew, M.A., Connell, M.M., 1987. Occupational lead exposure and blood pressure. Br. J. Ind. Med. 44, 744–748.

Perlstein, T., Weuve, J., Schwartz, J., Sparrow, D., Wright, R.O., Litonjua, A.A., et al., 2007. Cumulative community-level lead exposure and pulse pressure: the Normative Aging Study. Environ. Health Perspect. 115, 1154–1159.

Pocock, S.J., Shaper, A.G., Ashby, D., Delves, H.T., Clayton, B.E., 1988. The relationship between blood lead, blood pressure, stroke, and heart attacks in middle-aged British men. Environ. Health Perspect. 78, 23–30.

Rabinowitz, M., Bellinger, D., Leviton, A., Needleman, H., Schoenbaum, S., 1987. Pregnancy hypertension, blood pressure during labor, and blood lead levels. Hypertension 10, 447–451.

Robles, H.V., Romo, E., Sanchez-Mendozo, A., Rios, A., Soto, V., Avila-Casado, M.C., et al., 2007. Lead exposure effect on angiotensin II renal vasoconstriction. Hum. Exp. Toxicol. 26, 499–507.

Rodriguez-Iturbe, B., Sindhu, R.K., Quiroz, Y., Vaziri, N.D., 2005. Chronic exposure to low doses of lead results in renal infiltration of immune cells, NF-kB activation, and over expression of tubulointerstitial angiotensin. II. Antioxid. Redox Signaling 7, 1269–1274.

Rothenberg, S.J., Kondrashov, V., Manalo, M., Jiang, J., Cuellar, R., Garcia, M., et al., 2002. Increases in hypertension and blood pressure during pregnancy with increased bone lead levels. Am. J. Epidemiol. 156, 1079–1087.

Schober, S.E., Mirel, L.B., Graubard, B.I., Brody, D.J., Flegal, K.M., 2006. Blood lead levels and death from all causes, cardiovascular disease, and cancer: results from the NHANES III mortality study. Environ. Health Perspect. 114, 1538–1541.

Schwartz, J., 1988. The relationship between blood lead and blood pressure in the NHANES II survey. Environ. Health Perspect. 78, 15–22.

Schwartz, J., 1991. Lead, blood pressure, and cardiovascular disease in men and women. Environ. Health Perspect. 91, 71–75.

Schwartz, J., 1995. Lead, blood pressure, and cardiovascular disease in men. Arch. Environ. Health 50, 31–37.

Schwartz, B.S., Stewart, W.F., Todd, A.C., Simon, D., Links, J.M., 2000. Different associations of blood lead, meso 2,3-dimercaptosuccinic acid (DMSA)-chelatable lead, and tibial lead levels with blood pressure in 543 former organolead manufacturing workers. Arch. Environ. Health 55, 85–92.

Scinicariello, F., Yesupriya, A., Chang, M.H., Fowler, B.A., 2010. Modification by ALAD of the association between blood lead and blood pressure in the U.S. population: results from the Third National Health and Nutrition Examination Survey. Environ. Health Perspect. 118, 259–264.

Shinkai, Y., Yamamoto, C., Kaji, T., 2010. Lead induces the expression of ER chaperones ERP78 and GRP94 in vascular endothelial cells via the JNK-AP-1 pathway. Toxicol. Sci. 114, 378–386.

Silveira, E.A., Lizardo, J.H., Souza, L.P., Stefanon, I., Vassallo, D.V., 2010. Acute lead-induced vasoconstriction in the vascular bed of isolated perfused rat tails is endothelium dependent. Braz. J. Med. Biol. Res. 43, 492–499.

Staessen, J.A., Dolenc, P., Amery, A., Buchet, J.-P., Claeys, F., Fagard, R., et al., 1993. Environmental lead exposure does not increase blood pressure in the population: evidence from the Cadmibel study. J. Hypertens. 11 (Suppl. 2), S35–S41.

Staessen, J.A., Bulpitt, C.J., Fagard, R., Lauwerys, R.R., Roels, H., Thijs, L., et al., 1994. Meta-analysis. Hypertension caused by low-level lead exposure: myth or fact? J. Cardiovasc. Risk 1, 87–97.

Staessen, J.A., Roels, H., Lauwerys, R.R., Amery, A., 1995. Low-level lead exposure and blood pressure. J. Hum. Hypertens. 9, 303–328.

Staessen, J.A., Roels, A.H., Fagard, R., 1996. Lead exposure and conventional and ambulatory blood pressure: a prospective study. PheeCad Investigators. J. Am. Med. Assoc. 275, 1563–1570.

U.S. Agency for Toxic Substances and Disease Registry, 2007. Toxicologic Profile for Lead. U.S. Centers for Disease Control, Atlanta, GA.

U.S. Environmental Protection Agency, 1977. Air Quality Criteria for Lead. Report No. EPA-600/8-77-017. Office of Research and Development, Washington, DC.

U.S. Environmental Protection Agency, 1986. Air Quality Criteria for Lead, 4 vols. Report No. EPA/600/8-83/028bF. Environmental Criteria and Assessment Office, Washington, DC.

U.S. Environmental Protection Agency, 2006. Air Quality Criteria for Lead, vol. 1. Report No. EPA/600/R-05/144aF. National Center for Environmental Assessment, Washington, DC.

Vaziri, N., 2008. Mechanisms of lead-induced hypertension and cardiovascular disease. Am. J. Physiol. Heart Circ. Physiol. 295, H454–H465.

Vaziri, N.D., Ding, Y., Ni, Z., Gonick, H.C., 1997. Altered nitric oxide metabolism and increased oxygen free radical activity in lead-induced hypertension: effect of lazaroid therapy. Kidney Int. 52, 1042–1046.

Vaziri, N.D., Ding, Y., Ni, Z., 1999. Nitric oxide synthase expression in the course of lead-induced hypertension. Hypertension 34, 558–562.

Vaziri, N.D., Ding, Y., Ni, Z., 2001. Compensatory up-regulation of nitric-oxide synthase isoforms in lead-induced hypertension: reversal by a superoxide dismutase-mimetic drug. J. Pharmacol. Exp. Ther. 298, 679–685.

Vigdartchik, N.A., 1935. Lead intoxication in the etiology of hypertonia. J. Ind. Hyg. 17, 1–6.

Vupputuri, S., He, J., Muntner, P., Bazzano, L.A., Whelton, P.K., Batuman, V., 2003. Blood lead level is associated with elevated blood pressure in blacks. Hypertension 41, 463–468.

Weaver, V.M., Ellis, L.R., Lee, B.K., Todd, A.C., Shi, W., Ahn, K.D., et al., 2008. Associations between patella lead and blood pressure in lead workers. Am. J. Ind. Med. 51, 336–343.

Weiss, S.T., Munoz, A., Stein, A., Sparrow, D., Speizer, F.E., 1986. The relationship of blood lead to blood pressure in a longitudinal study of working men. Am. J. Epidemiol. 123, 800–808.

Weisskopf, M.G., Jain, N., Nie, H.L., Sparrow, D., Vokonas, P., Schwartz, J., et al., 2009. A prospective study of bone lead concentration and death from all causes, cardiovascular diseases, and cancer in the Department of Veteran Affairs—Normative Aging Study. Circulation 120, 1056–1064.

World Health Organization, 1995. Environmental Health Criteria 165: Inorganic Lead. International Programme on Chemical Safety, Geneva, Switzerland.

Zeller, I., Knoflach, M., Seubert, A., Kreutmayer, S., Stelzmüller, M., Wallnoefer, E., et al., 2010. Lead contributes to arterial intimal hyperplasia through nuclear factor erythroid-2 related factor-mediated endothelial interleukin 8 synthesis and subsequent invasion of smooth muscle cells. Arterioscler. Thromb. Vasc. Biol. 30, 1733–1740.

Zhang, L.-F., Peng, S.-Q., Wang, S., 2009. Decreased aortic contractile reaction to 5-hydroxytryptamine in rats with long-term hypertension induced by lead (Pb2+) exposure. Toxicol. Lett. 186, 78–83.

Chapter 14

Reproductive and Developmental Toxicity of Lead in Human Populations

14.1 INTRODUCTION

This chapter presents the toxicological evidence for lead as a potent reproductive and intrauterine developmental toxicant. Sections of the chapter deal with male and female gametotoxicity, embryotoxicity, fetotoxicity, and various gross indices of fetal injury such as spontaneous abortion, stillbirth, and neonatal survival. Prenatal developmental effects as they affect the whole organism are presented here, while prenatal effects on various systems and organs are discussed in system- and organ-specific chapters. For example, developmental neurotoxicity of lead as it occurs *in utero* is discussed in Chapter 12.

The bulk of the evidence for gross prenatal effects in humans such as survival to term and/or into the immediate postnatal period traces to early occupational exposures of women in various work settings and from the use of inorganic lead salts as abortifacients. This early history of acute or chronic lead poisoning is a history of extremely high lead exposures, whether in the workplace or from self-exposures among nonoccupational populations. This toxicological history traces to the nineteenth century (Lane, 1949; Oliver, 1911; Paul, 1860; Taussig, 1936) in terms of some sense of exposure levels associated with severe reproductive and developmental toxicity.

By the first decade of the twentieth century, recognition of reproductive and prenatal harm and subsequent responses greatly reduced the adverse effects of lead *in utero* during worker pregnancies through constraints on women of childbearing age in the lead workplace. In particular, the prevalence and incidence of such gross Pb toxicity outcomes as spontaneous abortions, stillbirths, and depressed early natal survival declined markedly.

It would be incorrect to assume that lead's adverse impacts on reproductive competence and prenatal health and development in human populations is mainly a matter of historical medical interest because current environmental

Trace Metals and other Contaminants in the Environment, Volume 10
ISSN: 1875-1121 DOI: 10.1016/B978-0-444-51554-4.00014-6

lead exposures of pregnant women or women of childbearing age now are well below those producing gross injury.

First, reproductive and developmental lead toxicity in both men and women remains a significant human health issue on several grounds, whether adequately characterized in terms of dose–response relationships or not. Occupational lead exposures remain as sources of toxicity in both industrialized and developing countries. These exposures of both male and female workers potentially persist at levels not greatly different from those reported in past decades.

Restricting women of childbearing age from the lead industry workforce for most of the twentieth century because of concerns about fetal exposures and toxic hazards to prenatal populations clearly reduced the prevalence and incidence of reproductive and developmental toxicity. Paternal exposures to lead in terms of reproductive competence indices has continued, based on occupational epidemiology data from several databases.

Second, U.S. legal rulings in the late twentieth century have challenged traditional workplace restrictions for younger women on grounds including economic and protective equity. Those rulings, in essence, require that reduced Pb exposure risk be adequate for workers regardless of gender.

A third cause of continued concern for the toxic potential of Pb is the inadequacy of U.S. occupational rules and regulations in recognizing the current health science and the human toxicological picture of lead's potency in both adult and pediatric human populations. U.S. occupational safety and health criteria and triggers for worker lead exposures in men, for example, specify PbB levels in the range of 50–60 μg/dl for removal from exposure and 40 μg/dl as the return to the workplace level. One reading of 60 μg/dl or two consecutive readings of 50 μg/dl mandate medical removal. These are levels above reported thresholds for reproductive toxicity in male workers. U.S. regulations also include a broad caveat and recommendation, not a requirement, that reproductive-age women in the workplace not exceed a PbB of 30 μg/dl. In essence, medical removal criteria and triggers for women in terms of regulatory requirements do not differ from those for male workers.

This incoherent and inconsistent body of regulations was correctly criticized by Bellinger (1994) as grossly unprotective for preventing fetal harm. Circulating Pb in pregnant women readily crosses the placental barrier and this access likely occurs within, and certainly by the end of, the first trimester. As noted later in this chapter, transplacental movement continues through gestation and to delivery. Available evidence indicates that fetal Pb closely tracks maternal circulating Pb, such that cord blood Pb is almost the same as maternal PbB at delivery. This relationship has been common regulatory scientific knowledge for decades.

A fourth concern about the need for continued appraisal of reproductive and developmental effects of Pb exposures in human populations is the incomplete and fragmentary picture for dose–response relationships in both humans and experimental test species. These dose–response gaps raise the question of which dose/exposure metric is most specific and sensitive for predicting reproductive and prenatal developmental harm. To date, the exposure metric has been PbB in terms of biomarkers. In other cases, environmental markers such as air Pb have been employed.

Two problems with PbB versus reproductive and fetal harm relationships are: (1) the sensitivity of current dose–responses in terms of both intensity and time course of Pb exposures and (2) the dynamic, labile nature of bone Ca resorption, associated bone Pb release, and Pb toxicokinetics through the course of pregnancy, especially in the third trimester.

This chapter presents the principal adverse outcomes for fetal health and development, arrayed in three categories. The first reflects the most pronounced and clinically indisputable adverse outcomes: spontaneous abortions, stillbirths, and first-year survival rates. Discussion of these outcomes is followed by a treatment of somewhat less severe effects: premature delivery and impaired fetal growth rates. The evidence for lead-associated congenital effects of an anatomical/structural nature, although limited for human populations, is also presented. The role of Pb as a male reproductive toxin, especially in those individuals having elevated occupational Pb exposures, is discussed. Studies of Pb effects on experimental animal reproduction and development are known and a selection of those results of general utility for human comparisons are presented, followed by the intrauterine toxicokinetics of Pb. Lead's ready entry into fetal tissue during gestation is the toxicological basis for subsequent adverse effects.

14.2 REPRODUCTIVE AND DEVELOPMENTAL Pb TOXICITY IN HUMAN POPULATIONS

14.2.1 Lead and Fetotoxic Effects

Lead's record as an injurious, even lethal fetal toxicant in humans dates in modern times to the late nineteenth and early twentieth centuries. Some of the associated statistics are presented in Table 14.1. Virtually all of this documented toxicological record for toxic exposures of pregnant women in the workplace is found in European records since more detailed registry information existed at those times in Europe than in the United States. The U.S. interest at the time was directed more to men working in the various lead industries, based on the widely disseminated research and writings of occupational physician Alice Hamilton (1914a,b, 1919, 1924, 1925).

Oliver's (1911) cumulative tabulations of the fetotoxic consequences of various workplace lead exposures in pregnant women (Table 14.1) depict two- to threefold higher spontaneous abortions, stillbirths, and neonatal deaths in married women in the lead trades versus housewives who did not

TABLE 14.1 Reproductive and Developmental Effects of High Lead Exposures in Human Populations: Spontaneous Abortion, Stillbirths, Infant Survival

Group	Pb Exposure	Endpoint	Results	References
Pregnant women, general population, 39, total studied, 22 in detail	Ingestion of Pb oleate (diachylon) as abortifacient	Abortions	12/22 women had miscarriages and several clinical Pb poisoning	Hall and Cantab (1905)
Pregnant Pb workers; pregnant non-Pb workers; housewives	Premarital, marital exposures in workplace, controls	Spontaneous abortions, stillbirths, Year 1 infant death rates	Abortions/stillbirths (per 1,000): Premarital Pb exposure, 86.0 Marital Pb exposure, 133.5 Mill workers, 47.6 Housewives, 43.2 Neonatal deaths (per 1,000): Premarital Pb exposure, 157 Marital Pb exposure, 271 Mill workers, 214 Housewives, 150	Oliver (1911)
Pregnant French Pb workers	Workplace Pb	Spontaneous abortion	61% (609/1,000) rate of spontaneous abortions	Cited by Taussig (1936) from French Department of Labor (1905)
Pregnant Pb workers	Air Pb = 750 μg/m^3	Spontaneous abortions, stillbirth rates	Increased number of abortions and stillbirths in pregnant Pb workers with elevated Pb-air	Lane (1949)
Japanese pregnant women, before and after Pb work (*N* = 104)	Workplace Pb	Spontaneous abortions	Pb work produced almost twofold increase in miscarriages	Nogaki (1957)

(*Continued*)

TABLE 14.1 Reproductive and Developmental Effects of High Lead Exposures in Human Populations: Spontaneous Abortion, Stillbirths, Infant Survival—(cont.)

Group	Pb Exposure	Endpoint	Results	References
Swedish smelter workers ($N = 551$)	Workplace Pb, other substances	Spontaneous abortion rate	Abortion rate=13.9% during pregnancy and smelter employment; 17% when also employed at smelter before pregnancy and living near smelter	Nordström et al. (1979)
Pregnancies in women living near an Australian lead smelter, tested after first trimester	Proximity to the smelter, using PbB as biomarker	Spontaneous abortion rate	No association of PbB with spontaneous abortions, but abortions higher in regions closer to smelter	McMichael et al. (1986)
Pregnant women in Pb smelter and control towns, former Yugoslavia ($N = 639$)	PbBs: mean 16.0 μg/dl, Pb-exposed; 5.2 μg/dl, controls	Spontaneous abortion rate	No association of PbB with abortion rate	Murphy et al. (1990)
Finnish pregnant women employed in various industries	Registry of occupational PbB measurements	Spontaneous abortion rate	No association between maternal Pb and abortion rates	Lindbohm et al. (1991a), Taskinen (1988)
Finnish pregnant women with paternal Pb exposures	Registry of occupational PbB measurements	Spontaneous abortion rate	No association between paternal PbBs and abortion rates	Lindbohm et al. (1991b)
Mexico City pregnant women followed from first trimester ($N = 668$)	PbB levels in nested case–control study design	Spontaneous abortion rate	Odds ratio of 1.8/5 μg/dl PbB increase; dose–response seen	Borja-Aburto et al. (1999)

work outside the home. French Department of Labor figures, discussed in the monograph of Taussig (1936), showed high fetal mortality rates, as recorded for 1905, including a 61% spontaneous abortion rate (609/1,000) among pregnant French workers in various Pb industries. Abortions induced by lead compounds were also recorded by Hall and Cantab (1905), who described the use of lead oleate ("diachylon") as an abortifacient. While the abortion rate with such substances was ca. 50% or higher using mainly the oleate, the high ingestion rate required to achieve such rates was accompanied by signs and symptoms of classical and severe Pb poisoning.

Little dose–response information in terms of environmental or biometric Pb levels exists for these earliest instances of occupational and induced fetal lethality, but by the mid-twentieth century some crude exposure rates could be deduced. Lane (1949) reported that workplace air Pb measuring 750 $\mu g/m^3$ was associated with significant increases in the numbers of spontaneous abortions and stillbirths among the pregnant workers. It is not possible to precisely translate such environmental levels to PbB in these particular workers given nonlinear modeling relationships and unknown particle size distributions. However, even crude calculations predict quite high PbB levels at these air Pb concentrations, based on later published relationships. EPA analysis of factory Pb workers (U.S. EPA, 1986, Ch. 11, Tables 11–37) included analysis of the data of King et al. (1979) for three factories having air Pb levels ranging from 35 to 1,200 μg Pb/m^3 and PbB from 25 to 90 $\mu g/dl$. For the Lane (1949) air Pb mean of 750 $\mu g/m^3$, the higher end of the PbB range, $\sim$90 $\mu g/dl$, would be expected as a gross estimate. Nogaki (1957) provides PbB data versus increases in fetal lethality rates. Pregnant women exposed to workplace Pb and with a PbB range of 110–337 $\mu g/dl$ had double the miscarriage rate over their prelead-employment rate. This group ($N = 104$) essentially served as their own controls.

Table 14.1, overall, shows little significant evidence of fetal or neonatal/early postnatal lethality with the lower workplace exposures of more recent periods or for nonoccupational exposures. Finnish population registry data linking occupational exposures to medical histories of pregnancies (Lindbohm et al., 1991a,b; Taskinen, 1988) for either pregnant women or paternal partners in more recent times failed to show relationships between Pb exposures and spontaneous abortion rates. Similarly, two cohorts of pregnant women enrolled in prospective lead studies in Australia and the former Yugoslavia provided no evidence of increased spontaneous abortion rates associated with increased PbB levels. These prospective results entailed assessments after the first trimester, while the negative data for the Australian cohort was for overall outcome. When analyzed for proximity to the smelter, the spontaneous abortion rate was higher in groups closer to the facility. Borja-Aburto et al. (1999) in their assessment of 668 pregnant women followed from the first trimester and using a case–control approach reported a dose–response relationship over the total cohort, and an odds ratio of 1.8 for each 5 $\mu g/dl$ PbB increase. These different outcomes for three groups of pregnant women exposed to ambient air levels impacted by lead smelters or other sources may reflect differences in use of the exposure biomarker or study designs.

Nonlethal, less severe toxicological impacts on fetal health in the United States are associated with lower levels of Pb exposures and these findings are given in Table 14.2. Fahim et al. (1976) evaluated preterm delivery and premature membrane rupture rates for groups of U.S. pregnant women living near or far from a lead smelter. Preterm delivery rate for women having a mean PbB of 17.5 μg/dl in the smelter area was fourfold (13% versus 3%) higher than the control group rate (PbB mean = 4.6 μg/dl). The premature membrane rupture rate was proportionately much higher, 17% versus <1%, in smelter-exposed versus control groups. Angell and Lavery (1982) examined preterm delivery and premature membrane rupture rates in pregnant women in the general population with reference to PbB levels. PbB was associated with increased preterm delivery rates down to a PbB level of 5 μg/dl and increased membrane rupture frequencies at PbB levels as low as 25 μg/dl.

Three reports of Pb effects from communities with differing Pb sources and included in international prospective Pb studies produced differing results for associating Pb exposure with preterm delivery rates (Table 14.2). The Port Pirie, Australia, prospective Pb study showed an association of maternal delivery PbB with preterm delivery rate in a dose–response fashion, with odds ratio estimates ranging to 4.4 for women with a PbB >13.5 μg/dl. Negative data for Pb effects on preterm delivery, however, were noted in two other prospective Pb studies, those in Cincinnati, OH (Bornschein et al., 1989), and the former Yugoslavia (Factor-Litvak et al., 1991). Differences may have to do with timing of PbB testing, levels of Pb exposure, etc. Savitz et al. (1990) evaluated pregnant women in the U.S. National Natality Survey with reference to histories of occupational Pb exposures. Pb exposure was associated with preterm delivery (odds ratio = 2.3). Another report noting positive data was that of Torres-Sánchez et al. (1999) evaluating Mexico City pregnant women ($N = 161$, preterm; $N = 459$, full term) in the general population. Despite a relatively modest difference in mean cord PbB for the preterm versus full-term groups, 9.8 versus 8.4 μg/dl, respectively, preterm delivery rate was statistically significantly associated with higher cord PbB.

Studies of the effects of Pb exposure on fetal growth rates measured as birth weights have produced conflicting results, as recorded in Table 14.3. These are from studies differing as to methods of classifying Pb exposure, exposure biomarkers, and endpoint design features. González-Cossio et al. (1997) evaluated 272 Mexico City mother–infant pairs using as exposure biomarkers maternal and cord PbB, tibial and patellar bone Pb. Only tibial Pb was significantly associated statistically with reduced birth weight, notably in the higher quartiles. Mothers and infants in the Norwegian birth registry, when analyzed with respect to any occupational Pb history, revealed that moderate to high Pb exposures were associated with risk of having a low-birth-weight infant (Irgens et al., 1998).

TABLE 14.2 Reproductive and Developmental Effects of Lead in Human Populations: Premature Delivery Rates

Group	Pb Exposure	Endpoint	Results	References
Pregnant women living near a Pb mining/smelting operation ($N = 254$) or in a control area	PbB means for preterm versus normal deliveries	Preterm delivery, premature membrane rupture	Lead smelter area preterm delivery = 13% versus control area = 3%; smelter area premature membrane rupture = 17% versus control area = <1%; Mean PbB for preterm delivery = 17.5 μg/dl, mean for membrane rupture = 14.2 μg/dl, and normal outcome mean = 4.6 μg/dl	Fahim et al. (1976)
Pregnant women in general population	PbB range down to 5 μg/dl	Preterm delivery, premature membrane ruptures	PbB associated with preterm delivery as low as 5 μg/dl, and with membrane rupture down to 25 μg/dl	Angell and Lavery (1982)
Pregnant women, the Port Pirie, Australia, prospective Pb study	Ranges of PbB versus effects	Preterm delivery	PbB is associated with preterm delivery: odds ratios of 2.1–4.4, PbB 7.7–10.6 and >13.5 μg/dl	McMichael et al. (1986)
Pregnant women in U.S. National Natality Survey	Occupational Pb exposures	Preterm delivery	Pb exposure is associated with preterm delivery, odds ratio = 2.3	Savitz et al. (1990)
Mexico City pregnant women, case–control study ($N = 459$ full term, $N = 161$ preterm)	Cord PbB both groups: 9.8 μg/dl, preterm; 8.4 μg/dl, full term	Preterm delivery	Increased PbB significantly related to preterm delivery	Torres-Sánchez et al. (1999)
Pregnant women in former Yugoslavia prospective Pb study	Cord PbB, or mid-pregnancy, maternal PbB	Preterm delivery	No association between either PbB measure and preterm delivery	Factor-Litvak et al. (1991)
Pregnant women in Cincinnati prospective Pb study	Mid-pregnancy or neonatal period (10 days) PbB	Preterm delivery	No association of either measure with preterm delivery	Bornschein et al. (1989)

TABLE 14.3 Reproductive and Developmental Effects of Lead in Human Populations: Fetal Growth Rates

Group	Pb Exposure	Endpoint	Results	References
Mexico City mother–infant pairs	Maternal venous and cord PbB, tibial and patellar bone Pb	Birth weight	After adjustments, tibia Pb was inversely significantly associated with birth weight, notably at higher quartiles	González-Cossio et al. (1997)
Pregnant women in the Cincinnati prospective Pb study	Mean PbB = 8.0 μg/dl	Birth weight	PbB was inversely related to birth weight on a cohort sample, going from 10 to 30 μg/dl; most effect seen for 30-year-old mothers	Bornschein et al. (1989)
Women and infants in the Norwegian birth registry	Occupational Pb exposure	Birth weight	Moderate–high Pb exposure associated with risk of having a low-birth-weight infant	Irgens et al. (1998)
Boston, MA, births of at least 20 weeks gestation (N = 5,183)	PbB range <1–35 μg/dl	Birth weight characterized as low weight (<2,500 g) and other indices	Increased PbB leads to relative risk increases; cord Pb ≥15 μg/dl associated with higher risks; no association with birth weight as a continuous variable	Bellinger et al. (1991)
Pregnant women and mother–infant pairs in the Port Pirie, Australia, prospective Pb study (N = 749), ≥20 weeks gestation	PbB mean near smelter = 10.1 μg/dl, control area, 7.0 μg/dl	Birth weight; small for gestational age; reduced intrauterine growth	No association between cord PbB or mid-pregnancy PbB and birth weight	McMichael et al. (1986)

(Continued)

TABLE 14.3 Reproductive and Developmental Effects of Lead in Human Populations: Fetal Growth Rates—(cont.)

Group	Pb Exposure	Endpoint	Results	References
Pregnant women and mother–infant pairs ($N = 401$) for smelter town versus unexposed individuals ($N = 506$)	PbB mean, mid-pregnancy, smelter area = 19.0 μg/dl; PbB mean for control area = 5.6 μg/dl	Birth weight	No association of PbB with birth weight	Factor-Litvak et al. (1991)

The international prospective Pb studies revealed conflicting results with respect to Pb impacts on birth weight (Table 14.3). The Port Pirie, Australia (McMichael et al., 1986), and former Yugoslavia (Factor-Litvak et al., 1991) birth cohorts revealed no association of maternal PbB with birth weights. By contrast, the Cincinnati prospective effort indicated an inverse link for maternal PbB and birth weight in going from 10 to 30 μg/dl (Bornschein et al., 1989). There was also a maternal age effect, the relationship being most apparent for 30-year-old mothers. The findings of Bellinger et al. (1991) for a large number of births in Boston, MA (N=5,183), where the cord PbB ranged from 1 to 35 μg/dl, showed that increases in PbB ≥15 μg/dl in newborns with weight <2,500 g was associated with higher risk of low birth weight. However, this inverse relationship was not evident with birth weight as a continuous variable.

The mixed results described above for birth weight likely trace to several factors. First, they differed with exposure biomarker, as was readily apparent in the results of González-Cossio et al. (1997). Only tibial Pb was linked to this endpoint, and neither of the blood markers showed associations. This comparative finding supports the growing acceptance of better measures of body lead burden than PbB, notably bone Pb. In Chapter 13, for example, many more instances of positive associations of bone Pb with cardiovascular effects were recorded than were shown for PbB. The findings of Bellinger et al. (1991) also indicated that the type of study design affects the results, while the timing of measurements of prenatal Pb exposures may be at issue in prospective studies.

Lead and Congenital Anomalies

Teratogenic substances can function in mammalian systems as either structural/anatomic or behavioral teratogens. The former define the conventional understanding of teratogenic effects, i.e., "birth defects," while the concept of behavioral teratogenesis is of more recent vintage. Higher exposures to teratogens are linked to visible birth defects, while it is generally held that behavioral teratogens arise from subtle effects at lower exposures. Bellinger (1994) has reviewed and discussed the topic of lead as a teratogen. Table 14.4 presents a number of studies that have examined congenital anomalies associated with Pb exposure in humans.

Lead's potency as a human structural teratogen has been infrequently studied and reported in the clinical and occupational health literature. The historical record for lead as a potent fetotoxic agent presented little data on the extent to which lethal outcomes in early worker pregnancies could be attributed to readily visible classes of birth defects. Organogenesis in the embryology of human reproduction occurs in weeks 3 through 8, 21–56 days. The relative paucity of data for Pb effects on organogenesis in the form of structural terata may be due to several factors operating in this gestational time window. First, transplacental Pb movement and transport to the embryo may occur after organogenesis. Alternatively, transplacental entry of Pb concurrent with organogenesis may in fact occur, but toxicological response mechanisms may favor systemic toxicity over teratogenesis in terms of relative sensitivity.

Pronounced birth defects associated with occupational Pb exposure have been infrequently reported in the literature, although one would expect the levels of Pb exposure in earlier workplace conditions to have maximized the likelihood for identifying overt human teratogenesis. Given the general absence of any systematic clinical and toxicological assessments of severe *in utero* effects in lead's early history, it is not possible to conclude that birth defects as adverse outcome measures would have competed with severe fetotoxicity and lethality in historically very high workplace exposures of pregnant workers.

Levine and Muenke (1991) reported the case of a pregnant lead worker who sustained brief, early, and quite elevated workplace exposure and delivered an infant with virtually all of the characteristic defects defining the VACTERL association seen with other human cases, where Pb was not explicitly identified as present. VACTERL refers to collective vertebral anomalies, anal artresia, cardiac defect, renal anomalies, and limb anomalies; only tracheoesophageal fistula was absent. A striking feature of these birth defects was their similarity to high-dose birth defects found in various test animal species, a urorecto/caudal pattern of malformations.

Several large studies involving paternal occupational Pb exposures flagged through occupational registries and associated congenital anomaly outcomes in their children matched through health registries identified diverse congenital anomalies, such as a fourfold increase in cleft lip and limited evidence for neural tube effects (Alexander et al., 1996; Irgens et al., 1998; Kristensen et al., 1993; Sallmén et al., 1992).

TABLE 14.4 Reproductive and Developmental Effects of Lead in Human Populations: Occurrence of Congenital Anomalies

Group	Pb Exposure	Endpoint	Results	References
Boston, MA, infants born at one hospital ($N = 4{,}354$)	Range of cord PbBs	Congenital anomalies (common anomalies = hemangiomas, lymphangiomas, skin tags and papillae, undescended testicles)	PbB was associated with total congenital anomalies but not individual forms	Needleman et al. (1984)
Fathers of children with a malformed child born in 1973–1982, case–control study	Occupational biological monitoring of fathers	Various congenital anomalies	Odds ratio for anomalies was increased (2.4, 1.9–3.2) with increased paternal exposure	Sallmén et al. (1992)
Fathers of children with congenital anomalies	Occupational Pb exposures of fathers	Various congenital anomalies	Increased Pb exposure associated with a fourfold increase in cleft lip in male children; no other effects	Kristensen et al. (1993)
Fathers of children with Pb exposure ($N = 2{,}021$ pregnancies)	Occupational Pb exposures of fathers in Pb smelters	Various congenital defects	Moderate and high Pb exposures associated with increased odds ratios, 1.9 and 2.7, respectively	Alexander et al. (1996)
English case–control study of 88,449 births over 25 years	Variable drinking water Pb levels in Fylde, England	Neural tube defects	Water Pb >10 μg/l associated with increased likelihood of neural tube defect	Bound et al. (1997)
Norwegian study of parental Pb and birth defects	Variable workplace Pb exposures for mothers or fathers	Birth defects	No linkage of Pb exposure with overall birth defects but limited evidence of neural tube effect	Irgens et al. (1998)
Ecological case–control study of MA mother–infant pairs drinking water Pb ($N = 14{,}130$ women total; women with 1,039 children having birth anomalies)	Variable municipal water Pb levels	Congenital anomaly and other adverse effects	Elevated water Pb (95th percentile of distribution) associated with higher risk of ear, face, neck, and cardiovascular anomalies	Aschengrau et al. (1993)

Environmental Pb exposures have been associated with birth anomalies in several nonoccupational populations, but the claims are made problematic by limited characterization of the individual Pb exposures. Aschengrau et al. (1993) evaluated the association of community water Pb exposures versus congenital anomalies using municipal water Pb levels as the environmental measure of potential exposures. Water Pb level, at the 95th percentile of water Pb distributions, was linked to higher risks of ear, face, neck, and cardiovascular anomalies. Bound et al. (1997) examined the extended temporal relationship (25 years) of available community drinking water Pb for a large number of residents of Fylde, England ($N = 88{,}449$ births), to the frequency of neural tube defects. They reported that water Pb levels >10 μg/l were associated with increased likelihood of neural tube defects. The study of Needleman et al. (1984) of a large number of Boston, MA, births ($N = 4{,}345$) along with associated cord blood Pb levels (range up to 35 μg/dl) showed PbB was associated with total common congenital anomalies such as hemangiomas, skin tags, and papillae, but not individual forms.

14.3 LEAD EFFECTS ON REPRODUCTIVE COMPETENCE

Lead exposure produces reproductive effects in both men and women, but reproductive gametotoxicity through several pathways has largely been studied and principally recognized in men. However, lead gametotoxicity in exposed women has been the topic of a number of recent studies.

14.3.1 Lead Gametotoxicity in Men

Three approaches to evaluating reproductive competence in Pb-exposed men, largely workplace exposures, have been elucidated: direct gametotoxicity through impaired spermatogenesis and quality, assessment of altered time to pregnancy, and reproductive history. These three aspects are set forth in Table 14.5. In virtually all cases, these toxic effects occurred at the relatively higher exposure levels seen in the workplace.

Effects on spermatogenesis and sperm quality in occupational Pb exposures, as noted in Table 14.5, have been documented for various types of lead facilities, e.g., battery plants and smelters, and in various countries (Alexander et al., 1996; Assennato et al., 1986; Bonde et al., 2002; Cullen et al., 1984; Lancranjan et al., 1975). The relatively early Lancranjan et al. (1975) study showed that two groups of Pb workers, one with a quite elevated mean PbB of 75 μg/dl defined as high exposure and a second with a mean PbB of 53 μg/dl defining moderate exposure, demonstrated decreased sperm counts and morphologically abnormal sperm forms. U.S. male Pb workers in the later study of Cullen et al. (1984) similarly showed that a high blood Pb range, 66–139 μg/dl, produced oligospermia, azoospermia, reduced testosterone, and thyroid dysfunction.

TABLE 14.5 Reproductive and Developmental Effects of Lead in Human Populations: Reproductive Effects in Men

Group	Pb Exposure	Endpoint	Results	References
A. Sperm quality				
Lead workers ($N = 150$)	High (mean PbB = 75 μg/dl), moderate (53 μg/dl), light (41 μg/dl) exposures, and controls (23 μg/dl)	Sperm count, motility, morphology	Decreased counts and morphologically abnormal sperm at two highest exposures	Lancranjan et al. (1975)
U.S. lead workers	PbB range 66–139 μg/dl	Abnormal spermatogenesis	Oligospermia, azoospermia, reduced testosterone, thyroid dysfunction	Cullen et al. (1984)
Italian battery workers ($N = 18$) versus control workers ($N = 18$)	PbB exposed workers = 61 μg/dl; controls = 18 μg/dl	Sperm counts, endocrine function biomarkers	Reduced sperm counts without endocrine dysfunction	Assennato et al. (1986)
Smelter workers ($N = 119$, semen samples; $N = 152$, PbB measures)	PbB range; Quartiles: <15, 15–24, 25–39, ≥40 μg/dl	Sperm characteristics, hormonal function	PbB associated with reduced sperm count, total motile sperm count, sperm concentration; no effect on reproductive hormonal function	Alexander et al. (1996)
Workers in Pb industry ($N = 503$)	PbB range versus dose–response	Sperm characteristics: lowest-observed-adverse-effect level	Threshold of 45 μg/dl for disturbances of spermatogenesis	Bonde et al. (2002)
Indian workers in battery or paint production plants	PbB mean = 68 μg/dl	Decreased sperm count, increased velocity and motility	DNA hyploidy and abnormalities	Naha and Manna (2007)

(*Continued*)

TABLE 14.5 Reproductive and Developmental Effects of Lead in Human Populations: Reproductive Effects in Men—(cont.)

Group	Pb Exposure	Endpoint	Results	References
Taiwanese battery workers	PbB mean = 40 μg/dl	Sperm quality, count	PbB >45 μg/dl increased head abnormalities; DNA denaturation; no differences in volume, motility, velocity	Hsu et al. (2009)
Polish lead workers	Two exposure groups: 25–40, 40–81 μg/dl PbB	Sperm count, quality	No differences in volume, count, morphology; higher exposure group, lower motility	Kasperczyk et al. (2008)
B. Time to pregnancy				
Male workers in European collaborative study (*N* = 1,108 (total), 638 with Pb exposure)	PbB in quartiles: <20, 20–29, 30–39, ≥40 μg/dl	Overall group: time to pregnancy	PbB was not associated with delays in pregnancy	Joffe et al. (2003)
Finnish couples with males exposed to Pb (*N* = 502)	PbB >10 μg/dl versus ≤10 μg/dl	Time to pregnancy	PbB in neither stratum associated with delayed time to pregnancy	Sallmén et al. (2000a)
Lead battery workers (*N* = 133 couples)	Variable PbB, including ≥30 μg/dl	Time to pregnancy	PbB increase increased time to pregnancy	Shiau et al. (2004)
C. Reproductive history				
New York State Pb workers matched with birth certificates (*N* = 4,256) versus controls (*N* = 5,148)	One or more PbBs >25 μg/dl defined exposure	Frequency of births	Pb exposure associated with fewer births; most evident with Pb work >5 years	Lin et al. (1996)

(*Continued*)

TABLE 14.5 Reproductive and Developmental Effects of Lead in Human Populations: Reproductive Effects in Men—(cont.)

Group	Pb Exposure	Endpoint	Results	References
Finnish workers exposed to Pb ($N = 2{,}111$) versus controls ($N = 681$)	PbB range for exposure: 10–>50 μg/dl	Pregnancy success	Impregnation failure rate increases with increased PbB; Risk = 1.9 for PbB >50 μg/dl	Sallmén et al. (2000b)
Danish battery plant workers ($N = 1{,}349$) versus controls ($N = 9{,}656$)	Variable Pb exposure in battery plants	Birth rates	No association between employment duration and birth rate	Bonde and Kolstad (1997)

Battery plant (Assennato et al., 1986) and smelter (Alexander et al., 1996) workers also had impaired spermatogenesis and sperm quality with increasing PbB, but both groups found negative results in terms of impaired endocrine function with elevated Pb exposures. These results, when contrasted with those of Cullen et al. (1984), indicated that the higher Pb exposures seen by Cullen are required for significantly impaired reproduction-associated endocrine function (Table 14.5). Bonde et al. (2002) determined the lowest adverse effect level for impaired spermatogenesis in a large group of Pb workers ($N = 503$). A threshold for such effects was determined to be 45 μg/dl.

Two categories of human reproductive success which have been evaluated for impacts of lead exposure in male workers are increases in time to pregnancy and adverse effects on reproductive histories (Table 14.5). The largest subject sample size examined for the association of Pb contact with time to pregnancy changes (total $N = 1{,}108$; exposure $N = 638$) was in Joffe et al. (2003). Pb exposure was stratified into PbB quartiles: <20, 20–29, 30–39, and ≥40 μg/dl. PbB was not associated with delays in pregnancy within any quartile. Similarly, Sallmén et al. (2000a) posted negative findings in the analysis using 502 Finnish couples, where men had lead exposure stratified into two broad groups: PbB <10 μg/dl and PbB ≥10 μg/dl. Impregnation failure rate increased with increased PbB (Sallmén et al., 2000b).

By contrast, Shiau et al. (2004) reported that lead battery plant workers ($N = 280$ pregnancies) exposed to Pb at levels up to and above 30 μg/dl showed an increased time to pregnancy with increased PbB. The difference here from negative studies likely included a relatively higher representation of high Pb exposures and possibly other differences.

Overall, Pb exposure appears to have some effect on reproductive histories for families with paternal Pb contact histories. Lin et al. (1996) reported that lead workers ($N = 4{,}256$) and controls ($N = 5{,}148$) in a large New York State occupational registry study showed reduction in frequency of births with increase in exposure, most evident with a Pb work history of more than 5 years. Similarly, the Finnish registry analyses of Sallmén et al. (2000a) included the finding that impregnation failure rate increased with increased PbB, failure rate risk being 1.9 for PbB >50 μg/dl (Sallmén et al., 2000b). By contrast, Bonde and Kolstad (1997) found no association between birth rate and Pb exposure among Danish battery plant workers ($N = 1{,}349$) versus controls ($N = 9{,}656$).

14.3.2 Gametotoxicity in Women

More recent studies support gametotoxicity from Pb exposure in women, as tabulated in Table 14.6. Such effects take the form of alterations in hormonal levels, fertility indices, and, in girls, indices of pubertal development.

Four studies looked at Pb effects on pubertal development in young girls. The most comprehensive of these pubertal studies were based on NHANES III subcohorts of young girls. Inverse associations were generally seen for conventional pubertal measures (Selevan et al., 2003; Wu et al., 2003), while Gollenberg et al. (2010) reported an inverse association of PbB and the hormone inhibin B.

14.4 EXPERIMENTAL ANIMAL EVIDENCE OF Pb EFFECTS ON REPRODUCTION AND DEVELOPMENT

Experimental animal data document that Pb is both a fetotoxin and a teratogen in various experimental test species albeit at relatively large doses. Lead's teratogenic potential, for example, has largely been characterized through parenteral administration of Pb compounds, with oral intake linked to weak or negative effects. In terms of relevance to human populations, data for nonhuman primates are of more direct interest than results from rodents or avian species. Much of the available data for experimental studies on the topic are somewhat older, reflecting the shift in emphasis in recent years to low-level, environmental exposure-related effects. However, such data are still of importance for reasons given earlier in this chapter.

14.4.1 Experimental Evidence of Pb-Associated Reproductive and Developmental Toxicity

Table 14.7 tabulates the various early studies of Pb effects on experimental animal reproduction and development, while Table 14.8 summarizes illustrative recent results, mainly in rats and mice.

Table 14.7 provides summary information about effects of Pb exposure on reproduction and development in three species of monkeys. High-dose

TABLE 14.6 Reproductive and Developmental Effects in Human Populations: Reproductive Effects in Women

Study Group	Pb Exposure	Endpoint	Results	References
NHANES III women aged 35–60 years	Group mean PbB = 2.8 μg/dl	Levels of FSH and LH	With increased PbB, both hormonal levels increased in postmenopausal women, those with removed ovaries; trend to increase in premenopausal women	Krieg (2007)
Taiwanese women with infertility, related difficulties, aged 23–44 years	PbB mean, infertility group = 3.55 μg/dl; PbB mean, prior delivery = 2.78 μg/dl	Odds ratio for infertility	Odds ratio for infertility = 2.94, PbB >2.5 μg/dl versus ≤2.5 μg/dl	Chang et al. (2006)
New York City 9-year-old girls	Pb median: 2.4 μg/dl	Puberty onset	No linkage between PbB and pubertal development	Wolff et al. (2008)
NHANES III female children, 8–16 years old	PbB mean = 2.5 μg/dl	Measures of pubertal development	Inverse association for some markers of pubertal development	Wu et al. (2003)
NHANES III female children, 8–18 years old	Geometric mean PbB <3 μg/dl; % >5 μg/dl differed with race, and highest % in Mexican-Americans, 12.8%	Measures of pubertal development	Inverse association with PbB, all measures in African-Americans, two of three in Mexican-Americans	Selevan et al. (2003)
NHANES III female children, 6–11 years old	Median PbB = 2.5 μg/dl, 5%>10 μg/dl	Reproductive hormone—LH, inhibin B levels	PbB inversely associated with inhibin B, no association with LH	Gollenberg et al. (2010)

TABLE 14.7 Selected Earlier Experimental Animal Studies of Lead Effects on Reproduction and Development: Reproductive and Developmental Toxicity and Teratogenesis

Experimental Animal	Pb Exposure	Test Endpoint	Results	References
Adult female Rhesus monkeys ($N = 10$, treated; $N = 7$, controls)	Daily Pb acetate in drinking water to give PbB range of 44–89 μg/dl	Disturbances in menstrual cycle characteristics	Pb exposure significantly reduced cycles per month, increased cycle length, more variable cycle length, and reduced duration of menses	Laughlin et al. (1987)
Adult female Rhesus monkeys ($N = 10$, treated; $N = 7$, controls)	Daily Pb acetate in drinking water to maintain PbB of 70 μg/dl	Circulating progesterone (P_4) for luteal function	Reduced P_4 with Pb exposure; maximum P_4 reduced. Luteal function suppressed	Franks et al. (1989)
Female adult *Macaca* monkeys ($N = 22$)	Pb injected, i.m.: 1 or 5 mg Pb/kg/day, gestation and/or lactation	Abortions, maternal death, fetal effects	High dose caused abortions and maternal death; lower dose caused maternal kidney effects and fetal cerebral injury	Tachon et al. (1983)
Pregnant squirrel monkeys ($N = 31$), fetuses, 19 mothers	PbB maternal mean = 37 μg/dl, range 22–87 μg/dl, oral exposure	Lead encephalopathic effects	One abortion, eight stillborns, three neonatal deaths, reduced cerebral weights, white matter hemorrhages	Lögdberg et al. (1988)

(Continued)

TABLE 14.7 Selected Earlier Experimental Animal Studies of Lead Effects on Reproduction and Development: Reproductive and Developmental Toxicity and Teratogenesis—(cont.)

Experimental Animal	Pb Exposure	Test Endpoint	Results	References
Pregnant Sprague–Dawley rats	IV Pb exposure at 25, 35, 50, 70 mg/kg on gestation days 8–17	Fetal toxicity	Pb caused fetal death and resorption on days 10–16. Evident at the 50 mg/kg dose; survivors had reduced body weight, crown-rump distances	McClain and Becker (1975)
Adult male albino rats	Pb acetate, drinking water, 18 weeks; 1,500, 3,500, 5,500 ppm Pb	Male fertility, litter number, weights of newborn and 11-day pups	No effect on male reproductive parameters	Piasek and Kostial (1987)
Adult female rats	Pb acetate, drinking water, 1,500, 3,500, 5,500 ppm Pb	Female reproductive performance	Significant adverse impacts on reproductive success	Piasek and Kostial (1985)
Pregnant rats	Oral dosing Pb acetate, up to 10 mg/kg/day: PbB 53–92 μg/dl	Fetal injury	Significant reduction of growth at 100 mg/kg	Miller et al. (1982)
Pregnant rats in a low Pb exposure regimen	Prolonged Pb exposure, weaning through lactation. Pb in water: 0.5, 5; Maternal 50, 250 ppm; PbB = 20 μg/dl	Fetal/offspring measures: weight, length measurement, physical development	High dose caused reduced fetal length; 50 and 250 ppm produced reduced postnatal weights	Kimmel et al. (1980)
Pregnant Sprague–Dawley rats	IV Pb exposure: 25, 35, 50, 70 mg/kg on different gestation days 8–17	Terata	Day 9 exposure optimal for a urorectalcaudal syndrome of malformations	McClain and Becker (1975)

(Continued)

TABLE 14.7 Selected Earlier Experimental Animal Studies of Lead Effects on Reproduction and Development: Reproductive and Developmental Toxicity and Teratogenesis—(cont.)

Experimental Animal	Pb Exposure	Test Endpoint	Results	References
Pregnant golden hamsters	Various Pb salts given parenterally	Terata	Day 8 exposure optimal for malformations of sacral and caudal vertebrae, absent or stunted tails	Ferm and Ferm (1971)

TABLE 14.8 Illustrative Recent Experimental Animal Studies of Lead Effects on Reproduction and Developmental Toxicity

Experimental Animal	Pb Exposure	Test Endpoint	Results	References
In vitro: fertilized buffalo oocytes in culture	Culture medium Pb = 0.05–10 μg Pb/dl	Indices of embryo development	Pb dose-dependent decline in blastocyst yield, increased embryo degeneration, asynchronous division	Nandi et al. (2010)
Wistar rats exposed to Pb *in utero*	Maternal Pb exposure: 0.025% Pb, gestation day 1–10, 11–20, 1–20	Fetal body weight and length	Significant decrease in fetal weight and length, gestation days 1–20 (greatest decrease) and 11–20 (next greatest)	Wang et al. (2009)
Mice, dosed and controls	Pb chloride, 77 or 100 ppm, drinking water	Experimental male reproductive toxicity	Inverse association of Pb dose and intact acrosomes	Oliveira et al. (2009)

(*Continued*)

TABLE 14.8 Illustrative Recent Experimental Animal Studies of Lead Effects on Reproduction and Developmental Toxicity—(cont.)

Experimental Animal	Pb Exposure	Test Endpoint	Results	References
Male rats, dosed and controls	Pb acetate, 8–24 mg/kg body weight	Spermatogenesis	Reduced spermatogenic cycle dose dependently	Rubio et al. (2006)
Male Sprague–Dawley rats, dosed and controls	1% Pb acetate, drinking water, 8 weeks	Oxidative stress marker levels	Increase in MDA, reduced activities of superoxide dismutase and catalase in plasma, testes	Salawu et al. (2009)
Female mice, F_1 and F_2 generations	Dams ingested 0.02–40 ppm Pb in diet; pup PbBs 0.7–13 μg/dl	Indices of sexual maturation	Delayed vaginal opening, age of estrus, plug formation, first parturition in F_1 animals; similar outcomes in F_2 animals	Iavicoli et al. (2006)
Female Charles Foster rats, dosed and controls	S.c. injection daily, 0.05 mg/kg body weight	Fertility rate or litter size	No changes with exposure, but implantation enzyme activities altered	Nampoothiri and Gupta (2008)
Female rats, dosed and controls	Pb acetate, 200 or 400 ppm, to 3 months postweaning	Reproduction and development measures	Litter size and pup weights reduced in dose-dependent manner	Teijon et al. (2006)

exposure of pregnant Macaca monkeys via i.m. injection at 1 or 5 mg/kg/day produced a range of lethal and nonlethal severe effects in both fetuses and mothers (Tachon et al., 1983). Higher dosing produced abortions and maternal death. Lower dosing induced maternal kidney effects and fetal cerebral injury. Oral exposures of pregnant squirrel monkeys producing a maternal PbB mean of 37 μg/dl induced stillborns, a spontaneous abortion, three neonatal deaths, and encephalopathic effects including reduced cerebral weights and white matter hemorrhages (Lögdberg et al., 1988).

Two studies of pregnant Rhesus monkeys orally exposed to lead acetate via drinking water to produce PbB at several levels (Table 14.7) revealed toxic effects on menstrual cycle characteristics and endocrine function. Laughlin et al. (1987) noted that lead exposure producing PbB in the 44–89 μg/dl range significantly reduced cycles per month, increased cycle length, and showed other effects. A PbB of 70 μg/dl suppressed luteal function and reduced circulating progesterone (Franks et al., 1989).

Lead exposures of male or pregnant female rats produced a range of effects at various dosing levels. Exposing male rats to soluble Pb in drinking water at 1,500, 3,500, or 5,500 ppm Pb showed no effects in terms of male reproductive parameters (Piasek and Kostial, 1987), but the same dosing model produced significant impacts on reproductive success (Piasek and Kostial, 1985). Miller et al. (1982) reported that oral dosing of pregnant rats up to 100 mg/kg/day produced significantly reduced fetal growth at the highest level. Kimmel et al. (1980) examined relatively low-dose oral exposures of rats to Pb and saw fetal effects down to a PbB of 20 μg/dl. These included body length changes, weight reductions, and delays in physical development.

Among more recent experimental animal studies, those illustrated in Table 14.8 show generally similar toxic impacts of Pb in male and female animals to those observed for somewhat higher dosing regimens.

14.4.2 Experimental Evidence of Pb Teratogenesis

Teratological effects can be induced in several animal species by Pb dosing at relatively high concentrations (Carpenter and Ferm, 1977; Ferm and Ferm, 1971; McClain and Becker, 1975). Dosing pregnant hamsters at gestation days 7–9 with soluble Pb (nitrate) at typically 31.3 mg/kg i.v. produced a specific set of malformations involving sacral and caudal vertebrae, producing absent or stunted tails. Similar findings with parenteral injection in pregnant Sprague–Dawley rats given 25, 35, 50, or 70 mg/kg i.v. injections on gestation days 8–17 were noted. Day 9 was optimal for such effects, urorectalcaudal syndrome of malformations.

Oral exposures of animal species largely fail to elicit the severe teratogenic responses seen with injection. This is likely due to the relative inability of oral Pb dosings to achieve high enough *in vivo* concentrations, and this in turn arises from the quite low absorption rates of ingested Pb in animals maintained on animal chows with low bioavailability of the element. This especially applies to soy meal-based animal diets. An oral uptake rate of 1% represents a huge reduction of effective dose compared to i.v. dosing at similar rates.

TABLE 14.9 Transplacental Transfer Toxicokinetics of Lead in Human Populations

Study Group	Pb Exposure	Measured Media	Results	References
Stillborn and spontaneously aborted fetal tissue	General population of pregnancies	Pb in blood, brain, bone, liver, kidneys, hearts	Pb in all examined tissues, increasing to term	Barltrop (1969)
First trimester aborted fetuses	General population of pregnancies	Pb in brain, kidney, liver	Pb levels measurable in first trimester aborted fetuses	Chaube et al. (1972)
Former Yugoslavia prospective Pb study ($N = 1{,}502$)	Smelter town PbB/cord PbB pairs versus control pairs	Cord PbB, maternal mid-pregnancy, and delivery PbBs	Maternal PbB at either time had high correlation with cord PbB, 0.8–0.9	Graziano et al. (1990)
Australian mother–infant pairs, European immigrants to Australia with varying stable Pb isotopes	Isotopic pregnancy and lactation Pb levels in maternal and fetal blood pre-/postdelivery	Maternal, cord, postnatal PbB levels	Pb is significantly mobilized from maternal bone, mainly in late gestation. Pb release persists into lactation	Gulson et al. (1999) and references therein
Australian mother–infant pairs, European immigrants to Australia	Isotopic pregnancy levels versus Ca nutrition	Maternal PbB	Low dietary Ca increase, bone Pb mobilization	Gulson et al. (2004)
Mexico City prospective Pb study pregnancies ($N = 615$)	Community Pb exposures	Maternal and cord PbB, plasma Pb, bone Pb	Plasma Pb was more linked to cord Pb than whole blood Pb; main input to plasma Pb, toxicokinetically, was bone Pb	Chuang et al. (2001)

14.5 TRANSPLACENTAL AND FETAL Pb TOXICOKINETICS

The principal features of transplacental and fetal Pb toxicokinetics provide an important part of the mechanistic features of lead's effects on reproduction and development in human populations and experimental animal test systems. See Table 14.9 for details.

Evidence noted earlier of ready transplacental and intrauterine movement of Pb in human reproduction and development included data showing Pb entering fetal tissue throughout much of gestation and the similarity and high correlation of maternal PbB at delivery and newborn cord PbB. The early toxicokinetic data of Barltrop (1969) and Chaube et al. (1972) showed Pb readily entering fetal circulation and tissues within the first trimester (Chaube et al., 1972) and certainly by the beginning of the second trimester (Barltrop, 1969).

A large number of studies have documented that fetal Pb at term is maximal and Pb uptake is highly correlated via cord PbB with maternal PbB. Graziano et al. (1990) noted that for mother–infant pairs in the former Yugoslavia prospective Pb study, such correlation was 0.8–0.9. Maternal PbB in this prospective study was measured at delivery or at mid-pregnancy.

Additional studies also confirm that transplacental transfer entails both exogenous and endogenous Pb, the latter arising from bone Ca mobilization with attendant release of co-deposited bone Pb (Gulson et al., 1999, 2004). The interplay of Pb presented to the conceptus via both exogenous and endogenous pathways is governed by maternal diets during pregnancy with reference to dietary calcium and calcium intakes in the final trimester when fetal skeletal mineral demands for Ca are maximal (Gulson et al., 2004).

A critical question in fetal Pb toxicokinetics is how one best measures dose–response relationships among various exposure biomarkers as well as relationships governing dose–toxic response relationships. It is now accepted that bone Pb is a better biomarker in constructing dose–toxic response relationships for a variety of toxic effects than are indicators such as PbB. This was demonstrated in Chapter 13 describing cardiovascular effects of Pb.

Chuang et al. (2001) examined the relative strength of associations of various maternal biomarkers for Pb with cord PbB in 615 Mexico City pregnancies. Plasma Pb was more closely correlated with cord Pb than was PbB, while the main input to plasma Pb toxicokinetically was bone Pb (Table 14.9).

REFERENCES

Alexander, B.H., Checkoway, H., Van Netten, C., Kaufman, J.D., Vaughan, T.L., Mueller, B.A., et al., 1996. Paternal occupational lead exposure and pregnancy outcome. Int. J. Occup. Environ. Health 2, 280–285.

Angell, N.F., Lavery, J.P., 1982. The relationship of blood lead levels to obstetric outcome. Am. J. Obstet. Gynecol. 142, 40–46.

Aschengrau, A., Zierler, S., Cohen, A., 1993. Quality of community drinking water and the occurrence of late adverse pregnancy outcomes. Arch. Environ. Health 48, 105–113.

Assennato, G., Paci, C., Baser, M.E., Molinini, R., Candela, R.G., Altamura, B.M., et al., 1986. Sperm count suppression without endocrine dysfunction in lead-exposed men. Arch. Environ. Health 41, 387–390.

Barltrop, D., 1969. Transfer of lead to the human foetus. In: Barltrop, D., Burland, W.L. (Eds.), Mineral Metabolism in Pediatrics. Davis, Philadelphia, PA, pp. 135–151.

Bellinger, D., 1994. Teratogen update: lead. Teratology 50, 367–373.

Bellinger, D., Leviton, A., Rabinowitz, M., Allred, E., Needleman, H., Schoenbaum, S., 1991. Weight gain and maturity in fetuses exposed to low levels of lead. Environ. Res. 54, 151–158.

Bonde, J.P., Joffe, M., Apostoli, P., Dale, A., Kiss, P., Spano, M., et al., 2002. Sperm count and chromatin structure in men exposed to inorganic lead: lowest adverse effect levels. Occup. Environ. Med. 59, 234–242.

Bonde, J.P.E., Kolstad, H., 1997. Fertility of Danish battery workers exposed to lead. Int. J. Epidemiol. 26, 1281–1288.

Borja-Aburto, V.H., Hertz-Picciotto, L., Lopez, M.R., Farias, P., Rios, C., Blanco, J., 1999. Blood lead levels measured prospectively and risk of spontaneous abortion. Am. J. Epidemiol. 150, 590–597.

Bornschein, R.L., Grote, J., Mitchell, T., Succop, P.A., Dietrich, K.N., Krafft, K.M., et al., 1989. Effects of prenatal lead exposure on infant size at birth. In: Smith, M.A., Grant, L.D., Sors, A.I. (Eds.), Lead Exposure and Child Development: An International Assessment. Kluwer Academic, Dordrecht, The Netherlands, pp. 307–319.

Bound, J.P., Harvey, P.W., Francis, B.J., Awwad, F., Gatrell, A.C., 1997. Involvement of deprivation and environmental lead in neural tube defects: a matched case–control study. Arch. Dis. Child. 76, 107–112.

Carpenter, S.J., Ferm, V.H., 1977. Embryopathic effects of lead in the hamster: a morphological analysis. Lab. Invest. 37, 369–385.

Chang, S.H., Cheng, B.H., Lee, S.L., Chuang, H.Y., Yang, C.Y., Sung, F.C., et al., 2006. Low blood lead concentration in association with infertility in women. Environ. Res. 101, 380–386.

Chaube, S., Swinyard, C.A., Nishimura, H., 1972. A quantitative study of human embryonic and fetal lead with considerations of maternal fetal lead gradients and the effects of lead on human reproduction. Teratology 5, 253.

Chuang, H.-Y., Schwartz, J., Gonzales-Cossio, T., Lugo, M.C., Palazuelos, E., Aro, A., et al., 2001. Interrelations of lead levels in bone, venous blood, and umbilical cord blood with exogenous lead exposure through maternal plasma lead in peripartum women. Environ. Health Perspect. 109, 527–532.

Cullen, M.R., Kayne, R.D., Robins, J.M., 1984. Endocrine and reproductive dysfunction in men associated with occupational inorganic lead intoxication. Arch. Environ. Health 39, 431–440.

Factor-Litvak, P., Graziano, J.H., Kline, J.K., Popovac, D., Mehmeti, A., Ahmedi, G., et al., 1991. A prospective study of birthweight and length of gestation in a population surrounding a lead smelter in Kosovo, Yugoslavia. Int. J. Epidemiol. 20, 722–728.

Fahim, M.S., Fahim, Z., Hall, D.G., 1976. Effects of subtoxic lead levels on pregnant women in the state of Missouri. Res. Commun. Chem. Pathol. Pharmacol. 13, 309–331.

Ferm, V.H., Ferm, D.W., 1971. The specificity of the teratogenic effect of lead in the golden hamster. Life Sci. 10, 35–39.

Franks, P.A., Laughlin, N.K., Dierschke, D.J., Bowman, R.E., Meller, P.A., 1989. Effects of lead on luteal function in rhesus monkeys. Biol. Reprod. 41, 1055–1062.

Gollenberg, A.L., Hediger, M.L., Lee, P.A., Himes, J.H., Buck Louis, G.M., 2010. Association between lead and cadmium and reproductive hormones in prepubertal U.S. girls. Environ. Health Perspect. 118, 1782–1787.

González-Cossio, T., Peterson, K.E., Sanin, L.-H., Fishbein, E., Palazuelos, E., Aro, A., et al., 1997. Decrease in birth weight in relation to maternal bone-lead burden. Pediatrics 100, 856–862.

Graziano, J.H., Popovac, D., Factor-Litvak, P., Shrout, P., Kline, J., Murphy, M.J., et al., 1990. Determinants of elevated blood lead during pregnancy in a population surrounding a lead smelter in Kosovo, Yugoslavia. Environ. Health Perspect. 89, 95–100.

Gulson, B.L., Mahaffey, K.R., Jameson, C.W., Patison, N., Law, A.J., Mizon, K.J., et al., 1999. Impact of diet on lead in blood and urine in female adults and relevance to mobilization of lead from bone stores. Environ. Health Perspect. 107, 257–263.

Gulson, B.L., Mizon, K.J., Korsch, M.J., Palmer, J.M., Donnelly, J.B., 2004. Blood lead changes during pregnancy and postpartum with calcium supplementation. Environ. Health Perspect. 112, 1499–1507.

Hall, A., Cantab, M.D., 1905. The increasing use of lead as an abortifacient: a series of 30 cases of plumbism. BMJ 1, 584–587.

Hamilton, A., 1914a. Lead Poisoning in the Smelting and Refining of Lead. Bulletin of the U.S. Bureau of Labor Statistics, No. 141. February 17. U.S. Department of Labor, Washington, DC.

Hamilton, A., 1914b. Lead poisoning in the United States. Am. J. Public Health IV, 477–488.

Hamilton, A., 1919. Lead poisoning in American industry. J. Ind. Hyg. 1, 8–21.

Hamilton, A., 1924. The prevalence and distribution of industrial lead poisoning. JAMA 83, 583–588.

Hamilton, A., 1925. Industrial Poisons in the United States. The MacMillan Co., New York.

Hsu, P.C., Chang, H.Y., Guo, Y.L., Liu, Y.C., Shih, T.S., 2009. Effect of smoking on blood lead levels in workers and role of reactive oxygen species in lead-induced sperm chromatin DNA damage. Fertil. Steril. 91, 1096–1103.

Iavicoli, I., Carelli, G., Stanek, E.J., Castellino, N., Li, Z., Calabrese, E.J., 2006. Low doses of dietary lead are associated with a profound reduction in the time to the onset of puberty in female mice. Reprod. Toxicol. 22, 586–590.

Irgens, Å., Krüger, K., Skorve, A.H., Irgens, L.M., 1998. Reproductive outcome in offspring of parents occupationally exposed to lead in Norway. Am. J. Ind. Med. 34, 431–437.

Joffe, M., Bisanti, L., Apostoli, P., Kiss, P., Dale, A., Roeleveld, N., et al., 2003. Time to pregnancy and occupational lead exposure. Occup. Environ. Med. 60, 752–758.

Kasperczyk, A., Kasperczyk, S., Horak, S., Ostalowska, A., Grucka-Mamczar, E., Romuk, E., et al., 2008. Assessment of semen function and lipid peroxidation among lead-exposed men. Toxicol. Appl. Pharmacol. 228, 378–384.

Kimmel, C.A., Grant, L.D., Sloan, C.S., Gladen, B.C., 1980. Chronic low-level lead toxicity in the rat. Toxicol. Appl. Pharmacol. 56, 28–41.

King, E., Conchie, A., Hiett, D., Milligan, B., 1979. Industrial lead absorption. Ann. Occup. Hyg. 22, 213–239.

Krieg, E.F., 2007. The relationships between blood lead levels and serum follicle stimulating hormone and luteinizing hormone in the Third National Health and Nutrition Examination Survey. Environ. Res. 104, 374–382.

Kristensen, P., Irgens, L.M., Daltveit, A.K., Andersen, A., 1993. Perinatal outcome among children of men exposed to lead and organic solvents in the printing industry. Am. J. Epidemiol. 137, 134–144.

Lancranjan, I., Popescu, H.I., Gavenescu, O., Klepsch, I., Serbanescu, M., 1975. Reproductive ability of workmen occupationally exposed to lead. Arch. Environ. Health 30, 396–401.

Lane, R.E., 1949. The care of the lead worker. Br. J. Ind. Med. 6, 125–143.

Laughlin, N.K., Bowman, R.E., Franks, P.A., Dierschke, D.J., 1987. Altered menstrual cycles in rhesus monkeys induced by lead. Fundam. Appl. Toxicol. 9, 722–729.

Levine, F., Muenke, M., 1991. VACTERL association with high prenatal lead exposure: similarities to animal models of lead teratogenicity. Pediatrics 87, 390–392.

Lin, S., Hwang, S.A., Marshall, E.G., Stone, R., Chen, J., 1996. Fertility rates among lead workers and professional bus drivers: a comparative study. Ann. Epidemiol. 6, 201–208.

Lindbohm, M.-L., Hemminki, K., Bonhomme, M.G., Anttila, A., Rantala, K., Heikkila, P., et al., 1991a. Effects of parental occupational exposure on spontaneous abortions. Am. J. Public Health 81, 1029–1033.

Lindbohm, M.-L., Sallmén, M., Anttila, A., Taskinen, H., Hemminki, K., 1991b. Paternal occupational lead exposure and spontaneous abortion. Scand. J. Work Environ. Health 17, 95–103.

Lögdberg, B., Brun, A., Berlin, M., Schütz, A., 1988. Congenital lead encephalopathy in monkeys. Acta Neuropathol. 77, 120–127.

McClain, R.M., Becker, B.A., 1975. Teratogenicity, fetal toxicity, and placental transfer of lead nitrate in rats. Toxicol. Appl. Pharmacol. 31, 72–82.

McMichael, A.J., Vimpani, G.V., Robertson, E.F., Baghurst, P.A., Clark, P.D., 1986. The Port Pirie cohort study: maternal blood lead and pregnancy outcome. J. Epidemiol. Commun. Health 40, 18–25.

Miller, C.D., Buck, W.B., Hembrough, F.B., Cunningham, W.L., 1982. Fetal rat development as influenced by maternal lead exposure. Vet. Hum. Toxicol. 24, 163–166.

Murphy, M.J., Graziano, J.H., Popovac, D., Kline, J.K., Mehmet, A., Factor-Litvak, P., et al., 1990. Past pregnancy outcomes among women living in the vicinity of a lead smelter in Kosovo, Yugoslavia. Am. J. Public Health 80, 30–35.

Naha, N., Manna, B., 2007. Mechanism of lead induced effects on human spermatozoa after occupational exposure. Kathmandu Univ. Med. J. 5, 85–94.

Nampoothiri, L.P., Gupta, S., 2008. Biochemical effects of gestational lead coexposure to lead and cadmium on reproductive performance, placenta, and ovary. J. Biochem. Mol. Toxicol. 22, 337–344.

Nandi, S., Gupta, P.S., Selvaraju, S., Roy, S.C., Ravindra, J.P., 2010. Effects of exposure to heavy metals on viability, maturation, fertilization, and embryonic development of buffalo (*Bubalus bubalis*) oocytes *in vitro*. Arch. Environ. Contam. Toxicol. 58, 194–204.

Needleman, H.L., Rabinowitz, M., Leviton, A., Linn, S., Schoenbaum, S., 1984. The relationship between prenatal exposure to lead and congenital anomalies. JAMA 251, 2956–2959.

Nogaki, K., 1957. On the action of lead on the body of lead refinery workers: particularly on the conception, pregnancy and parturition, in the case of females and on the vitality of their newborn. Igaku Kenkyu 27, 1314–1338.

Nordström, S., Beckman, L., Nordenson, I., 1979. Occupational and environmental risks in and around a smelter in northern Sweden. V. Spontaneous abortion among female employees and decreased birth weight in their offspring. Hereditas 90, 291–296.

Oliveira, H., Spano, M., Santos, C., Pereira, M.D., 2009. Lead chloride affects sperm motility acrosome reaction in mice. Cell Biol. Toxicol. 25, 341–353.

Oliver, T., 1911. A lecture on lead poisoning and the race. BMJ 1, 1096–1098.

Paul, C., 1860. Study on the effect of slow lead intoxication on the product of conception. Arch. Gen. Med. 15, 513–533.

Piasek, M., Kostial, K., 1985. Duration of exposure to lead and reproductive performance in rats. In: Lekkas, T.D. (Ed.), Heavy Metals in the Environment, vol. 1. CEP Consultants, Athens, Greece/Edinburgh, Scotland, pp. 463–465.

Piasek, M., Kostial, K., 1987. Effect of exposure to lead on reproduction in male rats. Bull. Environ. Contam. Toxicol. 39, 448–452.

Rubio, J., Riqueros, M.I., Gasco, M., Yucra, S., Miranda, S., Gonzales, G.F., 2006. *Lepidium meyenii* (Maca) reversed the lead acetate-induced damage on reproductive function in male rats. Food Chem. Toxicol. 4, 1114–1122.

Salawu, E.O., Adeeyo, O.A., Falokun, O.P., Yusuf, U.A., Oyerinde, A., Adeleke, A.A., 2009. Tomato (*Lycopersicon esculentum*) prevents lead-induced testicular toxicity. J. Hum. Reprod. Sci. 2, 30–34.

Sallmén, M., Lindbohm, M.-L., Anttila, A., Taskinen, H., Hemminki, K., 1992. Paternal occupational lead exposure and congenital malformations. J. Epidemiol. Community Health 45, 519–522.

Sallmén, M., Lindbohm, M.-L., Anttila, A., Taskinen, H., Hemminki, K., 2000a. Time to pregnancy among the wives of men occupationally exposed to lead. Epidemiology 11, 141–147.

Sallmén, M., Lindbohm, M.L., Nurminen, M., 2000b. Paternal exposure to lead and infertility. Epidemiology 11, 148–152.

Savitz, D.A., Whelan, E.A., Rowland, A.S., Klechner, R.C., 1990. Maternal employment and reproductive risk factors. Am. J. Epidemiol. 132, 933–945.

Selevan, S.G., Rice, D.C., Hogan, K.A., Euling, S.Y., Pfahles-Hutchens, A., Bethel, J., 2003. Blood lead concentration and delayed puberty in girls. N. Engl. J. Med. 348, 1527–1536.

Shiau, C.-Y., Wang, J.-D., Chen, P.-C., 2004. Decreased fecundity among male lead workers. Occup. Environ. Med. 61, 915–923.

Tachon, P., Laschi, A., Briffaux, J.P., Brain, G., Chambon, P., 1983. Lead poisoning in monkeys during pregnancy and lactation. Sci. Total Environ. 30, 221–229.

Taskinen, H., 1988. Spontaneous abortions among women occupationally exposed to lead. In: Hogstedt, C., Reuterwall, C. (Eds.), Progress in Occupational Epidemiology. Elsevier Science Publishers, New York, pp. 197–200.

Taussig, F.J., 1936. Abortion spontaneous and induced: medical and social aspects. The C.V. Mosby Co., St. Louis, MO, pp. 354–355.

Teijon, C., Olmo, R., Blanco, D., Romero, A., Teijon, J.M., 2006. Low doses of lead: effects on reproduction and development in rats. Biol. Trace Elem. Res. 111, 151–165.

Torres-Sánchez, L.E., Berkowitz, G., López-Carillo, L., Torres-Arreola, L., Rios, C., López-Cervantes, M., 1999. Intrauterine lead exposure and preterm birth. Environ. Res. 81, 297–301.

U.S. EPA (1986). U.S. Environmental Protection Agency. 1986. Air Quality Criteria for Lead. 4 Vols. Report No. EPA/600/8-83/028bF. Environmental Criteria and Assessment Office, Washington, DC.

Wang, Y.-Y., Sui, K.-X., Hong, L.I., Ma, H.-Y., 2009. The effects of lead exposure on placental NF-kappa B expression and the consequences for gestation. Reprod. Toxicol. 27, 190–195.

Wolff, M.S., Briton, J.A., Boguski, L., Hochman, S., Maloney, N., Serra, N., et al., 2008. Environmental exposures and puberty in inner-city girls. Environ. Res. 107, 393–400.

Wu, T., Buck, G.M., Mendola, P., 2003. Blood lead levels and sexual maturation in U.S. girls: the Third National Health and Nutrition Examination Survey, 1988–1994. Environ. Health Perspect. 111, 737–741.

Chapter 15

The Nephrotoxicity of Lead in Human Populations

This chapter describes the acute and chronic nephrotoxic effects of lead in human populations. These effects have long been recognized in chronic adult occupational lead exposures and in nonoccupational adult exposures arising from dietary Pb intakes, producing disorders such as gouty nephropathy. In acute childhood Pb exposure, severe kidney effects in the form of Fanconi syndrome were identified in the early pediatric literature. The syndrome often co-occurred with acute encephalopathy.

15.1 INTRODUCTION AND CONTEXT

Numerous reports on Pb-associated kidney injury in various subsets of human populations have been published over the years, and Pb nephropathy has been covered in various expert consensus reports for public agency or scientific organizations, such as those of the U.S. EPA (1977, 1986, 2006), the U.S. ATSDR (2007), the WHO (1995), and the NAS/NRC (1972, 1980, 1993). Individual monographs and critical reviews on the topic have also appeared (Ekong et al., 2006; Loghman-Adham, 1997; Wedeen, 1984, 1982).

Clinical lead nephropathy presents in several forms, depending on the severity and temporal nature of the Pb exposure. Historically, and arising from what were clearly very high exposures over many years, chronic Pb nephropathy was characterized histopathologically as interstitial and peritubular fibrosis, in which affected kidneys appeared as contracted organs with arteriosclerotic injury, fibrosis, glomerular atrophy, and hyaline degeneration. The Pb-associated disease in earlier worker patients was progressive with renal failure being a relatively common outcome (Wedeen, 1982).

More recent evidence in the occupational and environmental health literature identifies two general forms of nephropathological and nephrotoxicological responses to lead in lead workers and others. Acute nephrotoxic effects of lead present clinically and functionally with a different array of signs and symptoms than the chronic lead nephropathy syndrome, as discussed below.

Trace Metals and other Contaminants in the Environment, Volume 10
ISSN: 1875-1121 DOI: 10.1016/B978-0-444-51554-4.00015-8

Lead nephropathy as a toxicological topic has presented various interpretive clinical and empirical dilemmas. Some appear to have resolved to some extent. Some are mainly confined to nephropathic responses *per se*. Some are shared with other lead-associated toxic endpoints. A classical endpoint in Pb-induced nephropathy is a reduced estimated or measured glomerular filtration rate (GFR), typically employing creatinine clearance rates, in tandem with measurements of blood urea nitrogen (BUN) and serum creatinine levels. Such declines in GFR are proportional to the level of PbB in chronic injury, subsequent to any transitory hyperfiltration.

Some workers in this area earlier raised the issue of reverse causality, wherein impaired GFR not arising from lead-linked injury may produce a proportionately reduced rate of Pb removal from circulation and an increase in PbB. An analogous example of reverse causality once arose in the literature on the developmental neurotoxicity of lead exposure, noted in Chapter 12. There, the question was raised of impaired cognition in children leading to behaviors that elevated the risk of ingestion of lead through irregular or excessive oral exploratory behavior. Currently available evidence has refuted any likelihood of reverse causality in that circumstance. Evidence in the current literature, derived from well-done prospective studies of Pb nephropathy using various biomarkers and various time frames for predictive risk, has also largely refuted reverse causality in Pb nephrotoxicity.

A number of complications and empirical conundrums also arise in assessing lead worker cohorts with reference to endpoints such as morbidity and mortality associated with Pb nephropathy. There is the well-recognized and general healthy worker effect, where the healthiest and/or least adversely responsive workers exposed to toxics are most likely to remain at the job over time, potentially biasing the overall dose–toxic response relationships for various endpoints in humans to a less robust relationship. Survivor bias is related to this phenomenon in mortality studies. That is, those workers leaving employment within shorter time periods are not likely included in the statistical analyses of survival rates.

Characterizing lead nephropathy in those human populations with either occupational or environmental exposures has been plagued by a number of statistical issues. Worker study groups have often been few in number, and have remained small into more recent years. Worker Pb exposures were very high in earlier decades but have declined over the years so that body Pb burdens embodied in and determined by, for example, bone Pb content have similarly declined in younger workers. Statistically, one has to quantitatively accommodate this age–exposure–nephrotoxicity interaction in making sense of morbidity and/or mortality rates in the out years for these workers. General population exposures with regard to nephrotoxicity, by contrast, have entailed much larger sample sizes that have been available in recent years for study.

A much more quantitative understanding of the relative strengths and weaknesses of various biomarkers of human Pb exposures and early biomarkers of nephrotoxic effect has evolved, at least in terms of chronic effect thresholds. Until recently, Pb nephropathy was viewed as a problem in occupational hygiene having a rather high threshold, e.g., 60 μg/dl PbB, for onset of functional effects, and a threshold well above those other system and organ responses typically considered in risk and regulatory assessment. Such required high exposures and associated high toxicity thresholds for Pb nephropathy in the earlier clinical and epidemiological literature had less to do with this toxic effect being relatively insensitive in dose–response terms and more to do with the combination of comparatively crude methodologies for diagnosis and characterization, and the intrinsic characteristic of kidney function having a high reserve capacity. Consequently, lead nephropathic changes indexed using broad measures such as creatinine clearance identified injury which had progressed to a significant histopathological state.

More sensitive and reliable methods for identifying Pb nephropathy effects and more sophisticated epidemiological and statistical designs have evolved. These have been largely responsible for identifying currently recognized lower thresholds for nephrotoxicity and the type of nephrotoxic effects. In particular, concern increasingly extends to more of those adults and children having environmental, ambient exposures.

Using measured or estimated creatinine clearance has variability arising from subject variables such as muscle mass associated with it. Using such early biomarkers as cystatin C (a cysteine protease inhibitor having 120 amino acids and having a 3,000 Da molecular weight) is held to provide more stable results due to its constant generation rate and lower secretion than circulating creatinine. However, cystatin C is affected by age, sex, and race (Kottgen et al., 2008). There is also increasing use of early effect markers relative to kidney function: retinol-binding protein (RBP); β_2-microglobulin; and *N*-acetyl-β-D-glucosaminidase (NAG).

Unlike the neurotoxicological, developmental, and other toxic effects of lead, chronic lead nephropathy was historically viewed as a disease of adults, particularly lead workers, and children were generally considered to be at low risk. The exception to this view was kidney tubular injury in early, acute childhood Pb poisoning in the form of Fanconi Syndrome.

Lead nephropathy has also been complicated by the toxicological interactions of chronic kidney disease with adverse cardiovascular effects such as hypertension. Hypertension, as noted in this chapter, is a risk factor for Pb-associated and non-Pb-associated kidney disease, while mechanisms for inducing hypertension include the participation of kidney biochemistry and physiology via, for example, the renin–angiotensin pathway.

Finally, sections of this chapter demonstrate that clinical and scientific research on Pb nephropathy has produced parallels with other adverse effects in humans. That is, continued research produces continued declines in accepted thresholds for nephrotoxic effects of Pb over time. The scientific record now makes it clear that nephrotoxicity in terms of both tubular and glomerular injuries is produced at such low exposures that older segments of the general human population are at risk for nephrotoxicity by way of endogenous Pb releases from bone Pb reservoirs that became substantial in previous decades of higher environmental lead exposures and subsequent bone compartment uptake.

This chapter is organized into nine sections, reflecting both the evolution of the topic in terms of toxic impacts on various human population segments and the various categories of toxic harm. These include acute and chronic effects, lead nephropathy as a sequel of earlier lead poisoning, lead nephropathy occurring in occupational and nonoccupational/environmental exposures, dose–response relationships, genetic polymorphisms *vis-à-vis* Pb nephropathy in certain worker cohorts, Pb nephrotoxicity in children, and experimental animal data supporting interpretations of human nephrotoxic effects.

15.2 ACUTE NEPHROTOXIC EFFECTS OF LEAD IN DIVERSE HUMAN POPULATIONS

Lead exposures in diverse human populations produce both acute and chronic nephrotoxic effects. The chronic kidney disease association with occupational Pb exposures in the clinical literature, dating to the nineteenth century, was typically characterized as a glomerulonephritic disease histopathologically. This traced to the absence of evaluation of the temporal course of Pb-induced nephropathy, particularly the acute effects.

Table 15.1 presents illustrative studies of the principal features of acute lead nephropathy. The English toxicologist Oliver (1891) was one of the first to draw attention to distinguishing features of acute lead nephropathy. Acute nephropathy in that instance was evaluated in individuals with multiple acute Pb poisoning episodes sufficient to also produce encephalopathy and death. Kidney tubule injury was severe, with swelling and necrosis in renal proximal tubules. Somewhat lower exposures still occasioned acute effects in the kidney tubule, appearing as tubular dysfunction in adults (Clarkson and Kench, 1956; Pejić, 1928) and children (Chisolm, 1962, 1968) and histopathological evidence of morphological changes (Blackman, 1936; Cramér et al., 1974). Cramér et al. reported the formation of tubule cell intranuclear inclusion bodies, mitochondrial changes, and tubular cytomegaly early in lead worker employment. This trio of morphological and other changes has been confirmed in a number of high exposure studies in humans and experimental animals (Goyer et al., 1970).

Chisolm (1962, 1968) recorded the presence of the full Fanconi Syndrome of toxic proximal tubular injury in young children, i.e., aminoaciduria, glucosuria, and hyperphosphaturia, while other researchers have described elevated urinary levels of amino acids or glucose. The threshold in PbB associated with early tubular injury in acute childhood lead poisoning leading to Fanconi Syndrome was reported as typically >150 μg/dl, and lower levels of Pb,

TABLE 15.1 Acute Nephrotoxic Effects of Lead at Historically High Exposures in Diverse Human Populations

Group	Pb Exposure	Endpoint	Results	References
Individuals with multiple acute Pb poisoning episodes	Acute, high levels inducing encephalopathy, death	Nonspecific degenerative changes in kidney	Tubule cell damage: swelling and necrosis, notably in proximal tubules	Oliver (1891)
Variably aged subjects	Acute Pb exposures	Kidney injury sites	Proximal tubular injury	Pejić (1928)
Occupationally exposed adults	Elevated workplace exposures	Renal tubule dysfunction	Significant aminoaciduria	Clarkson and Kench (1956)
Environmentally exposed children	High Pb exposure, e.g., >150 μg/dl	Full Fanconi Syndrome	Aminoaciduria, glycosuria, hyperphosphaturia	Chisolm (1962, 1968)
Variably aged subjects	High exposures of variable duration	Early proximal tubular injury	Formation of tubular cell intranuclear inclusion bodies, mitochondrial changes, tubular cytomegaly	Blackman (1936), Cramér et al. (1974)

>60 μg/dl, have been associated with milder forms of tubular injury (Chisolm, 1962, 1968).

With continued Pb exposure in workers and other adults, tubular injury appears to give way to the classical pathological and functional signs of chronic kidney disease (Loghman-Adham, 1997).

15.3 LATE CHRONIC Pb NEPHROTOXICITY IN YOUNG ADULTS POISONED AS CHILDREN

The question of whether severe childhood Pb poisoning produces later chronic nephropathy remains a puzzle in light of inconsistent data from various countries and groups of poisoned subjects. Table 15.2 summarizes the topic. A series of reports in the earlier global literature on childhood lead poisoning included the well-known cluster of findings for children in Queensland, Australia, by Gibson et al. (1892) and Gibson (1904). A significant number of these Australian children surviving the poisoning episodes were eventually diagnosed as young adults to be suffering from chronic

TABLE 15.2 Late Chronic Nephrotoxic Effects of Lead in Adults Following Severe Childhood Pb Poisoning

Group	Pb Exposure	Endpoint	Results	References
Queensland, Australia, young adults with childhood Pb exposure ($N = 80$)	Deteriorated exterior Pb paint	Chronic nephropathy	35 (44%) died of nephropathy	Nye (1929)
Queensland, Australia, young adults with childhood Pb exposures ($N = 401$)	Deteriorated exterior Pb paint	Chronic nephropathy	Significantly increased rate of death from nephritis or hypertension; 165/401 (41%); all under age 40	Henderson (1954)
Queensland, Australia, young adults with childhood Pb exposures ($N = 32$)	Deteriorated exterior Pb paint	Chronic nephropathy	EDTA-mobilized Pb excretion significantly elevated in these patients versus other renal disease patients	Emmerson (1963)
Queensland, Australia, young adults with childhood Pb exposures	Deteriorated exterior Pb paint	Chronic nephropathy	Bone Pb mean = 94 ppm wet weight, approximately fourfold higher than typical level	Inglis et al. (1978)
U.S. young adults diagnosed in childhood for overt Pb toxicity in Boston, MA ($N = 139$)	Untreated Pb poisoning 20–35 years earlier	Clinical tests of renal function	No evidence of chronic kidney disease	Tepper (1963)
U.S. adolescents treated for childhood Pb poisoning ($N = 55$, age = 12–22 years)	Pb poisoning 11–16 years earlier; chelation therapy treatment	Clinical tests of renal function	No evidence of chronic kidney disease	Chisolm et al. (1976)
Chicago, IL, young adults with history of childhood Pb poisoning ($N = 62$)	Childhood PbB > 100 μg/dl, 17–23 years earlier with chelation therapy	Prospective study: serial clinical tests of kidney function versus results in control siblings	No differences in measures for poisoned or control subjects	Moel and Sachs (1992)
Boston, MA, adults with childhood Pb poisoning ($N = 35$) with matched controls ($N = 22$): 50-year follow-up	Documented Pb poisoning (1930–1942)	Clinical test of GFR	No differences in measures for poisoned or control subjects	Hu (1991)

nephropathy. An earlier group of these young adults ($N = 80$) studied by Nye (1929) sustained a dramatically high rate of kidney failure and mortality from the nephropathy, 35 of 80 patients or 44%. Henderson (1954), in a later case series report, found that of 401 Queensland young adults, 105 under the age of 40, 26%, died of hypertension and kidney failure.

Later studies by Emmerson (1963) and Inglis et al. (1978) provide valuable clues to the likely genesis of the nephrotoxic sequelae. Emmerson (1963) evaluated 32 patients who had both childhood Pb poisoning and presented with chronic nephropathy as young adults with respect to comparative EDTA-mobilized plumburesis levels versus patients with nephropathy without lead poisoning histories. Amounts of Pb excreted with EDTA challenge in the childhood poisoning group were significantly elevated versus the non-lead nephropathy patient group. Inglis et al. (1978), in an equally relevant study, reported that bone Pb levels in these Queensland childhood poisoning victims were a mean of 94 ppm Pb wet weight, fourfold higher than in a reference population.

Four reports (Table 15.2) have appeared over the years in U.S. follow-up studies of adolescents or adults with medical histories of severe childhood lead poisoning (Chisolm et al., 1976; Hu, 1991; Moel and Sachs, 1992; Tepper, 1963) for kidney function. These study groups differed significantly with respect to age at follow-up and childhood poisoning treatment history. Chisolm et al. (1976) described findings with 55 adolescents aged 11–16 years who had sustained lead poisoning and who had chelation therapy. Moel and Sachs (1992) investigated kidney function in 62 young adults who sustained lead poisoning 17–23 years earlier, with PbB levels >100 μg/dl. These individuals were also treated using chelation therapy at the time of diagnosis. Both Tepper (1963) and Hu (1991) described findings with renal function results for adults having untreated lead poisoning, Tepper describing 139 young Boston, MA, adults identified with childhood poisoning 20–35 years earlier and Hu reporting on 35 older adults poisoned as children 50 years before, 1930–1942.

Three of four U.S. studies recorded negative results for any association of childhood Pb poisoning with later development of chronic nephropathy, in conflict with results for the Australian kidney disease subjects, despite varying study characteristics and designs. One U.S. study (Hu, 1991) provided evidence, not at great odds with the Australian experience (see below). There are several likely reasons for the differences, such as the likely higher intensity of exposures, the nature of the exposures, and, in some cases, the role of effective treatment at the time of the poisonings in childhood. Queensland children were exposed to lead paint applied to exterior veranda surfaces (Gibson, 1904) where the combination of humidity, sunlight, and heat provided the environmental means for rapid deterioration and chalking of the lead paint to produce fine Pb-based particles. Bioavailability of this lead would have been very high (see Chapter 8).

Examination of Pb mobilization data through provocative chelation or chelation therapy plumburesis (Chisolm et al., 1976; Emmerson, 1963) showed large bone Pb stores in the Queensland young adults, while bone Pb concentrations in these nephropathy patients years after the lead paint exposures were also quite high. The mean bone Pb content reported by Inglis et al. (1978) for such patients rivals that reported for individuals with high occupational Pb exposures.

Other factors such as relatively small sample size and confounding through survivor bias and multieffect interactions may work for negative findings. The Hu (1991) study, for example, involved subjects from 50 to 60 years of age who had untreated childhood lead poisoning 50 years earlier. It does not appear that biasing from mortality among childhood poisoning victims due to lead-associated kidney failure prior to the study was adequately accounted for. A history of childhood poisoning 50 years before was significantly associated with hypertension, while creatinine clearance was observed to be supranormal, i.e., showing hyperfiltration. As noted below, hyperfiltration is held to be an early indicator of subsequent kidney disease. In the case of the Hu study, the Pb-associated hypertension (one mechanism for which traces to kidney dysfunction, noted by Campbell et al., 1985) and hyperfiltration collectively provided support for the Australian findings. In Moel and Sachs (1992) and Chisolm et al. (1976) studies, subjects were treated through chelation therapy at the time of their childhood poisoning diagnosis, which served to attenuate accumulation of high body lead burdens and would affect any comparison with untreated poisoning cases.

15.4 CHRONIC LEAD NEPHROPATHY IN ADULTS WITH OCCUPATIONAL EXPOSURES

Lead worker nephropathy has comprised a sizeable literature, at least in quantitative terms, within the earlier toxic nephropathy database. Illustrative studies are summarized in Table 15.3. Such nephropathy has typically, but not always, presented histopathologically as a focal interstitial nephritis and functionally as reduced GFR. Lead nephropathy has been documented in many countries, including the United States (Baker et al., 1979; Wedeen et al., 1975, 1979), Romania (Lilis et al., 1968), Brazil (Pinto de Almeida et al., 1987), Japan (Omae et al., 1990), Singapore (Chia et al., 1995a), and Belgium (Buchet et al., 1980).

Occupational Pb nephropathy as described in the older literature has been difficult to evaluate in terms of consistency across studies and for determining valid dose–response relationships. Studies have typically involved quite low sample sizes in terms of modern biostatistical and epidemiological criteria and in comparison to numbers in studies of Pb-associated nephropathy in general populations sustaining environmental exposures. The illustrative studies summarized in Table 15.3 are those which employed higher sample sizes. In dose–response terms and as briefly noted earlier, the relatively crude measures

TABLE 15.3 Illustrative Studies of Chronic Lead Nephropathy in Workers with High Occupational Lead Exposures

Group	Pb Exposure (μg/dl, μg/g)	Endpoints	Results	References
U.S. male lead workers (*N* = 140)	Workplace exposures	Chronic nephropathy; EDTA results for symptomatic Pb poisoning	Reduced GFR (*N* = 21) in 57 workers with positive EDTA Pb mobilization; 50% of a subset of workers with reduced GFR had focal interstitial nephritis	Wedeen et al. (1975, 1979)
Male workers in various Pb industries (*N* = 449)	Mean exposure period 12 years; PbB > 80 μg/dl in most smelter workers	BUN and creatinine levels	Positive association of BUN and creatinine with exposure duration	Lilis et al. (1980)
Male workers in Pb industries (*N* = 160, age 29–62 years)	Pb exposure: 4.5–31 years; PbB 16–280 μg/dl	BUN and creatinine clearance	Increased BUN, 28 workers; less clearance in eight workers	Baker et al. (1979)
Workers in Pb industries (*N* = 165)	Pb exposure range: 0.1–26 years; PbB mean = 37 μg/dl (9–60 μg/dl)	Creatinine clearance; uric acid clearance; β_2-microglobulin	No change in kidney markers with duration	Omae et al. (1990)
Pb workers (*N* = 137) and control workers (*N* = 153)	PbB > 60 μg/dl, time of exposure ≤12 years	Creatinine clearance, serum β_2-microglobulin	Higher frequency of abnormal microglobulin; eight workers had low creatinine	Chia et al. (1995a)
Pb workers (*N* = 128) and controls (*N* = 93)	Pb exposed mean = 33 μg/dl; control mean = 9 μg/dl	Biomarkers of kidney function	Urinary β_2-microglobulin increased with exposed workers	Chia et al. (1995b)
Pb workers (*N* = 160) and controls (*N* = 60)	Median exposure = 4.5 years; median PbB = 37 μg/dl	Biomarkers of lower-level renal effects	NAG; urinary NAG higher with exposure	dos Santos et al. (1994)

(Continued)

TABLE 15.3 Illustrative Studies of Chronic Lead Nephropathy in Workers with High Occupational Lead Exposures—(cont.)

Group	Pb Exposure (μg/dl, μg/g)	Endpoints	Results	References
Pb workers (*N* = 102)	Exposure period = 7–41 years; PbB range 42–141 μg/dl	Diagnosed kidney disease	17 of longest exposures had kidney failure; 13 had hypertension	Lilis et al. (1968)
Lead workers (*N* = 155) and controls (*N* = 126)	Exposed mean PbB = 48 μg/dl; control mean = 8 μg/dl	Renal biomarkers of lower kidney effects	Correlation of NAG and β_2-microglobulin with PbB	Vershoor et al. (1987)
Lead workers (*N* = 52) and controls (*N* = 44)	Exposed mean = 64 μg/dl; control mean = 26 μg/dl	Renal biomarkers	One-third of Pb workers had elevated serum creatinine versus 1/44 of controls	Pinto de Almeida et al. (1987)
U.S. lead smelter workers	Smelter Pb exposures: 1 year or more, 1940–1965; PbB mean, 1976: 56 μg/dl	Mortality rate from chronic kidney disease	Standardized mortality ratio = 2.71 with Pb exposure > 20 years	Steenland et al. (1992)
Korean current and former Pb workers (*N* = 803)	Workplace Pb exposures: PbB, tibial Pb, chelatable Pb. PbB mean = 32 μg/dl, tibial Pb mean = 37.2 μg/g bone mineral	Clinical measures: uric acid, BUN, serum creatinine, creatinine clearance, urinary NAG, RBP	Oldest third showed bone Pb and PbB association with uric acid	Weaver et al. (2003)
Korean current and former Pb workers (*N* = 537)	PbB, tibial Pb over 2-year prospective analyses	Serum creatinine, creatinine clearance	Males: serum creatinine decreased, clearance increased in one study, greatest with PbB decline; Females: increasing PbB with creatinine increase, may have been former Pb workers	Weaver et al. (2009)

in the older studies of occupational Pb nephrotoxicity provided a sense of a nephrotoxicity threshold at or around 60 μg/dl PbB. This threshold particularly applied to cases where such classical nephropathic features as interstitial nephritis were taken as endpoints (Baker et al., 1979; Wedeen et al., 1979).

One of the difficulties in developing a consistent picture for Pb nephropathy in both occupational and environmental populations, especially in older cohorts, is identifying an adequate biomarker for exposure. Cumulative markers of lead exposure, such as chelatable Pb or bone Pb in either cortical or trabecular bone regions, continue to gain favor as a more meaningful representation of dose in dose–response relationships for chronic, long-term toxic endpoints of Pb exposure like chronic kidney disease. As noted in various portions of this text, PbB is a measure of near-term Pb contact, weeks in duration or less, and partially any longer-term Pb burden accumulation. Both chelatable and bone Pb measurements enhance the robustness of the association of Pb exposure level and the severity of the associated Pb nephropathy. In addition, such measures help refute any claim of Pb nephropathy being a case of reverse causality.

Lesser or earlier manifestations of kidney injury appear to be associated with lower levels of Pb exposure, as would be expected in dose–response terms. Chia et al. (1995b) reported that urinary α_1-microglobulin was increased in Pb workers with a mean PbB of 33 μg/dl versus a control group mean of 9 μg/dl, while Dos Santos et al. (1994) found that urinary NAG, a biomarker for early kidney proximal tubular injury, was elevated and positively correlated with duration of Pb exposure.

Lead in the workplace has long been shown to produce chronic kidney disease in workers. Endpoints such as kidney failure and mortality have been investigated. One complicating factor in assessing Pb-associated kidney disease morbidity and mortality has been the changing picture for the severity of lead exposure; overall intensities of exposure in more recent years have been considerably less than the levels decades ago. This produced the need for statistical and epidemiological designs that accommodate interactions of age and length of Pb exposure with severity of kidney disease. Lilis et al. (1980) evaluated male workers in various Pb industries with reference to BUN and serum creatinine levels as markers of kidney disease with tenure of employment. These investigators reported a positive association of levels of BUN and creatinine with Pb exposure duration.

Weaver et al. (2003) showed a significant interaction of worker age and kidney disease in effect modification for the relationship of serum uric acid levels to either bone Pb or PbB in a mixed cohort of Korean current and former Pb workers ($N = 803$). While negative data for exposure-kidney endpoint associations were determined for the entire group, the oldest age tercile showed worsening kidney disease (increased uric acid) with increased exposure indexed as bone Pb and PbB, while the youngest subset had evidence of hyperfiltration responses.

Steenland et al. (1992) determined mortality rates from chronic kidney disease as a function of workplace Pb exposure duration in U.S. workers. They reported the highest standardized mortality ratio, 2.76, for those workers with more than 20 years of workplace exposures. The mean PbB for this cohort, noted in 1976, was 56 μg/dl.

15.5 LEAD NEPHROPATHY IN INDIVIDUALS WITH GOUT AND HYPERTENSION

Elevated risks for kidney disease from coexisting Pb exposures and gouty nephritis (saturnine gout) have been known since the nineteenth century (Yu, 1983). A major etiology in such Pb exposures was consumption of port wine heavily adulterated with Pb. Half of the patients presenting with Pb nephropathy were also afflicted with gout (Ball and Sorensen, 1969) in a study evaluating nephropathy patients with both Pb exposure and a clinical diagnosis of gout. These findings can be contrasted with the results of Batuman et al. (1981) for U.S. military veterans with diagnosed gout, where 50% of gout patients had both kidney failure and twice the amount of chelatable Pb mobilized from body stores. An additional risk group where both Pb poisoning and diagnosed gout were documented were illicit alcohol imbibers, notably consumers of "moonshine" living in the Southeastern United States. Kidney disease was a common feature of the clinical picture for those alcohol abusers (Morgan et al., 1966).

Individuals with hypertension and elevated Pb exposures appear to be at increased risk for Pb nephropathy (hypertensive nephrosclerosis), although the question of direction of any association remains problematic, in that hypertension aggravates the severity and course of kidney disease while Pb exposure has a hypertensive effect in the human cardiovascular system.

In the study of Beevers et al. (1976), Scottish subjects residing in an area with very elevated tap water Pb levels showed elevated PbB associated with elevated water Pb, while both serum uric acid and hypertension were correlated with PbB. U.S. military veterans described earlier were also examined for hypertension and chelatable lead with or without the presence of kidney failure. Chelant was administered as a single, 2-g dose over 3 days to patients with both hypertension and renal failure and to patients with hypertension but without kidney failure. Individuals with hypertension and kidney disease had the higher amounts of mobilized Pb.

Having manifest gout or hypertension does not appear to be necessary when considering uric acid formation and metabolism as a risk factor for Pb nephrotoxicity (Table 15.4). In the case of young adults of Queensland, Australia, with kidney disease and a clinical history of untreated childhood lead poisoning years earlier, vascular changes in kidneys coexisted with uric acid deposits (Inglis et al., 1978). A similar finding for Pb workers was noted by Cramér et al. (1974), with individuals showing both uric acid deposits and vascular changes in kidneys.

TABLE 15.4 Illustrative Studies of Chronic Lead Nephropathy in Human Populations with Diverse Risk Factors

Group	Lead Exposure (Various Units)	Endpoints	Results	References
A. Lead and gouty nephritis (saturnine gout)				
Nineteenth-century gout patients	Diet and beverage Pb exposures: port wine with high Pb levels	Clinical diagnosis of gout	Significant frequency of gout in patients drinking Pb-contaminated wine	Yu (1983)
Nephropathy patients	Variable exposures to Pb	Clinical diagnoses of gout	About 50% of patients with Pb nephropathy also present with gout	Ball and Sorensen (1969)
U.S. military veterans with gout	Pb exposure intensities unknown; current PbB comparable in normal function and kidney disease	EDTA Pb mobilization; renal function tests	50% of gout patients had renal failure; gout patients with kidney failure had almost twice the amount of chelatable Pb	Batuman et al. (1981)
Southeastern U. S. illicit alcohol drinkers	Elevated Pb content of illicit "moonshine" whiskey, reported up to ~5 mg/l	Symptomatic Pb poisoning along with gout	Kidney disease was often associated with presence of gout	Morgan et al. (1966)
B. Hypertensive nephrosclerosis				
Scottish subjects	PbB levels linked to water Pb	Serum uric acid, hypertension measurements	Both uric acid and hypertension correlate with PbB	Beevers et al. (1976)
U.S. military veterans with hypertension and with/without kidney failure	Chelatable Pb levels, 2 g chelant/3 days	Relative chelatable Pb in both groups	Individuals with hypertension and kidney disease had higher amounts of chelatable Pb	Batuman et al. (1983)
Patients with chronic Pb nephropathy	Severe childhood Pb exposures	Uric acid deposits and vascular changes in kidneys	Deposits and changes present in the absence of gout and hypertension	Inglis et al. (1978)

(*Continued*)

TABLE 15.4 Illustrative Studies of Chronic Lead Nephropathy in Human Populations with Diverse Risk Factors—(cont.)

Group	Lead Exposure (Various Units)	Endpoints	Results	References
Patients with chronic Pb nephropathy	Workplace Pb exposures	Uric acid deposits and vascular changes in kidneys	Deposits and changes present in the absence of gout and hypertension	Cramér et al. (1974)
C. Other chronic diseases				
Patients with existing chronic renal insufficiency followed 4 years ($N = 121$)	Environmental Pb exposures	Baseline serum creatinine 1.5 mg/dl; EDTA-chelatable Pb<600 μg/72 hours	Chelatable Pb over the period was significantly associated with increased serum creatinine (doubling) over 4 years	Yu et al. (2004)
Nonhypertensive NHANES III subjects ($N = 10{,}398$)	Environmental Pb exposure; PbB mean = 3.3 μg/dl	Diagnosed diabetes, chronic kidney disease	PbB was associated with a higher rate of kidney disease in diabetics	Muntner et al. (2003)

The question of long-term chronic disease states other than kidney disease amplifying any etiological role for Pb in kidney disease was probed by Muntner et al. (2003). A segment of the large U.S. NHANES III sampling cohort identified as normotensives ($N = 10{,}398$) was examined for the association of PbB with kidney disease in diabetics. These investigators reported that PbB was associated with a higher rate of kidney disease in diabetics versus nondiabetics. The mean PbB in this cohort from general ambient exposures was quite low, i.e., 3.3 μg/dl. This cohort is discussed in more detail in later sections.

15.6 NEPHROTOXIC EFFECTS OF Pb IN GENERAL HUMAN POPULATIONS

The literature on Pb nephrotoxicity in human populations exposed to relatively low, ambient levels of Pb has made up not only an extensive new literature but also a literature which has markedly improved our understanding of thresholds in Pb nephropathy, the prevalence and incidence of Pb nephrotoxic effects at low exposures, mechanisms of toxic action, etc. Furthermore, sampling and statistical issues are less problematic with general populations.

Sample sizes in the published database have been large, compared to the small and problematic numbers of Pb workers, and the extent and sophistication of statistical modeling for Pb nephrotoxic effects have been more involved. Demonstrable public health impacts of Pb nephrotoxicity at environmental levels raise the importance of this endpoint in terms of societal cost and both public health and health risk assessment policy.

The available environmental epidemiological literature on Pb nephrotoxicity in human populations has both cross-sectional and prospective studies. The latter have permitted analysis of reverse causality and the magnitude of risk predicted over time using baseline exposure biomarkers and changes in toxicity outcome marker over various periods.

General population cohorts for Pb–nephrotoxicity relationships have been studied both in the United States and internationally (Table 15.5). Major U.S. studies have been both cross-sectional and prospective in nature. Non-U.S. studies have largely been cross-sectional in nature, beginning with the first effort for environmental exposures via drinking water Pb in certain areas of Scotland (Campbell et al., 1977). Those researchers found that hyperuricemia was strongly associated with elevated PbB, the latter arising from tap water Pb levels >100 μg/l.

The prospective U.S. study of aging Boston, MA, residents produced a number of findings relating to Pb exposures and nephrotoxic effects (Table 15.5). Two prospective studies within the NAS cohort are of particular interest (Kim et al., 1996; Tsaih et al., 2004). Kim et al. examined a subset of the cohort longitudinally ($N = 459$ men) whose serial PbBs over 15 years (1979–1994) were measured along with changes in serum creatinine levels over the follow-up period. Mean baseline PbB was 9.9 μg/dl. A positive association was determined between ln-transformed PbB and serum creatinine. In these individuals, the highest PbB was $\leq$25 μg/dl and the β-coefficient was larger in subjects with PbB $\leq$ 10 μg/dl.

Tsaih et al. (2004) examined changes in serum creatinine versus baseline Pb exposure markers: PbB, tibia Pb, and patella Pb. Two measurements of serum creatinine were done, one at baseline and one at 6-year follow-up. A significant interaction was observed between PbB or tibial Pb and diabetes or hypertension versus changes in serum creatinine. Pb dose was not linked to creatinine changes in all participants.

Three of the NAS assessments entailed cross-sectional epidemiological analytical designs. Payton et al. (1994) reported that ln-transformed PbB was negatively associated with ln calculated creatinine clearance for 744 men evaluated between 1988 and 1991. Kim et al. (1996), in the cross-sectional portion of their NAS analyses, found a significant positive association of ln-transformed PbB with concurrent serum Pb. Wu et al. (2003) evaluated a subset of the cohort (total $N = 709$; 670 given full assessment) with respect to PbB, tibial Pb, and patellar Pb as exposure markers and both serum creatinine and estimated creatinine clearance rate as endpoints. They noted a

TABLE 15.5 Illustrative Studies of Nephrotoxic Effects of Lead in General Populations

Group	Pb Exposure (Various Units)	Endpoints	Results	References
Residents in Scottish households (household $N = 970$; subjects $N = 283$)	>100 μg Pb/l water	Hyperuricemia versus PbB	Hyperuricemia was strongly linked to PbB	Campbell et al. (1977)
Belgian Cadmibel Study ($N = 965$ men, $N = 1{,}016$ women)	PbB mean = 11.4 μg/dl, men; PbB mean = 7.5 μg/dl, women	Serum creatinine, β_2-microglobulin, measured and calculated creatinine clearances	Ln-transformed PbB was linked to reduced creatinine clearance in both genders	Staessen et al. (1992)
U.S. Boston, MA, Normative Aging Study (NAS, men 21–80 years old, $N = 744$, 1988–1991)	PbB mean = 8.1 μg/dl	Serum creatinine, measured and calculated creatinine clearances	PbB negatively associated with ln of creatinine clearance	Payton et al. (1994)
U.S. Boston, MA, NAS ($N = 459$ men)	PbB serially measured 1979–1994, 4–5 measurements, 10 μg/dl = PbB mean, baseline	Serum creatinine measured cross-sectionally and prospectively	Ln-transformed PbB was positively associated with concurrent serum creatinine; ln-transformed PbB positively linked to change in serum creatinine over follow-up in longitudinal study	Kim et al. (1996)
U.S. Boston, MA, NAS ($N = 709$ total, 670 for full assessment)	Pb exposure indexed 1991–1995: markers = PbB, tibial Pb, patella Pb. PbB mean = 6.2, tibial Pb = 22 μg/g mineral, patella Pb = 32 μg/g mineral	Serum creatinine, estimated creatinine clearance	Significant inverse association between patella Pb and creatinine clearance	Wu et al. (2003)

U.S. adults, male and female, in NHANES III: 1988–1994 ($N = 15{,}211$ total, $N = 4{,}813$ hypertensives)	PbB: normotensive mean = 3.3; hypertensive mean = 4.2	Dichotomous renal outcome measures, e.g., elevated serum creatinine and kidney disease; GFR < 60 ml/min/1.73 m^2, stratified into hypertensives and normotensives	Odds ratios increased for both kidney health measures, increased with PbB quartiles	Muntner et al. (2003)
NHANES subjects: 1999–2002, $N = 9{,}961$; NHANES III, 1988–1994, $N = 16{,}609$	Geometric mean (1999–2002) = 1.6 Geometric mean (1988–1994) = 2.8	As endpoints, the adjusted odds ratio for chronic kidney and peripheral artery diseases	For 1999–2002, highest quartile PbB (≥2.47 μg/dl) versus lowest (<1.06 μg/dl) were 2.72 and 1.92 (odds ratio) more likely to have chronic kidney and peripheral artery disease, respectively	Muntner et al. (2005)
NHANES subjects, 1999–2006, $N = 14{,}778$ adults≥20 years old	Geometric mean PbB = 1.6 (rounding); PbB quartiles used	Albuminuria (≥30 mg/g creatinine), reduced estimated GFR (<60 ml/min/ 1.73 m^2), combined endpoints; highest versus lowest quartile	Corresponding endpoints had odds ratio of 1.19, 1.56, and 2.39, respectively	Navas-Acien et al. (2009)
Swedish women ($N = 820$, age 53–64 years)	PbB mean = 2.2 μg/dl	Creatinine clearance and serum cystatin C measurements	Significant negative association of PbB with renal function markers	Åkesson et al. (2005)
Boston, MA, NAS participants ($N = 448$ men), 6% with diabetes, 26% hypertensive	PbB, tibial Pb, patellar Pb examined at baseline	Predictive changes, serum creatinine at baseline and 6 years later	Serum creatinine and PbB decreased over the period; interactions with diabetes and hypertension on creatinine changes	Tsaih et al. (2004)
Taiwanese chronic kidney disease patients ($N = 108$) without diabetes	Body Pb measured by chelation mobilization: <80 μg	Patients randomly assigned to EDTA chelation control groups: chelated 3 months, then 24 months periodically; patients followed to 1.25 times baseline serum creatinine and with changes in GFR	Patients without chelation showed increase in serum creatinine over baseline ($N = 14$) and/or mean decrease in GFR of 4.6 ml/min/1.73 m^2	Lin et al. (2006)

significant inverse association of patellar Pb with creatinine clearance. PbB mean was relatively low, 6.2 μg/dl, while tibial and patellar Pb means were 22 and 32 μg/g mineral, respectively.

The largest evaluation of low-level Pb exposure to kidney function relationships in a U.S. general population was that of Muntner et al. (2003) for the NHANES III, where 15,211 U.S. adult male and female normotensives and hypertensives were evaluated. Lead exposures were quite low, the population mean for normotensives being 3.3 μg/dl and for hypertensives, 4.2 μg/dl. The renal outcome measures were elevated serum creatinine and kidney disease indexed as reduced GFR (<60 ml/min/1.73 m^2). Subjects were stratified into normotensives and hypertensives and studied separately. The odds ratio for kidney dysfunction increased with quartile of PbB in hypertensives, the ratio in the highest PbB quartile being 2.07. Diabetics among the normotensives showed a higher frequency of kidney disease with higher PbB.

The Belgian Cadmibel Study (male $N = 965$, female $N = 1{,}016$) involved a large group of subjects in the Belgian general population having environmental Pb exposures (Staessen et al., 1992). PbB was the exposure marker, while levels of serum creatinine, β_2-microglobulin, and both estimated and measured creatinine clearance served as markers of kidney function. The PbB mean for men was 11.4 μg/dl, while that for women was 7.5 μg/dl. Ln-transformed PbB was associated with reduced creatinine clearance in both genders. A second large European study was that of Åkesson et al. (2005), who examined 820 Swedish women, aged 53–64 years, for associations between low-level PbB (mean = 2.2 μg/dl) and cadmium (various measures), and creatinine clearance rate or cystatin C measurements as indices of kidney function. They reported a significant negative association of PbB with these kidney function markers.

The very low mean PbBs in these women and those men and women in the Muntner et al. (2003) U.S. NHANES III provided good evidence that the relative threshold for Pb nephrotoxic effects with environmental exposures in the general population is an order of magnitude less than was observed in the older occupational Pb literature, ~60 μg/dl. Low body lead burdens indexed as chelatable Pb amounts of <80 μg/dl were also found to be a predictor of decreased kidney function (Lin et al., 2006). Taiwanese chronic kidney disease patients ($N = 108$) without diabetes were assigned to chelation and nonchelation groups. Those patients who were not periodically chelated over a 24-month testing period sustained a mean reduction in GFR of 4.6 ml/min/1.732 m^2 and showed increased serum creatinine.

15.7 GENETIC POLYMORPHISMS AND PB-ASSOCIATED OCCUPATIONAL NEPHROTOXICITY

A number of studies have been done exploring the potential associations of Pb exposure and Pb nephropathy modified by the ALAD genotype polymorphism. These studies are presented in Table 15.6. Five of the six cohorts

TABLE 15.6 ALAD Genetic Polymorphisms, Pb Exposures, and Occupational Nephropathy

Group	Pb Exposure (μg/dl)	Outcome Measures	Results	References
Korean Pb workers: ALAD 1-1, *N* = 716; ALAD 1-2, 2-2, *N* = 79	PbB mean, 1-1 allele = 31.7; variants, 34.2; other markers: chelatable Pb, tibial Pb, job tenure	BUN, serum uric acid, creatinine, creatinine clearance, NAG, RBP	ALAD 2 allele group had lower serum creatinine, higher creatinine clearance; ALAD 2 group showed hyperfiltration	Weaver et al. (2003, 2005)
Korean Pb workers: ALAD 1-1, *N* = 582; ALAD 1-2, 2-2, *N* = 63	PbB mean, 1-1 allele = 30.7; variants, 31.5; main bone Pb measure = patella	BUN, serum creatinine, measured/ estimated creatinine clearance, NAG-U, RBP-U	As in earlier studies, Pb measures linked to higher clearance rate, i.e., hyperfiltration in variant group	Weaver et al. (2006)
Vietnamese, Singaporean Pb workers; ALAD 1-1, *N* = 364; ALAD 1-2, 2-2, *N* = 40	PbB mean, 1-1 allele = 19.1; variants, 14.8	Urine albumin, β_2-micro-globulin, α_1-microglobulin, NAG, RBP	No associations; variant allele group had steeper dose–response slope, PbB and β_2-micro-globulin	Chia et al. (2006)
Chinese Pb workers; ALAD 1-1, *N* = 126; ALAD1-2, 2-2, *N* = 9	PbB mean, 1-1 allele = 41.0; variants, 62.5	Urine protein, β_2-micro-globulin, NAG	Variant allele showed higher association, PbB versus NAG, β_2-microglobulin	Gao et al. (2010)
Nonoccupational exposure cohort, aging males in Normative Aging Study: ALAD 1-1, *N* = 595; ALAD 1-2, *N* = 114	PbB mean, 1-1 allele = 6.3; variants, 5.8; Also, patellar, tibial Pb	Serum uric acid, creatinine, estimated creatinine clearance	No main associations; patella Pb with uric acid in variant allele group	Wu et al. (2003)

evaluated were Asian lead workers and one was from the ongoing Boston, MA, Normative Aging Study. Two of the worker groups were derived from the well-studied Korean Pb worker cohort, one involved Chinese and one mixed Vietnamese/Singaporean workers.

A consistent finding in the Korean worker cohort was a positive association of higher creatinine clearance in the variant group, consistent with a hyperfiltration response, an effect marker for early onset of chronic kidney disease. The Vietnamese/Singaporean findings were negative as to main effect but showed a higher slope in dose–response, β_2-microglobulin, and PbB. The nonworker Normative Aging cohort showed no main association, while the variant allele group showed a positive association of patellar Pb with uric acid.

15.8 EFFECTS OF Pb ON KIDNEY FUNCTION IN CHILDREN

Lead-associated nephropathy has typically been considered to be a toxic effect in adult men and women regardless of exposure source. Factors supporting this view, at least in the United States, were linked to the assumption of kidney toxicity of Pb at levels described in the older occupational and other Pb epidemiological literature with a relatively high threshold, e.g., $\geq$60–80 μg/dl. Such PbB levels are well above those leading to some level of medical intervention in children, i.e., $\geq$20 μg/dl. Various lead screening programs among high-risk children would readily identify and mark children with these high levels for medical intervention. Another factor has been the assumption, reasonably valid epidemiologically, that children are largely free of an array of chronic diseases such as hypertension that potentiate nephropathic associations for Pb. Finally, we would expect that any Pb–nephropathy relationships in children would be of the early-onset type, and one would seek associations of Pb with early outcome markers rather than clinically demonstrable chronic kidney disease. This has generally been the case for the six studies with children summarized in Table 15.7.

The earlier studies of Bernard et al. (1995) and Verberk et al. (1996) examined children where PbB was $\geq$10 μg/dl and the effect marker was either RBP (Bernard et al., 1995) or urinary NAG (Verberk et al., 1996). Both studies reported positive associations of the effect markers with elevated PbB.

Of particular importance are the studies of Fels et al. (1998), Staessen et al. (2001), and De Burbure et al. (2006). PbB levels were relatively low and all showed associations of exposure with early kidney dysfunction markers (Table 15.7). Staessen et al. (2001) noted that two groups of Belgian children having industrial suburban Pb exposures showed positive associations of PbB with both urinary β_2-microglobulin and serum cystatin C. The mean PbB levels for both industrial suburban groups were quite low, i.e., 1.8 and 2.7 μg/dl. De Burbure et al. (2006), evaluating up to 600 European children with regard to serum creatinine and cystatin C, reported an inverse association of increasing PbB and lower serum biomarkers of renal function, demonstrating hyperfiltration was operative in creatinine clearance rates (Table 15.7).

TABLE 15.7 Nephrotoxic Effects of Lead in Children

Group	Pb Exposure (μg/dl)	Endpoints	Results	References
Belgian children near a Pb smelter (*N* = 144) versus rural controls (*N* = 51)	Exposed PbB = 9.4–14.9 μg/dl	Urinary RBP	Increased urinary RBP levels in exposed children	Bernard et al. (1995)
Romanian children near a Pb smelter (*N* = 151, age 3–6)	Exposed PbB mean = 44 μg/dl; control mean = 16 μg/dl	Early biochemical indicator: urinary NAG	Exposed children showed a positive correlation of PbB with NAG	Verberk et al. (1996)
Belgian children (*N* = 200, 17 years old)	Two exposed PbB means: 1.8, 2.7 μg/dl; control mean 1.5 μg/dl	Urinary β_2-microglobulin, serum cystatin C	PbB was positively associated with both the urinary microglobulin and serum cystatin C	Staessen et al. (2001)
European children (*N* = variable with outcome, 300–600 children)	–	Serum creatinine, cystatin C	Inverse association of increasing PbB and lower renal markers (hyperfiltration)	De Burbure et al. (2006)
Polish children (*N* = 112 total: exposed *N* = 62; controls *N* = 50)	Exposed PbB mean = 13.3 μg/dl; control mean = 3.9 μg/dl	29 urinary or serum markers for nephron function	Increased β_2-microglobulin and Clara cell protein with increased PbB	Fels et al. (1998)
U.S. children (*N* = 769)	Mean PbB = 1.5 μg/dl; 99% < 10 μg/dl	Markers of GFR: serum creatinine, cystatin C	PbB levels > 2.9 μg/dl had lower GFR versus < 1.0 μg/dl PbB	Fadrowski et al. (2010)

15.9 EXPERIMENTAL ANIMAL STUDIES OF LEAD NEPHROPATHY

Inducing experimental Pb nephropathy in various test species using relatively modern protocols has comprised a considerable literature dating to at least the early 1960s. The dosing regimen in these studies of rats, rabbits, and dogs typically involved oral exposures through drinking water. Toxicity as early kidney dysfunction has most often been assessed and reported as morphological changes in the form of intranuclear inclusion bodies,

TABLE 15.8 Illustrative Experimental Animal Studies of Lead Nephropathy

Test Species	Pb Dosing System	Endpoint(s)	Results	References
Adult dogs	Pb acetate, 1× i.v., 0.2 or 10 mg/kg, stop-flow analysis	Pb-reuptake sites in the renal system	Proximal and distal tubular sites were identical; proximal tubular reuptake shown with citrate or bicarbonate infusion	Victery et al. (1979)
Rabbits	0.5% Pb acetate in diet ≤ 55 weeks	Morphological changes	Nuclear inclusion bodies and interstitial fibrosis were reported	Hass et al. (1964)
Rat	1% Pb in water, 9 weeks	Morphological changes	Nuclear inclusion bodies, increased mitochondrial swelling, and interstitial fibroses reported	Goyer (1971)
Rat	0, 0.5, 5, 25, 50, 250 ppm in water, ad lib	Morphological changes	Nuclear inclusion bodies, increased mitochondrial swelling, and proximal tubular cytomegaly in the higher exposure groups, more evident at 9 months	Fowler et al. (1980)
Rat	1% Pb acetate in water, up to 6 weeks	Kidney function up to 16 weeks postexposure: GFR and renal blood flow	Reductions in GFR and renal blood flow persisting to 16 weeks postexposure	Aviv et al. (1980)
Rat	Pb in drinking water: 100, 300, 500, 1,000, 2,000, 3,000, 5,500 ppm	GFR and effective renal plasma flow; adjustment for weight changes	Largely negative results at all dosing levels	O'Flaherty et al. (1986)

Male Wistar rats	0.5% or 2.0% Pb acetate in water, 2 months; 1% Pb acetate for 3 months	24-hour urinary glucose, total protein; LDH, lysozyme, β_2-microglobulin	At 2% dosing, increased microglobulin, glucose, total protein, LDH, lysozyme. At 1%, only β_2-microglobulin increased; no changes at 0.5%	Vyskocil et al. (1989)
Male Sprague–Dawley rats	Pb acetate in water (0.5%), 1–9 months, pair-fed controls; other dosings with/without chelation	Serum creatinine, BUN, creatinine clearance	At 6- or 9- month dosings, creatinine clearance was reduced and the blood measures increased; tubule-interstitial disease was severe; chelation did not abolish pathological changes, with some improvement in function	Khalil-Manesh et al. (1992a)
Male Sprague–Dawley rats	0.5% Pb acetate in water, 1–12 months, pair-fed controls	Renal pathology markers and GFR; early injury markers: NAG, GST, brush border antigens	Early effects: hypertrophy of tubules, nuclear inclusion bodies, increased GFR, increased urinary levels of early markers	Khalil-Manesh et al. (1992b)
Male Sprague–Dawley rats: controls, exposed had RK surgery	Pb acetate in water, 150 ppm—4 weeks; then RK + 12 weeks exposure: PbB = 26 μg/dl	Functional and pathological indices of kidney disease: creatinine clearance, arteriolar disease, peritubular capillary changes, tubulointerstitial integrity, macrophage infiltration	Exposure produced arteriolar damage, peritubular capillary loss, tubulointerstitial damage, macrophage infiltration	Roncal et al. (2007)

mitochondrial swelling, and interstitial fibrosis (Fowler et al., 1980; Goyer, 1971; Hass et al., 1964). More chronic forms of nephrotoxicity included reduced GFR and reduced renal blood flow, as occurred when rats were dosed with 1% Pb acetate for 6 weeks (Table 15.8; Aviv et al., 1980).

The relative utility of the rat as an animal model for human Pb nephropathy in the form of reduced GFR has been questioned by O'Flaherty et al. (1986). In their critique and comparison of the Aviv et al. (1980) findings versus their own experience, O'Flaherty et al. noted largely negative findings for the drinking water Pb regimen. These differences may arise from various factors, nutritional issues, animal strain differences, etc. However, Khalil-Manesh et al. (1992a,b) used a protocol that included pair feeding of controls in their assessments of Pb nephropathy in male Sprague–Dawley rats. They noted that clinical markers such as GFR and early tubular injury markers—urinary total protein, urinary glucose, NAG, GST, LDH—tracked Pb exposure. Interestingly, GFR in these rats was first increased, i.e., hyperfiltration was operative, and then decreased. This finding of hyperfiltration matches the finding in human cohorts (see above discussion).

The recent findings of Roncal et al. (2007), using male Sprague–Dawley rats and remnant kidney (RK) treatment of the animals to maximize renal functional changes, are of special interest. They noted that pathological manifestation of kidney damage in the form of arteriolar damage, peritubular capillary loss, tubulointerstitial damage, etc. occurred at an experimental PbB of 26 μg/dl. This blood level approximates those values noted in developmental neurobehavioral animal studies as linked to early neurotoxicity. The Roncal et al. data suggest that in terms of experimental dose–response relationships, those for Pb nephrotoxicity are not particularly insensitive when contrasted to findings in rats for neurobehavioral toxicity effects.

REFERENCES

Åkesson, A., Lundh, T., Vahter, M., Bjellerup, P., Lidfeldt, J., Nerbrand, C., et al., 2005. Tubular and glomerular kidney effects in Swedish women with low environmental cadmium exposure. Environ. Health Perspect. 113, 1627–1631.

Aviv, A., John, E., Bernstein, J., Goldsmith, D.I., Spitzer, A., 1980. Lead intoxication during development: Its late effects on kidney function and blood pressure. Kidney Int. 17, 430–437.

Baker Jr., E.L., Landrigan, P.J., Barbour, A.J., Cox, D.H., Folland, D.S., Ligo, R.N., et al., 1979. Occupational lead poisoning in the United States: Clinical and biochemical findings related to blood lead levels. Br. J. Ind. Med. 36, 314–322.

Ball, G.V., Sorensen, L.B., 1969. Pathogenesis of hyperuricemia in saturnine gout. N. Engl. J. Med. 280, 1199–1202.

Batuman, V., Landy, E., Maesaka, J.K., Wedeen, R.P., 1983. Contribution of lead to hypertension with renal impairment. N. Engl. J. Med. 309, 17–21.

Batuman, V., Maesaka, J.K., Haddad, B., Tepper, E., Landry, E., Wedeen, R.P., 1981. The role of lead in gout nephropathy. N. Engl. J. Med. 304, 520–523.

Beevers, D.G., Erskine, E., Robertson, M., Beattie, A.D., Campbell, B.C., Goldberg, A., et al., 1976. Blood-lead and hypertension. Lancet 2 (7975), 1–3.

Bernard, A.M., Vyskocil, A., Roels, H., Kriz, J., Kodl, M., Lauwerys, R., 1995. Renal effects in children living in the vicinity of a lead smelter. Environ. Res. 68, 91–95.

Blackman Jr., S.S., 1936. Intranuclear inclusion bodies in the kidney and liver caused by lead poisoning. Bull. Johns Hopkins Hosp. 58, 384–402.

Buchet, J.-P., Roels, H., Bernard, A., Lauwerys, R., 1980. Assessment of renal function of workers exposed to inorganic lead, cadmium or mercury vapor. J. Occup. Med. 22, 741–750.

Campbell, B.C., Beattie, A.D., Moore, M.R., Goldberg, A., Reid, A.G., 1977. Renal insufficiency associated with excessive lead exposure. BMJ 1 (6059), 482–485.

Campbell, B.C., Meredith, P.A., Scott, J.J.C., 1985. Lead exposure and changes in the renin–angiotensin–aldosterone system in man. Toxicol. Lett. 25, 25–32.

Chia, K.S., Jeyaratnam, J., Tam, C., Ong, H.Y., Ong, C.N., Lee, E., 1995a. Glomerular function of lead-exposed workers. Toxicol. Lett. 77, 319–328.

Chia, K.S., Jeyaratnam, J., Lee, J., Tan, C., Ong, H.Y., Ong, C.N., et al., 1995b. Lead-induced nephropathy: Relationship between various biological exposure indices and early markers of nephrotoxicity. Am. J. Ind. Med. 27, 883–895.

Chia, S.-E., Zhou, H.J., Yap, E., Tham, M.T., Dong, N.-V., Hong Tu, N.T., et al., 2006. Association of renal function and "delta"-aminolevulinic acid dehydratase polymorphism among Vietnamese and Singapore workers exposed to inorganic lead. Occup. Environ. Med. 63, 180–186.

Chisolm Jr., J.J., 1962. Aminoaciduria as a manifestation of renal tubular injury in lead intoxication and a comparison with patterns of aminoaciduria seen in other diseases. J. Pediatr. 60, 1–17.

Chisolm Jr., J.J., 1968. The use of chelating agents in the treatment of acute and chronic lead intoxication in childhood. J. Pediatr. 73, 1–38.

Chisolm Jr., J.J., Mellitts, E.D., Barrett, M.B., 1976. Interrelationships among blood lead concentration, quantitative daily ALA-U and urinary lead output following calcium EDTA. In: Nordberg, G.F. (Ed.), Proceedings of the Third Meeting of the Subcommittee on the Toxicology of Metals, Tokyo, Japan: November, 1974. Amsterdam, The Netherlands: Elsevier Publishing Co., pp. 416–433.

Clarkson, T.W., Kench, J.E., 1956. Urinary excretion of amino acids by men absorbing heavy metals. Biochem. J. 62, 361–372.

Cramér, K., Goyer, R.A., Jagenburg, R., Wilson, M.H., 1974. Renal ultrastructure, renal function, and parameters of lead toxicity in workers with different periods of lead exposure. Br. J. Ind. Med. 31, 113–127.

De Burbure, C., Buchet, J.-P., Leroyer, A., Nisse, C., Haguenoer, J.-M., Mutti, A., et al., 2006. Renal and neurologic effects of cadmium, lead, mercury, and arsenic in children: Evidence of early effects and multiple interactions at environmental exposure levels. Environ. Health Perspect. 114, 584–590.

Dos Santos, A.C., Colacciopo, S., Dal Bó, C.M., dos Santos, N.A., 1994. Occupational exposure to lead, kidney function tests, and blood pressure. Am. J. Ind. Med. 26, 635–643.

Ekong, E.B., Jaar, B.G., Weaver, V.M., 2006. Lead-related nephrotoxicity: A review of the epidemiological evidence. Kidney Int. 70, 2074–2084.

Emmerson, B.T., 1963. Chronic lead nephropathy: The diagnostic use of calcium EDTA and the association with gout. Australas. Ann. Med. 12, 310–324.

Fadrowski, J.J., Navas-Acien, A., Tellez-Plaza, M., Guallar, E., Weaver, V.M., Furth, S.L., 2010. Blood lead levels and kidney function in U.S. adolescents: The Third National Health and Nutrition Examination Survey. Arch. Intern. Med. 170, 75–82.

Fels, L.M., Wünsch, M., Baranowaski, J., Norska-Borowka, I., Price, R.G., Taylor, S.A., et al., 1998. Adverse effects of chronic low-level lead exposure on kidney function—a risk group study in children. Nephrol. Dial. Transplant 13, 2248–2256.

Fowler, B.A., Kimmel, C.A., Woods, J.S., McConnell, E.E., Grant, L.D., 1980. Chronic low-level lead toxicity in the rat. III. An integrated assessment of long-term toxicity with special reference to the kidney. Toxicol. Appl. Pharmacol. 56, 59–77.

Gao, A., Lux, X.T., Li, Q.Y., Tian, L., 2010. Effect of the delta-aminolevulinic acid dehydratase gene polymorphism on renal and neurobehavioral function in workers exposed to lead in China. Sci. Total Environ. 408, 4052–4055.

Gibson, J.L., 1904. A plea for painted railings and painted walls of rooms as the source of lead poisoning among Queensland children. Aust. Med. Gazette 23, 149–153.

Gibson, J.L., Love, W., Hardine, D., Bancroft, P., Turner, A.J., 1892. In: Huxtable, L.R. (Ed.), Transactions of the Third Intercolonial Medical Congress of Australasia. Charles Potter, Sydney, pp. 76–83.

Goyer, R.A., 1971. Lead toxicity: A problem in environmental pathology. Am. J. Pathol. 64, 167–181.

Goyer, R.A., Leonard, D.L., Moore, J.F., Rhyne, B., Krigman, M.R., 1970. Lead dosage and the role of the nuclear inclusion body: an experimental study. Arch. Environ. Health 20, 705–711.

Hass, G.M., Brown, D.V.L., Eisenstein, R., Hemmens, A., 1964. Relations between lead poisoning in rabbit and man. Am. J. Pathol. 45, 691–727.

Henderson, D.A., 1954. A follow-up of cases of plumbism in children. Australas. Ann. Med. 3, 219–224.

Hu, H., 1991. A 50-year follow-up of childhood plumbism: Hypertension, renal function, and hemoglobin level among survivors. Am. J. Dis. Child. 145, 681–687.

Inglis, J.A., Henderson, D.A., Emerson, B.T., 1978. The pathology and pathogenesis of chronic lead nephropathy occurring in Queensland. J. Pathol. 124, 65–76.

Khalil-Manesh, F., Gonick, H.C., Cohen, A.H., Alinovi, R., Bergamaschi, E., Mutti, A., et al., 1992a. Experimental model of lead nephropathy. I. Continuous high-dose lead administration. Kidney Int. 41, 1192–1203.

Khalil-Manesh, F., Gonick, H.C., Cohen, A., Bergamaschi, E., Mutti, A., 1992b. Experimental model of lead nephropathy. II. Effect of removal from lead exposure and chelation treatment with dimercaptosuccinic acid (DMSA). Environ. Res. 58, 35–54.

Kim, R., Rotnitsky, A., Sparrow, D., Weiss, S.T., Wager, C., Hu, H., 1996. A longitudinal study of low-level lead exposure and impairment of renal function. The Normative Aging Study. JAMA 275, 1177–1181.

Kottgen, A., Selvin, E., Stevens, L.A., Levey, A.S., Van Lente, F., Coresh, J., 2008. Serum cystatin C in the United States: The Third National Health and Nutrition Examination Survey (NHANES III). Am. J. Kidney Dis. 51, 385–394.

Lilis, R., Gavrilescu, N., Nestorescu, B., Dumitriu, C., Roventa, A., 1968. Nephropathy in chronic lead poisoning. Br. J. Ind. Med. 25, 196–202.

Lilis, R., Fischbein, A., Valciukas, J.A., Blumberg, W., Selikoff, I.J., 1980. Kidney function and lead: relationships in several occupational groups with different levels of exposure. Am. J. Ind. Med. 1, 405–412.

Lin, J.L., Lin-Tan, D.T., Li, Y.J., Chen, K.H., Huang, Y.L., 2006. Low-level environmental exposure to lead and progressive chronic kidney diseases. Am. J. Med. 119, 1–9.

Loghman-Adham, M., 1997. Renal effects of environmental and occupational lead exposure. Environ. Health Perspect. 105, 928–939.

Moel, D.I., Sachs, H.K., 1992. Renal function 17 to 23 years after chelation therapy for childhood plumbism. Kidney Int. 42, 1226–1231.

Morgan, J.M., Hartley, M.W., Miller, R.E., 1966. Nephropathy in chronic lead poisoning. Arch. Intern. Med. 118, 1729.

Muntner, P., He, J., Vupputuri, S., Coresh, J., Batuman, V., 2003. Blood lead and chronic kidney disease in the general United States population: results from NHANES III. Kidney Int. 63, 1044–1050.

Muntner, P., Menke, A., DeSalvo, K.B., Rabito, F.A., Batuman, V., 2005. Continued decline in blood lead levels among adults in the United States—The National Health and Nutrition Examination Surveys. Arch. Intern. Med. 165, 2155–2161.

National Academy of Sciences/National Research Council, 1972. LEAD. Airborne Lead in Perspective. National Academy Press, Washington, DC.

National Academy of Sciences/ National Research Council, 1980. Lead in the Human Environment. National Academy Press, Washington, DC.

National Academy of Sciences/National Research Council, 1993. Measuring Lead Exposure in Infants, Children, and Other Sensitive Populations. National Academy Press, Washington, DC.

Navas-Acien, A., Tellez-Plaza, M., Guallar, E., Muntner, P., Silbergeld, E., Jaar, B., et al., 2009. Blood cadmium and lead and chronic kidney disease in U.S. adults: A joint analysis. Am. J. Epidemiol. 170, 1156–1164.

Nye, L.J.J., 1929. An investigation of the extraordinary incidence of chronic nephritis in young people in Queensland. Med. J. Aust. 2, 145–159.

O'Flaherty, E.J., Adams, W.D., Hammond, P.B., Taylor, E., 1986. Resistance of the rat to development of lead-induced renal functional deficits. J. Toxicol. Environ. Health A 18, 61–75.

Oliver, T., 1891. Lead Poisoning in its Acute and Chronic Forms. Young J. Pentland, London, UK.

Omae, K., Sakurai, H., Higashi, T., Muto, T., Ichikawa, M., 1990. No adverse effects of lead on renal function in lead-exposed workers. Ind. Health 28, 77–83.

Payton, M., Hu, H., Sparrow, D., Weiss, S.T., 1994. Low-level lead exposure and renal function in the Normative Aging Study. Am. J. Epidemiol. 140, 821–829.

Pejić, S., 1928. The nature of the primary renal lesion produced by lead. Ann. Intern. Med 1, 577–604.

Pinto de Almeida, A.R., Carvalho, F.M., Spinola, A.G., Rocha, H., 1987. Renal dysfunction in Brazilian lead workers. Am. J. Nephrol. 7, 455–458.

Roncal, C., Mu, W., Reungjwi, S., Kim, K.M., Henderson, G.N., Ouyang, X., et al., 2007. Lead, at low levels, accelerates arteriolopathy and tubulointerstitial injury in chronic kidney disease. Am. J. Physiol. Renal Physiol. 293, F1391–F1396.

Staessen, J.A., Lauwerys, R.R., Buchet, J.-P., Bulpitt, C.J., Rondia, D., Van Renterghem, Y., et al., 1992. Impairment of renal function with increasing blood lead concentrations in the general population. N. Engl. J. Med. 327, 151–156.

Staessen, J.A., Nawroth, T., Den Hond, E., Thijs, L., Fagard, R., Hoppenbrouwers, K., et al., 2001. Renal function, cytogenetic measurements, and sexual developments in adolescents in relation to environmental pollutants: a feasibility study of biomarkers. Lancet 357, 1660–1669.

Steenland, K., Selevan, S., Landrigan, P., 1992. The mortality of lead smelter workers: an update. Am. J. Public Health 82, 1641–1644.

Tepper, L.B., 1963. Renal function subsequent to childhood plumbism. Arch. Environ. Health 7, 76–85.

Tsaih, S.-W., Korrick, S., Schwartz, J., Amarasiriwardena, C., Aro, A., Hu, H., 2004. Lead, diabetes, hypertension, and renal function: the Normative Aging Study. Environ. Health Perspect. 112, 1178–1182.

U.S. Agency for Toxic Substances and Disease Registry, 2007. Toxicologic Profile for Lead. U.S. Centers for Disease Control, Atlanta, GA.

U.S. Environmental Protection Agency, 1977. Air Quality Criteria for Lead. Report No. EPA-600/8-77-017. Office of Research and Development, Washington, DC.

U.S. Environmental Protection Agency, 1986. Air Quality Criteria for Lead, 4 vols. EPA 600/8-83/028bF. Environmental Protection Agency, Research Triangle Park, NC.

U.S. Environmental Protection Agency, 2006. Air Quality Criteria for Lead, vol. 1. Report No. EPA/600/R-05/144aF. Environmental Protection Agency, Research Triangle Park, NC.

Verberk, M.M., Willems, T.E.P., Verplanke, A.J.W., De Wolff, F.A., 1996. Environmental lead and renal effects in children. Arch. Environ. Health 51, 83–87.

Vershoor, M., Wibowo, A., Herber, R., van Hemmen, J., Zielhuis, R., 1987. Influence of occupational low-level lead exposure on renal parameters. Am. J. Ind. Med. 12, 341–351.

Victery, W., Vander, A.J., Mouw, D.R., 1979. Renal handling of lead in dogs: stop-flow analysis. Am. J. Physiol. 237, F408–F414.

Vyskocil, A., Panci, J., Tusl, M., Ettlerova, E., Semecky, V., Gasparova, L., et al., 1989. Dose-related proximal tubular dysfunction in male rats chronically exposed to lead. J. Appl. Toxicol. 9, 395–399.

Weaver, V.M., Lee, B.-K., Ahn, K.D., Lee, G.S., Todd, A.C., Stewart, W.F., et al., 2003. Association of lead biomarkers with renal function in Korean lead workers. Occup. Environ. Med. 60, 551–562.

Weaver, V.M., Schwartz, B.S., Jaar, B.G., Ahn, K.-D., Todd, A.C., Lee, S.-S., et al., 2005. Associations of uric acid with polymorphisms in the "delta-" aminolevulinic dehydratase, vitamin D receptor, and nitric oxide synthase genes in Korean lead workers. Environ. Health Perspect. 113, 1509–1515.

Weaver, V.M., Lee, B.K., Todd, A.C., Ahn, K.D., Shi, W., Jaar, B.G., et al., 2006. Effect modification by delta-aminolevulinic acid dehydratase, vitamin D receptor, and nitric oxide synthase gene polymorphisms on associations between patella lead and renal function in lead workers. Environ. Res. 102, 61–69.

Weaver, V.M., Griswold, M., Todd, A.C., Jaar, B.D., Ahn, K.D., Thompson, C.B., et al., 2009. Longitudinal associations between lead dose and renal function in lead workers. Environ. Res. 109, 101–107.

Wedeen, R.P., 1982. Lead nephrotoxicity. In: Porter, G. (Ed.), Nephrotoxic Mechanisms of Drugs and Environmental Toxins. Plenum Publishing Corp., New York, pp. 255–265.

Wedeen, R.P., 1984. Poison in the Pot: The Legacy of Lead. Southern Illinois University Press, Carbondale, IL.

Wedeen, R.P., Maesaka, J.K., Weiner, B., Lipat, G.A., Lyons, M.M., Vitale, L.F., et al., 1975. Occupational lead nephropathy. Am. J. Med. 59, 630–641.

Wedeen, R.P., Mallik, D.K., Batuman, V., 1979. Detection and treatment of occupational nephropathy. Arch. Intern. Med. 139, 53–57.

World Health Organization, 1995. Inorganic Lead. Environmental Health Criteria 165. International Programme on Chemical Safety, Geneva, Switzerland.

Wu, M.T., Kelsey, K., Schwartz, J., Sparrow, D., Weiss, S., Hu, H., 2003. A δ-aminolevulinic acid dehydratase (ALAD) may modify the relationship of low level lead exposure to uricemia and renal function: the Normative Aging Study. Environ. Health Perspect. 111, 335–341.

Yu, T., 1983. Lead nephropathy and gout. Am. J. Kidney Dis. 2, 555–558.

Yu, C.-C., Lin, J.-L., Lin-Tan, D.-T., 2004. Environmental exposure to lead and progression of chronic renal diseases. J. Am. Soc. Nephrol. 15, 1016–1022.

Chapter 16

Hematological Effects of Lead in Human Populations

This chapter discusses the hematological effects of lead in both human populations and various experimental animal test systems. Lead hematotoxicity, in common with effects such as neurotoxicity, nephrotoxicity, and cardiovascular impairments, has long been recognized as accompanying chronic childhood and adult lead exposures within various Pb exposure settings.

16.1 INTRODUCTION AND CONTEXT

Hematotoxic effects of lead have a prominent place in the clinical and occupational health history of the element, having been causally associated with chronic anemia arising from decreased erythropoiesis, impaired production of hemoglobin, and the presence of basophilic stippling in the gingival area, arising from deranged pyrimidine metabolism. The topic has generated many hundreds of individual research articles over the decades and has been heavily reviewed and critiqued in such expert consensus treatises as those of the U.S. EPA (1977, 1986, 2006), the U.S. CDC (1978, 1985, 1991, 2005), the U.S. ATSDR (1988, 2007), the NAS/NRC (1972, 1980, 1993), and the WHO (1995).

Much of the literature on Pb-associated hematotoxicity is of some vintage, having been the object of considerable research in the 1970s and 1980s. More recent studies have been directed at lingering dose−response and mechanistic issues.

Lead hematotoxicity is not only a toxic endpoint of Pb exposure in humans and animals but also provides biomarkers of Pb exposures via early toxic effects on the heme and erythropoietic pyrimidine biosynthesis pathways. Although heme biosynthesis is largely considered here in terms of heme production for utilization in hemoglobin, it is a critical cofactor for diverse metabolic processes, including in the cytochromes and for the 1-hydroxylase enzyme system responsible for generating the hormonal form

Trace Metals and other Contaminants in the Environment, Volume 10
ISSN: 1875-1121 DOI: 10.1016/B978-0-444-51554-4.00016-X

of vitamin D. These biomarkers of early effect have defined various segments of the Pb-hematotoxicity dose–response curve.

Hematotoxic effects of lead have triggered debates as to the nature and severity of lead hematotoxicity versus lead toxicity associated with brain, cardiovascular, and kidney injuries. Part of the debate goes to the basic question of what comprises a toxicologically adverse effect, a functionally adverse effect, and a biochemically adverse effect. Effects of Pb on heme formation from precursors are produced through effects on mitochondrial function.

A second area of contention springs from the question of reversibility. Effects of Pb which are assumed to be reversible when Pb exposure ceases are often assumed to be of less concern than those which are persistent and likely irreversible. Endpoints such as developmental neurotoxic effects on cognitive function in child populations are widely considered to be part of the latter. Reversible effects of Pb with exposure removal, however, are arguably of more theoretical than actual interest. Common Pb exposure scenarios where removal from exposure is highly unlikely, for example, children during the age of peak vulnerability in inner-city housing with deteriorating Pb paint, can have the effect function as a *de facto* persisting toxic endpoint for all practical purposes.

The relative significance of any hematotoxic effects of Pb at currently encountered low-level environmental Pb exposures in current human populations remains unclear, especially for those early effects which are Pb exposure biomarkers. The U.S. CDC (1991) recognized in its 1991 Statement on childhood Pb poisoning that EP (ZPP), a long-useful biomarker for childhood Pb exposure, was problematic at levels of 25 μg/dl and below, recommending only the direct measurement of PbB to identify elevated and toxic PbB values when screening childhood risk populations.

Hematotoxic effects of Pb are a consequence of Pb exposure in sensitive human populations where they appear to be modulated through differing genetic susceptibilities among various subgroups. Genetic susceptibility is generally accepted as affecting Pb toxicokinetics, but the net effect of genotypic influences foretelling less or more toxicity remains unclear.

This chapter is organized to first provide a historical look at the role of lead hematotoxicity in recorded classical clinical Pb poisoning. This section is followed by those dealing with the major components of Pb hematotoxicity.

16.2 HEMATOTOXIC EFFECTS OF LEAD IN CLINICAL LEAD POISONING

It is surprising, given the long history of Pb hematotoxicity as a topic in the classical poisoning literature, that the characteristics and etiology of hematotoxicity (e.g., unconfounded lead anemia) are still topics of discussion (U.S.

CDC, 1978, 1991, 2005; U.S. EPA, 1977, 1986, 2006). The purely lead-associated anemia is taken to be mildly hypochromic or normochromic and normocytic. With severe exposures, one occasionally sees basophilic stippling from deranged erythropoietic activity and accumulation of pyrimidine nucleotides. Reticulocytosis is seen subsequent to reduced erythrocyte survival time. Lead toxicity induces the anemia via the combined effects of reduced hemoglobin production and increased erythrocyte destruction.

One problem in young children is the confounding of iron-deficiency anemia with a Pb-induced anemia, at least on a group basis, requiring more extensive diagnostic laboratory measures. Consequently, hemoglobin (Hb) and hematocrit (Hct) measurements are combined with measures such as percent transferrin saturation. At present, it is generally accepted that iron deficiency in the presence of childhood lead poisoning works to exacerbate the lead-toxic component and vice versa (Chisolm, 1981; Marcus and Schwartz, 1987; Yip et al., 1981). The interaction appears to work at several levels, beginning with enhanced oral Pb uptake and elevated Pb retention in the presence of Fe deficiency. Pb-derived anemia has been well documented in cases where iron deficiency is not a confounder in older children (Roels et al., 1976) and in Pb workers (Lilis et al., 1978). Here, these results support a clear role for Pb-based anemia.

Earlier studies of Pb-associated anemia are included in Table 16.1 (Chisolm, 1971; Moore, 1988; U.S. CDC, 1978; Waldron, 1966). Identification of anemia in those Pb-exposed young children, where some level of Fe deficiency would co-occur, was based in the older literature on an Hb measurement of <10 g/dl (Chisolm, 1971), while the corresponding marker for exposed adults was <13 g/dl. Reduced Hb occurred because of significant reductions in both heme and globin production. Effects of Pb on heme production are presented in a later section, while impaired globin synthesis has been the subject of several studies, albeit with a small number of individuals (Wada et al.,1973; White and Harvey, 1972).

The relationship of Pb exposure and anemia in terms of dose–response relationships and thresholds indexed by various exposure markers has been recorded. In children, the thresholds in the older literature for onset of Pb-derived anemia with reference to Hb reduction are lower, i.e., <40 μg/dl than it is in adults, ≥40–50 μg/dl. At levels above these thresholds, studies typically identified an inverse relationship with PbB levels in children (Pueschel et al., 1972) and Pb workers (Baker et al., 1979). Several Pb worker epidemiological studies evaluated the percentage population response for a selected Hb reduction at varying levels of PbB. In the U.S. study of smelter workers by Baker et al. (1979), employing an Hb level <14.0 g/dl, 5% of workers had a Hb reduction at 40–59 μg/dl PbB, 14% at 60–79 μg/dl PbB, and 36% at ≥80 μg/dl PbB. A similar analysis by Grandjean (1979), however, showed much higher frequencies at a somewhat different Hb cutoff of <14.4 g/dl: 17% showed an Hb reduction at <25 μg/dl PbB, 26% at

TABLE 16.1 Hematological Effects of Pb in Human Populations: Lead-Induced Anemia via Reduced Hb Production[a]

Study Groups	Pb Exposure	Endpoint(s)	Results	References
Various earlier studies of Pb workers and exposed children	>40–100 μg/dl	Chronic Pb poisoning: Hb, children, <10 g/dl; workers, <13 g/dl	Pb-induced reduction in Hb, from both heme and globin reduction	Waldron (1966), Chisolm (1971), U.S. CDC (1978), Moore (1988)
Pb-poisoned children (*N* = 40)	PbB, 30–120 μg/dl	Reduced Hb level	Negative correlation between Hb and PbB	Pueschel et al. (1972)
Kosovo prospective Pb study children (*N* = 311) in high- and low-Pb areas	Mean high PbB: 30–40 μg/dl; mean low PbB: 6–9 μg/dl	Hb measurements at 4.5, 6.5, 9.5, 12 years of age	No significant association of Hb reduction with PbB	Graziano et al., 2004, Factor-Litvak et al. (1999)
Children (*N* = 86), age 1–6 years	General Pb exposure population: range, 2–84 μg/dl	Hb measurements versus PbB	No Pb-associated Hb reduction	Liebelt et al. (1999)
Lead workers (*N* = 202)	Wide PbB range: <25 to >60 μg/dl	PbB tercile versus % with lower Hb, <14.4 g/dl	17% <25 μg/dl; 26% 26–60 μg/dl; 45% >60 μg/dl	Grandjean (1979)
Kosovo prospective Pb study of pregnant women (*N* = 1,502) in high- and low-Pb areas	PbB range = 2–43 μg/dl; mean low PbB = 5 μg/dl; mean high PbB = 17 μg/dl	Hb versus PbB	No Pb-associated Hb reduction	Graziano et al. (1990)
Male workers in vinyl chloride stabilizer production (*N* = 1573)	Mean PbB = 13 μg/dl	Hb versus PbB	Positive association of Hb with PbB	Makino et al. (1997)
Male carpenters (*N* = 119)	Mean PbB = 8 μg/dl	Hb versus PbB or patella bone Pb	No Hb reduction versus PbB, reduction in Hb versus patella Pb	Hu et al. (1994)

[a]*Selected studies. Anemia variably defined over time as seen in the citations.*

26–60 μg/dl PbB, and 45% at >60 μg/dl PbB. The basis for the much higher figures determined by Grandjean (1979) is not clear since the difference in the Hb cutoff figures is not great, 14.0 versus 14.4 g/dl, and both study groups would be expected to show any "healthy worker" effects.

16.3 RECENT STUDIES OF ANEMIA IN HUMAN POPULATIONS VIA THE HB–Pb EXPOSURE RELATIONSHIP

More recent reports have produced mixed results for the absence or presence of dose–response relationships for Hb reduction with increased PbB levels (Table 16.1). A number of the newer investigations had the benefit of much better epidemiological methods: relatively large sample size, more extensive statistical controls for confounding, and methods for determining the relative utility of various Pb exposure markers in documenting hematotoxic effects. The former Yugoslavia/Kosovo prospective Pb study evaluated the relationship of Hb reduction with increasing PbB in smelter area children relative to a control community (N=311 for combined groups) (Graziano et al., 2004; Factor-Litvak et al., 1999). The high Pb exposure group had a mean PbB of 30–40 μg/dl, while the lower contamination site subjects had a mean of 6–9 μg/dl. Hb measurements were taken at four ages: 4.5, 6.5, 9.5, and 12 years. No significant association between reduced Hb and increased PbB was found. Liebelt et al. (1999), using a smaller group of children, aged 1–6 years, similarly reported no link between Hb reduction and increasing PbB.

One of the largest assessments of adults having environmental Pb exposures in terms of hematotoxic responses is that of Graziano et al. (1990), as part of the former Yugoslavia/Kosovo prospective Pb study. A total of 1,502 pregnant women with either high, smelter area exposures or lower, control town exposures were evaluated. The mean PbB level for the smelter area group was comparatively low for similarly situated study cohorts. The investigators reported the lack of any statistical association of Hb changes with increased Pb exposure. Similarly, the occupational Pb exposure report of Hu et al. (1994) for a group of carpenters whose mean PbB was 8 μg/dl indicated no association of Hb changes with PbB exposure.

By contrast, a large group of male workers in vinyl chloride stabilizer production ($N = 1{,}573$) showed a positive statistical association of Hb change with increased PbB (Makino et al., 1997). The mean PbB was relatively low compared to typical workplace exposures for Pb material workers, 13 μg/dl. The anomalous nature of this positive association is difficult to characterize. While it may reflect a biphasic, positive response at low exposures followed by decrements in Hb with higher exposures, such a response was absent in the environmental exposure study of Graziano et al. (1990). Here, the equally large study cohort ($N = 1{,}502$) of pregnant women would be expected to show a stronger hematological response to Pb exposure in either direction, based on available literature (Wildt et al., 1987). The positive association may also indicate the effect of a particular chemical form of lead or responses to multiple agents.

Hu et al. (1994) employed patellar bone Pb levels as the exposure biomarker to evaluate associations of Hb changes with indicators of body Pb burden. Unlike the negative results with PbB, they showed a significant inverse association of Hb with bone Pb. This finding is further support for the observation in studies of other systems that bone Pb is a more sensitive and reliable biomarker of Pb exposures for purposes of assessing long-term chronic toxicity risks. Bone Pb is of special note as an integrating marker of long-term risk where PbB is known to be too toxicokinetically mobile, particularly when PbB values are relatively low while bone Pb deposits remain elevated.

16.4 LEAD-INDUCED ANEMIA VIA ALTERED ERYTHROCYTE PRODUCTION AND DESTRUCTION

Markedly reduced erythrocyte survival is a contributing factor to the overall Pb-associated anemia in lead workers and lead-poisoned children (Table 16.2). It is difficult to precisely apportion anemia causes to hemoglobin reduction or to altered erythrocyte survival over the course of acute or chronic Pb poisoning at high exposures within individual children or across populations of children. Investigators have used erythropoietic markers such as altered erythrocyte volume or erythrocyte number—via hematocrit measurement—or erythropoietin production. An added factor destabilizing erythrocyte structure is the accumulation of pyrimidine nucleotide residues owing to potent inhibition of the pyrimidine metabolism enzyme, pyrimidine-5′-nucleotidase (P5N). This inhibitory process gives rise to basophilic stippling at relatively high PbB values.

The recognition of a hemolytic or erythrocyte survival component to the anemia of Pb exposure occurred quite early in the literature, and the topic has been reviewed in past years (Angle and McIntire, 1982; Moore, 1988; U. S. EPA 1977, 1986; Valentine and Paglia, 1980; Waldron, 1966). These earlier findings reflected the dose–response relationships for erythrocyte production and function at relatively high Pb exposures, notably for Pb workers, $\geq 40-80$ μg/dl.

Leiken and Eng (1963), in a typical study, reported that three of seven Pb-poisoned children showed evidence of shortened erythrocyte survival. Hernberg et al. (1967) employed H^3-labeled difluorophosphonate and Berk et al. (1970) used C^{14}-labeled glucose to demonstrate reduced erythrocyte survival via tracer kinetics. Earlier studies also documented that cell survival is impaired through the joint operation of increased erythrocyte fragility and increased osmotic resistance of cell membranes. A commonly accepted

TABLE 16.2 Hematological Effects of Pb in Human Populations: Lead-Induced Anemia via Altered Erythrocyte Production and Destruction[a]

Study Group(s)	Pb Exposure	Endpoint(s)	Results	References
Various groups of Pb-exposed adults and children in earlier literature	Significantly elevated PbB, >40–80 μg/dl	Anemia	Hemolytic anemia from shortened erythrocyte survival and various mechanisms	Waldron (1966), Moore (1988), Valentine and Paglia (1980), Angle and McIntire (1982)
Lead workers	Significantly elevated range of PbB levels	Erythrocyte survival time using tritium-labeled difluorophosphonate	Exposed workers had shortened erythrocyte survival times	Hernberg et al. (1967)
Children ($N=7$)	Diagnosis of Pb poisoning	Erythrocyte survival time	Three of seven children had shortened cell survival	Leiken and Eng (1963)
Lead workers	Significantly elevated range of PbB levels	Erythrocyte fragility	Exposed workers had increased mechanical fragility	Various earlier studies discussed in Waldron (1964)
Children ($N=579$) exposed to Pb smelter emissions	PbB range 11–164 μg/dl; 40% >40 μg/dl	Hct <35% probability versus group age	1-year-old 10% anemia probability, 20 μg/dl; 3-year-old 10% probability at 50 μg/dl; 5-year-old 10% probability at 75 μg/dl	Schwartz et al. (1990)
Battery workers ($N=183$)	Mean PbB, 5 μg/dl	Hct measurement versus PbB	Hct reduction with PbB increase	Gennart et al. (1992)
Male Pb workers ($N=160$)	PbB mean = 18–27 μg/dl	Hct measurement versus PbB	Hct reduction with PbB increase	Poulos et al. (1986)

(Continued)

TABLE 16.2 Hematological Effects of Pb in Human Populations: Lead-Induced Anemia via Altered Erythrocyte Production and Destruction[a]—(cont.)

Study Group(s)	Pb Exposure	Endpoint(s)	Results	References
Battery workers ($N = 30$) in a longitudinal analysis, 11 years	PbB mean = 30–60 μg/dl	Hct and cell count versus PbB	Significant association between PbB increase and increase in Hct and cell count	Hsiao et al. (2001)

[a]*Altered cell fragility and osmotic resistance. Hemolytic anemia from cell destruction.*

reason for these endpoints is potent Pb inhibition of erythrocyte membrane Na^+/K^+-ATPase. This enzyme controls cation transport.

Changes in erythrocyte volumes and numbers are reflected clinically in the Hct measurement and cell count measurements (Table 16.2). Schwartz et al. (1990) evaluated the relationship of the percentage of exposed child population with hematotoxicity indexed as a reduced Hct, <35%. Using a logistic statistical model, 1-year-old children had a 10% probability of an Hct <35% at 20 μg/dl PbB; those 3 and 5 years of age had this 10% probability at 50 and 75 μg/dl PbB, respectively. This group of children sustained very high Pb exposures from failed Pb emissions control equipment at a lead–zinc smelter in and near their residential neighborhoods. PbB ranged from 11 to 164 μg/dl, and 40% of the 579 children had PbB levels >40 μg/dl. There was no adjustment for concurrent Fe deficiency in this set of analyses.

Lead workers typically but not invariably showed reduced Hct with elevated Pb exposures (Gennart et al., 1992; Horiguchi et al., 1991; Hsiao et al., 2001; Hu et al., 1994; Makino et al., 1997; Poulos et al., 1986). Poulos et al. (1986) reported that male lead workers with a range of mean PbBs at 18–27 μg/dl showed reduced Hct with increasing PbB. Similarly, Pb battery workers with a mean PbB of 51 μg/dl evaluated by Gennart et al. (1992) showed reduced Hct with elevations in PbB. Hu et al. (1994) reported no link between Hct and PbB in the carpenters noted earlier but an inverse association when patellar bone Pb was used. As with Hb as the endpoint, bone Pb appears the more reliable marker of Pb exposures.

Two studies reported, by contrast, positive associations of Hct and/or cell number with PbB level. The results of Makino et al. (1997) using Hct parallel the positive link noted earlier for Hb changes. As also noted earlier, this anomalous finding is difficult to rationalize as simply a biphasic response at low Pb exposures as seen in the relatively low PbB mean. A similar positive association of Hct with PbB was reported by Hsiao et al. (2001) for PbB in the range 30–60 μg/dl, considerably above the mean value for Makino subjects of 13 μg/dl. This latter study was an 11-year assessment of the same cohort.

16.4.1 Pb Effects on Erythropoietin Biosynthesis and Function in Human Populations

Erythropoietin production, typically indexed as serum erythropoietin level, is a principal factor in erythropoiesis activity. It is significantly affected by lead in children (Graziano et al., 2004; Liebelt et al., 1999), in lead workers (Grandjean et al., 1989; Osterode et al., 1999; Romeo et al., 1996), in other workers (Sakata et al., 2007), and in adult women having environmental Pb exposure (Graziano et al., 1990). The results of illustrative studies are summarized in Table 16.3.

The study results of particular interest for environmentally exposed children are those of Graziano et al. (1990, 2004) derived from the Pb prospective study done in Kosovo, former Yugoslavia. These studies were done with large sample sizes, with age-based and other adjustments, such as hemoglobin level. The Kosovo children ($N = 311$, aged 4.5–15 years) were from both a smelter-contaminated community and a contamination-free town treated as a control. The contamination-zone children had PbB mean of 31–39 μg/dl, while those of control had 6–9 μg/dl. Levels of serum erythropoietin were analyzed for relationships to PbB at various age points. There were significant positive linkages (net stimulation) of serum levels with increasing PbB at ages 4.5 and 6.5 years but no association in older children.

A second study, that of Liebelt et al. (1999), found negative association of erythropoietin with PbB in 86 children, aged from 1 to 16 years. This difference from the former study probably relates to smaller sample size, a different age range and age distribution, and less control for age and other factors.

A subset of the Kosovo cohort of pregnant women ($N = 48$) having PbB in a range 2–40 μg/dl showed a negative association of erythropoietin with PbB. Lead workers, at Pb exposures producing typical PbB values in the range ≥30–40 μg/dl, typically showed negative associations (Osterode et al., 1999; Romeo et al., 1996). In addition, Grandjean et al. (1989) observed, besides reduced erythropoietin, reduced erythropoiesis reserve capacity when a quantity of blood was drawn from workers versus controls. These workers presented with a mean of 44.5 μg/dl PbB versus controls with a mean of 7 μg/dl.

TABLE 16.3 Hematological Effects of Pb in Human Populations: Lead Effects on Erythropoietic Formation and Erythropoietic Pyrimidine Metabolism

Study Group(s)	Pb Exposure(s)	Endpoint(s)	Results	References
A. Erythrocyte biosynthesis				
Lead battery workers ($N = 25$) and age-matched controls ($N = 25$)	Worker PbB mean = 44.5 μg/dl; control mean = 7 μg/dl	Levels of serum erythropoietin after blood volume removal (0.45 l)	Workers had reduced erythropoietin and improved erythropoiesis and reserve capacity	Grandjean et al. (1989)
Male lead workers ($N = 141$)	Lower exposure PbB mean = 30 μg/dl; higher exposure mean = 65 μg/dl	Level of serum erythropoietin versus PbB	Erythropoietin negatively associated with PbB	Romeo et al. (1996)
Children in the former Yugoslavia prospective Pb study ($N = 311$) aged 4.5–12 years	Mean PbBs for control town: 6–9 μg/dl; mean for exposed children: 31–39 μg/dl	Level of serum erythropoietin versus PbB	Significant positive association of erythropoietin versus PbB at ages 4.5 and 6.5 but insignificant linkage at older ages	Graziano et al. (2004)
Nepalese tricycle cab drivers ($N = 27$)	Mean PbB = 6.4 μg/dl	Level of serum erythropoietin versus PbB	Significant negative association of erythropoietin and PbB	Sakata et al. (2007)
Children with variable community exposures ($N = 86$) aged 1–6 years	PbB median = 18 μg/dl; PbB range = 2–84 μg/dl	Level of serum erythropoietin versus PbB	Negative association of erythropoietin versus PbB	Liebelt et al. (1999)
B. P5N and NAD-S inhibition				
Lead-exposed children ($N = 21$) aged 2–5 years	PbB range: 7–80 μg/dl	P5N activity versus PbB	Significant negative association of P5N inhibition with PbB; no evidence of threshold	Angle and McIntire (1978)

(*Continued*)

TABLE 16.3 Hematological Effects of Pb in Human Populations: Lead Effects on Erythropoietic Formation and Erythropoietic Pyrimidine Metabolism—(cont.)

Study Group(s)	Pb Exposure(s)	Endpoint(s)	Results	References
Lead-exposed children ($N = 42$) aged 1–5 years	PbB range: <10–72 μg/dl	P5N activity versus PbB	Significant negative association of log P5N activity and PbB; log–log positive correlation of cytidine phosphate and PbB, no evidence of threshold	Angle et al. (1982)
Male Pb workers ($N = 76$) and mixed gender controls ($N = 13$)	PbB up to 82 μg/dl	NADS activity versus PbB	Inverse relationship of NADS activity with PbB up to 82 μg/dl	Morita et al. (1997)
Pb workers ($N = 9$) showing overt Pb poisoning	PbB range: 80–160 μg/dl	P5N activity versus PbB	Significant association of P5N inhibition with PbB	Paglia et al. (1977)
Lead workers ($N = 278$)	PbB range: 5–74 μg/dl	P5N activity versus PbB	Significant negative association of P5N inhibition with PbB; level of inhibition similar to that of δ-ALA-D	Mohammed-Brahim et al. (1985)
Korean lead battery workers ($N = 66$) and 26 control workers	Mean PbB = 46 μg/dl, 39/66 >40 μg/dl	P5N activity versus PbB	Significant negative association of P5N inhibition with PbB	Kim et al. (2002)

The threshold for Pb erythropoietic effects in the form of reduced serum erythropoietin may be relatively low in dose–response terms. Sakata et al. (2007) reported that Nepalese (Katmandu) tricycle cab drivers with a mean PbB of 6.4 μg/dl had statistically significant lower erythropoietin levels with increasing PbB ($p < 0.001$).

16.4.2 Inhibition by Pb of P5N and NAD-S Activity and Erythropoietic Pyrimidine Metabolism

Markedly elevated Pb exposure is accompanied by erythrocytic basophilic stippling in tandem with an anemia with a hemolytic component that resembles a genetic disorder involving deranged pyrimidine metabolism from a deficiency of the enzyme P5N (Paglia and Valentine, 1975). Lead robustly inhibits P5N. Inhibition of the enzyme impairs phosphorolysis of the pyrimidine nucleotides cytidine and uridine phosphates. This ultimately leads to production of ribosomal fragments and altered erythrocyte stability and presents the ribosomal fragments histologically as basophilic stippling. Various dose–response aspects of P5N, adenine dinucleotide synthetase (NAD-S), and pyrimidine nucleotide alterations are summarized in Table 16.3.

Earlier studies in the epidemiological literature established the principal dose–response features of the P5N and PbB relationship. Paglia et al. (1977) noted significant inverse association of P5N activity with PbB in Pb workers with overt poisoning and the PbB range quite high: 80–160 μg/dl. In a larger worker cohort ($N = 278$), the observed significant inverse association of P5N activity and PbB indicated a level of PbB inhibition comparable to that of δ-ALA-D activity (Mohammed-Brahim et al., 1985). In Korean lead battery workers, Kim et al. (2002) likewise reported significant inhibition of P5N with elevated PbB.

Studies by Angle et al. (1982) also showed a robust inhibition of P5N by Pb in young children. Angle and McIntire (1978) and Angle et al. (1982) reported significant inhibition of P5N by Pb in children up to 5 years of age, without any evidence of a threshold at the low end of the PbB range (~5–10 μg/dl). In addition, Angle et al. (1982) reported a log–log positive correlation of cytidine phosphate levels with PbB.

NAD-S is an enzyme that has been identified in the recent literature as being inhibited by Pb at occupational exposures where PbB levels ranged up to 82 μg/dl (Morita et al., 1997).

16.4.3 Oxidative Stress Processes in Pb Hematotoxicity

Relatively high occupational Pb exposures clearly induce erythrocyte oxidative stress processes: generation of ROS, lipid peroxidation (indexed by MDA levels), and changes in activities of enzymes mediating antioxidant responses (Kasperczyk et al., 2009; Patil et al., 2006; Quintanar-Escorza et al., 2007). One factor operative here may be the potential of ALA accumulation to aid formation of ROS (Ahamad and Siddiqi, 2007).

Lead battery workers in India were found to have increased plasma MDA and ceruloplasmin levels and a significant positive correlation with PbB (Patil et al., 2006). Mexican lead battery recycling workers showed increased erythrocyte MDA levels. The MDA versus PbB correlation persisted to those workers who had PbB < 12 μg/dl (Quintanar-Escorza et al., 2007). Polish lead workers stratified as to presence of hypertension showed elevated MDA,

SOD, and glutathione activities in normotensives and hypertensives, but MDA values were significantly higher in hypertensives (Kasperczyk et al., 2009).

16.5 LEAD EFFECTS ON HEME BIOSYNTHESIS

This section deals with the long-recognized effects of Pb exposure on heme biosynthesis, specifically with reference to accumulation of heme intermediates in various organs, tissues, and biomarker media. The topic has been extensively reviewed in individual and consensus tracts (NAS/NRC, 1972, 1993; U.S. ATSDR, 2007; U.S. CDC, 1978, 1985, 1991, 2005; U.S. EPA, 1977, 1986, 2006; Waldron 1966). Lead impairs heme biosynthesis at various steps with a particularly significant effect in accumulation of protoporphyrin IX (EP, FEP, ZPP), analyzed as either the metal-free (FEP) or zinc (ZPP) complex. Other measured effects in the biosynthetic path include potent inhibition of the enzyme δ-ALA-D, moderate inhibition of the enzyme coproporphyrinogen oxidase, and relatively potent inhibition of the enzyme ferrochelatase. Associated occurrences following inhibitory effects on these enzymes include accumulation of δ-ALA in plasma (δ-ALA-P), various organs and tissues, and urine (δ-ALA-U), accumulation of coproporphyrin in urine (CP-U), and stimulation of the enzyme δ-ALA synthetase (δ-ALA-S) in derepressive, feedback response to accumulation of δ-ALA and EP.

In human and other mammalian populations, heme is the functional cofactor for critical diverse metabolic processes in the body such as oxygen transport by hemoglobin and myoglobin, the operation of various cytochromes in oxidative processes ranging from oxidative phosphorylation and detoxification of organic xenobiotics to enzymatically mediated formation of the hormonal form of vitamin D in kidney.

Heme biosynthesis in mammalian systems is a tightly regulated process with feedback loop and derepressive internal controls. Heme formation begins with formation of δ-ALA from glycine and succinyl coenzyme A catalyzed by the enzyme δ-ALA-S as the rate-limiting step in the overall biosynthesis. Two units of δ-ALA are condensed to form porphobilinogen (PBG), followed by formation of coproporphyrin, with further transformations leading to EP and, finally, heme when divalent Fe ion is inserted into EP. Sites of biosynthesis along this pathway vary. For example, the mitochondrion is the site of δ-ALA, EP, and heme formation, while a number of steps occur in the cytosolic portion downstream from formation of PBG via δ-ALA-D. Since reduced heme formation due to Pb inhibition of ferrochelatase in controlling ferrous Fe insertion and/or inhibiting ferrous ion transport to EP for insertion occurs in the mitochondrion, this negative impact has been considered as an evidence of general toxic mitochondrial injury.

16.5.1 Pb-Induced Stimulation of δ-ALA-S Activity

The stimulation of δ-ALA-S activity in response to Pb inhibition at steps downstream in heme biosynthesis, particularly Pb inhibition of δ-ALA-D, has been recorded in Pb workers (Meredith et al., 1978). Such stimulation, in tandem with δ-ALA-D activity inhibition, leads to the accumulation of the substrate δ-ALA in plasma, urine, and various mammalian organs and tissues (Moore, 1988). The threshold for enzymatic stimulation of δ-ALA-S in response to Pb impairment of heme biosynthesis is relatively high in terms of either δ-ALA accumulation or PbB level. Studies in leukocytes of Pb workers suggest a PbB threshold of ~40 μg/dl, while a PbB range of 18–40 μg/dl leads to a doubling of circulating δ-ALA.

16.5.2 Inhibition of δ-ALA-D by Pb

The enzyme δ-ALA-D (PBG-S; E.C. 4.2.1.24) is a Zn-requiring allosteric metalloenzyme which also requires a thiolic group at the active site. Its role is to catalyze the condensation of two units of δ-ALA to form PBG. It is extremely sensitive to inhibition by Pb owing to tight thiolic binding of Pb relative to Zn. Inhibition is reversible via thiolic group regeneration through the use of such agents as dithiothreitol (Granick et al., 1973) or zinc (Finelli et al., 1975). This capability for regeneration and the known genetic polymorphism of enzyme alleles (see next section) has been incorporated into methods for assessing Pb effects on the enzyme (see, for example, Chisolm et al., 1971). One employs an activity ratio of inhibited to full activity rather than direct measurement of δ-ALA-D activity.

Pb's inhibition of a Zn-requiring enzyme such as δ-ALA-D raises several questions about this ternary interaction with respect to environmental and occupational exposures to both metals versus enzyme effects and potential consequences for an adverse Pb effect on the heme biosynthesis pathway in groups with Zn deficiencies. Zn, as an essential trace nutrient, is under homeostatic control in terms of its biokinetics and metabolic disposition (Zinc. WHO, 1996). Consequently, one expects that bioavailability of Zn to zinc-metalloenzymes is regulated, unlike Pb, and has limited access to the enzyme active site in various species and various organs and tissues within species. Early studies (Meredith and Moore, 1980) have shown that a range of serum Zn levels have little impact on Pb-induced inhibition. One consequence is that smelter workers exposed to elevated levels of both Pb and Zn in Pb–Zn operations would probably not be expected to have elevated Zn exposures offsetting Pb inhibition of δ-ALA-D and overall heme biosynthesis. Roth and Kirchgessner (1981) in their studies of experimentally produced Zn deficiency showed lowered enzyme activity. However, Zn deficiency in human populations is relatively rare (Zinc. WHO, 1996).

The dose–response relationship in Pb inhibition of δ-ALA-D in various human populations has been the subject of many studies, dating from 1970 to recent years (Chisolm et al., 1985; Gurer-Orhan et al., 2004; Hernberg and Nikkanen, 1970; Hernberg et al., 1970; Lauwerys et al., 1978; Makino et al., 1997). Selected examples of such studies are given in Table 16.4.

TABLE 16.4 Hematological Effects of Pb in Human Populations: Lead Effects on Heme Biosynthesis, δ-ALAD Activity, ALA and CP Accumulations

Study Group(s)	Pb Exposure(s)	Endpoint(s)	Results	References
A. Inhibition of δ-ALA-D				
Adult Pb workers ($N=158$), male and female	PbB range 5–95 μg/dl	Dose–response relationship of PbB versus δ-ALA-D	Log of δ-ALA-D negatively correlated with PbB over entire PbB range	Hernberg et al. (1970)
Adults from general population ($N=25$)	PbB<16 μg/dl	Dose–response relationship of PbB versus δ-ALA-D	Similar relationship within this lower PbB range	Hernberg and Nikkanen (1970)
Adults from general population ($N=26$)	PbB range 12–56 μg/dl	Dose–response relationship of δ-ALA-D in both erythrocytes and kidney tissue	Inhibition by Pb occurs with δ-ALA-D from both sources	Secchio et al. (1974)
Belgian pregnant women and newborns ($N=500$)	Community exposure PbB range	Dose–response relationship, PbB versus δ-ALA-D	Inverse relationship for log δ-ALA-D and PbB over the lower environmental exposures	Lauwerys et al. (1978)
Lead-exposed children ($N=57$, median age = 54 months)	PbB range 3–40 μg/dl	Activated/nonactivated enzyme activity versus PbB	Ratio of activated/nonactivated enzymes is better correlate for effect versus PbB; threshold for Pb effect ~5 μg/dl	Chisolm et al. (1985)
Group of Japanese Pb workers	Workplace PbB range up to 100 μg/dl	Dose–response relationship, PbB versus δ-ALA-D	δ-ALA-D inhibition had a threshold of 5 μg/dl; plateaued for maximum inhibition above 40 μg/dl	Sakai and Morita (1996)
Japanese Pb workers ($N=58$ males) and 13 controls, male and female	Workplace Pb contact, total group range up to 82 μg/dl	Dose–response relationship, PbB versus δ-ALA-D	50% δ-ALA-D inhibition at 20 μg/dl PbB; minimal activity PbB ≥40 μg/dl	Morita et al. (1997)

(*Continued*)

TABLE 16.4 Hematological Effects of Pb in Human Populations: Lead Effects on Heme Biosynthesis, δ-ALAD Activity, ALA and CP Accumulations—(cont.)

Study Group(s)	Pb Exposure(s)	Endpoint(s)	Results	References
B. Plasma (ALA-P) and urinary ALA (ALA-U)				
Adult Pb workers ($N = 150$)	PbB range 7–92 μg/dl	Dose–response relationship ALA-U versus PbB; entire group and two subgroups < or >40 μg/dl	Entire group shows log ALA-U versus PbB relationship; two subsets show two linear correlation slopes	Selander and Cramér (1970)
Lead-poisoned children ($N = 51$) aged 1–5 years	PbB range 25–75 μg/dl	Dose–response relationship, ALA-U versus PbB	Exponential positive relationship across whole PbB range	Figures 4.6, NAS/NRC (1972); Unpublished data of Chisolm et al.
Variably workplace-exposed Japanese men and women	PbB range 10–99 μg/dl	Dose–response relationship, ALA-U versus ALA-P or versus PbB	ALA-U exponentially related to PbB; ALA-U doubles with 14 μg/dl PbB rise; ALA-U is linearly related to ALA-plasma	Oishi et al. (1996)
Japanese Pb workers	PbB range up to 100 μg/dl	Dose–response relationship, ALA-P versus PbB	High correlation of ALA-P with PbB; PbB threshold to ALA-P increase, 5 μg/dl	Sakai and Morita (1996)
C. Urinary coproporphyrin (CP-U)				
Lead-poisoned children ($N = 6$)	PbB range ≥60–80 μg/dl	CP-U relationship to PbB	Logarithmic positive relationship CP-U versus PbB; μg CP-U/24 hours = 450–3,420	Chisolm and Harrison (1956)
Adult lead workers	Variable	CP-U relationship to PbB	Logarithmic positive relationship CP-U versus PbB; μg CP-U/24 hours = 150–1,220	Haeger-Aronsen (1960)

There is a statistically significant negative association of the log of enzyme activity or the activity ratio (see above) with PbB level in both Pb workers and young children. The statistical depictions of this logarithmic relationship across studies are quite similar and suggest no effect of gender or age on the relationship. Various studies indicate that the threshold to inhibition in terms of PbB level is apparently ≤5 μg/dl (Hernberg and Nikkanen, 1970).

Roels and Lauwerys (1987) also found similar dose−response slopes for the logarithmic relationship, but this clearly does not translate to similar rates of δ-ALA accumulation in various biological compartments involved in the heme biosynthetic pathway (see below). Overall, a 50% inhibition of the enzyme appears to occur at a PbB ~ 20 μg/dl (Hernberg et al., 1970; Morita et al., 1997; Roels and Lauwerys, 1987), while there is virtually complete inhibition >40 μg/dl.

16.5.3 Genetic Polymorphism of δ-ALA-D in Human Populations

The enzyme δ-ALA-D is genetically polymorphic as to biosynthesis and activity in various ethnographic and geographic groups (Table 16.5). The two common, codominantly expressed alleles, ALAD-1 and ALAD-2, give rise to three genotypes: 1-1, 1-2, and 2-2 (Astrin et al., 1987; Battistuzzi et al., 1981; Doss et al., 1982; Potluri et al., 1987; Wetmur, 1994; Wetmur et al., 1991a). The enzyme is encoded by the gene expressing the genotypes on chromosome 9q34 (Potluri et al., 1987). The ethnological distribution of the minor heterozygote carrier, 1-2, or homozygote, 2-2, genotype of the enzyme is highest in those with Caucasian ancestries (10−20%), intermediate in Asian groups (5−10%), and virtually nonexistent in Afrocentric populations.

Important for differential behavior in terms of Pb toxicokinetics and Pb's effects on the heme biosynthetic pathway is that a single G→C transversion mutation of nucleotide 177 in the ALAD-2 allele provides for acquisition of a neutral asparagine at residue 59 instead of a positively charged lysine, decreasing the electropositive (increasing the electronegative) character of the binding region for Pb, leading to stronger binding of Pb (Wetmur et al., 1991b). This leads to the generally observed higher PbB levels in ALAD-2 carriers or homozygotes in children (Astrin et al., 1987; Perez-Bravo et al., 2004; Wetmur et al., 1991a), various worker populations with significantly elevated PbB levels (Fleming et al., 1998; Schwartz et al., 1995, 1997, 2000; Scinicariello et al., 2007; Wetmur et al., 1991a), and general populations with environmental Pb exposures (Miyaki et al., 2009).

It is not surprising that both major and minor genotypes for the enzyme expression are linked to circulating Pb, given that the major Pb binding site in the erythrocyte appears to be δ-ALA-D, rather than sites on hemoglobin (Bergdahl et al., 1997).

TABLE 16.5 Genetic Polymorphism of δ-ALA-D Genotypes and Relationship to Pb Distribution and Adverse Effects

Study Group	Pb Exposure (μg/dl)	Endpoint	Results	References
New York City children ($N = 1278$) in Pb screening, high EP	Mean PbB = 20.3	PbB levels by genotype	PbB mean, 1-1 genotype: 19.5 μg/dl; PbB mean, 1-2/2-2 genotypes: 27.1 μg/dl; statistically significant differences	Wetmur et al. (1991a), Wetmur (1994)
Chilean children (Antofagasta, $N = 93$) near Pb mining area	PbB geometric mean = 12.1–16.4	PbB levels by genotype and site distance	PbB geometric mean <200 m = 16.4 in ALA-D 1-2; mean = 12.1; PbBs higher closer to smelter in all genotypes. ALA-D 1-2 more frequent closer to contamination	Perez-Bravo et al. (2004)
Male Pb workers ($N = 202$)	Mean PbB = 40	PbB versus genotype, 1-1, 1-2/2-2	PbB = 38.4, 1-1; PbB = 47.0, 1-2/2-2; statistically significant	Wetmur et al. (1991a)
Korean lead battery workers, three factories ($N = 307$)	Mean PbB = 29.1 μg/dl	PbB, ZnPP, exposure duration versus genotype: 1-1, 1-2/2-2	PbB and ZPP for two genotypes not significantly different; number of 1-1 genotype workers have less exposure duration versus 1-2 or 2-2 genotype workers	Schwartz et al. (1995)
Korean Pb workers at battery plants ($N = 798$)	PbB mean = 32.0	PbB, Hb, tibial bone Pb, chelatable Pb versus genotype	1-2 genotype workers had higher PbB, similar Hb, and lower tibial Pb and chelatable Pb than 1-1 genotype	Schwartz et al. (2000)

(Continued)

TABLE 16.5 Genetic Polymorphism of δ-ALA-D Genotypes and Relationship to Pb Distribution and Adverse Effects—(cont.)

Study Group	Pb Exposure (μg/dl)	Endpoint	Results	References
Vietnamese and Singaporean Pb workers ($N = 459$ males and females)	Geometric mean PbB, all workers = 19.0	PbB, ALA-U, measures of renal function versus worker genotypes 1-1, 1-2/2-2	Genotype 1-1 workers had significantly higher PbB and ALA-U than 1-2/2-2 group; 1-2/2-2 genotype more sensitive for renal effects	Chia et al. (1995)
U.S. carpenters (subset $N = 122$, total $N = 691$)	Mean PbB = 8.0	PbB, patellar Pb, tibial Pb, BUN, uric acid, Cr	No effect of genotype on PbB, patellar Pb, tibial Pb but BUN and uric acid higher in 1-2/2-2 genotype	Smith et al. (1995)
Taiwanese general population ($N = 660$)	Mean PbB = 6.6	PbB versus genotype	No significant difference in PbB versus genotype	Hsieh et al. (2000)
Adult Japanese ($N = 101$) without occupational exposure	Mean PbB = 3.4	Relationship of PbB and biochemical measures to genotype	Genotype 1-2/2-2 individuals had higher PbB levels but no statistically significant differences in other laboratory measures	Miyaki et al. (2009)
Multiple studies of workers, general populations, children	Variable	Meta-analysis of effects of δ-ALA-D genotypes for various measures	Meta-analysis showed (a) 1-2/2-2 genotype had higher PbB, higher Hb, and lower ZPP and (b) inconclusive effects on other measures	Scinicariello et al. (2007)

The relationship of PbB to ALA-D genotype appears more clearly evident at higher Pb exposures (PbB levels) for workers (Schwartz et al., 1995; Smith et al., 1995) and children (Perez-Bravo et al., 2004; Wetmur et al., 1991a). Exceptions do exist, such as in the data of Miyaki et al. (2009) who reported that in Japanese individuals without occupational Pb exposure and a low mean PbB of 3.4 μg/dl, the ALAD 1-2 individuals still had significantly higher PbB levels ($p = 0.02$).

The more recent literature on the relationship of Pb-B, δ-ALAD genotype, various other markers of Pb exposures, and various toxic endpoints is a complex, seemingly inconsistent body of information across the various studies and some of the information appears anomalous (Table 16.5). Some issues of concern are the relationship of isozyme genotype to general indicators of exposure intensity such as employment duration and intensity, other biomarkers of Pb exposure such as patellar or tibial bone Pb levels, and biomarkers of early effect such as ZPP, plasma ALA, urinary ALA, and clinical markers of effect such as those for renal function.

Higher PbB levels in carrier (ALAD 1-2) individuals with tighter binding of Pb to the ALAD 1-2 form either do not produce any higher toxic risk to carriers or this binding may reduce risk. In several reports, evidence in carrier (ALAD 1-2) PbB data seems to suggest that having a genotype-associated higher PbB reduces toxicity risk. Korean lead battery workers with both higher employment longevity and elevated workplace Pb exposures were those having the ALAD 1-2/2-2 variants, suggesting the carriers displayed less circulating Pb toxicity risk (Schwartz et al., 1995), while these workers with the minor genotypes also presented with less chelatable Pb from overall body Pb burdens (Schwartz et al., 2000). While the findings with the battery workers appear to be an example of the "healthy worker" effect, whereby those with less intense adverse responses to workplace substance(s) remain and those more sensitive leave that employment, a similar scenario appears to apply with children's exposures.

Perez-Bravo et al. (2004) reported that Chilean children living near Pb contamination area from past mining operations and who had the carrier genotype had higher PbB than the 1-1 group and the typically seen inverse relationship of PbB with distance from the contaminated area existed. However, children living closer to the exposure source and having higher PbB were apparently more likely to be a carrier than those children living farther away. As in the case of the Korean battery workers, there may be selectivity for the carrier genotype over time owing to lower toxic responses in this subset.

The relationship of genotype to Pb burdens of bone presents a null picture in the literature (Kelada et al., 2001; Scinicariello et al., 2007), while chelatable Pb is lower in the heterozygotes (Schwartz et al., 1997, 2000).

The relationship of levels of heme intermediates in biological indicator media appears to be a complex one with respect to δ-ALAD genetic polymorphism. For plasma and urinary δ-ALA versus genotype, results from various studies do not allow conclusions, in part because of sample sizes and in part because of the range of exposures indexed as PbB from low, environmental exposures to the higher exposures seen in workers. A general consistent finding is that of lowered but not always significantly lowered ZPP and nonsignificantly increased Hb levels based on the meta-analysis by Scinicariello et al. (2007) using a number of relevant published data sets. Those studies with higher levels of PbB, mainly for Pb workers and particularly around or higher than 40 μg/dl PbB, showed statistically significant lowered ZPP. This particular finding is consistent with the δ-ALAD 1-2 subjects having body Pb burdens less bioavailable for impairing steps in the heme biosynthesis pathway.

The influence of genotype on various biomarkers for renal function and cardiovascular effects from Pb exposure is apparent with interactions for serum creatinine, while the meta-analysis showed little effect of genotype on SBP or DBP.

16.5.4 Effects of Pb Exposure on δ-ALA and CP Levels in Bioindicator Media

The combined effects of Pb on the above enzymatic steps within the heme biosynthetic pathway lead to accumulation of ALA-P, ALA-U, and in various organs and tissues of humans and experimental systems (Table 16.4). The latter are discussed in a later section. The dose−response relationships for ALA-U, the commonly tested bioindicator medium, take the form of a positive association between the log of ALA-U and PbB (Chisolm et al., as cited in NAS/NRC, 1972; Oishi et al., 1996; Sakai and Morita, 1996; Selander and Cramér, 1970). The threshold for this relationship is typically taken as 40 μg/dl, but Selander and Cramér (1970) showed two dose−response relationships for two worker subsets, one >40 μg/dl and the other at or below this value.

Sakai and Morita (1996) reported that there is a high correlation of plasma δ-ALA with PbB for Japanese workers ranging up to 100 μg/dl, and the threshold for ALA-P increase begins at the very low PbB value of 5 μg/dl. Oishi et al. (1996) noted that ALA-U, in addition to showing a logarithmic association with PbB, doubles in concentration with a PbB increase of 14 μg/dl. Meredith et al. (1978) reported that the relationship of ALA-U to ALA-P was logarithmic as was that of ALA-P to PbB.

Lead exposure leads to an increase in δ-ALA at body sites other than hematopoietic tissue. Virtually all of the data, for obvious reasons, are gathered from experimental animal studies.

Accumulation of urinary coproporphyrin has also been reported, principally in highly exposed children and lead workers and mainly in the older literature (Chisolm and Harrison, 1956; Haeger-Aronsen, 1960). CP accumulation, from inhibition by lead of coproporphyrin utilization in subsequent formation of protoporphyrin, is a relatively late responder to Pb exposure in both children and workers, and the association of urinary CP with PbB principally is seen at PbB values ≥40 μg/dl. It has rarely been used in recent years since the evolution of better, much more sensitive biomarkers of both exposure and early effects.

16.5.5 Protoporphyrin (EP, FEP, ZPP) Accumulation in Human Populations Exposed to Pb

The final step in heme biosynthesis, insertion of ferrous Fe into protoporphyrin IX to form heme, is inhibited by Pb. While the basis of this is widely held to be the inhibition of the enzyme modulating the process, ferrochelatase, one cannot rule out other mechanisms such as impaired ferrous Fe formation and/or intra-mitochondrial transport. As was typical with other steps in the heme pathway exploited diagnostically for ascertaining toxic Pb exposures in exposed populations, the association of EP accumulation with PbB is influenced by a number of factors, notably the presence of Fe deficiency anemia. With Fe deficiency, notably in children 6 years or younger, EP accumulation regardless of Pb exposure is significant. Various investigators have used different methods to quantify the impact of Pb on EP in the presence of EP increases from Fe deficiency (Marcus and Schwartz, 1987).

The dose–response relationships for EP change with Pb exposure, including thresholds, are logarithmic and appear to show that children are more sensitive than adults, while women are somewhat more sensitive than men (Table 16.6). For children, the dose–response relationship persists down to a threshold in blood lead on the order of 15–20 μg/dl, and in adults across gender, 25–35 μg/dl. The dose–response relationship of EP and PbB is affected by the time course of EP's accumulation with increase in Pb exposure indexed through PbB.

Increased exposure produces a relatively rapid rise in PbB compared to the corresponding rise in erythrocyte EP. That is, EP accumulation in erythropoiesis is a lagging indicator of Pb exposure and toxicity because accumulation occurs only in that population of cells formed subsequent to onset of exposure, i.e., a rate of 0.8% per day based on an erythrocyte full turnover rate of 120 days. The decline in accumulated EP will occur slowly with downward change in active Pb exposure. The presence of elevated EP with elevated PbB and the absence of Fe deficiency are more indicative of chronic than shorter or intermittent exposures.

In typical high-risk children, where both significantly elevated Pb exposure and marked Fe deficiency co-occur (see, for example, discussions in Marcus and Schwartz, 1987; Piomelli et al., 1982; U.S. CDC, 1985, 1991), the two factors have affected childhood Pb screening programs. In earlier Pb exposure screening periods with PbB average values of 25–30 μg/dl and higher, the recommended practice was to screen first for EP elevation and then follow up any positive results with a PbB measurement (U.S. CDC, 1978, 1985). This approach subsequently served to identify Fe deficiency in the absence of elevated PbB (false positives for PbB). At PbB levels

TABLE 16.6 Hematological Effects of Pb in Human Populations: Lead Effects on Heme Biosynthesis and Heme Accumulation

Study Group(s)	Pb Exposure(s)	Endpoint(s)	Results	References
New York City children (*N* = 2,004, aged 2–12 years)	PbB range 2–98 μg/dl; 1852 ≤30 μg/dl	EP versus PbB dose–responses, various models	Exponential positive relationship of EP to PbB threshold of 15–18 μg/dl	Piomelli et al. (1982)
NHANES II children (*N* = 1,677, aged 2–6 years)	U.S. national exposures via PbB measures	Blood levels of EP adjusted for iron status, percentage transferrin saturation	Dose–response curves for EP increased with PbB based on percentage transferrin saturation	Marcus and Schwartz (1987)
Lead workers (*N* = 79)	Workplace PbB range	EP versus PbB dose–responses	Exponential positive relationship of EP to PbB threshold of 25–30 μg/dl	Grandjean and Lintrup (1978)
Male and female subjects having environmental (*N* = 1,088 men, 511 women) or occupational (*N* = 200 male, 40 females) Pb exposures	Worker PbB: 10–80 μg/dl	EP levels versus gender, PbB	Exponential positive relationship of EP to PbB; female EP higher than males; control subjects mean EP = 25 μg/dl	Wildt et al. (1987)
Boston, MA prospective Pb study children (*N* = 232)	PbB levels <15 and ≥15 μg/dl	EP levels versus PbB dose–responses	EP was not associated with PbB < 15 μg/dl; fourfold increase in elevated EP ≥ 15 μg/dl	Rabinowitz et al. (1986)
Children in Cincinnati, OH prospective Pb study (*N* = 165), aged 3–36 months	Variable PbB versus age	Ratio of Zn-bound to total EP	At 33 months, all EP is in the Zn form, versus a minimum at 3 months	Hammond et al. (1985)

(*Continued*)

TABLE 16.6 Hematological Effects of Pb in Human Populations: Lead Effects on Heme Biosynthesis and Heme Accumulation—(cont.)

Study Group(s)	Pb Exposure(s)	Endpoint(s)	Results	References
U.S. hospital pediatric admissions ($N = 4,908$) 0–17 years, 2001–2002	PbB range 2.2–3.3 μg/dl; median, 3 μg/dl	EP and PbB comparisons versus 10 years ago	Exponential relationship of EP to PbB minimal	Soldin et al. (2003)
Ecuadoran children ($N = 78$) with 44 females, 34 males	Cottage ceramics industry family Pb exposures; PbB mean, 26.4 μg/dl; mean ZPP/heme ratio = 152 μmol/mol	Relationship of PbB to ZPP/heme ratio	Significant association of PbB to ZPP/heme ratio; higher association of PbB with log ZPP	Counter et al. (2007)

<25 μg/dl, the association of log EP with PbB is poor, so co-occurrence of Fe deficiency and relatively lower PbB levels in screening programs led to the problem of false negative values and under-detection of elevated PbB. Because of this limitation, avoiding any under-screening of PbB by direct PbB measurement was implemented (U.S. CDC, 1991, 2005).

Several reports provide information on the relationship of PbB to EP in children. Soldin et al. (2003) compared EP (as ZPP) and PbB using a database of 4,908 admissions, 0–17 years of age in 2001–2002, and compared the measurements to those of 10 years earlier. PbB had declined to a quite low range of mean, 2.2–3.3 μg/dl, with a median of 3 μg/dl. ZPP means ranged between 21.1 and 21.6 μg/dl, with median 21–27 μg/dl. These low PbBs were linked to EP through a complex power function model. Counter et al. (2007) evaluated the relationship of PbB to the ZPP/heme ratio in 78 Ecuadorian children having a PbB mean of 26.4 μg/dl and a mean ratio of 152 μmol/mol Hb. While there was a statistically significant association of PbB to the ratio, an even stronger association was calculated for log ZPP.

As noted earlier, heme serves as the prosthetic group for various proteins carrying out diverse metabolic functions, much of this information having been assessed in animal testings. Examples of this are described in various critical reviews and consensus documents (NAS/NRC, 1993; U.S. EPA, 1986). Children with relatively high Pb exposures, producing PbB in the range of 33–120 μg/dl, had significant reductions in the hormonal form of vitamin D, 1,25-$(OH)_2$-D (Rosen et al., 1980), which is formed via hydroxylation of the prohormone vitamin D by a heme-containing cytochrome P-450 oxidase complex in the kidney tubule, the 1-hydroxylase system (Rosen and Chesney, 1983). The impaired hydroxylation, then, potentially proceeds from initial reduced heme production by impaired ferrochelatase activity and/or ferrous Fe transport for insertion into EP (Fowler et al., 1980).

16.6 EXPERIMENTAL ANIMAL STUDIES OF THE HEMATOTOXIC EFFECTS OF Pb

Experimental studies of lead's effects on blood-forming and related systems comprise a large literature, much of it now of some vintage. Such studies clarified the nature and extent of the quantitative scope of Pb's hematotoxicity. This extended to the impacts of Pb exposures on heme biosynthesis across organs and tissues and the interactions of heme effects with toxic effects in other systems. Illustrative examples of such studies are presented in Table 16.7, summarizing data gathered in avians, rodents, rabbits, and nonhuman primates.

Effects of experimental Pb exposures have been investigated with respect to δ-ALA-D activity inhibition using a wide range of species. Dorward and Yagminas (1994) reported that chronic oral dosing of adult cynomolgus monkeys at 1.5 mg/kg/day for 8 years produced erythrocyte enzyme kinetics quite similar to those in humans. Redig et al. (1991) noted in their kinetic analyses that oral dosing in red-tailed hawks at 0.82 mg Pb/kg/day produced significant inhibition by the first week, with inhibition persisting for weeks after dosings ended. Using the activity ratio method for quantifying the extent of enzyme inhibition showed significant negative relationship of enzyme activity to PbB in orally dosed zebra finches (Scheuhammer, 1987), ducks (Murase et al., 1993), and mice (Tomokuni et al., 1989).

Multispecies studies have examined the relative extent of δ-ALA-D activity inhibition in blood versus effects in other organs and tissues. Dieter and Finley (1979) evaluated enzyme activity inhibition in brain and liver tissues versus erythrocytes for toxic effects of lead pellets ingested by waterfowl. Mallard ducks fed a lead pellet produced a PbB of 100 μg/dl by 4 weeks, and enzyme activity inhibition at sacrifice was 50% in the cerebellum and liver at tissue Pb contents of <0.5 and 2.5 ppm, respectively. Silbergeld and Lamon (1982) found that adult rats subjected to chronic Pb exposure had significantly reduced δ-ALA-D activity in blood, kidney, liver, and spleen but not in brain tissue. These investigators also found the activity of δ-ALA-S to be altered with lead exposure, but the direction of change differed with tissue. Chronically exposed rats had little change in synthetase activity in brain and kidney, a marked increase in spleen and a reduction in liver. Santos et al. (1999) also used adult rats to show significant reductions in δ-ALA-D activities in both erythrocytes and liver.

TABLE 16.7 Selected *In Vivo* Experimental Studies of Hematological Effects of Lead: Pb Effects on Heme Biosynthesis and Other Toxic Effects

Test Species	Dosing Regimen	Endpoints	Results	References
A. Effects on δ-ALA-D activity				
Mallard ducks	Lead pellet ingestion, PbB at 4 weeks, 100 μg/dl	Pb level and δ-ALA-D activity in blood and brain regions	Cerebellar δ-ALA-D activity reduced 50% at Pb<0.5 ppm; erythrocyte activity reduced 75%; liver activity reduced ~50% at Pb 2.5 ppm	Dieter and Finley (1979)
Neonatal mice	Dosing 1.0 mg/ml in water, $t=0$ to 17 days old	Brain δ-ALA-D activity	Statistically significant reductions in activity at all time points	Gerber et al. (1978)
Adult cynomolgus monkeys	Chronic oral exposure, 1.5 mg Pb/kg body weight, 8 years	Erythrocyte δ-ALA-D enzyme kinetics in monkeys with Pb exposure	Effects of Pb on monkey δ-ALA-D kinetics similar to humans	Dorward and Yagminas (1994)
Male Wistar rats ($N=47$) in four groups	Pb acetate (160 mg/l), 8 weeks, in semiliquid diet	δ-ALA-D activity in erythrocytes and liver	Reduction in δ-ALA-D in both erythrocytes and liver	Santos et al. (1999)
Adult rats	Chronic Pb exposure	Levels of ALA and activities of δ-ALA-D and δ-ALA-S with Pb exposure	Activities of δ-ALA-D in kidney, liver, and spleen reduced but not in brain; δ-ALA-S activity increased in spleen and reduced in liver	Silbergeld and Lamon (1982)
Adult female rabbits	Oral Pb dosing: 17 μM Pb acetate for 5 days	Levels of δ-ALA-D and δ-ALA-S activity versus PbB in various tissues	δ-ALA-D activity reduced in blood, bone marrow, and liver; δ-ALA-S activity unchanged in liver and bone marrow	Zareba and Chmielnicka (1992)

Red-tailed hawks	Oral Pb dosing: various levels of mg Pb/kg/day, 3 or 11 weeks	Levels of δ-ALA-D and δ-ALA-S activity versus Pb dosing	At 0.82 mg Pb/kg/day, significant reduction in enzyme activity by first week	Redig et al. (1991)
Zebra finches	Oral Pb dosing, 5–100 ppm range in diet	Level of δ-ALA-D activity in erythrocytes versus PbB	Log δ-ALA activity significantly negatively linked to PbB; activity ratio (activated/nonactivated enzyme activity) was significantly positively linked to PbB	Scheuhammer (1987)
Ducks	Dosing: 7 Pb pellets, 3 mm diameter	δ-ALA-D activity ratio (activated versus inactivated) versus PbB	Significant positive correlation between activity ratio and PbB; activity ratio declined with PbB after EDTA treatment	Murase et al. (1993)
Mice	Dosing: Pb in drinking water, 200 or 500 ppm, 14 or 30 days	δ-ALA-D activity versus dosing regimen	Erythrocyte δ-ALA-D activity was inhibited 90% versus 20–40% in liver	Tomokuni et al. (1989)
B. Levels of ALA and CP				
Adult rats	Dosing: chronic i.p. administration	δ-ALA levels in various tissues	9- to 15-fold increase in δ-ALA over controls seen in kidney and spleen with only moderate increase in brain	Silbergeld et al. (1982)
Adult rats	Dosing: 100–2,000 ppm Pb in drinking water, up to 18 months, with and without δ-ALA infusion	δ-ALA-P and δ-ALA-U levels; δ-ALA reabsorption rates by the kidney	ALA-U increased with Pb dosing and ALA-P; ALA reabsorption rate across Pb range unaffected	O'Flaherty et al. (1980, 1981)
Adult mice	Dosing: Pb in drinking water, 200 or 500 ppm, 14 or 30 days	δ-ALA-U and CP isomers I and III	δ-ALA-U increased up to 12-fold at higher dosings and time; no change in CP	Tomokuni et al. (1989)

(*Continued*)

TABLE 16.7 Selected *In Vivo* Experimental Studies of Hematological Effects of Lead: Pb Effects on Heme Biosynthesis and Other Toxic Effects—(cont.)

Test Species	Dosing Regimen	Endpoints	Results	References
Adult female rabbits	Oral Pb dosing: 17 μM Pb acetate, 5 days	Levels of δ-ALA-U and CP-U with exposure	Pb produced significant increases in δ-ALA and CP levels in urine	Zareba and Chmielnicka (1992)
C. Levels of EP/ZPP and changes in heme and related factors				
Adult female rabbits	Oral Pb dosing: 17 μM Pb acetate, 5 days	Level of EP with exposure	Pb produced a significant increase in EP	Zareba and Chmielnicka (1992)
Red-tailed hawks	Oral Pb dosing: various levels of Pb/kg/day, 3 or 11 weeks	Level of FEP activity versus Pb dosing	By 3 weeks at 0.87 mg/kg/day, Pb increased the free EP and ZPP; Hb decreased but only at higher dosing and 11 weeks	Redig et al. (1991)

Adult mice	Dosing: Pb in drinking water, 200 or 500 ppm for 14 or 30 days	EP levels, changes in Hb	No changes observed under any of the conditions	Tomokuni et al. (1989)
Sprague–Dawley rats: suckling ($N = 200$) and adult males ($N = 70$) and females ($N = 55$)	Dosing: Pb in drinking water, 3,000 ppm; animals had no surgical intervention, sham operation, or castration	FEP and Hb levels with Pb and different treatments	Significant declines in FEP and Hb; relative FEP response highest in young rats , followed by adult females	Roels et al. (1978)
D. P5N activity and nucleotide levels				
Neonatal rats, 21 days old	Acute dosing: Pb acetate	Dose–response relationship, P5N versus PbB in erythrocytes	Inhibition of P5N was linearly inversely related to PbB, 77–108 μg/dl	Konantakieti et al. (1986)
Adult mice	Dosing: Pb in drinking water, 200 or 500 ppm for 14 or 30 days	Dose–response relationship P5N versus PbB in erythrocytes and bone marrow	Inhibition of P5N activity significantly increased: 25% in marrow cells, 45% in erythrocytes	Tomokuni et al. (1989)
Rabbits	Dosing: oral Pb dosing 30 mg/kg/day, 60 days	Inhibition level of P5N and levels of pyrimidine nucleotides	Exposure increased levels of the nucleotides and increased inhibition of P5N	Swanson et al. (1980)

Female rabbits orally dosed for 5 days with 17 μM Pb showed reduced enzyme activity in erythrocytes, bone marrow, and liver (Zareba and Chmielnicka, 1992). These workers also noted an increased activity of the δ-ALA-S enzyme in these rabbits. Gerber et al. (1978) reported that oral dosing of neonatal mice with Pb at 0–17 days of age showed statistically significant reductions in brain enzyme activity at all time points.

The effects of Pb on experimental animal heme biosynthesis included altered levels of ALA and CP. The chronic dosing of adult rats by Silbergeld et al. (1982) led to a 9- to 15-fold increase in δ-ALA over controls in kidney and spleen with only a moderate increase in brain. Tomokuni et al. (1989), using oral dosings of adult mice, noted somewhat similar results to that of Silbergeld et al. (1982) that ALA-U was increased up to 12-fold at higher dosings and over longer periods. These results were paralleled by Zareba and Chmielnicka (1992) in their rabbit studies for ALA-U and CP-U. O'Flaherty et al. (1980, 1981) noted that very high Pb exposures were required to induce heme biosynthetic disturbances in their rat-exposure model, while ALA-U increased with sufficient Pb dosing in tandem with plasma increases of the intermediate. They also noted that the rate of reabsorption of ALA was unaffected.

Female rabbits (Zareba and Chmielnicka, 1992), red-tailed hawks (Redig et al., 1991), and adult rats (Roels et al., 1978) with relatively high oral Pb dosings had elevated ZPP, but the dosing level in adult mice was not sufficient to produce increases over controls. Neonatal rats acutely dosed with Pb acetate showed inhibition of P5N activity but at relatively high systemic levels in the PbB range 77–108 μg/dl (Konantakieti et al., 1986). Inhibition of P5N activity was also noted with oral Pb in the mouse (erythrocytes and marrow cells) and rabbit studies of Tomokuni et al. (1989) and Swanson et al. (1980), respectively.

REFERENCES

Ahamad, M., Siddiqi, M.K., 2007. Low level lead exposure and oxidative stress: current opinions. Clin. Chim. Acta 383, 57–64.

Angle, C.R., McIntire, M.S., 1978. Low level lead and inhibition of erythrocyte pyrimidine nucleotidase. Environ. Res. 17, 296–302.

Angle, C.R., McIntire, M.S., 1982. Children, the barometer of environmental lead. Adv. Pediatr. 29, 3–31.

Angle, C.R., McIntire, M.S., Swanson, M.S., Stohs, S.J., 1982. Erythrocyte nucleotides in children-increased blood lead and cytidine triphosphate. Pediatr. Res. 16, 331–334.

Astrin, K.H., Bishop, D.F., Wetmur, J.G., Kaul, B., Davidow, B., Desnick, R.J., 1987. δ-Aminolevulinic acid dehydratase isozymes and lead toxicity. Ann. N.Y. Acad. Sci. 514, 23–29.

Baker Jr., E.L., Landrigan, P.J., Barbour, A.G., Cox, D.H., Folland, D.S., Ligo, R.N., et al., 1979. Occupational lead poisoning in the United States: clinical and biochemical findings related to blood lead levels. Br. J. Ind. Med. 36, 314–322.

Battistuzzi, G., Petrucci, R., Silvagni, L., Urbani, F.R., Caiola, S., 1981. Delta-aminolevulinate dehydrase: a new genetic polymorphism in man. Ann. Hum. Genet. 45, 223–229.

Bergdahl, I.A., Grubb, A., Schütz, A., Desnick, R.J., Wetmur, J.G., Sassa, S., et al., 1997. Lead binding to δ-aminolevulinic acid dehydratase (ALAD) in human erythrocytes. Pharmacol. Toxicol. 81, 153–158.

Berk, P.D., Tschudy, D.P., Shepley, L.A., Waggoner, J.G., Berlin, N.I., 1970. Hematologic and biochemical studies in a case of lead poisoning. Am. J. Med. 48, 137–144.

Chia, K.S., Jeyaratnam, J., Lee, J., Tan, C., Ong, H.Y., Ong, C.N., et al., 1995. Lead-induced nephropathy: relationship between various biological exposure indices and early markers of nephrotoxicity. Am. J. Ind. Med. 27, 883–895.

Chisolm Jr., J.J., 1971. Increased lead absorption: toxicological considerations. Pediatrics 48, 349–352.

Chisolm Jr., J.J., 1981. Dose–effect relationships for lead in young children: evidence in children for interactions among lead, zinc, and iron. In: Lynam, D.R., Piantinida, L.G., Cole, J.F. (Eds.), Environmental Lead. Academic Press, New York, pp. 1–7.

Chisolm Jr., J.J., Harrison, H.E., 1956. The exposure of children to lead. Pediatrics 18, 943–958.

Chisolm Jr., J.J., Thomas, D.J., Hamill, T.G., 1985. Erythrocyte porphobilinogen synthase activity as an indicator of lead exposure in children. Clin. Chem. 31, 601–605.

Counter, S.A., Buchanan, L.H., Ortega, F., Rifai, N., Shannon, M.W., 2007. Comparative analysis of zinc protoporphyrin and blood lead levels in lead-exposed Andean children. Clin. Biochem. 40, 787–792.

Dieter, M.P., Finley, M.T., 1979. Delta-aminolevulinic acid dehydratase activity in blood, brain, and liver of lead-dosed ducks. Environ. Res. 19, 127–135.

Dorward, A., Yagminas, A.P., 1994. Activity of erythrocyte δ-aminolevulinic acid dehydratase in the female cynomolgus monkey (*Macaca fascicularis*): kinetic analysis in control and lead-exposed animals. Comp. Biochem. Physiol. B. Biochem. Mol. Biol. 108, 241–252.

Doss, M., Becker, U., Sixel, F., Geisse, S., Solcher, H., Schneider, J., 1982. Persistent protoporphyrinemia in hereditary porphobilinogen synthase (δ-aminolevulinic acid dehydrase) deficiency under low lead exposure: a new molecular basis for the pathogenesis of lead intoxication. Klin. Wochenscr. 60, 599–606.

Factor-Litvak, P., Wasserman, G., Kline, J.K., Graziano, J., 1999. The Yugoslavia prospective study of environmental lead exposures. Environ. Health Perspect. 107, 9–15.

Finelli, V.N., Klauder, D.S., Karaffa, M.A., Petering, H.G., 1975. Interactions of zinc and lead on δ-aminolevulinate dehydratase. Biochem. Biophys. Res. Commun. 65, 303–311.

Fleming, D.E.B., Chettle, D.R., Wetmur, J.G., Desnick, R.J., Robin, J.-P., Boulay, D., et al., 1998. Effect of the δ-aminolevulinate dehydratase polymorphism on the accumulation of lead in bone and blood in lead smelter workers. Environ. Res. 77, 49–61.

Fowler, B.A., Kimmel, C.A., Woods, J.S., McConnell, E.E., Grant, L.D., 1980. Chronic low-level lead toxicity in the rat. III. An integrated assessment of long-term toxicity with special reference to the kidney. Toxicol. Appl. Pharmacol. 56, 59–77.

Gennart, J.P., Bernard, A., Lauwerys, R., 1992. Assessment of thyroid, testes, kidney and autonomic nervous system function in lead-exposed workers. Int. Arch. Occup. Environ. Health 64, 49–57.

Gerber, G.B., Maes, J., Gilliavod, N., Casale, G., 1978. Brain biochemistry of infant mice and rats exposed to lead. Toxicol. Lett. 2, 51–63.

Grandjean, P., 1979. Occupational lead exposure in Denmark: screening with the hematofluorometer. Br. J. Ind. Med. 36, 52–58.

Grandjean, P., Lintrup, J., 1978. Erythrocyte–Zn–protoporphyrin as an indicator of lead exposure. Scand. J. Clin. Lab. Invest. 38, 669–675.

Grandjean, P., Hollnagel, H., Hedegaard, L., Christensen, J.M., Larsen, S., 1989. Blood lead–blood pressure relations: alcohol intake and hemoglobin as confounders. Am. J. Epidemiol. 129, 732–734.

Granick, J.L., Sassa, S., Granick, S., Levere, R.D., Kappas, A., 1973. Studies in lead poisoning: II. Correlation between the ratio of activated to inactivated δ-aminolevulinic acid dehydratase of whole blood and the whole blood level. Biochem. Med. 8, 149–150.

Graziano, J.H., Popovac, D., Factor-Litvak, P., Shrout, P., Kline, J., Murphy, M.J., et al., 1990. Determinants of elevated blood lead during pregnancy in a population surrounding a lead smelter in Kosovo, Yugoslavia. Environ. Health Perspect. 89, 95–100.

Graziano, J., Slavkovich, V., Liu, X., Factor-Litvak, P., Todd, A., 2004. A prospective study of prenatal and childhood lead exposure and erythropoietin production. J. Occup. Environ. Med. 46, 924–929.

Gurer-Orhan, H., Sabir, H.D., Ozgunes, H., 2004. Correlation between clinical indicators of lead poisoning and oxidative stress parameters in controls and lead-exposed workers. Toxicology 195, 147–154.

Haeger-Aronsen, B., 1960. Studies on urinary excretion of δ-aminolevulinic acid and other haem precursors in lead workers and lead-intoxicated rabbits. Scand. J. Clin. Lab. Invest. 12 (Suppl. 47), 1–128.

Hammond, P.B., Bornschein, R.L., Succop, P., 1985. Dose–effect and dose–response relationships of blood lead to erythrocyte protoporphyrin in young children. Environ. Res. 38, 187–196.

Hernberg, S., Nikkanen, J., 1970. Enzyme inhibition by lead under normal urban conditions. Lancet 1 (7637), 63–64.

Hernberg, S., Nurminen, M., Hasan, J., 1967. Nonrandom shortening of red cell survival times in men exposed to lead. Environ. Res. 1, 247–261.

Hernberg, S., Nikkanen, J., Mellin, G., Lilius, H., 1970. δ-Aminolevulinic acid dehydrase as a measure of lead exposure. Arch. Environ. Health 21, 140–145.

Horiguchi, S., Matsumura, S., Fukumoto, K., Karai, I., Endo, G., Teramoto, K., et al., 1991. Erythrocyte deformability in workers exposed to lead. Osaka City Med. J. 37, 149–155.

Hsiao, C.Y., Wu, H.D., Lai, J.S., Kuo, H.W., 2001. A longitudinal study of the effects of long-term exposure to lead among lead battery factory workers in Taiwan (1989–1999). Sci. Total Environ. 279, 151–158.

Hsieh, L.L., Liou, S.H., Chen, Y.H., Tsai, L.C., Yang, T., Wu, T.N., 2000. Association between aminolevulinate dehydratase genotype and blood lead levels in Taiwan. J. Occup. Environ. Med. 42, 151–155.

Hu, H., Watanabe, H., Payton, M., Korrick, S., Rotnitsky, A., 1994. The relationship between bone lead and hemoglobin. J. Am. Med. Assoc. 272, 1512–1517.

Kasperczyk, S., Kasperczyk, J., Ostalowska, A., Zalejska-Fiolka, J., Wielkoszynski, T., Swietochowska, E., et al., 2009. The role of the antioxidant enzymes in erythrocytes in the development of arterial hypertension among humans exposed to lead. Biol. Trace Elem. Res. 130, 95–106.

Kelada, S.N., Shelton, E., Kaufmann, R.B., Khoury, M.J., 2001. δ-Aminolevulinic acid dehydratase genotype and lead toxicity: a HuGE review. Am. J. Epidemiol. 154, 1–13.

Kim, R., Rotnitsky, A., Sparrow, D., Park, D.U., Yang, J.S., Park, I.J., et al., 2002. Evaluation of lead exposure in workers at secondary lead smelters in South Korea: with focus on activity of erythrocyte pyrimidine 5′-nucleotidase (P5N). Sci. Total Environ. 286, 181–189.

Konantakieti, C., Beuthin, F.C., Louis-Ferdinand, R.T., 1986. Erythrocyte pyrimidine 5′-nucleotidase inhibition by acute lead exposure in neonatal rats. J. Biochem. Toxicol. 1, 51–59.

Lauwerys, R., Buchet, J.P., Roels, H., Hubermont, G., 1978. Placental transfer of lead, mercury, cadmium, and carbon monoxide in women. I. Comparison of the frequency distributions of the biological indices in maternal and umbilical cord blood. Environ. Res. 15, 278–289.

Leiken, S., Eng, G., 1963. Erythrokinetic studies of the anemia of lead poisoning. Pediatrics 31, 996–1002.

Liebelt, E.L., Schonfeld, D.J., Gallagher, P., 1999. Elevated blood lead levels in children are associated with lower erythropoietin concentrations. J. Pediatr. 134, 107–109.

Lilis, R., Eisinger, J., Blumberg, W., Fishbein, A., Selikoff, I.J., 1978. Hemoglobin, serum iron, and zinc protoporphyrin in lead-exposed workers. Environ. Health Perspect. 25, 97–102.

Makino, S., Shimizu, Y., Takata, T., 1997. A study on the relationship between blood lead levels and anemia indicators in workers exposed to low levels of lead. Ind. Health 35, 537–541.

Marcus, A.H., Schwartz, J., 1987. Dose–response curves for erythrocyte protoporphyrin vs. blood lead: effects of iron status. Environ. Res. 44, 221–227.

Meredith, P.A., Moore, M.R., 1980. The *in vivo* effects of zinc on erythrocyte delta-aminolevulinic acid dehydratase in man. Int. Arch. Occup. Environ. Health 45, 163–168.

Meredith, P.A., Moore, M.R., Campbell, B.C., Thompson, G.G., Goldberg, A., 1978. Delta-aminolevulinic acid metabolism in normal and lead-exposed humans. Toxicology 9, 1–9.

Miyaki, K., Lwin, H., Masaki, K., Song, Y.X., Takahashi, Y., Muramatsu, M., et al., 2009. Association between a polymorphism of aminolevulinate dehydratase (ALAD) gene and blood lead levels in Japanese subjects. Int. J. Environ. Res. Publ. Health 6, 999–1009.

Mohammed-Brahim, B., Buchet, J.P., Lauwerys, R., 1985. Erythrocyte pyrimidine 5′-nucleotidase activity in workers exposed to lead, mercury or cadmium. Int. Arch. Occup. Environ. Health 55, 247–252.

Moore, M.R., 1988. Hematological effects of lead. Sci. Total Environ. 71, 419–431.

Morita, Y., Sakai, T., Araki, S., Araki, T., Masuyama, Y., 1997. Nicotinamide adenine dinucleotide synthetase activity in erythrocytes as a tool for the biological monitoring of lead exposure. Int. Arch. Occup. Environ. Health 70, 195–198.

Murase, T., Horiba, N., Goto, I., Yamato, O., Ikeda, T., Sato, K., et al., 1993. Erythrocyte ALA-D activity in experimentally lead-poisoned ducks and its change during treatment with disodium calcium EDTA. Res. Vet. Sci. 55, 252–257.

National Academy of Sciences: National Research Council, 1972. Lead: Airborne Lead in Perspective. National Academy Press, Washington, DC.

National Academy of Sciences: National Research Council, 1980. Lead in the Human Environment. National Academy Press, Washington, DC.

National Academy of Sciences: National Research Council, 1993. Measuring Lead Exposure in Infants, Children, and Other Sensitive Populations. National Academy Press, Washington, DC.

O'Flaherty, E.J., Hammond, P.B., Lerner, S.I., Hanenson, I.B., Roda, S.M.B., 1980. The renal handling of δ-aminolevulinic acid in the rat and in the human. Toxicol. Appl. Pharmacol. 55, 423–432.

O'Flaherty, E.J., Hammond, P.B., Taylor, E., 1981. Renal absorption and secretion of δ-aminolevulinic acid in the rat. Fund. Appl. Toxicol. 1, 278–281.

Oishi, H., Nakashima, M., Tokai, T., Tomokuni, K., 1996. Chronic lead exposure may inhibit endothelium-dependent hyperpolarizing factor in rats. J. Cardiovasc. Pharmacol. 28, 558–563.

Osterode, W., Barnas, D., Geissler, K., 1999. Dose dependent reduction of erythroid progenitor cells and inappropriate erythropoietin response in exposure to lead: new aspects of anaemia induced by lead. Occup. Environ. Med. 56, 106–109.

Paglia, D.E., Valentine, W.N., 1975. Characteristics of a pyrimidine-specific 5′-nucleotidase in human erythrocytes. J. Biol. Chem. 250, 7973–7979.

Paglia, D.E., Valentine, W.N., Fink, K., 1977. Lead poisoning: further observations on erythrocyte pyrimidine-nucleotidase deficiency and intracellular accumulation of pyrimidine nucleotides. J. Clin. Invest. 60, 1362–1366.

Patil, A.J., Bhagwat, V.R., Patil, J.A., Dongre, N.N., Ambekar, J.G., Jailkhani, R., et al., 2006. Effect of lead (Pb) exposure on the activity of superoxide dismutase and catalase in battery manufacturing workers (BMW) of Western Maharashtra (India) with reference to heme biosynthesis. Int. J. Environ. Res. Publ. Health 3, 329–337.

Perez-Bravo, F., Ruz, M., Moran-Jimenez, M.J., Olivares, M., Rebolledo, A., Codoceo, J., et al., 2004. Association between aminolevulinate dehydrase genotypes and blood lead levels in children from a lead-contaminated area in Antofagasta, Chile. Arch. Environ. Contam. Toxicol. 47, 276–280.

Piomelli, S., Seaman, C., Zullo, D., Curran, A., Davidow, B., 1982. Threshold for lead damage to heme synthesis in urban children. Proc. Natl. Acad. Sci. U.S.A. 79, 3335–3339.

Potluri, V.R., Astrin, K.H., Wetmur, J.G., Bishop, D.F., Desnick, R.J., 1987. Human delta-aminolevulinate dehydratase: chromosomal localization to 9q34 *in situ* hybridization. Hum. Genet. 76, 236–239.

Poulos, L., Qammas, S., Athanaselis, S., Maravelias, C., Koutselinas, A., 1986. Statistically significant hematopoietic effects of low blood lead levels. Arch. Environ. Health 41, 384–386.

Pueschel, S.M., Kopito, L., Schwachman, H., 1972. Children with an increased lead burden. A screening and follow-up study. J. Am. Med. Assoc. 222, 462–466.

Quintanar-Escorza, M.A., Gonzáles-Martinez, Navarro, L., Maldonado, M., Arévalo, B., Calderón-Salinas, J.V., 2007. Intracellular free calcium concentration and calcium transport in human erythrocytes of lead-exposed workers. Toxicol. Appl. Pharmacol. 220, 1–8.

Rabinowitz, M.B., Leviton, A., Needleman, H.L., 1986. Occurrence of elevated protoporphyrin levels in relation to lead burden in infants. Environ. Res. 39, 253–257.

Redig, P.T., Lawler, E.M., Schwartz, S., Dunette, J.L., Stephenson, B., Duke, G.E., 1991. Effects of chronic exposure to sublethal concentrations of lead acetate on heme synthesis and immune function in red-tailed hawks. Arch. Environ. Contam. Toxicol. 21, 72–77.

Roels, H., Lauwerys, R., 1987. Evaluation of dose–effect and dose–response relationships for lead exposure in different Belgian population groups (fetus, child, adult men, women). Trace Elem. Med. 4, 80–87.

Roels, H., Buchet, J.-P., Lauwerys, R., Hubermont, G., Bruaux, P., Claeys-Thoreau, F., et al., 1976. Impact of air pollution by lead on the heme biosynthetic pathway in school-age children. Arch. Environ. Health 31, 310–316.

Roels, H.A., Buchet, J.-P., Bernard, A., Hubermont, G., Lauwerys, R.R., Masson, P., 1978. Investigations of factors influencing exposure and response to lead, mercury, and cadmium in man and in animals. Environ. Health Perspect. 25, 91–96.

Romeo, R., Aprea, C., Boccalon, P., Orsi, D., Porcelli, B., Sartorelli, P., 1996. Serum erythropoietin and blood lead concentrations. Int. Arch. Occup. Environ. Health 69, 73–75.

Rosen, J.F., Chesney, R.W., 1983. Circulating calcitriol concentrations in health and disease. J. Pediatr. 103, 1–7.

Rosen, J.F., Chesney, R.W., Hamstra, A.J., DeLuca, H.F., Mahaffey, K.R., 1980. Reduction in 1,25-dihydroxyvitamin D in children with increased lead absorption. N. Engl. J. Med. 302, 1128–1131.

Roth, H.-P., Kirchgessner, M., 1981. [Ger.] Dependence of delta-aminolevulinic acid dehydratase upon alimentary zinc. Z. Tierphysiol. Tiernaehr. Futtermittelkd. 46, 59–63.

Sakai, T., Morita, Y., 1996. δ-Aminolevulinic acid in plasma or whole blood as a sensitive indicator of lead effects, and its relation to the other heme-related parameters. Int. Arch. Occup. Environ. Health 68, 126–132.

Sakata, S., Shimizu, S., Ogoshi, K., Hirai, K., Ohno, Y., Kishi, T., et al., 2007. Inverse relationship between serum erythropoietin and blood lead levels in Kathmandu tricycle taxi drivers. Int. Arch. Occup. Environ. Health 80, 342–345.

Santos, J.L., Fontanellas, A., Moran, M.J., Enriquez de Salamanca, R., 1999. Nonsynergic effect of ethanol and lead on heme metabolism in rats. Ecotoxicol. Environ. Saf. 43, 98–102.

Scheuhammer, A.M., 1987. Erythrocyte δ-aminolevulinic acid dehydratase in birds. II. The effects of lead exposure *in vivo*. Toxicology 45, 165–175.

Schwartz, J., Landrigan, P.J., Baker Jr., E.L., Orenstein, W.A., von Lindern, I., 1990. Lead-induced anemia: dose–response relationships and evidence for a threshold. Am. J. Publ. Health 80, 165–168.

Schwartz, B.S., Lee, B.-K., Stewart, W., Ahn, K.-D., Springer, K., Kelsey, K., 1995. Associations of δ-aminolevulinic acid dehydratase genotype with plant, exposure duration, and blood lead and zinc protoporphyrin levels in Korean lead workers. Am J. Epidemiol. 142, 738–745.

Schwartz, B.S., Lee, B.-K., Stewart, W., Sithisarankul, P., Strickland, P.T., Ahn, K.-D., et al., 1997. δ-Aminolevulinic acid dehydratase genotype modifies four hour urinary lead excretion after oral administration of dimercaptosuccinic acid. Occup. Environ. Med 54, 241–246.

Schwartz, B.S., Lee, B.-K., Lee, G.-S., Stewart, W.F., Simon, D., Kelsey, K., et al., 2000. Associations of blood lead, dimercaptosuccinic acid-chelatable lead, and tibia lead with polymorphisms in the vitamin D receptor and δ-aminolevulinic acid dehydratase genes. Environ. Health Perspect. 108, 949–954.

Scinicariello, F., Murray, H.E., Moffett, D.B., Abadin, H.G., Sexton, M.J., Fowler, B.A., 2007. Lead and delta-aminolevulinic acid dehydratase polymorphism: Where does it lead? A meta-analysis. Environ. Health Perspect. 115, 35–41.

Secchio, C., Erba, L., Cambiaghi, G., 1974. Delta-aminolevulinic acid dehydratase activity of erythrocytes and liver tissue in man: relationship to lead exposure. Arch. Environ. Health 28, 130–132.

Selander, S., Cramér, K., 1970. Interrelationships between lead in blood, lead in urine, and ALA in urine during lead work. Br. J. Ind. Med. 27, 28–39.

Silbergeld, E.K., Lamon, J.M., 1982. Effects of altered porphyrin synthesis on brain neurochemistry. Neurobehav. Toxicol. Teratol. 4, 635–642.

Silbergeld, E.K., Hruska, R.E., Bradley, D., Lamon, J.M., Frykholm, B.C., 1982. Neurotoxic aspects of porphyrinopathies: lead and succinylacetone. Environ. Res. 29, 459–471.

Smith, C.M., Wang, X., Hu, H., Kelsey, K.T., 1995. A polymorphism in the δ-aminolevulinic acid dehydratase gene may modify the pharmacokinetics and toxicity of lead. Environ. Health Perspect. 103, 248–253.

Soldin, O.P., Pezzullo, J.C., Hanak, B., Miller, M., Soldin, S.J., 2003. Changing trends in the epidemiology of pediatric lead exposure: interrelationship of blood lead and ZPP concentrations and a comparison to the U. S. population. Ther. Drug Monit. 25, 415–420.

Swanson, M.S., Angle, C.R., Stohs, S.T., Rovang, K.S., 1980. Erythrocyte nucleotides from control and lead-treated rabbits. Res. Commun. Chem. Pathol. Pharmacol. 27, 353–361.

Tomokuni, K., Ichiba, M., Hirai, Y., 1989. Effect of lead exposure on some biological indices related to porphyrin metabolism and the activity of erythrocyte pyrimidine 5′-nucleotidase in the mice. Arch. Toxicol. 63, 23–28.

U.S. Agency for Toxic Substances and Disease Registry, 1988. The Nature and Extent of Lead Poisoning in Children in the United States: A Report to Congress. Submitted to Congress, July 12, 1988. U.S. Public Health Service, Atlanta, GA.

U.S. Agency for Toxic Substances and Disease Registry, 2007. Toxicologic Profile for Lead. U.S. Public Health Service, Atlanta, GA.

U.S. Centers for Disease Control, 1978. Preventing Lead Poisoning in Young Children. A Statement by the Centers for Disease Control. U.S. Department of Health and Human Services, Atlanta, GA, April.

U.S. Centers for Disease Control, 1985. Preventing Lead Poisoning in Young Children. A Statement by the Centers for Disease Control. U.S. Department of Health and Human Services, Atlanta, GA.

U.S. Centers for Disease Control, 1991. Preventing Lead Poisoning in Young Children. A Statement by the Centers for Disease Control. U.S. Department of Health and Human Services, Atlanta, GA.

U.S. Centers for Disease Control, 2005. Preventing Lead Poisoning in Young Children. A Statement by the Centers for Disease Control. U.S. Department of Health and Human Services, Atlanta, GA.

U.S. Environmental Protection Agency, 1977. Air Quality Criteria for Lead. Report No. EPA-600/8-77-017. Office of Research and Development, Washington, DC.

U.S. Environmental Protection Agency, 1986. Air Quality Criteria for Lead, 4 vols. EPA 600/8-83/028bF. Environmental Criteria and Assessment Office, Washington, DC.

U.S. Environmental Protection Agency, 2006. Air Quality Criteria for Lead, vol. 1. Report No. EPA/600/R-05/144aF. National Center for Environmental Assessment, Washington, DC.

Valentine, W.N., Paglia, D.E., 1980. Erythrocyte disorders of purine and pyrimidine metabolism. Hemoglobin 4, 669–681.

Wada, O., Yano, Y., Ono, T., Toyokawa, K., 1973. The diagnosis of different degrees of lead absorption: in special reference to choice and evaluation of various parameters indicative of an increased lead absorption. Ind. Health 11, 55–67.

Waldron, H.A., 1964. The effect of lead on the fragility of the red cell incubated *in vitro*. J. Clin. Pathol. 17, 405–406.

Waldron, H.A., 1966. The anaemia of lead poisoning: a review. Br. J. Ind. Med. 23, 83–100.

Wetmur, J.G., 1994. Influence of the common human delta-aminolevulinate dehydratase polymorphism on lead body burden. Environ. Health Perspect. 102 (Suppl. 3), 215–219.

Wetmur, J.G., Lehnert, G., Desnick, R.J., 1991a. The δ-aminolevulinate dehydratase polymorphism: higher blood lead levels in lead workers and environmentally exposed children with the 1-2 and 2-2 isozymes. Environ. Res. 56, 109–119.

Wetmur, J.G., Kaya, A.H., Plewinska, M., Desnick, R.J., 1991b. Molecular characterization of the human delta-aminolevulinate dehydratase 2 (ALAD2) allele: implications for molecular screening of individuals for genetic susceptibility to lead poisoning. Am. J. Hum. Genet. 49, 757–763.

White, J.M., Harvey, D.R., 1972. Defective synthesis of α and β globin chains in lead poisoning. Nature 236, 71–73.

Wildt, K., Berlin, M., Isberg, P.E., 1987. Monitoring of zinc protoporphyrin levels in blood following occupational lead exposure. Am. J. Ind. Med. 12, 385–398.

World Health Organization, 1995. Environmental Health Criteria 165: Inorganic Lead. International Programme on Chemical Safety, Geneva, Switzerland.

World Health Organization. 1996. Zinc: Trace Elements in Human Nutrition and Health. Geneva, Switzerland, pp. 72–104 (Chapter 5).

Yip, R., Norris, T.N., Anderson, A.S., 1981. Iron status of children with elevated blood lead concentrations. J. Pediatr. 98, 922–925.

Zareba, G., Chmielnicka, J., 1992. Disturbances in heme biosynthesis in rabbits after administration *per os* of low doses of tin or lead. Biol. Trace Elem. Res. 34, 115–122.

Chapter 17

Carcinogenic and Genotoxic Effects of Lead in Human Populations

The focus on lead as an environmental and public health hazard has principally centered on noncancer endpoints in various exposed human and experimental populations. There is, however, an extensive database for lead carcinogenicity, derived largely from Pb exposures of adult lead workers.

17.1 INTRODUCTION AND PUBLIC HEALTH CONTEXT

Carcinogenic and genotoxic endpoints have not received as much attention as have evaluations of developmental neurotoxicity, cardiovascular effects, nephrotoxicity, immunotoxicity, hematotoxicity, and reproduction and development in terms of comparative human health risk assessment. Reasons include: (1) the nature of the cancer epidemiology literature and its limits; (2) the seeming relative insensitivity of cancer endpoints in terms of dose–response relationships; (3) much of the epidemiological data being derived from highly exposed lead workers; and (4) the relative magnitude of impacts of these endpoints in terms of numbers of affected individuals internationally.

One of the most problematic factors in evaluating Pb carcinogenesis in humans is the relative absence of a good exposure biomarker and particularly the absence of an exposure marker for an effect with a considerable latency period such as certain Pb-related cancers in humans. Added to this is the problem of markedly changing Pb exposures in human populations from previous decades to the present, declines which have been described quantitatively in earlier chapters.

Blood lead as measured in those older segments of human populations with occupational Pb histories can potentially integrate cumulative Pb exposures, but only where there were serial PbB measurements done accurately and proficiently. In many cases, single exposure measures, e.g., PbB, were only used in the course of studies on workplace carcinogenic responses. Until the late 1970s, little regulatory oversight of measurements or laboratory

Trace Metals and other Contaminants in the Environment, Volume 10
ISSN: 1875-1121 DOI: 10.1016/B978-0-444-51554-4.00017-1

requirements existed. Serial PbB measures for use in a cumulative index in adult subjects have historically only been available for occupational exposures. Single exposure measurements were typically used for environmental assessments of nonoccupational populations like those in U.S. NHANES III.

Bone Pb is considered both a more reliable indicator of body lead burden than PbB and a measure of integrated Pb exposures, and should theoretically serve as a better exposure biomarker for assessing cancer prevalence and incidence for occupational or general populations. Use of bone Pb over the more time-dependent PbB measurement, in fact, has been shown to produce significant associations of Pb exposure with such endpoints as cardiovascular toxicity in those studies where links with PbB were absent or marginally significant. See earlier chapters for such discussions. As of this writing, however, none of the cohort ($N = 29$) and case–control ($N = 16$) studies employed bone Pb as the exposure marker, and the nature of the epidemiological assessments are such that it would be difficult to begin such measurements *de novo* in surviving individuals enrolled in these studies.

A second problem with the cancer epidemiology of workplace Pb exposures is the requirement for large cohort sample sizes and relatively sophisticated statistical analyses, in that prevalences and incidences of any likely Pb-specific cancers are low in general populations and still relatively low in occupational cohorts with Pb exposures. A third problem is coexposures to carcinogens which often either occur geochemically and metallurgically with Pb, such as with the mining, milling, smelting, and refining of arsenic and cadmium, or with lifestyle practices, e.g., cigarette smoking.

Fourth, considerable confusion remains in the literature about whether Pb induces carcinogenicity through primary alteration of DNA or operates through epigenetic mechanisms. In the latter case, the evidence would indicate that the dose–response relationships for lead and cancer prevalence/incidence, at least in those worker cohorts evaluated so far, are relatively insensitive. On the other hand, regulatory regimens for Pb which prohibit exposures of populations to even quite low levels in diverse environmental media, and based on, for example, the more sensitive endpoints of cognitive or behavioral impairments in very young or even older children, would be expected to prevent a discernible incidence of Pb-related cancers.

The current expert consensus view on lead carcinogenesis is that certain lead compounds are probable carcinogens in humans and established carcinogens in animal models. The various discussions and rationales for this ranking are presented in expert consensus treatises including the recently updated (Vol. 87) monograph on inorganic and organic lead compounds by the International Agency for Research on Cancer (IARC) (IARC, 2006), the 11th and 12th Report on Carcinogens by the National Toxicology Program (NTP) (NTP, 2004, 2011), the U.S. EPA (2006), and the U.S. ATSDR (2007). A number of reviews of the nature of Pb carcinogenicity have appeared, including Silbergeld et al. (2000) and Silbergeld (2003).

Genotoxicity studies of Pb's potential as a human carcinogen have produced a considerable literature, but results are inconsistent, as are their interpretations. This chapter addresses more recent studies of Pb genotoxicity, where evaluations have been done within a greater understanding of genotoxicity and a broader context of mechanistic understanding, i.e., mode(s) of action (MOA). A particular reference here is to the ability of the various genotoxic outcomes to actually predict eventual carcinogenesis.

17.2 Pb-ASSOCIATED CANCER IN OCCUPATIONAL POPULATIONS

Virtually all of the available data for Pb-associated cancer in human populations are derived from worker exposures, with few studies done on general subgroups of populations having typically much lower environmental exposures. General population studies have been confined to the U.S. NHANES II and NHANES III for the periods 1976–1980 and 1988–1994, respectively. Furthermore, both the cohort and case–control studies of occupational exposures concerned poorly characterized exposures, exposures that varied over working lifetimes. Earlier decades involved much higher exposures than do more recent periods. In some cases, no environmental or biological exposure measures were determined; rather, worker cohorts were stratified by work categories expected to entail ranges of Pb exposures. In the case of general population exposures, studies predating the U.S. NHANES efforts produced data which were too problematic because of poor or little exposure characterization and those are not discussed further in this chapter.

Different occupational epidemiological reports differed as to sites of cancers, robustness of cancer incidences, and relative statistical significance for both cohort and case–control investigations. A significant portion of these differences arose from the heterogeneous nature of metal exposures in operations such as lead and lead–zinc smelters. Increased lung cancer incidences associated with Pb worker exposures hinged on the extent to which other known carcinogens, such as arsenic and/or cadmium in the case of lead, lead–zinc, or mixed metal operations, were present. Some increases in cancer incidences were seen in the form of trends, while others were determined to be statistically significant. Some studies attempted to clarify the temporal variability of Pb exposure intensity by stratifying subjects in terms of the level of biomarker measurement or the like. Some epidemiological studies carried out multiple analyses, with cohort and nested case–control studies.

A number of the epidemiological reports for Pb-associated cancer in worker population as cohorts or in case–control analyses indicated the lung and stomach as relatively more common cancer sites than other organs or systems. The relative role of these cancers in the various reports is not surprising, given the fact that workplace inhalation and ingestion of lead compounds comprised the principal intake and uptake sites from extractive industry substances, especially for poorly soluble mineral forms in contact with physiological media. These findings are summarized in Tables 17.1 and 17.2. Increased lung cancer rates have principally been reported from around the world for both lead battery workers (Fanning, 1988; Wong and Harris, 2000) and smelter workers (Ades and Kazantzis, 1988; Carta et al., 2005; Englyst et al., 2001; Lundström et al., 1997, 2006; Steenland et al., 1992; Wong and Harris, 2000) or workers classified in various industrial categories from which other registry information such as PbB levels can be retrieved for epidemiological assessments, as was noted for the Finnish adult worker population (Anttila et al., 1995) or for population subsets (IARC, 2006).

Other reports were more limited, confined to operations such as glass production using cohort (Sankila et al., 1990) or case–control (Wingren and Axelson, 1993) epidemiological designs. Several studies looked at lead chromate production workers (Sheffet et al., 1982). However, the relatively more robust role of the known human carcinogen chromate in any lung cancer incidence and mortality outcomes in these workers vitiates meaningful analysis of a role for lead.

Table 17.1 depicts nine reports on findings of increased lung cancer incidences and mortalities in both battery and smelter workers with elevated Pb exposures and one report derived from Pb exposure monitoring analysis. These cohorts and other subjects were from Britain, Finland, Italy, Sweden, and the United States. Pb exposures were indexed by actual environmental measurements in air or measurements of Pb in blood and urine. A job-exposure matrix approach was employed in part of one study using a nested case–control approach.

Mortality rates for lung cancer clustered around a relatively tight range except for Swedish smelter workers, the smelter workers showing much higher rates for cohort and case–control approaches. These data are presented separately. Standardized mortality ratio (SMR) estimates, as seen in Table 17.1 for the non-Swedish cohorts, typically ranged to <200. The Swedish smelter analyses comprise much more robust SMR and related analyses, and this has been attributed to a significant contribution from other workplace carcinogens, notably inorganic arsenic. The results for Swedish cohort or case–control studies (Table 17.1) are contained in the reports of Lundström et al. (1997), Englyst et al. (2001), and Lundström et al. (2006). Lundström et al. (1997) reported the total cohort's lung cancer deaths at 39, with an SMR of 280 (95% CI 200–380), and the high exposure group's lung cancer mortality at 19, SMR = 280 (95% CI 180–450). Accounting for a $\geq$15-year latency period, Standardized incidence ratio (SIR) values were similarly quite high (Table 17.1). Both total cohort and high-exposure subcohort mortality estimates were considerably higher than for other Pb exposure cohort outcomes. Pb exposures varied considerably over time, the 1950 mean PbB (62 μg/dl) being almost twice the figure for 1987 (33 μg/dl).

Englyst et al. (2001) attempted to account for the role of smoking and of arsenic, a potent human carcinogen co-occurring as a contaminant in the

TABLE 17.1 Carcinogenic and Genotoxic Effects of Pb in Human Populations: Lung Cancer in Lead Workers

Study Group	Study Details	Pb Exposure	Findings	References
U.S. battery plant workers	*N* = 4,518, men >1 year-work, 1947–1970; follow-up, 1947–1995	Urinary Pb, 2,275 workers; PbB, 1,863 men; mean: 63 μg/dl	Lung cancer deaths = 210; SMR = 113.9 (99.0–130.4), expected = U.S. male mortality rate	Wong and Harris (2000)
British battery plant workers	Proportional mortality odds ratio (PMOR) study: *N* = 2,073 males; frequency = matched, 1926–1985	Job-exposure matrix: high, *N* = 867; low, *N* = 1,206	Lung cancer PMOR = 0.93	Fanning (1988)
U.S. smelter workers	Cohort study: *N* = 2,300, six facilities, >1 year 1947–1970; follow-up, 1947–1995	Urinary (*N* = 2,275) and blood Pb (*N* = 1,863). Mean PbB = 80 μg/dl (*N* = 254)	Lung cancer deaths = 107, SMR = 121.5 (99.5–146.8)	Wong and Harris (2000)
U.S. smelter workers	Cohort study: *N* = 1,990 male smelter workers, >1 year employed, 1940–1965; subcohort with higher exposure, *N* = 1,436	PbB mean = 56 μg/dl; air Pb mean 3.1 mg/m^3; air As mean = 14 μg/dl	Lung cancer cases = 72; SMR = 118 (92–148)	Steenland et al. (1992)
Swedish smelter workers	Cohort study: *N* = 3,979 workers >1 year employed, 1928–1979; subcohort *N* = 1,992 workers in Pb department; mortality 1955–1987	PbB mean, 1950: 62 μg/dl; 1987: 33 μg/dl	Total cohort lung cancers = 39; SMR = 280 (200–380); high-exposure group lung cancers = 19; SMR = 280 (180–450)	Lundström et al. (1997)
	–		–	
	≥15-year latency period		Total cohort lung cancers = 42; SIR: 290 (210–400). Highest exposed group cancers = 23; SIR: 340 (220–520)	
	–		–	
	Lead-only workers or other lead-exposed departments		Lung cancers = 14; SIR: 310 (170–520)	

(*Continued*)

TABLE 17.1 Carcinogenic and Genotoxic Effects of Pb in Human Populations: Lung Cancer in Lead Workers—(cont.)

Study Group	Study Details	Pb Exposure	Findings	References
Swedish smelter workers	3,979 workers, primary Cu–Pb smelter, follow-up: 1958–1987. Subcohort 1: 710 workers in Pb department and elsewhere; Subcohort 2: 383 workers only employed in Pb department	Cumulative blood Pb index: PbB × work duration	Subcohort 1: lung cancer cases = 10; SIR = 240 (120–450). Subcohort 2: lung cancer cases = 5; SIR = 360 (120–830)	Englyst et al. (2001)
Swedish smelter workers	Nested case–control study; $N = 3,979$ primary smelter workers; workers with lung cancer ($N = 46$) compared with 141 age-matched male controls	Follow-up of earlier reports	Cancer cases showed a higher smoking rate versus controls. OR = 4.0 (1.6–10.1, $p = 0.003$). Air As, but not air Pb indices higher among the smokers with lung cancer; smoking and cumulative workplace As exposure accounted for the cancers	Lundström et al. (2006)

Italian (Sardinian) smelter workers	Cohort study, $N = 918$, followed 1972–2001	Environmental Pb and PbB available 1985–2001	Lung cancer mortality = 18, significant upward trend with Pb exposure. SMR = 196 (102–368, highest exposure)	Carta et al. (2005)
British smelter workers	Cohort and nested case–control studies: 4,173 Zn–Pb–Cd smelter workers, >1 year employed from 1943 to 1969, born before 1940	Mean PbB in Cd operation: 28 μg/dl; furnace PbB: 59 μg/dl; sinter: 56 μg/dl	Lung cancers and SMRs versus work years: 20–29: 44, SMR = 190 (138–256); 30–39: 28, SMR = 142 (94–205); ≥40: 8, SMR = 292 (126–575)	Ades and Kazantzis (1988)
	–	–		
	Nested case–control: $N = 174$ lung cancer cases, $N = 2,717$ controls frequency-matched on age, work start, etc.	Job-exposure matrix		
Finnish workers in various industrial operations	Cohort plus case–control design	PbB in national monitoring program; N, men = 18,329; N, women = 2,412. Highest PbB used, 1973–1988	SIR at mid-exposure range = 140 (100–190) for lung cancer; highest exposures = no risk increase	Anttila et al. (1995)

TABLE 17.2 Carcinogenic and Genotoxic Effects of Pb in Human Populations: Stomach Cancer in Lead Workers

Study Group	Study Details	Pb Exposure	Findings	References
U.S. battery factory workers	$N = 4{,}518$ men, >1-year work, 1947–1970; follow-up, 1947–1995	Urinary Pb, 2,275 men; PbB mean = 63 μg/dl, $N = 1{,}863$ men	Stomach cancer deaths = 45; SMR = 152.8 (111.5–201.5)	Wong and Harris (2000)
British battery factory workers	PMOR study: $N = 2{,}073$ males, frequency-matched by 10-year age group, 1926–1985	Job-exposure match	Stomach cancer deaths = 31; PMOR = 1.34	Fanning (1988)
U.S. smelter workers	Cohort study: $N = 2{,}300$, six facilities, >1 year, 1947–1970; follow-up, 1947–1995	Urinary ($N = 2{,}275$) and blood Pb ($N = 1{,}863$); mean PbB = 80 μg/dl ($N = 254$)	Stomach cancer deaths = 15; SMR = 133.4 (74.6–220.0)	Wong and Harris (2000)
U.S. smelter workers	Cohort study: $N = 1{,}990$ male smelter workers, >1-year employed, 1940–1965; subcohort with higher exposure ($N = 1{,}436$)	Mean PbB = 56 μg/dl; mean air Pb = 3.1 mg/m^3; mean air As = 14 μg/m^3	Stomach cancer deaths = 15; SMR = 136 (75–225)	Steenland et al. (1992)

workplace for this cohort and subcohorts. Lundström et al. (2006), using a nested case–control approach, evaluated 41 of these smelter workers with lung cancer ($N = 46$) versus 141 age-matched male controls. The cancer cases had a higher smoking rate (OR = 4.0, $p = 0.003$), while air arsenic but not air Pb was higher among the cancer cases who smoked. Smoking and workplace As appeared to principally account for the lung cancers.

Four Pb worker cohorts studied for incidence and mortality linked to stomach cancer yielded evidence of increased mortality from stomach cancer, and this is summarized in Table 17.2. The studies consisted of two battery factory (Fanning, 1988; Wong and Harris, 2000) and two Pb smelter (Steenland et al., 1992; Wong and Harris, 2000) cohorts. The proportional mortality odds ratio (PMOR) for the 2,073 male British battery workers studied by Fanning (1988) was 1.34 (based on 31 stomach cancer deaths). Pb exposure was assessed using a job-exposure matrix, with frequency matching by 10-year worker age bands, 1926–1985. U.S. battery workers (male $N = 4{,}518$) in the Wong and Harris report presented with an SMR of 152.8 based on 45 stomach cancer deaths.

U.S. smelter workers in two reports provided mortality estimates of 136 (Steenland et al., 1992) and 133.4 (Wong and Harris, 2000), respectively. Mean PbB values ranged from 56 to 80 μg/dl across the two studies. Smoking and the presence of other carcinogenic materials would be comparatively less of a complicating question for stomach cancer than for lung cancer.

Some occupational epidemiological studies have reported increased mortalities or incidences as "all-site" findings or sites other than lung or stomach. These findings are summarized in Table 17.3. Gerhardsson et al. (1995) reported that male Swedish secondary smelter workers have increased incidences and mortalities for all-site cancers relative to the national cancer rates for men. The SIR was 127 (CI = 91–174) and SMR 165 (CI = 109–244). Exposures were indexed via repeat PbB levels gathered from 1969 at a 2–3 months frequency. Anttila et al. (1995) employed PbB and occupational categorizations for 20,700 Finnish residents matched up with the national cancer registry to arrive at all-site RR of 1.4 (CI = 1.1–1.8). At least one PbB in the 1973–1983 span was required for inclusion.

Although not commonly found across the bulk of occupational epidemiological studies of Pb worker cancers, some reports have highlighted findings of increased cancers at certain other sites. Anttila et al. (1996) employed a nested case–control approach using the cohort described in Anttila et al. (1995, see above) to identify 26 CNS cancers, 16 of which were gliomas. For a subset of workers with PbB ≥28 μg/dl, the OR using maximum terciles or quartiles and highest PbB as marker was 2.0 for CNS cancers and 11.0 for gliomas. For the maximum stratum of cumulative exposure, lifetime mean exposure, and workplace exposure duration markers, the corresponding ORs were 12, 23, and 6.9. All tests for trend were significant ($p < 0.05$). One study reported increased kidney cancers (Steenland et al., 1992) for U.S. smelter workers, with the SMR for the subcohort with the heaviest Pb exposure at 239 (CI = 103–471) and that for the total cohort at 193.

Two meta-analyses of cancer risks among Pb-exposed workers in different Pb facilities have been carried out (Fu and Boffetta, 1995; Steenland and Boffetta, 2000). Risk estimates are presented in Table 17.4. A number of the cohorts in both analyses were the same, but the analyses dealt with the data sets in different ways. Steenland and Boffetta examined a total of eight Pb worker cohorts where the level of exposure documentation was relatively good for seven cohort and one nested case–control studies. Site-specific cancer risk estimates were for lung, stomach, kidney, and brain (gliomas). The strongest risks were estimated for stomach cancer, based on a combined tally of 181 deaths and with a combined RR of 1.34 (CI = 1.14–1.57). Lung cancer RR estimates differed, as noted in Table 17.4, depending on whether one study's relatively high mortality rate may have been due to coexposure to airborne arsenic. Inclusion of data from Lundström et al. (1997) yielded an RR of 1.30 (CI = 1.15–1.46), with the figure declining to 1.14 (CI = 1.04–1.73) when this cohort's findings were excluded. Kidney, bladder, and

TABLE 17.3 Carcinogenic and Genotoxic Effects of Pb in Human Populations: All Sites, CNS/Brain, and Kidney Cancers in Lead Workers

Study Group	Study Details	Pb Exposure	Findings	References
A. All-site cancers				
Swedish secondary smelter workers	Mortality and incidence; $N = 664$ males workers, worked ≥3 months, 1942–1987; expected rates from country	Repeated PbB levels from 1969, 2–3 months frequency	Increased mortality all-site cancers: 26 deaths, mainly before 1969; SMR = 1.65 (1.09–2.44). Increased incidence: SIR = 1.27 (0.91–1.74)	Gerhardsson et al. (1995)
Finnish residents with PbB and registry matching	Cohort and case–control designs: $N = 20{,}700$ workers in all industries; match-up with Finnish cancer registry	At least one PbB, 1973–1983	All cancers: RR = 1.4 (1.1, 1.8)	U.S. EPA (2006), Anttila et al. (1995)
B. Brain/CNS cancers				
Finnish residents with PbB and registry matching	Nested case–control study; from Anttila et al. (1995) database, 26 cases of CNS cancer, 16 gliomas	–	For PbB ≥28 μg/dl, OR ~2.0 for CNS cancers, OR = 11 (1.0–630); trend *p*-value = 0.037	Anttila et al. (1996)
C. Kidney cancers				
U.S. Pb smelter workers	Cohort study: $N = 1{,}990$ male smelter workers >1-year employment, 1940–1965; subcohort with higher exposure ($N = 1{,}436$)	Mean PbB = 56 μg/dl; air Pb = 3.1 mg/m^3	SMR: 1–5 years = 1.97; 5–20 years = 1.82; 20+ years = 2.05	Steenland et al. (1992)

brain (glioma) cancer risks were not materially different from the reference risk estimates.

The earlier meta-analysis (Fu and Buffetta, 1995) evaluated up to 15 cohort exposure categories. The analyses examined all work categories or just battery and smelter workers. Tables 5 and 6 of Fu and Buffetta (1995)

TABLE 17.4 Carcinogenic and Genotoxic Effects of Pb in Human Populations: Meta-analyses of Cancers in Highly Exposed Worker Populations

Lung Cancer	Stomach Cancer	Renal Cancer (Kidney, Bladder)	Brain Cancer (Gliomas)	References
Combined RR = 1.3 (1.15–1.46, 675 deaths)[a]; excluding Lundström et al. (1997), RR = 1.14 (1.04–1.73)[b]	Combined RR = 1.34 (1.14–1.57, 181 deaths)	Combined RR = 1.01 (0.72–1.42, 40 deaths)	Combined RR = 1.06 (0.81–1.40, 69 deaths)	Steenland and Boffetta (2000)
Case–control + cohort studies (*N* = 15 studies): RR = 1.24 (1.16–1.33) Battery + smelter studies: RR = 1.44 (1.29–1.62)	Case–control + cohort studies (*N* = 10): RR = 1.33 (1.18–1.49) Battery + smelter studies: RR = 1.50 (1.23–1.83)	Case–control + cohort studies (*N* = 5, kidney; *N* = 5, bladder) Kidney RR = 1.19 (0.96–1.48); Bladder RR = 1.41 (1.16–1.71) Battery + smelter studies: kidney RR = 1.26 (0.70–2.26)	–	Fu and Buffetta (1995)

[a]*Includes cohort risk estimates of Lundström et al. (1997).*
[b]*Excludes study of Lundström et al. (1997).*

tabulate the risk data for these two groups. At the time of this earlier analysis, the Lundström et al.'s lung cancer estimates had not appeared. Twelve studies were used in the case of lung cancer, using a random effects model derived from significant heterogeneity across studies. The lung cancer RR for the cohorts consisting of battery and smelter workers was 1.50, the RR for stomach cancer in these workers was 1.50, the RR for bladder was 1.41, and the RR for kidney was 1.26. No estimates were tabulated in the 1995 report for brain cancers for any cohort.

17.3 CASE–CONTROL STUDIES OF CANCER AND Pb IN GENERAL POPULATIONS

General human population analyses of the relationship of Pb to cancer have consisted of analyses of either case–control studies from data sets, where exposures may have been occupational in origin, or survey cohort data, where

Pb exposures were environmental in origin and considerably below those in occupational settings. The case–control studies involved populations from the United States, Canada, Finland, China, and Germany and described outcomes for multicancer sites, as well as stomach, kidney, brain, and other cancers.

Male Canadian residents aged 35–70 years, principally residents of Montreal, were evaluated with respect to multiple cancer site risks based on histories of either any kind of, or a substantial amount of, Pb exposure over the 1979–1985 time frame (Siemiatycki et al., 1991). Individuals with substantial Pb exposure histories presented with various ORs, depending on the chemical forms of Pb at issue. The OR for stomach cancer in individuals exposed to Pb dust was 21.6, the OR for lung cancer with lead oxide exposure was 2.2, while "all lead compounds," under substantial exposures, produced ORs of 1.5 and 1.8 for lung and stomach cancers, respectively. Two U.S.-based evaluations were done using occupational disease surveillance data; the case–control studies produced outcomes for stomach and brain/CNS cancers. The major investigation, by Cocco et al. (1998a), evaluated U.S. deaths from stomach cancer, 1984–1996. OR estimates differed with gender and race: African-American women, 1.76; White women, 1.53; African-American men, 1.15; and Caucasian men, 0.92.

Two reports, those of Partanen et al. (1991) and Pesch et al. (2000), involving residents of Finland and Germany, respectively, concerned OR estimates for kidney cancer. Partanen et al. evaluated 338 incident cases among Finnish men and women in their case–control study for the 1977–1978 period. An OR of 2.89 was calculated for individuals exposed to either lead or lead compounds. The study found an adjusted elevated but nonsignificant risk. The German case–control study, for men and women and incidences in the 1991–1995 period, yielded different ORs depending on gender and Pb exposure intensity. For men with high Pb exposures within the British job-matrix classification, the OR was 1.2, while for women, high exposure produced an OR of 1.0.

Brain/CNS cancer risks have been reported in case–control studies by Cocco et al. (1998b) and Hu et al. (1999) for U.S. and Chinese cases, respectively. The reported brain cancer risks for U.S. Caucasian men and women with lead exposure histories were 2.1 and 12.4, respectively. Chinese cases for 1989–1996 in men and women produced ORs for meningiomas of 7.20 and 5.69, respectively.

17.4 GENERAL POPULATION EXPOSURES TO ENVIRONMENTAL LEAD

Four reports on environmental, nonoccupational lead linkages to cancer mortality risk have appeared using U.S. NHANES data sets, two involving NHANES II covering the period 1976–1980 (Jemal et al., 2002; Lustberg and Silbergeld, 2002) and two dealing with NHANES III, covering

TABLE 17.5 Carcinogenic and Genotoxic Effects of Pb in Human Populations: Pb and Cancer in General Populations

Study Group	Study Details	Pb Exposure	Findings	References
U.S. NHANES II (1976–1980) participants: White subgroup, $N = 3,592$	Matchings: Deaths through 1992 with National Death Index and Social Security master death file: deaths from all-site cancers	Single PbB level at survey entry; group median = 12 μg/dl	Nonsignificant increased risk of mortality from any cancer versus PbB quartile; highest PbB quartile RR was 2.0, men and 1.6, women; lowest exposure quartile had small numbers of deaths; dose–response relationship only in women	Jemal et al. (2002)
U.S. NHANES II cohort of Jemal et al. (2002) for non-Whites, total $N = 4,190$. Follow-up mortality $N = 929$, cancer $N = 240$	More confounder adjustments than in Jemal et al. (2002) (see Tables 2 and 3 in Lustberg and Silbergeld, 2002)	Single PbB level at entry, median PbB = 13 μg/dl	All-site cancer mortality RR = 1.68 (95% CI 1.02–2.78) at 20–29 μg/dl versus <10 μg/dl	Lustberg and Silbergeld (2002)
U.S. NHANES III cohort (1988–1994): subcohort ≥40 years old, $N = 9,686$	Mortality followed through 2/2000; Deaths = 543 from cancer	Single PbB level at entry; PbB terciles (median)=<5 (2.6 μg/dl), 5–9 (6.3 μg/dl), ≥10 (11.8 μg/dl)	All-site cancer mortality RR = 1.69 (1.14–2.52) at ≥10 μg/dl; RR = 1.44 (1.12–1.86) at 5–9 μg/dl	Schober et al. (2006)
U.S. NHANES III cohort (1988–1994): adults ≥20 years old, $N = 13,946$	All-cause and cause-specific mortality versus PbB tercile	Single PbB level at entry; PbB terciles: ≥3.6 μg/dl, 1.9–3.6 μg/dl, <1.9 μg/dl	No association of any PbB tercile with cancer mortality risks	Menke et al. (2006)

1988–1994. Summary data are presented in Table 17.5. The NHANES II analyses differed in a number of ways. As seen in Tables 2 and 3 of Lustberg and Silbergeld (2002), these researchers did more confounder adjustments, and extended analysis to individuals other than just Whites.

Both analyses employed the same exposure database, single PbB measurements for individuals entering the study. Median PbB levels for the subgroups were similar, 12 and 13 μg/dl for the Jemal et al. and Lustberg and Silbergeld groups, respectively. The NHANES design is complex, as noted elsewhere in this monograph, and a complex cancer risk design is attached to it. For example, in the Jemal et al. approach, Cox proportional hazard regression models were used to ascertain dose–response relationships across (log-transformed) quartiles for all-sites cancers and site-specific cancers.

Increases in risk for cancer mortality in the Jemal et al. report in terms of all-site and site-specific groups were noted at the highest quartile of PbB, but statistical significance for a dose–response pattern was only apparent for women in terms of all cancer sites. Lustberg and Silbergeld (2002) reported increases in all-site and lung cancer mortality risks across three ranges of PbB: <10, 10–19, and 20–29 μg/dl; however, no cancer outcome estimates achieved statistical significance. Capping the PbB range at 30 μg/dl in the latter paper (to limit the statistical skewing of estimates by occupational exposures) would potentially also have attenuated the strength of the association in the highest PbB band. The association of smoking with PbB also introduced a confounding factor for the lung cancer estimates.

Schober et al. (2006) and Menke et al. (2006) analyzed the U.S. NHANES III database for cancer mortality risks using somewhat different statistical designs. Schober et al. only included those individuals ≥40 years of age ($N = 9{,}686$; mortality followed through February 2000), while Menke et al. employed a much broader age band projecting to an age threshold of 20 years or older ($N = 13{,}946$). As with the NHANES II data analyses above, both studies based Pb exposure on the single PbB measurements gathered at study entry. Mortality risks were stratified via PbB terciles. Schober et al. employed as terciles in PbB: <5, 5–9, and ≥10 μg/dl. Menke et al. stratified into <1.9, 1.9–3.6, and ≥3.6 μg/dl.

Schober et al. (2006) reported for the highest tercile, RR = 1.69 (CI = 1.14–2.52), and for the mid-tercile, 5–9 μg/dl, 1.44 (CI = 1.12–1.86). By contrast, Menke et al. found no association of mortality risks with any tercile of PbB. The outcome differences likely traced in part to the survey cohorts in the analyses. For example, the age band differences were considerable.

17.5 EXPERIMENTAL ANIMAL STUDIES OF Pb CARCINOGENICITY

A number of studies on lead and cancer using various animal models of Pb carcinogenicity and genotoxicity have been published. A significant number have been described in older reports where methodologies relating to the topic were limited in histopathology, genotoxicity, and statistical methods. A number of these older data are presented in compendia such as the 1986

EPA lead criteria document (U.S. EPA, 1986, Ch. 12). Where possible, newer data are the focus of this section of the chapter and these data are also reviewed in U.S. EPA (2006), U.S. ATSDR (2007), NIH/NTP (2004), and IARC (2006). Mammalian test systems employed with respect to Pb carcinogenicity have principally been confined to rats and mice, with isolated reports of data from other species such as rabbits and nonhuman primates. Both male and female animals of both species were used and produced positive responses, indicating absence of any gender- or species-specific toxicokinetic or metabolic factor operating.

While a number of the earlier experimental animal lead and cancer studies entailed parenteral dosings (see IARC, 2006) via intravenous and intraperitoneal administration, the most relevant routes of dosing would be ingestion or inhalation. Table 17.6 presents illustrative data for tumorigenic results of Pb dosings of male and female rats and mice given various lead compounds orally via diet or drinking water. Various studies in the table and in the overall experimental animal literature on Pb and cancer differ in sizes of the dosing groups, with some raising questions whether enough animals were used to minimize false negatives, i.e., whether the study had sufficient statistical power. Specifically, as the dosing number increases, smaller incidence rates can be detected. A dosing count of 50 animals per group will detect an incidence rate of at least 8%, much lower than that with 25 animals per group.

Mice administered soluble Pb salts in drinking water or diet showed various tumorigenic responses. Waalkes et al. (1995) found that male offspring of pregnant mice given Pb in drinking water through gestation and 4 weeks postpartum showed proliferative lesions in a significant dose–response fashion plus increased incidence of renal adenomas and occasional renal carcinomas. Waalkes et al. (2004) later reported that adult male mice of the wild type, in a study involving animals of the wild type or having the metallothionein-I/II knockout gene, showed a dose–response-related incidence of renal proliferative lesions and tumors when dosed for 112 weeks with Pb (acetate) in drinking water. Adult female mice, given Pb (acetate) in drinking water at 50 or 1,000 ppm, showed lymphocytic leukemic responses (Blakley, 1987).

Eight rat studies for animals exposed to Pb in diet or drinking water with or without coexposures or interventions are also summarized in Table 17.6. All of the selected rat studies showed significant renal tumorigenesis, with or without other exposures or interventions, and regardless of rat strain: Sprague–Dawley, Wistar, and Fischer. For example, in the reports of Kasprzak et al. (1985) and Bogden et al. (1991), Pb dosing alone produced significant percentages of renal tumors, but the rate increased even further when dietary Ca was increased. Some of the earlier studies showed tumor occurrence at sites other than the kidney. Oyasu et al. (1970) reported both cerebral gliomas and kidney tumors, while Zawirska (1981) noted lung and brain tumors (gliomas), as well as those of the kidney.

TABLE 17.6 Carcinogenic and Genotoxic Effects of Pb in Human Populations: Selected Experimental Animal Carcinogenesis Studies of Pb

Test Animal	Dosing Design	Results	References
A. Mice			
Female Swiss albino mice, 8 weeks old, $N = 42$–46/group	Pb acetate in drinking water at 0, 50, 1,000 μg/ml	Lymphocytic leukemia induced by Pb exposure	Blakley (1987)
Pregnant C57Bl/6NCr mice, $N = 10$–15/group	Pb acetate in drinking water: 0, 500, 750, 1,000 ppm; dosing from gestation to birth, continuing 4 weeks postpartum; pups observed 112 weeks ± 500 ppm Na barbital	Male offspring showed significant dose–response relationship in proliferative lesions and incidence of renal adenomas; chronic nephropathy not present	Waalkes et al. (1995)
Adult male wild-type or metallothionein-I/II knockout mice	Pb acetate in drinking water: 0, 1,000, 2,000, 4,000 ppm; started at 8 weeks, dosed for 112 weeks	Wild-type mice showed a dose–response relationship in renal proliferative lesions and tumors	Waalkes et al. (2004)
B. Rats			
Male Sprague–Dawley rats, 5–8 weeks old	Pb subacetate in diet, 1% ± 2 acetyl-aminofluorine (2-AAF), 12–17 months	Pb groups (±2-AAF) showed significant increase in cerebral gliomas and kidney tumors	Oyasu et al. (1970)
Male Wistar rats, 6–8 weeks old, $N = 11$–13	Pb subacetate in diet, 1.5%, 23 or 48 weeks; with/without unilateral nephrectomy	Intact animals fed Pb showed 9/11 kidney tumor incidence at 48 weeks; 64% adenomas, 36% carcinomas	Ito (1973)
Male Sprague–Dawley rats, $N = 28$–30/group	Pb subacetate in diet, 1%±varying levels of Ca, dosed 79 weeks	Pb exposure alone produced significant rate of renal tumors, 13/29 (45%); increasing Ca increased tumor rate	Kasprzak et al. (1985)

(Continued)

TABLE 17.6 Carcinogenic and Genotoxic Effects of Pb in Human Populations: Selected Experimental Animal Carcinogenesis Studies of Pb—(cont.)

Test Animal	Dosing Design	Results	References
Male, female Wistar rats, $N = 47$ each sex; 31 weeks of age	Pb acetate in diet, 3 mg/day, 60–504 days, controls; observed to 800 days	Pb exposure produced 102 tumors: 12 rats with kidney adenomas, 15 with lung adenomas, 10 with brain gliomas, and tumors at other sites: dosing time determined kidney tumor incidence	Zawirska (1981)
Male, female Wistar rats, $N = 15–20$ males, 19–26 females, age 2–2.5 months	Pb acetate in diet ± sulfathiazole; 3 mg Pb acetate/day, 18 months + 7 additional observation months	Pb acetate alone produced 14 renal tumors in males, with 5 carcinomas, and 1 carcinoma in females	Waszynski (1977)
Male Sprague–Dawley rats, $N = 7–16$/group	Pb acetate (26 or 2,600 ppm) in drinking water, total of 76 weeks; at 28 weeks, sodium nitrite and ethyl urea added in some groups	The 2,600 ppm Pb group alone showed 13/16 tumor rate, as tubule carcinomas	Koller et al. (1985)
Male, female Fischer 344 rats, $N = 24$ each sex	Pb acetate (500, 2,000, 8,000 ppm) in diet, up to 725 days, ± other carcinogens	The 2,000 and 8,000 ppm Pb dosings alone produced significant increases in malignant kidney tumors; no interactions seen with other carcinogens	Fears et al. (1989)
Male Wistar rats ($N = 48$), 5 weeks old, in six groups	Pb in drinking water: 0, 1.0, 100 μg Pb/ml plus high (4%) or low (0.2%) Ca in diet	All rats fed 4% Ca and 100 μg/ml Pb had renal hyperplasia or invasive transitional cell carcinoma, but none at low Ca	Bogden et al. (1991)

17.6 HUMAN GENOTOXICITY STUDIES OF LEAD IN OCCUPATIONAL AND GENERAL POPULATIONS

Recent years have seen a spate of human genotoxicity studies attempting both to rationalize the mixed findings for Pb carcinogenesis in human populations and to exploit a number of the major advances in genotoxicity methodologies. Data from mainly newer studies using more specific and sensitive techniques are of particular interest. For logistical reasons, genotoxicity studies in workers and the general population entail small sample sizes, in contrast to high counts of individuals in cohort and case–control studies using as gross outcomes cancer incidences and mortality.

Genotoxic responses, i.e., those associated with genotoxic chemical carcinogens, are those that produce DNA damage directly or indirectly. Direct damage occurs by mechanisms such as damage to DNA via adduct formation or broken DNA strands. Indirect mechanisms include inhibition of synthesis and repair, and interactions with binding proteins and tumor suppressor proteins. Mechanisms for nongenotoxic carcinogens can operate via some combination of promoting carcinogenesis by genotoxic substances, impaired control of gene expression, and dysregulation of transcription and signal transduction.

For Pb, a number of recent studies have indicated that the element is genotoxic, although exact MOA are difficult to define. One issue is the predictive relationship of genotoxic markers to eventual actual cancer incidence or to cancer-associated mortality. One may have clearly positive associations of genotoxic biomarkers with carcinogen or genotoxin exposures but still have a problematic predictor for cancer occurrence.

At present, it appears that for carcinogens in general and carcinogens encountered in various occupational settings, chromosomal aberrations (CAs) indexed as either chromosome-type (CSAs) or chromatid-type (CTAs) aberrations are the most robust markers (Hagmar et al., 2004; Rossner et al., 2005) of carcinogens. These testings employ peripheral lymphocytes as the cellular and subcellular surrogate for like effects in organs and tissues. CTAs arise from the action of chemical clastogens and UV light via base alterations, cross-links, and single-strand breaks. Double-strand breaks are associated with lesioning agents such as ionizing radiation.

Genotoxicity when applied to human populations raises a number of dose–response issues. There is the question of cytotoxicity versus genotoxicity when using large treatment concentrations in *in vitro* or *in vivo* experimental systems, and then the question of relevance of findings for highly exposed Pb workers to general populations exposed to much lower environmental levels of Pb. Illustrative findings in the more recent literature for genotoxic endpoints, i.e., clastogenic effects, DNA damage, and micronuclei (MN) formation rates, are presented in Table 17.7.

DNA damage assessment in Pb-exposed workers with more current techniques uses the Comet assay via single cell gel electrophoresis (SCGE), and

TABLE 17.7 Carcinogenic and Genotoxic Effects of Pb in Human Populations: Selected Human Genotoxicity Studies

Study Group	Design Details	Pb Exposure	Results	References
A. DNA Damage: workers				
Chinese secondary Pb smelter workers ($N = 46$) versus 28 controls	SCGE/Comet assay	PbB medians <13 to >37 μg/dl	Statistically significant increase in tail length ($p < 0.05$) in dose–response fashion	Ye et al. (1999)
Colombian battery plant workers ($N = 43$) versus 13 controls	SCGE/Comet assay	Mean = 98.5 μg/dl workers; 5.4 μg/dl controls	Significant increase in tail length, $p < 0.05$	DeRestropo et al. (2000)
Italian battery plant workers ($N = 37$) versus 29 controls	SCGE/Comet assay	Mean = 39.6 μg/dl workers; 4.4 μg/dl controls	Significant dose–response for increase in tail moment, $p = 0.011$	Fracasso et al. (2002)
Polish battery plant workers ($N = 44$) versus 40 controls	SCGE/Comet assay	Mean = 50 μg/dl workers; 5.6 μg/dl controls	Increased % of cells with increased tail length ($p < 0.05$)	Palus et al. (2003)
Indian secondary Pb smelter workers ($N = 45$) versus 36 controls	SCGE/Comet assay	Mean = 24.8 μg/dl workers; 2.8 μg/dl controls	Significant increase in % cells with increased tail length, $p < 0.05$; PbB and exposure duration correlated to DNA damage	Devi et al. (2003)
Taiwanese battery plant workers ($N = 23$, high Pb; $N = 34$, low Pb) versus 30 controls	DNA–protein cross-links	PbB: high = 32.5 μg/dl; controls 4.2 μg/dl	Percent cross-links significantly increased in high Pb group ($p < 0.05$)	Wu et al. (2002)

(Continued)

TABLE 17.7 Carcinogenic and Genotoxic Effects of Pb in Human Populations: Selected Human Genotoxicity Studies—(cont.)

Study Group	Design Details	Pb Exposure	Results	References
B. MN percentage: workers				
Bulgarian battery plant workers ($N = 103$) versus 78 controls	Frequency (%) of MN cells	Mean PbB: workers = 56 μg/dl; controls = 19 μg/dl	Significant increase (43%, $p < 0.001$) in MN cells in workers	Vaglenov et al. (2001)
Polish battery plant workers ($N = 44$) versus 52 controls	MN assay	PbB: 28.2–65.5 μg/dl workers; 1.7–18 μg/dl controls	Higher % of MN cells in Pb workers, 18.6%, versus 6.6%, controls ($p < 0.01$)	Palus et al. (2003)
C. SCE and CAs: worker, general groups				
Healthy Czech Republic residents ($N = 11{,}834$) with cytogenetic records, 1975–2000	Lymphocyte CA evaluation: cancer cases in the national registry ($N = 485$) versus the cytogenetic records, various occupational backgrounds	–	Significant association between overall cancer incidence and CSAs but not CTAs; stomach cancer strongly linked with total CAs	Rossner et al. (2005)
Healthy Nordic and Italian individuals: ($N = 1{,}573$, CA; 877 CTA/CSA)	Healthy subjects, cytogenetically tested and then followed for 17 years (median)	–	High levels of CAs associated with increased total cancer incidence in Nordic and total cancer mortalities in Italians	Hagmar et al. (2004)
Mexican male painters ($N = 25$) and matched controls	SCE in peripheral blood lymphocytes	PbB: 50.4 μg/dl workers; 5.6 μg/dl controls	Significant increase in SCE ($p < 0.05$) versus controls	Pinto et al. (2000)

Polish battery plant workers ($N = 30$) versus 43 controls	SCE in lymphocytes	–	Significant increase in SCE ($p < 0.05$) versus controls	Palus et al. (2003)
Turkish battery plant workers ($N = 71$) versus 20 controls	SCE in lymphocytes	PbB: 34.5 μg/dl workers; 10.4 μg/dl controls	Significant increase in SCE ($p < 0.05$) for workers with PbB >50 μg/dl	Duydu and Süzen (2003)
Male Thai police officers	SCE levels	PbB mean = 43.5 μg/dl; stratified into two PbB groups; cutoff = 49.7 μg/dl	Higher PbB group showed higher SCE levels	Wiwanitkit et al. (2008)
Taiwanese battery plant workers ($N = 23$ high exposure, $N = 34$ low exposure) versus 30 controls	SCE in lymphocytes	PbB: high exposure = 32.5 μg/dl; controls = 4.2 μg/dl	Significant increase in SCE of high-exposure workers ($p < 0.01$) versus controls	Wu et al. (2002)
Chinese battery plant workers ($N = 21$ total: seven highest exposures) versus seven controls	SCE in lymphocytes	PbB: highest exposure = 86.9 μg/dl; control = 7.8 μg/dl	Significant increase in SCE ($p < 0.001$) of highest exposed versus controls	Huang et al. (1988)
Italian battery plant workers ($N = 11$) also serving as preemployment controls prospectively	CAs	Postemployment PbB = 45 μg/dl; preemployment PbB = 34 μg/dl	Significant increase in CAs ($p < 0.05$)	Forni et al. (1976)
Chinese battery plant workers ($N = 21$ total, seven highest exposures) versus seven controls	CAs	PbB highest exposure = 86.9 μg/dl; control 7.8 μg/dl	Significant increase in CAs ($p < 0.01$) versus controls	Huang et al. (1988)

(*Continued*)

TABLE 17.7 Carcinogenic and Genotoxic Effects of Pb in Human Populations: Selected Human Genotoxicity Studies—(cont.)

Study Group	Design Details	Pb Exposure	Results	References
D. Genotoxicity in general populations				
Flemish women ($N = 99$) living near Antwerp	HPRT gene variant frequencies versus urbanization and PbB	–	HPRT gene mutation rates correlated with PbB levels	Van Larebeke et al. (2004)
Residents of Bremen, Germany ($N = 141$)	Oxidative DNA damage	Median PbB = 4.6 μg/dl	No increase in damage	Merzinich et al. (2001)
Children in Milan, Italy, near smelter ($N = 19$) versus 12 controls	SCE	PbB near smelter = 29.3–62.7 μg/dl; PbB controls = 10.0–21.0 μg/dl	No change in SCE	Dalprá et al. (1983)
Children from Lagunera, Mexico: 6–11 years old, $N = 65$	DNA baseline and repair damage with/without peroxide challenge	PbB = 65% >10 μg/dl	No association of PbB and DNA repair ability	Méndez-Gómez et al. (2008)
Polish children, 5–14 years old, $N = 74$	SCE and MN levels	PbB mean = 7.7 μg/dl; range 2.7–23.0	No association of PbB with SCE but association with MN levels ($p = 0.007$)	Mielzynska et al. (2006)
Epigenetic study: subset of the NAS cohort, Boston, MA; $N = 517$	DNA methylation markers: Alu, long-interspersed nuclear element (LINE-1)	PbB = 4.1 μg/dl; patellar Pb = 27.4 ppm; tibial Pb = 20.5	Patellar Pb inversely associated with LINE-1; no association with either methylation marker for tibia Pb, PbB	Wright et al. (2010)
Epigenetic study: mother–infant pairs, Mexican early life exposure study, $N = 103$; cord blood samples	Alu, LINE-1 DNA methylation markers in umbilical cord blood	Maternal tibia Pb = 10.5 ppm; patella Pb = 12.9 ppm	Association of patella Pb and LINE-1; tibial Pb with Alu. No link of cord blood Pb with either methylation marker	Pilsner et al. (2009)

characterizes such morphological features as the % frequency of cells with increased tail length/moment. Sample sizes are small and controls are employed. Despite limited numbers of individuals, the exposure differences in terms of PbB versus control measurements are great, while the outcomes summarized in Table 17.7 have typically produced statistically significant differences in the Comet assays. This approach has been employed in worker exposures in China, Taiwan, India, Poland, and Italy. Some studies have employed multiple methods, e.g., DNA damage, MN formation, and clastogenic effects.

Collectively, DNA damage findings were observed across a wide range of Pb exposures and worker groups, so that one could discern positive outcomes at relatively moderate PbB measures, at least in terms of typically elevated exposures. Regarding the highest relative exposures, ≥40 μg/dl, Colombian battery plant workers, with a mean PbB of 98.5 μg/dl, showed a statistically significant increase in the results versus controls with a mean PbB about 20-fold lower, i.e., 5.4 μg/dl (DeRestropo et al., 2000). Palus et al. (2003) noted that a group of Polish battery plant workers having a mean PbB of 50.4 μg/dl, versus controls with PbB an order of magnitude lower (5.6 μg/dl), showed a statistically significant increase in the percentage of cells with increased tail length. Ye et al. (1999) saw statistically significant increases, in dose–response fashion, in tail length in samples for the Comet assay in Chinese secondary smelter workers whose PbB medians ranged from <13 to >37 μg/dl.

Lower worker exposures were still associated with positive findings in the Comet assay for DNA damage. Italian battery plant workers, mean PbB 39.6 μg/dl, showed statistically significant increases in tail moment versus controls (mean 4.4 μg/dl). Indian secondary smelter workers (Devi et al., 2003), with mean PbB 24.8 μg/dl, showed correlations of Comet assay results with both PbB and duration of Pb exposure. Taiwanese battery plant employees (Wu et al., 2002), with mean PbB value of 32.5 μg/dl in the higher Pb exposure group, showed a significant relationship of PbB with the percent of DNA–protein cross-links.

A second test of genotoxicity in Pb workers quantified the frequency (%) of cells with MN as a function of Pb exposure. Bulgarian battery plant workers studied by Vaglenov et al. (2001) showed a statistically significant increase in MN cells versus controls, with workers having a mean PbB of 56 μg/dl and controls 19 μg/dl. Similarly, Palus et al. (2003) noted that Polish battery plant workers with a PbB range of 28.2–65.5 μg/dl had an almost threefold higher percentage of MN (18.6% versus 6.6% in controls), a statistically significant difference ($p < 0.01$).

Chromosomal damage indices in Pb worker and general population peripheral lymphocytes, indexed as CAs and sister chromatid exchanges (SCEs), have been noted by a number of researchers for workers from around the world. Mexican painters (Pinto et al., 2000) and battery plant workers from Poland (Palus et al., 2003), Taiwan (Wu et al., 2002), China (Huang et al., 1988), and Turkey (Duydu and Süzen, 2003) showed statistically significant increases in SCE in Pb workers versus controls. PbB mean values ranged greatly, down to mean values <40 μg/dl. Italian (Forni et al., 1976) and Chinese (Huang et al., 1988) investigators noted findings of significantly increased CAs in battery factory workers across a range of PbB levels. These are illustrative of more recent findings indicating that this particular biomarker, when taken with its apparent higher predictive power of subsequent cancer induction versus other damage markers, suggests that CAs are significantly linked to Pb and also significantly associated with carcinogenic endpoints (Hagmar et al., 2004; Rossner et al., 2005).

Five studies, two involving children, probed whether Pb exposure is associated with genotoxic responses in general populations. The adult studies were based on residents of Antwerp, Belgium (Van Larebeke et al., 2004), Bremen, Germany (Merzinich et al., 2001), and Milan, Italy (Dalprá et al., 1983), while the child studies occurred in Mexico (Méndez-Gómez et al., 2008) and Poland (Mielzynska et al., 2006). The Flemish subjects in the Antwerp study presented with HPRT (hypoxanthine phosphoribosyl transferase) gene mutation rates that correlated positively with PbB levels. The German subjects in Bremen, however, showed no responses to ambient Pb exposures when using oxidative DNA damage as the injury marker. Milanese residents living near a Pb smelter showed no change in SCE rates with PbB up to 62.7 μg/dl. In general, except for the Flemish results, which require confirmation, genotoxic markers in adult general populations having relatively low or moderate Pb exposures are little changed across PbB concentrations.

Mexican children (Lagunera, 6–11 years old, $N = 65$) with PbB >10 μg/dl = 65% showed no link of PbB with DNA damage or repair ability (Méndez-Gómez et al., 2008). Mielzynska et al. (2006) reported that Polish children 5–14 years of age ($N = 74$) displayed an association of PbB with MN level ($p = 0.007$).

Two epigenetic studies have recently been described, one for Mexican mother–infant pairs (Pilsner et al., 2009) and one for U.S. elderly subjects, selected from the Normative Aging Study (Wright et al., 2010). In the Wright et al. study, there was no association of tibia Pb or PbB with either the Alu or LINE-1 methylation markers. Patellar Pb was inversely associated with LINE-1. These are summarized in Table 17.7. There was no linkage of cord PbB with either test of DNA methylation in the Mexican children, but an association was noted for maternal patellar Pb and the LINE-1 marker.

17.7 GENOTOXICITY STUDIES OF LEAD EXPOSURE IN *IN VIVO* EXPERIMENTAL ANIMAL SYSTEMS

The relationships of Pb exposures of experimental animals to *in vivo* genotoxic responses have been investigated under a variety of experimental conditions: animal species, mode of administration, chemical form of the lead compound, and diverse endpoints. Table 17.8 summarizes findings for

TABLE 17.8 Carcinogenic and Genotoxic Effects of Pb in Human Populations: Selected *In Vivo* Experimental Genotoxicity Studies

Test Animal	Dosing/Design	Results	References
A. In vivo studies: oral or inhalation dosing			
CD-1 male mice	DNA damage: SCGE/Comet assay. Pb inhalation, 6,800 μg/m^3: 60 minutes × 2/week, 4 weeks	Weakly positive DNA damage seen in liver, kidney, nasal cavity, brain, bone marrow cells; negative data for testicular cells	Valverde et al. (2002)
Swiss male mice, 9–12 weeks old	CAs: bone marrow and spermatocytes; Pb by gavage, 200 or 400 mg/kg Pb acetate. Ca at 40 or 80 mg/kg 2 weeks post-Pb exposures	Chromosomal damage at both Pb dosing levels in bone marrow and spermatocytes; Ca prevented CA	Aboul-Ela (2002)
Kunming female mice, multigenerations: second, third	DNA damage: SCGE assay; drinking water: 1 μg/ml *in utero* to sexual maturity	Positive evidence, DNA damage for second, third generations	Yuan and Tang (2001)
Swiss Albino mice, 4 weeks old	Lymphocyte single-strand breaks; Pb (nitrate) by gavage: 24, 48, 72 hours; 1 or 2 weeks	Non-dose-dependent increased rate of single-strand breaks	Devi et al. (2000)
Male C57B1 mice	CAs, bone marrow; 0.5% Pb, one month, low or normal diet Ca	Increased CAs in low but not regular diet plus Pb feeding	DeKnudt and Gerber (1979)
Swiss albino mice, 7–8 weeks old	CAs, bone marrow; co-exposed with zirconium salt, 110 or 220 mg/kg	Pb (nitrate) produced dose-related CAs in marrow; giving Pb and zirconium salt together increased CAs	Dhir et al. (1992a)
NMRI mice	Dominant lethal mutations. Pb (chloride): 1.3 mg/ml drinking water	No mutations linked to Pb	Kristensen et al. (1993)

(*Continued*)

TABLE 17.8 Carcinogenic and Genotoxic Effects of Pb in Human Populations: Selected *In Vivo* Experimental Genotoxicity Studies—(cont.)

Test Animal	Dosing/Design	Results	References
Wistar male rats, 30 days old	CAs; Pb (acetate), 10 mg/kg, 5×/ week, 4 weeks, co-administered cypermethrin	Pb acetate induced increase in aneuploidy and % damaged cells; both agents together increased structural aberrations	Nehez et al. (2000)
Sprague–Dawley rats, unilateral nephrectomy	DNA damage, SCGE, and MN formation. Pb: 78 mg/kg orally × 3	Positive evidence of DNA damage and MN formation	Robbiano et al. (1999)
B. In vivo studies: parenteral dosing			
Male Swiss mice, 9–12 weeks old	Chromosomal damage, SCE in bone marrow, spermatocytes; i.p. Pb acetate, 25–400 mg/kg, single or up to 7 daily injections	Dose-dependent chromosomal damage at single or repeated doses in bone marrow; SCE changes at 50, 100 mg/kg	Fahmy (1999)
Swiss albino mice, 6–8 weeks old	MN formation in bone marrow; 0.63–80 mg/kg Pb (nitrate), 12, 24, or 36 hours	Pb induced MN but not in dose-dependent fashion; more in males	Jagetia and Aruna (1998)
Swiss albino mice, 6–8 weeks old	SCE in bone marrow; 10, 20, 40 mg/kg Pb (nitrate), i.p., 24 hours; with/without ascorbic acid or fruit extract	Pb alone induced increased SCE dose-dependently; extract and ascorbic acid reduced damage	Dhir et al. (1993)
Swiss albino mice, 8 weeks old	Chromosome damage in bone marrow; 5, 10, 20 mg/kg Pb (nitrate), i.p., 24 hours; given with/ without ferric chloride	Pb increased chromosomal damage in dose–response fashion; iron did or did not alter damage based on dose timing	Dhir et al. (1992b)

(*Continued*)

TABLE 17.8 Carcinogenic and Genotoxic Effects of Pb in Human Populations: Selected *In Vivo* Experimental Genotoxicity Studies—(cont.)

Test Animal	Dosing/Design	Results	References
Algerian mice, $N = 6$, dosed groups or controls	SCE levels in bone marrow; PbOac, i.p. injections, 5 × or 10 ×, 0.46 mg Pb acetate/kg; alone or with Cd or Zn	Pb alone, increased SCE; further increase with higher time interval; further SCE increase with Cd or Zn	Tapioso et al. (2009)
Female and male Sprague–Dawley rats	CAs in bone marrow; 104 mg/kg Pb, i.p.	Pb increased CAs in females but not males	Tachi et al. (1985)

investigative reports in the more recent literature, using test methods that are more sophisticated, sensitive, and specific for genotoxic responses than approaches in the older literature. Table 17.8 focuses first on oral or inhalation exposures, the routes principally relevant to human Pb exposure, followed by parenteral administration data. Various animal species have been used, but exposure results in various strains of rats and mice comprised the majority of the relevant literature. Various genotoxic endpoints, using methods also employed in human studies, have been sought including DNA damage using the Comet/SCGE assay or single-strand breaks, frequencies of MN cells, clastogenic effects, and gene mutations.

Several strains of mice given Pb orally at various stages, including *in utero*, postnatal, and next-generation treatments, have shown genotoxic evidence of DNA, chromosomal, and other damage (Table 17.8). Yuan and Tang (2001) noted that female Kunming mice exposed to water Pb from *in utero* and continued through second and third generations showed DNA damage for both generations, using the Comet assay. Adult Swiss Albino mice administered Pb by gavage gave evidence of either chromosomal damage at dosings of 200 or 400 mg Pb (acetate)/kg in bone marrow or spermatocytes (Aboul-Ela, 2002), or DNA damage in the form of increased rate of single-strand breaks (Devi et al., 2000).

Genotoxicity results appear to be affected by dietary cofactors or co-exposures. The study of Aboul-Ela (2002) showed that any clastogenic effects were prevented by increased dietary Ca, a finding consistent with the older results noted by DeKnudt and Gerber (1979). The latter showed there were CAs in male C57B1 mice bone marrow in the low diet Ca but not the high Ca groups. Adult Swiss Albino mice given both zirconium salt and Pb nitrate in diet (0.5% Pb; 110, 220 mg/kg Zr) produced dose-related CAs in bone marrow at a higher rate than with Pb alone (Dhir et al., 1992a). No studies appeared to be positive for dominant lethal mutations in mice exposed to lead (Kristensen et al., 1993).

Valverde et al. (2002) showed that adult CD-1 mice exposed to Pb by inhalation (6,800 $\mu g/m^3$; 60 minutes × 2/weeks, 4 weeks) and examined using the Comet assay showed weakly positive DNA damage in liver, kidney, and other tissues.

Rat studies of experimental Pb genotoxicity showed Pb-associated genotoxic effects. Adult Sprague–Dawley rats dosed with either Pb alone (acetate, 10 mg/kg, 4 weeks) or with Pb plus cypermethrin showed increased aneuploidy and CAs with Pb alone and higher rates with dosing by both agents. DNA damage (Comet assay) and increased % MN cells were recorded in Sprague–Dawley rats fed Pb at 78 mg/kg.

Parenteral administration of Pb has also been associated with genotoxicity in mice and rats, mainly in the former. Effects were largely in the form of chromosomal damage. Adult mice, Swiss Albino strain, given injections of Pb at various dosing rates or frequency presented with chromosomal damage mainly in bone marrow and in the form of increased SCE rates (Dhir et al., 1992b, 1993; Fahmy, 1999; Jagetia and Aruna, 1998). Female and male Sprague–Dawley rats administered Pb i.p. showed increased CAs in females but not males (Tachi et al., 1985).

17.8 *IN VITRO* STUDIES OF Pb GENOTOXICITY USING CELLS IN CULTURE

A large part of the literature on the experimental evaluation of Pb genotoxicity has involved use of human and animal cell culture systems under various Pb contact conditions. The full range of genotoxicity testing methods for DNA damage, mutagenicity, and chromosomal damage and their results were described (Table 17.9). Mixed results in terms of positive or negative findings for both human and animal cell systems in culture were noted.

17.8.1 Human Cells in Culture

Use of various human cultured cell systems produced positive results for DNA damage in some reports (Robbiano et al., 1999; Wozniak and Blasiak, 2003) but not in others (Costa et al., 1996; Hartwig et al., 1990; Hartwig, 1994; Snyder and Lachmann, 1989). As a function of human cell or derived cell type, human Burkitt's lymphoma and HeLa cells produced negative data for DNA damage, while results were positive for primary fibroblasts, lymphocytes, and kidney cells. Overall, little evidence has been forthcoming that Pb contact conclusively produces mutagenic effects in human cell culture systems.

TABLE 17.9 Carcinogenic and Genotoxic Effects of Pb in Human Populations: Selected *In Vitro* Experimental Genotoxicity Studies

Test System	Dosing/Design	Results	References
A. Human cells in culture			
Human primary lymphocytes in RPMI 1640 without serum	DNA strand breaks and DNA–protein cross-links, Comet assay; Pb (acetate), 1–100 μM for 1 hour	Single-strand breaks maximum at 1 μM; double-strand breaks higher at all Pb levels; slight increase in DNA–protein cross-links	Wozniak and Blasiak (2003)
Human kidney cells	DNA strand breaks; Pb, 1.8 mM	Increased DNA strand breaks	Robbiano et al. (1999)
Human foreskin fibroblasts: DMEM + 10% FCS	Anchorage independence, 500–2,000 μM, 24 hours: catalase inhibitor, 3-aminotriazole, 80 mM	Dose-dependent increases in anchorage independence	Hwua and Yang (1998)
Human foreskin fibroblasts: DMEM + 10% FCS	HPRT mutagenicity assay; Pb (acetate): 500–2,000 μM, 24 hours	No mutagenicity observed	Hwua and Yang (1998)
Human Burkitt's lymphoma cells: EBV transformed in RPMI 1640 + 10% FCS	DNA–protein cross-links; Pb (nitrate): 1–10 mM for 6 hours	No DNA–protein cross-links induced by Pb	Costa et al. (1996)
Human keratinocytes, MEM + 2% FBS	HPRT mutagenesis assay; 0.1–100 μM/ml for 2–24 hours	Only indirect evidence cited as indicating mutagenesis	Ye (1993)
HeLa cells: AMEM + 5% FBS	DNA single-strand breaks; Pb (acetate), 500 μM, 20–25 hours	Pb did not induce single-strand breaks	Hartwig et al. (1990)
HeLa cells: HEPES/glucose buffer	DNA strand breaks; Pb (acetate), 100 μM, 0.5–4 hours	Pb did not induce strand breaks	Snyder and Lachmann (1989)

(*Continued*)

TABLE 17.9 Carcinogenic and Genotoxic Effects of Pb in Human Populations: Selected *In Vitro* Experimental Genotoxicity Studies—(cont.)

Test System	Dosing/Design	Results	References
B. Animal cells in culture			
Chinese hamster V79 cells: DMEM + 10% FCS	MN formation; Pb (acetate): 0.01–10 μM for 18 hours	Concentration-dependent modest increase in MN	Bonacker et al. (2005)
Chinese hamster V79 cells	MN formation; Pb (acetate): 0.05 μM	Increased MN formation	Thier et al. (2003)
Chinese hamster ovary cells: HBSS, then Ham's F12 + 5% FBS	Mutagenicity: gpt assay; Pb (chloride): 0.1–1.0 μM, 1 hour	Pb produced point mutations; more deletions at 0.5–1.0 μM	Ariza and Williams (1999)
Primary rat kidney cells	DNA strand breaks; Pb (acetate): 560 μM	Increased strand breaks	Robbiano et al. (1999)
Chinese hamster ovary AA8 cells: DMEM + 10% NCS + crown ethers (binding, uptake)	CAs; Pb (nitrate): 0.05–1 μM, 3–12 hours	No chromosomal damage	Cai and Arenaz (1998)
Chinese hamster ovary-derived AS52 cells: HBSS, then Ham's F12 + 5% FBS	Mutagenicity: gpt assay; Pb (chloride): 0.1–1.0 μM for 1 hour; xanthene oxidase inhibitor added	At 0.8, 1.0 μM, Pb was mutagenic; inhibitor reduced mutagenesis	Ariza et al. (1998)
Chinese hamster ovary K1 cells	Mutagenicity: Hprt locus; Pb (acetate): 0.5 mM	Pb at this dose was mutagenic	Yang et al. (1996)
Chinese hamster ovary-derived AS52 cells: HBSS, then Ham's F12 + 5% FBS	Mutagenicity: gpt assay; Pb (chloride): 0.1–1.0 μM for 1 hour	Pb as chloride produced TG-resistant mutants in dose-dependent fashion	Ariza and Williams (1996)
Chinese hamster ovary AA8 cells: AMEM + 10% FBS	CAs; Pb, 500–2,000 μM, 24 hours as either glutamate or nitrate	Pb glutamate induced damage at 1 mM; Pb nitrate did not induce damage	Wise et al. (1994)

(*Continued*)

TABLE 17.9 Carcinogenic and Genotoxic Effects of Pb in Human Populations: Selected *In Vitro* Experimental Genotoxicity Studies—(cont.)

Test System	Dosing/Design	Results	References
Chinese hamster ovary cells: EMEM + 10% FBS	CAs; SCE and MN formation; Pb (nitrate) 3–30 μM, 2 hours + 16 hours recovery	Pb did not induce CAs or MN formation; Pb did induce dose-dependent SCE increase	Lin et al. (1994)
Chinese hamster G-12-V79 cells with one copy gpt gene: Ham's F12 + 5% FBS	DNA strand breaks defined as nick translation; Pb (acetate): 1,700 μM, 5 days; mutagenesis-gpt: Pb (nitrate) 5–2,000 μM, 5 days	Pb did not induce single-strand breaks; Pb not mutagenic	Roy and Rossman (1992)
Chinese hamster V79 cells: AMEM + 10% FBS	Mutagenesis: HPRT; Pb (acetate) 0.5–5 μM, 44 hours; SCE: 1–10 μM Pb (acetate), 26 hours	Pb acetate was not mutagenic; Pb acetate alone did not induce SCE	Hartwig et al. (1990)

17.8.2 Experimental Animal Cells in Culture

Various animal cell types have been subjected to Pb treatments under various cell culture conditions. The most common experimental cell systems have been from primary or derived Chinese hamster cells, particularly ovary cells (CHO). Results have been both positive and negative across the various types of genotoxicity assessments. Lead was not mutagenic nor did it produce altered SCE rates in Chinese hamster V79 cells in the findings of Hartwig et al. (1990), but in a second study, Pb produced mutations in ovary cells over the Pb range of 0.01–10.0 μM (Ariza and Williams, 1999), produced mutagenesis in CHO-derived A522 cells at Pb levels up to 1.0 μM (Ariza et al., 1998), produced mutagenic effects in CHO-derived K1 cells (Yang et al., 1996), and produced TG-resistant mutants in a dose-dependent fashion (Ariza et al., 1998).

No chromosomal damage was reported by Lin et al. (1994) or Cai and Arenaz (1998). Increased MN from Pb contact were noted by Thier et al. (2003) and Bonacker et al. (2005). DNA damage was not observed by Roy and Rossman (1992) but was seen by Robbiano et al. (1999).

REFERENCES

Aboul-Ela, E.I., 2002. The protective effect of calcium against genotoxicity of lead acetate administration on bone marrow and spermatocyte cells of mice *in vivo*. Mutat. Res. 516, 1–9.

Ades, A.E., Kazantzis, G., 1988. Lung cancer in a non-ferrous smelter: the role of cadmium. Br. J. Ind. Med. 45, 435–442.

Anttila, A., Heikkila, P., Pukkala, E., Nykyri, E., Kauppinen, J., Hernberg, S., et al., 1995. Excess lung cancer among workers exposed to lead. Scand. J. Work Environ. Health 21, 460–469.

Anttila, A., Heikkila, P., Nykyri, E., Kauppinen, T., Pukkala, E., Hernberg, S., et al., 1996. Risk of nervous system cancer among workers exposed to lead. J. Occup. Environ. Med. 38, 131–136.

Ariza, M.E., Williams, M.V., 1996. Mutagenesis of AS52 cells by low concentrations of lead (II) and mercury (II). Environ. Mol. Mutagen. 27, 30–33.

Ariza, M.E., Williams, M.V., 1999. Lead and mercury mutagenesis: type of mutation dependent upon metal concentration. J. Biochem. Mol. Toxicol. 13, 107–112.

Ariza, M.E., Bijur, G.N., Williams, M.V., 1998. Lead and mercury mutagenesis: role of H_2O_2 superoxide dismutase, and xanthine oxidase. Environ. Mol. Mutagen. 31, 352–361.

Blakley, B.R., 1987. The effect of lead on chemical- and viral-induced tumor production in mice. J. Appl. Toxicol. 7, 167–172.

Bogden, J.D., Gertner, S.B., Kemp, F.W., McLeod, R., Bruening, K.S., Chung, H.R., 1991. Dietary lead and calcium: effects on blood pressure and renal hyperplasia in Wistar rats. J. Nutr. 121, 718–728.

Bonacker, D., Stoiber, T., Bohm, K.J., Prots, L., Wang, M., Unger, E., et al., 2005. Genotoxicity of inorganic lead salts and disturbance of microtubule function. Environ. Mol. Mutagen. 45, 346–353.

Cai, M.Y., Arenaz, P., 1998. Antimutagenic effect of crown ethers on heavy metal-induced sister chromatid exchanges. Mutagenesis 13, 27–32.

Carta, P., Aru, G., Nurchis, P., Cadeddu, C., Polizzi, M., Nieddu, V., et al., 2005. Study of mortality from specific causes in lead and zinc foundry workers in Sardinia. G. Ital. Med. Lav. Ergon. 27 (Suppl. 1), 43–45.

Cocco, P., Dosemeci, M., Heineman, E.F., 1998a. Brain cancer and occupational exposure to lead. J. Occup. Environ. Med. 40, 937–942.

Cocco, P., Dosemeci, M., Heineman, E.F., 1998b. Occupational risk factors for cancer of the central nervous system: a case–control study of death certificates from 24 U.S. States. Am. J. Ind. Med. 33, 247–255.

Costa, M., Zhitkovich, A., Gargas, M., Paustenbach, D., Finley, B., Kuykendall, J., et al., 1996. Interlaboratory validation of a new assay for DNA–protein crosslinks. Mutat. Res. 369, 13–21.

Dalprá, L., Tibiletti, M.G., Nocera, G., Giulotto, P., Auriti, L., Carnelli, V., et al., 1983. SCE analysis in children exposed to lead emission from a smelting plant. Mutat. Res. 120, 249–256.

Devi, K.D., Banu, B.S., Grover, P., Jamil, K., 2000. Genotoxic effect of lead nitrate on mice using SCGE (comet assay). Toxicology 145, 195–201.

Devi, K.D., Rozati, R., Saleha Banu, B., Hanumanth, R.P., Grover, P., 2003. DNA damage to workers exposed to lead using comet assay. Toxicology 187, 183–193.

DeKnudt, G., Gerber, G.B., 1979. Chromosomal aberrations in bone-marrow cells of mice given a normal or calcium-deficient diet supplemented with various heavy metals. Mutat. Res. 68, 163–168.

DeRestropo, H.G., Sicard, D., Torres, M.M., 2000. DNA damage and repair in cells of lead exposed people. Am. J. Ind. Med. 38, 330–334.

Dhir, H., Ghosh, S., Sharma, A., Talukder, G., 1992a. Interaction between IV metals—lead and zirconium—in bone marrow cells of *Mus musculis in vivo*. Biometals 5, 81–86.

Dhir, H., Sharma, A., Talukder, G., 1992b. Modifying effect of iron on lead-induced clastogenicity in mouse bone marrow cells. Biol. Trace Elem. Res. 34, 279–286.

Dhir, H., Roy, A.K., Sharma, A., 1993. Relative efficiency of *Phyllanthus emblica* fruit extract and ascorbic acid in modifying lead and aluminum-induced sister-chromatid exchanges in mouse bone marrow. Environ. Mol. Mutagen 21, 229–236.

Duydu, Y., Süzen, H.S., 2003. Influence of δ-aminolevulinic acid dehydratase (ALAD) polymorphism on the frequency of sister chromatid exchange (SCE) and the number of high-frequency cells (HFCs) in lymphocytes from lead-exposed workers. Mutat. Res. 540, 79–88.

Englyst, V., Lundström, N.G., Gerhardsson, L., Rylander, L., Nordberg, G., 2001. Lung cancer risks among lead smelter workers also exposed to arsenic. Sci. Total Environ 273, 77–82.

Fahmy, M.A., 1999. Lead acetate genotoxicity in mice. Cytologia 64, 357–365.

Fanning, D., 1988. A mortality study of lead workers, 1926–1985. Arch. Environ. Health 43, 247–251.

Fears, T.R., Elashoff, M., Schneiderman, M.A., 1989. The statistical analysis of a carcinogenic mixture experiment III. Carcinogens with different target organ systems. Aflotoxin B1, *N*-butyl-*N*-(4-hydroxybutyl) nitrosamine, lead acetate and thiouracil. Toxicol. Environ. Health 5, 1–23.

Forni, A., Cambiaghi, G., Secchi, G.C., 1976. Initial occupational exposure to lead: chromosome and biochemical findings. Arch. Environ. Health 31, 73–78.

Fracasso, M.E., Perbellini, L., Solda, S., Talamini, G., Franceschetti, P., 2002. Lead induced DNA strand breaks in lymphocytes of exposed workers: role of reactive oxygen species and protein kinase C. Mutat. Res. 515, 159–169.

Fu, H., Boffetta, P., 1995. Cancer and occupational exposure to inorganic lead compounds: a meta-analysis of published data. Occup. Environ. Med. 52, 73–81.

Gerhardsson, L., Hagmar, L., Rylander, L., Skerfving, S., 1995. Mortality and cancer incidence among secondary lead smelter workers. Occup. Environ. Med. 52, 667–672.

Hagmar, L., Stromberg, U., Bonassi, S., Hansteen, I.-L., Knudsen, L.-E., Lindholm, C., et al., 2004. Impact of types of lymphocyte chromosomal aberrations on human cancer risk: results from Nordic and Italian cohorts. Cancer Res. 64, 2258–2263.

Hartwig, A., 1994. Role of DNA repair inhibition in lead- and cadmium-induced genotoxicity: a review. Environ. Health Perspect. 102 (Suppl. 3), 45–50.

Hartwig, A., Schlepegrell, R., Beyersmann, D., 1990. Indirect mechanism of lead-induced genotoxicity in cultured mammalian cells. Mutat. Res. 241, 75–82.

Hu, J., La Vecchia, C., Negri, E., Chatenoud, L., Bosetti, C., Jia, X., et al., 1999. Diet and brain cancer in adults: a case–control study in northeast China. Int. J. Cancer 81, 20–23.

Huang, J., He, F., Wu, Y., Zhang, S., 1988. Observations on renal function in workers exposed to lead. Sci. Total Environ 71, 535–537.

Hwua, Y.S., Yang, J.L., 1998. Effect of 3-aminotriazole on anchorage independence and mutagenicity in cadmium- and lead-treated diploid human fibroblasts. Carcinogenesis 19, 881–889.

International Agency for Research on Cancer, 2006. IARC Monographs on the Evaluation of Carcinogenic Risks to Humans: Vol. 87. Inorganic and Organic Lead Compounds. World Health Organization, Lyon, France.

Ito, N., 1973. Experimental studies on tumors. Acta Pathol. Jpn. 23, 87–109.

Jagetia, G.C., Aruna, R., 1998. Effect of various concentrations of lead nitrate on the induction of micronuclei in mouse bone marrow. Mutat. Res. 415, 131–137.

Jemal, A., Graubard, B.I., Devesa, S.S., Flegal, K.M., 2002. The association of blood lead levels and cancer mortality among whites in the United States. Environ. Health Perspect. 110, 325–329.

Kasprzak, K.S., Hoover, K.L., Poirier, L.A., 1985. Effects of dietary calcium acetate on lead subacetate carcinogenicity in kidneys of male Sprague–Dawley rats. Carcinogenesis 6, 279–282.

Koller, L.D., Kerkvliet, N.I., Exon, J.H., 1985. Neoplasia induced in rats fed lead acetate, ethyl urea and sodium nitrite. Toxicol. Pathol. 13, 50–57.

Kristensen, P., Eilertsen, E., Einarsdóttir, E., Øvrebø, S., Haugen, A., 1993. Effect modification by inorganic lead in the dominant lethal assay. Mutat. Res. 302, 33–38.

Lin, R.H., Lee, C.H., Chen, W.K., Lin-Shiau, S.Y., 1994. Studies on cytotoxic and genotoxic effects of cadmium nitrate and lead nitrate in Chinese hamster ovary cells. Environ. Mol. Mutagen 23, 143–149.

Lundström, N.-G., Nordberg, G., Englyst, V., Gerhardsson, L., Hagmar, L., Jin, T., et al., 1997. Cumulative lead exposure in relation to mortality and lung cancer morbidity in a cohort of primary smelter workers. Scand. J. Work Environ. Health 23, 24–30.

Lundström, N.-G., Englyst, V., Gerhardsson, L., Jin, T., Nordberg, G., 2006. Lung cancer development in primary smelter workers: a nested case-referent study. J. Occup. Environ. Med. 48, 376–380.

Lustberg, M., Silbergeld, E., 2002. Blood lead levels and mortality. Arch. Intern. Med. 162, 2443–2449.

Méndez-Gómez, J., Garcia-Vargas, G.G., López-Carrillo, L., Calderón-Aranda, E.S., Gómez, A., Vera, E., et al., 2008. Genotoxic effects of environmental exposure to arsenic and lead on children in region Lagunera, Mexico. Ann. N.Y. Acad. Sci. 1140, 358–367.

Menke, A., Muntner, P., Batuman, V., Silbergeld, E.K., Guallar, E., 2006. Blood lead below 0.48 micromol/L and mortality among U.S. adults. Circulation 114, 1388–1394.

Merzinich, H., Hartwig, A., Ahrens, W., Beyersmann, D., Schlepegrell, R., Scholze, M., et al., 2001. Biomonitoring on carcinogenic metals and oxidative DNA damage in a cross-sectional study. Cancer Epidemiol. Biomarkers Prev. 10, 515–522.

Mielzynska, D., Siwinska, E., Kapka, L., Szyfter, K., Knudsen, L.E., Merlo, D.F., 2006. The influence of environmental exposure to complex mixtures including PAHs and lead on genotoxic effects in children living in Upper Silesia, Poland. Mutagenesis 21, 295–304.

National Toxicology Program, 2004. Lead and lead compounds. In: Report on Carcinogens: 11th Edition, Carcinogen Profiles. National Institute of Environmental Health Sciences, Research Triangle Park, NC, pp. III-158–III-162.

National Toxicology Program, 2011. Lead and lead compounds, In: Report on Carcinogens. 12th Edition, Carcinogen Profiles. National Institute of Environmental Health Sciences, Research Triangle Park, NC, pp. 251–256.

Nehez, M., Lorencz, R., Desi, I., 2000. Simultaneous action of cypermethrin and two environmental pollutant metals, cadmium and lead, on bone marrow cell chromosomes of rats in subchronic administration. Ecotoxicol. Environ. Saf. 45, 55–60.

Oyasu, R., Battifora, H.A., Clasen, R.A., McDonald, J.H., Hass, G.M., 1970. Induction of cerebral gliomas in rats with dietary lead subacetate and 2-acetylaminofluorene. Cancer Res. 30, 1248–1261.

Palus, J., Rydzynski, K., Dziubaltowska, E., Wyszynska, K., Natarajan, A.T., Nilsson, R., 2003. Genotoxic effects of occupational exposure to lead and cadmium. Mutat. Res. 540, 19–28.

Partanen, T., Heikkila, P., Hernberg, S., Kauppinen, T., Moneta, G., Ojajarvi, A., 1991. Renal cell cancer and occupational exposure to chemical agents. Scand. J. Work Environ. Health 17, 231–239.

Pesch, B., Haerting, J., Ranft, U., Klimpel, A., Oelschlagel, B., Schill, W., 2000. Occupational risk factors for renal cell carcinoma: agent-specific results from a case–control study in Germany. Int. J. Epidemiol. 29, 1014–1024.

Pilsner, J.R., Hu, H., Ettinger, A., Sanchez, B.N., Wright, R.O., Cantonwine, D., et al., 2009. Influence of prenatal lead exposure on genomic methylation of cord blood DNA. Environ. Health Perspect. 117, 1466–1476.

Pinto, D., Ceballos, J.M., Garcia, G., Guzman, P., Del Razo, L.M., Vera, E., et al., 2000. Increased cytogenetic damage in outdoor painters. Mutat. Res. 467, 105–111.

Robbiano, L., Carrozinno, R., Porta Puglia, C., Corbu, C., Brambilla, G., 1999. Correlation between induction of DNA fragmentation and micronuclei formation in kidney cells from rats and humans and tissue-specific carcinogenic activity. Toxicol. Appl. Pharmacol. 161, 153–159.

Rossner, P., Boffetta, P., Ceppi, M., Bonassi, S., Smerhovsky, Z., Landa, K., et al., 2005. Chromosomal aberrations in lymphocytes of healthy subjects and risk of cancer. Environ. Health Perspect. 113, 517–520.

Roy, N.K., Rossman, T.G., 1992. Mutagenesis and comutagenesis by lead compounds. Mutat. Res. 298, 97–103.

Sankila, R., Karjalainen, S., Pukkala, E., Oksanen, H., Hakulinen, T., Teppo, L., et al., 1990. Cancer risk among glass factory workers: an excess of lung cancer? Br. J. Ind. Med. 47, 815–818.

Schober, S.E., Mirel, L.B., Graubard, B.I., Brody, D.J., Flegal, K.M., 2006. Blood lead levels and death from all causes, cardiovascular disease, and cancer: results from the NHANES III mortality study. Environ. Health Perspect. 114, 1538–1541.

Sheffet, A., Thind, I., Miller, A.M., Louria, D.B., 1982. Cancer mortality in a pigment plant utilizing lead and zinc chromates. Arch. Environ. Health 37, 44–52.

Siemiatycki, J., Gerin, M., Dewar, R., Nadon, L., Lakhani, R., Begin, D., et al., 1991. Associations between occupational circumstances and cancer. In: Siemiatycki, J. (Ed.), Risk Factors for Cancer in the Work Place. CRC Press, Boca Raton, FL, pp. 141–145.

Silbergeld, E.K., 2003. Facilitative mechanisms of lead as a carcinogen. Mutat. Res. 533, 121–133.

Silbergeld, E.K., Waalkes, M., Rice, J.M., 2000. Lead as a carcinogen: experimental evidence and mechanisms of action. Am. J. Ind. Med. 38, 316–323.

Snyder, R.D., Lachmann, P.J., 1989. Thiol involvement in the inhibition of DNA repair by metals in mammalian cells. J. Mol. Toxicol. 2, 117–128.

Steenland, K., Boffetta, P., 2000. Lead and cancer in humans: where are we now? Am. J. Ind. Med. 38, 295–299.

Steenland, K., Selevan, S., Landrigan, P., 1992. The mortality of lead smelter workers: an update. Am. J. Public Health 82, 1641–1644.

Tachi, K., Nishimae, S., Saito, K., 1985. Cytogenetic effects of lead acetate on rat bone marrow cells. Arch. Environ. Health 40, 144–147.

Tapioso, J.T., Marques, C.C., Mathias, M.D., Ramalhinho, M.D., 2009. Induction of micronuclei and sister chromatid exchange in bone-marrow cells and abnormalities in sperm of Algerian mice exposed to cadmium, lead and zinc. Mut. Res. 678, 59–64.

Thier, R., Bonacker, D., Böhm, K.J., Wang, M., Unger, E., 2003. Interaction of metal salts with cytoskeletal motor protein systems. Toxicol. Lett. 140–141, 75–81.

U.S. Agency for Toxic Substances and Disease Registry, 2007. Toxicologic Profile for Lead. U.S. Department of Health and Human Services, Atlanta, GA.

U.S. Environmental Protection Agency, 2006. Air Quality Criteria for Lead. Vol. 1. Report No. EPA/600/R-05/144aF. National Center for Environmental Assessment, Washington, DC.

Vaglenov, A., Creus, A., Laltchev, S., Petkova, V., Pavlova, S., Marcos, R., 2001. Occupational exposure to lead and induction of genetic damage. Environ. Health Perspect. 109, 295–298.

Valverde, M., Fortoul, T.I., Diaz-Barriga, F., Majia, J., Del Castillo, E.R., 2002. Genotoxicity induced in CD-1 mice by inhaled lead: differential organ response. Mutagenesis 17, 55–61.

Van Larebeke, N., Koppen, G., Nelen, V., Schoeters, G., Van Loon, H., Albering, H., et al., Flemish Environment and Health Study Group 2004. Differences in HPRT mutant frequency among middle-aged Flemish women in association with area of residence and blood lead levels. Biomarkers 9, 71–84.

Waalkes, M.P., Diwan, B.A., Ward, J.M., Devor, D.E., Goyer, R.A., 1995. Renal tubular tumors and atypical hyperplasias in B6C3F1 mice exposed to lead acetate during gestation and lactation occur with minimum chronic nephropathy. Cancer Res. 55, 5265–5271.

Waalkes, M.P., Liu, J., Goyer, R.A., Diwan, B.A., 2004. Metallothionein-I/II double knockout mice are hypersensitive to lead-induced kidney carcinogenesis: role of inclusion body formation. Cancer Res. 64, 7766–7772.

Waszynski, E., 1977. Nonneoplastic and neoplastic changes in the kidneys and other organs of rodents fed lead acetate and sulfathiazole chronically. Pathol. Pol. 28, 101–111.

Wingren, G., Axelson, O., 1993. Epidemiologic studies of occupational cancer as related to complex mixtures of trace elements in the art glass industry. Scand. J. Work Environ. Health 19 (Suppl. 1), 95–100.

Wise Sr., J.P., Stearns, D.M., Wetterhahn, K.E., Patierno, S.R., 1994. Cell-enhanced dissolution of carcinogenic lead chromate particles: the role of individual dissolution products in clastogenesis. Carcinogenesis 15, 2249–2254.

Wiwanitkit, V., Suwansaksri, J., Suphan, S., 2008. White blood cell sister chromatid exchange among a sample of Thai subjects exposed to lead: Lead-induced genotoxicity. Toxicol. Environ. Chem. 90, 765–768.

Wong, O., Harris, F., 2000. Cancer mortality study of employees at lead battery plants and lead smelters, 1947–1995. Am. J. Ind. Med. 38, 255–270.

Wozniak, K., Blasiak, J., 2003. *In vitro* genotoxicity of lead acetate: induction of single and double DNA strand breaks and DNA–protein cross-links. Mutat. Res. 535, 127–139.

Wright, R.O., Schwartz, J., Wright, R.J., Bollati, V., Tarantini, L., Park, S.K., et al., 2010. Biomarkers of lead exposure and DNA methylation within retrotransposons. Environ. Health Perspect. 118, 790–795.

Wu, F.-Y., Chang, P.-W., Wu, C.-C., Kuo, H.-W., 2002. Correlations of blood lead with DNA–protein cross-links and sister chromatid exchanges in lead workers. Cancer Epidemiol. Biomarkers Prev. 11, 287–290.

Yang, J.-L., Yeh, S.-C., Chang, C.-Y., 1996. Lead acetate mutagenicity and mutation spectrum in the hypoxanthine guanine phosphoribosyl transferase gene of Chinese hamster ovary K1 cells. Mol. Carcinog. 17, 181–191.

Ye S.-H. 1993. Hypoxanthine phosphoribosyl transferase assay of lead mutagenicity on keratinocytes. Zhongguo Yaoli Xuebao 14, 145–147. Cited in Chapter 5, U.S. EPA, 2006.

Ye, X.B., Fu, H., Zhu, J.L., Ni, W.M., Lu, Y.W., Kuang, X.Y., et al., 1999. A study on oxidative stress in lead-exposed workers. J. Toxicol. Environ. Health A 57, 161–172.

Yuan, X., Tang, C., 2001. The accumulation effect of lead on DNA damage in mice blood cells of three generations and the protection of selenium. J. Environ. Sci. Health A. Tox. Hazard Subst. Environ. Eng. 36, 501–508.

Zawirska, B., 1981. The role of the kidneys in disorders of porphyrin metabolism during carcinogenesis induced with lead acetate. Environ. Res. 24, 391–408.

Chapter 18

Immunotoxic Effects of Lead in Human Populations

18.1 INTRODUCTION

Effects of Pb on the human immune system comprise a relatively newer literature, paralleling growth in the cellular and molecular biology of immunological phenomena. However, some of the frank manifestations of Pb immunotoxicity, historically in lead workers and taking such forms as impaired resistance to infections, have been known for decades. One can ascertain relative growth in the two literatures by examining how much of the environmental immunotoxicity database has appeared in the last 15 years versus what was available before 1995. The 1986 U.S. EPA lead criteria document described immunotoxic effects of lead largely as they were gathered from experimental animal systems (Chapter 12.8 of that document). That earlier document and its cited literature provided the general functional taxonomy of humoral and cell-mediated immunity and their corresponding subdivisions, as understood by the early to mid-1980s.

Immunotoxic effects of lead overlap with findings for additional endpoints in other sections of this book. For example, by the 1970s, it was known that immunology had a role in the neoplastic and tumorigenic effects of Pb in experimental animal systems when functioning as a tumor promoter or in oncogenesis stimulation. Gainer (1973, 1974) showed that CD-1 mice exposed to Pb had enhanced oncogenic response to Rauscher leukemia virus via immunosuppression, with splenomegaly and enhanced splenic virus titers. There was the long-known interactive effect of Pb enhancing the chemically induced tumorigenesis from benzo[a]pyrene given to Syrian hamsters (Kobayashi and Okamoto, 1974).

Immunotoxic effects of Pb also played a role earlier in the clinical history of the element through categories of effects which clearly subsumed a significant immunological component. For example, early individual and consensus compendia on lead toxicity and that of other chemical substances often mentioned "respiratory effects." A number of these effects had to do with altered host resistance to various inhaled pathogens and toxicants from

Trace Metals and other Contaminants in the Environment, Volume 10
ISSN: 1875-1121 DOI: 10.1016/B978-0-444-51554-4.00018-3

Pb-associated humoral and cell-mediated immunological effects. Removal of inhaled dust particles and pathogens via alveolar macrophages, for example, is impaired in the presence of inhaled Pb and is accompanied by mitochondrial and endoplasmic reticular damage to alveolar macrophages and type-1 alveolar epithelial cells (Bingham et al., 1968).

Lead is now known to affect both forms of immunological expression in human populations and experimental systems: humoral and cell-mediated immunology. It does so in complex ways that make for adverse responses identified at increasingly lower exposures. One critical aspect of Pb immunotoxicity in terms of this robust set of dose–response relationships is that it does not impart direct toxic effects which can be discerned histochemically or ultrastructurally but produces effects by disrupting the regular function of immunological components.

The Pb immunotoxicity focus in this chapter, in keeping with the intent of the book itself, is first on subsets of human populations who show Pb effects on health working through immunotoxic pathways and then on experimental data that support the adverse impacts demonstrated epidemiologically. This chapter is not intended to be a comprehensive compendium of Pb effects at the biomolecular and cellular immunology level but a discussion of the most relevant endpoints visibly affecting public health.

This chapter is organized into three major sections. The first concerns the relatively general immunotoxic outcomes in human populations and experimental animals, where Pb exposures have been studied for effects on host resistance, hypersensitivity, and potential autoimmunity. These general functional effects at the organismal level are a complex integration of both humoral and cell-mediated immunological responses. These latter two topics comprise the remaining two sections of the chapter.

Humoral and cell-mediated immunotoxic endpoints from Pb exposures are interconnected in complex ways. A division of labels is not intended to imply totally separate categories of immunological activity but is used for ease of characterizing eventual impacts on human populations and experimental test systems via recognized signs and symptoms. Each of the major sections presents data on human subjects followed by information for experimental systems examined first *in vivo* and then *in vitro*.

18.2 EFFECTS OF Pb ON HOST RESISTANCE, POTENTIAL HYPERSENSITIVITY, AND POTENTIAL AUTOIMMUNITY

18.2.1 Studies in Human Populations

The earliest evidence of immunotoxic effects of Pb in human populations arose for effects seen with the heaviest Pb exposures, those in lead workers. Resistance to viral, bacterial, and parasitic infections in workers, while the crudest marker of immunological effects, was the most readily discernible

TABLE 18.1 **Effects of Lead on Human Host Resistance, Hypersensitivity, and Potential Autoimmunity**

Subjects	Pb Exposure	Immunotoxic Endpoint	Results	References
Lead workers	PbB range: 22–89 μg/dl	Respiratory ailments, associated effects on immune responses	More colds and influenza infections per year than controls; secretory IgA suppressed	Ewers et al. (1982)
Urban children (Boston, MA)	PbB levels ≥10 μg/dl versus <10 μg/dl	Respiratory tract illness, otitis media; other infectious diseases	Increased relative risks versus PbB <10 μg/dl	Rabinowitz et al. (1990)
Urban children (Chicago, IL, *N* = 104)	Elevated PbB: 100–471 μg/dl	*S. enteritidis* infections versus PbB	High PbB levels associated with *S. enteritidis* had prolonged diarrhea	Sachs (1978)
Southeastern Michigan children screened for PbB at 1–3 years of age (*N* = 4,634; 69.5% African-American)	8.6% ≥10 μg/dl; 39.0% ≥5 μg/dl. Mean PbB: 4.7 μg/dl; median PbB: 4.0 μg/dl	Period prevalence and incidence of asthmas: African-Americans and Caucasians	No dose–response associations of asthma with PbB in African-Americans; African-Americans at more risk than Caucasians. Trend toward increased risk ≥5 μg/dl in Caucasians	Joseph et al. (2005)

effect to the occupational health community, integrating humoral and cell-mediated immunological mechanisms. Table 18.1 presents illustrative studies of the search for these Pb effects, studies done mainly in children.

Rabinowitz et al. (1990), in their prospective study of developmental neurotoxicological effects of low-level lead exposures in Boston, MA, children of relatively affluent families, assessed the incidence of infectious diseases such as respiratory illnesses and otitis media versus PbB level. Children stratified into the low Pb group, PbB <10 μg/dl, had lower relative risk than did those with PbB ≥10 μg/dl. This study took minimal cognizance of covariates and confounding variables such as race, socioeconomic and demographic variables, and immunological assays. The earlier study of Sachs (1978) examined lead-poisoned Chicago inner-city children having a relatively high range of PbB: 100−471 μg/dl. The higher end of this poisoning range showed infection with *Salmonella enteritidis*.

The potential relationship of Pb exposures to childhood asthma via immunotoxic responses to Pb has long intrigued researchers with its marked similarities in terms of socioeconomic and demographic risk elements. For example, elevated PbB increases the production of IgE in children (Sun et al., 2003), while IgE levels are increased in atopic and nonatopic individuals with asthma (Romanet-Manent et al., 2002). Such factors, however, would be difficult to disentangle in terms of causality and dose−response relationships. Joseph et al. (2005) prospectively evaluated the relationship of childhood asthma, defined using two criteria, and the level of PbB in a cohort of 1- to 3-year-old children who were enrolled in a southeastern Michigan health system. Characteristics included annual income, gender, race, PbB stratified as to <5, ≥5 μg/dl, and period incidences for either definition of asthma.

An increased, nonsignificant hazard ratio for asthma was observed in Caucasian children, while African-American children had a higher risk of asthma than the Caucasian subjects, regardless of PbB group. The authors concluded that asthma risks overall were not linked to PbB level, but urged further studies given the likely operation of risk factors such as race, asthma phenotype, genetic predispositions, and concurrent socioeconomic and environmental exposures.

Lead workers in the report of Ewers et al. (1982), with a PbB range of 22−89 μg/dl, presented with more colds and flu infections annually than did controls. IgA levels were also suppressed in these workers.

18.2.2 Studies of Pb Effects on Experimental Animal Resistance to Disease

A number of experimental animal studies, some in the older literature, have described effects of lead treatment on resistance to various intra- and extracellular pathogens. In some cases, endotoxin-producing microbiota have been evaluated. Illustrative reports are summarized in Table 18.2. The majority of these experimental efforts have been carried out with various mouse strains, but some testings have been done in rats and fowl. Host resistance endpoints have involved bacterial, viral, and parasitic diseases, and tumor formation and promotion.

Various strains of mice exposed to Pb orally, by inhalation or by parenteral administration, showed markedly increased morbidity and mortality

TABLE 18.2 Illustrative Studies of Effects of Pb on Experimental Animal Host Resistance, Hypersensitivity, and Autoimmunity

Test Species	Dosing Regimen	Endpoint Measures	Results	References
A. Host resistance				
Swiss-Webster mice	I.p., 30 days, 100 or 250 μg/0.5 ml, Pb as nitrate	Mortality rates versus controls, inoculation with *Salmonella typhimurium*	54% mortality for 100 μg/0.5 ml	Hemphill et al. (1971)
CBA/J mice	Oral Pb in diet: 16, 80, 400, 2,000 ppm	Mortality rates versus controls: treated with *L. monocytogenes*	100% mortality in 80–2,000 ppm groups versus 0% in 0 or 16 ppm	Lawrence (1981)
WMRI mice	Inhaled Pb, 13–14 $\mu g/m^3$ Pb aerosol	Clearance rate of *S. marcescens* versus Pb exposure	Clearance of *S. marcescens* significantly reduced	Schlipköter and Frieler (1979)
Male CD-1 mice	Pb in drinking water, 0.004–0.1 M, starting 4 weeks of age	Various viruses given i.p. 3–6 weeks after, including Rauscher leukemia virus and encephalomyocarditis virus	Increased encephalomyocarditis virus mortality from Pb dosing: up to 100% at 0.1 M Pb; mouse interferon protective effect not reduced by Pb	Gainer (1974)
Swiss mice	Pb gavage: 62.5–500 mg/kg body weight, 28 days	Semliki forest virus challenge after start of dosing	Significant mortality at 250 mg/kg body weight	Gupta et al. (2002)
C57 BL/C mice	Pb in drinking water, 130–1,300 ppm, chronic exposure	Moloney sarcoma virus challenge and tumor development	Growth of primary tumors enhanced by Pb dosing	Kirkvliet and Boecher-Steppan (1982)

(*Continued*)

TABLE 18.2 **Illustrative Studies of Effects of Pb on Experimental Animal Host Resistance, Hypersensitivity, and Autoimmunity—(cont.)**

Test Species	Dosing Regimen	Endpoint Measures	Results	References
Rats	Pb i.v., single, 20 mg/kg	Challenged with *E. coli* or *S. epidermidis*	With *E. coli*, mortality 96%; 80% mortality with *S. epidermidis*	Cook et al. (1975)
B. Autoimmunity				
Strains of New Zealand mice predisposed to lupus erythematosus	3 ×, i.p. injection, Pb acetate, 0.1 ml, 3 weeks	Sera assays for IgG1, IgG2A; mortality and morbidity rates recorded	Pb-enhanced lupus manifestations and mortality but in gender-variable ways	Hudson et al. (2003)
Avian embryos	Single injections (5 days old) of Pb, 5 or 10 μg	At 4 weeks, immunized with bovine serum albumin; at 6 or 8 weeks, sensitized with self-antigen thyroglobulin	IgG levels in males significantly increased at higher dose; males induced to produce auto-anti-thyroglobulin IgG. No changes in delayed hypersensitivity	Bunn et al. (2000, 2001)

rates when infected with various bacteria. Swiss-Webster adult mice given i.p. injections (30 days, 100 or 250 μg/dose, nitrate) showed as much as 54% mortality at the lower dose (Hemphill et al., 1971), while CBA/J mice fed Pb at levels of 0, 16, 80, 400, and 2,000 ppm for 4 weeks presented with 100% mortality at all doses except that at 16 ppm Pb, which produced 0% mortality when infected with the intracellular bacterium *Listeria monocytogenes* (Lawrence, 1981).

Inhalation of Pb as an aerosol at a concentration of 13–14 μg/m^3 air significantly reduced the clearance rate of *Serratia marcescens* in the findings of Schlipköter and Frieler (1979), while rats dosed with single 20 mg Pb/kg injections showed 96% and 80% mortality rates after infection with *E. coli* or *S. epidermidis*, respectively (Cook et al., 1975).

Recent findings shed light on how Pb may work to alter host responses to at least bacterial pathogens. Torres et al. (2004) found that effective responses in 6 to 12-weeks-old toll-like receptors (TLR-2)-deficient (−/−) and TLR-2 control (+/+) mice and MyD88-1 (−/−) mice to *L. monocytogenes* infection required rapid activation of the innate immune system via TLR-2, given that MyD88-1-animals deficient in an adapter protein involved in TLR-2 signaling died from the infection, and TLR-2-deficient animals are much more vulnerable to this pathogen than are wild-type mice. Levels of TLR-2-deficient animal macrophages and dendritic cell macrophages stimulated by the pathogen were significantly reduced, as were levels of NO and interleukin (IL)-12.

Several reports described enhanced mortality rates when various Pb-treated mouse strains were infected with viruses. The findings of Gainer (1974) indicated that Pb exposures of male CD-1 mice via drinking water (0.004–0.1 M Pb) for 4 weeks and more yielded increased mortality versus controls when infection occurred 3–6 weeks post-Pb dosings. Over this exposure range, animal mortalities with encephalomyocarditis virus ranged up to 100% at the highest (0.1 M Pb) dosing. Pb dosing by gavage of adult Swiss mice at 62.5–500 mg/kg/day for 4 weeks produced significant mortality at the 250 mg/kg rate subsequent to infection with the Semliki forest virus (Gupta et al., 2002). Tumor development in mice challenged with Moloney sarcoma virus was enhanced with chronic oral Pb dosing (130–1,300 ppm Pb in drinking water) (Kirkvliet and Boecher-Steppan, 1982). A primary cell-mediated response in spleen and lymph nodes of tumor-bearing mice was augmented significantly by Pb. Little effect was reported on T-cell mediated responses, but there was a reduction in macrophage activity along with a significantly reduced number of macrophages.

Youssef (1996) found that the mortality rate in chickens infected with a virulent Newcastle disease virus was significantly enhanced with Pb dosing (20 or 40 mg/kg/day at 1–56 days). A vaccine for the virus decreased the mitogenic response in lymphocytes.

Several investigations have assessed the association of Pb exposures with autoimmunity, using experimental animals with a propensity to develop autoimmunological responses. Hudson et al. (2003) found that Pb given i.p. (3 weeks, as acetate, 1.3 mM Pb) to four strains of New Zealand mice genetically predisposed to lupus development showed variable responses in Pb-induced autoimmune response. The level of lupus exacerbation differed with strain and gender. Bunn et al. (2000) reported that chicken embryos dosed at embryonic day 5 with Pb at 5 or 10 μg Pb, then immunized after hatching (4 weeks) with bovine serum albumin, then at 6–8 weeks with self-antigen thyroglobulin showed that 10 μg Pb exposure increased IgG levels in males, and males were induced to produce auto anti-thyroglobulin IgG. No changes in delayed type hypersensitivity (DTH) were noted.

18.3 Pb EFFECTS ON HUMORAL IMMUNOLOGICAL RESPONSES IN HUMAN POPULATIONS AND EXPERIMENTAL SYSTEMS

Humoral immunological responses with respect to Pb exposures in human populations have typically been less robust than cell-mediated immunomodulation processes (McCabe, 1994; McCabe et al., 1991). For example, Pb can induce functional changes in the immune system while antibody responses remain relatively unaltered, although they are different from controls in humans and animals. Humoral immunological biomarkers have typically employed total serum Ig (IgG, IgM, IgA, IgE) levels, but such measures, at least in experimental test species, were not found to offer very robust endpoints (Tryphonas, 2001). One factor underpinning modest humoral responses to some toxicant exposures is the immune system's large reserve capacity, such that lymphocyte numbers or shifts in lymphocyte subsets may not co-occur with immunological changes (Tryphonas, 2001). With animal test species, infectivity tests are highly associated with and highly predictive of chemical-induced immunosuppression. One Pb effect of particular concern is the direction of B-cell maturation, B cells being important to humoral responses. T-cell-driven maturation of B cells involves skewing class switching, and this complex effect prevails with relatively modest Pb exposures.

The earlier literature produced inconsistent results with respect to null effects or increases in Ig groups, tracing at least in part to different exposure intensities, exposure duration, and timing of exposures at particularly sensitive developmental points (see U.S. EPA, 1986, 2006, for summary information).

18.3.1 Humoral Immunological Responses in Humans

This section principally concerns two subgroups within human populations having enhanced Pb-associated immunotoxic risk because of increased toxic vulnerability, i.e., young children, or adults having enhanced risk by virtue of greatly increased Pb exposure levels in the workplace (versus the general adult population with largely environmental Pb contact).

International studies of young children having a range of Pb exposures have demonstrated various Pb-linked changes in immunoglobulin levels. The result summaries of these children, from China, France, Germany, and the United States, are described in Table 18.3. PbB levels across the child cohorts ranged considerably, but three group means were <10 μg/dl: Parisian neonates had a cord blood mean of 7 μg/dl (Annesi-Maesano et al., 2003), German children 7–10 years of age had a mean PbB of 3 μg/dl (Karmaus et al., 2005), and U.S. children living near mining and smelting operations had a mean PbB of 7 μg/dl (Sarasua et al., 2000). Two child cohorts presented with mean PbBs ranging above 10 μg/dl: Chinese children aged 3–5 years (Sun et al., 2003) and U.S. children aged 9 months to 6 years (Lutz et al., 1999).

All studies reported elevations of IgE in children except the study of Sarasua et al. (2000), where no positive results were noted. The German children (Karmaus et al., 2005) had increases in previous 12-month infection rates with increases in PbB. In addition, the numbers of T cells, cytotoxic T cells ($CD8^+$), and B cells ($CD19^+$) were all significantly reduced. IgE elevations did not follow a dose–response relationship. One difficulty here and in other cross-sectional studies where clear dose–response is not evident is ascertaining whether such immune responses existed before or after Pb exposure, muddling a causal role.

The neonate study of Annesi-Maesano et al. (2003) is particularly significant to the issue of a robust developmental immunotoxicity of Pb, in that the neonate hair levels were correlated with cord blood IgE. Newborn hair Pb would not be subject to the contamination problems often seen in older children (see the earlier discussion on Pb measurement media), while simultaneously giving a full gestational accounting of Pb exposure versus cord blood. On the other hand, the strength of correlation was more significant in newborns whose mothers were nonallergic, indicating that familial allergenicity may muddle the Pb–immune response relationship. The second youngest cohort of children, 6–35 months old whose exposures arose from extractive industry contamination, was evaluated by Sarasua et al. (2000). All three Ig profiles that were reported—IgA, IgG, and IgM—were elevated with PbB in this data set. Older children and adults, by contrast, had little immunological response.

Chinese children with a relatively wide PbB range, up to 40 μg/dl, were stratified for purposes of humoral immunoresponses into PbB <10 and ≥10 μg/dl. Ig levels were altered across the groups but in different ways. IgE was higher in the higher PbB stratum but only in females, while IgG and IgM were lower, IgG being significantly lower statistically. Lutz et al. (1999) in their U.S. urban cohort found a statistically significant relationship

TABLE 18.3 Effects of Pb on Humoral Immunity in Human Populations

Subjects	Pb Exposure	Endpoint Measures	Results	References
Hesse, Germany children ($N = 331$, aged 7–10 years)	Geometric mean PbB 2.7 μg/dl, quartile groupings	Serum Ig levels: IgA, IgE, IgG, IgM	Only IgE was elevated with increased PbB ($p < 0.05$)	Karmaus et al. (2005)
Hangzhou, China children ($N = 38$ high PbB; $N = 35$ low PbB; aged 3–6 years)	PbB ≥10 μg/dl versus <10 μg/dl	Serum Ig levels: IgG, IgM, IgE	IgE was significantly correlated with PbB in female children; high PbB group versus controls for females significant at $p < 0.05$	Sun et al. (2003)
Paris, France, newborns; two sets of mother–newborn pairs: 137, 237: combined cohort $N = 374$	Newborn hair Pb levels	Cord blood IgE	IgE was statistically associated with hair Pb; *in utero* Pb exposures linked to humoral responses	Annesi-Maesano et al. (2003)
U.S. children living in mining communities	Mean PbB, youngest children = 7 μg/dl; mean for 36 to 71-month-olds = 6 μg/dl	Serum IgA, IgE, IgM, circulating B lymphocytes	Children <3 years old and with PbB >15 μg/dl showed increases in all three globulins	Sarasua et al. (2000)
U.S. children (Green County, MO) $N = 279$, aged 9 months to 6 years	PbB range = 1–45 μg/dl	Serum IgE and cell-mediated immunological markers	IgE was positively correlated statistically with PbB	Lutz et al. (1999)
Korean lead battery workers ($N = 606$)	PbB mean of age group means = 22 μg/dl	Serum IgE levels and cellular immunity markers	IgE is significantly positively correlated with PbB; IgE higher in PbB group ≥30 μg/dl	Heo et al. (2004)

(*Continued*)

TABLE 18.3 Effects of Pb on Humoral Immunity in Human Populations—(cont.)

Subjects	Pb Exposure	Endpoint Measures	Results	References
Iranian workers variably exposed to Pb	PbB ≥25 μg/dl	Serum IgA, IgG, IgM	IgG was reduced	Ayatollahi (2002)
U.S. adults residing in mining communities or control areas	PbB geometrical mean = 4.3 μg/dl	Serum IgA, IgG, IgM	No alterations with increased PbB	Sarasua et al. (2000)
Turkish lead workers (worker $N=25$, control $N=25$)	PbB mean = 75 μg/dl	Serum IgG, IgM, C_3, C_4 complement	With increased PbB, significant reduction of IgG, IgM, C_3, C_4	Başaran and Undeğer (2000)
U.S. smelter workers (worker $N=145$, control $N=84$)	Mean PbB = 39 μg/dl	Serum IgA, IgG, IgM, salivary IgA, C_3 complement, cellular markers	IgG was reduced; no differences in other measures	Pinkerton et al. (1998)
Turkish lead battery workers ($N=25$, control $N=25$)	PbB elevated, >40 μg/dl	Serum IgG, IgM, C_3, C_4 complement	Significant reduction in these markers ($p<0.05$) with increased PbB	Ündĕger et al. (1996)

between PbB and IgE only in children aged 9 months to 6 years, the association being a direct one.

A number of more recent studies have described humoral and cell-mediated immunotoxic responses in lead workers in a variety of industrial settings. Responses in terms of Ig levels versus workplace Pb exposures have been observed internationally, covering various racial and ethnic worker populations in the United States, Taiwan, Turkey, Korea, and Iran. Illustrative reports are summarized in Table 18.3. Heo et al. (2004) noted that Korean Pb battery workers with PbB >30 μg/dl showed a statistically significant direct relationship to serum IgE levels versus workers in the lower PbB group. U.S. lead workers having a median PbB of 39 μg/dl, along with a control group, were evaluated by Pinkerton et al. (1998) with respect to immunological markers of Pb exposures. No differences were noted in serum Ig measures except for a reduction in IgG. Both the number and percentage of B cells were increased for the elevated PbB group versus control workers.

Several studies in Turkish lead battery workers (Başaran and Undeğer, 2000; Ündĕger et al., 1996) indicated Pb effects on immune function in workers with a mean PbB of 75 μg/dl. Significant reductions were observed in serum levels of IgG, IgM, and C3, C4 complement. It should be noted that intensity of Pb exposure was generally greater than in other worker study groups and may account for more adverse humoral immunotoxic responses. The mean PbB levels in the reports of Heo et al. (2004) and Pinkerton et al. (1998) were 22 and 39 μg/dl, respectively. Ayatollahi (2002) reported that Iranian workers with diverse forms of Pb exposures such as battery workers, car painters, and printers had reduced serum Ig levels, particularly IgG.

18.3.2 Experimental Animal Studies of Humoral Immunotoxicity of Pb

Reports of Pb effects on humoral immunity in experimental animals have principally involved strains of mice, but studies using rats have also appeared. Table 18.4 presents experimental *in vivo* immunotoxicity data.

TABLE18.4 *In Vivo* **Effects of Pb on Humoral Immunity in Experimental Animals**

Animal Species	Dosing Details	Endpoints	Results	References
Several mice strains	Pb (acetate) in drinking water (10 mM, 8 weeks)	Plaque-forming cells with sheep RBC challenge	BALB/c mice showed elevated responses but not others	Mudzinski et al. (1986)
Female Fischer 344 rats, 9 weeks old	Pb (acetate) in drinking water during breeding, pregnancy; Pb: 0, 100, 250, 500 ppm	Immune function in female offspring, 13 weeks	Serum levels of IgE were increased in rats at 100 ppm exposure	Miller et al. (1998)
Pregnant BALB/c mice, with or without cross-fostered litters	Pb (acetate) in maternal drinking water, day 15 of gestation; Pb: 0, 0.08, 0.4, 1 mM; cross-fostering: 0 or 0.1 mM Pb acetate	2-Week neonate plasma IgE levels before, after birth	Neonates with *in utero* or lactational Pb exposure had higher IgE levels	Snyder et al. (2000)
Adult BALB/c female mice	Subcutaneous dosing: 50 μg Pb, 3× per week, 3 weeks	Plasma IgE levels	IgE levels were increased	Heo et al. (1996)

Mudzinski et al. (1986) examined the occurrence of plaque-forming cells with sheep erythrocyte challenge, using several strains of mice exposed to Pb (acetate) in drinking water (10 mM, 8 weeks). There was elevated plaque formation response with the BALB/c mouse strain but not with the others. Snyder et al. (2000) evaluated immunotoxic effects of Pb in neonate mice when dams were exposed to Pb in drinking water from gestation day 15 at four dosing levels: 0, 0.08, 0.40, or 1 mM Pb. Cross-fostered litters were exposed at 0.0 or 0.1 mM Pb. Testing at 2 weeks postnatally showed that *in utero* and lactational Pb exposures produced higher IgE concentrations in the pups.

Similarly with adult female BALB/c mice, subcutaneous dosing (50 μg Pb, 3× per week, 3 weeks) elevated plasma IgE levels (Heo et al., 1996). With rats, 9-week-old Fischer 344 animals given Pb (acetate) in drinking water at 0, 100, 250, or 500 ppm during breeding and pregnancy produced female offspring whose sera at 13 weeks showed increases in IgE at the 100 ppm dosing level (Miller et al., 1998).

Both data of Snyder et al. (2000) and Miller et al. (1998) showed that the prenatal developmental window is a time for robust expression of IgE effects from Pb exposure. Relatively robust humoral responses prenatally have developmental immunological parallels with cell-mediated immunotoxic responses to Pb, as noted below.

18.4 CELL-MEDIATED IMMUNOTOXIC RESPONSES IN HUMANS AND EXPERIMENTAL SYSTEMS

18.4.1 Responses in Humans

Studies in humans of cell-mediated immunotoxic responses to Pb comprise a growing literature. Illustrative reports are set forth in Table 18.5. Such effects have focused on young children with Pb exposures or lead workers.

Overall, lead appears to particularly target T cells, dendritic cells, macrophages, and neutrophils. Innate natural killer (NK) cells appear to have less sensitivity to Pb.

Pb exposures in adult humans fail to affect lymphocytic T- and B-cell activation in some cohorts (Pinkerton et al., 1998) or reduce proliferation in others (Mishra et al., 2003). Typically, impaired activation appears linked to higher PbB values. In one case, at lower PbB exposure, Fischbein et al. (1993) reported that decreased lymphocyte abundance occurred with decreased abundance of $CD3^+$ and $CD4^+$ cells with PbB in the range of 15–55 μg/dl. A continuing area of research interest addresses the question of Pb impacts on cellular abundances of T cells versus responses shifting to Th2 from Th1-dependent activities. For example, Heo et al. (2007) noted that Pb attenuates interferon-γ (IFN-γ). Such cell phenotype shiftings have predominantly been elucidated in experimental studies. Limited epidemiological data exist to go beyond the inverse relationship in, for example, children

TABLE 18.5 Effects of Lead on Cell-Mediated Immunity in Human Populations

Subjects	Pb Exposure	Endpoint(s)	Results	References
Hesse, Germany children ($N = 331$); age 7–10 years; 96% 7–8 years old	Geometric mean PbB = 2.7 μg/dl; PbB quartiles	Abundances of various T and B cells versus PbB level	Decreased abundance of T cells ($CD4^+$), cytotoxic T cells ($CD8^+$) and B cells; effect greatest at the 2.2–2.8 μg/dl quartile	Karmaus et al. (2005)
U.S. children (Springfield, Green County, MO, $N = 279$), aged 9 months to 6 years	PbB range: 1–45 μg/dl	Levels of cytokine IL4, lymphocyte TNF (soluble CD27), soluble CD25 (FL2 receptor)	No changes in the cell-mediated immunomarkers versus PbB	Lutz et al. (1999)
U.S. children (near mining, smelting sites): $N = 1{,}561$ children plus adults; $N = 372$ children aged 6–35 months	PbB means: 7 μg/dl, 6–35 months; 6 μg/dl, 36–71 months; 4 μg/dl, 6–15 years	Levels of peripheral blood lymphocyte immunocell phenotypes: T cells, B cells, NK cells, CD4/CD8 subsets	Circulating B cells elevated with PbB >15 μg/dl in children <3 years old	Sarasua et al. (2000)
Zhejiang Province, China, children ($N = 73$) 3–6 years old	PbB range: 3–40 μg/dl	Abundance of selected lymphocyte cell types: cytotoxic $CD8^+$ T cells, $CD3^+$ T cells, $CD4^+/CD8^+$ T helper cells. $CD4^+$ T helper cells, B cells	With children's PbBs ≥10 μg/dl, lower abundance of $CD3^+/CD4^+$ and $CD4^+/CD8^+$ and higher % of $CD3^+/CD8^+$ cells	Zhao et al. (2004)
Lagunera, Mexico, children near a primary Pb smelter	65% PbB >10 μg/dl (range: 3–49 μg/dl)	Monocyte NO and superoxide production, direct activation via IFN-γ + lipopolysaccharide (LPS); indirect activation effects on Th-1 responses	NO in indirectly activated monocytes negatively associated with PbB; superoxide production positively associated; gender effects noted	Pineda-Zavaleta et al. (2004)

U.S. adults (Springfield, Green County, MO) near mining/smelting operations ($N = 433$)	Median PbB = 4.3 μg/dl	Lymphocyte B and multiple T-cell phenotypes	No changes in B or T cells with PbB	Sarasua et al. (2000)
Adult U.S. Pb workers ($N = 145$) and controls ($N = 84$)	Median worker PbB = 39 μg/dl	Various T-cell phenotypes and B cells: $CD3^+$, $CD4^+$,cytotoxic $CD8^+$, $CD4^+CD8^+$, T-memory cells, NK cells	No marked differences, exposed versus controls, except % and numbers of T-memory cells, B-cells positively associated with cumulative Pb exposure	Pinkerton et al. (1998)
Adult Turkish Pb battery workers ($N = 25$) and controls ($N = 25$)	Worker PbB mean = 75 μg/dl	Peripheral lymphocyte subsets: T, $TCD4^+$, $TCD8^+$, NK, and B cells	Significant depression in T-helper cells	Başaran and Undeğer (2000)
Adult Taiwanese Pb battery workers versus teacher control group	Variable PbB levels	Percentage of monocytes and T cells	Higher percent of monocytes and lower percent of T8 cells in workers; T4/T8 ratio higher in workers	Kuo et al. (2001)
Adult Japanese Pb workers producing Pb stearate ($N = 71$) and controls without exposure ($N = 28$)	Mean worker PbB = 19 μg/dl	Number and percentage of $CD3^+$ and $CD45RO^+$ memory T and $CD8^+$ cells	Significant reduction in number of memory T cells and increase in % of $CD8^+$ cells	Sata et al. (1998)
U.S. adult firearms instructors ($N = 87$)	Mean PbB, high exposure = 31 μg/dl	Peripheral blood lymphocyte subsets of cells: T, Th, Tc, THC, Tm, NK	With PbB >25 μg/dl, percent and number of $CD3^+$, $CD4^+$ cells reduced	Fischbein et al. (1993)

to lowered T-cell abundances, increased B-cell abundance and IgG levels (Karmaus et al., 2005; Sarasua et al., 2000; Zhao et al., 2004).

Macrophages, to include the blood-based monocytes, are sensitive targets of Pb immunotoxic responses, the end results of which are such impacts as hyperinflammatory injury. The global systemic distribution of macrophages and their responses to Pb would foretell multiorgan and multisystem toxicities, operating through increased levels of ROS, lowered NO, increased tumor necrosis factor (TNF)-α, increased prostaglandin E2, and reduced phagocytic activity (Kasten-Jolly et al., 2010).

Dendritic cells may be involved with any impacts of lead on acquired immunity. Data are largely from cell culture studies, wherein Gao et al. (2007) reported Pb-delayed dendritic cell maturation and function with Pb addition, such as a significantly reduced ratio of IL-12/IL-10.

However, both epidemiological and experimental animal data indicate that Pb principally targets macrophages and T lymphocytes. The restriction of Pb toxicity to immunocellular function rather than direct cellular injury, the more typical case with other organs and systems, complicates elucidating the mechanisms of immunotoxic action of Pb and identifying dose–response relationships. For example, Pb impacts T-helper-1 cells in the form of DTH and affects production of cytokines. In the case of macrophages, Pb induces a hyperinflammatory response.

A number of pediatric epidemiological reports have documented that relatively low PbB level is inversely linked to T-cell abundances (and increases in B-cell abundances). Infants and toddlers (6–35 months old) with a mean PbB of 7 μg/dl showed reduced T-cell ($CD3^+$) abundance and elevated B-cell levels with rising PbB (Sarasua et al., 2000). No changes were observed for T-helper, cytotoxic T, or NK cells. Three studies, those of Karmaus et al. (2005), Zhao et al. (2004), and Lutz et al. (1999), described somewhat contradictory results for older children up to 10 years of age (Table 18.5). Reports of Lutz et al. (1999) and Zhao et al. (2004) were more consistent about types of cell-mediated immunotoxicity effects observed. Both studies showed no change with PbB for total T-cell abundances and no changes in B-cell levels.

Several factors may be operative in diverging endpoint results. The studies of Karmaus et al. (2005), besides providing the most detailed quantification of T-cell subsets, involved the lowest measured PbB values. They reported a mean PbB of 3 μg/dl and a range of 1–5 μg/dl. They also reported reduced abundances of T and B cells, and cytotoxic T cells, and no changes in T-helper, T-memory, and NK cells. Sarasua et al. (2000) showed some differences and similarities with the findings of Karmaus et al. (2005). A potential second factor with these two reports is the age range, the former cohort consisting of infants and toddlers and the latter of children 7 years and older.

An additional cohort of children from a Mexican primary smelter community, exposed to both Pb and arsenic (As), presented with reduced monocyte NO and increased superoxide anion production via either direct IFN-γ activation with lipopolysaccharide or by indirect means (Pineda-Zavaleta et al., 2004). These children had relatively high PbB levels, with 65% $\geq$10 μg/dl, ranging up to 49 μg/dl. Superoxide anion levels were directly associated with PbB but inversely linked to urinary As. These children showed a gender association with the effects, particularly for both NO and superoxide anion formation.

Illustrative cases from studies of Pb immunotoxicity in lead workers are also given in Table 18.5. As with the child cohorts, they are an international representation of effects: the United States, Turkey, Korea, and Japan. As with the child cohorts, endpoints quantified with regard to Pb immunotoxicity mainly consisted of relative abundances of lymphocyte subsets of cells. Pb exposure intensities ranged considerably across the worker cohorts, while all were considerably above exposures seen with children. In general, Pb workers had reductions in abundances of T cells, e.g., Kuo et al. (2001), Başaran and Undeğer (2000), Fischbein et al. (1993), Sata et al. (1998), while Pinkerton et al. (1998) noted increased abundances in T-memory cells and no changes in T, T-helper, or cytotoxic T cells.

Other forms of immunotoxic expression with occupational Pb exposures have been described. Brazilian Pb workers (Queiroz et al., 1993) had significantly reduced peripheral blood neutrophil populations versus controls, using both chemotaxis and nitroblue tetrazolium reduction. These impairments were seen in workers with PbB $<$60 μg/dl. Mishra et al. (2003) examined immunotoxic effects of Pb in three groups of workers versus controls: lead battery plant workers, jewelry makers, and three-wheeler drivers. Phytohemagglutinin-induced lymphocyte proliferation was inhibited versus controls, but no dose–response relationship was seen. T-cell mitogen-stimulated monocytes also showed an increase in IFN-γ but NK-cell level was unaffected. There was a positive correlation of IFN-γ with PbB.

As has been seen with other toxic effects of Pb, there appear to be temporal and developmental windows for immunotoxic response. The developing immune system is 3–10 times more sensitive than the developed one. Sarasua et al. (2000) observed children 6–35 months of age who had statistically increased IgA, IgM, and B cell abundances.

18.4.2 Studies of Pb Effects on Cell-Mediated Immunity in Experimental Animals

Recent studies of Pb-associated immunotoxicity in experimental animals have been quite helpful in elucidating the relevance of developmental periods and other host factors on the expression of immunotoxicity and also helpful in revealing the full range of immunological responses not readily accessible in available pediatric or occupational epidemiological efforts. Selected reports across test species are summarized in Table 18.6.

TABLE 18.6 *In Vivo* Effects of Pb on Cell-Mediated Immunity in Experimental Animals

Animal Species	Dosing Details	Endpoint(s)	Results	References
Female Fischer 344 rat offspring, 13 weeks old; dams bred at 9 weeks	Pregnant dams exposed to 0, 100, 250, 500 ppm Pb (acetate) beginning 9 weeks old; offspring Pb-exposed only *in utero*	Pb modulated immune, macrophage cytokine and effector functions	TNF-α, NO production elevated at 250 ppm; decreased DTH reactions in 250 ppm group and IFN-γ in the 500 ppm group	Miller et al. (1998)
Female Fischer 344 rat offspring tested at 13 weeks old; offspring from chelant treatment alone or with Pb	Pb given to dams 2 weeks before mating to birth: Pb in drinking water, up to 500 ppm; DMSA chelant given 6–21 days of gestation via gavage (60 mg/kg/day)	Various measures of Pb-impacted immune functions	Pb alone reduced Th-1 responses: DTH, IFN-γ, enhanced Th-2 responses; IL-4 production and TNF-α; DMSA reversed Pb-based immunological alterations; DMSA alone produced lowered DTH in offspring when adults	Chen et al. (1999)
Pregnant Sprague–Dawley rats, offspring	Pb (acetate) in drinking water early (3–9 days) or later ingestion (15–21 days); Pb=500 ppm	Differential immunotoxic outcomes tested: DTH, IL-10 production, monocyte numbers, thymic weights	Late gestational Pb exposure produces worst immunotoxicity; females more affected: depressed DTH, elevated IL-10 production, monocyte numbers increased; male offspring had increased IL-12 and decreased IL-10 production	Bunn et al. (2001)
Pregnant Sprague–Dawley rats, offspring	Pb (acetate) in drinking water: from breeding until parturition, 250 ppm; two isocaloric diets: 10%, 20% casein	Immunotoxicity testing in dams, 8 weeks postpartum; offspring tested at 13 weeks old	Dams showed no Pb effect; offspring showed IFN-γ and DTH reduced with Pb; high protein group showed elevated IL-4 and TNF-α	Chen et al. (2004)

BALB/c mice	Injection with Pb (chloride)	V beta expression, Th1, Th2 frequency	Pb induces oligoclonal response: increased V beta 5^+, -7^+ and V beta 13^+CD^+splenic T cells; skewed frequency of Pb-induced Th2 cells over Th1 cells	Heo et al. (1997)
BALB/c mice	S.c. injection with Pb (chloride)	Measure of cytokine production *in vivo*	Enhanced production of IL-4 by stimulated splenic T cells; IFN-γ production inhibited; plasma IL-4 levels increased	Heo et al. (1996)
BALB/c neonate mice, dams	Pb (acetate) in drinking water, 0.5 mM; dams postpartum to weaning; pups after weaning	Pb with *L. monocytogenes* dosing, 22 days old	Pups Pb-exposed plus the infective agent showed prolonged sickness behavior versus controls; serum IL-1β, IL-6 enhanced with exposure; reduced thymic $CD4^+$, $CD8^+$, $CD4^+CD8^+$, and CD4(−) CD8(−) T cells	Dyatlov and Lawrence (2002)
TO mice, six per group	I.p. Pb, 2 weeks, with/without α-tocopherol	Effects of Pb $\pm$ α-tocopherol on T-cell mitogens, production of NO	Pb produced splenomegaly with $CD11^{b+}$ cell influx; no NO production; Pb inhibited mitogenic responses; Pb plus tocopherol partially reduced the Pb-produced changes	Fernandez-Cabezudo et al. (2003)
Female BALB/c mice, 8 weeks old	Single foot-pad injection: 5, 25, 40, 50 μg Pb chloride	Immunomodulation testing using PLNA or reporter antigen-PLNA	Pb strongly stimulated T- and B-cell proliferation and antigen-secreting cells; skewed toward T-helper type 2 responses	Carey et al. (2006)

(*Continued*)

TABLE 18.6 *In Vivo* Effects of Pb on Cell-Mediated Immunity in Experimental Animals—(cont.)

Animal Species	Dosing Details	Endpoint(s)	Results	References
Fertilized Cornell K strain white Leghorn chicken eggs	Pb (acetate) injected on E12, 200 μg per egg; on E8, embryos given testerone or vehicle	Embryo testing, E20: thymocytes tested for cell surface markers: CD3, CD4, CD8, TCR1, TCR2	Testosterone plus Pb embryo group produced $CD4^+$ $CD8^+$ double positive cells versus Pb alone	Hussain et al. (2005)
Cornell K strain white Leghorn juvenile chickens	Embryonic Pb exposure (E12): 400 μg Pb per egg	DTH and cytokine IFN-γ production by thymocytes	Decreased thymocyte production of IFN-γ cytokine and DTH	Lee and Dietert (2003)
Adult rabbits	Pb inhalation (oxide) at 30 μg/m^3, 4 days, 3 hours daily	Immunotoxicity testings in lung culture at 0, 24, 72 hours postsacrifice using macrophage functions	No effects on cell viability or number. Functional effects; Pb increased H_2O_2 and superoxide anion production across time points; TNF-α activity depressed, TNF-α release was increased	Zelikoff et al. (1993)

Key issues for immunotoxic responses in experimental animals that are useful to the human immunology of Pb include the extent to which the stage of development, gender, level of exposure intensity, and the windows for Pb exposure at whatever stage of development affect immunotoxic endpoints. In addition, cytokine production and distribution, along with switching rates among T-cell subtypes, can be probed. These factors relate to the mechanisms by which Pb triggers immunotoxic responses.

Miller et al. (1998) reported the developmental immunotoxicological effects of Pb presented only *in utero* to Fischer 344 rat offspring tested at 13 weeks. Dams were exposed to lead at 0, 100, 250, or 500 ppm Pb (acetate). Macrophage cytokines and effector functions were specifically addressed as endpoints. TNF-α and NO production were elevated and DTH was reduced at 250 ppm, while IFN-γ was only reduced at the highest exposure, 500 ppm. Dams showed no changes in immune endpoints at any Pb exposure. As noted by the authors, the observed changes with Pb dosing were developmentally driven and may involve biasing of T-helper cell subsets in terms of the balance in type-1 versus type-2 T-helper cell responses.

A related *in utero* immunotoxic animal model using pregnant Fischer 344 rats exposed to the above Pb dosing range was that of Chen et al. (1999). Here, Pb dosing was accompanied by treatment with a common Pb chelating agent, DMSA. The chelant was given by gavage days 6–21 of gestation. Pb alone reduced T-helper type-1 responses, i.e., reduced IFN-γ and DTH. Enhanced T-helper type-2 responses appeared in the form of IL-4 and TNF-α production. The chelant reversed Pb-induced immunotoxic alterations, while by itself, it lowered DTH in offspring when tested as adults. Offspring of Sprague–Dawley dams, exposed to Pb either late (15 days) or early (3–9 days) in gestation, showed that late gestational exposure produced worse immunotoxicity and mainly females were affected (Bunn et al., 2001). There was depressed DTH, increased IL-10 production, and increased numbers of monocytes. Male offspring, by contrast, had increased IL-12 and decreased IL-10 production. An effect of casein level in diet (10%, 20%) was examined in the related study of Chen et al. (2004). No effects were noted in dams, but the offspring showed reduced IFN-γ and DTH reduction at 13 weeks, while the higher dietary protein loading also indicated elevated IL-4 and TNF-α.

Three studies have noted embryonic Pb immunotoxicity in Cornell K strain white Leghorn chickens. This testing animal model has the virtue of relatively high-throughput exposure studies at a multiplicity of embryonic time windows (Hussain et al., 2005; Lee and Dietert, 2003; Lee et al., 2001). Lead exposure at E12 with 400 μg per egg resulted in decreased production of IFN-γ cytokine and reduced DTH. Chickens exposed to Pb at 5–400 μg per egg and at E5, E7, E9, or E12 embryological time points showed suppressed IFN-γ and NO production at the first three exposure time points at 5–6 weeks of age. DTH was unaltered. Hussain et al. (2005) evaluated the joint effects of Pb and testosterone administration in the chickens on embryo day 20, using 200 μg per egg injections. The Pb was given on day 12, while testosterone was given earlier, day 8. Cell surface markers in T cells showed that joint dosing produced $CD4^{+}CD8^{+}$ double positive cells versus Pb alone.

Neonate BALB/c mice exposed to drinking water Pb after weaning or before weaning, and postpartum to weaning, demonstrated prolonged sickness behavior after infection with *L. monocytogenes* at 22 days of age. Increased serum levels of IL-1β and IL-6 were determined, while there was a reduction in $CD4^{+}CD8^{+}$ and CD4(−)CD8(−) T cells (Dyatlov and Lawrence, 2002).

Adult rabbits inhaling Pb oxide aerosol at 30 $\mu g/m^3$ (4 days, 3 hours daily) showed no effects on macrophage viability or number at 0, 24, or 72 hours in isolated lung tissue after sacrifice. Pb exposure did increase H_2O_2 and superoxide anion production across all time points. Depression in TNF-α activity and increase in TNF-α release were noted (Zelikoff et al., 1993).

Carey at al. (2006) used adult BALB/c mice given foot pad injections of Pb at 5, 25, 40, or 50 μg and the popliteal lymph node assay (PLNA) or a variant, reporter antigen-PLNA, to show strongly induced T- and B-cell proliferation and antigen-secreting cells. The results skewed toward T-helper type-2 responses.

REFERENCES

Annesi-Maesano, L., Pollitt, R., King, G., Bousquet, J., Hellier, G., Sahuquillo, J., et al., 2003. *In utero* exposure to lead and cord blood total IgE. Is there a connection? Allergy 58, 589–594.

Ayatollahi, M., 2002. Study of the impact of blood lead level on humoral immunity in humans. Toxicol. Ind. Health 18, 39–44.

Başaran, N., Undeğer, U., 2000. Effects of lead on immune parameters in occupationally exposed workers. Am. J. Ind. Med. 38, 349–354.

Bingham, E., Pfizer, E.A., Barkley, W., Redford, E.P., 1968. Alveolar macrophages: reduced numbers in rats after prolonged inhalation of lead sesquioxide. Science 162, 1297–1299.

Bunn, T.L., Marsh, J.A., Dietert, R.R., 2000. Gender differences in developmental immunotoxicity to lead in a chicken: analysis following a single early low-level exposure *in ovo*. J. Toxicol. Environ. Health A 61, 677–693.

Bunn, T.L., Parsons, P.J., Kao, E., Dietert, R.R., 2001. Gender-based profiles of developmental immunotoxicity to lead in the rat: assessment in juveniles and adults. J. Toxicol. Environ. Health A 64, 223–240.

Carey, J.B., Allshire, A., Van Pelt, F.N., 2006. Immune modulation by cadmium and lead in the acute reporter antigen-popliteal lymph node 3 assay. Toxicol. Sci. 91, 113–122.

Chen, S., Golemboski, K.A., Sanders, F.S., Dietert, R.R., 1999. Persistent effect of *in utero*-meso-2,3-dimercaptosuccinic acid (DMSA) on immune function and lead-induced immunotoxicity. Toxicology 132, 67–79.

Chen, S.C., Golemboski, K.A., Piepenbrink, M., Dietert, R.R., 2004. Developmental immunotoxicity of lead in the rat: influence of maternal diet. J. Toxicol. Environ. Health A 67, 495–511.

Cook, J.A., Hoffmann, E.O., Di Luzio, N.R., 1975. Influence of lead and cadmium on the susceptibility of rats to bacterial challenge. Proc. Soc. Exp. Biol. Med. 150, 741–747.

Dyatlov, V.A., Lawrence, D.A., 2002. Neonatal lead exposure potentiates sickness behavior induced by *Listeria monocytogenes* infection of mice. Brain Behav. Immun. 16, 477–492.

Ewers, U., Stiller-Winkler, R., Idel, H., 1982. Serum immunoglobulin, complement C3, and salivary IgA levels in lead workers. Environ. Res. 29, 351–357.

Fernandez-Cabezudo, M.J., Hasan, M.Y., Mustafa, N., El-Sharkawy, R.T., Fahim, M.A., Al-Ramadi, B.K., 2003. Alpha tocopherol protects against immunosuppressive and immunotoxic effects of lead. Free Radic. Res. 37, 437–445.

Fischbein, A., Tsang, P., Luo, J.-C., Roboz, J.P., Jiang, J.D., Bekesi, J.G., 1993. Phenotypic aberrations of the $CD3^+$ and $CD4^+$ cells and functional impairments of lymphocytes at low-level occupational exposure to lead. Clin. Immunol. Immunopathol. 66, 163–168.

Gainer, J.H., 1973. Activation of the Rauscher leukemia virus by metals. J. Natl. Cancer Inst. 51, 603–613.

Gainer, J.H., 1974. Lead aggravates viral disease and represses the antiviral activity of interferon inducers. Environ. Health Perspect. 7, 113–119.

Gao, D., Mondal, T.K., Lawrence, D.A., 2007. Lead effects on development and function of bone-marrow derived dendritic cells promote Th2 immune responses. Toxicol. Appl. Pharmacol. 222, 69–79.

Gupta, P., Husain, M.M., Shankar, R., Seth, P.K., Maheshwari, R.K., 2002. Lead exposure enhances virus multiplication and pathogenesis in mice. Vet. Hum. Toxicol. 44, 205–210.

Hemphill, F.E., Kaeberle, M.L., Buck, W.B., 1971. Lead suppression of mouse resistance to *Salmonella typhimurium*. Science 172, 1031–1032.

Heo, Y., Parsons, P.J., Lawrence, D.A., 1996. Lead differentially modifies cytokine production *in vitro* and *in vivo*. Toxicol. Appl. Pharmacol. 138, 149–157.

Heo, Y., Lee, W.T., Lawrence, D.A., 1997. *In vivo* the environmental pollutants lead and mercury induce oligoclonal T cell responses skewed toward type-2 reactivities. Cell. Immunol. 179, 185–195.

Heo, Y., Lee, B.-K., Ahn, K.-D., Lawrence, D.A., 2004. Serum IgE elevation correlates with blood lead levels in battery manufacturing workers. Hum. Exp. Toxicol. 23, 209–213.

Heo Y., Mondal T.K., Gao D.H., Kasten-Jolly J., Kishikawa H., Lawrence D.A. 2007. Posttranscriptional inhibition of interferon-gamma production by lead. Toxicol. Sci. 96: 92–100.

Hudson, C.A., Cao, L., Kasten-Jolly, J., Kirkwood, J.N., Lawrence, D.A., 2003. Susceptibility of lupus-prone NZM mouse strains to lead exacerbation of systemic lupus erythematosus symptoms. J. Toxicol. Environ. Health A 66, 895–918.

Hussain, I., Piepenbrink, M.S., Dietert, R.R., 2005. Impact of *in ovo*-administered lead and testosterone on developing female thymocytes. J. Toxicol. Environ. Health A 68, 1309–1319.

Joseph, C.L.M., Havstad, S., Ownby, D.R., Peterson, E.L., Maliarik, M., McCabe, J.M.J., et al., 2005. Blood lead levels and risk of asthma. Environ. Health Perspect. 113, 900–904.

Karmaus, W., Brooks, K.R., Nebe, T., Witten, J., Obi-Osius, N., Kruse, H., 2005. Immune function biomarkers in children exposed to lead and organochlorine compounds: a cross-sectional study. Environ. Health 4 (5). 1–10. doi:10.1186/1476-069X-4-5.

Kasten-Jolly, J., Heo, Y., Lawrence, D.A., 2010. Impact of developmental lead exposure on splenic factors. Toxicol. Appl. Pharmacol. 247, 105–115.

Kirkvliet, N.I., Boecher-Steppan, L., 1982. Immunotoxicology studies on lead. Effects of exposure on tumor growth and cell-mediated tumor immunity after syngeneic or allogenic stimulation. Immunopharmacology 4, 213–224.

Kobayashi, N., Okamoto, T., 1974. Effects of lead oxide on the induction of lung tumors in Syrian hamsters. J. Natl. Cancer Inst. 52, 1605–1610.

Kuo, H.-W., Hsiao, T.-Y., Lai, J.-S., 2001. Immunological effects of long-term lead exposure among Taiwanese workers. Arch. Toxicol. 75, 569–573.

Lawrence, D.A., 1981. *In-vivo* and *in-vitro* effects of lead on humoral and cell-mediated immunity. Infect. Immun. 31, 136–143.

Lee, J.-E., Dietert, R.R., 2003. Developmental immunotoxicity of lead: impact on thymic function. Birth defects. Res. A Clin. Mol. Teratol. 67, 861–867.

Lee, J.-E., Chen, S., Golemboski, K.A., Parsons, P.J., Dietert, R.R., 2001. Developmental windows of differential lead-induced immunotoxicity in chickens. Toxicology 156, 161–170.

Lutz, P.M., Wilson, T.J., Ireland, A.L., Gorman, J.S., Gale, N.L., Johnson, J.C., et al., 1999. Elevated immunoglobulin E (IgE) levels in children with exposure to environmental lead. Toxicology 134, 63–78.

McCabe Jr., M.J., 1994. Mechanisms and consequences of immunomodulation by lead. In: Dean, J.H., Luster, M.I., Munson, A.E., Kimber, I. (Eds.), Immunotoxicology and Immunopharmacology, third ed. Raven Press, Ltd., New York, pp. 143–162.

McCabe Jr., M.M., Dias, J.A., Lawrence, D.A., 1991. Lead influences translational or posttranslational regulation of IgA expression and increases invariant chain expression in mouse B cells. J. Biochem. Toxicol. 6, 269–276.

Miller, T.E., Golemboski, K.A., Ha, R.S., Bunn, T., Sanders, F.S., Dietert, R.R., 1998. Developmental exposure to lead causes persistent immunotoxicity in Fischer 344 rats. Toxicol. Sci. 42, 129–135.

Mishra, K.P., Singh, V.K., Rani, R., Yadav, V.S., Chandran, V., Srivastava, S.P., et al., 2003. Effect of lead exposure on the immune response of some occupationally exposed individuals. Toxicology 188, 251–259.

Mudzinski, S.P., Rudofsky, U.H., Mitchell, D.G., Lawrence, D.A., 1986. Analysis of lead effects on *in vivo* antibody-mediated immunity in several mouse strains. Toxicol. Appl. Pharmacol. 83, 321–330.

Pineda-Zavaleta, A.P., Garcia-Vargas, G., Borja-Aburto, V.H., Acosta-Saavedea, L.C., Vera Aguilar, E., Gómez-Muñoz, A., et al., 2004. Nitric oxide and superoxide anion production in monocytes from children exposed to arsenic and lead in region Lagunera, Mexico. Toxicol. Appl. Pharmacol. 198, 283–290.

Pinkerton, L.E., Biagini, R.E., Ward, E.M., Hull, R.D., Deddens, J.A., Boeniger, M.F., et al., 1998. Immunologic findings among lead-exposed workers. Am. J. Ind. Med. 33, 400–408.

Queiroz, M.L.S., Almeida, M., Gallao, M.I., Höehr, N.F., 1993. Defective neutrophil function in workers occupationally exposed to lead. Pharmacol. Toxicol. 72, 73–77.

Rabinowitz, M., Allred, E.N., Bellinger, D.C., Leviton, A., Needleman, H.L., 1990. Lead and childhood propensity to infectious and allergic disorders: Is there an association? Bull. Environ. Contam. Toxicol. 44, 657–660.

Romanet-Manent, S., Charpin, D., Magnan, A., Lenteaume, A., Vervloet, D., 2002. Allergic vs. nonallergic asthma: What makes the difference? Allergy 57, 607–613.

Sachs, H.K., 1978. Intercurrent infection in lead poisoning. Am. J. Dis. Child. 132, 315–316.

Sarasua, S.M., Vogt, R.F., Henderson, L.O., Jones, P.A., Lybarger, J.A., 2000. Serum immunoglobulins and lymphocyte subset distributions in children and adults living in communities assessed for lead and cadmium exposure. J. Toxicol. Environ. Health A 60, 1–15.

Sata, F., Araki, S., Tanigawa, T., Morita, Y., Sakurai, S., Nakata, A., et al., 1998. Changes in T cell subpopulations in lead workers. Environ. Res. 76, 61–64.

Schlipköter, H.-W., Frieler, L., 1979. The influence of short-term lead exposure on the bacterial clearance of the lung. Zentralbl. Bakteriol. Parasitenkd. Infektionskr. Hyg. Abt. Orig. Reihe B 168, 256–265.

Snyder, J.E., Filipov, N.M., Parsons, P.J., Lawrence, D.A., 2000. The efficiency of maternal transfer of lead and its influence on plasma IgE and splenic cellularity of mice. Toxicol. Sci. 57, 87–94.

Sun, L., Hu, J., Zhao, Z., Li, L., Cheng, H., 2003. Influence of exposure to environmental lead on serum immunoglobulin in preschool children. Environ. Res. 92, 124–128.

Torres, D., Barrier, M., Bihl, F., Quesniaux, V.J., Maillet, I., Akira, S., et al., 2004. Toll-like receptor 2 is required for optimal control of Listeria monocytogenes infection. Infect. Immun. 72, 2131–2139.

Tryphonas, H., 2001. Approaches to detecting immunotoxic effects of environmental contaminants in humans. Environ. Health Perspect. 109 (Suppl.), 877–884.

U.S. Environmental Protection Agency, 1986. Air Quality Criteria for Lead, 4 vols. Report No. EPA/600/8-028bF. Environmental Criteria and Assessment Office, Washington, DC.

U.S. Environmental Protection Agency, 2006. Air Quality Criteria for Lead, vol.1. Report No. EPA/600/R-144aF. National Center for Environmental Assessment, Washington, DC.

Ündeğer, Ü., Başaran, N., Canpinar, H., Kansu, E., 1996. Immune alterations in lead-exposed workers. Toxicology 109, 167–172.

Youssef, S.A.H., 1996. Effect of subclinical lead toxicity on the immune response system of chickens to Newcastle's disease virus vaccine. Res. Vet. Sci. 60, 13–16.

Zelikoff, J.T., Parsons, E., Schlesinger, R.B., 1993. Inhalation of particulate lead oxide disrupts pulmonary macrophage-mediated functions important for host defense and tumor surveillance in the lung. Environ. Res. 62, 207–222.

Zhao, Z.Y., Li, R., Sun, L., Li, Z.Y., Yang, R.L., 2004. Effect of lead exposure on the immune function of lymphocytes and erythrocytes in preschool children. J. Zhejiang Univ. Sci. 5, 1001–1004.

Chapter 19

Effects of Lead on Other Organs and Systems in Human Populations

This chapter summarizes toxicological and epidemiological information for a cluster of adverse effects on organs and systems not individually assessed in previous chapters. Material addressed here does not mean that it is of lesser toxicological or public health consequence. These topics have either been studied only to a limited extent or have been studied and found to comprise relatively insensitive toxic endpoints in dose–response terms. In other cases, some effects are incorporated into toxic response discussions presented elsewhere. For example, some of the endocrine system toxic endpoints induced by Pb are included in Chapter 14.

19.1 EFFECTS OF LEAD ON THE ENDOCRINE SYSTEM OF HUMANS AND EXPERIMENTAL ANIMALS

Endocrinological toxicity due to Pb has not been evaluated in any systematic or holistic fashion and available data array into several areas of study: the human reproductive system, the thyroid endocrine system, and the vitamin D–calcium relationship.

Table 19.1 summarizes data for the thyroid endocrine effects of Pb, while Tables 19.2 and 19.3 present effects on reproductive hormone functions of Pb exposures in humans and animals. Table 19.4 presents the main effects of Pb on the GI and hepatic systems.

19.1.1 Effects of Pb on Thyroid Function in Humans and Experimental Animals

Toxic effects on thyroid hormonal function due to Pb exposure typically occur at relatively high Pb exposure levels usually associated with lead workers (Table 19.1). While more recent reports took advantage of more sophisticated methods for quantifying relatively low-level changes in hormonal levels, there appears to have been little downward shift of

Trace Metals and other Contaminants in the Environment, Volume 10
ISSN: 1875-1121 DOI: 10.1016/B978-0-444-51554-4.00019-5

TABLE 19.1 Effects of Lead on Thyroid Function in Human Populations

Subjects	Pb Exposure	Endpoints Studied/ Reported	Reported Results	References
U.S. children (11 months–7 years old, $N = 68$)	Mean PbB = 25 μg/dl	Total (TT4) and free (FT4) thyroxine	No effect of Pb on either measure	Siegel et al. (1989)
Lead workers ($N = 58$)	Variable PbB levels	TT4, FT4	No effect versus PbB reported	Refowitz (1984)
Secondary smelter workers ($N = 88$, workers and controls)	Median PbB = 31 μg/dl; bone Pb also measured	FT3, FT4, TSH levels	No effect of Pb on thyroid using any measure of either Pb biomarker	Erfurth et al. (2001)
Canadian (Trail, BC) primary smelter workers ($N = 151$)	Mean PbB = 24 μg/dl; quartiles of PbB current and 10-year means: <15, 15–24, 25–39, ≥40 μg/dl	FT4, TT4, TSH levels	No effect of Pb on any measure using either PbB index	Schumacher et al. (1998)
Battery plant workers ($N = 176$)	Mean PbB = 56 μg/dl; range 5–134 μg/dl	TT3, FT4, TT4, TSH levels	Reduced levels of TT3, TT4 and no effect on TSH	Tuppurainen et al. (1988)
General population ($N = 82$)	Mean 15 μg/dl	TT4, TTR levels	Pb exposure duration significantly linked to decreasing levels of TT4 and FT4; no effects on TTR reported	Zheng et al. (2001)

dose–response relationships. This is in contrast to new findings in neurodevelopmental toxicity, immunotoxicity, genotoxicity, nephrotoxicity, and cardiovascular outcomes. Toxic effects of Pb on the thyroid endocrine system have mainly been studied in adults, either lead workers (Erfurth et al., 2001; Refowitz, 1984; Schumacher et al., 1998; Tuppurainen et al., 1988) or adults in the general population (Zheng et al., 2001).

Siegel et al. (1989) described assessments of Pb and thyroid function in children. A small group of children ($N = 68$) aged 11 months to 7 years were evaluated by Siegel et al. for evidence of any association of PbB with changes in serum total or free thyroxine at a clinical facility. Mean PbB was moderately high, i.e., 25 μg/dl. Using only simple statistics, no associations of PbB with free or total serum thyroxine were detected. These data would suggest that any pediatric link for Pb and thyroid function does not extend to this low range of PbB, but alternatively, that a combination of larger sample size and more rigorous statistical designs is required, along with the absence of any complicating concurrent health problems.

The twin problems of small sample size and potential influence of co-occurring health problems may also attach to the Zheng et al. (2001) findings, where adult patients with diverse neurological and neurotrauma problems ($N = 82$) were tested for any connections of PbB with serum and cerebrospinal fluid (CSF) total thyroxine and transthyretin (TTR) levels. Here, Pb exposures were modest, with the mean PbB reported as 15 μg/dl (rounding). The sole significant finding was a statistically significant association of PbB increase with declining TTR levels.

Four relatively recent reports concern various categories of Pb workers evaluated with respect to Pb exposure and thyroid function. Schumacher et al. (1998) studied 151 primary smelter workers at a lead–zinc operation in Trail, British Columbia, Canada. Mean PbB was 24 μg/dl and workers were arrayed into quartiles based on PbB measurement: <15, 15–24, 25–39, and ≥40 μg/dl. Endocrine endpoints were levels of free (FT4) and total (TT4) thyroxine, total triiodothyronine (TT3), and thyroid stimulating hormone (TSH). No relationship was found with any endpoint for either mean or quartiled PbB values. This particular operation's concurrent worker PbB levels were relatively low compared to historic primary smelter exposures and reflected major changes in metallurgical processes and emission controls beginning in 1996 (Hilts, 2003). It is not apparent from the findings whether the long-term Pb exposure index, bone Pb, would have shown positive results in those long-term workers at the plant before marked reductions in lead releases.

Tuppurainen et al. (1988) evaluated a cohort of battery plant workers ($N = 176$) presenting with a relatively high PbB (mean = 56, range = 134 μg/dl), using the same battery of thyroid endpoint measures as did Schumacher et al. Increasing Pb exposure duration in the workers was significantly associated with reductions in serum FT4 and TT4 levels, but no significant links were noted with PbB measures. Workers with more than 7.6 years of battery plant experience showed the strongest association.

Erfurth et al. (2001) reported no effects of PbB (mean = 31 μg/dl) on serum T3, T4, or TSH in their study of secondary smelter workers ($N = 88$), paralleling the findings with primary smelter workers. Here, too, Pb exposures in terms of PbB were lower than typical in such lead workers. Dursun and Tutus (1999) described results with workers manufacturing lead powders in terms of Pb and thyroid function tests. Increases in T3 and T4 but no changes in TSH levels were reported with increasing PbB.

Overall, Pb appeared to produce inconsistent effects across the various studies. In some cases, the endpoint measures were reduced with increased exposures, increases were seen elsewhere, and in still other cases, no effects were observed. It may well be, at least for thyroid endocrinological endpoints, that Pb in blood is not a particularly useful biomarker. Tuppurainen et al.'s finding that exposure duration was more significantly linked to thyroid effects than PbB suggests that a better exposure biomarker for thyroid toxicity relationships may be bone Pb measures.

19.1.2 Effects of Lead on Male and Female Sex Hormones in Humans and Experimental Animals

Current data, mainly gathered in experimental animal models, indicate that sex hormone levels and functions are affected by Pb at multiple sites in males and females within a hypothalamic–pituitary–gonadal axis, with Pb exerting potential impacts on testosterone (TES), estradiol (E_2), luteinizing hormone (LH), insulin growth factor 1 (IGF_1), follicle-stimulating hormone (FSH), prolactin (PRL), and gonadotropin-releasing hormone (GnRH). There are some data that indicate that such an axis exists and functions in humans as well, although available information is largely confined to male lead workers with relatively high Pb exposures. Little has appeared in the lead epidemiological literature describing Pb effects on women's sex hormone levels or functions, e.g., stability of the menstrual cycle.

A number of more recent studies indicate that Pb appears to function as a disrupter of endocrine action, based on data from both male and female animals and results of studies in male Pb workers. The nature and extent of such disruptions, however, are not consistent. The search for endpoints in human populations has measured levels of various hormones in accessible biological media, such as serum or plasma, obtained from mainly smelter or lead battery plant employees. PbB levels have ranged considerably across these investigations (Table 19.2), with typically observed mean or median PbB values of 30–40 μg/dl. These studies complement, at a lower part of the dose–response curve, the earlier findings of Braunstein et al. (1978) and Cullen et al. (1984) where Pb exposures were much greater, sufficient to produce clinical evidence of lead poisoning.

Ng et al. (1991) reported positive association of serum FSH and LH in battery workers with rising PbB, along with reduced TES and no change in PRL. This group had a relatively moderately elevated mean PbB of 35 μg/dl. Two groups of secondary smelter workers with PbB mean values not materially different produced somewhat inconsistent results. Gustafson et al. (1989) found no effects on PRL or TES, but reductions in levels of FSH and LH, for their cohort of smelter workers having a mean PbB of 39 μg/dl. A median PbB of 31 μg/dl in the smelter worker cohort of Erfurth et al. (2001) failed to yield any effects on any of the hormonal measures versus controls. Alexander et al. (1998) examined primary smelter workers and found an inverse relationship of serum TES level with PbB (range = 5–58 μg/dl) but no changes in the other endpoints (FSH, LH).

TABLE 19.2 Effects of Lead on Male Sex Hormone Levels in Lead Workers

Subjects	Pb Exposure	Measures[a]	Results	References
Secondary smelter workers ($N = 88$, exposed + controls)	Median PbB = 31 μg/dl	FSH, LH, PRL, TES	No material effects of Pb on any reported measure	Erfurth et al. (2001)
Primary smelter workers ($N = 152$, exposed + controls)	Range, 5–58 μg/dl	FSH, LH, TES	Testosterone levels were reduced with increasing serum Pb levels	Alexander et al. (1998)
Secondary smelter workers ($N = 42$, exposed + controls)	Mean PbB = 39 μg/dl	FSH, LH, PRL, TES	No effects on PRL, TES; reduction in FSH and LH	Gustafson et al. (1989)
Lead battery production workers ($N = 171$, exposed + controls	Mean PbB = 35 μg/dl	FSH, LH, PRL, TES	Mixed results: increased FSH, LH levels; reduced testosterone; no change in PRL	Ng et al. (1991)

[a]*FSH, follicle-stimulating hormone; LH, luteinizing hormone; PRL, prolactin; TES, testosterone.*

The inconsistencies in hormonal changes in these lead workers have no ready explanation. In the case of TES, all studies adjusted for age, given its inverse relationship to TES concentration. Sample sizes were generally small, particularly those of Erfurth et al. and Gustafson et al. for secondary smelter employees. A threshold to such effects in terms of PbB is also not apparent from a collective look at the data.

Experimental data are much more revealing with regard to sex hormone changes *vis-a-vis* Pb exposures, particularly findings in nonhuman primates sustaining chronic exposures at realistic Pb concentrations for relevance to human effects. Illustrative studies are tabulated in Table 19.3.

By and large, chronic Pb dosings of female nonhuman primates produce various effects on sex hormonal levels and stable periodicities in menstrual cycle and menstrual flow. These effects are dependent on dosing level, periods of exposure, developmental time windows employed for dosings, etc. The most sensitive endpoint in the monkey was reduced corpus luteal production of progesterone over the PbB range of 10–20 μg/dl (Foster et al., 1996). In exposed rats, delay in onset of puberty and reduced serum concentrations of IGF_1, LH, and E_2 were noted. PbB concentrations in rat pups whose developmental Pb exposures were either during gestation, lactation, or both declined to low levels, 1–3 μg/dl, before pubertal onset (Dearth et al., 2002). This suggested a persisting disturbance of *in utero* and preweaning hormonal function by Pb. Furthermore, no dosing group displayed a

TABLE 19.3 Experimental Animal Studies of Pb Effects on Sex Hormone Levels and Functions

Test Species	Testing Regimen	Pb Exposures	Endpoints	Results	References
A. Female animals					
Cynomolgus monkeys	Chronic Pb exposures (acetate), birth to 15–20 years, oral dosing, 50–2,000 μg/kg/day	PbB: 10–15 μg/dl, low group; 25–30 μg/dl in higher groups	20-δ-OH-progesterone, progesterone, and E_2 levels; menstrual cycle	Reduced progesterone, no change in E_2 or other measures	Foster et al. (1996)
Cynomolgus monkeys	Chronic Pb exposures (acetate), daily at 1.5 mg/kg and shorter dosing windows	PbB<40	LH, FSH, E_2 levels; progesterone, menstrual cycle	Significant reductions in LH, FSH, and E_2 during menstrual cycle; no change in progesterone or cycle	Foster (1992)
Rhesus monkeys	Pb (acetate) in drinking water (2–8 mg/kg, 33 months)	Exposed PbB mean = 69 μg/dl	Progesterone levels, menstrual cycle, menstrual flow period	Reduced progesterone levels; longer, more variable cycles; shorter flow	Franks et al. (1989)
Squirrel monkeys	Pb (acetate), 0.1% or less in diet, maternal dosing, 5–8.5 weeks pregnancy→d1	Mean maternal PbB = 37 μg/dl across dose groups	Changes in placental weight and pathology	Reduced placental weight, lesions such as hemorrhages	Lögdberg et al. (1988)

Fischer 344 rat	12/mg/ml Pb (acetate); gavage; 30 days prior to breeding to weaning	Overall maternal PbB mean = 40 μg/dl; pups, up to 38 μg/dl	Puberty onset, serum levels of IGF and E_2	Delay in puberty; significantly reduced IGF_1, E_2 levels	Dearth et al. (2002)
B. Male animals					
Cynomolgus monkeys	Pb (acetate) daily oral dosing: 0–1,500 μg/kg; period ranging from lifetime to infancy or postinfancy exposures	PbB = 3–26 μg/dl, lifetime group	LH response to GnRH stimulation in various groups	Reduced LH response in lifetime group	Foster et al. (1993)
Cynomolgus monkeys	Pb (acetate) daily, birth to 10 years, or other periods	Mean PbB = 35 μg/dl	Circulating levels of FSH, LH, TES	No effects of Pb on FSH, LH, TES at these dosing regimens	Foster et al. (1998)
Sprague–Dawley rats	0.6% Pb (acetate) in drinking water at various times: postnatal, 24–74 days for pubertal exposure +postpubertal exposures	PbB>250 μg/dl	TES in males	Reduced TES	Ronis et al. (1996)

maximum PbB value higher than 38 μg/dl, the maximum for animals sustaining Pb exposures through the prenatal period and through lactation.

Male monkeys exposed to Pb during their lifetime or early infancy showed reduced LH responses to GnRH stimulation. The lifetime exposure group PbB values ranged from 3 to 26 μg/dl over time (Foster et al., 1993). Foster et al. (1998) also noted no effects of Pb at a later age in terms of circulating levels of FSH, LH, and TES. Male rats exposed to Pb during prenatal and lactational periods produced much higher PbB values with continued dosing up to 85 days of age, the highest oral dosing level producing 214 μg/dl.

19.1.3 Effects of Pb on Stress and Other Hormones in Humans

A new area of research with very important implications for pediatric plumbism at very low doses has to do with Pb effects on the hypothalamic–pituitary–adrenal (HPA) axis in humans and experimental systems. Several studies in young children relevant to the topic have been conducted by Gump et al. (2005, 2008). The association between increased pediatric cardiovascular reactivity in young children with increasing PbB (Gump et al., 2005) was noted in the discussion on cardiovascular effects. Later, Gump et al. (2008) reported that in their longitudinal child cohort, prenatal and early postnatal PbB levels were independently associated with salivary cortisol levels once an acute stress (cold pressor test, using hand immersion in ice water) was applied when these children were an average of 9.5 years old. Resting cortisol levels were not affected by PbB. PbB levels were quartile-ranked for both cord blood and postnatal (2.6 years of age) samples. Increasing PbB levels were significantly associated with heightened salivary cortisol responses. In the prenatal PbB group, the maximum PbB value was 6.3 μg/dl while the maximum postnatal PbB was 13.1 μg/dl. Only six children of a total of 169 had PbB values ≥10 μg/dl. These data indicate that HPA stimulation and pediatric cortisol levels in response to acute stress are quite sensitive to effects of Pb.

Several studies in the older literature address effects of Pb on calcitropic endocrine functions in humans. Earlier discussion addressed effects of Pb on children's 1,25-$(OH)_2$-D_3 levels, where relatively high PbB concentrations in children (>30 μg/dl) depressed vitamin D levels. Calcium repletion even in the presence of elevated PbB values appears to abolish the reduction. Mixed results for levels of parathyroid hormone (PTH), calcitonin (CAL), 1,25-$(OH)_2$-D_3, and 25-hydroxy vitamin D (25-OH-D_3) have been noted for lead workers with high PbB values compared to adults in the general population. Kristal-Boneh et al. (1998) saw increases in the two hormones reported, serum PTH and 1,25-$(OH)_2$-D, with increasing PbB (mean = 43 μg/dl). Mason et al. (1990), by contrast, saw an increase in the dihydroxy vitamin D but no change in 25-OH-D_3.

19.1.4 Experimental Pb Effects on Other Hormonal Systems

Effects of Pb on other hormonal systems have been largely absent from the epidemiological literature and one is confined to results from experimental animal systems. Ronis et al. (2001) noted a Pb-induced reduction in pituitary response to hypothalamic stimulation as the underlying basis for reduced bone length and strength in rats exposed via combined prenatal and postnatal dosings. Subsequently, Dearth et al. (2002) found that Pb appeared to disturb secretion of IGF_1, with this hormone then decreasing LH-releasing hormone in the hypothalamus.

A more recent area of research on experimental Pb-induced hormonal disturbance is that of stress hormonal alterations. Cory-Slechta et al. (2004) reported a persistent effect of maternal Pb dosing on corticosteroid concentrations in offspring reaching adulthood. Corticosteroid levels remained elevated in both sexes, with females showing an amplified effect when there was joint use of a toxic stress of Pb exposure and a physical stress via physical restraint. Here again, a developmental endocrinological effect is seen to be both more sensitive and more persistent, particularly with prenatal intoxication. A second experimental regimen shown to be positive for association of stress responses with early (neonatal) Pb exposure in animals was the finding of Yu et al. (1996) that Pb reduced endurance in the cold water swimming test.

19.2 EFFECTS OF Pb ON THE GI TRACT IN HUMANS AND ANIMALS

Lead "colic" has long been recognized among clinicians as a symptomatic correlate of clinical-level lead exposures in infants, toddlers, and highly exposed lead workers. The earlier literature has been described in several expert consensus documents (NAS/NRC, 1972; U.S. CDC, 1978; U.S. EPA, 1977, 1986). Typically, the clinical GI effects of Pb that are clustered under the label of "lead colic" include anorexia, abdominal pain, constipation, nausea, and vomiting. These may erupt along with acute encephalopathy or be a persistent feature of chronic neuropathy in infants and toddlers, while Pb workers are typically plagued with chronic GI disturbances.

GI toxicity of Pb, while of considerable vintage in terms of a clinical literature, has largely been ignored as the focus of more current interest is in the lead epidemiology or experimental toxicity literature. This has to do with the broad view that GI effects are relatively insensitive in terms of either dose–response relationships or with regard to being a useful early effect biomarker. Part of this lack of quantitative relationships is that Pb colic can express itself across a wide range of exposures and has little to offer the clinician in terms of anything pathognomonic for lead poisoning. Pb toxicity is difficult to distinguish etiologically from other sources of abdominal effects, e.g., the presence of peptic ulcer or pancreatitis in adults.

Available dose–response data for GI effects in humans principally arise from findings with workers exposed to lead (Table 19.4). These occupational groups include smelter workers (Dahlgren, 1978; Lee et al., 2000), battery plant employees (Awad El Karim et al., 1986; Matte et al., 1989), demolition

TABLE 19.4 Effects of Pb on the GI and Hepatic Systems in Human Populations

Subjects	Pb Exposure	Endpoints Measured	Reported Results	References
A. GI system				
Workers in secondary smelters, battery production, PVC stabilization production ($N = 95$)	Median PbB = 46 μg/dl	GI symptoms: appetite loss, constipation, abdominal pain	No association of PbB with GI effects	Lee et al. (2000)
Battery production and repair workers ($N = 63$)	PbB divided into: PbB <60 μg/dl and PbB ≥60 μg/dl	GI symptom battery: impaired appetite, nausea, abdominal pain	No significant difference between the two PbB groups	Matte et al. (1989)
Demolition workers at two PbB ranges ($N = 119$ total): 50–70 ($N = 87$); >70 ($N = 19$)	PbB ranges: 50–70 μg/dl; >70 μg/dl	GI symptom set: abdominal pain, constipation	Statistically significant increase in prevalence of symptoms at higher PbBs versus PbB <50 μg/dl	Holness and Nethercott (1988)
Battery production workers	PbB range: 55–81 μg/dl, battery workers; 7–33 μg/dl, control workers	GI symptoms: "abdominal colic," constipation	Increased prevalence of GI symptoms in the battery workers	Awad El Karim et al. (1986)
Smelter workers ($N = 34$)	70 μg/dl, mean in workers with chronic GI symptoms	Nausea, abdominal pain, abnormal bowel movements	15 workers at high PbB had chronic GI effects but no dose–response seen	Dahlgren (1978)
New York City lead cable splicers ($N = 90$)	PbB means: with symptoms, 30 μg/dl; without symptoms, 27 μg/dl	Lead colic	21% of workers reported GI symptoms; PbB difference not significant	Fischbein et al. (1980)

(Continued)

TABLE 19.4 Effects of Pb on the GI and Hepatic Systems in Human Populations—(cont.)

Subjects	Pb Exposure	Endpoints Measured	Reported Results	References
B. Hepatic system				
Pb-chelatable children ($N = 26$, aged 2–9 years)	Mean PbB, 46 μg/dl	Urinary levels of 6-β-OH-cortisol/cortisol ratio	Lower urinary results in both biomarkers versus control children; significant correlation with PbB and chelatable Pb	Saenger et al. (1984)
Adult Thailand volunteers ($N = 118$) from general population	Urinary Pbs	Urinary levels of 7-OH-coumarin metabolite of coumarin	Females had a significant association of Pb–U with 7-OH-coumarin	Satarug et al. (2004)
Lead battery workers ($N = 30$) longitudinally studied	Mean PbBs: 60 μg/dl at start; 30 μg/dl at finish	Nonspecific liver injury indexed by enzyme levels	No reduction in ALT	Hsiao et al. (2001)
Workers exposed to Pb in various jobs: garage, gas pump, printing, construction ($N = 100$)	Mean PbB = 78 μg/dl, workers; mean PbB = 20 μg/dl, controls	Liver enzyme levels: ALT, AST, γ-GT)	No differences in ALT, AST, or γ-GT levels between PbB groups	Al-Neamy et al. (2001)

workers (Holness and Nethercott, 1988), and cable splicers (Fischbein et al., 1980). Results in terms of symptomatology versus exposure levels are mixed and do not reveal any clear dose–response relationship. Lee et al. and Matte et al. both reported no GI effects in workers having either a median PbB of 46 μg/dl (Lee et al.) or workers in two groups having PbB <60 or ≥60 μg/dl (Matte group). Those studies searched for appetite loss, constipation, nausea, and abdominal pain.

The remaining four studies summarized in Table 19.4 did show various effects with workplace exposure. These studies generally consisted of workers presenting with higher PbB levels but did not allow for a clear threshold or for dose–response relationships to be estimated. In the aggregate, it appears that a PbB level of 50 μg/dl or somewhat higher can be linked to at least some GI effects. One can also conclude, within limits of the current available evidence, that GI effects collectively are still to be considered insensitive measures of Pb toxicity in humans relative to developmental neurotoxicological, cardiovascular, nephrotoxicological, and immunotoxicological dose–responses.

19.3 EFFECTS OF Pb ON THE HEPATIC SYSTEM IN HUMANS AND ANIMALS

Hepatic toxicity of Pb has been infrequently studied, and available data are of some vintage. Two types of effects have been reported in the literature. Evidence of a generalized injury indexed by elevations in circulating levels of enzymes in serum has been reported, as have disturbances in cytochrome P450 subtypes with Pb exposure, comprising a second category of adverse responses. The generalized injury appears to be a relatively insensitive endpoint for Pb toxicity, a conclusion supported by several more recent studies in the occupational literature. Hsiao et al. (2001) noted no effects of PbB on serum alanine aminotransferase (ALT) activity in a longitudinal study of lead battery workers where the starting mean PbB was 60 μg/dl and the ending mean was 30 μg/dl. Similarly, no changes with elevated PbB concentrations in diversely exposed workers (e.g., printers, construction workers) were observed for ALT, aspartate aminotransferase (AST), or γ-glutamyl transferase (γ-GT) activities. The mean PbB of the exposure group was quite high, 78 μg/dl, versus a control mean of 20 μg/dl. No dose–response relationships were estimated nor could a threshold PbB be identified from the relatively meager available information.

The effects of Pb on the induction and functions of the hepatic cytochromes were addressed to some extent as part of the earlier discussion on multiorgan heme impairments by elevated Pb exposure. Hepatic cytochrome P450 isozymes, of necessity, have principally been studied using experimental animals. Few human data are available, and in those cases, investigators indirectly evaluated Pb impairments via changes in metabolite levels arising from substrates known to be transformed by P450 systems. In a small group of children ($N = 26$, age 2–9 years of age) with an average PbB of 46 μg/dl, Saenger et al. (1984) reported that Pb reduced the urinary level of the cortisol metabolite 6-β-hydroxycortisol and the metabolite/cortisol ratio, this metabolite arising from hepatic CYP3A4. The ratio was significantly correlated with PbB, urinary Pb, and EDTA-mobilization Pb.

Satarug et al. (2004) reported that Thai women among the general population showed a significant inverse relationship between urinary Pb level and the urinary excretion of the 7-hydroxy metabolite of coumarin, a transformation mediated by the CYP2A6 phenotype. The investigators reported blood Pb as serum, not whole blood, concentrations. It is difficult to convert the values to whole blood without estimates that are too high for the general population under discussion.

REFERENCES

Alexander, B.H., Checkoway, H., Faustman, E.M., Van Netten, C., Muller, C.H., Ewers, T.G., 1998. Contrasting associations of blood and semen lead concentrations with semen quality among lead smelter workers. Am. J. Ind. Med. 34, 464–469.

Al-Neamy, F.R., Almehdi, A.M., Alwash, R., Pasha, M.A.H., Ibrahim, A., Bener, A., 2001. Occupational lead exposure and amino acid profiles and liver function tests in industrial workers. Int. J. Environ. Health Res. 11, 181–188.

Awad El Karim, M.A., Hamed, A.S., Elhaimi, Y.A., Osman, Y., 1986. Effects of exposure to lead among lead-acid battery factory workers in Sudan. Arch. Environ. Health 41, 261–265.

Braunstein, G.D., Dahlgren, J., Loriaux, D.L., 1978. Hypogonadism in chronically lead-poisoned men. Infertility 1, 33–51.

Cory-Slechta, D.A., Virgolini, M.B., Thiruchelvam, M., Weston, D.D., Bauter, M.R., 2004. Maternal stress modulates the effects of developmental lead exposure. Environ. Health Perspect. 112, 717–730.

Cullen, M.R., Kayne, R.D., Robins, J.M., 1984. Endocrine and reproductive dysfunction in men associated with occupational inorganic lead intoxication. Arch. Environ. Health 39, 431–440.

Dahlgren, J., 1978. Abdominal pain in lead workers. Arch. Environ. Health 33, 156–159.

Dearth, R.K., Hiney, J.K., Srivastava, V., Burdick, S.B., Bratton, G.R., Dees, W.L., 2002. Effects of lead (Pb) exposure during gestation and lactation on female pubertal development in the rat. Reprod. Toxicol. 16, 343–352.

Dursun, N., Tutus, A., 1999. Chronic occupational lead exposure and thyroid function. J. Trace Elem. Exp. Med. 12, 45–49.

Erfurth, E.M., Gerhardsson, L., Nilsson, A., Rylander, L., Schütz, A., Skerfving, S., et al., 2001. Effects of lead on the endocrine system in lead smelter workers. Arch. Environ. Health 56, 449–455.

Fischbein, A., Thornton, J., Blumberg, W.E., Bernstein, J., Valciukas, J.A., Moses, M., et al., 1980. Health status of cable splicers with low-level exposure to lead: results of a clinical survey. Am. J. Public Health 70, 697–700.

Foster, W.G., 1992. Reproductive toxicity of chronic lead exposure in the female cynomolgus monkey. Reprod. Toxicol. 6, 123–131.

Foster, W.G., McMahon, A., Younglai, E.V., Hughes, E.G., Rice, D.C., 1993. Reproductive endocrine effects of chronic lead exposure in the male monkey. Reprod. Toxicol. 7, 203–209.

Foster, W.G., McMahon, A., Rice, D.C., 1996. Subclinical changes in luteal function in monkeys with moderate blood lead levels. J. Appl. Toxicol. 16, 159–163.

Foster, W.G., Singh, A., McMahon, A., Rice, D.C., 1998. Chronic lead exposure effects in the monkey (*Macaca fascicularis*) testis. Ultrastruct. Pathol. 22, 63–71.

Franks, P.A., Laughlin, W.K., Dierschle, D.J., Bowman, R.E., Meller, P.A., 1989. Effects of lead on luteal function in rhesus monkeys. Biol. Reprod. 41, 1055–1062.

Gump, B.B., Reihman, J., Stewart, P., Lonky, E., Darvill, T., Matthews, K.A., et al., 2005. Prenatal and early childhood blood lead levels and cardiovascular functioning in 9.5-year-old children. Neurotoxicol. Teratol. 27, 655–665.

Gump, B.B., Stewart, P., Reihman, J., Lonky, E., Darvill, T., Parsons, P.J., 2008. Low-level prenatal and postnatal blood lead exposure and adrenocortical responses to acute stress in children. Environ. Health Perspect. 116, 249–255.

Gustafson, Å., Hedner, P., Schütz, A., Skerfving, S., 1989. Occupational lead exposure and pituitary function. Int. Arch. Occup. Environ. Health 61, 277–281.

Hilts, S.R., 2003. Effect of smelter emission reductions on children's blood lead levels. Sci. Total Environ 303, 51–58.

Holness, D.L., Nethercott, J.R., 1988. Acute lead intoxication in a group of demolition workers. Appl. Ind. Hyg. 3, 338–341.

Hsiao, C.Y., Wu, H.D., Lai, J.S., Kuo, H.W., 2001. A longitudinal study of the effects of long-term exposure to lead among lead battery factory workers in Taiwan (1989–1999). Sci. Total Environ 279, 151–158.

Kristal-Boneh, E., Froom, P., Yerushalmi, N., Harari, G., Ribak, J., 1998. Calcitropic hormones and occupational lead exposure. Am. J. Epidemiol. 147, 458–463.

Lee, B.-K., Ahn, K.-D., Lee, S.-S., Lee, G.-S., Kim, Y.-B., Schwartz, B.S., 2000. A comparison of different lead biomarkers in their associations with lead-related symptoms. Int. Arch. Occup. Environ. Health 73, 298–304.

Lögdberg, B., Brun, A., Berlin, M., Schütz, A., 1988. Congenital lead encephalopathy in monkeys. Acta Neuropathol. 77, 120–127.

Mason, H.J., Somervaille, L.J., Wright, A.L., Chettle, D.R., Scott, M.C., 1990. Effect of occupational lead exposure on serum 1,25-dihydroxyvitamin D levels. Hum. Exp. Toxicol. 9, 29–34.

Matte, T.D., Figueroa, J.P., Burr, G., Flesch, J.P., Keenlyside, R.A., Baker, E.L., 1989. Lead exposure among lead-acid battery workers in Jamaica. Am. J. Ind. Med. 16, 167–177.

National Academy of Sciences: National Research Council, 1972. Lead: Airborne Lead in Perspective. National Academy Press, Washington, DC.

Ng, T.P., Goh, H.H., Ng, Y.L., Ong, H.Y., Ong, C.N., Chia, K.S., et al., 1991. Male endocrine functions in workers with moderate exposure to lead. Br. J. Ind. 48, 485–491.

Refowitz, R.M., 1984. Thyroid function and lead: no clear relationship. J. Occup. Med. 26, 579–583.

Ronis, M.J.J., Badjer, T.M., Shena, S.J., Roberson, P.K., Shaikh, F., 1996. Reproductive toxicity and growth effects in rats exposed to lead at different periods during development. Toxicol. Appl. Pharmacol. 136, 361–371.

Ronis, M.J.J., Aronson, J., Gao, G.G., Hogue, W., Skinner, R.A., Badger, T.M., et al., 2001. Skeletal effects of developmental lead exposure in rats. Toxicol. Sci. 62, 321–329.

Saenger, P., Markowitz, M.E., Rosen, J.F., 1984. Depressed excretion of 6-β-hydroxycortisol in lead-toxic children. J. Clin. Endocrinol. Metab. 58, 363–367.

Satarug, S., Nishijo, M., Ujjin, P., Vanavanitkun, Y., Baker, J.R., Moore, M.R., 2004. Evidence for concurrent effects of exposure to environmental cadmium and lead on hepatic CYP2A6 phenotype and renal function biomarkers in nonsmokers. Environ. Health Perspect. 112, 1512–1518.

Schumacher, C., Brodkin, C.A., Alexander, B., Cullen, M., Rainey, P.M., van Netten, C., et al., 1998. Thyroid function in lead smelter workers: absence of subacute or cumulative effects with moderate lead burdens. Int. Arch. Occup. Environ. Health 71, 453–458.

Siegel, M., Forsyth, B., Siegel, L., Cullen, M.R., 1989. The effect of lead on thyroid function in children. Environ. Res. 49, 190–196.

Tuppurainen, M., Wagar, G., Kurppa, K., Sakari, W., Wambugu, A., Froseth, B., et al., 1988. Thyroid function as assessed by routine laboratory tests of workers with long-term lead exposure. Scand. J. Work Environ. Health 14, 175–180.

U.S. Centers for Disease Control, 1978. Preventing Lead Poisoning in Young Children. A Statement by the Centers for Disease Control. U.S. Department of Health, Education, and Welfare, Atlanta, GA.

U.S. Environmental Protection Agency, 1977. Air Quality Criteria for Lead. Report No. EPA-600/8-77-017. Office of Research and Development, Washington, DC.

U.S. Environmental Protection Agency, 1986. Air Quality Criteria for Lead, 4 vols. EPA 600/8-83/028bF. Environmental Criteria and Assessment Office, Washington, DC.

Yu, S.Y., Mizinga, K.M., Nonavinakare, V.K., Soliman, K.F., 1996. Decreased endurance to cold water swimming and delayed sexual maturity in the rat following neonatal lead exposure. Toxicol. Lett. 85, 135–141.

Zheng, W., Lu, Y.M., Lu, G.Y., Zhao, Q., Cheung, O., Blaner, W.S., 2001. Transthyretin, thyroxine, and retinol-binding protein in human spinal fluid: effect of lead exposure. Toxicol. Sci. 61, 107–114.

Part 4

Human Health Risk Assessment

Chapter 20

Human Health Risk Assessment for Lead: Introduction, Context, Rationale

Part 4 consists of five chapters that provide quantification of the health risks of lead exposure in segments of human populations at significant risk of adverse health effects from such exposures. They sequentially present and quantify risk expressions using a reasonable and generally acceptable model of human health risk assessment for those contaminants such as environmental lead that *have already been emitted into the human environment*. Such a risk assessment model proceeds through merging and quantitative integration of the previous parts of this book and attempts to answer the question of how much of a threat to human health lead exposures pose.

The previous parts of this monograph have already provided some quantitative looks at lead and public health within more specific topics. Part 1 presented a measure of the entirety of lead contamination in the human environment, while Part 2 offered descriptions of the nature and extent of lead exposures in the United States and internationally. Part 3, in providing the range of Pb toxic effects in human populations, also referred to various expressions of Pb exposures reported as linked to these various effects epidemiologically. The need to translate those earlier parts and chapters into a coherent expression of threat of harm to human health remains.

20.1 INTRODUCTION

Health risk assessment in humans has been variably defined by different disciplines and subdisciplines (NAS/NRC, 1983, 2009). These definitions broadly settle on a meaning which seeks to describe, *within the available evidence*, a process for the quantitative determination of toxicity risks in proportion to the extent of actual exposure among those likely to sustain such exposures from either having predisposing characteristics or being more likely to be at elevated risk. The definition is as sparse as it is limited. For example, risks of harm to health by various environmental contaminants, in environmental reality, range considerably as to: (1) the nature of the effects

Trace Metals and other Contaminants in the Environment, Volume 10
ISSN: 1875-1121 DOI: 10.1016/B978-0-444-51554-4.00020-1

such as being carcinogenic or noncarcinogenic, (2) the nature and size of the human populations at risk for such toxic responses, and (3) variability and uncertainty attending the risk assessments.

The relatively recent emergence on the regulatory and public health policy scene of the theory and practice of risk assessment has been accompanied by an associated vocabulary and use of diverse methods, methods which have been variably validated. Responses to this form of evaluating likelihood of environmental harm from contaminant exposures by the scientific and public health world have been mixed. For example, regulated stakeholders and/or their agents are tempted to advance the view that risk assessment can remove much of claimed guesswork from regulatory actions, thereby allowing (but hardly guaranteeing) less adverse economic impacts on those same stakeholders.

The idea of using the tools of risk assessment to underpin legislative, policy, and regulatory actions, when represented by their practitioners as science in the service of policy, would potentially appeal to legislators, health officials, and regulatory staffs. Some environmental regulatory agencies dealing with risk assessment practices and methodologies may view quantitative approaches as a considerable improvement over using qualitative or semiquantitative decision making.

Such takes on the topic have not been universal, however, and skepticism is not without merit (Michaels, 2008; Montague, 2004; Silbergeld, 1993). Those holding a more skeptical view assert that risk assessment concepts and methodologies necessarily entail their own cluster of uncertainties, i.e., one set of preexisting uncertainties is replaced by a second, perhaps more complex set of uncertainties. In such cases, risk assessment remains *risky* assessment. Furthermore, persisting uncertainty in the risk assessment process, whatever the specifics of methods, will still require extensive use of professional judgments, judgments that are ideally informed and guided by the intervention of strong multidisciplinary backgrounds. Skepticism also attends the view that protracted risk evaluation approaches merely serve to drag out regulatory timetables.

The use of judgments by practitioners in public health risk assessment is a particularly thorny issue. Not only are there potential disciplinary differences in judgments in risk assessment but there are also differences in any particular biases that may be brought to the table. Problems of scientific bias or favoritism among scientists have been broadly recognized in past years, as seen in the 1980 essay, *On Bias: Does Where You Stand Really Depend on Where You Sit?* by Philip Handler, the then President of the NAS/NRC (Handler, 1980). The nature and scope of the issue has become increasingly difficult to recognize or characterize in the contemporary scientific community by its societal clients, constituents, and even by scientists themselves.

Scientific review committees and other panels charged with producing bias-neutral products for public sponsors have typically relied, and continue to rely, on the mechanism of aggregate balance and overall consensus. That is, any individual biases or views on scientific topics among members of some public deliberative body would be assumed to cancel out, eventually leading to a neutral product. The operative problem in recent years has been one of assuring full disclosure of all such biases and potential conflicts of interest. Elliott (2011) has recently described the special difficulties of bias recognition when research activity funding is not disclosed.

Human health risk assessment of the type introduced above is not the only risk assessment modality operative in various areas. A second approach is that of the Precautionary Principle (see, for example, Percival 2006). The Precautionary Principle figures in health risk assessment to some extent but here it is presented for discussion within the final topic of this work, lead regulation. While this approach has been defined in various ways and often appears to be misunderstood, it basically describes a proactive, *a priori*, and more conservative methodology in terms of the consumer sector. This approach attempts to prevent or minimize risk of any public health and safety hazards before commercial introduction of a product occurs. Substances enter or do not enter the commercial domain based on evaluations generated beforehand.

These two forms of approaching public health risk figure in how different regulatory and political jurisdictions have conceptualized risk. While the human health risk assessment model first described by the NAS/NRC (1983) that has evolved for use by U.S. public agencies is compartmentalized and formulaic, public bodies such as the EU and its member states rely on the Precautionary Principle. The Precautionary Principle, in terms of its regulatory functioning within the EU, operates through the currently implemented and evolving EU program, where there is a process of testing, evaluation, and registration of new marketable substances, the REACH (Registration, Evaluation, Authorization, and Restriction of Chemicals) program.

The use of human health risk assessment versus use of the Precautionary Principle through REACH depends, to some extent, on the nature of the risk and associated regulation and the stage of some economically useful substance's life cycle. The Precautionary Principle by definition is most relevant and applicable to *ante hoc* control over new substances prior to their marketing and entry into the environment or for assessment of new uses for existing substances already released into human environments. Sequenced and compartmentalized risk assessment methods extant in the United States by definition are *post hoc* tools for dealing with ongoing or potential risks from substances already in the human environment.

These two modalities also interact differently with public health policy. Conventional risk assessment helps inform efforts at secondary disease prevention or disease risk prevention by first quantifying the extent of contact with existing toxic substances and then the extent of harm to human and other populations from such toxic substances. The Precautionary Principle focuses on, at least conceptually, primary prevention of disease by attempting to avoid human contact with toxic substances to begin with.

The emergence of a global economy and global aspects of environmental impacts from players in that economy would suggest that nations or world regions are not fully at liberty to promote or practice what they might wish in terms of environmental risk assessment and management. For example, implementing various elements of the EU's Precautionary Principle has the obvious effect of extending to and requiring its use by non-EU entities wishing to do business in EU member states since they must first abide by REACH criteria and protocols.

This chapter and the four that follow confine themselves to conventional human health risk assessment, insofar as environmental Pb has been released or continues to be released into the environment in enormous quantities, and conventional risk assessment is the *post hoc* tool for dealing with these past or ongoing releases. To help link this part with the preceding material, Table 20.1 sets forth the connections.

TABLE 20.1 Components of Human Health Risk Assessment for Pb and Their Databases in This Book

Risk Component/ Compartment	Chapters with Supporting Data	Added Elements of Confirmation
Pb hazard characterization in humans, animals, *in vitro* systems	Chapters 11–19	1. Adherence to Bradford Hill criteria 2. Official positions of consensus-based expert treatises 3. Official statements of public agencies nationally, internationally
Pb dose–toxic response relationships	Dose: Chapters 7–10 Toxic response: Chapters 11–19	1. Adherence to Bradford Hill criteria 2. Official positions of consensus-based expert treatises 3. Official statements of public agencies nationally, internationally
Case-specific Pb exposure determinations	Illustrative examples in Chapter 23 using documented cases	Using accepted scientific methods to determine Pb levels, Pb risk groups, Pb exposure pathways
Pb risk characterization	Illustrative examples in Chapter 24 using documented examples	Use of all of the above elements of scientific confirmation

20.2 THE COMPONENTS OF CONVENTIONAL HUMAN HEALTH RISK ASSESSMENT

The extent to which one can do quantitative risk assessments for a contaminant hinges largely on the available database, information which arrays itself into any of the four generic components of a typical risk assessment: hazard characterization, dose–response relationships, case-specific exposure characterization, and the ultimate product, overall risk characterization for some specific contamination and health impact scenario. Hazard characterization and subsequent assessment components can additionally partition into those cases involving carcinogenic/genotoxic effects and those viewed as noncarcinogens. The thoroughness of hazard characterization rests on the nature and the quantity of the supporting scientific information. The more complete and reliable the information base, the more extensive and quantitative the risk assessment can be. Hazard characterization and dose–response relationships are general for a particular substance, while exposure assessment and final risk characterization require specific information as input.

The size and quality of the available database for an environmental pollutant will vary greatly across substances and will also vary within the four components of the typical risk assessment. The variety of adverse health risks of a substance may be qualitatively well known, for example, but dose–response relationships may be poorly quantifiable because of either limits of inadequate exposure measurement data or absence of good biomarkers of adverse effect or absence of information on the full span of the dose–response curve. Hazard characterization and dose–response relationships may both be understood as general descriptors, but case-specific or scenario-specific exposure data may be lacking, requiring judgment about alternative approaches (e.g., default values).

Epidemiological studies involving various subsets of human populations ranked by either biological/other host characteristics or nature of exposures are of primary interest, followed by experimental test system results using animal models. Ranked next are data supporting MOA, i.e., mechanistic, underpinnings for the epidemiological and experimental results. MOA rationales are particularly critical to how carcinogens express effects.

The term "dose–response" in environmental epidemiology is typically understood to mean that quantitative relationship in which adverse effect severity and multiplicity increase in proportion to the intensity of exposure or "dose" indexed externally (intake/uptake quantities) or internally (exposure biomarkers). In the case of experimental animal exposures, reference is often to the administered dose, but biomarkers can also be available. "Dose–response" has also been employed to denote impacts at some selected effect level of a toxicant in terms of increasing affected fractions of some population as exposure increases. In this case, a dose–population response label is more precise.

Many environmental contaminants are ranked as to health risk in terms of dose or exposure markers largely quantified in terms of some inhaled, ingested, or dermally assimilated amount of contaminant. These quantities comprise online databases such as the U.S. EPA's IRIS (U.S. EPA, 2009), where much of the database is given over to derivation of parameters such as the RfD or RfC. The former quantity refers to the amount of a substance ingested daily on a body-weight basis that is taken to be without risk of toxic harm. The latter is the amount of a substance in inhaled air also likely to be free of toxic risk. RfDs and RfCs are derived and applicable to lifetime exposures and thus applicable to risk population exposures where lifetime contact with the contaminant is assumed to occur. Such approaches, however, are not necessary for or applicable to lead for reasons discussed below.

20.3 ENVIRONMENTAL LEAD: ISSUES FOR HEALTH RISK ASSESSMENT

The elements of human health risk assessment for Pb are presented in Table 20.1. Lead is one of those environmental contaminants and human toxicants that produce both (a) actual or expressed toxicity as medically diagnosed in Pb-intoxicated individuals or in groups of individuals and (b) some likely, probable risk of toxicity in those mainly Pb-exposed groups. The interplay in the lead case between any diagnosed individual toxic harm from Pb and informing a clinician's diagnostic decision using an epidemiologically derived database for Pb toxicity is neither unique nor outside conventional practice.

In actuality, it is typically the case that a diagnosis of harm from a substance in a young child would draw reflexively on assembled databases from epidemiological studies. With Pb, the database of epidemiological studies is huge. Furthermore, such criteria as having supporting epidemiological data as a prerequisite for diagnosis and any follow-up clinical management are increasingly being applied at the forensic interface. U.S. courts increasingly hold that an epidemiological database can be an important factor in judging diagnostic probity arising from such judicial precedents as the 1993 U.S. Supreme Court Daubert decision (Daubert v. Merrill Dow Pharmaceutical, Inc.: 509 U.S. 579, 585 (1993)). See, for example, Berger (2005) for impacts of the Daubert decision on toxic tort actions.

Stratification of lead's toxicity in humans into diagnosable disease or probabilistic risks of disease is not simply defined by where exposures and toxic effects align across the full spectrum of lead's dose–response relationships. That is, one cannot infer that only historic clinical effects at high Pb exposures comprise diagnosable disease and that subclinical effects always require population or epidemiological approaches. See the historical scientific perspective on these topics by Mushak (1992).

20.3.1 Lead as an Exemplary Environmental Model for Human Health Risk Assessment

Lead would appear to be a particularly valuable, even unique, model for assessing human health risk assessments for toxic environmental contaminants. There are several factors establishing that status. The first is the amount of available data and the nature of that data. Second, we can identify subsets of human populations at particular risk for both exposure to and toxic effects of lead. Third, the available data for Pb and public health readily partition into the typical components of risk assessment. Finally, all data components present with acceptable quality in the nature and extent of the information.

20.3.2 Lead and Human Health: Volume of Evidence

Lead's current status as a public health issue is defined by a larger database than exists today for any other environmental contaminant. This huge base of knowledge led the U.S. Surgeon General in 1991 to identify childhood lead poisoning as the principal environmental health issue for U.S. children (Sullivan, 1991). That database consists of numerous epidemiological assessments, experimental animal testing results, and numerous mechanistic studies of diverse experimental stripes. This makes lead ideally suited to be a qualitative and quantitative model for useful quantitative application of the elements of human health risk assessment. Lead's status in terms of its wide scope of demographic and socioeconomic impact also makes it a good model for risk assessment as a societal case study.

First, we now have a rather detailed grasp of the many and varied adverse impacts of lead exposure, based on the enormous information base for lead's hazard characterization. Lead produces a range of adverse effects in humans, beginning with very subtle but still significant effects in "asymptomatic" children at PbB levels ranging from <10 to ~ 25 μg/dl. These include: (1) decreases in IQ and academic achievement and increases in a range of neurobehavioral problems such as attention and performance deficits, neuroelectrophysiological and social behavioral indices; (2) immunotoxicity; (3) cardiotoxicity; (4) nephrotoxicity; and (5) alterations in hematological indices. Higher levels of exposure produce additional signs and symptoms of lead poisoning in both adults and children, such as anemia, chronic and acute lead encephalopathy, and in certain cases added adverse effects on the kidney, cardiovascular effects, and continued, more severe neurotoxicity and immunotoxicity.

20.3.3 Human Populations at Higher Risk for Pb exposure and Toxicity

Lead's toxic effects are expressed relatively early in certain subsets of human populations, and these subsets can be defined in terms of certain particular Pb exposure sensitivities or sensitivities arising from biological or physiological vulnerabilities. The lead worker, of course, defines a risk population simply on the basis of greater intensities and durations of Pb exposure arising from their employment. There are some intrinsic behavioral characteristics that define this population subset, but the principal extrinsic characteristic is the higher exposure to Pb. Pb worker cohorts with employment histories of longer duration, however, may demonstrate the "healthy worker" effect, with actual toxicological hyporesponsiveness to workplace Pb contact.

Pb exposure relationships for other population subsets are behaviorally or biologically defined. Two known principal population risk groups are the fetus and the very young child. The fetus, during much of its gestational existence, sustains Pb exposure with little biological constraint on Pb movement from the mother and the mother's Pb exposure sources. Infants and toddlers, however, actually have amplified Pb exposures that occur through several behavioral mechanisms. First, infants and toddlers orally explore their environment once they are sufficiently mobile. This co-occurs with extensive and normal hand-to-mouth activity that allows ingestion of Pb in interior or exterior dusts, in soils, and/or in other nonfood items.

Older literature erroneously characterized such normal hand–mouth activity producing Pb exposure as mainly existing in children showing evidence of pica, the abnormal ingestion of any form of nonfood material. Pica was therefore considered the behavioral factor that permitted lead exposure and attending poisoning, not the presence of high Pb hazards *per se*. Clinical definitions of pica in quantitative terms are still inconclusive, but generally the term applies to ingestion rates greatly above the amounts characterizing normal behaviors for nonfood materials. Mushak (1998) noted that one index of pica is statistical—the ingestion rates at the 95th percentile determined from a representative group of study children. Second, lead exposures and their adverse risks are associated with a number of internal exposure biomarkers that provide the dose/exposure–response relationships for Pb as well as for case-specific exposure and risk characterizations.

Dose–response relationships for lead, in terms of PbB as the exposure biomarker or dose versus some adverse effect, are probably better known than those for virtually any other environmental contaminant and certainly for any other metal or metal-related pollutant. Furthermore, these relationships are buttressed by a vast scientific and public health literature derived from clinical and epidemiological studies, experimental animal testings, and a large variety of sophisticated mechanistic toxicological studies *in vivo* and *in vitro*.

Exposure assessments for lead for both specific and general population cases have an equally huge database associated with them. For example, exposure assessments of environmental contamination at various hazardous waste sites are increasingly important for lead, given that lead occurs at a high frequency as a contaminant at these sites, more so than most other substances.

In addition to a large database on lead exposures assembled empirically over the decades and centuries, the more recent lead literature includes a number of biokinetic models to ascertain body lead burdens and active exposure biomarkers such as lead in whole blood or Pb in bone. These predictive models of systemic lead exposure are of differing complexity and utility in varied exposure settings. Historically, they can be defined as classical compartment models, a hybrid of the compartmental and PB-PK models, or the PB-PK model type.

Lead is a multimedia pollutant, i.e., it provides exposures through diverse environmental media. The specific characteristics of each environmental contributor, such as Pb concentrations and intake amounts of some environmental medium, help determine the extent to which different individuals and populations sustain actual exposures. This also means that while some Pb sources can be characterized on a national, international, or other macroscale as being significant, the actual sources and pathways in specific cases for substances such as lead or other elements require specific evaluation. Some segments of human populations may sustain exposures from Pb in more than one environmental medium, in which case one employs methods to sort out potential contributions or the relative total contributions to total exposures.

Risk assessment efforts for lead exposure and poisoning risk have been described piecemeal in the recent literature covering areas such as extractive industry wastes in the U.S. Intermountain West and industrial or process waste risk assessments more broadly directed throughout America. Parts of this fragmentation of risk assessment effort are regulatory and legislative in origin, topics discussed in Part 5.

20.4 ORGANIZATION OF THE RISK CHAPTERS

The organization of this portion of the book broadly follows the sequence of risk assessment components recommended in the 1983 NAS/NRC report on risk assessment in the Federal government. That report's generic organization is often relied on by public agencies such as the U.S. EPA in its protocols and documents dealing with various mandated environmental health activities for contaminants such as lead.

20.4.1 Human Health Hazard Characterization of Environmental Pb

Chapter 21 presents the principal elements of lead health hazard characterization, an assessment of toxic effects in humans and experimental systems done in broader ranging terms. Also included is Pb's characterization using the Bradford Hill criteria for establishing causal relationships in dose–responses. The quantitative aspects of human lead toxicity are set forth in Chapter 22, presenting the most relevant features of lead's dose–toxic response relationships in various human populations.

Lead induces numerous adverse effects involving multiple tissues, organs, and systems, but not all of them are of equal relevance or gravity for human health nor are all of them well characterized for risk assessment purposes. Chapter 21, in much the same fashion as the rest of this monograph, does not offer an indiscriminate compendium of the universality of adverse effects but rather is a selective treatment of those toxic effects of current principal concern to lead toxicology and public health, documented as to data reliability.

Chapter 21 achieves this collective human health hazard characterization for lead via a ranking of the component human adverse health effects with respect to thoroughness and quality of the data, the relative sensitivity or ranges of sensitivities of the toxic responses with reference to typical ambient or occupational Pb exposures, the relative gravity of the toxic response in terms of both injury and persistence/irreversibility, and reliable supporting evidence from experimental and mechanistic data.

Human health hazard characterization for lead viewed over an extended time, as seen in other sections, has relied on temporal shifts in the definitions of what can be characterized as an adverse effect, i.e., evidence of organ or tissue damage. Health hazard characterization in Chapter 21 is discussed as the evidence for such assessment currently exists. The current database is obviously more extensive than that of past decades, but one cannot exclude the possibility or even the probability that the database will grow with continued research, particularly given the availability of increasingly more sophisticated tools. Table 20.1 notes that health hazard characterization in Chapter 21 mainly draws from the preceding Pb toxicity chapters.

20.4.2 Human Dose–Response Relationships for Pb

The quantitative relationships between how much Pb exposure occurs and what multiplicity and severity of toxic effects ensue from such exposures in sensitive segments of human populations are set forth in Chapter 22. Here, dose–response data are stratified into various risk populations. A generically obvious category of vulnerability in humans is defined in terms of Pb exposure intensity, i.e., the lead worker. However, occupational Pb poisoning *per se* is not the main focus of this tract beyond subsuming the category under the broad label of adult exposures. For example, much of the available data for lead's potential as a human carcinogen derives from findings in occupational Pb exposure cohorts.

20.4.3 Exposure Characterizations for Pb in Human Populations

Chapter 23 presents a focused discussion of the main elements of exposure assessment. Exposure characterization has both generalized and scenario-specific elements. General aspects characterize the overall approaches to exposure measurements regardless of case requirements. Two approaches are those that entail measurement data and those which rely on exposure simulations via modeling results (see, for example, Mushak, 1998). Environmental Pb has an extensive history in terms of both approaches; their specifics have been set forth in earlier chapters. While hazard characterization and dose–responses attach to Pb impacts in human populations as universal, generic parameters, exposure characterization employs scenario- or case-specific data to integrate the first two aspects of risk assessment with exposure specifics.

20.4.4 Health-Based Risk Characterizations for Pb in Human Populations

This melding of components is input for risk characterization, wherein one projects the likelihood of health-based risk in human population segments, given measured or simulated exposures and relevant dose–response relationships.

Chapter 24 presents some general discussions of the health risk characterization process in combination with illustrative uses of Pb risk characterization approaches for recent regulatory or advisory efforts for lead in various media. These examples employ the best characterized and epidemiologically grounded toxic endpoints combined with various reasonably well-grounded exposure approaches. The scope of the illustrative applications also varies considerably, ranging from tightly defined geographic sites to entire countries (United States) or continents (Europe and elsewhere).

An adequate treatment of health risk characterizations for Pb in humans requires that all three input components be evaluated. Reliable dose–toxic response data also require reliable exposure characterizations to complete the risk assessment. Conversely, well-characterized exposures would further require well-established dose–response relationships for the particular toxic contaminant. Absence of hazard assessment data prevents the evaluation of toxic responses in any dose–response relationships regardless of the quality of the dose or exposure characterizations.

20.5 INTERACTIVE, FEEDBACK RELATIONSHIP OF LEAD SCIENCE AND HUMAN HEALTH RISK ASSESSMENT FOR LEAD

An earlier discussion in this book concerned the interactive relationship of lead research and lead regulatory and health policy (Mushak, 1991; NAS/NRC, 1993). This part on health risk assessment for lead demonstrates an analogous relationship for research and risk assessment as depicted in Figure 20.1. New scientific research data on lead toxicology and epidemiology are used to quantify lead risk assessment. Remaining research gaps in

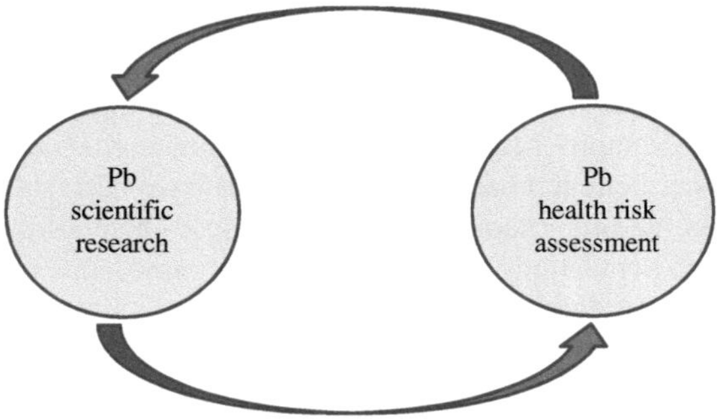

FIGURE 20.1 Interactive, feedback loop relationship for lead science and health risk assessment.

risk assessment then identify the need for more scientific research, the new data then continuing the loop.

REFERENCES

Berger, M.A., 2005. What has a decade of *Daubert* wrought? Am. J. Public Health 95, S59–S65.

Elliott, K.C., 2011. Is A Little Pollution Good For You? Incorporating Societal Values into Environmental Research. Oxford University Press, New York (Chapter 4).

Handler, P., 1980. Chairman's Essay: On Bias—Does Where You Stand Really Depend on Where You Sit? In: The National Research Council/1980: Issues and Current Studies. National Academy of Sciences, National Academy of Engineering, Institute of Medicine. National Academy Press, Washington, DC, pp. 1–17 (Publ. 1981).

Michaels, D., 2008. Doubt Is Their Product: How Industry's Assault on Science Threatens Your Health. Oxford University Press, New York.

Montague, P., 2004. Reducing the harms associated with risk assessment. Environ. Impact Assess. Rev. 24, 733–748.

Mushak, P., 1991. U.S. Agency for Toxic Substances and Disease Registry's report to Congress on childhood lead poisoning in America: review and update. In: Proc. First National Conference on Laboratory Issues in Childhood Lead Poisoning Prevention, Columbia, MD, pp. 79–104.

Mushak, P., 1992. Perspective. Defining lead as the premiere environmental health issue for children in America: criteria and their quantitative application. Environ. Res. 59, 281–309.

Mushak, P., 1998. Uses and limits of empirical data in measuring and modeling human lead exposure. Environ. Health Perspect. 106 (Suppl. 6), 1467–1484.

National Academy of Sciences: National Research Council, 1983. Risk Assessment in the Federal Government: Managing the Process. National Academy Press, Washington, DC.

National Academy of Sciences: National Research Council, 1993. Measuring Lead Exposure in Infants, Children, and Other Sensitive Populations. National Academy Press, Washington, DC.

National Academy of Sciences: National Research Council, 2009. Science and Decisions. Advancing Risk Assessment. National Academy Press, Washington, DC.

Percival, R.V., 2006. Who's afraid of the precautionary principle? Pace Environ. Law Rev. 23, 21–61.

Silbergeld, E.K., 1993. Risk assessment: the perspective and experience of U.S. environmentalists. Environ. Health Perspect. 101, 100–104.

Sullivan, L., 1991. Speech of Secretary of US Department of Health and Human Services Louis Sullivan. First Annual Conference on Childhood Lead Poisoning. October 7. Washington, DC.

U.S. Environmental Protection Agency, 2009. IRIS Substance File: Integrated Risk Assessment Information System. National Center for Environmental Assessment, Cincinnati, OH.

Chapter 21

Hazard Characterization for Lead in Human Populations

A vast scientific and medical record exists for lead toxicity and public health down through the decades. Part 3 of this monograph presented a focused representation of this literature. Critical and summarizing appraisals of the human health effects of lead were also noted as the topic of various governmental and other scientific consensus documents reflecting informed scientific and medical opinion. These include U.S. ATSDR (2007), U.S. EPA (1977, 1986a, 2006), U.S. CDC (1991a, 2005), NAS/NRC (1972, 1980, 1993), and WHO (1987, 1995, 2000).

21.1 GENERAL ISSUES FOR HAZARD CHARACTERIZATION: LEAD AS THE PROTOTYPIC MODEL FOR HUMAN HEALTH RISK ASSESSMENT

The earliest clinical and public health history relevant to hazard characterization for inorganic lead was one where the toxicant was only considered to produce severe poisonings. Lead poisoning in this early record for Pb was viewed as grave only if the poisoning was fatal (McKhann and Vogt, 1926; Thomas and Blackfan, 1914). Furthermore, there was little differentiating among risk groups. Lead was generally thought equally toxic to children and adults until the mid- to late nineteenth century, given that no database resembling quantitative toxicology or epidemiology for Pb existed at that time. As noted by this author's perspective article on lead (Mushak, 1992), overt, readily apparent clinical and toxic effects of lead were taken to be the sole effects of lead because those were the only effects that could be identified using the relatively crude clinical and scientific means then available.

Later, as seen below, death was not the only toxic outcome of concern for lead poisoning. Later clinical appraisals included severe and persisting brain damage in the form of outcomes such as manifestation of mental retardation. Eventually, serious but not catastrophic clinical sequelae of severe poisonings among survivors were identified, particularly in childhood lead

Trace Metals and other Contaminants in the Environment, Volume 10
ISSN: 1875-1121 DOI: 10.1016/B978-0-444-51554-4.00021-3

poisonings (Byers and Lord, 1943). The increasing awareness about lead as a public health issue in subsequent decades and on to the present served to focus concerns about dose–toxic response relationships on lower exposure levels, not higher. That is to say, no recent accounts of the scientific and medical/public health impacts of lead have identified credible findings that lead's adverse impacts were less severe than earlier thought.

Progress in defining more conservatively and quantitatively the full toxicological impact of environmental lead has been neither rapid nor smooth. There were several reasons for this. Periodic challenges to new research for Pb hazard characterization came from various nonscientific sources, particularly from industrial and economic sectors having a stake in an industrially significant commodity. Specifically, these extra-scientific elements included complex jostling for advantage and juggling of economic interests, political realities, contending interests within the health communities, public awareness, the public's technical or health literacy, and the trade-offs of costs and benefits (see Graebner, 1987; Mushak, 1992).

Progress in defining ever more subtle adverse effects at ever lower lead exposures was also dependent on broader scientific developments. Specifically, progress had to await methods sensitive and specific enough to look for such effects. This factor in turn mandated that resources be found to acquire the methodologies and then to do the research. Independent research on the twin topics of lead exposure and toxicity was established and proliferated with the creation and growth of funding mechanisms not linked to any particular affected industries or other economic interests, beginning in the late 1960s and early 1970s (Graebner, 1987; Mushak, 1992). Second, a critical mass of research had to accumulate to induce and sustain continued research and findings. The interplay of funded research and its findings with societal interests in and provision of resources for such research are embodied in the closed, feedback-loop schematic presented in Mushak (1991) and NAS/NRC (1993). First, research findings engage the interest of the public health and health agency sectors, and these then provide the resources for further research, the additional research providing more data of interest to the sponsoring entities.

The more subtle the toxicity of an element, the more involved the technical tools required to identify and quantify the toxic effects. Studies of subclinical effects in particular required interdisciplinary efforts; those efforts required a research evaluation and funding mechanism that went beyond simply funding a single investigator. The case of developmental neurotoxicity of Pb is instructive. The development of the multi-investigator program project and center grant approach by components of the U.S. National Institutes of Health made it possible to do "mission" research. In turn, these organized efforts attracted the expertise of epidemiologists, biostatisticians, neurotoxicologists, biochemical toxicologists, and analytical toxicology laboratories to measure increasingly lower levels of lead in indicator media and to study subclinical, subtle neurotoxic effects of lead.

Not all of the affected segments of the lead-and-health issue progressed at a similar pace. For example, societal or professional responses to the need to recognize and deal with lead poisoning outside the workplace, such as lead poisoning in children, traditionally were through medical intervention, using what this author would characterize as the medical intervention model (Mushak, 1994). The medical intervention model can be defined as one in which the physician or other medical caregiver responds to one or more identified case(s) of lead poisoning using traditional methods for identification and diagnosis, and traditional standards for intervention and medical care. Lead poisoning of children in past decades, continuing into the 1960s and even 1970s, was deemed important only if it was functionally detrimental, clinically apparent, and amenable to medical intervention on an individual postpoisoning treatment basis. Even overtly poisoned individuals first had to be identified before the medical model mechanisms were triggered.

This response model neither has become inappropriate nor has it been abandoned. Quite the opposite; it has been refined and supplemented by the public health or epidemiological model. The epidemiological model, by definition, addresses the scientific evidence for population-level adverse health impacts. This is not to say that the medical intervention model operated to identify childhood lead poisonings most of the time. Lead poisoning has few signs or symptoms that are specific to just lead poisoning when using individual effects in isolation. Other substances can mimic lead's toxic effects. For example, other conditions can make children listless and produce gastrointestinal dysfunction or anemia. However, few toxicants would produce collective toxicity involving all organs and systems as does lead. What the medical model did was to trigger a prescriptive protocol for what to do once lead poisoning was teased out, e.g., via a blood lead or biochemical indicator test. Lead-poisoned children were still those who presented with either acute CNS damage or chronic effects in such targets as the nervous and blood-forming systems in the form of mental deficits and frank anemia, respectively.

More recently, the scientific and public health communities have accepted that lead's adverse effects can occur at increasingly lower lead exposures and that these outcomes do not easily lend themselves to identification as lead toxicity by the classical medical intervention model. For example, practically all of the longitudinal environmental epidemiological studies appearing in the peer-reviewed literature over the past 15–20 years are statistical statements about the probability of children in a lead-exposed community having blood lead levels above the level of concern for adverse effects. Longitudinal epidemiological data for low-level Pb neurotoxicity from around the world, like that generated in the United States (Boston, MA; Cincinnati, OH), the former Yugoslavia, Mexico, and Australia (Port Pirie), have produced statistical or aggregate statements about how groups of exposed subjects respond to graded levels of lead exposure. That is, the associated dose–response relationships are population based and population defined. However, such evidence, as noted earlier, is of major value in informing a clinical assessment for Pb toxicity in individual diagnostic settings for specific children.

Population dose–response data obviously point regulatory or health policy to remedies that differ from the traditional physician-centered approach. An overt poisoning case will entail hospitalization, chelation therapy, and medical management of the signs and symptoms of the poisoning. In addition, the patient would ideally be discharged to a lead-free or lead-safe setting. Options where there is widespread but subtle poisoning within an entire community entail community-level actions, coupled perhaps with systematic environmental cleanup interventions. The latter are part of the rationale behind the more recent statements of the U.S. CDC (1991a, 2005).

Something akin to this community-level approach is already a part of regulatory practice for the U.S. EPA and state environmental regulators. Soil lead cleanups in communities are often prescribed when a certain fraction (5%) of the community's children are determined by a blood lead survey or are estimated by EPA's IEUBK computerized modeling of blood lead to exceed the current U.S. CDC blood lead guidance and action level of 10 μg/dl. The upshot of this cleanup intervention is that the presence of elevated community lead in contact media for children, in tandem with measured or modeled blood lead distributions that largely fall below typical medical intervention levels, is adequate to produce an environmental intervention rather than a medical one. Reducing exposure with the goal of preventing more exposure in more children would also label such an approach a preventive medical model or a model of primary poisoning prevention since unabated toxic levels of lead in dusts and soils would make it likely that additional clinical and subclinical cases of lead poisoning would occur.

The assessment of difficult-to-discern effects in terms of their societal and monetized resource costs and the monetized benefits that accrue from prevention require macro- or population-level economic analyses. The various longitudinal studies on lead and IQ effects as well as lead's effects on other body systems and organs indicate that such effects are apt to have population-level costs, especially in lost cognitive acuity in a global economy that is increasingly competitive in high-value sectors that require high-functioning cognition. Other sectors of the global economy that are service based and that require tedious, repetitive activities place a premium on attention to detail. Such requirements leave individuals with attention deficits from lead exposure, a universally recognized outcome, at an economic and functional disadvantage.

Such analyses have been carried out either for source-specific lead exposure reduction in the United States (U.S. CDC, 1991b; U.S. EPA/Schwartz, 1985; U.S. EPA/Levin, 1986b) or for all lead sources and pathways (Gross et al., 2002; Salkever et al., 1995; Schwartz, 1994). Collectively, these assessments show that there are major monetized benefits of reducing lead exposures on a nationwide basis. Such benefits would not be assessed under conventional older models of harm to health nor could they be quantified.

21.2 THE CHRONOLOGY OF HUMAN HEALTH HAZARD CHARACTERIZATION FOR Pb

The time-dependent human health hazard characterization for lead is best summarized by chronological blocks of time and those adverse health effects that were considered of medical or scientific importance. Associated with the various levels of adverse health effect severity are body lead burdens or dose. The combination of these dose ranges, whether measured at the time or retroactively imputed as likely lead doses given current knowledge, defines the dose–response relationship, also called the dose–effect severity relationship. This topic as the third component of health risk assessment is addressed in Chapter 22.

In brief, this relationship indicates that as dose increases above some threshold, effects get worse and more effects occur. This relationship has been a dictum of toxicology for centuries and still applies across the whole dose–response spectrum for lead. Presented here is the chronology of dose–response thresholds in PbB from the onset of adverse health effects in terms of minimal amount of exposure assumed to produce such toxicity.

The chronology for thresholds to toxicity severity over time when such effects were considered clinically or epidemiologically adequate evidence of harm to lead poisoning victims and indexed in terms of some PbB threshold was noted by Mushak (1992). The earliest time band for our purposes was that linked to the most draconian definitions of what constituted lead poisoning, especially in children. With the passage of time and evolution in defining lead as a health issue, the lead level for "official" definitions of toxicity or adverse effects by the medical community or governmental health agencies continued to decline markedly.

21.3 CLINICAL OR OVERT TOXIC EFFECTS OF LEAD IN HUMANS

Earlier critical assessments of the state of the Pb poisoning problem divided the worst clinical effects defined in the 1930s and early 1940s into chronic lead encephalopathy and acute encephalopathy (NAS/NRC, 1972). We now recognize that both forms of the worst damage to the CNS, especially to the developing brains of the preschool child, represented a combination of exposures and responses. With lead paint poisoning being the source of the most severe U.S. childhood poisoning episodes in the 1920s and onward to the present, a high chronic exposure baseline of adverse but largely unrecognized poisoning in the form of anemia, general misbehavior, and attention problems would be augmented by intermittent massive exposures via paint chip or paint flake ingestion producing the classic acute lead encephalopathy. This stage was sequentially listlessness, severe irritability, convulsions proceeding to coma, and, in the most severe cases, brain swelling, and death.

Other mixes of chronic and acute severe lead poisonings would be expressed in other organs. The Fanconi Syndrome is acute kidney injury that may subside rather rapidly with early medical intervention. Some evidence suggests that there may be nephropathic sequelae, but this has not been a consistent finding. Reports in the older literature have periodically noted cardiovascular injury.

Severe adverse effects in other organs and systems with mainly chronic or subchronic poisoning of the younger child included gastrointestinal injury in the form of enteric and related nerve tissue injury causing reduced motility and severe pain. Frank anemia with marked reduction in hematocrit, hemoglobin, and marked increase in markers of impaired working of the blood-forming system was also observed.

The chronological course in recognizing adult poisonings in certain occupations likewise began with rather severe definitions of poisoning. While inorganic lead was potentially lethal in acute adult exposures, severe chronic lead poisoning from the workplace or through idiosyncratic intoxication, such as Pb-contaminated illicit alcohol from lead-bearing stills, was the more common poisoning scenario.

Chronic adult poisoning at the highest exposures typically consisted of a quartet of disease symptoms: peripheral neuropathy with mainly motor nerve damage, leading to foot and wrist drop and a general inability to walk; severe anemia; GI effects (lead colic); and chronic lead nephropathy, which in its most severe form led to end-stage renal disease in a number of reports. GI tract effects were invariably present, commonly noted in the earlier occupational medical literature as the "Monday Morning Colic" since weekend respites from exposure would somewhat ameliorate the severe GI pain and discomfort phase of GI tract injury, if doing little to reduce paralysis of GI motility and the resulting constipation.

As recently as the early 1940s, the pediatric medical community generally assumed that lead-poisoned children surviving the acute stage had no persisting physical damage. However, a classic study of childhood lead poisoning by Byers and Lord in 1943 in the United States showed that in fact there were severe persisting neurological and neurobehavioral sequelae in terms of mental retardation, destructive or disruptive behaviors, and general social maladjustment. The Byers and Lord study was slow to be widely accepted since childhood lead poisoning in those and later years was assumed to be a problem of low-income people. Popular thought held that if caregivers would simply look after their children appropriately, lead exposure and associated severe poisoning would not occur. This social and socioeconomic bias by the medical profession was an impediment to timely responses and remedies to an environmental problem that afflicted a population segment where the form of the lead exposure—deteriorating lead paint—could manifest itself more severely and overtly than occurred in more affluent segments where children were not in contact with peeling paint.

Despite the Byers and Lord study, several decades passed in which the clinical notion persisted that acute or severe lead poisoning was the main manifestation of the disease that one had to worry about.

Other drawbacks in the history of lead poisoning could be seen slowing public health and scientific responses to the problem. The evolution of publicly funded, competitive research grant systems for independent pursuit of lead toxicity research was noted earlier. Until then, money and other resources to carry out clinical and population researches were not available outside lead industry sources (Graebner, 1987). Second, no systematic diagnostic or exposure identification protocols existed by which intervention could occur before full-blown, life-threatening lead encephalopathy occurred. One drawback was the absence of a reliable and sensitive way to routinely measure body lead burden to permit both a differential diagnosis and signal the relative urgency for medical intervention. This latter gap was closed by the 1930s to early 1940s with the establishment of a colorimetric method for blood lead measurement. The method was first developed in health laboratories and later perfected and promoted by the American Public Health Association (NAS/NRC, 1972).

Early blood lead measurements were used to assemble dose–response relationships for clinical toxic effects, and this was the case during the 1950s to the 1970s. With the availability of an accepted measure of body dose, one could be alerted to the likely degree of poisoning. With time and in more recent years, the clinical and scientific definitions of significant adverse effects of lead have been associated with lower and lower blood lead thresholds.

21.4 THE TRANSITION PERIOD FOR CHARACTERIZING Pb TOXICITY IN HUMANS

Between the time of the Byers and Lord findings in the early 1940s and the 1979 study of Needleman et al. showing that subclinical effects were occurring in children who earlier would not have been judged lead poisoned, there were a number of reports showing that more moderate clinical or borderline clinical toxic health effects could be detected at blood lead levels below those producing encephalopathy. Some neurological deficits, diminished mental acuity, and evidence diagnostic of anemia were identified in children in the range 50 or 60 μg/dl up to 80 μg/dl PbB, where early symptoms of encephalopathy might be in place. Consistent with this, the advisory record from the U.S. Public Health Service in the 1950s and 1960s first shows some official confidence in advising 80 μg/dl as the body burden which must be reached before harm occurs, followed by a revising downward in the 1960s to 60 μg/dl.

By the early 1970s, with the recommendation of the U.S. Surgeon General for screening high-risk children, the threshold level for toxic effects was revised down to 40 μg/dl. The effect outcome underpinning this advisory was measurable decline in hemoglobin associated with lead's effects on heme biosynthesis. The U.S. CDC in the 1970s issued two advisories on childhood lead poisoning, in 1975 and 1978, which targeted a PbB of 30 μg/dl as the lowest level showing a statistically measured hemoglobin drop on a group basis. Other group-basis effects seen in the research literature at this time were increased erythrocyte ZPP at a threshold in children of 15–20 μg/dl PbB and reduced nerve conduction velocity in children at about 30 μg/dl PbB.

21.5 THE CURRENT ERA OF HUMAN Pb TOXICITY: SUBCLINICAL Pb TOXICITY IN ITS NUMEROUS DIMENSIONS

The landmark Needleman et al. (1979), study marked the advent of current views about the neurobehavioral and other effects of body lead at lower and lower levels. The Needleman report made it clear that, because the kinds of adverse individual health effects on the brains of children associated with lead were not unique to lead but had other causes as well, further assessment of lead as a health problem at low environmental exposures would require multidisciplinary and sophisticated test instruments to show lead–neurotoxic effect relationships.

With knowledge of these constraints, the studies over the last 30 years or so have been carried out with increasing statistical, epidemiological, and toxicological sophistication. This pattern typifies not only human lead exposure studies but also the many hundreds of studies done each year on experimental animal models of low-dose lead toxicity and explorations of mechanisms of low-dose toxicity in diverse cellular, subcellular, and biomolecular systems.

Neurobehavioral studies, as a public health template for lead toxicity, have been carried out around the world and have been of two types: retrospective or cross-sectional and longitudinal or prospective studies. The retrospective study has more limits than a prospective one, in that the role of lead as an etiological agent is easier to define in prospective studies. Typically, groups are studied in the latter with much better control for confounding factors that could lower or obliterate any lead–effect relationship. The former appeared much earlier and varied in their quality. However, about a dozen or so retrospective studies from Europe and the United States show a lead–neurobehavioral effect relationship. When analyzed through inclusion in a meta-analysis, the robustness of the lead–IQ effect is amplified.

These cross-sectional studies in general showed that for every 10-unit rise in blood lead on a child group basis, two or more IQ points are lost. These mean declines, while modest, would be reflecting much greater changes in the tails of the IQ distribution when the full-scale impact is calculated.

The prospective studies and their results are the best evidence to date for low-level lead effects on IQ, attention deficits, and indicators for later educational and social performance. These studies show a more robust lead–IQ relationship than is apparent from cross-sectional data, and the robustness is linked to the segment of the dose/exposure curve, as is apparent in the internationally pooled analysis by Lanphear et al. (2005). Equally important, these studies show lead effects persisting to older ages in children and there may well be irreversible effects.

21.6 Pb POISONING IN EXPERIMENTAL MODELS OF LEAD TOXICITY

It is a given in toxicology that seeing the same toxic effects for a substance in test animals or other test systems as in humans is strong supporting evidence for a causal relationship. The causal relationship is strengthened when the dose–response relationships in test species overlap those seen in humans. It is also a dictum of mechanistic toxicology that a documented mechanism in test systems that explains a particular toxicant's mode of action in humans is a further proof of a cause–effect relationship.

Multiple supporting bodies of evidence exist for lead, especially at low exposures. Numerous studies in the nonhuman primate show subtle neurobehavioral effects of the type seen in children. Furthermore, many mechanism studies are able to explain the very high neurotoxic potency of lead in children, specifically effects in the developing nervous system's neuronal development and function.

Pb toxicity is seen in nonhuman species exposed to the element in the wild or under controlled experimental conditions. The most uncontrolled severe poisonings seen in animals occur in livestock and certain wildlife species with either high acute or high chronic exposures. Livestock and horses sustain severe neuropathological effects from grazing on plants growing in contaminated soils. Peripheral neuropathy and pronounced hematotoxicity occur, progressing to generalized systemic collapse and even death. Migratory waterfowl are typically acutely poisoned, as noted earlier, from lead shot ingestion. Gizzard paralysis occurs first. Generalized paralysis occurs with liver and kidney damage along with marked impairment of the blood-forming system. Lethal levels of lead exposure can be reached in waterfowl. Other birds are lost to predators because of impaired survival skills.

Raptors, notably hawks, have occasionally been lead poisoned by preying on urban birds such as pigeons that have high lead body burdens. This was a more severe problem when gasoline lead consumption produced very high dust and soil lead levels alongside urban roads.

Dogs and other pets that spend time outside can be lead poisoned, but usually the poisoning is not as severe. This type of poisoning involves impaired functioning of the heme biosynthesis pathway and occasionally the nervous system.

These comments deal with environmental levels likely to produce toxic responses at the indicated species level. The focus has been on the probability of risk groups undergoing ecotoxic responses, mainly for comparative purposes. We do not deal with complex effects that may be expressed at low lead exposures in systems and populations, or interplays between relatively larger groupings that occur or may occur because certain populations are reduced, directly or indirectly, by changes in environmental compartment levels of lead. Lead readily accumulates in sediments so that there may be different sedimentary populations that relate to that lead accumulation more detrimentally than others. This topic of subtle, large-scale ecotoxic effects as an aspect of concern largely unknown in the lead literature is presented in WHO's 1989 report on the ecotoxicity of lead (WHO, 1989) and sections of the current EPA Pb criteria document (U.S. EPA, 2006).

21.7 IDENTIFYING RELEVANT TOXIC EFFECTS IN HUMANS FOR LEAD HEALTH HAZARD CHARACTERIZATION

While the toxicological literature on Pb is huge, not all aspects of that literature are of sufficient scientific quality or highly relevant in terms of important criteria for human health risk assessment. These criteria are qualitatively outlined in Table 21.1.

The suite of adverse human health effects noted in Table 21.1 occurs within various affected human population subsets and within the criteria set forth above. They are included on the basis of the available evidence, including evidence for their importance as hazardous endpoints, scaled as to severity, duration of adverse impacts, and affected numbers in human populations likely to sustain toxic harm.

On this basis, developmental neurotoxicological effects, along with any anatomical and other evidence of such neurotoxicity, seen in *in utero* and with early postnatal exposures, appear first.

Reproductive and developmental effects—while occurring before or accounting for developmental neurotoxicity—along with cardiovascular, nephrotoxic, and immunotoxic effects appear next on the basis of potency of effects and sizes of the population affected. These are followed by carcinogenic/genotoxic, hematological, and miscellaneous effects generally associated with higher Pb exposures in human populations across various stages of development.

TABLE 21.1 Demonstrable Human Health Hazard Characterizations for Lead

Health Hazard	Range of Dose Responses	Identified Risk Groups	Types of Evidence Available: Epidemiological	Experimental	Mechanistic
1. Developmental neurotoxicity a. Cognitive deficits b. Academic achievement deficits c. Executive functioning d. Social malbehavior	Yes	Fetus, infants, toddlers, older children for CNS effects	Yes	Yes	Yes
2. Adult neurotoxicity: peripheral nervous system; some CNS outcomes	Unclear	Occupational exposures	Apparent	Yes	?
3. Reproductive/developmental	Overall, yes	Occupational, higher environmental exposure populations	Yes	Yes	Yes
4. Cardiovascular effects	Yes, to include lower exposures	Environmental and occupational populations	Yes	Yes	Yes
5. Nephrotoxicity	Yes, to include lower exposures	Environmental and occupational populations	Yes	Yes	Yes
6. Immunotoxicity	Yes, to include lower exposures	Possibly fetus; infants, older children; occupational groups	Yes	Yes	Yes
7. Carcinogenesis/genotoxicity	Mainly elevated workplace Pb exposures	Worker groups, some general population data	Yes	Yes	Yes
8. Hematotoxicity	Yes	Young children; worker groups	Yes	Yes	Yes

21.8 CRITERIA FOR Pb-BASED CAUSALITY OF TOXIC EFFECTS

The remainder of this chapter discusses the evidence for the various causal relationships between Pb exposure and toxicity and disease. As set forth in Table 21.2, adverse human health effects of lead are placed in causal context using the nine proofs-of-causality criteria first enunciated by British medical statistician Sir Austin Bradford Hill in 1965 and subsequently labeled "Hill's Criteria of Causation." These criteria are relevant and of particular value in the environmental epidemiology of Pb, given the many contentious arguments over "association" versus "cause" that have been lodged against Pb's human health impacts, particularly in earlier eras of lead health research and in the evolution of Pb as a major public health factor.

The nine criteria noted by Hill (1965) include the following:

1. *Consistency*: The existence of consistent findings as endpoints expressed among different groups and in different places by different researchers strengthens effect causality.
2. *Specificity*: The likelihood of causality in the relationship exists if specific populations have effects or disease rates at specific sites with no alternative explanations.
3. *Temporality*: Some effect occurs subsequent to the exposure or other suspected cause or after an expected delay.
4. *Biological gradient*: Basically, dose–response relationships exist. The toxic response can be direct, inverse, or simply inducible by some substance's presence.
5. *Coherence*: Agreement between epidemiological and experimental outcomes increases the relationship being causal.
6. *Strength*: Associations which are not robust do not necessarily disprove a causal relationship.
7. *Plausibility*: A plausible mechanism for cause and effect is helpful if currently known.
8. *Experiment*: Experimental data may be adequate in the face of limited epidemiological findings.
9. *Analogy*: The effect(s) of similar factors or agents may be known.

It should be noted that Hill's *Criteria of Causation* are scientifically intended to address *general causality* in support of mainly epidemiological evidence. Within legal frameworks, there is also the matter of establishing *specific causality* in the case of, say, personal injury litigation. This latter form of causation is typically sought through the diagnostic conclusions of someone qualified to render medical or other professional diagnostic opinions for specific cases. General causality addresses the question, "*Could* lead cause this particular injury in people?", based on the weight of the scientific evidence, while specific causality addresses the question, "*Did* lead cause this particular injury in this particular person?" These two forms of

TABLE 21.2 Pb Causality in Human Health Hazards Using Hill's Criteria of Causation[a,b]

Adverse Effect	Hill Criteria								
	Consistency	Specificity[c]	Temporality	Biological Gradient (Dose–Response)	Coherence	Strength	Plausibility	Experimental	Analogy
Developmental neurotoxicity: fetus, young children	Yes	Yes: collective effects	Yes	Yes	Yes	Yes	Yes	Yes	Yes
Reproductive–developmental effects	Yes	Yes: collective effects	Yes	Partial	Yes	Yes	Yes	Yes	Yes
Cardiovascular effects	Yes	Yes: collective effects	Yes	Yes	Yes	Yes	Yes	Yes	Yes
Nephrotoxic effects	Yes	Yes: collective effects	Yes	Yes	Yes	Yes	Yes	Yes	Yes
Immunotoxic effects	Yes	Yes: collective effects	Yes	Yes	Yes	Yes	Yes	Yes	Yes
Carcinogenicity/ genotoxicity	Yes	Yes: collective effects	Yes	Partial, at higher exposure	Yes	Yes	Yes	Yes	Yes
Hematotoxicity	Yes	Yes: collective effects	Yes	Yes	Yes	Yes	Yes	Yes	Yes

[a] *Hill (1965).*
[b] *Criteria defined in text.*
[c] *Individual effects in isolation not pathognomonic for lead poisoning; collectively, most or all indicated effects in the same affected individuals are more specific to lead.*

causality are scientifically and medically related. As noted earlier, general causality in concert with epidemiology is assumed to provide the template by which diagnostic decisions are informed and guided.

Virtually all of the questions that can be asked of lead using the Hill causality criteria can be answered in the affirmative, allowing a conclusion of general causality of the various adverse health effects long or newly connected with lead exposures in various elevated risk segments of human populations. Table 21.2 offers a number of the responses to the nine causality proofs of Hill that merit discussion.

Consider the element of specificity of the toxic insult (Table 21.2). It has been noted that lead does not produce a *single* signature injury in humans, particularly children, which would allow differentiating it from other causal factors, e.g., U.S. EPA, (2006). Many causes can be invoked for headaches, or for irritability, or anemia, or impaired mental acuity. However, lead arguably produces a *de facto collective* signature injury, particularly when concurrency or sequencing of multiple endpoints is synchronous with one or more elevations in lead exposure biomarkers such as blood and/or bone lead.

Few nonpathogenic, chemical contaminants simultaneously or sequentially induce in young children developmental neurotoxicity, impaired growth and development, hematotoxicity, early nephrotoxicity, or immunotoxicity in concert with elevated Pb exposure biomarkers. Similarly, few such contaminants induce simultaneously or sequentially in men or women peripheral neurotoxicity, cardiovascular toxicity, hematotoxicity, various stages of nephrotoxicity, various stages of immunotoxicity, and miscellaneous endpoints such as GI effects in concert with elevated Pb exposure biomarkers.

Consistency across studies by various investigators in various parts of the world has been seen with the various toxic endpoints in both Tables 21.1 and 21.2. This has particularly been true for the various Pb developmental neurotoxicity studies in the various forms indicated in Table 21.1 and as illustrated in the prospective neuroepidemiological studies from around the globe.

The temporality criterion among the Hill causality proofs can be readily answered in the affirmative. For example, one aspect of temporality that was once of concern was the potential existence of reverse causality, i.e., is elevated blood lead concurrent with a developmental neurotoxic endpoint because the neurotoxic impairment enhances lead exposure rather than the reverse? The outcomes of multiple prospective epidemiological studies of lead neurotoxicity in children effectively ruled out the likelihood of reverse causality.

The biological gradient (dose–response) criterion can also be readily answered in the affirmative. Two points in appraising the enumerated toxic effects are that: (1) the persistence of the dose–response relationships span relatively wide exposure ranges in many cases, e.g., neurotoxicity, hematotoxicity, reproductive and developmental effects, immunotoxicity and (2) a number extend to relatively low exposures. None of the indicated toxic endpoints in Table 21.2 fail to show any dose–response relationships within or across studies.

The coherence element among the causality criteria is readily answered in the affirmative. The toxic effects in Table 21.2 have epidemiological, experimental, and mechanistic data attached to their database. In fact, the relative amounts of these three categories of evidence across effects are better illustrated with lead than with any other contaminant.

Criteria 6–8 are interconnected with the coherence criterion and are similarly answered in the affirmative. Criterion 9, effect analogies in the epidemiological and toxicological literature, can generally be said to exist with other toxic metals, e.g., cadmium and inorganic/methyl mercury.

The nine Hill causality criteria provide results that collectively support a firm conclusion that Pb *causes* multiple toxic effects in various subsets of exposed human populations.

REFERENCES

Byers, R.K., Lord, E.E., 1943. Late effects of lead poisoning on mental development. Am. J. Dis. Child. 66, 471–494.

Graebner, W., 1987. Hegemony through science: information engineering and lead toxicology, 1925–1965. In: Rosner, D., Markowitz, G. (Eds.), Dying for Work: Workers' Safety and Health in Twentieth-Century America. Indiana University Press, Indianapolis, IN, pp. 140–159.

Gross, S.D., Matte, T.D., Schwartz, J., Jackson, R.J., 2002. Economic gains resulting from the reduction in children's exposure to lead in the United States. Environ. Health Perspect. 110, 563–569.

Hill, A.B., 1965. The environment and disease: association or causation? Proc. R. Soc. Med. 58, 295–300.

Lanphear, B.P., Hornung, R., Khoury, J., Yolton, K., Baghurst, P., Bellinger, D.C., et al., 2005. Low-level environmental lead exposure and children's intellectual function: an international pooled analysis. Environ. Health Perspect. 113, 894–899.

McKhann, C.F., 1926. Lead poisoning in children, with notes on therapy. Am. J. Dis. Child. 32, 386–392.

Mushak, P., 1991. U.S. Agency for Toxic Substances and Disease Registry's report to Congress on childhood lead poisoning in America: review and update. In: Proc. First National Conference on Laboratory Issues in Childhood Lead Poisoning Prevention, Columbia, MD, pp. 79–104.

Mushak, P., 1992. Perspective: defining lead as the premiere environmental health issue for children in America: criteria and their quantitative application. Environ. Res. 59, 281–309.

Mushak, P., 1994. The 1993 NAS Lead Report: Adverse Health Effects and Critical Sources of Exposure in Children. National Conference: Building a Lead-Safe Future, Washington, DC, May 15–18.

National Academy of Sciences: National Research Council, 1972. Lead: Airborne Lead in Perspective. National Academy Press, Washington, DC.

National Academy of Sciences: National Research Council, 1980. Lead in the Human Environment. National Academy Press, Washington, DC.
National Academy of Sciences: National Research Council, 1993. Measuring Lead Exposure in Infants, Children, and Other Sensitive Populations. National Academy Press, Washington, DC.
Needleman, H.L., Gunnoe, C., Leviton, A., Reed, R., Peresie, H., Maher, C., et al., 1979. Deficits in psychologic and classroom performance of children with elevated dentine lead levels. N. Engl. J. Med. 300, 689–695.
Salkever, D., 1995. Updated estimates of earning benefits from reduced exposure of children to environmental lead. Environ. Res. 70, 1–6.
Schwartz, J., 1994. Societal benefits of reducing lead exposure. Environ. Res. 66, 105–124.
Thomas, H., Blackfan, K., 1914. Recurrent meningitis due to lead, in a child of five years. Am. J. Dis. Child. 8, 377–380.
U.S. Agency for Toxic Substances and Disease Registry, 2007. Toxicological Profile for Lead. U.S. Department of Health and Human Services, Atlanta, GA.
U.S. Centers for Disease Control, 1991a. Preventing Lead Poisoning in Children: A Statement by the Centers for Disease Control. U.S. Department of Health and Human Services, Atlanta, GA.
U.S. Centers for Disease Control, 1991b. Strategic Plan for the Elimination of Childhood Lead Poisoning. U.S. Department of Health and Human Services, Atlanta, GA.
U.S. Centers for Disease Control, 2005. Preventing Lead Poisoning in Young Children: A Statement by the Centers for Disease Control. U.S. Department of Health and Human Services, Atlanta, GA.
U.S. Environmental Protection Agency, 1977. Air Quality Criteria for Lead. Report No. EPA-600/8-77-017. Office of Research and Development, Washington, DC.
U.S. Environmental Protection Agency, 1985. Costs and Benefits of Reducing Lead in Gasoline. Final Regulatory Impact Analysis. (Schwartz, J., principal author). Report No. EPA-230-05-85-006. Office of Policy Analysis, Washington, DC.
U.S. Environmental Protection Agency, 1986a. Air Quality Criteria for Lead, 4 vols. Report No. EPA/600/8-83/028bF. Environmental Criteria and Assessment Office, Washington, DC.
U.S. Environmental Protection Agency, 1986b. Reducing Lead in Drinking Water: A Benefit Analysis. (Levin, R., principal author). Report No. EPA-230-09-86-019. Office of Policy Planning and Assessment, Washington, DC.
U.S. Environmental Protection Agency, 2006. Air Quality Criteria for Lead, vol. 1. Report No. EPA/600/R-05/144aF. National Center for Environmental Assessment, Washington, DC.
World Health Organization, 1987. Air Quality Guidelines for Europe. Lead. WHO Regional Publications: European Series No. 23. Regional Office for Europe, Copenhagen, Denmark.
World Health Organization, 1989. Environmental Health Criteria 85: Lead - Environmental Aspects. International Programme on Chemical Safety, Geneva, Switzerland.
World Health Organization, 1995. Environmental Health Criteria 165: Inorganic Lead. International Programme on Chemical Safety, Geneva, Switzerland.
World Health Organization, 2000. Air Quality Guidelines for Europe. Lead, second ed. WHO Regional Publications: European Series No. 91. Regional Office for Europe, Copenhagen, Denmark.

Chapter 22

Dose–Response Relationships for Toxic Effects of Lead in Human Populations

The second element of health risk assessment for lead is the identification of dose–response relationships for the various adverse health effects noted in the previous chapter. Dose–response relationships, like health hazard characterizations, are general in nature and independent of site-specific lead exposure scenarios.

22.1 INTRODUCTION AND CONTEXT

This chapter quantifies relationships between levels of Pb exposure and intensity or multiplicity of toxic effects in various human populations and experimental testing models. It combines evidence of Pb exposures across a range of values with established toxicological and/or epidemiological findings for lead's toxic responses across this exposure range.

The chapter is organized into sections that first summarize the dose or exposure component of dose–toxic responses for Pb, presenting the empirical evidence for comparative advantages and limits of various Pb exposure biomarkers. This exposure/dose section draws on data in the more detailed, earlier chapters on this topic.

The second portion of this chapter presents the dose–toxic response relationships for Pb for risk group-specific lead exposures as published in reliable scientific literature. These relationships are stratified on the basis of affected human risk population subsets, e.g., young children, adults, lead workers, older adults. In addition, individuals within these larger risk categories may be at added differential risk owing to intrinsic factors such as genetic susceptibilities. These are followed by data selected to illustrate dose–toxic response relationships observed in various experimental test systems, particularly data that help buttress an understanding of the range and depth of lead dose–responses in humans.

The dose–toxic response relationships for environmental lead in humans, taken collectively, produce a picture in which there is not a readily apparent

Trace Metals and other Contaminants in the Environment, Volume 10
ISSN: 1875-1121 DOI: 10.1016/B978-0-444-51554-4.00022-5

threshold in dose–response terms for the most sensitive toxic endpoints. Absence of a threshold for Pb developmental neurotoxicity, in particular, is recognized and accepted in official statements of the U.S. EPA (2006), the U.S. ATSDR (2007), the U.S. CDC (2005), and the NAS/NRC (1993). Other toxic endpoints are increasingly being linked with lower Pb effect threshold levels. Increasing numbers of findings of dose–response relationships are being reported for not only long-known sensitive endpoints such as neurocognitive and other neurobehavioral effects but also other effects at exposure marker levels (PbB) below the U.S. CDC guidance value of 10 μg/dl and even <5 μg/dl.

While theoretical thresholds may remain for lead and its toxic endpoints between current empirically generated data and zero, that threshold can only lie within an extremely narrow range of a few units of PbB above zero. What is more, these thresholds are approaching conventional performance limits for laboratory measurement of lead in whole blood.

A second characteristic of lead's dose–toxic response relationships is that they continue to decline from previously studied accepted thresholds as new evidence emerges. Some newer findings in young children, in particular, are striking in the extent to which early toxic responses in non-neural systems begin at exposures well below those long thought to be stable thresholds for such effects. For example, Fadrowski et al. (2010) reported that one can observe impaired kidney function in children with a mean PbB of 1.5 μg/dl (99% PbB <10 μg/dl) using a more sensitive and stable nephrotoxicity indicator, the protein cystatin C level, versus the conventional creatinine level.

Viewed over the totality of clinical and epidemiological history, some of the most sensitive developmental neurotoxic effects of lead are linked to biomarker threshold values one to two orders of magnitude lower than what they were a century or more ago (see Figure 22.1 for a PbB-based multiple

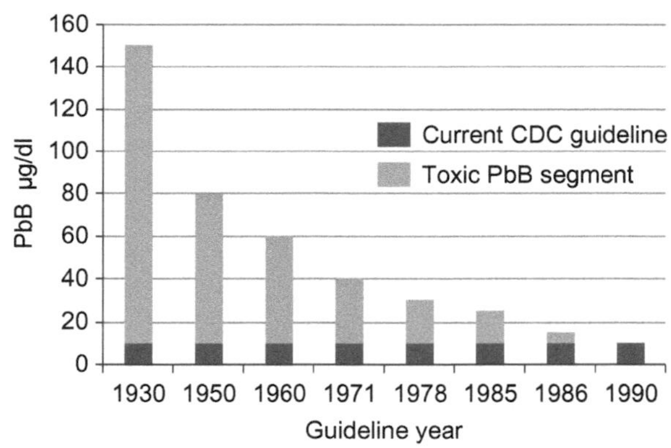

FIGURE 22.1 Chronology of guidelines for declines in lead toxicity thresholds in young children. *Source: Mushak (1992a). © Elsevier Inc. Reprinted with permission.*

effect threshold chronology). For example, current early toxic effects seen at ≤5 μg/dl or ≤10 μg/dl are markedly below the imputed thresholds for lethal outcome estimated around or above the 100–150 μg/dl mark in the first half of the twentieth century.

22.2 THE DOSE/EXPOSURE COMPONENT OF LEAD'S DOSE–RESPONSE RELATIONSHIPS

As noted, dose–toxic response relationships differ with respect to which segments of human populations are affected. Some segments of human populations are much more sensitive to lead contact than others, and these are termed risk, or increased risk, subpopulations. Given qualitative and quantitative differences across risk populations in terms of dose–responses, these segments are discussed separately. Since the basis of differential dose–response relationships in these populations is both exposure driven and toxicology/biology driven, each factor defining higher risk is discussed.

Refer to earlier chapters for detailed presentations of lead exposure assessment methodologies and characteristics. Summary discussions are presented here. "Exposure" is often quantified as the amount of some bioactive substance entering receiving body compartments of "exposed" individuals. Toxicologists and health scientists also speak of exposure as meaning systemic entry or "delivered dose," i.e., the amount of lead absorbed into the central circulating body compartment, the bloodstream, and then distributed to systems, organs, and tissues where effects from the substance can occur. Mechanistic toxicologists speak even more narrowly of lead entering and affecting target organs, cells, or organelles, and then expressing various toxic responses by means of site-specific mechanistic rationales. For example, delivered dose effects can be imparted by a mechanism where the element's presence is required for persistence or where effects can occur and persist tardively and postkinetically, even when the substance is no longer present at the site(s).

An adequate characterization of the dose/exposure portion of reported dose–responses for lead requires discussion of the merits and limits of those exposure biomarkers which have traditionally been used and others which are relatively new and gaining favor. These two broad groups of exposure assessment instruments are further stratified in terms of temporal and toxicokinetic relationships for lead's associated toxic responses and in terms of accessibility and acceptability for routine use in the research and medical communities.

Earlier sections provided the above characterizations in some detail and are only summarized here.

A third division of the discussion comprises those exposure measures which are measured and those which are modeled, i.e., those simulated or projected, using diverse methods in each category. Chapters 8 and 9 discuss the specifics in more detail.

22.3 BLOOD Pb AS THE DOSE/EXPOSURE MEASURE IN Pb DOSE–TOXIC RESPONSE RELATIONSHIPS

To date, the great majority of published and accepted dose–response relationships for lead in various human populations at increased risk have entailed blood lead values for the "dose" portion.

22.3.1 Advantages and Limits to Use of PbB in Dose–Toxic Response Relationships

The relative accessibility of whole blood Pb to various sampling methods and measurements, such as capillary and/or venous puncture collection and routine quantitative measurements using diverse spectroscopic and electrochemical techniques, largely accounts for the great majority of available dose–response relationship data for Pb in human populations (see, for example, Mushak, 1992b; U.S. CDC, 1991, 2005). Field or clinic sampling of whole blood has typically preferred venous puncture, although proficient technologists can perform capillary samplings with minimal complications from external contamination or sample dilution by tissue serum from excess digital pressure (Schlenker et al., 1994).

Routine analytical methods typically include micro (graphite furnace) atomic absorption spectrometry and electrochemical approaches such as anodic-stripping voltammetry. More complex, expensive and nonroutine/research approaches are inductively coupled plasma-mass spectrometry and definitive methods such as thermal ionization-mass spectrometry. These methods have the requisite sensitivity, specificity, and record of reliability for quantification across the range of environmental exposures that humans presently encounter. Combining current instrumental methods with careful quality assurance and quality control protocols permits adequate proficiency for even low Pb concentrations, values of $\leq 1-2$ μg/dl.

It is generally accepted that lead in whole blood is distributed unevenly between the erythrocytes and plasma or serum, with the former binding virtually all of the total blood lead content based on long-known data tracing mainly to the 1970s (see previous chapters). Little difference in lead concentration is typically seen between serum and plasma, about 1% of the total amount in blood. At PbB levels ≥ 25 μg/dl, the equilibrium concentration in plasma as a fraction of total PbB rises in a curvilinear fashion (DeSilva, 1981; Manton and Cook, 1984).

As noted earlier, plasma Pb, although the more precise and meaningful toxicokinetic measure for interorgan Pb distribution and eventual dose–response relationships, has a myriad of analytical problems associated with its measurement (Mushak, 1998; U.S. EPA, 2006). For example, plasma lead content even in high exposures is quite low, so that contaminating Pb because of external contamination or Pb passage from hemolyzed cells to plasma (Everson and Patterson, 1980; Mushak, 1998) can create huge errors in plasma Pb relative to true values. It is likely that using plasma/serum Pb values will be confined to research studies.

Using PbB measurement as a reliable and routine index of lead exposure by clinical laboratories dates to the late 1930s and 1940s, when colorimetric methods became established. Two colorimetric methods that remained in use until current methods arrived are the APHA (American Public Health Association, 1955) and U.S. PHS (Keenan et al., 1963) methods. These required about 5 ml of whole blood.

The conventional view had long been that lead in the human erythrocyte is largely bound to hemoglobin (85%), particularly Hb-A_1 and Hb-A_2 as noted, for example, by Moore (1988). The balance was assumed bound to membrane sites. Recent studies and observed kinetic behavior of lead in blood, however, have implicated other major sites of erythrocyte lead binding, one being the cytosolic vestigial heme synthesis pathway enzyme in the cell, δ-ALA-D (Bergdahl et al., 1998), and perhaps, at high lead levels, a lead-binding protein that resembles metal-inducible metallothionein. This newer information on lead storage in blood cells has a number of implications for lead binding in diverse populations. These ALA-D binding sites for Pb are affected by genetic diversity in human populations as also are the ranges and distribution of PbB values across screening and study populations. Given that genetic polymorphisms affect Pb binding in blood, they also affect the subsequent dose–toxic response relationships. That is, genetic variability enhances the variability of PbB between subjects with the same overall external lead contact. The isozyme ALA-D variants and Pb binding differ with racial makeup of the population, Caucasians showing about 80% of ALA-D 1-1, 18–19% ALA-D 1-2, and 1–2% of the homozygotic isozyme 2-2. Afrocentric populations show little of the variant allele expression and associated isozyme variability, while those of Asian descent appear intermediate in the fractions of isozyme variants.

Epidemiological data suggest that children (Astrin et al., 1987; Wetmur, 1994) and adult lead workers (Bergdahl et al., 1998; Ziemsen et al., 1986) with the ALA-D-2 variant have higher PbB levels, while lead workers with the variant accumulate higher levels of Pb in their bones with age (Smith et al., 1995a). Differences in lead–ALA-D isozyme variant interactions also appear to influence the severity of toxic effects, but in inconsistent ways (Bergdahl et al., 1997a, b; Schwartz et al. 1997; Smith et al. 1995b). A number of studies have attempted to characterize the impact of ALA-D-1 and ALA-D-2 allele forms on lead binding in blood of adults. These studies have looked at lead workers or other occupational groups, where the overall lead exposure would be quite elevated compared to the general adult population. Lower lead exposures in adult worker groups show less ALA-D polymorphic linkage, at least with regard to PbB.

Do ALA-D variants affect the degree of toxic expression of lead in exposed subjects, particularly by raising or lowering the PbB threshold in various dose–toxic response relationships? The results of studies of Pb workers present a mixed picture, as noted in Chapter 16. In some cases, higher PbB is linked with lower toxic endpoint values; in others, the reverse (Bergdahl et al., 1997; Schwartz et al., 1995, 1997; Smith et al., 1995a, b).

22.3.2 Genetic Hemoglobinopathies, Blood Lead, and Dose–Responses Using Blood Lead

Does the co-existence of elevated lead exposures and PbB with hemoglobinopathies such as sickle cell anemia trait or the thalassemias alter the dose portion of dose–toxic responses versus their absence? Specifically, do they alter lead toxicokinetics?

The available information is mixed and inconclusive. The largest study of children in whom this relationship would likely be present was the early investigation of Adebonojo (1974) for 1559 urban African-American children residing in Philadelphia, PA. Children 6 months to 18 years of age were tested concurrently for PbB and the presence of sickle cell trait, sickle cell disease, and G6PD deficiency. This author noted that the presence or absence of disease (HbSS) or sickle cell trait (HbAS) did not appear to affect mean blood lead levels nor did the presence or absence of G6PD deficiency. However, the author's use of mean values for all children within an age band, with its associated high standard deviation, would have obscured any relationship at more elevated PbB values. Stratification of PbB by deciles and statistical testing might have revealed differences.

22.3.3 Erythrocyte Stability, Pb Toxicokinetics, and PbB Use in Dose–Response Assessments

Some of the genetic disorders described above are associated with chronic anemias of apparent hemolytic origin. Since the overwhelming fraction of lead in blood is bound in the erythrocytes, erythrocyte survival might affect lead's persistence in, and removal from, the blood compartment. However, the extent to which such alterations in the relationship of erythrocytes to lead biokinetics can be captured in single PbB measurements has not been investigated in any detail.

Erythrocyte stability and survival are also affected by lead even in the absence of any genetic disorders predisposing to reduced erythrocyte survival and stability. Shortened erythrocyte life span, associated increase in reticulocytosis, and lead-impaired heme and globin synthesis collectively work to produce a lead-associated anemia, largely at the relatively high values of PbB encountered in occupational Pb contact (see, for example, U.S. EPA, 1986, Ch. 12). Genetically induced anemia in individuals also at risk for Pb-induced anemia would, in theory, further aggravate the reliability of the dose portion of dose–toxic response relationships across the spectrum of toxicity.

Lead biokinetics can potentially be affected by erythrocyte instability in ways such as increased erythrocyte fragility and brittleness due to lead's inhibitory effects on blood cell membrane stability, ionic pump activity, and ionic flux homeostasis. These abnormalities occur at chronic Pb exposures producing PbB of 50–60 μg/dl or higher, where numerous toxic endpoints have been triggered.

22.3.4 Other Erythrocyte Lead-Binding Proteins

It is known that some lead workers are more sensitive to lead exposure in terms of symptoms of lead poisoning than others even when PbB levels are comparable. Raghavan and Gonnick (1977) reported that when lead workers had high levels of a particular erythrocyte lead-binding protein, they were asymptomatic. Further lead worker erythrocyte studies (Raghavan et al., 1980, 1981) led these investigators to conclude that the intra-erythrocyte protein, when present at relatively high levels, protected workers against toxic effects. Church et al. (1993) reported this protein behaved most like metallothionein, a metal-inducible, low-molecular-weight metal-binding protein. The protein is present in other tissues, kidney, and brain, based on animal data.

Lolin and O'Gorman (1988) studied several groups of workers classified by the length of exposure and the intensity of lead exposure, with reference to the presence or absence of lead poisoning symptoms. The lead-binding protein was absent in controls (meaning it was inducible), was variably induced in workers having a PbB of 40 μg/dl or higher, and was lower in workers showing lead toxicity. This erythrocyte protein was also correlated with ALA-D activity, both being low in symptomatic workers. This relationship provides a mechanistic rationale for at least part of the binding protein's protective role, i.e., it sequesters lead that otherwise would impair the heme biosynthesis pathway, including the level of activity of ALA-D.

22.3.5 Other Genetic Determinants of PbB in Humans

The VDR gene polymorphism has been reported to affect PbB levels and produce certain cardiovascular effects in lead workers (Lee et al., 2001; Schwartz et al., 2000). Korean workers with the B allele were reported to have a higher PbB than those with the bb variant receptor gene, while the genotype of VDR in these same workers predicted blood pressure. The B allele worker group had higher SBP and DBP and higher prevalence of hypertension (Lee et al., 2001).

22.3.6 Temporal Characteristics of Pb in Whole Blood

A previous chapter on Pb toxicokinetics covered the topic in detail and only summary comments are given here.

Lead in whole blood is considered to mainly reflect short-term exposures, presenting as a half-life of PbB decline with reduced exposures on the order of 30 days or so. Conversely, PbB in many subsets of human populations can rise relatively rapidly, based on specifics of the increased Pb contact (Mushak, 1993, 1998; Tola et al., 1973; U.S. ATSDR, 1988, 2007; U.S. EPA, 1986, Ch. 10, 2006). There are potentially some variable fractions of longer-term contributions to total PbB levels, depending on the Pb exposure profile and history of a given individual or exposure group. A toxicokinetic scenario wherein PbB is essentially all attributable to recent exposure would be that of a child (or children) in infancy or toddlerhood with a history of moderate or low chronic exposures over the longer term.

An intermediate Pb toxicokinetic scenario reflecting exposures captured in the short term and in bone compartment accumulations would be exposures of older children with significant histories of earlier lead exposures. One case in which the main contributor of Pb to PbB is accumulated bone Pb would be bone Pb resorption to blood in retired Pb workers.

With regard to kinetic compartments, the short term, i.e., rapid blood Pb, compartment is depicted in kinetic curves as the rapid Pb decay component, even in the case of lead workers where the fast decay component is evident within weeks or several months. Slower kinetic compartments are represented by Pb from the two bone compartments, trabecular bone Pb kinetics with a shorter half-life, and the longest-lived bone compartment, cortical bone (Nilsson et al., 1991).

The transitory nature of PbB behavior would logically be of most concern in dose–response expressions where exposures are short term or intermittent or in cases where PbBs are measured at single or isolated points in time. In these instances, the extent to which the single measurement may underestimate true exposure risks would be unknown and one cannot be assured how representative of overall exposures such single measurements are. Another complication is the seasonality of PbB measurements, in that levels are generally considered to be maximal in late summer (~30% higher than the winter minimum) and minimal in late fall or winter. This long-recognized variability gave rise to the label of childhood lead poisoning as "the Summer Disease," when both PbB values and hospital admissions for lead poisonings were maximal (Hunter, 1978). This toxicokinetic phenomenon is still considered to be a result of both hormonal metabolic changes and increased contact with lead-bearing dust and soil.

In those cases where Pb exposures are or are likely to be relatively stable, one can still use PbB as a measure of more extended exposures. This is the case for occupational Pb exposures and perhaps persisting exposures in children. In the former case, an integrated measure of worker serial PbB measurements, i.e., concentration × time, has been reported to be highly correlated with bone Pb values (Roels et al., 1995).

22.4 BONE Pb AS THE DOSE/EXPOSURE MEASURE IN Pb DOSE–TOXIC RESPONSE RELATIONSHIPS

As noted in earlier chapters, Pb lodged in bone has drawn considerable interest as a Pb exposure biomarker for those human populations and experimental animal species where such deposition would be significant and diagnostic of longer-term Pb exposure histories. That interest reflects the bodies of new evidence characterizing bone Pb's toxicokinetic mobility. Historically, bone Pb had been viewed as an inert body storage compartment for the element with little likelihood of any subsequent exposure hazard. Supporting this earlier view were early tissue Pb concentration studies showing that bone Pb accumulates with age, up to 90% of total body burden, particularly in individuals with occupational Pb exposure histories (Barry, 1975; Barry and Mossman, 1970).

This assumption of kinetic and toxicological inertness began to change with studies showing existence of: (1) a huge bone pool of Pb exchanging Pb with other compartments (Rabinowitz et al., 1977), (2) bone Pb as the main source of chelatable Pb in children (Saenger et al., 1982) and in adult Pb workers (Hansen et al., 1981), (3) the persisting elevated PbB levels and persisting toxicity in long-retired lead workers only consistent with bone Pb resorption in retirement (Alessio et al., 1976), (4) the evidence of bone Pb releases using stable Pb isotopic techniques (Gulson et al., 1997), (5) the increase in PbB in postmenopausal women with osteoporosis (Silbergeld et al., 1988), and (6) the increasing number of epidemiological studies showing various relationships of accumulated Pb levels in various bone types using newer, noninvasive approaches to toxic endpoints (Hu et al., 1998; Schwartz et al., 2001).

Lead accumulates in different types of human bone over a lifetime of active exposure, up until 50–60 years of age, followed by decline. In postmenopausal women, bone Pb eventually declines. The toxicokinetic determinants of accumulation differ in terms of bone type, anatomical sites of bone, the nature of Pb exposures, nutritional status, and various host-specific factors that affect quantitative *in vivo* movement of Pb such as gender, stage of development, physiological stresses such as pregnancy, physiological changes such as menopause and osteoporosis, and presence of genetic polymorphism in Pb binding analogous to the above discussion for blood lead.

Discussing PbB and bone Pb as biomarkers of ongoing or immediately toxic exposure requires that these measures be kept distinct from measures of body lead burden. These measures have been the source of considerable confusion in the environmental health literature. Body lead burden, the total amount of body Pb regardless of relative immediate mobilizability, may not be as sensitive a measure of concurrent exposure risk as a measure such as PbB in cases where circulating Pb *in vivo* is available to have immediate toxic impacts in target tissues and where the time for toxic response is short. PbB can rise considerably and induce toxic effects in sensitive organs and tissues without materially altering the total body lead burden for reasons including the large size of the latter versus the former. However, mobilized bone Pb is the more relevant exposure marker for long-term exposures of workers or the general adult population.

There are differing relationships between PbB and bone Pb within the context of being internal biomarkers. A good conceptual depiction of temporal relationships between the two measures appears in Figure 4.5 of U.S. EPA (2006), Ch. 4.

The role of bone as the principal body depository for Pb makes it the principal biomarker of body lead burden. By comparison, the total Pb content of whole blood is considerably less (Rabinowitz et al., 1976). Given current evidence, however, the kinetic and metabolic mobility of bone Pb is sufficient to qualify this Pb reservoir as a biomarker of Pb exposure and a biomarker relevant for much longer periods than PbB. Some recent epidemiological studies, like those for toxic effects described in the toxicity chapters of this book, indicate associations of bone Pb with long-term chronic adverse effects in human populations while PbB provided negative results.

Both cortical and trabecular bones accumulate Pb but do so within different kinetic parameters. Cortical bone in human Pb exposure subjects is associated with the longest half-life of Pb decline, ranging in the older literature up to a decade or more. Trabecular bone is much more mobile, having a half-life of decline as short as a year or less. The much shorter half-life of Pb mobilization from trabecular (spongy) bone back to the circulating compartment indicates that this provides the highest relative fraction of all bone Pb releases (Hu et al., 1998).

22.5 DOSE–TOXIC RESPONSE RELATIONSHIPS FOR CHILDREN BASED ON BLOOD LEAD

Tabulated dose–responses in this chapter are presented for subsets of human populations based on degrees of both exposure and toxicity risks. Children, particularly as infants and toddlers, are at highest risk due to both exposure and toxicological vulnerability factors. Long recognition of Pb as a developmental toxicant based on an extensive body of scientific data would particularly argue for singling out developing systems in human populations as at greatly increased risk of toxic harm.

Table 22.1 sets forth dose–toxic response relationships for young children in terms of PbB thresholds for adverse effects as exposures increase. Here, thresholds are taken to mean those lowest PbB levels that have been reliably associated in the international literature with toxic effects. Identifying some empirical threshold on the basis of the low point of ranges in regression or other statistical analyses should not be taken to mean an actual threshold exists (see above on lack of thresholds for Pb in humans).

TABLE 22.1 Full-Range Dose–Toxic Response Relationships for Lead in Children[a–c]

Adverse Health Effect	Lowest Dose Level(s)	Comments	Illustrative References
Developmental toxicity	≤10	Fetal effects reported in some studies	
Early nephrotoxicity	≤10	Based on certain biomarkers	
Early immunotoxicity	~10	Both humoral (IgE) and cell (T-cell abundance) immunity affected	
Early hematotoxic effects	10–20	P5N inhibition; ALA-D inhibition; EP accumulation	
Early peripheral neuropathy	20–30	Nerve conduction velocity studies	Chapters this book, U.S. CDC (1991, 2005), NAS/NRC (1993), U.S. EPA (2006), U.S. ATSDR (1988, 2007)
Later hematotoxicity	30	Plasma, urine ALA increases	
Vitamin D biosynthesis, function	30	1,25-$(OH)_2$ vitamin D impairments	
Later hematotoxicity	40–45	Hemoglobin formation impaired via heme and globin effects	
GI and clinical hematotoxicity	50–60	Lead "colic;" anemia	
Nephrotoxicity	60–80	Late onset nephropathy	
Chronic encephalopathy	80–100	Various impaired clinical measures of mental functioning: e. g., mental retardation; U.S. CDC—PbB ≥70 is an "acute" medical emergency	
Acute effects and death	100+	Acute encephalopathy; cerebral vasculopathy; acute nephropathy	

[a]*Dose metric = PbB.*
[b]*Dose measures = central tendency or range.*
[c]*Adverse effects as selected in Chapter 21.*

Information in Table 22.1 is drawn from data meeting the following criteria: (1) exposure (PbB) measurements have been reliably done, (2) associations with PbB have been determined using the more reliable toxic endpoints, (3) dose–toxic responses have been confirmed internationally by multiple investigator teams, (4) evaluations of dose–responses have included aggregated data analyses, such as meta-analyses and pooled analyses, and (5) where multiple forms of expression for the dose measure are presented and all are comparably plausible, that form providing the lowest-exposure relationship is stated.

The full spectrum of toxic effects ranges from subtle disturbances in multiple organs and systems at the low end of exposure, <10 μg/dl, to death or profound brain damage at the high, extreme concentrations >100 μg/dl. Both are increasingly likely at PbB values above 100 μg/dl. For ease of presentation, Table 22.1 notes the specific new effects in terms of a lowest PbB value associated with those toxic endpoints. However, as PbB increases in the table, the number of effects and the severity of effects first identified at lower associations will increase. Finally, the lowest levels of PbB empirically linked to adverse health effects in children are not thresholds below which the specified effects do not occur.

Historically, toxic endpoints linked to clinical lead poisoning were accepted as beginning from the highest entry, ">100 μg/dl," in the early twentieth century, declining to those toxic endpoints with PbB thresholds of 40–60 μg/dl as recently as the early 1970s. In more recent years, the focus among the health science, medical, and regulatory communities has been on the low end of the dose–response spectrum for lead in children. These lower exposures are more relevant for nonoccupational risk populations and follow the large growth of evidence in the scientific literature for absence of toxicity thresholds for lead.

Several topics allied with dose–toxic response relationships for humans at risk, particularly in pediatric populations, merit comment. The first is the scope of definition of a lead-associated adverse health effect. That definition first requires an acceptable definition of "health." In the 1940s, the World Health Organization (1948) defined the term as being much more than simply the absence of manifest disease or some infirmity, declaring, health "... is a state of complete physical, mental, and social well-being ..." This relatively holistic definition of what comprises "health" provides informing criteria for what impacts of Pb exposure are adverse health effects, i.e., any effect which impairs complete physical, mental, and social well-being. One extension of the definition would be the ability of a toxicant (lead) not only to induce toxicity but also to reduce the capacity of organs and systems to respond to toxic stresses, such as by reducing the intrinsic reserve capacity of critical functions. For example, Pb-induced nephrotoxicity also impacts reserve capacities for kidney function.

A second topic of importance to low-level Pb effects is that of dose–population response relationships, relationships where the endpoint is the prevalence or incidence of some sentinel or index effect, rather than the typically understood dose–toxic response relationship. The former is statistical, while the latter is toxicological. As PbB increases across a target population, the fraction of exposed individuals presenting with some adverse effect at a defined level increases.

The epidemiological studies that establish the basic validity of dose–toxic response relationships at low exposures also establish the validity of general causality of various adverse effects by lead. Equally important, such studies also establish the validity of a physician choosing to include lead-related effects when articulating a diagnosis of specific causality of some effect in a given patient.

The third conceptual point is that of the societal impacts of low-level toxic effects expressed as seemingly modest changes when viewed in isolation. As noted in U.S. EPA (1986, 2006), U.S. CDC (1991), and multiple references in those, small changes translate to large impacts across entire populations of affected individuals, e.g., child populations sustaining Pb-associated IQ decrements.

The PbB dose–responses presented in Table 22.1, particularly for subclinical endpoints at lower exposures relevant to populations today, typically reflect the results of numerous studies by many researchers. Overall, the dose/exposure parts of the dose–responses in the table were determined as mean values, often geometric mean measures of central tendency. This repeated confirmation and commonality in findings offsets potential uncertainties such as may occur where data have been derived from cross-sectional and/or isolated PbB measurements. In the aggregate, data in this table are the distillation of different statistical and epidemiological designs, including meta-analyses, pooled analyses, and large longitudinal evaluations.

Viewed in the aggregate, dose–response relationships depict adverse effects from Pb exposures ranging from peaks of exposure carrying risks of lethal effects to PbB values below 5 μg/dl.

Virtually all of the dose–response relationships in Table 22.1 reflect greater sensitivity to Pb in children than parallel information depicted in this chapter for adults. Given the recognized greater sensitivity to Pb of young children when compared to adults or even older children, this is fully predictable. However, as noted in Table 22.4, some early toxic endpoints in adult groups are also identified with PbB thresholds <10 μg/dl.

Table 22.2 presents an expanded depiction of dose–toxic response relationships in children at PbB levels at or below 10 μg/dl. This tabulation particularly focuses on developmental neurotoxic effects in young children, ranging from IQ decrements to deficits in school achievement and neuromuscular function (see Chapter 12 for details). Other systems are affected and include the renal, hematological, and immunological systems. For example, early kidney effects in the form of reduced GFR using cystatin C as the biomarker were noted by Fadrowski et al. (2010), with effects seen in children with a mean PbB of 1.5 μg/dl and >99% with PbB <10 μg/dl. Immunotoxicity in children due to low Pb exposures is discussed in Chapter 18.

The U.S. CDC in its 2005 Statement and the U.S. EPA (2006) noted that it is established that toxic effects of Pb occur in young children at levels

TABLE 22.2 Dose–Toxic Response Relationships for Lead in Children at PbB Levels <10 μg/dl[a,b]

Adverse Health Effect	Lowest PbB Dose Level (μg/dl)	Comments	References
Developmental neurotoxicity: IQ deficits	0–7.5	Slope of PbB–IQ deficit curve higher <10 μg/dl than ≥10 μg/dl	Chapters this book, e.g., recent studies summarized in Chapter 21, U.S. CDC (2005), U.S. EPA (2006), U.S. ATSDR (2007), pooled and meta-analyses
Developmental neurotoxicity: academic achievement; reading, math scores; multiple large-sample data sets	~2	Data based on, for example, pooled analysis from multiple neurocognitive studies; concurrent PbB shows strongest dose–response	
Impaired executive function	Mean PbB = 6.5; 80% <10 μg/dl	–	
Immunotoxicity: elevated IgE	Geometric mean PbB = 2.7	–	
Immunotoxicity: lowered T-cell abundance	Geometric mean PbB = 2.7	–	
Pubertal development alterations by Pb	Geometric mean = 2.5 μg/dl	NHANES III female children, 8–16 years old; inversely associated with PbB	
Altered hormonal levels in female children, 6–11 years old	Median PbB = 2.5 μg/dl	PbB inversely associated with inhibin B	
Altered early nephrotoxicity markers: β_2 microglobulin, serum cystatin-C	Two child exposure cohort PbB mean values: 1.8, 2.7 μg/dl	PbB positively associated with both markers	
Lowered GFR using serum creatinine, cystatin-C	PbB mean = 1.5 μg/dl; 99% <10 μg/dl		

[a]*Dose measures = central tendency or range.*
[b]*Adverse effects as selected in Chapter 21.*

<10 μg/dl. The U.S. CDC guidance level was retained at 10 μg/dl PbB, however. Some have misinterpreted this value as a type of preservation threshold, which it is not. U.S. CDC Statements, rather, have been and are relatively flexible prescriptions for prioritizing the management of exposures in child populations most at risk for Pb poisoning. Iterations of the U.S. CDC Statements beginning in 1975 have made two characterizations of the statements explicit. First, there was and is every expectation that the nature and advisory utility of the statements would change as the science changed. For example, the CDC's April 1978 Statement (U.S. CDC, 1978), which was used in the derivation of the 1978 NAAQS primary standard of 1.5 μg/dl stated that (p. 1, Introduction):

"The CDC recognizes that there will doubtless be further development in this field which may alter or redefine our current understanding."

The 1985 Statement (p. 1, Introduction, U.S. CDC, 1985) noted:

"The revised recommendations in this 1985 Statement reflect current knowledge concerning screening, diagnosis, treatment, follow up, and environmental intervention for children with elevated blood lead levels."

The second aspect of all statements, including the 2005 version, was the expectation that the statements would eventually effect primary lead poisoning prevention for children in the United States and elsewhere. Furthermore, this ultimate goal of primary prevention has been presented in print to the public health and clinical communities for more than 30 years. We clearly see this in the evolution of the prescriptive language of successive statements which simply translates as, "it is much easier to prevent than to treat." For example, the 2005 Statement (p. ix, Preface) notes:

"The data demonstrating that no 'safe' threshold for BLLs in young children has been identified highlights the importance of preventing childhood exposures to lead. It confirms the need for a systematic and society wide effort to control or eliminate lead hazards in children's environments before they are exposed. This emphasis on primary prevention, although not entirely new, is highlighted here and is clearly the foremost action supported by the data presented in *A Review of Evidence of Adverse Health Effects Associated with Blood Lead Levels <10 μg/dL in Children.*"

The CDC Statements of 1978 and 1985 were presented by CDC as tabulated Pb toxicity risk advisories for *secondary prevention of childhood lead poisoning*. That is, the focus was on identification of the frequency and nature of lead exposures in U.S. children where the mechanisms are *post hoc* in nature: children are identified as having toxic lead exposures after they have sustained such exposures.

These statements also held that responses addressing low but toxic Pb levels required primary prevention strategies, not just secondary prevention via identifying *post hoc* poisonings through screenings.

22.6 DOSE–TOXIC RESPONSE RELATIONSHIPS FOR ADULTS BASED ON BLOOD LEAD

A minimum of two human risk population categories can be identified among Pb-exposed adults. One group is identified by virtue of relatively high exposures, i.e., lead workers at primary and secondary smelters, primary Pb refineries, lead battery production facilities and lead fabricators, or others occupationally exposed to Pb. A second is characterizable because of systemic, somatic changes which enhance internal or endogenous Pb exposures. This category would include pregnant and nursing women with high body lead burdens in bone, and women during menopause sustaining osteoporotic changes in bone metabolism and mineralization processes.

People exposed to lead in the workplace clearly have the highest exposures overall, and primary and secondary smelter workers probably have the highest exposures in the occupational category. Certain non-lead production workers also have had high exposures, such as those in the construction and remodeling industries. This includes those who remove old lead paint from bridges and other surfaces. A detailed discussion quantifying such workplace exposures is outside the scope of this chapter, but those workers, ranked in terms of Pb toxicity, are discussed in earlier chapters. It is critical to keep in mind that environmental exposures that affect typical risk populations add to workplace exposures.

Table 22.3 presents the full-spectrum of dose–toxic response relationships for lead toxicity in adults in terms of lowest reported adverse effect level thresholds. As with the earlier tables, only the lowest values for PbB in the associations specific for the indicated range are identified. It is also understood with these adult tabulations that as PbB values rise above lowest levels of determined associations, those toxic effects increase in severity and multiplicity. Toxicity criteria for reliability and validity parallel those enumerated for the earlier childhood tables. A number of dose–responses scaled to dose/exposure are identified in the table. In many cases, including hematotoxicity, peripheral neurotoxicity, and nephrotoxicity, effects are qualitatively similar but occur at higher empirically measured thresholds.

Certain other effects for adults in this table are apparent. First, acute encephalopathy is relatively rare in adults when identified at all. Peripheral neuropathy is typically assumed to be more frequent in adults than is brain injury, but that traditional stance is changing. Hematotoxic measures are generally but not totally associated in adults with higher PbB thresholds. However, there are a number of low-level Pb effects in adults sustaining nonoccupational exposures at PbB thresholds not greatly different than in children.

TABLE 22.3 Full-Range Dose–Toxic Response Relationships for Lead in Adults

Adverse Health Effects	Lowest Dose Level (PbB, μg/dl)	Comments	References
Elevated BP	~10	More consistently, systolic BP	U.S. EPA (2006), U.S. ATSDR (2007), meta-analyses and other data in Chapters 12–19, this book
Elevated serum creatinine	~10	Hyperfiltration as early nephrotoxic effect	
Elevated EP, women	15	Group basis, lowest end of range	
Elevated EP, men	20–25	Group basis, lowest end of range	
Potential cognitive impairment	20–25	Small group studies	
Peripheral (sensory) nervous system effects	~30	Nerve conduction velocity measurement	
Kidney tubular injury	~30	Impaired reabsorption	
Genotoxicity	40–50?	Micronuclear, DNA damage/ mitosis; chromosomal effects in Pb workers	
Later hematotoxicity	40–60	Anemia indexed as lower hemoglobin	
Later nephrotoxicity	60+	Interstitial nephritic effects	
GI effects	60+	"Monday morning colic" among lead workers	
Later peripheral neuropathic effects	60–80+	Wrist drop, foot drop neuromotor injury	

Table 22.4, like Table 22.2 for children, is an expanded depiction of low-level PbB effects at or below 10 μg/dl. Of particular public health importance, early cardiotoxic and nephrotoxic effects have been identified in meta-analyses and well-designed large studies as occurring at PbB $\leq$ 10 μg/dl. Details are in the health effect chapters (Chapters 12 to 19).

TABLE 22.4 Dose–Toxic Response Relationships for Lead in Adults at PbB Levels <10 μg/dl[a,b]

Adverse Health Effects	Lowest Dose Level (μg/dl)	Comments	References
Increases in hypertension, SBP, DBP	Range of means: 3.0–5.4	Black males, females in NHANES III: 1988–1994; PbB significantly linked to SBP, DBP; linear association in Black females	NAS/NRC (1993), U.S. EPA (1986, 2006), U.S. ATSDR (2007), meta-analyses and other studies in recent reports noted in Chapter 21
Increased DBP, use of antihypertensive medicine	Male geometric mean = 3.7; female geometric mean = 2.6	1995 English Health Survey: females, significant association of DBP with PbB, use of antihypertensives	
Increased hypertension, SBP, DBP	Geometric mean = 1.64	NHANES 1999–2002: Mexican-Americans and Blacks showed significant increases in OR versus PbB quartile	
Increased risk of cardiovascular mortality	Study of Schober et al., PbB terciles: <5, 5–9, ≥10; study of Menke et al., terciles <1.93, 1.93–3.63, 3.63–10	NHANES III, PbB in higher two groups linked to higher cardiovascular mortality in Schober et al. (2006); HR trend significant for all groups in Menke et al. (2006)	
Increased risk of PAD	Lowest quartile, 1.4; highest quartile, >2.9	Fourth quartile significantly linked with PAD	
Increased risk of chronic kidney disease, PAD	Geometric mean (1999–2002) = 1.6; geometric mean (1988–1994) = 2.8	Increased OR in NHANES subject of CKD and PAD with PbB for 1999–2002 period	
Increased risk of albuminuria, reduced estimated GFR and combined endpoints	Geometric mean = 1.6; PbB quartiles used	NHANES subjects, 1999–2006; ORs for endpoints at highest quartile were elevated versus lowest one	NAS/NRC (1993), U.S. EPA (1986, 2006), U.S. ATSDR (2007), meta-analyses and other studies in recent reports noted in Chapter 21
Reduced creatinine and serum cystatin-C	Mean PbB = 2.2	Swedish women; significant negative association of PbB with renal function markers	

SBP, systolic blood pressure; DBP, diastolic blood pressure; PAD, peripheral artery disease; CKD, chronic kidney disease.
[a]*Dose measure=central tendency or range.*
[b]*Adverse effects noted in Chapter 21.*

22.7 DOSE–TOXIC RESPONSE RELATIONSHIPS FOR ADULTS BASED ON BONE LEAD

Data on bone Pb and various toxic endpoints in humans have generally been generated for adults, and for adult populations having either occupational exposures or principally environmental contacts with the element. Bone Pb in children, for reasons presented in earlier chapters, does not lend itself and has not lent itself to serving as a stable dose/exposure metric versus PbB in the very young child.

The available literature, as described in the adverse effects chapters, allows one to conclude that bone Pb is a better dose metric for long-term, chronic toxic effects in adults than are single measurements of PbB. In fact, that literature shows associations of trabecular and/or cortical bone Pb with a number of toxic endpoints when no links are shown with PbB levels. This comparative set of outcomes is a strong argument for using bone Pb or combined bone Pb and PbB as exposure metrics.

While individual epidemiological studies show associations of bone Pb with toxicity, the available evidence does not allow identifying any clear set of thresholds or gradients in dose–response for bone Pb versus diverse toxic endpoints, particularly for neurological, cardiovascular, and nephrotoxic effects (Hu et al., 2007; Navas-Acien et al., 2009; Shih et al., 2007). One difficulty with such studies is their requirement for considerable laboratory expertise in the form of XRF bone Pb noninvasive measurements and using expensive equipment. This logistical reality still limits such studies to research project-type evaluations rather than to routine use. For example, NHANES II, NHANES III, and post-NHANES III epidemiological evaluations were limited to the single PbB measurements gathered within the survey designs, when bone Pb measurements might actually have been more revealing, had they been done.

While dose–response relationships for bone Pb and toxic endpoints are limited in human environmental exposures, those which have been recorded (see effect chapters) suggest a number of associations. There are comparatively more data for bone studies in lead workers. Here, there are still uncertainties recognized earlier—more intense workplace Pb exposures that prevailed in many lead industries in past decades and the difficulty of translating worker bone Pb values to environmental levels of the toxicant. Schwartz and Hu (2007) recommended that a safe threshold for prevention of chronic toxic endpoints in Pb workers be set at 15 μg/g mineral in tibial bone, but this figure requires further validation. Nonetheless, worker protections based on bone Pb have considerable promise.

REFERENCES

Adebonojo, F.O., 1974. Hematological status of urban Black children in Philadelphia. Clin. Pediatr. 13, 874–888.

Alessio, L., Bertazzi, P.A., Monelli, O., Toffoletto, F., 1976. Free erythrocyte protoporphyrin as an indicator of the biological effect of lead in adult males. III. Behavior of free erythrocyte protoporphyrin in workers with past lead exposure. Int. Arch. Occup. Environ. Health 38, 77–86.

American Public Health Association, 1955. Methods for Determining Lead in Air and in Biological Materials. Committee on Chemical Procedures of the Occupational Health Section, New York.

Astrin, K.H., Bishop, D.F., Wetmur, J.G., Kaul, B., Davidow, B., Desnick, R.J., 1987. δ-Aminolevulinic acid dehydratase isozymes and lead toxicity. Ann. N.Y. Acad. Sci. 514, 23–29.

Barry, P.S.I., 1975. A comparison of concentrations of lead in human tissues. Br. J. Ind. Med. 32, 119–139.

Barry, P.S.I., Mossman, D.B., 1970. Lead concentrations in human tissues. Br. J. Ind. Med. 27, 339–351.

Bergdahl, I.A., Gerhardsson, L., Schütz, A., Desnick, R.J., Wetmur, J.G., Skerfving, S., 1997a. Delta-aminolevulinic acid dehydratase polymorphism: influence on lead levels and kidney function in humans. Arch. Environ. Health 52, 91–96.

Bergdahl, I.A., Grubb, A., Schütz, A., Desnick, R.J., Wetmur, J.G., Sassa, S., et al., 1997b. Lead binding to δ-aminolevulinic acid dehydratase (ALAD) in human erythrocytes. Pharmacol. Toxicol. 81, 153–158.

Bergdahl, I.A., Sheveleva, M., Schütz, A., Artamonova, V.G., Skerfving, S., 1998. Plasma and blood lead in humans: capacity-limited binding to δ-aminolevulinic acid dehydratase and other lead-binding components. Toxicol. Sci. 46, 247–253.

Church, H.J., Day, J.P., Braithwaite, R.A., Brown, S.S., 1993. Binding of lead to a metallothionein-like protein in human erythrocytes. J. Inorg. Biochem. 49, 55–68.

DeSilva, P.E., 1981. Determination of lead in plasma and its relationship to lead in erythrocytes. Br. J. Ind. Med. 38, 209–217.

Everson, J., Patterson, C.C., 1980. "Ultra-clean" isotope dilution/mass spectrometric analyses for lead in human blood plasma indicate that most reported values are artificially high. Clin. Chem. 26, 1603–1607.

Fadrowski, J.J., Navas-Acien, A., Tellez-Plaza, M., Guallar, E., Weaver, V.M., Furth, S.L., 2010. Blood lead level and kidney function in US adolescents: The Third National Health and Nutrition Examination Survey. Arch. Intern. Med. 170, 75–82.

Gulson, B.L., Jameson, C.W., Mahaffey, K.R., Mizon, K.J., Korsch, M.J., Vimpani, G., 1997. Pregnancy increases mobilization of lead from maternal skeleton. J. Lab. Clin. Med. 130, 51–62.

Hansen, J.P.B., Døssing, M., Paulev, P.-E., 1981. Chelatable lead body burden (by calcium-disodium EDTA) and blood lead concentration in man. J. Occup. Med. 23, 39–43.

Hu, H., Rabinowitz, M., Smith, D., 1998. Bone lead as a biological marker in epidemiological studies of chronic toxicity: conceptual paradigms. Environ. Health Perspect. 106, 1–8.

Hu, H., Shih, R., Rothenberg, S., Schwartz, B.S., 2007. The epidemiology of lead toxicity in adults: measuring dose and consideration of other methodologic issues. Environ. Health Perspect. 115, 455–462.

Hunter, J.M., 1978. The summer disease: some field evidence on seasonality in childhood in childhood lead poisoning. Soc. Sci. Med. 12, 85–94.

Keenan, R.G., Byers, D.H., Saltzman, B.E., Hyslop, F.L., 1963. The "USPHS" method for determining lead in air and in biological materials. Am. Ind. Hyg. Assoc. J. 24, 481–491.

Lee, B.-K., Lee, G.-S., Stewart, F.S., Ahn, K.-D., Simon, D., Kelsey, K.T., et al., 2001. Associations of blood pressure and hypertension with lead dose measures and polymorphisms in the vitamin D receptor and δ-aminolevulinic acid dehydratase genes. Environ. Health Perspect. 109, 383–389.

Lolin, Y., O'Gorman, P., 1988. An intra-erythrocytic low molecular weight lead-binding protein in acute and chronic lead exposure and its protective role in lead toxicity. Ann. Clin. Biochem. 25, 688–697.

Manton, W.I., Cook, J.D., 1984. High accuracy (stable isotope dilution) measurements of lead in serum and cerebrospinal fluid. Br. J. Ind. Med. 41, 313–319.

Menke, A., Muntner, P., Batuman, V., Silbergeld, E.K., Guallar, E., 2006. Blood lead below 0.48 micromol/L (10 microg/dL) and mortality among U.S. adults. Circulation 114, 1388–1394.

Moore, M.R., 1988. Hematological effects of lead. Sci. Total Environ. 71, 419–431.

Mushak, P., 1992a. Perspective. Defining lead as the premiere environmental health issue for children in American: Criteria and their quantitative application. Environ. Res. 59, 281–309.

Mushak, P., 1992b. The monitoring of human lead exposure. In: Needleman, H.L. (Ed.), Human Lead Exposure. CRC Press, Boca Raton, FL, pp. 45–64.

Mushak, P., 1993. New directions in the toxicokinetics of human lead exposure. Neurotoxicology 14, 29–42.

Mushak, P., 1998. Uses and limits of empirical data in measuring and modeling human lead exposure. Environ. Health Perspect. 106 (Suppl. 6), 1467–1484.

National Academy of Sciences: National Research Council, 1993. Measuring Lead Exposure in Infants, Children, and Other Sensitive Populations. National Academy Press, Washington, DC.

Navas-Acien, A., Tellez-Plaza, M., Guallar, E., Muntner, P., Silbergeld, E., Jaar, B., et al., 2009. Blood cadmium and lead and chronic kidney disease in US adults: a joint analysis. Am. J. Epidemiol. 170, 1156–1164.

Nilsson, U., Attewell, R., Christoffersson, J.-O., Schütz, A., Ahlgren, L., Skerfving, S., et al., 1991. Kinetics of lead in bone and blood after end of occupational exposure. Pharmacol. Toxicol. 69, 477–484.

Rabinowitz, M.B., Wetherill, G., Kopple, J.D., 1976. Kinetic analysis of lead metabolism in healthy humans. J. Clin. Invest. 58, 260–270.

Rabinowitz, M.B., Wetherill, G., Kopple, J.D., 1977. Magnitude of lead intake from respiration by normal man. J. Lab. Clin. Med. 90, 238–248.

Raghavan, S.R.V., Culver, B.D., Gonick, H.C., 1980. Erythrocyte lead-binding protein after occupational exposure. Environ. Res. 22, 264–270.

Raghavan, S.R.V., Culver, B.D., Gonick, H.C., 1981. Erythrocyte lead-binding protein after occupational exposure. II. Influence of lead inhibition of membrane Na^+, K^+-adenosine triphosphatase. J. Toxicol. Environ. Health 7, 561–567.

Raghavan, S.R.V., Gonick, H.C., 1977. Isolation of low-molecular-weight lead-binding protein from human erythrocytes. Proc. Soc. Exp. Biol. Med. 155, 164–167.

Roels, H., Konings, J., Green, S., Bradley, D., Chettle, D., Lauwerys, R.R., 1995. Time-integrated blood lead concentration is a valid surrogate for estimating the cumulative lead dose assessed by tibial lead measurement. Environ. Res. 69, 75–82.

Saenger, P., Rosen, J.F., Markowitz, M.E., 1982. Diagnostic significance of edetate disodium calcium testing in children with increased lead absorption. Am. J. Dis. Child. 136, 312–315.

Schlenker, T.L., Fritz, C.J., Mark, D., Layde, M., Linke, G., Murphy, A., et al., 1994. Screening for pediatric lead poisoning. Comparability of simultaneously drawn capillary and venous blood samples. JAMA 271, 1346–1348.

Schober, S.E., Mirel, L.B., Graubard, B.I., Brody, D.J., Flegal, K.M., 2006. Blood lead levels and death from all causes, cardiovascular disease, and cancer: Results from the NHANES III mortality study. Environ. Health Perspect. 114, 1538–1541.

Schwartz, B.S., Hu, H., 2007. Adult lead exposures. Time for change. Environ. Health Perspect. 115, 451–454.

Schwartz, B.S., Lee, B.-K., Stewart, W., Ahn, K.-D., Springer, K., Kelsey, K., 1995. Associations of δ-aminolevulinic acid dehydratase genotype with plant, exposure duration, and blood lead and zinc protoporphyrin levels in Korean lead workers. Am J. Epidemiol. 142, 738–745.

Schwartz, B.S., Lee, B.-K., Stewart, W., Sithisarankul, P., Strickland, P.T., Ahn, K.-D., Kelsey, K., 1997. δ-aminolevulinic acid dehydratase genotype modifies four hour urinary lead excretion after oral administration of dimercaptosuccinic acid. Occup. Environ. Med. 54, 241–246.

Schwartz, B.S., Lee, B.-K., Lee, G.-S., Stewart, W.F., Simon, D., Kelsey, K., Todd, A.C., 2000. Associations of blood lead, dimercaptosuccinic acid-chelatable lead, and tibia lead with polymorphisms in the vitamin D receptor and δ-aminolevulinic acid dehydratase genes. Environ. Health Perspect. 108, 949–954.

Schwartz, B.S., Lee, B.-K., Lee, G.-S., Stewart, W.F., Lee, S.S., Hwang, K.Y., et al., 2001. Associations of blood lead, dimercaptosuccinic acid-chelatable lead, and tibia lead with neurobehavioral test scores in South Korean lead workers. Am. J. Epidemiol. 153, 453–464.

Shih, R.A., Hu, H., Weisskopf, M.G., Schwartz, B.S., 2007. Cumulative lead dose and cognitive function in adults: a review of studies that measured both blood lead and bone lead. Environ. Health Perspect. 115, 483–492.

Silbergeld, E.K., Schwartz, J., Mahaffey, K., 1988. Lead and osteoporosis: mobilization of lead from bone in postmenopausal women. Environ. Res. 47, 79–94.

Smith, C.M., Wang, X., Hu, H., Kelsey, K.T., 1995a. A polymorphism in the δ-aminolevulinic acid dehydratase gene may modify the pharmacokinetics and toxicity of lead. Environ. Health Perspect. 103, 248–253.

Smith, C.M., Hu, H., Wang, X., Kelsey, K.T., 1995b. ALA-D genotype is not associated with HT or HB levels among workers exposed to low levels of lead. Med. Lav. 86, 229–235.

Tola, S., Hernberg, S., Asp, S., Nikkanen, J., 1973. Parameters indicative of absorption and biological effect in new lead exposure: a prospective study. Brit. J. Ind. Med. 30, 134–141.

U.S. Agency for Toxic Substances and Disease Registry, 1988. The Nature and Extent of Lead Poisoning in Children in the United States. U.S. Public Health Service, Atlanta, GA, Submitted to Congress, July 12, 1988.

U.S. Agency for Toxic Substances and Disease Registry, 2007. Toxicologic Profile for Lead. U. S. Centers for Disease Control, Atlanta, GA.

U.S. Centers for Disease Control, 1978. Preventing Lead Poisoning in Young Children. A statement by the Centers for Disease Control. U.S. Department of Health and Human Services, Atlanta, GA, April.

U.S. Centers for Disease Control, 1985. Preventing Lead Poisoning in Young Children. A statement by the Centers for Disease Control. U.S. Department of Health and Human Services, Atlanta, GA.

U.S. Centers for Disease Control, 1991. Preventing Lead Poisoning in Young Children. A Statement by the Centers for Disease Control. U.S. Department of Health and Human Services, Atlanta, GA.

U.S. Centers for Disease Control, 2005. Preventing Lead Poisoning in Young Children. A Statement by the Centers for Disease Control. U.S. Department of Health and Human Services, Atlanta, GA.

U.S. Environmental Protection Agency, 1986. Air Quality Criteria for Lead, 4 vols. Report No. EPA 600/8-83/028bF. Environmental Protection Agency, Research Triangle Park, NC.

U.S. Environmental Protection Agency, 2006. Air Quality Criteria for Lead, vol. 1, Report No. EPA/600/R-05/144aF. Environmental Protection Agency, Research Triangle Park, NC.

Wetmur, J.G., 1994. Influence of the common human delta-aminolevulinate dehydratase polymorphism on lead body burden. Environ. Health Perspect. 102 (Suppl. 3), 215–219.

World Health Organization, 1948. WHO Definition of Health. Preamble to the Constitution of the World Health Organization as adopted by the International Health Conference, New York, 19–22 June, 1946; signed on 22 July 1946 by the representatives of 61 States (Official Records of the World Health Organization, no. 2, p. 100) and entered into force on 7 April 1948.

Ziemsen, B., Angerer, J., Lehnert, G., Benkmann, H.-G., Goedde, H.W., 1986. Polymorphism of delta-aminolevulinic acid dehydratase in lead-exposed workers. Int. Arch. Occup. Environ. Health 58 (3), 245–247.

Chapter 23

Exposure Characterizations for Lead in Specific Human Populations

23.1 INTRODUCTION AND RATIONALE

Part 2 and its chapters presented the topic of human lead exposure in global and categorical terms, addressing the technical areas of lead intakes, uptakes (absorption), toxicokinetics, integration of toxicokinetics into *in vivo* disposition in a manner allowing quantitative assessments of lead exposure, etc. In contrast to these broadly descriptive aspects of human Pb exposure, the applied health discipline of quantitative risk assessment requires prescriptive approaches for site-specific, case-specific, and environmental scenario-specific lead exposure characterizations. Data from such specific exposure characterizations are combined with available data for lead dose–response relationships to arrive at some quantitative risk characterization indexed as some endpoint for human health risk.

Typically, human health risk characterization for lead is carried out in a regulatory context and for environmental Pb exposures in various communities. The process invariably entails characterizing risk to young children and other risk groups residing in those communities. "Community," in turn, can be disaggregated to include as few as one individual per exposure unit, siblings in an exposure unit, a neighborhood block, etc. In certain cases, Pb exposures can be assessed for much greater geographic and demographic descriptors, including populations within national boundaries. U.S. national surveys such as NHANES II, III, IV, later NHANES and Hispanic HANES, are illustrative of the latter.

Lead as an environmental contaminant within various regulatory scenarios is unique among the suite of contaminant substances usually encountered at waste sites or the like, generated through cumulative anthropogenic activities. Exposure and subsequent risk characterization occurs first through determination of levels of PbB in risk groups and then relating these levels to some health risk metric. The health risk can be determined in two ways: by quantifying the distribution fraction of PbB values in excess of some

Trace Metals and other Contaminants in the Environment, Volume 10
ISSN: 1875-1121 DOI: 10.1016/B978-0-444-51554-4.00023-7

health risk threshold value or by doing a more detailed computation of the degree of deficits in some health metric versus determined PbB levels. Chapter 24 addresses the latter portion of this process.

By contrast, virtually all other contaminants identified at hazardous or other waste sites are evaluated for their potential human health risk by evaluating aggregated intakes or uptakes of the substance and comparing these values with some risk criterion level or cancer slope factor metric for noncarcinogenic and carcinogenic effects, respectively.

Two general forms of exposure assessment described in Part 2 apply for specific site or case exposure characterizations: measuring exposure biomarkers, typically PbB, or using toxicokinetic models to simulate likelihoods of specific levels of exposure, given sufficient data for environmental levels of Pb in relevant media contacted by human populations or in the absence of reliable empirical methods for measuring more remote exposures. Chapter 10 presented the environmental epidemiology of Pb as community-specific, regional, national, or international studies. That chapter principally described Pb exposure epidemiology using measured levels of an exposure biomarker, typically PbB. Chapter 9, by contrast, described various modeling approaches to predicting body levels of the element, with PbB as the modeling output. These two chapters have as their end effort the quantification of Pb exposures in risk groups as categories, with no additional analyses being done on a case-specific basis. Actual health-based risk associated with such measured or modeled PbB or other biomarkers were not estimated.

This chapter presents some specific case studies where Pb exposure risks in groups of the highest risk category, young children, were quantified in various ways using either measured or modeled PbB levels.

23.2 SOME KEY METHODOLOGICAL AND INTERPRETIVE ISSUES

23.2.1 Uses of Measured PbB Data

Single-shot, i.e., isolated blood lead, surveys have been employed in a number of lead-impacted U.S. communities. Some of these efforts have been carried out in areas where the U.S. EPA has done biokinetic modeling of lead exposure via the EPA IEUBK model. One needs to be careful about calibrating EPA's model or any other biokinetic model on haphazard, i.e., unstructured, PbB surveys. The merits and limits of such surveys have been critiqued elsewhere (Mushak, 1998). The problems are several: (1) such measures may be affected by various artifacts that are less influential in more structured sampling; and (2) such measures relate to bioavailability at some specific point in time and with a certain current land-use practice and population distribution. Demographic or socioeconomic characteristics of affected populations could change over time with movements in and out of areas of interest. Site land use could change; leaded media will change in terms of lead release over time; and interaction of child populations with the media could change in the future.

The IEUBK model is currently the standard risk assessment tool by which Federal and state lead risk assessments are carried out at Pb exposure sites with young children potentially exposed. Reliance on predictive modeling rather than requiring sole use of PbB screening of impacted children is driven in large part because of those concerns.

A follow-on issue is that of interpreting data from a community's risk residents where both PbB measurements and PbB modeling using the IEUBK computerized model have been done, and where results may be different. Does one assume automatically that the PbB measurements are valid and useful for calibration of predictive models or does one further scrutinize the PbB measurements? A good illustration of the problems of reconciling PbB measurement data with modeled PbB is to be found in sequential analyses of a PbB survey done in a Superfund community in Illinois by the Illinois Department of Public Health (IDPH, 1994) and its contractor, under U.S. ATSDR sponsorship. The authors of the IDPH report concluded, based on their statistical analysis, that there was no association between the PbB levels and lead in soil arising from prior contamination by a now-defunct lead smelter. However, a detailed later analysis by the U.S. EPA (Marcus, 1995 as reviewed in Mushak, 1998), using the inferential statistical technique of structural equation modeling, demonstrated a clear association. Furthermore, it noted that EPA's results from use of the IEUBK model for this area were not materially at odds with measured PbB values.

There has been a tendency in some regulatory quarters to calibrate the uptake input parameter in the EPA biokinetic model at a particular site by using any available PbB data, ignoring the need to first establish the reliability of blood lead information itself and the need to determine that any model versus measurement differences only exist because of the uptake variable (Mushak, 1998). That is, fine-tuning is often done in the absence of any pathway analysis that would quantitatively ascertain whether there is child contact with leaded soils, as well as whether there is an operative pathway between soils and inside dusts, etc.

23.2.2 Predictive PbB Modeling Parameters: The Key Roles of Pb Intake and Absorption Rate in Determining Childhood Pb Exposure and Health Risk

It has generally been assumed that children, usually undefined as to actual age, have a higher lead absorption rate than adults (NAS/NRC, 1993; U.S. CDC, 1991). The strength of the evidence for this observation across the ages of childhood, and its implications for lead health risk assessment, has been examined by Mushak (1998) along with some recent data impinging on the topic (U.S. EPA, 2006). Overall, from available data, it is reasonable to say that infants up to about 18–24 months of age absorb lead at a rate of about 40–50%.

How soon after this age does this uptake rate decline and when in childhood does uptake equal that for adults? More recent Australian data (Gulson et al., 1997), though based on few subjects, suggest that lead uptake rates in children 6–11 years old appear to approach those of adults, but where in this age span any decline that is most significant was not determined. O'Flaherty (1998) assumed, without empirical evidence, that the 5-year-old child has a lead uptake rate from diet of half that of younger children, 25% for purposes of modeling. Mushak (1998) analyzed epidemiological data of Tong et al. (1996) to determine that an uptake figure of 50% persists to around 3 years of age.

It is still unclear what the nature of the enhanced lead uptake is in infants and toddlers versus adults and perhaps older children. In the early infant, developmental factors are arguably at work. At what point do purely dietary factors begin to affect uptake of lead in preschool children? It is significant that one can simulate in adult volunteers these high lead uptakes of infancy with fasting conditions. That is, there may be nothing intrinsic about the adult gut lead uptake process that appears to operate at all times to produce a consistently low, i.e., 10%, lead uptake rate. The infant and toddler uptake rate of 50% may or may not apply only at feedings or during fasting periods as well. These questions have serious risk assessment implications.

A second parameter affecting Pb exposure biomarkers such as PbB is the daily intake rate of Pb. This factor was discussed earlier and one only needs to keep in mind that both daily intake and uptake rates can vary on a site-specific and case-specific basis. As also pointed out earlier, daily intake rates of Pb in young children, notably by mouth, are the product of Pb concentration and mass of ingested material containing Pb.

23.2.3 Sampling Issues for Assessing Site-Specific or Case-Specific Human Pb Exposure Characterizations in Risk Assessment

Both measured and modeled Pb exposures in risk groups have been reported in more recent years. A critical methodological component of either of these approaches is the sample collection protocol. The choice of sampling protocol for PbB measurement in young children or other risk groups (pregnant women, women of childbearing age, female employees working at Pb-contaminated sites) is determined by the hypothesis-testing questions posed. The statistical and epidemiological design for a simple study of distributions of PbB values within a community is different from approaches intended to show spatial or other geographic associations of PbB level with potential lead sources and/or lead pathways for risk group exposures. In the latter cases, one would evaluate such parameters as PbB samplings of children in residences whose locations differ with distance or directionality (downwind or upwind from a stationary source) or, as was the case in the past with leaded gasoline use, mobile sources such as major urban traffic corridors. Earlier chapters presented these factors in considerable detail.

A critical component of evaluating risk groups' exposures to Pb is the sampling of case-specific environmental media known to provide pathways from Pb source to human receptor(s). Using environmental data with measured PbB values provides for regression analysis modeling. Use of environmental data is also desirable and often necessary in order to use biokinetic models, such as the IEUBK model.

A second important factor in PbB measurement is the temporality of the PbB sampling. Childhood lead poisoning has long been known clinically as the "summer disease" because prevalences of elevated PbB and prevalences of medical interventions such as hospitalizations of young children peaked in late summer (Hunter, 1978; NAS/NRC, 1993). Part of this phenomenon traces to enhanced opportunities for increased Pb exposures, e.g., outside play during summer vacations, and to changes in vitamin D with changes in solar irradiation (NAS/NRC, 1993; U.S. EPA, 1986, 2006). The upshot of this seasonality factor is that PbB measurements made in the late fall and in winter will be lower by 30% or more than those carried out in the late summer.

23.3 ILLUSTRATIVE CASE STUDIES OF SITE-SPECIFIC Pb EXPOSURE CHARACTERIZATIONS IN RISK ASSESSMENT

The nature of the site in terms of lead exposure could be urban and dispersed, e.g., a city neighborhood with residential units having variably deteriorating interior and exterior Pb paint and associated dusts and contaminated soils, or a rural community with specific neighborhoods impacted by point source Pb contamination, e.g., an operating or defunct primary Pb smelter. Site exposure assessments are often conducted within the regulatory purview of the U.S. EPA or other public agencies such as state offices responsible for hazardous waste sites. Results of PbB findings help determine whether health risk is sufficiently high to require remedial intervention, typically expressed as a "clean-up" level in cases where the environmental medium mainly affected is soil and/or household dusts.

This section presents three case examples of site-based lead exposure characterizations for subsequent risk assessment purposes. The case studies are based on research where environmental media-specific levels of Pb were determined as part of the epidemiological and statistical designs. For purposes of this section, such two-part case studies are desirable for, first, quantifying baseline exposure characterizations and eventual baseline human health risk characterizations and, second, for identification of which subsequent media-specific regulatory interventions are likely to result in reduced Pb exposures and reduced health risk.

The first example summarizes exposure characterizations of individuals living in a well-known and studied, geographically broad locale impacted by a Pb point source in Idaho that had long released Pb to the environment. The second example examines a series of site-specific exposure assessments pooled to produce predictive relationships of PbB in children 3 years of age or younger to such lead sources as dusts and soils. This second illustration presented exposure characterizations for specific locations that were of two general site types: urban lead exposure settings in old city neighborhoods with lead paint and industrial emission histories, and Pb exposure endpoints reported for mining and other extractive industry sites. The third illustration presents results of case studies of U.S. populations selected by the U.S. EPA as part of regulatory action for air Pb.

23.3.1 The Bunker Hill/Silver Valley, ID, Superfund Site

The point source for the single-site example dispersed lead and other inorganic metals via multiple pathways from a large-volume lead, lead–zinc, and silver mining–smelter operation in Idaho, known as the Silver Valley/Bunker Hill metallurgical complex. This facility operated continuously in several configurations for about a century until the early 1980s, with some intermittent operations later. Multiple studies of various environmental and historical segments of the facility's impacts were carried out over the years, beginning in the early 1970s (von Lindern et al., 2003a,b; Yankel et al., 1977). A number of exposure assessments for the Silver Valley site over the decades since the 1970s were done both inside and outside of regulatory actions. As a result, this "megasite," a term introduced in a report of the NAS/NRC (2005) dealing with large regulated mining sites, has produced extensive databases for contamination and human risk population exposures applicable to extractive industrial settings in the United States and elsewhere.

This megasite is spatially subsumed within its wider geographic feature, the Coeur d'Alene River Basin (CDARB), a geographic and hydrological drainage entity of 3,700 square miles. A 53-mile stretch of the Coeur d' Alene River was evaluated for human and ecological toxic impacts, from the Idaho/Montana border on the East through the geographically centered contaminated zone segment and proceeding to Lake Coeur d'Alene on the West. There were about 11,000 residents within the entire Basin, of whom two-thirds resided within the Superfund area.

In 1983, the industrial part of the mining–milling–smelting complex was added to the NPL for Superfund evaluation and intervention by the U.S. EPA. This part consisted of a 21-square-mile area containing the facilities and five proximate communities: Kellogg, Page, Pinehurst, Smelterville, and Wardner. Two operable units for regulatory and administrative record purposes were defined in this Superfund "Box." Operable Unit 1 defined the populated community parts of the Box area. Operable Unit 2 applied to the unpopulated portions of the Box which also had extractive industry-related contamination from Pb and other toxicant releases.

Subsequent to the designation, a 1983 detailed lead study was carried out and since 1985, annual PbB screenings focused on children 9 years old and younger have been done by the Panhandle Health District/Shoshone County, ID, with cooperation of the Idaho Department of Health and Welfare (IDHW).

In 1998, the U.S. EPA decided to expand the geographical reach of the original superfund designation and NPL listing. The U.S. EPA, in its 2002 Record of Decision (ROD), formally identified and promulgated remedial actions for Operable Unit 3 (OU-3) for contaminated areas in the CDARB outside the Bunker Hill Box site, but excluding Lake Coeur d'Alene itself (U.S. EPA, 2002). A component of the Administrative Record for the ROD was a detailed environmental epidemiological study of lead exposures via PbB measurements and predictive modelings in the late 1990s.

Additional evaluations at this site have included the 5-year review of selected remedy effectiveness. Of notable concern in such a diffusely contaminated megasite is the question of whether Pb contamination has reoccurred through releases from unremediated portions of the site.

The first of a series of Pb exposure assessments of young children within the overall CDARB was carried out in the 1974–1975 period by the IDHW (Yankel et al., 1977) as part of an environmental epidemiological assessment of potential health threats from failed smelter emission controls in the form of a major baghouse fire in the early 1970s. The fire disabled the particulate emission control system and permitted a large increase in monthly emissions of Pb into the atmosphere, up to a monthly average of 73 tons of Pb (U.S. ATSDR, 2000) for about a year. The epidemiological effort later included the U.S. CDC (Landrigan et al., 1976) in evaluating the smelter area children for such health endpoints as anemia and peripheral neuropathy (reduced nerve conduction velocities). Prevalences of toxic levels of PbB, then defined as ≥40 μg/dl, rose to 99% in Smelterville, adjacent to the smelting complex.

The 1974–1975 Silver Valley study data are summarized in Table 23.1. Data are presented as they were gathered in 1974, before emission control systems were restored. Geometric mean PbB values varied with distance of child residences from the smelter emission points but varied only moderately with age of the children. Unlike other Pb exposure scenarios, those based on smelter emissions indicated that excessive exposures persisted into later years. The Silver Valley findings are consistent with findings elsewhere (Neri et al., 1978). Urban children exposed to lead paint more typically show peak PbB values at 2–3 years of age followed by a marked decline. In Area I, with homes within 1 mile of the facility, PbB varied from 75 μg/dl

TABLE 23.1 Blood Lead Levels (μg/dl, Geometric Mean) for Children in the Silver Valley Study, 1974–1975[a]

Geographic Area[b]	Age (years)					
	1	2	3	4	5	9
I	69[c]	72	75	75	68	57
II	50	51	55	46	49	40
III	33	36	36	35	35	32
IV	31	35	34	31	31	30
V	27	35	29	29	29	24

[a]Source: *Selected data in Yankel et al. (1977); Children N=1,149, 1–9 years old, August 1974.*
[b]*Area I, homes within 1 mile of smelter; Area II, 1–2.5 miles; Area III, 2.5–6 miles; Area IV, 6–15 miles; Area V, 15–20 miles.*
[c]*Erroneously noted in Yankel et al. (1977) as 59 μg/dl.*

for 3- and 4-year-olds to 57 μg/dl for 9-year-olds. Area II, 1–2.5 miles from the facility, was also significantly impacted. PbB levels for this group ranged from 55 μg/dl for 3-year-olds to 40 μg/dl for 9-year-olds. Childhood PbB levels for Areas III–V, ranging spatially from 2.5 to 20 miles from the complex, differed only moderately among the areas. Every age group in the two closest areas presented with geometric mean PbB levels that reached or exceeded the then-existing upper PbB limit.

The relationships of area-specific PbB levels for Silver Valley children with environmental Pb levels were also explored by the authors. As noted in Table 23.2, the expected decline in PbB with distance from the facility is accompanied by lower mean air Pb and soil Pb values. In terms of extremes in air Pb, measurements proximate to the smelter stacks are an order of magnitude higher than those seen in Area V, 15–20 miles distant from stack emissions. Area II, 1–2.5 miles from emissions, is also associated in the table with high geometric mean PbB and both air and soil Pb content.

The air levels in Table 23.2 were the last recordings of active smelter emissions in the peer-reviewed literature, since the smelting activity ceased continuous operation in 1981 for a variety of economic and regulatory reasons, as noted in NAS/NRC (2005). From this point forward, the considerable number of epidemiological assessments done within and outside the Bunker Hill Superfund Box area was for a defunct operation, with the various analyses and findings reported over the years generally focused on contaminants continuing to provide exposure and health risks through their persisting presence in the human environment in this area, notably dusts and soils.

The 1983 Pb study in the Bunker Hill Box area (Table 23.3) showed that relatively elevated PbB levels remained in Bunker Hill site communities,

TABLE 23.2 Geometric Mean PbB and Other Descriptive Statistics for Silver Valley, ID, Children in Areas I–V, 1974[a,b]

Geographic Area	PbB (μg/dl)[c,d]	Estimated Pb–Air ($\mu g/m^3$)[c]	Estimated Pb–Soil (ppm Pb)[e]
I	66	18.0	7,750
II	48	14.0	3,200
III	34	6.7	1,125
IV	32	3.1	1,320
V	28	1.5	2,250

[a]Source: *Data in Yankel et al. (1977); further analyses in U.S. EPA (1986). Area definitions per Table 23.1.*
[b]*Data only for 1974.*
[c]*Estimated by U.S. EPA (1986), Table 11.27.*
[d]*Rounding.*
[e]*Yankel et al. (1977), interpolations from Figure 3.*

being highest in the closest town, Smelterville, with arithmetic and geometric means of ~20–21 μg/dl. Pinehurst showed a geometric mean of ~11 μg/dl.

Table 23.3 presents, in addition to 1983 results, Box community PbB descriptive statistics for the five towns for selected years from 1988 to 1999. In one of the years, there was only combined community data (1983), while in 1988 only four communities were evaluated. Sitewide data were included for 1997 onward.

Over these years, the combination of yard-specific and community remediations and public health interventions (education, awareness, hygiene guidance, and serial biomonitoring) led to marked reductions in mean PbB concentrations, reductions which significantly exceeded the overall national picture of PbB decline owing to food and gasoline Pb reductions (U.S. EPA Region 10 ROD, 2002). Smelterville, the community closest to the facility, yielded a geometric mean PbB of 19.6 μg/dl in 1983, 8.8 μg/dl in 1990, 6.2 μg/dl in 1995, and 3.6 μg/dl in 1999. This decline, about 18 μg/dl, occurred over the total period following the shutdown of smelter operations in 1981 and did not simply reflect reductions of inputs to the communities from active Pb stack and fugitive emissions. Although the magnitudes of decline varied, depending on the original PbB statistics, the 1999 means were not materially different across communities.

Sitewide arithmetic and geometric mean PbB levels derived for the testing year 1999 were 4.7 and 3.9 μg/dl, respectively. Each of the remaining communities within the Superfund Box was reported to show PbB declines over time. A factor affecting the rate of community-specific PbB declines

TABLE 23.3 Mean PbB Levels at the Bunker Hill, ID, Superfund Box Site by Year and Site Community[a–d]

Year	Community	PbB Range (μg/dl)	PbB Mean (μg/dl)	
			Arithmetic	Geometric
1983	Smelterville	6–35	21.4	19.6
	Area 2	5–45	17.1	15.5
	Pinehurst	1–40	12.2	10.8
1988	Kellogg	4–39	9.2	8.0
	Page	4–26	10.3	8.5
	Smelterville	4–55	14.2	11.6
	Wardner	4–18	8.5	7.8
1990	Kellogg	4–25	9.3	8.3
	Page	4–21	11.0	9.4
	Pinehurst	4–20	7.4	6.7
	Smelterville	4–30	9.9	8.8
	Wardner	4–15	9.1	8.5
1992	Kellogg	4–26	8.1	6.9
	Page	4–10	6.1	5.8
	Pinehurst	4–15	6.0	5.5
	Smelterville	4–30	8.3	7.4
	Wardner	4–15	7.2	6.8
1995	Kellogg	1–30	6.4	5.2
	Page	2–12	6.1	5.3
	Pinehurst	1–15	4.6	4.0
	Smelterville	2–17	7.2	6.2
	Wardner	3–10	5.5	5.1
1997	Kellogg	1–22	5.9	5.0
	Page	2–9	6.3	5.7
	Pinehurst	1–17	4.2	3.5
	Smelterville	2–10	5.6	5.2
	Wardner	1–10	4.9	3.9
	Sitewide	1–22	5.4	4.5

(Continued)

TABLE 23.3 Mean PbB Levels at the Bunker Hill, ID, Superfund Box Site by Year and Site Community[a–d]—(cont.)

Year	Community	PbB Range (μg/dl)	PbB Mean (μg/dl)	
			Arithmetic	Geometric
1999	Kellogg	1–14	4.5	3.7
	Page	1–8	4.1	3.5
	Pinehurst	1–17	5.0	4.2
	Smelterville	1–17	4.3	3.6
	Wardner	1–12	5.4	4.5
	Sitewide	1–17	4.7	3.9

[a]*Smelter closed, 1981.*
[b]*Superfund designation, 1983.*
[c]*Selected years from all years in IDHW (2000) (TerraGraphics Environmental Engineering).*
[d]*Area 2 = combined Kellogg, Wardner, Page communities.*

shown in the table was how the contaminant remediation was prioritized across communities. In 1999, Smelterville yard and community soil Pb remediation was complete, while rates of completion for Kellogg, Wardner, Page, and Pinehurst at that time were 20%, 44%, 17%, and 5%, respectively.

Data in Table 23.3 are for children up to 9 years old. Children's PbB levels are at a peak around 1–3 years of age and decline in older children in communities with those residual interior dust and soil Pb contamination problems remaining in defunct smelter communities like these. The Pb exposure pathways here differ from what existed in these same towns when the Bunker Hill complex was operating and emitting large amounts of Pb via the air fallout pathways. The latter exposure routes produce less change with age among exposed children (see Table 23.1). We would expect that the younger age band subjects would show more significant PbB changes than older children where oral contact activity is less pronounced.

Another index of efficacy of remediation and public health intervention practices was the decline in the maximum reported PbB level, the latter providing some measure of the extent to which the approaches also benefited those children in the upper tail of the log-normal distribution. The maximum PbB reported for Smelterville declined from 55 μg/dl in 1988 to 10 μg/dl in 1997 (Table 23.3), the year remediation was completed in this community.

A number of correlation and regression analyses of differing complexity have been described for the Bunker Hill residents, principally for the site's childhood exposures. Table 23.4 describes two types of simpler analyses, the correlational matrix and a multiple regression model for log-transformed Pb source variables and geometric mean levels, notably the raw dust model

TABLE 23.4 Correlation and Multiple Regression Analyses for Bunker Hill, ID, Superfund Box Children[a,b]

Correlation Matrix: PbB, All Communities, All Years[c,d]		
Parameter	**PbB (μg/dl)**	**ln PbB (μg/dl)**
Child age	−0.244	−0.259
Dust Pb (ppm)	0.068	0.080
ln Dust Pb	0.241	0.256
Yard soil Pb (ppm)	0.237	0.230
ln Yard soil Pb	0.226	0.211
Community geometric mean soil Pb (ppm)	0.373	0.384

Multiple Regression Model: Log-Transformed Raw Dust			
	$R^2 = 0.231$	$p < 0.0001$	
Parameter	**Estimate**	**Standardized Estimate**	**Probability**
Age	−0.44803	−0.25541	0.0001
ln Dust Pb	0.83723	0.15677	0.0001
ln Soil Pb	0.21461	0.06466	0.0080
Geometric mean, soil Pb; 200 ft radius	0.73100	0.12938	0.0001
Geometric mean, community soil Pb	1.76000	0.19709	0.0001

[a]Source: *Selected data in vol. II, IDHW (2000) (TerraGraphics Environmental Engineering).*
[b]*Period of time 1988–1998.*
[c]*Principal media Pb sources: soil Pb, house dust Pb.*
[d]$p \leq 0.001$ *all entries except PbB versus dust Pb,* $p \leq 0.05$.

form. As commonly observed, age is inversely linked to children's PbB levels in Table 23.4, the relationship driven by both behavioral and anthropometric factors. Other significant correlational parameters for PbB and log-normal (ln) PbB are household dust, ln household dust, yard soil, ln yard soils, and geometric mean city soil. All correlations were significant at $p < 0.001$.

Multiple regression estimates for dust and the two expressions for soil Pb presented in Table 23.4 ($R^2 = 0.231$; $p < 0.0001$) show log-transformed dust and soil Pb estimates and standardized estimates as significant predictors of

PbB, as well as geometric mean soil Pb at a 200-ft radius and community-wide soil Pb.

An important part of the statistical and epidemiological assessments of Bunker Hill children was the impact of treatment modality on children's PbB levels. This was done using a mixed model repeat measures analysis comparing two exposure control methods over the period 1989–1998. Treatments consisted of either yard soil Pb reduction or public health interventions without soil removal by the Panhandle Health District, using the various methods described earlier. Table 23.5 notes comparative data for selected years plus data for all years combined, and compares control pre- and post-PbB with exposed pre- and post-PbB values for the indicated year.

In the earlier years, where the fractional extent of yard soil removals was relatively low, the public health interventions were predictably seen to be more effective. In 1990, for example, post-PbB means within the site for all towns were 2.5-fold lower for residence interventions than for soil removals, 12.0 and 4.6 μg/dl, respectively. With increases in the fraction of soil Pb abatements in the site communities, the differences became less. The data in Table 23.5 for testing year 1998 showed a decline of 4.9 μg/dl versus controls using nurse visit intervention and 2.6 μg/dl versus control children using yard soil Pb removal.

It is critical to keep in mind the nature of the treatment modalities over the long term versus the short term. While Table 23.5 comparisons would indicate that intensive intervention methods such as nurse visits, hygiene guidance, education, and awareness are associated with higher declines in postremediation PbB, these are transitory methods and methods that assume stability in the populations with children. The typically highly mobile residents in the Silver Valley benefit only to the extent they understand and accept the methods identified within the public health intervention approach. New residents would need to be identified and the process begun all over again. By contrast, soil Pb removal is relatively permanent, absent any recontaminating processes such as soil- and sediment-mobilizing flooding and/or high winds re-entraining fugitive dusts.

The U.S. EPA IEUBK model was extensively used in assessing Pb exposures in the Superfund Box area and with subsequently designated areas outside the smelter area. An earlier computational version of the computerized model was used in earlier years, with a more current version used from about the mid-1990s to recent years. The model has been employed in determining mean PbB outputs (simulated *in vivo* exposures) using multiple media-specific Pb intakes and uptakes, particularly for those Pb sources difficult to quantify via multiple regression analysis models. The model can also be used in estimating health risk probabilities and fractions of exposed children above and below some selected risk threshold. These latter uses are presented in the following chapter. Estimates from model use are of value in

TABLE 23.5 Mixed Model Repeat Measures Analyses of PbB Changes Versus Exposure Control Methods for Bunker Hill, ID, Superfund Box Children: 1988–1998[a–d]

	Control Homes				Intervention Homes				Remediated Soil Homes			
Year	Pre	Post	Difference	*p*-Value	Pre	Post	Difference	*p*-Value	Pre	Post	Difference	*p*-Value
1990	10.2	8.9	1.3	0.0137	28.0	16.0	12.0	0.0091	15.3	10.7	4.6	0.0001
1992	6.5	7.1	−0.6	0.0321	24.3	9.0	15.3	0.0001	9.7	10.4	−0.7	0.4032
1994	5.3	5.7	−0.4	0.0385	15.2	11.2	4.0	0.0001	6.4	8.1	−1.6	0.0072
1996	5.2	4.8	0.3	0.0992	14.3	10.7	3.7	0.0001	9.4	7.9	1.5	0.015
1998	4.8	4.5	0.3	0.1398	13.4	9.3	4.1	0.0001	7.3	5.7	1.6	0.0207
All years 1989–1998	6.7	6.2	0.5	0.0001	16.4	11.5	4.9	0.0001	10.6	8.0	2.6	0.0001

[a]Source: *Selected data in IDHW (2000) (TerraGraphics Environmental Engineering).*
[b]*PbB change: Post- versus preindicated action.*
[c]*Remediation = high home soil Pb removal.*
[d]*Intervention = actions including PbB monitoring, nursing follow-up, education, awareness.*

determining remediation strategies where inputs subsequent to some Pb reduction scenario are tested for resulting outputs as PbB means.

Model use in the many analyses done inside and outside the Box areas but within the CDARB also involved adjustments to the IEUBK model default settings based on relative concordance with measured PbB values. Unlike many site cases where the only empirical data set for comparison with this model's simulated PbB outputs is an isolated PbB screening, the Box communities have been the beneficiaries of annual extensive PbB monitoring by the Panhandle Health District, with more than 5,000 measurements systematically and reliably gathered. This large body of measurement data makes the calibration exercise for the model much more reliable.

As noted earlier, the principal parameter typically requiring site-specific adjustment is the uptake or bioavailability rate. This factor is mainly driven by geochemical and chemical properties of the ingested media. Another parameter potentially requiring adjustment is the daily intake rate. However, intake is a variable that is subject (child) dependent and not environmental medium dependent, and would not vary within a given child population as much as geochemically variable Pb media. With the latter, ingestion could include process tailings, fugitive dust particulate, weathered flotation pond fine particulate, etc. These media differ in ease of Pb mobilization from them.

The IEUBK model permits batch estimates for simulated PbB analyses of multiple samples like those at the Bunker Hill site and also permits batch comparisons of the estimates with corresponding PbB measurements. One can do serial runs for the latter comparisons varying one or more site-dependent input parameters in the exposure module. Table 23.6 has summary batch comparisons selected from reported empirical data gathered at the Box communities over the years. Annual comparison estimates have been reported and selected years are given in the table. The results are for a site-adjusted bioavailability of 18%, using a site-specific proportioning of Pb between dust and soil Pb, combining both yard and community-wide soil contributions reported in the IDHW (2000) report. Children in Table 23.6 are those 2 years of age or younger.

Reported sitewide PbB means with parameter adjustments for modeling show that, over the 11 years of testing media Pb, 1988–1998, there was close concordance between measured and modeled PbB means. The most smelter-impacted community, Smelterville, also showed reasonable agreement overall as did the community of Kellogg. The observed Sitewide PbB means declined to 6.1 versus 5.3 μg/dl as the estimated value.

The IDHW carried out an investigation of Pb contamination outside the Box communities in the late 1990s. The results formed part of the 2000 ATSDR environmental health assessment for those CDARB communities outside the Box geographic confines (U.S. ATSDR, 2000) and also provided support for the U.S. EPA extending the reach of the original Superfund designation to communities to the East (Upper Basin) and West (Lower Basin)

TABLE 23.6 IEUBK Model—Predicted Versus Observed Child PbB levels for the Bunker Hill, ID, Superfund Box Site: 1988–1988[a–d]

	Sitewide PbB (μg/dl)		Smelterville PbB (μg/dl)		Kellogg PbB (μg/dl)	
Year	Observed	Predicted	Observed	Predicted	Observed	Predicted
1989	13.4	15.6	13.7	13.4	13.6	16.6
1990	10.9	9.7	13.3	15.1	11.4	10.8
1992	8.2	8.4	9.6	9.7	8.7	9.1
1994	7.7	6.8	8.4	9.5	8.6	7.7
1996	8.1	6.5	10.4	3.4	8.3	7.6
1998	6.1	5.3	7.0	4.6	6.8	5.8

[a]Source: *Selected data from IDHW (2000) (TerraGraphics Environmental Engineering).*
[b]*Children 2 years of age in indicated year, arithmetic means.*
[c]*Sitewide = all communities.*
[d]*Selection of 18% bioavailability; 42% Pb as dust Pb; soil Pb in ratio 27:19:12 for community, neighborhood, yard soils. Table 4.30, IDHW (2000).*

in 1998 as OU-3 (U.S. EPA Region 10 ROD, 2002). Eight non-Box CDARB communities or areas were included in OU-3: Mullen, Burke/Nine Mile, Wallace, Silverton, Osburn, Side Gulches, Kingston, and Lower Basin.

Measurements of PbB among children in these stretches of the Basin were carried out for 765 individuals, 21% of the estimated census of 3,651 people. There were 47 children 0–5 years of age included out of 231 total for that age band, 21%. Age-stratified childhood data included two sets of measurements for pre-6-year-old children: 9 to 60 months old and 9 to 24 months old. Results are tabulated in Table 23.7 for the 1996 epidemiological assessment period.

Overall, the eight geographic areas representing the Basin outside the Superfund Box had mean PbB values for both age bands that were considerably below age-parallel groups tested over the years within the five Superfund Box communities. Arithmetic and geometric mean values in all eight sites were <10 μg/dl. The lowest arithmetic mean across sites was 4.8 μg/dl for the Side Gulches, while the corresponding lowest geometric mean was 4.3 μg/dl for Osburn. The maximum PbB in the younger child group across all eight areas was a single measurement in Silverton of 23 μg/dl, followed by a figure of 20 μg/dl for the Burke/Nine Mile site. Maxima in the remaining six locales ranged from 9 to 18 μg/dl. Table 23.8 presents the PbB mean summary data for all non-Box areas in the Basin for the years 1996 through 2001. While the values show scatter, the figures for 2001 are below those for 1996.

TABLE 23.7 Mean PbB Levels for CdA River Basin Children in Two Age Groups in Areas Outside the Superfund Box Region[a–c]

	Arithmetic Mean		Geometric Mean	
Community/ Area	9–24 Months Old	9–60 Months Old	9–24 Months Old	9–60 Months Old
Mullan	7.8	6.8	7.4	5.8
Burke/Nine Mile	9.1	7.9	6.6	6.3
Wallace	8.1	7.0	7.0	6.0
Silverton	7.5	6.0	6.1	5.0
Osburn	6.2	5.0	5.7	4.3
Side Gulches	5.2	4.2	4.8	3.8
Kingston	7.7	6.7	6.3	5.4
Lower Basin	8.7	8.2	7.1	6.2

[a]Source: *Selected tabular data from U.S. EPA Region 10 ROD (2002): Part 3 Tables 3.7-3a, 3.7-4a.*
[b]*Children in groups 9–24 months, 9–60 months.*
[c]*With repeat observations.*

TABLE 23.8 Annual PbB Levels for 2-Year-Old Children in CdA Basin Communities Outside the Bunker Hill, ID, Superfund Box: 1996–2001[a–c]

Year	Arithmetic Mean PbB (μg/dl)	Geometric Mean PbB (μg/dl)
1996	5.7	4.6
1997	–[d]	–[d]
1998	6.6	5.5
1999	9.0	7.1
2000	6.4	5.5
2001	4.4	3.7

[a]Source: *Selected data from U.S. EPA Region 10, ROD (2002), OU-3.*
[b]*U.S. EPA Region 10, ROD (2002), Chapter 12, Table 12.19.*
[c]*PbB data for all OU-3 communities outside the Box.*
[d]*Insufficient sample size.*

A number of statistical assessments relating data for PbB in the young children and older children/adult groups with various environmental Pb measures were carried out for this portion of the Basin (U.S. ATSDR, 2000). Using stepwise logistical and linear regression techniques, the most robust association with log-transformed children's PbB values was the log daily exterior floor mat Pb loading rate. This indicated that interior dust exposure via tracked-in leaded particulate from soil reservoirs was a principal Pb pathway from contaminated soils. Consistent with this were additional, non-environmental Pb measures that showed highly significant linkages with PbB. These included play areas for children such as dirt patches or sandboxes, frequency of hand–mouth activity, and (inversely) time spent at day-care centers as a marker for time away from residential Pb exposures.

The IEUBK predictive model was used extensively in these non-Box Basin studies. As for the Box evaluations, the biokinetic modeling was necessary for evaluating responses of PbB simulations to environmentally variable media Pb inputs and calibrations for intake–uptake parameters and for determining risk levels linked to various Pb input scenarios. Table 23.9 indicates the concordance between measured and predicted PbB levels for the non-Box subjects and environmental data sets using either default or indicated bioavailability and dust/soil ratios. The site-specific partitioning values

TABLE 23.9 Observed Versus Predicted PbB Levels for Children 9–24 Months Old for All Basin Community Children Outside the Bunker Hill, ID, Superfund Box[a–d]

Parameter	Observed PbB (μg/dl)	Predicted PbB (μg/dl) (Model Defaults)	Predicted PbB (μg/dl) 40:30:30 Dust–Soil–Soil
Minimum (μg/dl)	2	2	2
Maximum (μg/dl)	23	27	17
Arithmetic mean (μg/dl)	7.5	9.2	6.1
Geometric mean (μg/dl)	6.2	7.9	5.5
Geometric standard deviation	1.82	2.08	1.91

[a]Source: *Selected data from U.S. EPA Region 10, ROD (2002) OU-3.*
[b]*40:30:30 = 40% house dust Pb, 30% yard soil Pb, 30% community soil Pb; default in IEUBK model is 55% dust Pb, 45% soil Pb.*
[c]*With repeat observations.*
[d]*Prediction model = U.S. EPA IEUBK model.*

provide a closer match with observed values than do the results using the default dust/soil ratio.

23.3.2 The Lanphear et al. Pooled Analyses of Soil and Dust Pb Exposures in U.S. Children

Lanphear et al. (1998) carried out a pooled analysis of 12 of these studies that met specific statistical and epidemiological criteria for factors such as completeness of measurements, and selected subject characteristics from this pooled analysis are summarized in Table 23.10. Five data sets had been gathered for inner-city or other urban settings and seven for various locales impacted by mining/milling/smelter wastes containing Pb, and these are depicted in the table. Three of the five inner-city/urban child cohorts were from prospective studies, while most of the mining/smelter communities were evaluated through cross-sectional analyses.

Children in this pooled analysis were in the infancy age band and had an average age of 16 months. Consequently, their Pb exposures were largely those characterizable as one or more interior home environments with limited external, i.e., public, contact with Pb in settings such as daycare centers. In these exposure cases, interior dust Pb and bulk surficial soil (plus external surface dust Pb) are known to be principal sources and pathways for infant–toddler residential Pb exposures. Dust Pb exposures with lead loadings as the metric rather than Pb concentration appear to better capture the statistical size of any exposure relationships (Lanphear et al., 1995; Milar and Mushak, 1982).

In terms of exposure groupings, inner-city urban children such as those in the Cincinnati cohorts presented with much higher mean PbB values than either the remaining urban groups or the means for the mining/smelting sites, about twice or more higher. Mean PbB levels across the seven mining/smelting sites ranged from 3.2 to 4.9 μg/dl. Dust Pb loadings for the urbanized subjects markedly exceeded mean values for the mining/smelting sites, reflecting to some extent relatively higher inputs to interior dusts from deteriorated lead paint surfaces as well as tracked-in soil Pb and other exterior media Pb.

Table 23.11 presents selected summary data from the pooled analysis, showing estimated PbB as a function of both exterior Pb and interior dust Pb loadings using the statistical model described in a published appendix to the pooled analysis. Dust Pb loadings at even quite low levels and with moderate soil and other exterior Pb inputs show potentially significant mean PbB levels in terms of health risk. Health risk characterizations for these pooled study subjects are discussed in some detail in the next chapter. At the exterior Pb level corresponding to the U.S. national median soil Pb of 72 ppm, dust Pb loadings over the range 1–100 $\mu g/ft^2$ Pb correspond to a mean PbB of 2.8–7.3 μg/dl. At dust loadings of 40–100 $\mu g/ft^2$ Pb and soil Pb of 2,000 ppm, mean PbB ranges from 8.7 to 10.5 μg/dl.

TABLE 23.10 Descriptive Data for Dust Pb Pooled Analyses of Urban and Extractive Industry Community Children Exposed to Lead[a–d]

Subgroup	*N*	PbB (μg/dl) (Geometric Mean)	Dust Pb Loading, μg/m^3 (Geometric Mean)	Exterior Pb (ppm) (Geometric Mean)	Mean Age (months)
Urban child cohorts					
Boston Prospective Study	175	4.3	2.3	247	13.5
Cincinnati Prospective Study	285	11.2	293.4	472	13.6
Cincinnati Soil Study	99	10.4	20.4	966	20.0
Rochester Prospective Study	274	2.9	8.3	914	6.1
Rochester Dust Pb Study	205	6.3	17.8	690	20.4
Mining/milling/smelting site child cohorts					
Butte, MT	118	3.6	2.5	520	21.1
Bingham Creek, UT	335	3.2	1.9	97	23.7
Leadville, CO	108	4.9	4.7	755	20.2
Magna, UT	64	4.5	8.8	247	21.3
Sandy, UT	46	3.2	6.1	416	22.3
Midvale, UT	86	4.6	3.7	327	19.6
Palmerton, PA	45	4.7	5.9	582	20.8
All studies	1,861[e]	5.1	13.5	509	16.3

[a]Source: *Adapted from Lanphear et al. (1998).*
[b]*Urban = inner-city and other urban locales impacted by paint Pb.*
[c]*Extractive industry = western U.S. mining, milling, smelting communities.*
[d]*Either cross-sectional or longitudinal designs employed.*
[e]*Total* N *from Lanphear et al. (1998), Table 3.*

Table 23.11 indicates that current U.S. Federal Pb standards and practices regarding media Pb are insufficient to adequately protect infants. The current Pb hazard rule standard for floor Pb is 40 μg/ft^2, while a common clean-up

TABLE 23.11 Predicted Children's PbB Levels (μg/dl) as a Function of Floor Dust Pb Loadings and Exterior Soils[a–d]

Dust Lead Loading	Exterior Pb (ppm)					
(μg/ft²)	10	72[e]	100	500	1,000	2,000
1	2.3	2.8	2.9	3.5	3.8	4.1
5	3.2	4.0	4.1	4.9	5.3	5.7
10	3.7	4.6	4.7	5.6	6.1	6.5
20	4.2	5.3	5.4	6.5	7.0	7.6
40	4.9	6.1	6.3	7.5	8.1	8.7
70	5.5	6.8	7.0	8.4	9.1	9.8
100	5.9	7.3	7.6	9.0	9.7	10.5

[a]Source: *Adapted from Lanphear et al. (1998).*
[b]*Geometric mean PbB, children <3 years of age.*
[c]*Based on studies for seven U.S. mining–smelting and five urban sites.*
[d]*Loadings: μg Pb/ft².*
[e]*Median U.S. housing survey level, 1989–1990.*

level of soil Pb at residential portions of Superfund sites is 500 ppm. The table indicates a geometric mean PbB of 7.5 μg/dl at these model inputs. This value and other health risk characterization aspects of the pooled study are discussed further in the next chapter.

23.3.3 The U.S. EPA Full-Scale Pb Exposure Assessment for Three U.S. Geographic Case Studies

The U.S. EPA carried out a series of measured and modeled lead exposure estimates for three specific case studies in support of a full-scale risk assessment for representative lead exposure scenarios (U.S. EPA, 2007). These addressed and supported the risk basis in young children for a revision downward in the primary and secondary National Ambient Air Quality Standards (NAAQS) for air Pb (U.S. EPA, 2008).

As part of the process, the specific sites were a primary smelter community (Herculaneum, MO), a secondary smelter community (Troy, AL), and a generic U.S. urban area. Three lead modeling and measurement exercises were carried out to determine levels in ambient air, soil, and indoor dusts. Using these estimates as inputs to Pb exposure modeling, PbB values were calculated as median or 95th percentile levels.

Table 23.12 presents the EPA full-scale risk-associated environmental media methodologies employed in these exercises. For ambient air Pb, available air Pb monitors in various locales were used for the U.S. general urban

TABLE 23.12 Media Pb Concentration Determinations for the U.S. EPA/OAQPS Risk Assessments Based on Three Case Studies[a–d]

Spatial Template	General Urban Case Study: Single Generic Study Area	Primary Pb Smelter Case Study: Combination of U.S. Census Blocks and Block Groups to 10 km Radius	Secondary Pb Smelter Case Study U.S. Census Blocks to a 6 km Radius
Media Pb			
Ambient air: measurement approach	Use of available ambient air Pb monitor data from monitors in various locales	Dispersion modeling	Dispersion modeling
Methodology performance assessment	Using above monitor data	Comparison to Pb TSP monitor data from this site	Comparison to Pb-TSP monitor data from this site
Outdoor soil Pb: measurement approach	Nationally representative residential soil Pb values	Soil sampling data near facility; regression modeling farther away	Predictive model to estimate soil Pb, using a surrogate location
Methodology performance assessment	Using above empirical data on soil Pb	Estimates using surrogate data	Modeled estimate compared to surrogate data, other point source sites
Interior dust Pb: measurement approach	Used two approaches: a hybrid compartment model and empirical, air Pb-only regression model	Site-specific regression model near facility; air plus soil pooled analysis regression model	Statistical (air-only) regression model
Methodology performance assessment	Case study estimates compared to literature estimates and national survey	Site-specific sampling data; case estimates compared to literature estimates	Study estimates compared to literature data and national survey

[a]*Primary smelter case study: Herculaneum, MO, smelter operation.*
[b]*Secondary smelter case study: Troy, AL, smelter operation.*
[c]*General urban case study: no specific geographic locations employed.*
[d]*Adapted from Table 3.1, U.S. EPA (2007).*

TABLE 23.13 Estimated Annual Ambient Air Pb Levels (μg/m³) for Three Case Studies[a–d]

	Estimated Air Pb at Different Air Pb Standards or Conditions			
Statistic	**Current Conditions**	**1.5 μg/m³, Pb NAAQS**	**0.5 μg/m³, Monthly Maxima**	**0.2 μg/m³, Monthly Maxima**
General urban case				
Single study area	High end: 0.11 Mean: 0.06 (rounding)	0.60	0.13	0.05
Primary Pb smelter case				
Maximum	–	0.74	0.33 (rounding)	0.13
Median	–	0.04	0.02	0.01
95th Percentile	–	0.15	0.07 (rounding)	0.03
5th Percentile	0.01 (rounding)	0.01 (rounding)	0.01 (rounding)	0.00 (rounding)
Secondary smelter case				
Maximum	0.13	–	0.07	0.03
Median	0.00 (rounding)	–	0.00 (rounding)	0.00 (rounding)
95th Percentile	0.02	–	0.01	0.00 (rounding)
5th Percentile	0.00 (rounding)	–	<0.001	<0.001

[a]*Primary smelter case study: Herculaneum, MO, smelter.*
[b]*Secondary smelter case study: Troy, AL, smelter.*
[c]*General urban case study: no specific geographic location.*
[d]*Adapted from Table 3.3, U.S. EPA (2007).*

case study, while dispersion modeling was done at the two smelter communities. Soil Pb estimates used representative published national soil levels for the urban scenario while existing measured soil Pb in the two smelter locales were used for those cases.

Two models were employed for interior dust Pb estimating: a hybrid compartment model and a regression model using only air Pb. The smelter community dust Pb estimates involved site-specific modeling.

Estimated annual ambient air Pb levels (μg Pb/m³) for the three core studies are depicted in Table 23.13, while interior dust Pb estimates are

TABLE 23.14 Estimated Indoor Dust Concentrations (ppm) for Three Case Studies[a–e]

		Estimated Dust Pb at Different Air Pb Standards or Conditions		
Statistics	**Current Conditions**	**1.5 μg/m³ Pb NAAQS**	**0.5 μg/m³, Monthly Maxima**	**0.2 μg/m³, Monthly Maxima**
General urban case study	High end: 157–198 Mean: 107–146	426–566	166–206	102–140
Primary Pb smelter case				
Maximum	–	1,944	1,077	557
Median	–	84	73	67
95th Percentile	–	219	173	149
5th Percentile	–	53	47	44
Secondary smelter case				
Maximum	166	–	120	84
Median	63	–	61	61
95th Percentile	73	–	67	63
5th Percentile	60	–	60	60

[a]*Primary smelter case study: Herculaneum, MO, smelter.*
[b]*Secondary smelter case study: Troy, AL, smelter.*
[c]*General urban case study: no specific geographic location.*
[d]*Adapted from Table 3.6, U.S. EPA (2007).*
[e]*Original table includes an NAAQS of 0.05 μg/m³.*

given in Table 23.14. The air Pb estimates in the smelter cases are presented as distributions for four air Pb scenarios: maximum, median, 95th percentile, and 5th percentile. For the "current conditions" case, the general urban example shows a mean Pb of 0.06 $\mu g/m^3$ and a high end of 0.11 $\mu g/m^3$. For the three potential standard-based cases in the table where the indicated air Pb is, for example, monthly maxima, the urban example air Pb ranges from 0.05 to 0.60 μg Pb/m^3. Median air Pb estimated levels for the primary smelter locale range from 0.01 to 0.04, while the secondary smelter town showed 0.00, 0.00, and 0.00 $\mu g/m^3$ as seen in Table 23.13.

The three estimates of case-specific interior dust levels are given in Table 23.14. The general urban area scenario shows a dust Pb mean for

TABLE 23.15 Model-Estimated Median PbB Values (μg/dl) for Three Case Studies[a–e]

Case Study/Air Pb Scenario	Concurrent PbB (μg/dl)			
	Recent Air[f]		Total Pb Exposure[g]	
	Low	High	Low	High
General urban case study[h]				
Current conditions	0.4	0.8	2.0	2.1
NAAQS, 1.5 /m^3	1.9	2.2	3.1	3.7
NAAQS, 0.5 μg/m^3, monthly max	0.5	0.9	2.0	2.2
NAAQS, 0.2 /m^3 max	0.2	0.5	1.7	1.9
Primary Pb smelter: 1.5 km radius subarea[h]				
NAAQS, 1.5 μg/m^3	–		4.6	
NAAQS, 0.5 μg/m^3, monthly max	–		3.2	
NAAQS, 0.2 μg/m^3, monthly max	–		2.3	
Secondary Pb smelter: 1.5 km radius subarea[h]				
Current conditions	0.1		1.3	
NAAQS, 0.5 μg/m^3, monthly max	<0.1		1.3	
NAAQS, 0.2 μg/m^3, monthly max	0.1		1.3	

[a] *Primary smelter case study: Herculaneum, MO, smelter.*
[b] *Secondary smelter case study: Troy, AL, smelter.*
[c] *General urban case study: no specific geographic location.*
[d] *Adapted from Table 3.7, U.S. EPA (2007).*
[e] *Concurrent PbB as the exposure metric.*
[f] *Inhalation of current air Pb+ingestion, recently deposited dust Pb.*
[g] *Recent and past air Pb inputs, other Pb inputs.*
[h] *Original table includes an NAAQS of 0.05 μg/m^3.*

current conditions in the range of 107–146 ppm Pb. Levels for the three indicated air Pb standard limits for the urban case are 426–566, 166–206, and 102–140 ppm. Dust Pb estimates in terms of distribution statistics for the two smelter locales are also given in the table.

Tables 23.15 and 23.16 present the IEUBK model-estimated child PbB distribution with regard to two air Pb impact scenarios: "recent air" Pb and "total Pb exposures", and the four air Pb standard ceilings, including the current conditions case. Further divisions were the stratification of estimates in the urban case into "low" and "high," as defined in Chapter 3 of U.S. EPA (2007).

TABLE 23.16 Model-Estimated 95th Percentile PbB Values (μg/dl) for Three Case Studies[a–e]

	Concurrent PbB (μg/dl)			
	Recent Air[f]		Total Pb Exposure[g]	
Case Study/Air Pb Scenario	Low	High	Low	High
General urban case study[h]				
Current conditions	1.1	2.8	5.1	7.2
NAAQS, 1.5 μg/m^3	4.6	7.5	7.6	11.5
NAAQS, 0.5 μg/m^3, monthly max	1.2	2.6	4.8	6.7
NAAQS, 0.2 μg/m^3 monthly max	0.5	1.8	4.2	6.4
Primary Pb Smelter: 1.5 km radius subarea[h]				
NAAQS, 1.5 μg/m^3		–		12.3
NAAQS, 0.5 μg/m^3, monthly max		–		8.5
NAAQS, 0.2 μg/m^3, monthly max		–		6.1
Secondary Pb smelter: 1.5 km radius subarea[h]				
Current conditions		0.5		3.3
NAAQS, 0.5 μg/m^3, monthly max		0.4		3.3
NAAQS, 0.2 μg/m^3, monthly max		<0.1		3.1

[a]*Primary smelter case study: Herculaneum, MO, smelter.*
[b]*Secondary smelter case study: Troy, AL, smelter.*
[c]*General urban case study: no specific geographic location.*
[d]*Adapted from Table 3.8, U.S. EPA (2007).*
[e]*Concurrent PbB as the exposure metric.*
[f]*Inhalation of current air Pb+ingestion, recently deposited dust Pb.*
[g]*Recent and past air Pb inputs, other Pb inputs.*
[h]*Original table includes an NAAQS of 0.05 μg/m^3.*

Table 23.16 gives the corresponding 95th percentile values for the three case studies. As expected, for the general urban case study, the then-extant air Pb standard as ceiling gave PbB high values for the total Pb exposure category of 7.6 and 11.5 μg/dl.

REFERENCES

Gulson, B.L., Mahaffey, K.R., Jameson, C.W., Vidal, M., Law, A.J., Mizon, K.J., et al., 1997. Dietary lead intakes for mother/child pairs and relevance for pharmacokinetic models. Environ. Health Perspect. 105, 1334–1342.

Hunter, J.M., 1978. The summer disease: some field evidence on seasonality in childhood in childhood lead poisoning. Soc. Sci. Med. 12, 85–94.

Idaho Department of Health and Welfare, 2000. 1999 Five Year Review Report Bunker Hill Superfund Site. Vol. II. TerraGraphics Environmental Engineering, Inc. Division of Environmental Quality, Boise, ID.

Illinois Department of Public Health, 1994. Madison County Lead Exposure Study. Granite City, IL, February, Illinois Department of Public Health.

Landrigan, P.J., Baker Jr., E.L., Feldman, R.G., Cox, D.H., Eden, K.V., Orenstein, W.A., et al., 1976. Increased lead absorption with anemia and slowed nerve conduction in children near a lead smelter. J. Pediatr. 89, 904–910.

Lanphear, B.P., Emond, M., Jacobs, D.E., Weitzman, M.L., Tanner, M., Winter, N.L., et al., 1995. A side-by-side comparison of dust collection methods for sampling lead-contaminated house dust. Environ. Res. 68, 114–123.

Lanphear, B.P., Matte, T.D., Rogers, J., Clickner, R.P., Dietz, B., Bornschein, R.L., et al., 1998. The contribution of lead-contaminated house dust and residential soil to children's blood lead levels. A pooled analysis of 12 epidemiological studies. Environ. Res. 79, 51–68.

Marcus, A.H., 1995. Statistical analysis of data from the Madison County lead study and implications for remediation of lead-contaminated soil. Attachment 4: U.S. EPA Region V Decision Document/Explanation of Significant Differences for the NL Industries/Taracorp Site, Granite City, IL. Environmental Criteria and Assessment Office, Research Triangle Park, NC.

Milar, C.R., Mushak, P., 1982. Lead-contaminated house dust: hazard, measurement, and decontamination. In: Chisolm Jr., J.J., O'Hara, D.M. (Eds.), Lead Absorption in Children: Management, Clinical, and Environmental Aspects. Urban & Schwarzenberg, Baltimore, MD, pp. 143–152.

Mushak, P., 1998. Uses and limits of empirical data in measuring and modeling human lead exposure. Environ. Health Perspect. 106 (Suppl. 6), 1467–1484.

National Academy of Sciences: National Research Council, 1993. Measuring Lead Exposure in Infants, Children, and Other Sensitive Populations. National Academy Press, Washington, DC.

National Academy of Sciences: National Research Council, 2005. Superfund and Mining Megasites. Lessons from the Coeur d'Alene River Basin. National Academy Press, Washington, DC.

Neri, L.C., Johansen, H.L., Schmitt, N., Pagen, R.T., Hewitt, D., 1978. Blood lead levels in children in two British Columbia communities. Trace Subst. Environ. Health XII, 403–410.

O'Flaherty, E.J., 1998. A physiologically based kinetic model for lead in children and adults. Environ. Health Perspect. 106 (Suppl. 6), 1495–1503.

Tong, S., Baghurst, P., McMichael, A., Sawyer, M., Mudge, J., 1996. Lifetime exposure to environmental lead and children's intelligence at 11–13 years: the Port Pirie cohort study. BMJ 312, 1569–1575.

U.S. Agency for Toxic Substances and Disease Registry, 2000. Coeur d'Alene River Basin Environmental Health Exposure Assessment. Final Report. Division of Health Studies, Atlanta, GA.

U.S. Centers for Disease Control, 1991. Preventing Lead Poisoning in Young Children. A Statement by the Centers for Disease Control. U.S. Centers for Disease Control and Prevention, Atlanta, GA.

U.S. Environmental Protection Agency, 1986. Air Quality Criteria for Lead, 4 vols. Report No. EPA 600/8-83/028bF. Environmental Criteria and Assessment Office, Washington, DC.

U.S. Environmental Protection Agency, 2002. Record of Decision. The Bunker Hill Mining and Metallurgical Complex Operable Unit 3. U.S. EPA, Seattle, WA, Region 10, September.

U.S. Environmental Protection Agency, 2006. Air Quality Criteria for Lead, vol. 1. Report No. EPA/600/R-05/144aF. National Center for Environmental Assessment, Washington, DC.

U.S. Environmental Protection Agency, 2007. Lead: Human Exposure and Health Risk Assessments for Selected Case Studies. Report No. EPA-452/R-07-014a, vol. I. Office of Air Quality Planning and Standards, Washington, DC.

U.S. Environmental Protection Agency, 2008. Office of the Administrator. National Ambient Air Quality Standards for Lead. Final Rule, 73 FR 66964-67062, November 12.

von Lindern, I.H., Spalinger, S.M., Bero, B.N., Petrosyan, V., von Braun, M.C., 2003a. The influence of soil remediation on lead in house dust. Sci. Total Environ. 303, 59–78.

von Lindern, I., Spalinger, S., Petrosyan, V., von Braun, M.C., 2003b. Assessing remedial effectiveness through the blood lead soil/dust relationship at the Bunker Hill Superfund site in the Silver Valley of Idaho. Sci. Total Environ. 303, 139–170.

Yankel, A.J., von Lindern, I.H., Walter, S.D., 1977. The Silver Valley lead study: the relationship between childhood blood lead levels and environmental exposure. J. Air Pollut. Control Assoc. 27, 763–767.

Chapter 24

Health Risk Characterization of Lead Effects in Human Populations

24.1 INTRODUCTION, CONTEXT, AND RATIONALE

The final step in the risk assessment methodology employed here for environmental contaminants is quantification of the extent of risks to health in various human populations. This step combines the general elements of hazard characterization and dose–toxic response relationships for lead with case-specific data quantifying the extent of toxic exposure to arrive at some quantitative value for risk. Various regulatory entities differ in the specifics of how this methodological template is applied but retain these broad features.

For those environmental hazards having both carcinogenic and noncarcinogenic, systemic toxicological impacts, risk characterizations are done to address both cancer and noncancer risks. Expressions of cancer risk take various forms in these cases. One form would be lifetime or annualized cancer risks over population morbidity and mortality baselines. This would typify cancer risks in skin or internal organs from exposures to inorganic arsenic in drinking water. Risk managers are typically concerned with cancer rates within the range of $1 \times E-4$ to $1 \times E-6$.

Lead has been shown to be carcinogenic in animals and is considered a probable carcinogen in humans. The current view of Pb carcinogenicity *vis-a-vis* systemic toxicity was set forth in Chapter 17. In terms of overall human health risks, carcinogenic endpoints for lead in terms of quantifiable cancer risk metrics in populations-at-large are not as reliably drawn as are noncarcinogenic effects. Noncarcinogenic adverse endpoints are the focus for adverse health effects from Pb exposure in humans.

As briefly noted earlier, lead is unique in that it has both (1) a voluminous record of clinical and subclinical poisonings in humans, and ranges of multiple toxicities in experimental animals and ecological populations, and (2) an extensive epidemiological record for adverse health risks in humans.

Trace Metals and other Contaminants in the Environment, Volume 10
ISSN: 1875-1121 DOI: 10.1016/B978-0-444-51554-4.00024-9

Put differently, lead can not only poison someone, but it has already poisoned many.

Communities where increased environmental lead exposures and various toxic endpoints have been explicitly incorporated into a dose–response framework currently highlight health risk characterization using case-specific evidence. Data generated in such communities typically first permit determination of distributions of risk group PbB values, specifically prevalences and incidences of PbB above an accepted health risk threshold. These PbB statistics are then incorporated into health risks using dose–toxic response relationships.

Many substances have been quantified as to poisoning threats, principally via formulaic risk characterizations using such methods as low-dose extrapolations, uncertainty factors across species and dosing specifics, benchmark dose (BMD)–response determinations above some background prevalence such as 5% or 10%. Many substances are characterized for health risk based on level of exposure relative to some reference value taken to be without risk of toxic harm over a lifetime, e.g., the U.S. EPA RfD for ingested substances or the agency's RfC for inhaled substances. Given the nature of high-risk exposures to Pb in humans, the apparent absence of determinable "safe" exposure in high-risk populations, and the temporal confines of such risk, the lifetime approach using RfDs or concentrations is not employed with lead.

Risk characterization may theoretically be applied to a lead-poisoning patient or to an assemblage of individuals with common Pb exposure rates and quantifiable for their group poisoning risks. In the former case, the clinician makes a diagnosis that is based on all the evidence presented for purposes of diagnosis, to include available evidence for lead's general causal relationships to health hazards. In the latter, one typically examines probabilistically how risk is fractionally distributed across a risk population or how incremental changes in the exposure metric are epidemiologically shown to produce changes in the health risk metrics.

Both of these quantification approaches are presently being for group risk estimates. The U.S. EPA or various state risk assessors determine via the IEUBK model what percentages of children exposed to lead at a contamination site exceed some preselected threshold value indexed as a geometric mean PbB. The latter was extensively employed in the recent significant downward change in the U.S. primary and secondary NAAQS for lead from 1.5 to 0.15 μg Pb/m^3 (U.S. EPA, 2008).

24.2 METHODOLOGICAL AND INTERPRETIVE ISSUES FOR LEAD HEALTH RISK CHARACTERIZATION IN HUMANS

The choice of data to be analyzed for selection of health hazard characterization and applicable dose–responses entails a series of methodological and interpretive factors. Given adequate supporting evidence, one would select the most sensitive toxic endpoint to represent hazard characterization and the most sensitive dose–response relationships to apply to subsequent health risk characterizations. The simple rationale here is that preventing the earliest harm to the most sensitive populations (such as very young children in the case of Pb) at the lowest exposures would obviously prevent harm expressed as less-sensitive toxic endpoints, i.e., effects occurring at higher PbB thresholds, expressed in less sensitive individuals, e.g., lead workers.

It is fortunate in the case of lead risk characterization that there is a confluence of the most sensitive endpoint and dose–response data with a huge body of scientific evidence cementing general causality within the framework of Bradford Hill's proofs of causality (see Chapter 21, Table 21.2). While the bulk of that evidence applies to developmental neurotoxicity, it is striking from the tables in Chapter 22 the extent to which there is multi-effect dose–response convergence below 10 μg/dl for both children and adults.

There are several ways one can quantify human health risk characterization for humans at risk through lead exposure. The first and simplest examines the prevalences or incidences of blood lead levels above some health risk threshold, with frequencies of exceedance identifying those at more risk (compared to those with PbB values below the risk threshold). Expressions of health risk in terms of elevated PbB occurrences (e.g., ≥10 μg/dl) do not simultaneously provide quantitative estimates of organ- or system-specific toxic harm, such as actual loss of IQ points or increases in SBP or DBP. A health risk threshold indexed in terms of a PbB level, however, represents the synthesis of numerous empirical dose–toxic response relationships, as developed and discussed in previous chapters.

Estimations of exposure frequencies above some health risk threshold in terms of PbB can be done using empirical data or through lead exposure modeling approaches, such as the U.S. EPA IEUBK model for childhood lead exposure simulations. Both approaches have their advantages and limitations, and a comparative look at the topic was provided by Mushak (1998).

As employed by regulators, PbB cutoff values are assumed to cover one or more unacceptable toxic endpoints and are chosen to minimize unacceptable frequencies of effects in the target population. However, this minimizing rarely permits the prospect of 100% avoidance of toxic health risk within some risk population, so the approach entails both risk assessment and risk management components. As illustration, the latter can take the form of the inclusion of a stated fraction of a targeted risk population to be protected.

One current regulatory practice in characterizing lead's health risks in U.S. communities with lead contamination sources is to (1) quantify toxic risk for lead exposures using the IEUBK model in terms of a central tendency estimate of exposure (geometric mean), (2) use percentile distributions to identify the fraction of a risk population (i.e., children) to be protected at the identified PbB health risk threshold value, and (3) use modeling to examine responses in PbB statistics to various exposure reduction interventions. Current U.S. EPA risk management policy for protecting children at hazardous waste sites, using the IEUBK lead model, is to keep 95% of exposed children below a PbB level of 10 μg/dl.

A second approach applied at a number of Pb contamination sites is the identification and quantification, in a dose–response fashion, of site-specific associations of toxic endpoints with lead contamination. At the Silver Valley, ID Superfund Site's Box communities, dose–toxic response relationships were documented for childhood anemia and pediatric nerve conduction velocity impairments from lead exposures (Landrigan et al., 1976; Schwartz et al., 1988, 1990).

A third approach to quantifying health hazard risk is one in which incremental changes in an exposure biomarker, like blood lead, is linked to incremental changes in some health metric, e.g., reduction in IQ. In this process, aggregated empirical data for both the lead exposure metric and the health harm metric are employed to generate a numerical slope relationship. The most quantitative use of this approach meta-analytically has been that of Lanphear et al. (2005) for longitudinally derived epidemiological relationships of elevated PbB to IQ decrements in preschool or elementary school children described in international studies.

This is a two-component approach, in common with dose–response relationships *per se*, in that one has to derive both the most robust dose/exposure metric and the health risk markers for the relevant range of risk populations anticipated. Such approaches as that of Lanphear et al. (2005) not only establish the basic dose–response relationships for developmental neurotoxicity of Pb and the level of exposure but also allow (1) evaluations of choices among dose metrics for use in the final dose–response expressions and (2) choice of health risk statistical models for subsets of the overall dose–response relationships in the entire cohorts. This approach, furthermore, has been employed with both modeled and measured PbB values and using various statistical modeling forms to express the most robust forms of the dose–response relationship (U.S. EPA, 2007, 2008).

24.3 CASE STUDIES OF LEAD HEALTH RISK CHARACTERIZATIONS FOR HUMAN RISK GROUPS

This chapter presents several case examples of health risk characterization for lead in various U.S. lead exposure scenarios. Two of these follow case examples described in the previous chapter for selected actual exposure scenarios derived from U.S.-based measurement or modeled data. In these cases, risk is characterized at some accepted value for an operational, i.e., empirical threshold for lead toxicity, keeping in mind that there has, in fact, been no blood lead value determined to be free of toxic risk. One scenario revisits the large database available for the Silver Valley, ID Superfund "megasite" (IDHW, 2000; U.S. EPA, 2002).

The second revisits the multistudy pooled analysis of exposure modeling as compiled for interior dust Pb and soil/exterior Pb exposures (Lanphear et al., 1998). This pooled analysis presented predicted exceedance percentages above selected toxicity thresholds.

The third presents some of the salient features of a full-scale health risk characterization described by the U.S. EPA OAQPS (U.S. EPA, 2007), carried out as part of the risk assessment modalities employed as part of the revisions of the U.S. lead NAAQS promulgated by the agency in November 2008 (U.S. EPA, 2008). This third case is considerably more complex in its computational and statistical design features and used more modeling approaches than the others. One particular segment of this U.S. EPA–OAQPS set of health risk assessments is for the overall U.S. urban population, and this is the segment mainly selected for illustrative discussion. Two other case example exercises were carried out and reported by the agency's OAQPS: a primary smelter community case study and a case study using a secondary smelter community as its focus (U.S. EPA, 2007).

24.3.1 Health Risk Characterization of Pb at the Silver Valley, ID/Superfund Site

As noted in Chapter 23, this large mining/milling/smelting lead–silver–zinc complex has been evaluated in terms of two population and geographic areas: (1) exposure populations in communities within the main Bunker Hill Superfund Box comprising 21 square miles and (2) later extending the confines of the site to a larger area of the Coeur d'Alene River Basin and evaluating that latter segment of the overall impact zone. Health risks have been determined at the Idaho site in the following two ways: establishing that lead-associated toxicity existed in Box community children and determining through measurement and modeling the percentages of children's PbBs above a health risk threshold, 10 μg/dl.

Landrigan et al. (1976) reported an inverse association of peroneal nerve conduction velocity and PbB level in 202 smelter community children of Bunker Hill. A threshold to the effect was subsequently determined by Schwartz et al. (1988). Schwartz et al. (1990) evaluated dose–response data for anemia that included hematocrit measurements. Data from 579 children, aged 1–5 years, who presented with relatively high blood Pb measurements were used: range 11–164 μg/dl, 40% >40 μg/dl. Using a logistic model, these workers estimated that a 10% probability of anemia (Hct < 35%) in children at 1, 3, and 5 years, respectively, occurred at corresponding PbB values of 20, 50, and 75 μg/dl.

A number of studies of Bunker Hill site children in terms of PbB screening programs began in the 1980s and have continued. Much of this information appeared in the previous chapter. This chapter confines itself to PbB frequencies at selected health risk thresholds that have been widely accepted, most frequently PbB concentrations ≥10 μg/dl. These various PbB screening results have been summarized in Tables 24.1–24.4.

Mean PbB concentrations and prevalences of elevated PbB levels ≥10 μg/dl in the Idaho Box communities as a function of age and year are summarized in Table 24.1. The age range of particular interest as shown in Table 24.1 is that of 2–6 years, the age bracket deemed by the U.S. CDC to be that with particularly high risk of Pb exposure and toxicity. In general, the younger the children in this overall age bracket, the higher the frequency of PbB values ≥10 μg/dl. Within this age range for the first year of systematic screening, 1988, exceedance frequencies ranged up to 62.1%. This high frequency is not unexpected, given that organized lead remediation efforts in and around the Box communities had yet to begin. Nine-year-old children showed the lowest frequency of elevated PbB, 28.9%, to be expected, given the attenuation of various behavioral and other risk factors in children of that age in similar exposure settings.

Six illustrative years of the total described for annual community screenings 1988–1998 are depicted: 1988, 1990, 1992, 1994, 1996, and 1998. From 1988 onward, there was an increasingly higher fraction of remediated yards of the total of residential yards available for cleanup. Among 2- and 3-year-old children, in particular, the impact of increased remediation with time on elevated PbB frequencies is shown in Table 24.1 to be evident by 1992. For these ages, there was an approximate 50% decline in PbB exceedance rates for the risk threshold, from the low 60s to the low 30s. By 1998, there were three- to fivefold declines in those frequencies.

Surprisingly, 6- and 9-year-old children showed the greatest relative declines in frequencies ≥10 μg/dl over the period 1988–1998, 22- and 15-fold, respectively. The likely reason for this is that pervasive lead contamination in smelter communities is such that children of all ages are affected; with abatement of such contamination, the lower risk behaviors in older children, e.g., lower mouthing activity, are expressed more prominently, accounting for steeper decline rates.

Frequencies of PbB exceedances at three health risk threshold levels set forth for variably aged children over the period 1988–1998 are shown in Table 24.2. The table has contracted the annual screening data to the years indicated. Frequencies of PbB levels ≥25 μg/dl, a threshold risk level requiring some form of medical intervention and one well above the current risk threshold of ≥10 μg/dl, were recorded as above zero in the earlier years of Bunker Hill screenings, particularly among 2-year-old children. Among older children and for children in more recent screenings, few measurable frequencies at this risk threshold have been recorded. The greatest clustering of

TABLE 24.1 Means and Prevalences of PbB ≥10 μg/dl for Children in the Bunker Hill, ID, Superfund Box Site by Age Group and Year[a–d]

	1988		1990		1992		1994		1996		1998	
Child Age Group	Mean	% ≥10 μg/dl	Mean	% ≥10 μg/dl	Mean	% ≥10 μg/dl	Mean	% ≥10 μg/dl	Mean	% ≥10 μg/dl	Mean	% ≥10 μg/dl
2-year-olds	11.7	60.9	11.1	54.3	8.2	31.4	7.7	28.1	8.1	26.1	6.2	20.5
3-year-olds	12.5	62.1	9.9	43.8	8.1	32.5	6.6	17.1	6.3	21.4	5.9	11.8
4-year-olds	8.6	36.8	10.2	44.4	8.1	32.0	6.0	14.6	7.3	20.7	5.3	9.3
5-year-olds	11.7	56.7	9.4	39.5	7.3	23.3	6.3	17.4	5.7	8.1	4.7	10.8
6-year-olds	10.0	54.2	8.0	33.3	7.3	31.6	5.5	11.8	5.3	11.5	4.4	2.5
9-year-olds	7.6	28.9	6.4	12.2	6.3	16.7	4.7	8.7	3.6	0.0	3.2	2.0

[a]*Selected data from all years in IDHW (2000) (TerraGraphics Environmental Engineering).*
[b]*PbB data from annual screenings.*
[c]*Children in all Box communities.*
[d]*Arithmetic mean.*

TABLE 24.2 Prevalences of PbB (%) ≥10, ≥15, and ≥25 μg/dl[a–c]

Child Age Group	2-Year-Olds	3-Year-Olds	4-Year-Olds	5-Year-Olds	6-Year-Olds	9-Year-Olds
1988						
≥10 μg/dl	60.9	62.1	36.8	56.7	54.3	28.9
≥15 μg/dl	30.4	17.2	15.8	23.3	20.8	5.3
≥25 μg/dl	4.3	10.3	0.0	3.3	0.0	0.0
1990						
≥10 μg/dl	54.3	43.8	44.4	39.5	33.3	12.2
≥15 μg/dl	22.9	18.8	19.4	11.6	6.3	0.0
≥25 μg/dl	5.7	0.0	0.0	0.0	0.0	0.0
1992						
≥10 μg/dl	31.4	32.5	32.0	23.3	31.6	16.7
≥15 μg/dl	9.8	7.5	8.0	17.4	7.9	2.1
≥25 μg/dl	0.0	2.5	0.0	0.0	0.0	0.0
1994						
≥10 μg/dl	28.1	17.1	14.6	17.4	11.8	8.7
≥15 μg/dl	7.0	4.9	2.1	2.2	2.0	2.2
≥25 μg/dl	0.0	0.0	0.0	0.0	2.0	0.0
1996						
≥10 μg/dl	26.1	21.4	20.7	8.1	11.5	0.0
≥15 μg/dl	8.7	3.6	3.4	2.7	0.0	0.0
≥25 μg/dl	2.2	0.0	1.7	0.0	0.0	0.0
1998						
≥10 μg/dl	20.5	11.8	9.3	10.8	2.5	2.0
≥15 μg/dl	2.6	2.9	0.0	0.0	0.0	0.0
≥25 μg/dl	0.0	0.0	0.0	0.0	0.0	0.0

[a]*Selected data from all years noted in IDHW (2000) (TerraGraphics Environmental Engineering).*
[b]*PbB data from annual screenings.*
[c]*Children in all Box communities.*

TABLE 24.3 Observed Versus Predicted Prevalences of PbB ≥10 μg/dl for 2-Year-Old Children in the Bunker Hill, ID, Superfund Box Site, by Year[a–d]

	Sitewide % ≥10 μg/dl		Smelterville % ≥10 μg/dl		Kellogg % ≥10 μg/dl	
Year	Observed	Predicted	Observed	Predicted	Observed	Predicted
1988	46.2	49.0	71.0	54.8	41.4	54.6
1990	37.3	22.3	30.8	57.7	41.1	30.3
1992	27.2	9.4	31.5	18.5	38.1	12.5
1994	17.1	5.2	20.8	10.4	20.2	6.3
1996	12.6	4.1	11.8	3.9	17.6	6.3
1998	7.8	1.7	14.0	0.0	9.2	1.9

[a]*Selected data from all years in IDHW (2000) (TerraGraphics Environmental Engineering).*
[b]*Using dust–soil partitions and GSDs per Table 4-30, IDHW (2000).*
[c]*Sitewide and two communities: Kellogg and Smelterville.*
[d]*IEUBK model: defaults 55% dust Pb, 45% yard soil Pb.*

TABLE 24.4 Observed Versus Predicted Prevalences of PbB ≥10 and ≥15 μg/dl for 2-Year-Old Children in the Coeur d'Alene Basin Outside the Bunker Hill, ID, Box[a–d]

		Predicted PbB μg/dl	
Parameter	Observed PbB μg/dl	Model Default	40:30:30
% ≥10 μg/dl	23	37	17
% ≥15 μg/dl	12	19	6
Geometric standard deviation	1.82	2.08	1.91
Geometric mean PbB (μg/dl)	6.2	7.7	5.5

[a]*Selected data from U.S. EPA: Region 10 (2002), ROD, OU-3: Part 3.*
[b]*40:30:30 = 40% house dust Pb, 30% yard soil Pb, 30% community soil Pb.*
[c]*IEUBK default: 55% house dust Pb, 45% yard soil sitewide; includes eight communities; repeated observation data.*
[d]*IEUBK model run in batch mode, Version 0.99d.*

elevated PbB frequencies within the overall ≥10 μg/dl fraction is in the 10–15 μg/dl interval, with some further significant frequencies in the 15–25 μg/dl span.

Two-year-old children showed significant frequencies of PbB ≥15 μg/dl through much of the 11-year screening period covered as shown in Table 24.2, particularly 1988–1992. Relatively marked declines in frequencies in all threshold PbB levels were observed as the fraction of remediated residential areas, the principal source of childhood exposures, increased.

The multiple lead exposure studies carried out at the Bunker Hill communities employed both measured and modeled descriptive statistics, including the frequency of PbB levels exceeding health risk thresholds. Table 24.3 compares the measured PbB results with those simulated using the U.S. EPA IEUBK model for three geographic areas, the entire site and two communities within the Bunker Hill site: Kellogg, ID, and Smelterville, ID. Two-year-old children were compared in terms of measured and predicted PbB levels for 11 years, 1988–1998, for those three areas. Sitewide percentage frequencies of measured PbB ≥10 μg/dl versus predicted values were generally higher across the entire time period and in some cases significantly lower for modeled data. Predicted levels were up to three- to fourfold lower. In the earliest year, 1988, the two figures were approximately equal.

Similarly, measured versus predicted values for the towns of Smelterville, ID, and Kellogg, ID, indicated that, in most cases, the model underestimated measured PbB percentage frequencies by considerable margins. In 1996, for example, the measured percentage frequencies of PbB ≥10 μg/dl exceeded the modeled, predicted values by about threefold. The values for 1988 show a mixed picture, with the observed frequency exceeding that predicted for Smelterville, and the opposite being true in Kellogg, ID, children. It is not obvious what the basis for the shift in relative measured versus simulated PbB levels and exceedance frequencies would be, particularly over the years of PbB screenings and the various media-specific lead remediation actions that were taken. While the underprediction of the model versus measurement indicates higher lead intakes and uptakes, it is not clear why discordance in these modeling parameters would be variable over time.

Risk characterizations for individuals outside the Superfund site but residing within the Coeur d'Alene River valley are also in the form of percentages of PbB measurements exceeding health risk thresholds. Table 24.4 provides summary data for non-Box percentage frequencies for both observed and predicted PbB levels in children of 2 years of age or younger at or above two health risk thresholds: 10 and 15 μg/dl. Also included in the table are the sitewide geometric mean and geometric standard deviation values. The modeled PbB concentrations are for partitions of combined soil/dust contamination media into the soil and dust components. Screening data for PbB were gathered as part of the non-Box Basin assessments done in the late 1990s.

The observed percentages of PbB measurements in these children at or exceeding 10 and 15 μg/dl as the health risk thresholds were 23% and 12%, respectively. The corresponding geometric mean PbB in this cohort was 6.2 μg/dl. Of the 23% frequency figure for the threshold value of ≥10 μg/dl, about half or 11% of the total exceedance was in the 10–14 μg/dl concentration interval, while the remainder occurred at values ≥15 μg/dl.

Predicted PbB levels for these children ranged considerably across two selections in the model parameters for apportioning dust and soil. The selection which included community soil Pb (the 40:30:30 distribution in Table 24.4) gave a closer estimate (17%) to the measured frequency for the percentage frequency of PbB estimates at or above the health risk threshold ≥10 μg/dl. For frequency percentages ≥15 μg/dl, the estimate using the same 40:30:30 distribution underestimated the measured frequency by two-fold, 6% versus 12%. Use of the IEUBK model's default Pb apportionment figure for dust and soil, by contrast, markedly exceeded the measured frequencies for ≥10 and ≥15 μg/dl, 37% and 19%, respectively.

24.3.2 Health Risk Characterization for Pb at Multiple U.S. Communities Providing Pooled Soil and Dust Pb Loading Data

The Lanphear et al.'s (1998) analysis of 12 pooled studies of dust and soil Pb relationships to young children's PbB value was described in the previous chapter in terms of modeled estimates of PbB mean values, given changes in interior dust Pb loadings (μg/ft^2) and exterior (soil plus exterior surface Pb) Pb concentrations (ppm). Here, the analysis is extended to estimates of percentage of PbB exceeding two health risk thresholds, ≥10 and ≥15 μg/dl. These health risk threshold exceedances parallel the approach used in the case study for the smelter Superfund site in the Silver Valley of Idaho.

Table 24.5 summarizes the percent probability estimates for the two PbB thresholds for children <3 years of age. For example, at a dust Pb loading of 100 μg/ft^2 and a yard soil concentration of 500 ppm, the frequency of PbB values ≥10 μg/dl is 43%. A peak value is encountered at this dust loading and 2,000 ppm exterior Pb, 54%. The frequency for the higher PbB risk threshold, ≥15 μg/dl, at the 100 μg/ft^2 dust loading and 500 ppm exterior soil/surface Pb level is approximately half that of the lower threshold figure, 18%.

24.3.3 U.S. EPA's Health Risk Characterizations for Air Pb and Standard Setting Using Case Studies

The most comprehensive and quantitative illustrations of the use of human health risk assessments in regulatory scientific frameworks for environmental

TABLE 24.5 Percent Probability Estimates of PbB ≥10 and 15 μg/dl in Young Children Exposed to Various Dust Pb Loadings and Exterior Pb Levels[a–d]

	Exterior Pb Level (ppm)											
	10		72		100		500		1,000		2,000	
Dust Pb Loading (μg/ft²)	≥10	≥15	≥10	≥15	≥10	≥15	≥10	≥15	≥10	≥15	≥10	≥15
1	0.33	0.03	1.0	0.11	1.2	0.13	2.7	0.37	3.7	0.55	4.9	0.82
5	1.8	0.22	4.4	0.70	5.0	0.84	9.3	1.9	12	2.7	1.5	3.7
10	3.3	0.48	7.4	1.4	8.3	1.7	14	3.5	18	4.8	22	6.3
20	5.7	0.99	12	2.7	13	3.1	21	6.1	26	8.0	30	8.5
40	9.4	1.9	18	4.7	20	5.4	30	10	35	13	40	16
70	13	3.2	24	7.2	26	8.2	37	14	43	18	48	21
100	17	4.3	28	9.3	31	10	43	18	48	21	54	26

[a]Source: *Adapted from Lanphear et al. (1998).*
[b]*Children <3 years old.*
[c]*Loading: μg Pb/ft².*
[d]*Based on studies of seven U.S. mining–smelting and five urban sites.*

Pb are given in the U.S. EPA's (2007) risk assessment document prepared by the agency's OAQPS. This effort was part of the agency's risk analyses in support of the review and eventual proposed air Pb rulemaking for the U.S. NAAQS and was published in October 2007 (U.S. EPA, 2007). These full-scale exposure and health risk methodologies appeared in two volumes.

As briefly noted earlier, the U.S. EPA method differs conceptually and methodologically from the approaches defined in terms of relative frequencies of risk population PbB levels exceeding some designated risk threshold, e.g., $\geq$10 μg/dl. Here, three U.S. case studies were set forth as typifying Pb exposures for associated segments of risk groups, two site-identified and one geographically generalized: (1) populations impacted by Pb emissions from a primary smelter, (2) populations impacted by Pb emissions from a secondary smelter, and (3) a general U.S. urban population exposed to a mix of Pb emission sources. The following two major components define this approach: a detailed exposure assessment and a detailed human health risk assessment. The outcome of this method is the quantitative linkage of incremental changes in PbB with incremental changes in IQ as the health risk metric derived from the core adverse effect, the developmental neurotoxicity of Pb in the form of cognitive deficits (reduced IQ). The focus here, as it is with other risk approaches for Pb, is the quantification of adverse health impacts on young children. U.S. EPA's health risk characterization proceeds via the generation of a distribution of IQ loss estimates derived from estimated PbB levels for two exposure metrics (concurrent and lifetime-average PbB). The previous chapter presented the blood Pb portion of this two-part exercise.

The overall schematic for quantified health risk estimates in the analysis of U.S. EPA (2007) entailed combining concentration–response functions with blood lead distributional statistics generated for each of the three case studies to produce distributions of IQ loss estimates for each study population. Before the quantitative analyses of health risk were done via using differing concentration–response functions, the health risk portion in U.S. EPA's full-scale health risk assessment was evaluated to produce several statistical modeling and assessment steps for the risk metric, IQ point loss, in young children sustaining developmental neurotoxicity effects at various PbB estimates.

The U.S. EPA risk assessment based its concentration–response function derivations on the international pooled analysis of Lanphear et al. (2005). Specifically, log-linear concentration–response functions for IQ loss were developed for two dose metrics: concurrent PbB and lifetime average PbB. The pooled analysis determined the concurrent PbB metric to be more robust versus the lifetime average exposure indicator.

In its health risk portion, the U.S. EPA developed the following three concentration–response functions from the Lanphear et al.'s (2005) statistical models: (1) log-linear function with cut point, (2) log-linear function with low-exposure linearization, and (3) two-piece linear function.

TABLE 24.6 Estimated Health Risks (IQ Point Loss) for Modeled Median PbB Values (μg/dl) and Concentration–Response Functions for Three Case Studies[a–e]

	IQ Point Losses			
	Recent Air[f]		Total Pb Exposure[g]	
Case Study/Air Pb Scenario	Low	High	Low	High
General urban case study				
Current conditions—95th percentile, air Pb	0.4	2.0	1.8	5.4
NAAQS, 1.5 $\mu g/m^3$, max quarterly	2.1	4.3	3.5	7.0
NAAQS, 0.5 $\mu g/m^3$, max monthly	0.5	2.1	1.9	5.4
NAAQS, 0.2 $\mu g/m^3$, max monthly	0.3	1.7	1.7	5.2
Primary Pb smelter: 1.5-km-radius subarea				
NAAQS, 1.5 $\mu g/m^3$, max quarterly	–	–	3.7	6.8
NAAQS, 0.5 $\mu g/m^3$, max monthly	–	–	2.6	5.8
NAAQS, 0.2 $\mu g/m^3$, max monthly	–	–	1.9	2.0
Secondary Pb smelter: 1.5-km-radius subarea				
Current conditions	<0.1	0.4	0.8	3.8
NAAQS, 0.5 $\mu g/m^3$, max monthly	<0.1	0.1	0.7	3.7
NAAQS, 0.2 $\mu g/m^3$, max monthly	<0.1	0.1	0.6	3.6

[a]*Primary smelter case study: Herculaneum, MO, smelter.*
[b]*Secondary smelter case study: Troy, AL, smelter.*
[c]*General urban case study, no specific geographic location.*
[d]*Adapted from Table 4-2, U.S. EPA (2007).*
[e]*Concurrent PbB as exposure metric.*
[f]*Inhalation of current air Pb + recently deposited dust Pb.*
[g]*Has both recent and past Pb inputs.*

Tables 24.6 and 24.7 tabulate distributions of IQ point losses for the three case studies, for various air Pb impacts within each case study, and for various existing or potentially NAAQS-regulated air Pb scenarios. In all cases, the dose function is concurrent PbB level. Table 24.6 derives medians in IQ point losses, while 95th percentile values for IQ point losses are summarized in Table 24.7.

Median IQ point losses are significant for all the three case studies when one estimates for the high end of the total Pb exposure groups. For the

TABLE 24.7 Estimated Health Risk (IQ Point Loss) for Modeled 95th Percentile PbB and Concentration–Response Functions for Three Case Studies[a–e]

	IQ Point Losses			
	Recent Air[f]		Total Pb Exposure[g]	
Case Study/Air Pb Scenario	Low	High	Low	High
General urban case study				
Current conditions—95th percentile, air Pb	1.0	3.4	4.2	8.8
NAAQS, 1.5 μg/m^3, max quarterly	3.3	6.0	5.5	10.5
NAAQS, 0.5 μg/m^3, max monthly	1.0	3.6	4.2	9.1
NAAQS, 0.2 μg/m^3, max monthly	0.4	2.3	3.9	8.4
Primary Pb smelter: 1.5-km-radius subarea				
NAAQS, 1.5 μg/m^3, max quarterly	–	–	6.8	9.5
NAAQS, 0.5 μg/m^3, max monthly	–	–	5.8	8.5
NAAQS, 0.2 μg/m^3, max monthly	–	–	4.9	7.6
Secondary Pb smelter: 1.5-km-radius subarea				
Current conditions	0.5	1.4	3.2	6.3
NAAQS, 0.5 μg/m^3, max monthly	0.4	0.5	3.2	6.3
NAAQS, 0.2 μg/m^3, max monthly	<0.1	0.2	3.1	6.1

[a]*Primary smelter case study: Herculaneum, MO, smelter.*
[b]*Secondary smelter case study: Troy, AL, smelter.*
[c]*General urban case study, no specific geographic location.*
[d]*Adapted from Table 4-3, U.S. EPA (2007).*
[e]*Concurrent PbB as exposure metric.*
[f]*Inhalation of current air Pb+recently deposited dust Pb.*
[g]*Has both recent and past Pb inputs, other Pb inputs.*

general urban example, the high end ranges from 5.2 to 7.0 lost IQ points as a function of selected ambient air Pb standards or current conditions. The low-end range is 1.7–3.5 IQ points lost for the total exposure group. The high-end portion of the recent air group within the general urban case study presents IQ point losses at differing standards ranging from 1.7 to 4.3 IQ point losses. Median IQ point loss estimates were not made for the primary smelter case's recent air portion, while the total Pb exposure group medians ranged from 2.0 to 6.8 IQ points lost in the high group, and a range of medians of 1.9–3.7 IQ points lost for the low group. As would be expected, the lowest ranges of median IQ point losses are generated for the relatively lower air Pb scenarios and modeling approaches in the secondary smelter case study.

In all three case studies, median IQ point losses predictably track closely and directly the level of permissible Pb in ambient air via the various air Pb NAAQS scenarios. This applies to all groups: recent air or the total Pb exposures.

Estimated health risks for children in the form of IQ point losses using the 95th percentiles of modeled PbB levels are presented in Table 24.7. For the total Pb exposure group, the low and high end of IQ point losses across the three case studies represented significant decrements in IQ. The high end of total Pb exposure within the general urban case study ranged in IQ losses from 8.4 to 10.5 IQ points, depending on the air standard scenario, and for the low end of total Pb exposures, the range was 3.9–5.5, approximately half that for the high end of the PbB range. Recent air estimates for the high group ranged from 2.3 to 6.0 IQ points lost and from 0.4 to 3.3 IQ points lost in the low group, depending on selected ceilings for air Pb as a function of NAAQS. The primary smelter case did not estimate recent air Pb point losses, while the total Pb exposure's high group showed a range of 7.6–9.5 IQ points lost. For the high group, the point loss range varied with NAAQS scenario from 4.9 to 6.8 IQ points.

The secondary smelter example, as seen in Table 24.7, showed much lower IQ point losses for all the groups, categories, and NAAQS scenarios relative to the primary smelter and general urban case studies. In all three case studies, 95th percentile IQ point losses, in much the same way as with the median values noted above, predictably track closely and directly the level of permissible Pb in ambient air via the various air Pb NAAQS scenarios. This applies to all groups: recent air, recent and past air, or the total Pb exposures.

REFERENCES

Idaho Department of Health and Welfare, 2000. 1999 Five Year Review Report Bunker Hill Superfund Site, vol. II. Division of Environmental Quality, TerraGraphics Environmental Engineering, Inc., Boise, ID.

Landrigan, P.J., Baker Jr., E.L., Feldman, R.G., Cox, D.H., Eden, K.V., Orenstein, W.A., et al., 1976. Increased lead absorption with anemia and slowed nerve conduction in children near a lead smelter. J. Pediatr. 89, 904–910.

Lanphear, B.P., Matte, T.D., Rogers, J., Clickner, R.P., Dietz, B., Bornschein, R.L., et al., 1998. The contribution of lead-contaminated house dust and residential soil to children's blood lead levels. A pooled analysis of 12 epidemiological studies. Environ. Res. 79, 51–68.

Lanphear, B.P., Hornung, R., Khoury, J., Yalton, K., Baghurst, P., Bellinger, D.C., et al., 2005. Low-level environmental lead exposure and children's intellectual function: an international pooled analysis. Environ. Health Perspect. 113, 894–899.

Mushak, P., 1998. Uses and limits of empirical data in measuring and modeling human lead exposure. Environ. Health Perspect. 106 (Suppl. 6), 1467–1484.

Schwartz, J., Landrigan, P.J., Feldman, R.G., Silbergeld, E.K., Baker Jr., E.L., von Lindern, I.H., 1988. Threshold effect in lead-induced peripheral neuropathy. J. Pediatr. 112, 12–17.

Schwartz, J., Landrigan, P.J., Baker Jr., E.L., Orenstein, W.A., von Lindern, I.H., 1990. Lead-induced anemia: dose–response relationships and evidence for a threshold. Am. J. Public Health 80, 165–168.

U.S. Environmental Protection Agency, 2002. Record of Decision. The Bunker Hill Mining and Metallurgical Complex Operable Unit 3. U.S. EPA Region 10, Seattle, WA, September.

U.S. Environmental Protection Agency, 2007. Lead: Human Exposure and Health Risk Assessments for Selected Case Studies, Report No. EPA-452/R-07-014a, vol. I. Office of Air Quality Planning and Standards, Washington, DC.

U.S. Environmental Protection Agency, 2008. Office of the Administrator. National Ambient Air Quality Standards for Lead. Final Rule. 73 FR 66964-67062, November 12, 2008.

Part 5

Regulatory Approaches to Control

Chapter 25

Legislative Aspects of Lead Regulation and Regulatory Policies

This chapter of the monograph is an introduction to the topic of environmental lead regulation and some scientific and regulatory scientific issues *vis-a-vis* Pb regulation. It also provides a brief history of legislative highlights in regulating lead in the human environment.

25.1 INTRODUCTION

Viewed within a broad perspective, the elements and pathways for regulating human lead exposure have unequally evolved in tandem with regulation of the element to control ecological Pb exposures. This dichotomized, hierarchical model of lead control began to yield more equal attention to both human and ecological populations in the 1970s and 1980s with the appearance of legislative mandates in the United States and elsewhere, including the National Environmental Policy Act (NEPA) of 1969 and the creation of such agencies as the U.S. EPA. Despite overtures at parity in regulatory protection, the statutory and administrative regulatory focus has largely remained on reducing human exposure risks, regardless of adequacy for ecological protection.

The rationale for this, as indicated in the standard-setting language (U.S. EPA, 1978), appears to be the arbitrary assumption that humans are the most Pb-sensitive species and that protection of humans conveys protection to ecological segments of the biosphere. This apparent anthropocentric bias appeared to be predicated on little hard evidence about relative Pb toxicological sensitivity. The most recent U.S. EPA lead criteria document (U.S. EPA, 2006) does take considerable account of the desirability of full evaluation of the adequacy of any primary Pb standards (protecting humans) in guiding promulgation of secondary standards (protecting ecological compartments). Reasons for the human populations focus of these chapters were noted earlier; the focus is not a judgment of relative importance in the above context.

Part 5 discusses regulatory matters stratified by United States and international perspectives, and each is given a separate section in the later chapters.

Trace Metals and other Contaminants in the Environment, Volume 10
ISSN: 1875-1121 DOI: 10.1016/B978-0-444-51554-4.00025-0

Some aspects of Pb regulation are universal in nature and traverse national or world regional boundaries. Others are framed and implemented by the legal and regulatory instruments available to specific national or regional jurisdictions. The current preference in the European Union is for regulation of new substances within the concept of the Precautionary Principle. Already emitted lead as a contaminant is regulated by directives and regulations analogous to those in the United States and Canada. This approach is distinct from enforcement mechanisms relied on in the U.S. regulatory community. In terms of the U.S. picture, regulation and regulatory policies for Pb are dispersed in diverse ways across all segments of the Federal government—legislative, enforcement, and judicial.

25.2 SCIENTIFIC AND REGULATORY ISSUES FOR U.S. REGULATORY INITIATIVES AND POLICIES FOR Pb IN THE HUMAN ENVIRONMENT

Lead is a multimedia pollutant and consequently has impacts on human populations via their contact with diverse Pb-contaminated media. Multiple media intakes and uptakes provide a cumulative, integrated internal body level of Pb and integrated expressions of Pb in the body are what are quantitatively linked to Pb poisoning risks. While Pb is a multimedia contaminant, not all media in all places or at all times are equally contaminated or provide comparable human exposures. The relative significance of Pb in some particular medium or medium type differs from case to case and from exposure scenario to exposure scenario.

While Pb exposures in human populations are cumulative as to sources of Pb intakes, regulation of such exposures is commonly fragmented and piecemeal. This is because of the legislative and executive nature of pollutant regulation in the United States and elsewhere. Different governmental agencies regulate Pb in different environmental media. Historically, little or no cognizance was taken of the cumulative, integrative nature of multimedia lead exposures and aggregate toxicity risk in human populations.

Mushak and Schroeder (1980) addressed this disparity between single-medium legislative mandate and multimedia pollutant science in two reports to the U.S. National Commission on Air Quality (Mushak and Schroeder, 1980; Schroeder and Mushak, 1980). In the report dealing with the legislative history of multimedia pollutants, Schroeder and Mushak (1980) noted the origins of the media-specific fragmentation of regulatory control for environmental contaminants and used a set of case studies to describe the effects of such piecemeal control for lead.

A second scientific factor that complicates effective Pb regulation and population exposure reduction efforts is the environmental medium-specific temporality and persistence of Pb in environmental media. Pb emitted from a stationary source to the atmosphere results in relatively transitory elevations in localized air concentrations of Pb relative to dry or wet depositional events and/or long-range transport. Where there is also variability to emissions for any of a number of reasons and over whatever time metric, how that variability is captured will considerably affect the nature and extent of Pb regulation. A regulation for air lead that is indexed for compliance as an annual mathematical average permits more frequent excursions in elevated Pb emissions and more opportunities for subject exposures than, say, quarterly maximum averages, while quarterly maximum averages permit more emission excursions upward than monthly maximum averages.

An additional complication in this regard is the propensity for more Pb accumulation in some environmental media than in others. Ambient air levels of Pb with relatively stable Pb input and removal rates would not greatly increase over time, while levels in exterior or interior dusts would be expected to accumulate indefinitely, in the absence of any interventions reducing accumulation, concurrent with reducing inputs to the medium. This factor would be of particular concern for risk groups impacted by the sources and pathways of an operating smelter generating dust Pb fallout onto contact surfaces or by interior surfaces containing deteriorating lead paint and thereby producing interior dust accumulations. Soil as a Pb medium is typically the largest accumulator of post-depositional or dispersed Pb in terms of total Pb mass, but not necessarily Pb concentration. Soil's retaining capacity as a long-term depository medium is high, decades and even centuries in duration when indexed as half-life, while sheer bulk of soil mass permits huge Pb input burdens.

Regulating Pb also requires recognizing how regulatory efficacy is to be achieved in terms of the time frame of target population exposures. Implementing regulatory controls for Pb in children to reduce childhood PbB levels via whatever statistical metric would provide more real-time responses in terms of exposure reductions. Using such controls on the basis of PbB values in older adults would be less effective for remediating adult exposures since a significant fraction of older adult PbBs would be endogenous Pb releases from lifetime Pb accumulations in their bones. Such Pb remediation would, however, serve to reduce future population exposures.

25.3 INTERACTIVE FEEDBACK-LOOP RELATIONSHIPS OF LEAD SCIENCE, RISK ASSESSMENT, AND REGULATION

Figure 1.1 depicts the three components of an overall interactive feedback-loop relationship that is tripartite in nature, involving scientific research, health risk assessment, and the regulation of lead in the human environment. Scientific research informs both risk assessment and regulatory initiatives. Subsequent needs for adequate risk assessment and stronger, more effective regulation then dictate that there be more research. Adequate regulation is also informed by adequate human health risk assessments, subsequent regulatory data gaps then dictating the need for more reliable risk assessments. Such risk assessment data are then gathered from more scientific research.

25.4 HISTORICAL HIGHLIGHTS OF GENERAL AND SPECIFIC U.S. LEGISLATIVE INITIATIVES FOR LEAD REGULATION

This section presents a brief history and context for the legislative initiatives that laid the groundwork for more recent legislative actions on Pb. Later chapters detail the regulation of Pb by various U.S. and international agencies on a media-specific basis. This portion has a summary of early environmental legislation and a summary of more recent environmental initiatives from the United States. Congress for general or specific pollution controls directly focused on lead. It includes a discussion of some of the interactions between Congress and various economic and public health stakeholders, including industrial producers and users of lead on the one hand and various public groups on the other. Such interactions have occurred via Congressional committees, subcommittees, and individual members of the House and Senate.

25.4.1 Early Congressional Initiatives for Environmental Protection

The earliest legislative actions for environmental control of lead releases and human exposures were general in nature, applicable to contaminants generically or by broad class rather than by specific substance. The U.S. legislative initiatives at the Federal or lower legislative levels specifically enacted for lead control and public health protection arose considerably later, tracking closely the growth of the scientific and public health history of lead. The U.S. Congress did not enact environmental legislation to control broadly defined "pollution" in environmental media until the first quarter of the twentieth century. These laws and their implementation appeared to have had marginal success. Part of the problem was the array of obstacles and drawbacks within the enabling language. This conclusion is borne out by the persisting severe air, water, and other pollution that continued well after 1925.

Early legislation was general for all pollutants and mostly addressed water pollution. Lead contamination was rarely specifically addressed by implementation programs from state or Federal agencies. This absence of any legislative initiatives for lead at this time was in marked contrast to international, non-U.S. activities, including organized movements to ban uses of lead paint in interiors of buildings around the world.

The earliest Congressional legislation broadly dealing with any definition of pollution abatement and control was the 1899 Refuse Act, commonly known as the Rivers and Harbors Act. Its purpose was actually to simply assure unimpeded navigation of waterways by banning solid refuse that might interfere with vessel movement. This act was explicitly intended to address gross waterway contamination by debris; street runoff and sewage effluents were exempted. Its value to water quality through regulation of pollutants in water came much later, in the 1960s, when certain provisions of the 1899 law were interpreted by the U.S. Supreme Court in a 1960 case to mean that the law was applicable to pollutants as well as to navigational hazards. As a consequence, permits were required for pollutant discharges. Subsequent court decisions followed this precedent.

Congressional initiatives in the early twentieth century were reactive rather than proactive, to the extent that comprehensive pollution prevention initiatives did not hold legislative interest and progress was fragmentary and episodic. In 1912, Congress authorized the U.S. Public Health Service to evaluate health effects of navigable water pollutants but not much else. The 1924 Oil Pollution Control Act was passed to control oil spills in America's coastal waters, with enforcement by the U.S. Secretary of War. How well enforcement worked or whether there were fewer oil spills is not evident.

A cluster of federal laws followed: the 1948 Federal Water Protection and Control Act, a 1951 extension of that act, and a permanent act in 1956; the 1958 Fish and Wildlife Coordination Act; and the 1961 Federal Water Quality Control Act. The 1948 act, using a legislative blueprint that served for quite a bit of future environmental legislation, gave the primary responsibility for keeping waterways clean, along with funds for treatment plants and water quality research, to the states. Procedural hurdles were many and arguably served to thwart effective water quality protection. For example, interstate waters could be protected in litigation, but those suits required a hearing, multiple notices to the polluter(s), and permission from the states to proceed. The 1956 legislation omitted the state consent requirement but used a Federal–state conference, a public hearing, and a court suit.

The 1961 water quality legislation increased the federal presence in pollution control and transferred authority to the then Department of Health, Education and Welfare. As noted by D'Itri and D'Itri (1977), there were centralized control shortcomings in this legislation. The subsequent 1965 Water Quality Act and the 1966 Clean Water Restoration Act were improvements. The 1965 law created an enforcement mechanism, the Federal Water Pollution Control Administration (FWPCA), and required standards and state timetables for submitting control plans to the FWPCA. In its early history, FWPCA was passed around among several agencies; it found a permanent home at the U.S. EPA in 1970.

State-based pollution control was not very effective. A 1970 finding from the Office of Enforcement and Standards Compliance indicated that even though states had the authority to enforce water quality regulations, most elected not to do so or were negligent in meeting mandated deadlines.

A contentious issue was the presence on state pollution control boards of individuals who worked for regulated entities. The relative lack of state enthusiasm for enforcing the early laws, coupled with the structural complexity of federal enforcement, meant that water quality protection initiatives were extremely limited or inefficient in practice. The latter was shown in the Interior Department's fruitless attempts to enforce an action against a New Jersey chemical company in 1970.

25.4.2 More Recent Congressional Initiatives for Regulation

The more recent era of environmental awareness and a public demand for cleaner air and water produced a flurry of laws and associated regulations from the late 1960s onward. These are detailed in the four remaining chapters of this book. Only running comments are presented here. A number of these address pollution control problems where lead was included either through general provisions, by subsequent amendments and administrative actions, or where litigation has forced implementation of categorical provisions specific to lead.

The NEPA of 1969, in an idealized view, systematized and provided a conceptual and structural framework for how the Federal government would deal comprehensively with environmental pollution in America and what would be expected of Federal agencies in implementing regulatory policies. Among its many provisions, NEPA contained several key ones: it sets a national policy to "... encourage productive and enjoyable harmony between man and his environment"; taking proactive stances to prevent environmental harm in the first place was another mandate.

Specific, tangible provisions of NEPA included establishing a Council on Environmental Quality and requiring environmental impact statements where necessary. These were not minor elements in the legislation. Environmental impact statements were to consider the environmental impacts of alternatives to desired actions and their impacts, however unsavory the economic or political dimensions. In this sense, environmental impact statements are an environmental "full disclosure law."

The Clean Air Act (CAA) of 1970 along with its early amendments addressed national air quality issues. After it passed, a number of references to air lead and air mercury were incorporated in the CAA. Section 108(a)(2) of the CAA as amended in 1974 and administrative action by the EPA Administrator in early 1976 led to designation of lead as a criteria pollutant. A criteria pollutant requires air quality criteria developed through an air quality criteria document which becomes the basis for setting primary and secondary NAAQS. As a simple distinction, ambient air pollutants and standards are required when emission sources are sufficiently diffuse and pervasive that their control requires centralized action. Furthermore, health and welfare issues are paramount in standard setting under the CAA and

Congress intended health issues to be immune from economic considerations in regulatory action taken under this section.

Between passage of the CAA and the 1976 administrative action declaring lead a criteria pollutant, EPA dealt with lead in gasoline as a regulatory issue under §211(c)(1) and other 211 subsections of the CAA. Under this section, the EPA Administrator has broad authority to regulate or prohibit manufacture and sale of any fuel additive if its emission products would produce pollution that endangers public health or welfare.

The 1990 CAA Amendments (CAAA) comprised the most sweeping legislative changes in the basic 1970 CAA, adding a host of changes after considerable debate and maneuvering by affected parties. A number of amendments that affected lead regulation occurred here. An important CAAA provision mandated the phaseout, rather than just the phasedown of lead in gasoline by 1995. Section 106 of the amendments (Title I, Part D, Subtitle 5 of the original CAA) deals with nonattainment provisions for lead. Solid waste combustion emissions of lead are covered in §305(a) of the amendments (§129(a)(4) in the original CAA).

Section 112(n)(1)(A) of the amended CAA directed EPA to evaluate the health hazards associated with emissions of hazardous air pollutants (HAPs) from electric utility power generating units that used fossil fuels. HAPS identified for further evaluation included lead, out of a screening list of 67 substances.

The Federal Water Pollution Control Act and the Clean Water Act (CWA) of 1972, with 1977 and 1987 Amendments, collectively addressed water pollution as it impacts both groundwater and surface water quality and discharge permits. Section 304(a)(1) of the CWA as amended in 1977 required EPA to publish criteria for water quality reflecting the best scientific knowledge of the effects of water pollutants on health and welfare. Section 307(a)(1) of the amended CWA lists 65 toxic water pollutants for which EPA must publish criteria documents. Under the act, the Natural Resources Defense Council sued EPA to produce the documents, and a number of water quality criteria documents have been published, including criteria for lead.

The CWA and its 1977 and 1987 amendments covered other topics: Title III, enforcement, dealt with effluent limitations, standards and implementation plans for specific pollutants, inspections, monitoring, and enforcement; Title I covered research and related programs; Title IV outlined treatment plant construction grants, permits, and licenses. The general provisions covering citizen suits, reports to Congress and state authorities under the CWA were present in Title V.

The 1974 SDWA and 1986 Amendments collectively protect the drinking water supplies of the United States that are both subject to regulation and used by the public. Both surface water and groundwater drinking supplies are covered. Lead figures prominently in the SDWA. In the 1986 amendments, for example, Part B, §1417 prohibits use of lead pipes, lead joint solder, and leaded fluxes under a 2-year phaseout timetable. Part F, §1461–1467 covers the removal of school water appliances containing lead, focusing on drinking water coolers.

A critical part of the SDWA has to do with setting MCLs and MCL goals (MCLGs) for lead, mercury, and other contaminants in regulated drinking water supplies. The CWA defines a regulated "public water supply" as one serving 25 or more people and having at least 15 service connections. MCLs are standards that are enforceable while MCLGs are targets that suppliers strive to meet for drinking water quality. The following two levels of drinking water protection are provided: (1) primary drinking water standards refer to regulated supplies and contamination control via MCLs or treatment techniques and target protection of human health and (2) secondary regulations, which are not enforceable, deal with aesthetic criteria such as taste or smell. As described in Chapter 28, mechanisms other than MCLs can be used for drinking water contaminant control. Drinking water regulations can also be implemented as action levels. Pb in U.S. tap water is currently managed in this way.

Other Congressional Legislation

Other congressional legislation also deals with lead pollution. The 1980 Comprehensive Environmental Response, Compensation, and Liability Act (CERCLA) or Superfund and its 1986 amendments, the Superfund Amendments and Reauthorization Act (SARA), dealt with lead in a number of ways. The 1986 SARA required that the U.S. PHS, through the Agency for Toxic Substances and Disease Registry (ATSDR), prepare a report to Congress on the nature and extent of childhood lead poisoning in America. The ATSDR sent its report to Congress on the topic in July 1988 (U.S. ATSDR, 1988; Mushak et al., 1989; Mushak and Crocetti, 1989, 1990; Crocetti et al., 1990 a, b). Another act involving lead is the Toxic Substances Control Act (TSCA) of 1976.

25.4.3 U.S. Federal Legislation Specific to Lead

Federal laws specific just to lead have been enacted. In 1971, for example, Congress passed the LBPPPA directing the HUD to determine the nature and extent of the lead paint poisoning problem, particularly among young children, and the most effective ways to remove lead paint from residential surfaces that provide exposure paths for children. In 1973, Congress amended the LBPPPA to require HUD to eliminate, as far as practicable, the hazards of lead-based paint in housing with the hazard. In 1976, the LBPPPA was extended and revised. In early 1988, Congress passed the LBPPP section (§556) of the Housing and Community Development Act of 1987, directing

HUD to do four things: develop and issue regulations for the lead paint hazard as defined by Congress; launch a 5-year effort to identify lead paint in all public housing; establish a demonstration project to include abatement methods; and create a plan to identify and abate all privately owned housing in the United States. In 1988 and 1989, sections of other acts dealt with the HUD directives spelled out earlier.

A more recent piece of U.S. legislation with a lead focus was Title X of the Housing and Community Development Act, referred to as the 1992 Residential Lead-Based Paint Hazard Reduction Act or simply Title X. Title X provides a number of steps to eliminate the most hazardous of lead-painted residences in public and certain classes of private housing. Its guiding purpose was to develop a national approach to creating "infrastructure" to eliminate hazards as rapidly as possible. Some provisions of the legislation as interpreted by HUD and EPA are controversial, particularly in the matter of interim controls of lead paint exposures in the most deteriorated living units.

A second series of actions springing from Title X mandates have had to do with the repair and/or renovation of private housing or other child-occupied structures. These regulations were recently implemented as the Lead Renovation, Repair, and Painting (LRRP) Rule.

25.4.4 Congressional Interactions with Various Stakeholders Affected by Legislative Initiatives for Lead

Various stakeholder interests, e.g., industrial and public interest groups, interact with political bodies extensively and intensively in the course of their legislating. These interactions have occurred with Pb. Some illustrations can be offered.

In 1982, the Subcommittee on Environment, Energy and Natural Resources of the House Committee on Government Operations held a hearing on the public health dangers of lead in gasoline. The committee was examining the Reagan Administration's proposal to relax or rescind lead phasedown standards for gasoline. Medical and scientific experts in the field addressed the consequences of childhood lead poisoning, the scientific evidence on lead health risks, and the role of gasoline lead in addition to toxic effects from other lead sources. They were joined as witnesses at the hearing by the parent of a lead-poisoned child, other health workers dealing with the problem of lead poisoning, representatives of advocacy groups, and the Inspector General of the EPA, who reported on his investigation of alleged misuse of discretionary authority concerning the lead standards by the EPA Administrator. Industry representatives did not appear.

When the Senate Committee on the Environment and Public Works looked at the Airborne Lead Reduction Act of 1984, a proposed amendment to the CAA that would ban the sale of leaded gasoline for general use, medical experts reviewed the medical research on health effects of lead exposure and its toxicity in children. At this hearing, the medical experts were joined by representatives of the lead additive manufacturing and consuming industries, oil companies, and farmer cooperatives.

Medical experts were back before the Senate body in 1990 as the Subcommittee on Toxic Substances, Environmental Oversight, Research and Development looked again at health effects in children and lead poisoning control strategies. Many of the government witnesses focused on cooperative EPA–HUD efforts to remove lead-based paint in housing. Scientists on the panel described new findings on the neurobehavioral problems associated with lead exposure, problems associated with prenatal exposure to lead, and research projects on the effects of lead in soil. Public advocacy group speakers criticized federal efforts at preventing lead poisoning in the face of growing knowledge about the hazards.

When the same Senate subcommittee convened later in 1990 to consider the Lead Ban Act of 1990 and the Lead Exposure Reduction Act of 1990, medical experts, the American Academy of Pediatrics, the National Education Association, and others spoke in favor of both measures.

In September 1990, the U.S. House Subcommittee on Transportation and Hazardous Materials held a hearing on the proposed Lead Pollution Prevention Act of 1990, which would restrict the amount of lead in various products, require battery recycling, and establish monitoring and abatement programs for lead-based paint in three Cabinet agencies and the EPA. Here, the U.S. EPA opposed the measure, based on the argument that it had adequate existing authority for lead abatement programs. At this hearing, lead poisoning expert Dr Herbert Needleman presented a meta-analysis of low-level lead exposure studies looking at IQ effects in children that was placed in the record. Educators, advocates, pediatricians, and the National Institute of Environmental Health Sciences had representatives speaking for the measure.

25.5 CASE STUDIES OF LEAD REGULATION: MEDIUM-SPECIFIC FEDERAL STATUTES FOR CONTROL OF A MULTIMEDIA CONTAMINANT

As part of reports to the U.S. National Commission on Air Quality on multimedia contaminants, their science (Mushak and Schroeder, 1980), and their legislative history (Schroeder and Mushak, 1980), the authors described seven case studies of attempts at environmental regulation of multimedia pollutant exposures via single-medium statutory treatment. The discussions of the problems associated with multimedia pollutants being regulated by single-medium statute control mechanisms included scientific issues of duplicative data gathering, research, and analysis, jurisdictional disputes, separate standard-setting proceedings, separate standard-setting procedures, and differing time limits within these. These case studies were based on seven general statutes then in force: (1) the CAA; (2) the CWA; (3) the SDWA; (4) the Food, Drug, and Cosmetic Act (FDCA); (5) the Federal Insecticide, Fungicide and Rodenticide Act (FIFRA); (6) the Resource Conservation and Recovery Act (RCRA); and (7) the TSCA. The focus with these case studies was the general nonoccupational population. With few exceptions, the earlier incarnations of Federal statutes addressing environmental contaminants were not drafted with the intent of addressing multimedia control of toxicants such as lead. Some exceptions showed up in the language of TSCA and RCRA. Four of the seven statutory case studies were assembled to address contamination of the major natural media: ambient air, through the CAA; ambient/drinking water, through the CWA and the SDWA; and terrestrial repositories/soil, through RCRA. Product-oriented statutes addressed three categories: pesticides, through FIFRA; foods and drugs, through the FDCA; and a wide range of manufactured and marketed products producing contaminants, through TSCA. These latter three categories and associated statutes can, therefore, be viewed as substance focused as well as medium focused.

Lead was a particular focus of the legislative and statutory analysis of contaminants having multimedia pathways of human exposures. While the media through which human risk populations are exposed to lead are largely secondary in toxicological and toxicokinetic terms once uptake to the body occurs, it is all important from a regulatory perspective. At the time of the Schroeder and Mushak analysis in 1980, lead and its compounds were regulated under the following sections: §211(c)(1)(A) of the CAA, which regulated the element as a constituent of lead fuel additives and their emission products; §108(a)(1) and §109 of the CAA, which regulates Pb in air as a criteria pollutant and therefore requires appropriate NAAQS; §301 and §304 of the CWA, requiring the setting of effluent standards for Pb and its compounds; §302 of the CWA requiring the Administrator of EPA to set water quality criteria; §311, requiring regulation of hazardous spills; §1412 of the SDWA, requiring the Administrator to establish national drinking water standards via promulgation of MCLs; and §§ 3001–3004 of RCRA, establishing disposal standards for lead waste products from the lead acid battery industry, the lead smelting industry, and other lead users. Lead regulation by the U.S. FDA was under §406 of the FDCA, requiring the Commissioner to establish tolerances for residues in commodities and foods; under an added section, §191.9 of the Federal Hazardous Substances Act (FHSA), requiring the FDA Commissioner to promulgate the regulation of lead in paints to an interim value of 0.5% and then a final value of 0.06%. Subsequent to the FDA Commissioner's promulgation of reduced lead in paints under FHSA, lead in paint was permanently brought under the control of the Consumer Product Safety Commission (CPSC) in 1977, under §30(d) of the Consumer Product Safety Act. This regulatory transfer was dictated by the LBPPPA of 1971 and 1976 amendments requiring the CPSC to set a safe lead paint content level by the end of 1976. Lead was banned at levels above 0.06% for residential and like uses (toys, etc.), effective from February 1978.

25.5.1 The Clean Air Act

With respect to research and data collection, the language of §103(f)(1)(A) of the CAA contained some rationales that airborne contaminants such as Pb, which could produce adverse effects on human health, can be researched for their relative contributions to health impacts. Quantification of relative contribution of the airborne components of a multimedia contaminant obviously requires quantification of other pathway-dependent contributions.

The CAA via §§108 and 109 mandated establishment and periodic review of U.S. NAAQS, with the first promulgation of a primary and secondary NAAQS for airborne Pb as a criteria pollutant occurring in 1978. §108 defines air quality criteria for the pollutant, e.g., lead, and §109 specifies the requirement for setting criteria-based standards. Primary NAAQS are to be set to protect public health with an adequate margin of safety. The notion of an adequate margin of safety was statutorily advanced to drive consideration of those adverse effects likely to be of major health concern and for which data were/are not sufficiently complete to precisely define human dose–response relationships. The secondary NAAQS for criteria pollutants such as airborne Pb are set for "welfare" effects as defined in §302 (h) of the Act, and include effects on vegetation, visibility, water, crops, man-made materials, animals, economic values, and personal comfort and well-being. Section 110 of the Act deals with SIPs, which is the actual mechanism by which Federally mandated CAA criteria pollutant regulations are implemented.

Other provisions of the CAA deal with various aspects of criteria pollutant regulation, among them §111(a) giving the EPA Administrator control over new source performance standards. This aspect is of obvious importance in the case of multimedia pollutants such as Pb since there is the potential difficulty from shifting what is regulated through shifting across media-specific regulation. Specifically, the control of new source performance standards via §111 is not permitted to generate additional health effects via another medium. Overall, §111 does not appear to permit the Agency to hold out for regulating multimedia pollutants such as airborne Pb by controlling Pb in other media that might be less expensive to control.

Other regulatory classes of multimedia pollutants are within the purview of the CAA. Section 112 gives the EPA Administrator the authority to control air pollutants such as Pb which are not identified as criteria pollutants but entail emissions from new or existing sources of pollutants, via National Emission Standards for Hazardous Air Pollutants (NESHAPS). Here, the Administrator must first determine that the HAPs can cause or are expected to cause increases in mortality or illness. Here, as for criteria pollutants, standards are to be set with an ample margin of safety. While the language of this section of the CAA would appear to bar economic costs of control as a factor in NESHAPS issuance, costs in a few cases seemed to figure in standard setting. For example, the vinyl chloride standard issued in 1975 was based on some consideration of the impact of setting a zero level for a "no threshold" pollutant on the viability of this industry. EPA proposed instead a standard in 1975 based on the then-extant best available control technology.

The various sections of the CAA, overall, allow consideration of the role of nonambient air levels of lead in quantifying rationales for setting primary and secondary NAAQS. The 1978 NAAQS value of 1.5 $\mu g/m^3$ (quarterly maximum arithmetic average) as the primary and secondary standards for Pb derived from multimedia analysis of contributions to overall biomarkers of exposure, i.e., PbB, and the details are presented in a later chapter on air Pb regulation.

25.5.2 The Clean Water Act

Section 104 of the CWA provides sufficient authority to the EPA Administrator in the areas of research and data collection for the Agency to conduct and promote coordination of research, investigations, experiments, surveys, and studies "... relating to the causes, effects ... reduction ... and elimination of pollution." [§104(a)(1)]. The CWA [§104(c)] also allows cooperative administrative efforts on research with agencies such as the "... Secretary of Health, Education, and Welfare" (now Health and Human Services). For multimedia pollutants like Pb, there is the statutory wherewithal to look at Pb contained in environmental media other than ambient waters.

The CWA has a number of provisions for setting standards as they apply to Pb as a multimedia pollutant. These provisions centered on effluent guidelines, new source performance standards, and substances characterizable as toxicants. Section 301 of the Act, pursuant to §304, required that all point sources of pollution meet technology-based limits for their discharges. Categories of pollutants included conventional contaminants such as Pb, biological oxygen demand, suspended solids, fecal coliform bacteria, and pH. Such substances and conditions are to be controlled by the "best achievable control technology" in the case of existing waste stream systems while new sources require the Administrator to promulgate "standards of performance." In all of the Agency's options for effluent regulations under the CWA, costs are considered in effluent reductions.

The EPA Administrator, in addition to addressing effluent limits within a category-by-category regulatory approach, has had authority under §307 to impose uniform toxic pollutant limitations across all sources of such discharges. Such limits, analogous to provisions in the CAA, must contain an "ample margin of safety." Multimedia pollutants such as Pb come within the uniform toxicant effluent regulations of the CWA in that there is explicit authority in §307 to consider what controls may be derived from other Federal statutes. One serious complication, however, to multistatute approaches for a multimedia pollutant like lead is the differing statutory and regulatory requirement for "margin of safety" across at least three statutes: the CAA, the CWA, and the SDWA. References are made to "adequate margin of safety" and "ample margin of safety." These statutes involve different pathways for Pb exposures and toxic responses for a given pollutant, and the distinctions between "ample" and "adequate" have produced variable intensities of regulatory control. However, court review has held that "adequate" and "ample" do not require separate treatment or consideration.

Some recognition of multimedia contamination by waterborne toxic pollutants also resides in EPA's promulgation of pretreatment standards. These standards apply where discharge is to public wastewater treatment plants rather than directly into navigable waterways. Pretreatment standards, however, parallel the same overall requirements as for direct discharges, although pretreatment can be relaxed somewhat if the treatment processes in a public facility significantly reduce toxicant loadings in treated effluent. The language of these pretreatment standards recognizes that removal of toxicants like Pb from a sewage stream creates a solid waste disposal problem, with the potential for contaminating dispersal sites.

25.5.3 The Safe Drinking Water Act

The SDWA directs EPA's Administrator to issue national primary drinking water regulations using advice on recommended MCLs, as set forth in §§1412(b)(1)(A) and 1412(3). The recommending scientific body has been the U.S. NAS. At that time, Pb was regulated in drinking water at an MCL value of 50 μg/l, followed in 1991 by EPA promulgating a tap water action level expressed as a distributional statistic, i.e., a not-to-exceed value of 15 μg/l in 90% of a community's samples of tap water levels. Exceedences at the 90th percentile action value are to be dealt with through water treatment technologies or more extensive actions such as lead water pipe removals.

In its language and rationales, the SDWA has the clearest recognition of multimedia contaminants, in that the NAS in its recommendations for MCLs had explicitly recognized the potential for cumulative exposures of toxicants such as Pb. Elements of this explicit recognition included: (1) identification of sensitive groups of human populations at particularly high risk for exposures and/or toxicity; (2) the exposure to contaminants such as Pb in other media than water, such as in food, ambient air, occupational sites, resulting in a cumulative body burden; (3) synergistic effects arising from multitoxicant interactions when multiple toxicants are present. As a separate aspect of multimedia Pb regulation, the SDWA would appear to link with the U.S. FDA via the FDCA, mainly through the mechanism of regulation of water additives. In 1979, however, an interagency agreement between U.S. EPA and U. S. FDA gave U.S. EPA exclusive control over public drinking water supplies, with the U.S. FDA retaining jurisdiction over bottled water and water used in food or food processing once present in the processing facility.

25.5.4 The Food, Drug, and Cosmetic Act

Although research and analysis responsibilities that particularly address the impacts of multimedia pollutants such as lead via the FDCA were not explicit in FDA's authority under the Act, this component of the FDA's regulation of additives, foods, and drugs is implied. In 1971, addressing multimedia FDCA contaminants by more than only the FDA took the form—under §§406 and 505 of the Act—of establishing the National Center for Toxicological Research, a joint effort through an interagency agreement with the Department of Health and Human Services. A second shared activity is with the U.S. EPA for regulating pesticide chemicals, where EPA sets tolerances for substances which the FDA enforces.

Under §406 of the Act, the U.S. FDA promulgates maximum levels as "tolerances" within the language of the Act, applicable to "any poisonous or deleterious substance added to any food." Tolerance-setting responsibility for pesticides and related chemicals rests with the U.S. EPA (§406a). A finding by the FDA Commissioner that a substance requires a tolerance under §406 requires setting that tolerance "... to such extent as he finds necessary for the protection of public health ..." This applies to addition of substances to raw agricultural commodities. This section and its mandated authority recognize the aggregated toxic exposure problem with multimedia pollutants such as lead. For example, §406a language dictates setting a tolerance by the U.S. EPA, e.g., for pesticides, such that the tolerance recognizes "... the other ways in which the consumer may be affected by the same pesticide chemical or by other related substances that are poisonous or deleterious."

The FDCA empowers the FDA Commissioner to regulate and prohibit the presence of environmental contaminants in food. In some cases, this regulatory responsibility is shared with the U.S. EPA. Here, maximum levels permitted in food are in the form of tolerances for "any poisonous or deleterious substance added to any food" that is required for food production and cannot be avoided by good manufacturing practice. The term "adulterated" in the language of the FDCA, §402(a)(1), refers to those cases where the regulated substance can be avoided or eliminated through good practices.

Under §406, the FDA Commissioner's promulgation of tolerance limits must be "to such extent as he finds necessary for the protection of human health ... and the other ways in which the consumer may be affected by the same or other poisonous or deleterious substances." This mandate language clearly dictates viewing lead in food as part of a multimedia contribution to total Pb intakes and recognizing toxicant–toxicant interactions. Similarly, the responsibility of the U.S. EPA for setting tolerance limits for pesticide residues in foods is mandated to take into account all the ways a consumer can encounter the pesticide via multimedia exposures.

25.5.5 The Federal Insecticide, Fungicide, and Rodenticide Act

FIFRA, with respect to research, data collection, and analysis with any eye to multimedia lead considerations, enables the U.S. EPA to do research to achieve the legislation's ends, while §§20(a) and 20(b) direct the U.S. EPA Administrator to cooperate with other Federal agencies in executing the Act's provisions. FIFRA requires the Agency to register all pesticides, to set standards for the use of restricted pesticides, and to initiate cancellation or suspension for those substances posing unreasonable adverse effects on the environment.

Section 3(c)(5) requires pesticides' entry to the market have their chemical composition support claims for use, meet labeling requirements, and perform the claimed function without "unreasonable adverse effects on the environment." This requirement that there not be unreasonable risks of harm to humans or ecosystems foretells analysis for costs versus benefits and is assumed to include the cumulative effects of multimedia exposures to the pesticide.

EPA's legal actions in the earlier years of the act's existence principally involved those pesticides showing some evidence of carcinogenic action. Carcinogenicity is one toxicological behavior of paramount importance in addressing cumulative loading into environmental compartments encountered by human populations. Cumulative contributions in human exposures certainly call forth a multimedia dimension. Pesticides have primarily been those with organic chemical structures. Lead-based pesticides were phased out from heavy use in orchard crop production, e.g., apple orchards, long ago. Lead as a larvicide included its application as lead arsenate, the arsenate having recognized high carcinogenic potency in exposed humans at particularly high risk.

25.5.6 The Resource Conservation and Recovery Act

This legislation began with a multimedia dimension to the extent that RCRA addresses solid waste and solid waste volumes that increased with growth and use of pollution control technologies for contaminants in air and water. On the other hand, solid wastes dispersed to landfills and other disposal sites can result in contaminants leaching into drinking water supplies.

RCRA provides (§8001) that the U.S. EPA engage with other agencies to coordinate research, investigations, and studies relating to adverse health and welfare effects of releases of toxic contaminants in solid wastes into the environment and to eliminate such effects. Furthermore, §1006 of RCRA requires that there be integration of all provisions of the act with like provisions in other media-specific legislation to avoid duplication.

The portions of RCRA dealing with hazardous wastes, one of two categories of wastes under the act, are of interest for lead contamination in multiple media; such hazardous material is covered in §§3001–4009 of the legislation. The act regulates generation, transportation, storage, and disposal of hazardous wastes and is quite broad in its scope. The various aspects of operating methods, techniques, and practices as detailed in the act must function (§3004) so as to "... protect human health and the environment."

Priority setting under RCRA is one mechanism for addressing multimedia pollutants such as lead and lead compounds in solid wastes, but the feasibility of that approach may be problematic, given the large listings of hazardous chemicals within all three bodies of legislation with potential multimedia dimensions: RCRA, the CWA, and the SDWA. RCRA itself lists lead acetate, lead phosphate, lead subacetate, lead compounds not otherwise specified.

25.5.7 The Toxic Substances Control Act

Section 10 of TSCA gives the U.S. EPA Administrator broad authority to coordinate research activities with other agencies and assigns the lead to EPA for distributing data on the toxicology of substances including lead which may have a multimedia dimension. Also, §4(e) of TSCA establishes a multiagency panel to establish priorities within the act and thereby gives attention to the multimedia impacts of actions under the legislation. The act contains provisions for testing substances prior to release, if the manufacture of the material will present "... an unreasonable risk of injury to health or the environment." The U.S. EPA Administrator is also empowered to restrict or ban the use of chemicals if, as set forth in §6 of TSCA, "... there is a reasonable basis to conclude that the manufacture ... will present an unreasonable risk of injury to health or the environment ..."

Under TSCA, the EPA Administrator is to take into account methods available via toxicant controls in other media that may be more cost-effective. Cross-media least-cost comparisons necessarily bring into play the multimedia nature of pollutants such as lead and lead compounds.

25.6 INTERNATIONAL CONTROL EFFORTS FOR LEAD

The history of lead regulation outside the United States, particularly in European countries, differs both qualitatively and quantitatively across countries and political systems. A significant part of this differential history traces to the tenure of the problem. Attempts at lead regulation in the United States were largely a twentieth-century effort, particularly in the second half of the century. By contrast, lead control in other political jurisdictions, particularly in Europe, has had a multicentury history, and the nature of that control has differed markedly from American approaches. Control characteristics reflect the differing political and societal composition of the different countries, and even within non-U.S. jurisdictions over time with the transition from principalities, kingdoms, colonies, and such to more modern governmental forms.

International positions on environmental toxicants like lead have also been translated into regulatory devices showing more variety than in the United States. For example, the WHO, through such collective mechanisms as environmental health criteria documents, issues guidelines in the form of recommended levels of lead and other toxicants in various environmental media, not regulations. Regulations are implemented downstream within the relevant ministerial jurisdictions internationally, e.g., the European Union. WHO guidelines have been promulgated for lead in air in Europe (WHO, 1987, 2000), for lead in drinking water for all member countries (WHO, 1993), for lead in food via the WHO/FAO provisional tolerable weekly intakes (FAO/WHO, 1993), and for total intakes and exposures via multimedia sources and pathways as contained in WHO's Environmental Health Criteria document (No. 165) for inorganic lead (WHO, 1995).

25.6.1 Some Historical Differences in Pb Control in the United States and Internationally

The role of lead in both commercial and public health histories is a much larger chronicle internationally than in the comparatively younger United States. This is particularly the case for European nations whose histories date to ancient empires and parts of empires, to include Greco-Roman times. Previous chapters presented the details of Pb's history as a contaminant and early poisoning records from exposure and these are not repeated here.

The earliest recordings of lead control in the United States or its founding colonies began in the eighteenth century. McCord, in a series of articles in 1953 and 1954 in the journal *Industrial Medicine and Surgery*, described lead exposure and poisoning as it existed in early America (McCord, 1953a,b, 1954). It is worthwhile to note that the first regulatory action against lead in either the colonies or the United States was enacted by the Massachusetts Bay Colony government in 1723. This action banned certain parts that were made with lead from rum distillery equipment, so that rum distilleries would not produce a lead-contaminated product. One impetus for this action, in the account by Benjamin Franklin (noted by McCord, 1953a), was the widespread lead poisoning among North Carolinians who consumed rum made in Massachusetts. Franklin also drew attention in correspondence (McCord, 1953a) to the problem of lead poisoning from hot lead type used in the printing trade and to the problem of poisonings from lead water pipes. That is, poisonings were recognized as occurring both within and outside the workplace.

Given the increased interest in the sources and nature of lead poisoning in Europe and the Colonies in the eighteenth century, one might expect that public awareness would have increased considerably, at least among the literate segments of the population. For reasons given earlier, this may have occurred to some extent, but the bulk of the information remained in the medical community.

25.6.2 Lead in the Workplace Versus the Ambient Environment: Differences in the International and American Perspectives

Regulating lead in the general human environment was preceded by associated controls on worker exposures. Occupational regulatory history for Pb was in place and in a structural framework in Europe and elsewhere first, before it was in place in the United States. In European countries, the focus on occupational exposures and their consequences for human lead toxicity in general began earlier and entered the occupational and general clinical literature earlier than in the United States. The extensive occupational medicine report by the French medical researcher Tanquerel des Planches, first published in 1839 in France as *Traite des Maladies de Plumbe ou Saturnines* and subsequently edited and translated into English by Dana in 1848, covered more than 1,200 cases, most of whom were exposed to lead in the workplace. As described earlier, this work was a compilation of lead poisoning cases and summary assessments of signs and symptoms from occupational lead exposure of workers in a variety of settings.

This report also began the practice of approaching lead poisoning in a systematic way, a considerable improvement over the fragmented and inconsistent earlier attempts at defining lead poisoning as a recognizable disease entity. Des Planches' publication and others appearing somewhat later in the nineteenth century did not by any means guarantee that technical information reached or was used by the general public. It did, in all likelihood, educate the industrialist and the occupational hygiene community to the nature of the hazards associated with lead production and lead use.

The early lead poisonings ascribed by Tanquerel des Planches to lead paint production and use probably gained the attention of those who would provide alternatives to lead paints as consumer products later in the nineteenth century. The nonleaded paint producers would market their substitutes by touting their low hazard to health (see NAS/NRC, 1993). Finally, the French treatise did provide relatively useful information on which occupational medicine could build as the century progressed.

Tanquerel des Planches' study group of more than 1,200 workers included 800 who were either involved in lead pigment manufacturing or were house painters. It was this awareness of occupational lead hazards in the first half of the nineteenth century that laid the groundwork for international regulatory action on general uses of leaded paints in the early twentieth century, as detailed in the next chapter (ILO, 1927). The history of lead paint as an occupational hazard in the nineteenth century and even earlier has been told by Stevenson (1949), as cited and discussed by Fee (1990). This source notes the recognition of lead paint poisoning and there were calls for its prohibition in commerce by various individuals and organizations.

Occupational lead exposures in the United States as recorded in the nineteenth century largely appeared to be disconnected from any community histories of environmental lead poisonings, lacked any governmental regulatory framework for Pb itself, and appeared to be more diverse than was the case in Europe. For example, the extractive lead industry and its mining, milling, and smelting sectors in the late nineteenth and early twentieth centuries in the United States employed a large workforce that also had widespread and largely unmanaged chronic lead poisoning (Derickson, 1987; McCord, 1953b).

In the rapidly developing western United States in the last quarter of the nineteenth century, the rapid and unregulated growth of extractive metal industries occurred with a rash of severe occupational lead poisonings among lead and other metals miners. The frequency of such poisonings was especially high among those mining lead, lead–silver, and lead–zinc ores. Lead in the western United States ore bodies was often present in geochemical forms with relatively high bioavailability to humans, i.e., a high uptake rate of lead into the bloodstream from receiving compartments such as the lung or the gastrointestinal tract (Mushak, 1991). This particularly included lead carbonate ore (cerussite) or ores having lead carbonate mixed with the less bioavailable form of lead ore, galena (lead sulfide). Chapter 11 provides a tally of occupational Pb poisonings in different parts of Utah from approximately 1870 to 1900 (McCord, 1953b).

Extractive industry operations in the West included milling and smelting as well as mining. Beyond miner exposures at the mining phase of these operations, workplace exposures would have occurred in the milling and smelting steps. The old jig and stamping method of ore separation would have produced dust exposures, while the flotation method yielded mill tailings and other process wastes which would have been toxic in the workplace. Similarly, early smelters would have produced both workplace and environmental contaminations. The magnitude of these aspects of occupational exposure was poorly recorded and remains unquantified.

Within the extractive industry world, lead poisoning appeared to have been well established as an occupational hazard despite absence of workplace regulations, at least among miners, in the American West. Derickson (1987) noted that in certain Western mining areas, and particularly in the Coeur d'Alene valley of the Idaho Panhandle, miners' unions responded to the hazards of lead–silver mining by establishing a hospital system for treating mine injuries and presumably lead poisoning.

Environmental damage associated with smelter lead emissions in the nineteenth century was recognized at least on a local level and would have produced some level of regional awareness. The NAS/NRA (2005) report on the scope, consequences, and remediation of the historical environmental lead contamination of the Coeur d'Alene River Basin in Idaho noted multiple legal actions for damages taken by farmers and other residents along the river against Bunker Hill mill and smelter owners in the nineteenth century.

25.7 THE PRECAUTIONARY PRINCIPLE AND ITS PLACE IN LEAD REGULATION GOING FORWARD

Lead already in the human environment has typically been controlled by legislative and executive means in the United States, Canada, and Europe through regulatory mechanisms focused on lead in various media. An earlier discussion touched on the issue of how regulation is to apply to lead and other substances proposed for market entry, i.e., use of proactive bars to entry of untested substances. A second and related question has to do with the status of scientific and public health uncertainty in guiding or impeding regulatory initiatives. These twin problems in regulation of contaminants such as lead have prompted the development of the Precautionary Principle in various parts of the regulatory world. Percival (2006) has presented a legal analysis of the principle with respect to environmental law and regulation.

The Precautionary Principle was explicitly framed in 1992 via the Rio Declaration by representatives of 178 nations as:

> "Where there are threats of serious or irreversible damage, lack of full scientific certainty shall not be used as a reason for postponing cost-effective measures to prevent environmental degradation." [Rio Declaration on Environment and Development, Principle 15, U.N. Doc. A/CONF. 151/26 (Aug. 12, 1992)].

The principle has been widely well received and has been formally endorsed by the European Union as part of the body of European Commission directives (CEC, 2000).

Of particular interest, lead emissions and their post-release, pervasive adverse effects on the broad biosphere have been argued as confirming the need for a proactive mechanism for primary preventive controls on toxic substance introduction and marketing without adequate certification as to health and safety requirements. Two examples can be used to illustrate consequences of the absence of regulatory actions or rulings on lead releases or related actions (Percival, 2006).

Percival (2006) noted that U.S. legislative initiatives on lead in diverse human exposure media sources in the 1970s and 1980s were enacted within a largely precautionary framework. An element of this precautionary, proactive stance was its pursuit in the face of scientific uncertainty. Notable, and mentioned earlier in this chapter, §§108 and 109 of the CAA anticipated and provided for scientific uncertainties in public health protection.

A second instance of the principle, that of a careful judicial stance in rulings on lead regulation-based lawsuits, arose in connection with EPA rulemaking on reducing air lead levels in 1978 (Ethyl Corp. v. EPA, D.C. Cir., 1978). The Federal DC Circuit Court, the venue for regulatory legal challenges, ruled against the Ethyl Corporation's challenge to TEL curbs by EPA earlier in the 1970s, noting:

"Where a statute is precautionary in nature, the evidence difficult to come by, uncertain, or conflicting because it is on the frontiers of scientific knowledge, the regulations designed to protect the public health, and the decision that of an expert administrator, we will not demand rigorous step-by-step proof of cause and effect. Such proof may be impossible to obtain if the precautionary purpose of the statute is to be served" (Ethyl Corp. v. EPA, 541 f. 2d 1, 6, 11, 55, D.C. Cir, 1976).

REFERENCES

Council of the European Communities, 2000. Communication from the Commission on the Precautionary Principle. Available at: <http://europa.eu.int/comm/dgs/health_consumer/library/pub/pub07en.pdf>.

Crocetti, A.F., Mushak, P., Schwartz, J., 1990a. Determination of numbers of lead-exposed U.S. children by areas of the United States: an integrated summary of a report to the U.S. Congress on childhood lead poisoning. Environ. Health Perspect 89, 109–120.

Crocetti, A.F., Mushak, P., Schwartz, J., 1990b. Determination of numbers of women of child-bearing age and pregnant women by areas of the United States: an integrated summary of a report to the U.S. Congress on childhood lead poisoning. Environ. Health Perspect 89, 121–124.

Derickson, A., 1987. To be his own benefactor: the founding of the Coeur d'Alene Miners' Union Hospital, 1891. In: Rosner, D., Markowitz, G. (Eds.), Dying for Work: Workers' Safety and Health in Twentieth-Century America. Indiana University Press, Bloomington, IN, pp. 3–18.

D'Itri, P.A., D'Itri, F.M., 1977. Mercury Contamination: A Human Tragedy. John Wiley & Sons, New York.

Fee, E., 1990. Public health in practice: an early confrontation with the 'silent epidemic' of childhood lead poisoning. J. Hist. Med. 45, 570–606.

Food and Agricultural Organization/World Health Organization, 1993. Evaluation of Certain Food Additives and Contaminants. Forty-First Report of the Joint FAO/WHO Expert Committee on Food Additives. Technical Report Series 837. World Health Organization, Geneva.

International Labour Office, 1927. White Lead. Data Collected by the International Labour Office in Regard to the Use of White Lead in the Painting Office. Studies and Reports: Series F, Industrial Hygiene, No. 11. Geneva, Switzerland.

McCord, C.P., 1953a. Lead and lead poisoning in early America: Benjamin Franklin and lead poisoning. Ind. Med. Surg. 22, 393–399.

McCord, C.P., 1953b. Lead and lead poisoning in early America: lead mines and lead poisoning. Ind. Med. Surg. 22, 534–539.

McCord, C.P., 1954. Lead and lead poisoning in early America: the lead pipe period. Ind. Med. Surg. 23, 27–31.

Mushak, P., 1991. Gastro-intestinal absorption of lead in children and adults: overview of biological and biophysico-chemical aspects. Chem. Speciation Bioavail. 3, 87–104.

Mushak, P., Crocetti, A.F., 1989. Determination of numbers of lead-exposed American children as a function of lead source: integrated summary of a report to the U.S. Congress on childhood lead poisoning. Environ. Res. 50, 210–229.

Mushak, P., Crocetti, A.F., 1990. Methods for reducing lead exposure in young children and other risk groups: an integrated summary of a report to the U.S. Congress on childhood lead poisoning. Environ. Health Perspect. 89, 125–135.

Mushak P., Schroeder C., 1980. Multiple Media Pollutants. Report to the National Commission on Air Quality. Contract No. 23a-AQ-6981. Discussed and cited in the 1981 National Commission on Air Quality report: To Breathe Clean Air, pp. 2.1.9, 2.2–4, 3.1–30, 3.1–34.

Mushak, P., Davis, J.M., Crocetti, A.F., Grant, L.D., 1989. Prenatal and postnatal effects of low-level lead exposure: integrated summary of a report to the U.S. Congress on childhood lead poisoning. Environ. Res. 50, 11–26.

National Academy of Sciences: National Research Council, 1993. Measuring Lead Exposure in Infants, Children, and Other Sensitive Populations. National Academy Press, Washington, DC.

National Academy of Sciences: National Research Council, 2005. Superfund and Mining Megasites. Lessons from the Coeur d'Alene River Basin. National Academy Press, Washington, DC.

Percival, R.V., 2006. Who's afraid of the precautionary principle? Pace Environ. Law Rev. 23, 21–61.

Schroeder C., Mushak P., 1980. Multiple Media Pollutant Exposures and Their Regulation. Legislative Treatment of Multi-Media Pollutant Exposures. Report to the National Commission on Air Quality. Contract No. 23a-AQ-6981. Discussed and cited in the 1981 National Commission on Air Quality report: To Breathe Clean Air, pp. 2.1.9, 2.2–4, 3.1–30, 3.1–34.

Stevenson L.G., 1949. A History of Lead Poisoning, Ph.D. Dissertation. The Johns Hopkins University, Baltimore, MD.

Tanquerel des Planches L., 1839. Lead Diseases Translated. edited Dana, S.L. 1848. Daniel Bixby and Co., Lowell, MA.

U.S. Agency for Toxic Substances and Disease Registry, 1988. The Nature and Extent of Lead Poisoning in Children in the United States: A Report to Congress. U.S. Department of Health and Human Services, Washington, DC.

U.S. Environmental Protection Agency, 1978. National Primary and Secondary Ambient Air Quality Standards for Lead. 43 FR 46246-46263, October 5, 1978.

U.S. Environmental Protection Agency, 2006. Air Quality Criteria for Lead, vol. 1. Report No. EPA/600/R-05/144aF. Environmental Protection Agency, Research Triangle Park, NC.

World Health Organization, 1987. Air Quality Guidelines for Europe. Lead. WHO Regional Publications: European Series No. 23. Regional Office for Europe, Copenhagen, Denmark.

World Health Organization, 1993. Guidelines for Drinking Water Quality, second ed., vol. 1. World Health Organization, Geneva, Switzerland, (Recommendations).

World Health Organization, 1995. Environmental Health Criteria 165: Inorganic Lead. International Programme on Chemical Safety, Geneva, Switzerland.

World Health Organization, 2000. Air Quality Guidelines for Europe. Lead, second ed. WHO Regional Publications: European Series No. 91, Regional Office for Europe, Copenhagen, Denmark.

Chapter 26

Lead Regulation and Regulatory Policies: Lead in Paint

26.1 A BRIEF HEALTH PERSPECTIVE ON LEAD IN PAINT

The interactive, feedback-loop relationships for lead among scientific research, risk assessments, and regulatory initiatives graphically depicted in previous chapters would predict that lead regulation in whatever environmental medium in the human environment arises from demonstrated threats to health shown by both scientific research and health-risk assessment. Consequently, discussions of lead in this and subsequent chapters require prefacing with brief perspectives on medium-specific lead exposures and associated human health hazards. This perspective differs from the previous detailed discussions in the health effects chapters in this book, Chapters 11–19.

Lead paint, as it enters the human environment and in its subsequent environmental behavior, is an illustrative model for regulation considerably different from other lead-containing media or pathways to human exposures. While lead enters regulated media such as ambient air, drinking water, ambient waters, and food through a mechanism involving lead release and contamination, lead in paint presents itself as a medium that is potentially a contaminant in its entirety. In the former instances, releases from contamination sources such as air and water Pb arise from some technological or industrial waste stream linked to, and following, lead production or consumption. These activities were presented in the early chapters.

Lead paint applied to a myriad of structural surfaces began environmental life as a marketed commodity for the specific purpose of applications to those structural surfaces. Lead paint enjoyed an extensive commercial history in its role as a utilitarian commodity. Lead released from lead paint surfaces because of both intrinsic properties such as surface chalking (once promoted as an economic plus) and extrinsic properties such as relative surface instability producing flaking and peeling could in theory eventually result in all or virtually all of the material being consumed via these modes of dispersal. For example, chalking lead paint surfaces release very fine-sized lead paint particles that then add to existing dusts, producing more hazardous dusts with an increased Pb level or loading.

Trace Metals and other Contaminants in the Environment, Volume 10
ISSN: 1875-1121 DOI: 10.1016/B978-0-444-51554-4.00026-2

There are two parts to lead paint as a public health issue: occupational health issues arising from Pb exposures of workers producing, using, maintaining, or removing lead paint, and pediatric health concerns from Pb exposures of young children once the substance is applied to surfaces in residential or other surroundings.

Lead paint was known as a major occupational poisoning hazard long before it became an environmental health hazard from application and ultimate deterioration of lead paint. In terms of history, lead paint production and use was known as an occupational hazard in Europe and elsewhere before its recognition in the United States. This body of information first emerged in the nineteenth century.

The limited literature on lead paint production or use in nineteenth-century America was rooted in the occupational health literature, with a particular focus on lead pigment workers and some concern for house and commercial painters (see, for example, Hamilton, 1911; Knerr, 1992). Of particular concern to industrial hygienists and occupational physicians and labor activists were the significant rates of clinical lead poisonings among workers producing white lead by the Dutch Process (Hamilton, 1911; Knerr, 1992).

This relatively crude and dusty process, entailing a highly bioavailable form of lead that was also readily inhaled, was particularly hazardous in terms of poisoning rates and likely poisoning of family members through bearing home dusty white lead on their clothing and on shoes. Comparatively, Hamilton (1911) reported that American white lead facilities were more hazardous than similar European operations, but she observed subsequent improvement in worker hygiene in the former over time.

The more significant aspect of lead paint as a public health matter, although of more recent vintage, attends the role of lead paint in childhood lead paint poisoning. While the earliest chronicles of childhood poisonings emerged in Australia, the majority of the problem subsequently resided in the United States.

The findings of Gibson et al. (1892) as first described in the 1890s heralded lead as a child environmental health problem. This Australian physician and his colleagues (Gibson, 1904; Gibson et al., 1892) were the first to establish lead paint poisoning in young Australian children due to ingestion of weathered (chalking) exterior paint used in Queensland. The extremely demanding climate in terms of lead paint stability produced rapid deterioration of the painted surface and ready dispersal of paint lead in the form of small flakes or, more typically, very small particles of lead released from deteriorating lead-painted surfaces such as veranda railings.

In the United States and elsewhere, lead exposures severe enough to produce clinical Pb poisoning typically involved lead paint. One critical factor in this source-poisoning relationship was the ubiquity of lead paint. Historically, not only did lead occupy a central economic position over the centuries via its myriad commercial forms, but also lead paint products were increasingly popular for surface coverings and were successful commercially. Lead paint had a number of technical virtues as a surface covering. It had the reputation of durability, and the leading brand touted the characteristic of a persisting "fresh" look. This appeared to be due to the constant weathering action on the surface and ongoing shedding of surface layers as fine paint dust particles to surrounding surfaces.

Historically, it is not surprising that deteriorated and deteriorating lead paint-covered surfaces in the older U.S. housing in Northeastern and Midwestern cities began to translate into childhood lead paint poisoning cases by the 1910s–1920s. McKnight et al. (1989), citing Mattiello (1942), reported that the first U.S. factory to produce white lead pigments was established in 1804 in Philadelphia, PA. Knerr (1992), in his corporate history of Eagle-Picher Industries (long a major lead pigment and other lead products producer), traced the evolution of the white lead pigment industry for paints in the Midwest to such locales as Cincinnati, OH. This city contained one of three white lead paint pigment production facilities west of the Allegheny Mountains, the Cincinnati Manufacturing Co., which opened in 1815. Two more plants opened soon after, with only one surviving into the 1820s. By 1850, there were 51 U.S. factories producing white lead paint pigment. This number burgeoned after the Civil War. Eagle White Lead Co., one of the corporate predecessors to Eagle-Picher Industries, was established in Cincinnati in 1867.

In the United States, the early decades of the twentieth century produced the first clinical inklings of what would prove to be the enormous childhood lead poisoning problem arising from lead in paint. This period witnessed an international effort to control or eliminate use of lead paint, an effort notable by the United States' absence. There were also continuing efforts to control and characterize lead exposures in the workplace.

Although Gibson et al. (1892) laid some of the groundwork for the clinical identification and treatment of childhood lead poisoning at the close of the nineteenth century, it took some time for the American medical community to recognize and accept the severity of the childhood lead poisoning problem, especially among the poor and disadvantaged populations living in deteriorated housing, much of it marginal rental housing. There were various reasons for this. First, physicians were not dealing with signs and symptoms peculiar to lead. Numerous causes of individual effects of lead such as severe nervous system toxicity, or kidney damage, or hematotoxic effects can exist from other causes (NAS/NRC, 1993; U.S. CDC, 1978, 1985, 1991, 2005). Since physicians in this early period of the problem had little routine means for measuring lead exposure or body lead burdens, ready testing for lead poisoning was not at hand.

Second, those principally afflicted were often but not exclusively outside the mainstream of pediatric practice. It is of no surprise that the earliest findings of childhood lead poisonings were in urban poor children in older and larger cities like Baltimore, Boston, and Philadelphia. This permitted a medical and public health perception that this form of lead intoxication was a contained medical problem. Closely related to that point was the fact that lead poisoning was not a reportable disease, locally or nationally. Consequently, there was no oversight or review of any increase in the problem beyond the individual clinician's practice experience if such poisonings were not published in medical journals. Third, the earliest definition of lead poisoning was a rather draconian one, the finding of coma and convulsions in children who subsequently died from the high exposures. There were few early warning signs of onset of potentially fatal lead paint poisoning and no screening framework by which to identify and intervene against such poisonings.

Finally, it was the nature of the lead paint problem, certainly in the United States, that the problem got worse and more noticeable over time with lead-painted-surface deterioration and involved a lead source that was pervasive in the environment of the youngest children, infants, and toddlers. Housing deterioration and increased paint lead access for children because of this availability increased over time and was amplified with the changed demographics of America's inner-city neighborhoods. At the same time, cribs and furniture with painted surfaces were usually present, and chewing and gnawing those surfaces and ingestion as typically happened would produce lead poisoning.

In the early decades of the twentieth century, an international effort to ban or otherwise control the production, sale, and use of leaded paints gained momentum. Building on the Tanquerel des Planches and later reports, the International Labor Office (ILO, 1927) developed a control mechanism for lead paint in the form of the White Lead Treaty, articulated by the 1921 White Lead Convention. The initial signatory list in the 1920s included a number of nations, but the United States was not among them. Subsequently, other nations signed the pact. The treaty had a number of shortcomings. It focused on interior paints, had no retroactivity (which would have mandated lead paint abatement), and had a rather high cutoff of 2% (20,000 parts-per-million (ppm)) lead content. The protocol was confined to only two lead pigments, the carbonate and the sulfate. Nonetheless, the interest and attention directed to lead paints in these early decades by labor organizations helped focus attention on the lead problem outside the medical and scientific communities.

26.2 LEAD PAINT REGULATION IN THE UNITED STATES

The exposure and toxic injury consequences resulting from lead paint as a public and child health problem first elicited highly fragmented and relatively localized regulatory responses. This reflected in part the perception that the lead paint problem was confined to certain U.S. communities and states and therefore was best dealt with by local initiatives. Table 26.1 provides summary information on some local, state, and Federal regulations for

TABLE 26.1 Significant U.S. Regulations or Guidelines for Lead in Paint

Statutory/ Regulatory Initiatives	Codified Law or Regulation	Years	Key Features
A. Early non-Federal actions			
City ordinances: Baltimore, MD	Regulation 17 and predecessors, Baltimore City Ordinance 384	1940s–1950s	Banned use of Pb paint for dwelling unit interiors
City ordinances: Philadelphia, PA	Amendments to various existing city ordinances	1950–1966	Identification of lead paint or other Pb sources for identified Pb poisoning cases; called for Pb paint or other Pb source removal; enforcement spotty and difficult
City ordinances for other U.S. cities	Various ordinances for Chicago; Cincinnati, OH; Jersey City, NJ; New Haven, CT; New York City; St. Louis, MO; Washington, D.C; Wilmington, DE	By the early 1970s	Various levels of required Pb paint removal
Massachusetts state statutes	[105 CMR 460.050] for 1990; earlier regulations in 1973	1973, 1990	Collectively, Pb paint abatement in older homes, strict liability for rental property owners, inspections when poisoning found; screening of children 9 months to 4 years
Rhode Island state statutes and amendments	R 23-24-6PB	1992, 1997 amendments	Voluntary provisions for Pb paint abatement, no civil or criminal penalties
Maryland state statutes and amendments	MD. ANN. CODE PUB. GEN. LAWS 1957 §§2-301, 2-307	1957	Financial assistance for Pb abatement; later statutes prohibited use of Pb paint for interiors; Pb poisoning made a reportable disease; financial penalties to violators

(*Continued*)

TABLE 26.1 Significant U.S. Regulations or Guidelines for Lead in Paint—(cont.)

Statutory/ Regulatory Initiatives	Codified Law or Regulation	Years	Key Features
Lead paint producer voluntary initiatives	American Standards Association Directive Z66.1-1955	1955	Voluntary producer compliance with 1% Pb content focused on painted surfaces likely chewed by children; multiple problems with this standard (see text)
B. Federal regulatory initiatives			
U.S. LBPPPA [PL 91-965], later amendments	84 Stat. 2078, U.S. C. 4821 et seq.	1971; 1972, 1973, 1976 amendments	Act prohibited lead-based paint (>1% Pb) in residential structures built, owned, or maintained by the Federal government; authorized funding for mass screening program
U.S. FDA briefly regulated Pb paint (3/1972) under Federal Hazardous Substances Act; Federal FDC Act	§2(f)1(A)74 STAT. 372, 374, Part 191; §701 e, f, g, 52 STAT. 1055-56; 70 STAT. 919 et seq.	1972–1974	The Act mandated Pb paint content 0.5% to 12/1974; then, in 1976, to 0.06%. Implemented statute and amendments
U.S. LBPPPA amended by the National Consumer Health Information and Health Promotion Act [PL 94-317]	90 STAT. 705-706	1976	Mandated lowering Pb in paint and painted children's products
U.S. CPSC began regulating Pb in new paints/painted surfaces under the Consumer Product Safety Act [PL 92-53]	86 STAT. 1231 and amendments; §30(d), 1978 Part 1303	1976	Banned new Pb paint with >0.06% Pb and new painted surfaces >0.06% on toys, furniture

(Continued)

TABLE 26.1 Significant U.S. Regulations or Guidelines for Lead in Paint—(cont.)

Statutory/ Regulatory Initiatives	Codified Law or Regulation	Years	Key Features
LBPPPA amended by the 1987 Housing and Community Development Act (§556) and 1988 Stewart B. McKinney Homeless Assistance Amendments Act [PL 100-628]	102 STAT. 3224	1987–1988	Mandated series of regulations by the U.S. HUD, including reports to Congress (1995) and interim and final guidance for existing Pb paint hazards
Title X, 1992 Housing and Community Development Act	§1017, Title X	1992 et seq.	Mandated a number of requirements involving both the U.S. HUD and the U.S. EPA. See text for details.
TSCA amendments—Lead: Renovation, Repair, and Painting Program	§402(c)(3) 40 C.F.R. Part 745	2008 to present	Mandated set of regulations covering pre-1978 U.S. dwellings that are owner occupied and COFs. See text for details.
Consumer Product Safety Improvement Act (CPSIN), [PL 110-314]	§101(f); U.S.C. §1278a; 16 C.F.R. §1303.1	2008 to present	Lowered the permissible levels of Pb in children's toys and other products and Pb in paint on toys and other products; lowered Pb content to 90 ppm total for paint (August 14, 2009); lowered Pb level in children's products in phases, to 100 ppm in 2011

U.S. lead paint control efforts. Regulation of lead paint exposures and control of poisonings, furthermore, focused on the owners of properties with a lead paint problem, i.e., numerous landlords dispersed throughout America's cities. In some cases, U.S. cities with a growing childhood lead poisoning problem confined their response efforts to epidemiological studies, case finding, and community education programs for poisoning prevalence and the various demographic and socioeconomic correlates thereof (Lin-Fu, 1982). Such efforts typified the lead poisoning responses in Chicago, IL, and in New York City.

Other cities, such as Philadelphia, PA, and Baltimore, MD, combined case finding and public education with attempts at abatement or other lead paint remediation via local enforcement mechanisms. Such efforts in the 1950s and 1960s gave way to Federally assisted massive screening and various levels of intervention in the 1970s and later with the enactment of the Federal LBPPPA in 1971 [42 USC 4821-4846 and subsequent amendments]. A good case study of this early, localized approach to lead paint would be the efforts at regulation in the city of Baltimore, MD, notably in the 1930s through 1960s. Fee (1990) has provided a useful chronology of this city's efforts at childhood lead paint poisoning control.

Illustrative examples of efforts at lead paint exposure control at the state level are those implemented in states in the Northeastern United States. State efforts at preventing or reducing childhood lead paint poisoning are of much more recent vintage than were those of U.S. cities. This is the case with laws in Massachusetts and Rhode Island. As with U.S. cities, not all attempts at lead paint poisoning control at the state level were equally effective or as stringently enforced across the board.

26.2.1 Early Lead Poisoning and Lead Paint Control Initiatives: The Baltimore, MD, Case

The first American clinical references to childhood lead poisoning were the studies of Thomas and Blackfan (1914) and Blackfan (1917) of Baltimore children treated at the Johns Hopkins Hospital. Later, other reports began to appear from different cities. At that time, lead poisoning in Baltimore and elsewhere was defined in rather stark terms, i.e., the presence of coma or convulsions and other severe symptoms, often leading to a fatal outcome. Children surviving any lead poisoning episode were considered to be free of further toxic harm. In Baltimore, these and later cases involved exposure to lead paint in older housing in the poorest sections of the city occupied by predominantly low-income African-Americans.

Fee's (1990) history of the lead paint problem in Baltimore during the 1930s–1960s illustrates quite well the interplay among a lead source—lead paint—a population affected by this source in the form of severe lead poisoning in children, the public health and political responses to this threat, and the role of the print and radio media in Baltimore in both informing and educating the local residents to an environmental health problem.

One underlying reason for Baltimore being the site of America's earliest attention to a widespread childhood lead poisoning problem was attributed by Fee to the city's commissioner of health, Huntington Williams, appointed in 1931. Williams had, in addition to a good grounding in medicine and public health, the support and encouragement of influential members of the community and its elected officials, especially Mayor Howard Jackson. Williams and his medical colleagues realized that lead paint was the problem behind the childhood lead poisoning then being reported in the city's main clinic. The commissioner apparently also recognized that this source of poisoning was associated with a number of local political, medical, and economic issues among various influential quarters in the city.

Among various alternatives for responding to the growing epidemic of lead paint poisoning on a city-wide basis, the health department under Williams elected to go the route of local media attention and both community awareness and education to deal with the problem. This approach would, understandably, be of only limited effectiveness. It was an approach that, according to Fee, was criticized by other public health experts as weak and unlikely to get at the root problem, which was the persisting presence of increasingly deteriorating lead paint in deteriorating housing. Nonetheless, public awareness and education were sought effectively by combined use of a weekly radio health program, health messages via the print media including the local newspapers, and production of a monthly department newsletter, *Baltimore Health News*. A popular pamphlet, "Lead Poisoning in Children—A Disease You Can Prevent," had a printing of 20,000 copies, distributed in well-baby clinics and other public locations.

Baltimore/Commissioner Williams' approach accepted the now-dubious premise that appropriate family and child exposure avoidance behavior in highly hazardous residences containing lead paint would be adequate to minimize risks to children exposed in the structures. The Baltimore approach likely assured that high-risk neighborhoods and much of the rest of the city would know about lead paint and the lead poisoning of young city residents and would have comprehended at least some ways in which lead exposures and associated poisoning could be minimized. It would not, however, guarantee that awareness and education programs would eradicate the problem.

The epidemiological course of the lead paint poisoning epidemic in Baltimore suggested that public awareness and risk population education are only partial solutions. Part of the rationale for Baltimore's approach was that the now-accepted normal mouthing activity and ingestion of lead paint materials by infants and toddlers orally exploring their environment were considered then to be "perverse" habits of these children. If these abnormal behaviors could be prevented by whatever means, the problem would abate. A second dictum of the city's logic was that, once parents and other care providers were informed of the hazards of lead paint in their environment, it was then principally their responsibility. This notion that education about the problem of lead poisoning rather than removing the source of the lead poisoning is an effective approach has had great staying power. It has persisted as an argument against comprehensive lead source abatement to the present.

The roles of the local media in public awareness and education during this period were several. Media outlets were regularly employed for passing along health information. Secondly, lead poisoning outbreaks gained considerable press attention. This was certainly the case in an outbreak of lead poisonings among Baltimore children connected with impoverished families' that burnt lead battery casings as fuel in cold weather during the depths of the Depression. Here, newspapers and local radio helped to spread the warnings in an effective way.

During the tenure of health commissioner Williams and his relatively active, *albeit* secondary, approach to lead poisoning prevention, poisonings continued to occur and in fact were getting worse. That could have meant an ineffective approach or simply that the more attention paid to the problem, the more the problem was recognized. For example, the city began the intensely publicized practice of providing cost-free lead poisoning screening throughout the city along with sample testings of paint suspected of containing lead. Fee (1990) provided evidence that convincingly shows this was at least partly the case for Baltimore's lead poisoning statistics versus the rest of America. Baltimore had a disproportionately high portion (about 24%) of the nation's lead poisoning cases but only 0.7% of her population. The risk group demographics of Baltimore resembled too closely those for many other older American cities to have produced true large differences in poisoning prevalences and incidences. It was also likely the case that the overall problem was getting worse as the city's old housing stock with the most hazardous lead paint problems continued to deteriorate and landlord neglect took a cumulative toll on housing quality. We cannot rule out, from the available information, that the growth of the problem might have been much worse except for the city's efforts.

Baltimore's health apparatus also used the media and public attention when implementing city ordinances to begin regulating the poisoning problem. A 1941 ordinance specified that when lead poisoning was identified in a unit, the owner would be required to remove the lead paint in it. This statute was not enforced, since opponents argued lead paint removal was expensive for landlords with marginal, low-rent housing. Furthermore, enforcement meant confronting powerful economic interests in the city. Limited ordinance enforcement was part of the reason the health apparatus turned to more public education, including a popular radio program, "Not Fit to Eat," which explained lead paint hazards and asked listeners, especially parents, to pass along the information to their family and friends. A second approach was to widely recommend the use of lead-free paint to cover deteriorating lead paint surfaces. This was of limited effectiveness since many tenants could not afford the paint, and landlords were reluctant to spend money on the paint. There was also no way to know whether the available paint was lead-free or not since this information was not on labels at the time.

A modestly improved 1948 ordinance required landlords to remove lead paint from units where a lead-poisoned child had already been identified. Any level of subsequent abatement was piecemeal, however, and this was commonly the case in a number of U.S. cities in this period. That is, only areas where paint was seen to be peeling and flaking were to be abated. Comparatively, landlord response rates were relatively high in Baltimore, although deteriorated paint residues such as postabatement peeling and flaking and interior lead paint dusts on floors, window wells, and window sills were ignored in the ordinance. In June 1951, Baltimore adopted Regulation 17 as part of Baltimore City Ordinance 384, wherein "No paint shall be used for interior painting of any dwelling or dwelling unit or any part thereof unless the paint is free of any lead pigment." This local initiative was proactive in trying to avoid the lead paint problem on new surfaces versus reactive approaches in the 1948 ordinance. Major problems with this 1951 ordinance remained in that the mere application of theoretically lead-free paint over poorly prepared lead-painted surfaces did little to avoid continued peeling and flaking of older layers. Second, no container labeling as to safety of contents *vis-a-vis* lead content was required. One did not know what new paints had or did not have lead.

In the 1950s and 1960s, the city would turn away from its principal reliance on public education to prevent lead poisoning and would turn more to regulating paint lead entering the local commercial pipeline, enforcing housing quality ordinances by holding landlords more accountable and liable, and using more screening data to identify hotspots of poisonings within the city. This did not mean that media attention and public attention ceased. Radio spot announcements and articles in local newspapers focused on these newer areas, in the process downplaying the need to hold caregivers responsible. For example, how to identify the probable presence of lead in paints by the color was publicized. The public was warned to avoid all yellow, orange, and green paints owing to a likely presence of lead pigments. The rates of lead poisoning in Baltimore continued to occupy media space, and as poisonings increased in the 1950s, this attention created concern and pressure from the public for the city to do more.

The continued growth in the prevalence of pediatric lead paint poisonings in Baltimore next prompted ordinances for paint can labeling, beginning with a modest 1958 ordinance requiring identification of lead at loadings of $>1\%$ in the contents of new containers and warnings about hazards in use, and a 1959 ordinance specifying that new paint can labels and their wording be of sufficient size to be clear and free of ambiguity.

In 1962, the policy of the city changed with the appointment of a new health commissioner to succeed the retiring Williams. Public education again became the key element in the city's approach, despite its repeated inability to eradicate the problem. Despite the shortcomings of local public awareness and education in Baltimore, most other American cities with similar demographics in the 1930s–1960s did not even offer that level of response to paint lead poisoning.

26.2.2 Other Local U.S. Efforts at Lead Paint Control Initiatives

The city of Philadelphia had a program from 1950 to 1970 of identifying lead sources in symptomatic lead poisoning cases (Sobolesky, 1982). In the period 1950–1966, the poisoning victim's environment was evaluated and orders issued through existing ordinance for lead source removal, typically lead paint, by the property owner. Enforcement was spotty and difficult, given Philadelphia's lack of regulations specifying lead paint or lead in other health hazards. In 1966, the City Code was amended to direct removal of lead sources from the environment of poisoned children, using the city's Court of Equity. Effective implementation of this 1966 ordinance was hobbled by lack of expertise, reliable abatement methodologies, and resources, a situation which improved with new funds from the Federal Model Cities Program.

In the 1970s, Philadelphia employed its own staff hired within the Model Cities initiative to inspect for the presence of lead paint and to develop or fine-tune lead paint abatement methodologies. Covering costs within the Federal program rather than from residents or landlords helped in its wide acceptance within the high lead-paint and lead paint poisoning zones within the city. In 1966, Chicago instituted the first mass screening program.

By the early 1970s (as tabulated by Gilsinn, (1972)), the following U.S. cities banned the use of lead-based paint on interior surfaces: Baltimore, MD; Chicago, IL; Cincinnati, OH; Jersey City, NJ; New Haven, CT; New York City, NY; Philadelphia, PA; St. Louis, MO; Washington, DC; and Wilmington, DE.

26.2.3 Illustrative U.S. Lead Paint Poisoning Prevention Efforts at the State Level

Lead paint screening and mandated attention to lead paint hazards where appropriate had also been developed in various states. In 1973, Massachusetts enacted a relatively stringent and actively enforced lead paint abatement law, the two core features of which were: (1) lead paint or other lead hazard abatement in any homes built before 1978 and occupied by children younger than 6 years of age and (2) property owners are legally responsible (strict liability) for lead toxic injury of lead-exposed children (Klein, 1977). Additional strictures within the law included comprehensive inspections of the interior and exterior of homes with lead-poisoned children; permanent removal or covering of lead paint on all chewable and accessible surfaces to a height of 5 ft and on all window surfaces subject to impact, i.e., frictional surfaces; removal of loose and peeling lead paint from areas;

potential for criminal charges for neglect of compliance; and notification of other tenants in multi-unit buildings and of mortgage and lien holders. In 1990, the state buttressed this lead paint control and poisoning liability machinery with a law [105 CMR 460.050] requiring yearly universal screening of children 9 months through 4 years of age for lead poisoning.

In 1992, Rhode Island enacted regulations for addressing lead paint poisonings of children in deteriorated housing [State of Rhode Island 1992 with 1997 Amendments: R23-24-6PB]. The effectiveness of the Rhode Island approach versus that of its neighbor, Massachusetts, was more limited in a number of ways (Brown, 2002; Sargent et al., 1999). A major difference with Massachusetts requirements was the voluntary nature of requirements for lead paint hazard abatement in that Rhode Island owners had to agree to do the abatement work. Strict liability, possibility of criminal charges, and notification provisions were not in the Rhode Island regulations, but these elements were in the Massachusetts regulations. Not surprisingly, lead paint poisoning statistics were lower in Massachusetts than in Rhode Island, while a cost–benefit comparative analysis of relative impacts of corresponding regulatory effectiveness in the two states showed that broad requirements and stricter enforcement in Massachusetts led to lower rates of childhood poisonings and lower societal and economic costs (Brown, 2002; Sargent et al., 1997, 1999).

Maryland, given the impetus provided by the efforts at lead paint poisoning case findings and attempts at lead paint regulation in Baltimore, enacted several sets of lead paint control regulations in the earlier years. In 1957, Maryland acknowledged that lead in older dwellings is a problem and set up a financial assistance program for persons who wished to do lead paint abatement [MD. ANN. CODE PUB. GEN. LAWS 1957: §§2-301, 2-307]. Other statutes prohibited the use of lead-based paint on interior surfaces in units with young children, required that physicians report lead poisoning cases, established an advisory council to study the problem, and required that owners violating orders to abate lead paint hazards deposit funds in a court-managed escrow system. Tenants involved in episodes of violation could not be evicted or financially penalized.

A number of other states enacted lead paint hazard control remedies in more recent years, and as of 1990, the U.S. HUD noted statutes dealing with lead-based paint in the following states besides those noted above: Arizona, Arkansas, California, Connecticut, Delaware, Illinois, Kentucky, Louisiana, Maine, New Hampshire, New Jersey, New York, South Carolina, and Wisconsin.

Lead Paint Manufacturer Initiatives in Reducing Childhood Lead Paint Exposures and Poisonings

In February 1955, the American Standards Association (ASA) issued its directive Z66.1-1955, which set forth a 1% lead in paint standard (with compliance to be voluntary on the part of lead paint producers) along with other declarations:

"A liquid coating material to be deemed suitable, from a health standpoint, for use on articles such as furniture, toys, etc., or for interior use in dwelling units where it may be chewed by children (a) shall not contain lead compounds of which the lead content (calculated as Pb) is in excess of 1 percent of the total weight of the contained solids (including pigments and drier); (b) shall not contain compounds of antimony, arsenic, cadmium, mercury, selenium, or barium (when soluble by stirring for 10 minutes with 5 percent hydrochloric acid at room temperature) introduced as such in the formulation of such coatings."

"Coatings complying with this standard may be marked: Conforms to American Standard Z66.1-1955, for use on surfaces which might be chewed by children."

The 1% ASA standard, while concurred in by some health scientists and clinicians, was not based on any obvious body of toxicological data that this value would actually result in toxicologically safe levels. This level of 1% or 10,000 ppm Pb would clearly be a toxic level of exposure. This 1% figure had, as one factor in the numerical choice, an accommodation for the industry's concerns that some level of lead needed to remain in paint to serve as a drier, allowing the paint film to dry within tolerable time limits. The validity of the 1% figure as the minimum level for a drier function remains questionable.

A number of other problems were attached to this voluntary standard, some obvious and others more subtle. By its wording, the ASA standard was directed to chewable interior or other painted surfaces, mainly residential interior woodwork, furniture, and toys. Other forms of lead paint, e.g., chips, flakes, dusts, from (nonchewable) interior walls and ceilings, exterior woodwork and walls, were not subsumed under the standard. The labeling for new 1% Pb paint containers was misleading, since chewable surfaces with lead paint levels below the new coat(s) with the 1% lead pigment/drier, but containing much higher Pb loadings or concentrations, would still provide a hazardous chewable surface once the topmost layer(s) of 1% Pb paint were chewed through. A subtle hazard was one identified by Chisolm and Harrison (1956) shortly after the ASA 1955 voluntary standard appeared. That is, ingestion of peeling or flaking paint with multiple layers of the 1% Pb paint would still lead to relatively high total Pb ingestion rates in children when indexed as a complete paint chip down to the substrate. Furthermore, this cumulative Pb hazard likely would be misidentified by paint chip laboratory analyses measuring Pb concentration rather than lead loadings per unit area (mg Pb/cm^2).

Compliance with this voluntary standard was more assumed than systematically verified among national paint sales figures in the years between 1955 and the final banning of lead in paint for residential uses or uses on furniture and toys above 0.06% Pb, effective February 1978. As a voluntary standard, few sanctions attached to any firms violating the 1% loadings of Pb in the products of a paint manufacturer.

26.2.4 U.S. Federal Guidelines and Regulations for Control of Childhood Lead Paint Poisoning

The U.S. government did not become significantly involved in the U.S. lead paint poisoning problem as a national public health issue until the early 1970s, and that involvement continues in a substantial way. Federal efforts took several forms: (1) rules and regulations implemented through multiple Federal agencies such as the U.S. Department of Housing and Urban Development (U.S. HUD), the U.S. Environmental Protection Agency (U.S. EPA), the U.S. Food and Drug Administration (U.S. FDA), and the U.S. Consumer Product Safety Commission (U.S. CPSC); (2) guidelines issued by both the U.S. CDC and the U.S. HUD; (3) the preparation of reports to Congress on childhood lead poisoning in its various dimensions; and (4) mandating public information initiatives by various agencies to educate the public at large with data gathered on the nature and scope of the U.S. lead paint poisoning problem.

Rules and regulations took the form of both banning lead paint for various residential uses in the United States and managing lead paint exposures via various pathways from existing lead-painted surfaces through mass screening and funding of community control interventions. Guidelines included those issued by U.S. HUD concerning methods and practices for reducing children's lead exposures in housing within its jurisdiction (U.S. HUD, 1990, 1995), and those issued by U.S. CDC in the form of periodic statements on childhood lead poisoning as prescriptive guidance to physicians and other health care providers for the control and management of childhood lead paint exposures and associated poisonings. The 1990 U.S. HUD report to Congress (U.S. HUD, 1990) set forth the agency's comprehensive and workable plan for the abatement of lead-based paint in privately owned housing, and included within the funded projects responding to Congress's request: (1) an overview of lead in the environment; (2) the contribution of lead-based paint to lead poisoning; (3) the extent of lead-based paint in the then-existing U.S. housing stock; (4) the cost analyses for lead-based paint testing and abatement; (5) an analysis of the then-current governmental and private-sector activities in this area; and (6) description of the overall comprehensive plan including updating HUD's lead-based paint regulations, public information in the form of brochures, notices, and audiovisual materials for the general public, research on testing, abatement, and health effects of abatement methodologies for paint lead and lead in soil, and sources of financial support for abatement in private housing.

A number of Federal statutes enacted over the past 40 years mandate a variety of regulatory interventions dealing with lead in paint. Two bodies of legislation were major players in mandating Federal roles in lead paint exposure and poisoning prevention. In 1971, Congress passed the LBPPPA [PL 91-965. 84 Stat 2078; U.S.C. 4821 et seq.] with later amendments. A more recent piece of U.S. legislation with a lead focus was Title X of the 1992 Housing and Community Development Act, referred to as the 1992 Residential Lead-Based Paint Hazard Reduction Act or simply Title X [P.L. 101-550; 42 U.S.C. 4851 et seq.].

The 1971 LBPPPA:

1. Required that the Secretary of Health, Education, and Welfare (now Health and Human Services) prohibit the use of lead-based paint in residential structures constructed or maintained by the Federal government or those units getting Federal assistance in any form; the LBPPPA defined lead paint as that paint having more than 1% Pb content.
2. Authorized a national program for cities and states to conduct mass screening programs to identify childhood lead poisoning, refer for medical treatment, investigate sources of the lead poisoning, and order lead paint abatement. This screening program began under the aegis of the U. S. CDC in the 1970s, was folded into block grants from the U.S. Department of Health and Human Services' Maternal and Child Health Services Program in 1981, and reverted, in reduced size, to the U.S. CDC in 1988 under the 1988 Lead Contamination Control Act.

In 1972, the U.S. HUD issued regulations barring use of lead-based paint in HUD-associated housing. The LBPPPA was amended in 1973 to lower the allowed level of Pb in paint to 0.5% until the end of December 1974 and 0.06% afterward unless the CPSC found that a higher level was safe. CPSC reported in 1974 that 0.5% was a safe level. In 1976, the LBPPPA was further amended to define lead paint as paint having more than 0.06% Pb unless the CPSC deemed a higher level safe and issued that as a finding. CPSC declined to issue such a finding and this level became the definition of lead paint by the LBPPPA.

The LBPPPA with its early amendments triggered parallel rulemaking setting lead limits in paint sold to the general public and banning sale of new furniture and toys painted with lead paint above an interim level of 0.5% Pb as solid content or dried paint film and an eventual level of 0.06% Pb. In March 1972, the U.S. FDA announced (37 FR 5229-5231; March 11, 1972) regulations under the Federal Hazardous Substances Act [2(f)(1)(A) 74 STAT. 372, 374; Part 191] and the Federal Food, Drug and Cosmetic (FDC) Act [§701 e, f, g; 52 Stat. 1055-56 and 70 Stat. 919 et seq.] for new sales.

In 1976, the LBPPPA was amended by the National Consumer Health Information and Health Promotion Act of 1976 [PL 94-317; 90 Stat. 705-706], mandating a determination of a safe level of lead. Subsequently, the

U.S. CPSC announced a ban on new lead paint or new toys and furniture with lead paint above 0.06% Pb (42 FR 44192-44202; September 1, 1977) effective February 28, 1978, using its authority under the Consumer Product Safety Act [PL 92-573; 86 Stat. 1231 and Amendments; §30 (d)), Part 1303].

Later amendments included those under the Lead-Based Paint Poisoning Prevention (LBPPP) Section 556 of the Housing and Community Development Act of 1987 and §1088 of the Stewart B. McKinney Homeless Assistance Amendments Act of 1988 [PL 100-628: 102 Stat 3224]. In addition, the 1989 appropriation act funding the U.S. HUD and independent Federal agencies contained more language on the LBPPP. The 1987 Amendments called for four particular regulatory or regulatory-related actions: (1) inclusion of intact paint in one of the definitions of a lead paint hazard; (2) changes in the lead-based paint requirements of the HUD public housing program; (3) an extensive research and demonstration project; and (4) several reports, including the 1990 Report to Congress of a "Comprehensive and Workable Plan" for lead paint abatement in nonpublic housing. One upshot of the amendments was mandating a construction cutoff date of 1978 and not 1973 or 1950 for targeted housing.

In 1990, the U.S. HUD issued interim guidance for lead paint identification and abatement methodologies for public and Indian housing under the above LBPPPA Amendments (55 FR 14556-14614). These guidelines were addressed to testing for lead-based paint, criteria for lead paint abatement contractor selections, specific lead paint abatement criteria and methodologies, cleanup postabatement, and waste disposal.

A final set of guidelines was issued in June 1995. These final guidelines replaced and expanded on the interim ones published in 1990 and refined and extended the activities called for into related activities such as renovations. However, the final guidelines principally target public and Indian housing but are readily applicable to the private housing sector.

Title X of the 1992 Housing and Community Development Act considerably expanded upon strategies and methodological approaches for dealing with lead paint hazards in the United States. Statutory and regulatory distinctions in the LBPPPA and its later amendments in the 1970s and 1980s with those in Title X reflect in large measure the growing health science and technological understanding of lead and lead paint as a public health and child health issue. That is, these two bodies of statutes and subsequent regulations demonstrated the interplay, or feedback loops, of science, risk assessment, and regulation illustrated earlier.

A good example of the evolution of statute, regulations, and guidelines as reflective of changing science in some 20 years between the enactments of the LBPPPA and Title X and subsequent issuance of corresponding mandated rules and regulations is the June 1995 final U.S. HUD Guidelines for lead paint testing and abatement, compared to HUD's interim guidelines issued in April 1990. The interim guidelines were issued within the directions provided by amendments to the 1971 LBPPPA and Fiscal Year 1990 appropriations legislation for the HUD and independent agencies, while the final guidelines were issued pursuant to §1017 of the 1992 Title X Act.

A second example of evolving science intertwining with U.S. legislation and associated mandated regulations is the explicit acknowledgment of the need to involve multiple Federal agencies in order to adequately control lead paint exposure and poisoning because of lead being a multimedia pollutant. The previous chapter discussed the general issue of regulating multimedia pollutants via single-media legislative approaches. A number of the regulations evolving from Title X mandates call for joint action. Two illustrations are the U.S. HUD's Final Rule (64 FR 50140-50199; September 15, 1999) on Requirements for Notification, Evaluation and Reduction of Lead-Based Paint Hazards [§§1012, 1013 of Title X; 24 C.F.R. Part 35 et seq.] and the U.S. EPA's Final Rule (66 FR 1206-1240; January 5, 2001) on Lead: Identification of Dangerous Levels of Lead [Title X, §§402, 403 of the TSCA; 40 C.F.R. Part 745].

Title X goes beyond the intent and scope of the earlier LBPPPA and its multiple amendments. The LBPPPA, drawing on the science of the time and the public health characterizations of the as-then-understood nature and extent of lead paint poisoning of U.S. children, was more absolutist and narrow in its approaches to lead paint control. The act's original definition of lead paint hazard, 1% lead, made no stratification of risk based on lead paint surface condition or location nor did it develop the multimedia or multipathway dimensions of the lead paint issue. Conceptually, the act operated from the notion of a *(paint) lead-free* child's environment rather than an environment that was *(paint) lead-safe.* Some hold that Title X requirements in terms of lead paint regulation are on their face more flexible while also being more comprehensive in terms of lead exposure mechanisms and pathways, while others hold that flexibility in the form of partial remediation versus rigidity in the form of total abatement presents its own problems of continued Pb exposure risks in the long term.

To the extent Title X's framing language goes to "any condition that causes exposure to lead ...," it incorporates the need to address lead-contaminated dust (interior dusts with significant inputs from lead paint) and bare, lead-contaminated soil (some fraction of which arises from exterior lead paint weathering or improper lead paint residue disposal) as well as lead paint. Furthermore, lead-based paint is a hazard if it is deteriorated or present on accessible (e.g., chewable), friction, or impact surfaces. Intact lead-based paint is a potential hazard until it becomes an immediate hazard through a variety of mechanisms. The notion of imminence or "immediate" over "potential" lead paint hazard does call for caution, however, in terms of underestimating risk for adverse exposures. Painting over a frictional surface, e.g., a newly painted window jamb, with existing lead paint underneath does not provide an "intact" surface within any assumed meaning of permanency.

As soon as the window is used, the newly painted surface is abraded as a necessary and predictable consequence, exposing the under layers of lead paint.

Title X, unlike the LBPPPA, provides for the use of a "risk assessment" within a narrow meaning of having lead-painted surface inspections in terms of the Act's hazard definitions. Here, risk assessment does not refer to the scientific definition of risk assessment as an applied discipline drawing upon multiple academic branches of science and requiring application of principles and practice as set forth in, for example, Part 4 of this monograph.

The June 1995 final HUD Guidelines for evaluation and control of lead-paint hazards (U.S. HUD, 1995), generated in response to the requirements of Title X, consists of six major divisions: (1) an overview discussion of childhood lead poisoning; (2) the conceptual and operational frameworks for Title X; (3) the current definitions of what lead-based paint hazards are; (4) the core operational part, the lead hazard control process; (5) a compendium of other issues, such as a more comprehensive look at overall health of children in these residences, and potential sources of exposure besides lead paint; and (6) innovation and cost-effectiveness issues. The operational part consists of 18 chapters and a set of appendices.

Part IV of the guidelines, the lead hazard evaluation and control/intervention portion, consists of planning, the lead hazard evaluation, compliance issues for certification and training of abatement/interim control contractors, control of the lead hazards, cleaning and clearance, and reevaluation. The planning step, within the meaning of Title X, concerns some gross assessments of the units or buildings to be evaluated. What is the age of the house? What is the condition and likely remaining life of the property? The hazard evaluation partitions into two forms of assessment activity, a lead paint inspection or a risk assessment. A "paint inspection" involves a systematic investigation of painted surfaces in interior and exterior areas, and in individual residential or commons areas. This approach includes the use of lead loadings in surface paints, using either portable X-ray fluorescence units or laboratory analysis of excised paint chip samples. "Risk assessments" within the specific meaning of Title X refer to the process of identifying and describing lead paint hazards, e.g., peeling, flaking, chipping paint containing lead, lead in interior leaded dusts, lead in soils, and the need for structural repairs, thereby identifying the nature, extent, and locations of defined lead paint hazards. Some environmental measures may also be taken. Often, both lead paint inspections and risk assessment efforts are done and the results combined.

The guidelines address lead hazard controls in terms of both interim measures and lead paint abatements, where the interim controls refer to measures intended to temporarily reduce exposures of residents to lead paint hazards, such as specialized cleaning, repairs, maintenance, painting, hazard monitoring over time, and various educational approaches to foster awareness. With respect to "abatement," it is important to make the distinction between the definition put forth in Title X and absolute abatement.

Title X language refers to two forms of abatement. In the case of private ownership, abatement does not mean removal of all lead paint in the structure or unit but is intended to have a "permanency" defined as being at least 20 years in duration. This can be done by containment or encapsulation of the lead paint surfaces as well as outright removal, removing lead-painted components and fixtures, and removing contaminated dusts. While this form of abatement is prescribed for general ownership units at large, Title X requires total lead paint removal in public and Indian housing. These major abatement approaches are typically done in the course of major renovations or rehabilitation of housing.

HUD Guidelines, in common with some other Federal guidelines dealing with lead poisoning issues, do not carry the force of law via enforceable standards. While the 1995 HUD Guidelines do not have legal sanctions for noncompliance, deviations from these guidelines get in the way of interested stakeholders acquiring Federally assisted housing insurance and financing.

The U.S. HUD promulgated its legally enforceable rules and regulations under Title X in September 1999. They approximate quite closely the 1995 U.S. HUD Guidelines. The regulations cover three areas for compliance by relevant entities involved with Federally owned properties or units receiving Federal assistance. In the aggregate, the 1999 regulations emphasize identification of lead hazards in such housing, notification to occupants of their existence, and mandate through detailed regulations the control of these hazards.

The scope of the rules and regulations within this 1999 U.S. HUD rule-making also spells out five joint actions by HUD and the U.S. EPA. EPA's obligations under these Title X-mandated joint activities are to apply to much of privately owned housing what these regulations provide for Federally owned or assisted housing.

The joint disclosure rule, issued under §1018 of Title X [42 U.S.C. 4852d], requires HUD and EPA to issue joint regulations for disclosure of known lead paint or lead paint hazards by individuals selling or leasing housing built before 1978. A joint final rule was promulgated on March 6, 1996. This provided for EPA to issue an information pamphlet, provided for availability of records pertaining to lead paint or paint hazards on surfaces in the property, and included disclosure language in contracts or leases. Importantly, with regard to limitations within these rules, buyers, sellers, landlords, or tenants are not required to initiate testing or hazard control activities.

The second category of joint endeavor covers EPA certification mandates and work practices standards under Title IV of TSCA and Title X. This regulation requires EPA to establish mechanisms for training and certification of workers involved with risk assessment, paint inspections, or various forms of lead paint or lead paint hazard abatements.

U.S. EPA was also required to issue standards for hazardous levels of lead in paint, dust, and soil under TSCA §403, discussed below with respect to EPA's Lead Hazard Rule issued in January 2001. U.S. HUD issued interim standards in the 1999 rulemaking and that final lead rule was issued in January 2001.

The U.S. EPA established a laboratory accreditation program under TSCA [§405(b) (15 U.S.C. 2685(b)]. Called the NLLAP, the activity recognizes and accredits laboratories which accurately analyze lead in paint, dust, and soil samples. Accreditation is acquired by participating in the Environmental Lead Proficiency Analytical Testing (ELPAT) program and agreeing to auditing. A continuously updated list of approved laboratories is provided to the general public. U.S. HUD is required to test using NLLAP firms for analyses of samples from Federally owned or assisted housing units.

A fifth area of collaboration involves EPA regulations recently promulgated for renovation and remodeling activities for owner-occupied housing or child-occupied facilities (COFs) (e.g., day-care facilities) and HUD participation via renovation and remodeling regulation for Federally owned or assisted housing. This Renovation, Repair Lead Paint Rule for owner-occupied residences and COFs was recently issued and is discussed below.

This HUD Final Rule also consolidated all of HUD's lead paint requirements into Part 35 of 24 CFR, subsumed under one of 12 subparts of Part 35. This provided simplicity in comprehending the material and streamlining the relevant texts.

In this Final Rule, U.S. HUD sets forth seven strategies for lead paint or paint hazard evaluation and hazard reduction, varying in relative stringency from least to most: (1) safe work practices during rehabilitation; (2) ongoing lead paint maintenance practices to assure continued intact surfaces; (3) visual assessment and paint stabilization; (4) risk assessment and interim controls (as defined above); (5) combined risk assessment and lead paint inspection with interim controls; (6) risk assessment and abatement of lead paint hazards; and (7) lead-based paint inspection and abatement via removal of all lead paint.

The U.S. EPA's Final Lead Rule "Lead: Identification of Dangerous Levels of Lead" [66 FR 1206-1240; January 5, 2001; 40 C.F.R. Part 745] basically applies various lead paint hazard regulations beyond those contained in the August 1996 (U.S. EPA) and September 1999 (HUD) regulations for U.S. Federally related housing units to U.S. private housing. It supports already issued joint rules from HUD and EPA. The rule applies to targeted housing, targeted housing being single or multiple units built before 1978 (with exceptions for elderly and disabled adult housing) and these regulations extend as well to COFs.

The substance of this rule provides a cluster of standards that (1) risk assessors use to identify hazards and evaluate clearance tests; (2) specify when certified professionals are needed for lead cleanup; (3) specify what training providers have to teach in their courses; (4) identify, via results of measurements, hazards that Federal agencies or property owners have to abate in old housing; or (5) when known, hazards that have to be disclosed to buyers or tenants by property owners or landlords.

A significant part of this rule is the set of definitions for hazards springing from lead in paint, in dusts, and in soils. Lead-based paint is that paint which contains $\geq$1% Pb/cm^2 surface area Pb or $\geq$0.5% Pb by weight in a collected sample analyzed in a laboratory. "Hazardous" lead paint, as distinct from just Pb paint *per se*, is that lead paint on a friction surface subject to abrasion and where the surface beneath the abradable surface has already accumulated dust Pb loading exceeding the rule-defined dust hazard Pb loading level. It is also that lead paint on an impact surface caused by impacts from adjacent components, and any chewable lead-based paint on which there are teeth marks. Surface areas of deteriorated lead paint are specified for triggering the requirements of occupant protection, clearance testing, and use of trained, certified workers: >2 ft^2 of deteriorated lead paint in a room, >20 ft^2 of exterior lead painted surface, or >10% of total surface areas on an interior or exterior component.

The dust lead standard is binary, in terms of specified hazard levels, and is expressed in units of loading, not concentration. Dust Pb "hazard" standards are defined as a floor level of $\geq$40 μg/ft^2 Pb and an interior window sill measurement $\geq$250 μg/ft^2. Dust Pb "clearance" standards, i.e., standards for surface wipe Pb loadings postabatement, are the same for floors and sills and $\geq$400 μg/ft^2 for window troughs (wells).

The soil Pb "hazard" standard is also dual: bare soil play areas frequented by young children have a hazard level of $\geq$400 ppm, while other yard areas with bare soil are governed by a higher standard, 1,200 ppm Pb.

Other actions within this rule include: (1) changes in existing regulations for interpreting results of test sampling for lead-paint hazards; (2) changes in practices to mandate testing of all deteriorated paint on surfaces in units with paint histories suspect for lead paint; (3) changes to practices for dust/soil sampling locations; (4) work practices for management of soil removal in abatement activities; (5) amendments to tribal and state program authorization requirements; and (6) changing the definition of "abatement" to apply to cases with lead-based paint hazards.

The U.S. EPA's principal set of regulations for renovations and repairs of pre-1978 U.S. private housing stock are embodied in the April 2008 Final Rule: Lead: Renovation, Repair, and Painting Program; Lead Hazard Information Pamphlet; Notice of Availability (73 FR 21692-21769; April 22, 2008; 40 C.F.R. Part 745). Issued under §402(c)(3) of the TSCA, this rule, the LRRP rule, applies to lead-based paint hazards created by renovation, repair, and painting activity associated with target housing (defined above) and COFs. TSCA §402(c)(3) requires EPA to revise the lead paint activities regulations within its purview and jurisdiction to apply to renovation and remodeling activities that create lead paint hazards. Scheduled to take effect in April 2010, certain of the regulations were undergoing changes in 2011 in response to comments from various affected stakeholders.

EPA's promulgation of this rule required field studies to quantify the lead paint lead exposure hazard scenarios for workers and any occupants during renovations and repairs that lead to and accompany hazardous disturbances of lead paint. These were carried out in four phases described in the rule. Two of the field evaluations addressed lead exposures of workers involved in repair and renovation of contemporary or historical aged housing, while two directly or indirectly focused on childhood lead exposures associated with the amounts of leaded dusts generated from repair and renovation disturbing lead paint-covered surfaces. Phase I concerned the lead dust study and the amounts of leaded dusts released in units associated with the following five activities: paint removal by abrasive sanding, demolitions and other major interior jobs, window replacement, carpet removal, and heating, ventilation and air conditioning (HVAC) repair or replacement.

Phase II was the Worker Characterization and Blood Lead Study, which from the title had to do with the PbB levels of renovation and repair workers, while Phase III of the field data gathering evaluated children's PbB levels associated with repairs and renovations in units, the Wisconsin Childhood Blood Lead Study. The final Phase IV segment concerned blood lead data gathering in workers doing repairs and renovations in old, historic homes. In the Dust Study, 12 interior and 12 exterior renovation activities were carried out in vacant units in Ohio.

The U.S. EPA determined from these field studies that renovation and remodeling activities create lead paint hazards as defined by existing agency lead paint control regulations. The agency in this Final Rule sets forth requirements for training, containment, cleaning, and cleaning verification that EPA asserts achieve its goal of minimizing exposures to lead paint hazards. Renovations must be done by certified firms, via a certification process spelled out by the agency. The rule spells out the work practice standards. These standards include warning signs and work area containment during repair and renovation, waste management covering hazardous debris and residues generated, cleaning steps, and cleanup verification steps.

Recent U.S. Federal Regulations for Lead Paint in Domestic Painting and Use on Toys

In August 2008, Congress enacted the Consumer Product Safety Improvement Act [PL 110-314; 15 U.S.C. §1278a], which lowered the permissible levels of lead paint for structural use, paint on furniture, paint on/in children's products, and lead paint on toys intended for children and sold in the U.S. Effective August 14, 2009, such paints cannot exceed 90 ppm total Pb content as determined by laboratory testings [16 CFR §1303.1].

26.3 INTERNATIONAL LEAD PAINT REGULATION

Outside the U.S., the first efforts to promulgate lead paint regulations and controls on associated hazards to human health traced to the controls placed on lead paint use by European countries in the late nineteenth century and in first decades of the twentieth century. Table 26.2 summarizes some of the significant regulatory efforts at lead paint control in both earlier and more recent periods. The earlier efforts were by individual countries, while the most concerted, multinational effort for both Europe and other countries was the body of constraints placed on uses of lead paint pigments and referred to as the White Lead Convention, convened in 1921 after a preliminary 1919 conference in Washington, DC, by the International Labour Office of the former League of Nations (ILO, 1927). White lead in this context referred to white lead paint pigments including basic lead carbonate and basic lead sulfate, materials used for white paint or as tinting base for colored paints.

The second multinational effort occurred within guidelines and ensuing enforceable directives issued by the European Parliament and various regulatory directorates within the European Union and its members. This ongoing multistate effort dates, of course, to the establishment of the European Union in the early 1970s.

The first formal regulatory initiatives on lead paint production or use in jurisdictions outside of either the U.S. or Europe took place in Great Britain's dominions: Australia, New Zealand, Union of South Africa, Rhodesia, and India. Controls on production and sale of lead paint in more recent years were implemented in other parts of Asia including China, Singapore, and Malaysia.

26.3.1 Illustrative International Efforts at Control of Lead Paint Production and Use: Individual Countries

Individual national efforts outside the United States to control lead paint production and/or use, particularly in Europe, arose out of the nineteenth century organized labor movement that accompanied the industrialization of Europe. In 1867, occupational concerns over white lead production and use resulted in the Congress of the Workers' International Association while 1896 witnessed the International Workers' Conference on the same topic. In 1901, the Assembly of the International Association for Labour Legislation was convened to deal with the topic of white lead.

Such organizing efforts appeared to result in a number of major European jurisdictions enacting various constraints on the production of white lead and its uses. Several conclusions can be drawn from the early record of white lead control initiatives in Europe and elsewhere. First, legislation and regulation for production and use of white lead and related paint pigments in the major European countries provided much of the evidence

TABLE 26.2 Significant International Regulations or Guidelines for Lead in Paint

Statutory Initiative	Codified Law or Regulation	Year(s)	Key Features
A. Early regulatory initiatives			
White Lead Convention: International Labour Office, League of Nations	Country-specific codification	1921 and continuing; Russian Federation ratified in 1991 for a total of 13 nations	Convention document consisted of 15 articles, detailed in the text. Banned interior Pb paint use for dwellings and other structures; focus was on painter safety
Early regulatory initiatives for individual countries: France, Great Britain, Germany	National laws	France: 1909, 1913, 1914, 1919	Banned Pb paint for both interior and exterior building surfaces, banned dry sanding, other hazardous acts by painters, not occupants; banned women and children from the Pb painting and manufacturing trades
		Great Britain: 1891, 1901, 1920	Similar occupational restrictions for Pb paint use and production to France; lead poisoning cases were eligible for workman's compensation
		Germany: through the 1920s	Similar occupational restrictions as in France and Great Britain, plus extensive controls on workplace Pb dusts
B. Regulation of Pb paint in the European Union			
European Council directives	67/548/EEC, 77/728/EEC, 88/379/EEC	1967–1988 and later	Required all Pb paint with >0.5% Pb to be sold in cans with specified warning label. In 1981, the Directive tightened label wording
European Council directive	76/769/EEC	1976 and later	Restricted Pb content of children's products
European Council directive	89/677/EEC	1989–1992	Sale of lead paint for commercial use was banned, beginning in 1992. Only certain specialty uses allowed

(*Continued*)

TABLE 26.2 Significant International Regulations or Guidelines for Lead in Paint—(cont.)

Statutory Initiative	Codified Law or Regulation	Year(s)	Key Features
European Council directive	88/378/EEC, general directive for EU toy sales; current regulations include the EN 71-3 toy safety directive for Pb in painted surfaces; Directive 2002/95/EC is for a limit in electronic toys and consumer products	1988 to recent amendments	Pb in paints on surfaces of toys is held to a minimum of 90 ppm as Pb for paint on toys and 100 ppm in electronic toys
European Council directive	82/605/EEC for Pb worker protections in producing various Pb products including Pb paints	1982 and later	Applies to both Pb metal and Pb ionic components; provisions include Pb monitoring protocols
European Council directive	77/312/EEC	1977 and later	Sets forth the general provisions for biological screening (PbB) for exposed children on a periodic basis

and guidance for worker protections embodied in the later White Lead Convention language established in 1921 (Hutton, 1923; ILO, 1927).

Second, prohibitions and controls on exposures to lead paint and lead paint pigment production and use in painting residential and other structures were for the benefit of the lead worker or painter in preventing lead poisoning. Little of the regulatory efforts toward lead paint was focused on the occupants of residences being painted and school pupils. Nonetheless, the control of lead paint exposures among European painters simultaneously provided some protections for residents. Similarly, the requirements for worker hygiene practices like washing and changing work clothes to street clothes before leaving work obviously served to reduce the incidence and intensity of secondary or take-home lead exposures of highly bioavailable and hazardous lead and lead paint dusts.

Review of the rules and regulations governing white lead production and use in various European countries, particularly prior to widespread adoption of the White Lead Convention's elements, indicates the appearance of some common elements of regulation. In the case of structural painters, for example, painting interiors with lead paints containing more than 2% lead was prohibited, as was dry sanding or burning or other practices generating hazardous worksite leaded dusts. In the case of white lead production, factories were prohibited from employing children <15–18 years of age or women, particularly those of childbearing age. Workplace atmospheres with high amounts of leaded dusts were prohibited. Some early regulations governing lead paint production and use in major European industrialized countries are summarized below.

France

In July 1909, France banned the use of white lead for painting both interior and exterior surfaces in buildings, enacted worker hygiene measures in October 1913 along with requirements calling for medical supervision, prohibited women and children from working in the lead trades in March 1914, and required that occupational lead poisoning be both a reportable and a compensable disease in October 1919. Some prohibited painter practices included dry sanding, pumice stone rubbing, and applying white lead to surfaces directly by hand, while requirements included clean clothing and washing facilities for dust removal. Consistent with an overall focus on workmen rather than occupants of buildings being painted, these French regulations did not apply white lead bans for building surface painting if painting was done by the owner–occupant.

Great Britain

The first law in Great Britain pertaining to white lead was enacted in 1891 (Hutton, 1923) and, like France, incorporated occupational controls on lead and lead paint, including white lead pigment production and use through a series of regulations enacted over the early twentieth century.

The 1901 British Factory and Workshop Act applied generally to the worker's employment but a number of provisions were subsumed within this general stream of regulations. The discharge of the act was under the aegis of the Chief Inspector of Factories. The 1901 act set forth a number of prohibitions and requirements similar to those in France, such as prohibiting employment of women and children under 18 years old in white lead production, while requiring adequate hygiene facilities and apparel. The ban on women and children was further augmented by legislation in 1920. Reporting requirements for occupational lead poisoning were also in the 1901 act. The 1906 Workmen's Compensation Act set forth a scheme for compensation for workplace lead poisoning.

Unlike France, where lead trades were generally dealt with collectively, the British approach individualized regulations to the particular category of lead trade. The first was the white lead control apparatus instituted in 1891. Companion regulations for "manufacture of paints and colors" appeared in 1907 while controls for manufacturing litharge and red lead appeared in 1911. British legislation and regulations banning lead paints used in interior painting of buildings, despite a generally heavy body of worker regulations for lead trades, were never successfully promulgated either prior to the White Lead Convention or afterward. A history of the deliberations and debates in Britain in these years is presented in ILO (1927). Hutton (1923) discussed lead paint regulations in Britain's Dominion nations.

Germany

Germany was the third country in Europe with an extensive regulatory apparatus for the production and use of white lead pigments and other products. A number of general occupational protections were implemented in the first two decades of the twentieth century, resembling to some degree those in place during this period in France and Great Britain. A special body of 11 regulations set forth controls on the painting industry as a specific category of the lead trades. They included a ban on employing children under 18 years, control measures on grinding and other uses of white lead so as to avoid leaded dust exposures, provision of worksite clothing, and means for washing dusts from the body.

Other European Countries

Other European jurisdictions with legislation and regulations for production and uses of lead paints included Austria, Belgium, Bulgaria, Greece, Hungary, Italy, the Netherlands, Russia, Sweden, and Switzerland.

Australia

Lead paint legislation and regulations specifically drawn for preventing toxic lead exposures and associated poisoning for nonoccupational settings, in contrast to the focus on lead workers and lead painters in Europe and in Asia, were promulgated for Queensland, Australia, in 1922 [Acts of the Parliament of Queensland, 1922]. These were specifically controls on residential or public use of lead paints potentially accessible to children. The legislation was in response to the childhood lead poisoning epidemic of several decades' standing in Queensland in the late nineteenth and early twentieth centuries (see earlier discussions).

Key provisions of this act, applicable to children under 14 years, included (1) banning paint containing more than 5% soluble lead (calculated as a percentage of lead oxide on the dry material free of varnish) for use on the outside of residence, hall, school, or other building, or on veranda railings, gates and fences; (2) banning such paint less than 4 ft from the floor or ground; and (3) banning dry rubbing (or sanding) lead paints exceeding 5% Pb content. While the intent of the local legislation was to avoid exposures to weathering or chalking lead paints on exterior surfaces of Queensland properties to avoid the rash of lead poisonings revealed by Gibson et al. (1892), the threats posed by interior lead paints or lead paint on toys and furniture were not addressed.

European Lead Paint Regulations in Recent Years by Individual Countries

Lead paint regulations in Europe in more recent years were subsumed in large measure within legislation and regulations of the European Union for those nations within the European Union. In the several decades between the early flurry of regulation and controls within the European Union, intermittent guidelines or regulations appeared. Furthermore, some countries took actions apart from joint regulatory promulgations from the European Union.

For example, in 1963, a voluntary agreement between the British Ministry of Health and the Paintmakers Association of Great Britain provided that paint cans containing any lead pigment or drier would carry cautionary wording for Pb levels >1% by weight in the dry paint film, "do not apply on surfaces which might be chewed by children." A later, 1968 action was the preparation of British Standard 4310, for labeling low lead paints as those having <0.5% Pb, in association with a label: "this product has a low lead content and complies with BS 4310." This standard was preempted by an EEC directive in 1977 (see discussion below).

Britain's Toy Safety Regulations [S.I. 1974], however, markedly lowered the permissible Pb content in toy paints to 0.25% by weight, on the belief that oral contact with toy surfaces by children occurs more frequently and intensely than for structural surfaces (Home Office, 1974).

Denmark in 2000 banned import and marketing of paints with lead used as driers (siccatives) except for lead carbonate and lead sulfate. Corrosion prevention and antifouling paints with lead contents of less than 250 and 1,250 ppm Pb, respectively, were permitted until further notice.

The Russian Federation imposed controls on interior paints containing Pb in 1991 and 1992, augmenting earlier regulations in the former Soviet Union from 1984. However, exterior lead paints appear to be available in the open Russian market [as discussed in SAICM/WHO-UNEP (2009)].

Recent Lead Paint Actions Outside of Europe and the United States

Regulatory actions for lead paint in recent years for individual jurisdictions outside of Europe are illustrated by those in Australia. In 1992 and 1997, the lead content of paints for domestic use in Australia was reduced to 0.25% and 0.1%, respectively. In 2008, Australia banned the import for sale in

Australia of 14 coating materials containing Pb in excess of 1% [Australian Uniform Paint Standard/Appendix I, Schedule I, 2008].

In Africa, South Africa banned household lead paint with >600 ppm Pb, effective March 2009. In China, India, and Thailand, despite regulations controlling lead paints, retail outlets were marketing Pb paints >600 ppm Pb in recent years (Clark et al., 2006).

26.3.2 The (International) White Lead Convention

The first multi-country effort to deal with leaded paints came at the White Lead Convention, a joint effort of the International Labour Office, various labor groups, occupational health organizations, and affected legislative and regulatory bodies. The organizing instrument for this effort was an initial conference after World War I (1919), followed by the Third International Labour Conference, an official convention in Geneva, Switzerland, in 1921, followed by ratification by the various governmental bodies who were signatories to the convention. In some cases, ratification proceeded quickly, due in part to the preexisting matrix of edicts, proclamations, regulations, and other enforceable instruments already in place in those countries. Other nations, despite an extensive regulatory structure and history for occupational lead exposures, did not respond rapidly, if at all.

The United States was not a signatory to the convention; reasons for this refusal have been recorded by various historians. Knerr (1992) attributed this to master painters resisting the proscriptions on lead paint, although there is equal or more evidence that various organizations of painters or painters' unions favored ratification (ILO, 1927).

There were two particularly salient purposes of the convention (Hutton, 1923; ILO, 1927). Although generally considered an international mechanism for banning uses of white lead in painting interiors of structures, the convention also provided an omnibus provision seeking to protect women and young children from employment or poisoning in numerous components of the international lead industry.

The declarations within the White Lead Convention document consisted of 15 Articles, seven of which deal with the nature and scope of provisions for the lead pigments, with the remainder dealing with timetables, legal aspects, and magisterial matters. Article 1 called for prohibiting the use of white lead or lead sulfate and all paints and other products containing these pigments for interior painting, except for painting railway stations or industrial facilities. Article 2 included the provision that governments define the limits of such forms of painting as artistic painting or "fine lining." Article 3 banned employment of children under 18 years and all women within the ranks of painters using white lead and other lead pigments, while Article 4 spelled out a timetable for implementing the provisions, 6 years after closure of the 1921 (Third) session of the International Labour Conference. Article 5 enumerated four parts for worker hygiene practices for painters using white lead or other lead pigments, while Article 6 mandated the enforcement of the convention's provisions by the various governmental entities. Article 7 provided for keeping mortality and morbidity statistics among painters within the jurisdictions of the signatories.

Article 8 called for ratification of the document to be registered with the Secretary-General of the League of Nations while Article 12 called for governments agreeing to the convention provisions to apply it to possessions, colonies, and protectorates.

According to the ILO (1927), as of September 30, 1926, 12 nations had registered ratifications: Austria, Belgium, Bulgaria, Chile, Czechoslovakia, Estonia, France, Latvia, Poland, Rumania, Spain, and Sweden. Four states—Greece, Hungary, Italy, and the Netherlands—approved ratification, while ratification was recommended in five states: Argentina, Cuba, Denmark, Germany, and Uruguay. In 1991, the Russian Federation became the final signatory.

26.3.3 The European Union: Post-White Lead Convention Multi-State Actions

The European Union has issued a number of directives and regulations having to do with such aspects of Pb paint as warning labels on paint cans, occupational exposures to Pb and Pb paint, screening children exposed to lead, banning Pb paint in member states, and more recently, controls on Pb in or on children's products.

The EEC in 1977 issued a directive that required all paints with Pb >0.5% by weight to be sold with the label reading: "contains lead. Should not be used on surfaces liable to be chewed or sucked by children." Containers of volume <125 ml could bear the warning: "Warning. Contains lead" (*Official Journal of the European Communities*, 1977). In 1981, a follow-up directive (*Official Journal of the European Communities*, 1981) revised the can warning to replace "should" with "must."

Lead paint for general painting use was barred from sale in EU member countries by an EU directive effective in 1992 [European Council Directive 89/677/EEC]. Only specialty uses were permitted.

Recent activities within the European Union concerning lead paint have had to do with the growing problem of lead paint on imported toys with paint levels exceeding permitted lead contents. In this regard, attempts at recent controls parallel those in the U.S. implemented by the U.S. CPSC (see the earlier section). The EU's Toy Directive 88/378/EEC governs regulatory matters for lead in toys marketed in EU member states. Lead in toy paints are particularly covered under EN 71-3 "Safety of Toys, Part 3: Migration of Certain Elements," with respect to propensity for mobilization and migration. Lead in toys, including that on painted surfaces, is held to a maximum Pb level of 90 μg/g (ppm). Testing Pb mobility in toy products employs protocols mandated under EN71-3:1994/AC: 2002. Finger paints have a Pb limit of 25 ppm, under EN 71-7: Safety of Toys Part 7: Finger Paints.

Total Pb content in electrical toys and consumer products is limited to 100 ppm by EU Directive 2002/95/EC, while total lead levels in nonelectrical products are left to individual members states for control (see, e.g., the above discussion of Danish regulations). For Denmark, the limit is 100 ppm via Danish Statutory Order No. 1012.

REFERENCES

Blackfan, K.D., 1917. Lead poisoning in children with especial reference to lead as a cause of convulsions. Am. J. Med. Sci. 153, 877–887.

Brown, M.J., 2002. Costs and benefits of enforcing housing policies to prevent childhood lead poisoning. Med. Decis. Making 22, 482–492.

Chisolm Jr., J.J., Harrison, H.E., 1956. The exposure of children to lead. Pediatrics 18, 943–958.

Clark, C.S., Rampal, K.G., Thuppil, V., Chen, C.K., Clark, R., Roda, S., 2006. The lead content of currently available new residential paints in several Asian countries. Environ. Res. 102, 9–12.

Fee, E., 1990. Public health in practice: an early confrontation with the 'silent epidemic' of childhood lead poisoning. J. Hist. Med. 45, 570–606.

Gibson, J.L., 1904. A plea for painted railings and painted walls of rooms as the source of lead poisoning among Queensland children. Aust. Med. Gazette 23, 149–153.

Gibson, J.L., Love, W., Hardine, D., Bancroft, P., Turner, A.J., 1892. In: Huxtable, L.R. (Ed.), Transactions of the Third Intercolonial Medical Congress of Australasia. Charles Potter, Sydney, pp. 76–83.

Gilsinn, J.F., 1972. Estimates of the Nature and Extent of Lead Paint Poisoning in the United States. National Bureau of Standards, Washington, DC, p. 11: Table 1

Hamilton, A., 1911. The white lead industry in the United States. In: Dubofsky, M. (Ed.), Bulletins of the U.S. Bureau of Labor and the U.S. Bureau of Labor statistics, 1895–1919: 95. University Publications of America, Bethedsa, MD, pp. 189–259.

Home Office, 1974. Toy Safety Regulations. Consumer Protection Statutory Instrument 1974/1367. HMSO, London.

Hutton, R., 1923. Lead Poisoning. A Compilation of Present Knowledge. Provincial Board of Health of Ontario, Toronto.

International Labour Office, 1927. White Lead. Data Collected by the International Labour Office in Regard to the Use of White Lead in the Painting Industry. Studies and Reports: Series F, Industrial Hygiene, No. 11. Geneva, Switzerland.

Klein, R.Z., 1977. Lead poisoning. Adv. Pediatr. 24, 103–130.

Knerr, D., 1992. Eagle-Picher Industries. Strategies for Survival in the Industrial Marketplace, 1840–1980. Ohio State University Press, Columbus, OH.

Lin-Fu, J.S., 1982. The evolution of childhood lead poisoning as a public health problem. In: Chisolm, J.J., Jr. O'Hara, D.M. (Eds.), Lead Absorption in Children. Management, Clinical, and Environmental Aspects. Urban & Schwarzenberg, Baltimore, MD, pp. 1–10.

Mattiello, J.J. (Ed.), 1942. Protective and Decorative Coatings, Vol. II. John Wiley & Sons, New York.

McKnight M.E., Byrd W.E., Roberts W.E., Logegren E.S. 1989. Methods for Measuring Lead Concentrations in Paint Films, Report No. NISTR 89-4209. National Institute of Standards and Technology, Washington, DC, p. 1.

National Academy of Sciences: National Research Council, 1993. Measuring Lead Exposure in Infants, Children, and Other Sensitive Populations. National Academy Press, Washington, DC.

Sargent, J.D., Bailey, A., Simon, P., Blake, M., Dalton, M., 1997. Census tract analysis of lead exposure in Rhode Island children. Environ. Res. 74, 159–168.

Sargent, J.D., Dalton, M., Demidenko, E., Simon, P., Klein, R.Z., 1999. The association between State housing policy and lead poisoning in children. Am. J. Publ. Health 89, 1690–1695.

Sobolesky, W.J., 1982. Some problems associated with the environmental phases of a childhood lead poisoning prevention program. In: Chisolm, J.J., Jr. O'Hara, D.M. (Eds.), Lead Absorption in Children. Management, Clinical, and Environmental Aspects. Urban & Schwarzenberg, Baltimore, MD, pp. 135–141.

Thomas, H., Blackfan, K., 1914. Recurrent meningitis due to lead, in a child of five years. Am. J. Dis. Child. 8, 377–380.

U.S. Centers for Disease Control, 1978. Preventing Lead Poisoning in Young Children. A Statement by the Centers for Disease Control. April. U.S. Department of Health and Human Services, Atlanta, GA.

U.S. Centers for Disease Control, 1985. Preventing Lead Poisoning in Young Children. A Statement by the Centers for Disease Control. U.S. Department of Health and Human Services, Atlanta, GA.

U.S. Centers for Disease Control, 1991. Preventing Lead Poisoning in Children. A Statement by the Centers for Disease Control. Atlanta, GA: U.S. Department of Health and Human Services.

U.S. Centers for Disease Control, 2005. Preventing Lead Poisoning in Young Children. A Statement by the Centers for Disease Control. Department of Health and Human Services, Atlanta, GA.

U.S. Department of Housing and Urban Development, 1990. Comprehensive and Workable Plan for the Abatement of Lead-Based Paint in Privately-Owned Housing. Report to Congress. Office of Policy Development and Research, Washington, DC.

U.S. Department of Housing and Urban Development, 1995. Guidelines for the Evaluation and Control of Lead-Based Paint Hazards in Housing. Office of Lead Hazard Control, Washington, DC.

U.S. Department of Housing and Urban Development, 1999. Requirements for Notification, Evaluation and Reduction of Lead-Based Paint Hazards in Federally-Owned Residential Property and Housing Receiving Federal Assistance. Final Rule (64 FR 50140-50199; September 15, 1999).

World Health Organization. United Nations Environmental Programme, 2009. Background Information in Relation to the Emerging Policy Issue of Lead in Paint. Report SAICM/ICCM.2/INF/38. International Conference on Chemicals Management, May 11–15, 2009. Strategic Approach to International Chemicals Management, Geneva, Switzerland.

Chapter 27

Regulation and Regulatory Policies for Lead in Ambient Air

27.1 A BRIEF PERSPECTIVE ON AMBIENT AIR Pb

The history of air lead regulation largely began with controlling leaded dusts generated in the workplace that might be inhaled or ingested later once deposited on worksite surfaces or brought home where family members might be exposed. Specifically, the previous chapter noted that European nations' controls on lead in paints addressed concerns about workmen generating leaded paint dusts through highly hazardous work activities, particularly dry sanding or rubbing lead painted surfaces and generating high levels of suspended paint dusts. Other than these interrelationships, the bulk of human exposures to air Pb are from ambient air contaminated by lead emissions from varied sources described in earlier chapters.

Ambient air lead comprises a significant source for human Pb exposures and is also a contaminated environmental compartment that arises from lead within waste streams emitted to the atmosphere. Unlike lead paint, air Pb's history has no identity as a lead product *per se* and is purely a contaminated medium encountered by various human risk populations.

The movement of Pb in air has a major impact on how the element is to be effectively regulated. Earlier chapters described the details of Pb behavior in the atmosphere. For example, ambient air Pb levels vary greatly with intensity of lead emissions, with the regularity of emissions, with distances between emission point and measurement or exposure point, with the type of lead emissions to ambient air (mobile or stationary source emissions), and with long-term temporal relationships among these emissions. Emissions of Pb into ambient air from vehicles burning leaded gasoline deposited the largest leaded particles close to the exhaust tailpipes, the smaller particles farther away from roadways, and the smallest, that is, submicron particles, extended distances through long-range transport mechanisms.

By contrast, stack emissions of lead to the atmosphere from point sources such as primary and secondary smelters impact areas much closer to these point sources, with the bulk of air Pb and fallout levels highest adjacent to the emitting stack(s), declining with increasing distance. Typically, Pb

Trace Metals and other Contaminants in the Environment, Volume 10
ISSN: 1875-1121 DOI: 10.1016/B978-0-444-51554-4.00027-4

increases in the air and dust/soil receiving compartments are maximal within 1–2 km of the source or less. How airborne Pb behaves as a function of its emission and atmospheric characteristics affects how public agencies can control air Pb and the types of regulatory strategies required.

In terms of the clinical and toxicological history of Pb, the element's presence in ambient air was largely ignored until recent decades. There were several reasons for this neglect. Measurement methods were relatively crude and insensitive for the range of air Pb measurements required, even in cases of relatively heavy ambient air contamination. Ambient air Pb, unlike paint Pb, expresses its adverse impacts across a broader segment of human populations. Collectively, ambient air Pb levels reaching human populations have not only served to complicate the science and technology of particulate aerometry but also made effective regulation of any toxicological hazards of air Pb difficult. The health impacts history of air Pb is short when compared to histories of other media-specific Pb forms traced back to ancient times—Pb in paints, foodstuffs, beverages, and waters.

Besides some of the technical complexities noted above, air Pb at toxic levels for human exposures is at microlevel and submicrolevel amounts. A brief reading of the levels of air Pb permissible in emissions to the human environment makes this apparent. This analytical reality meant that quantitative analysis of air Pb for purposes of health risk characterization and evaluation of dose–response relationships was not feasible. Second, the fact that lead entered ambient air where people live by either mobile (largely vehicular emissions) or point sources (e.g., smelters, incinerators, other pyrogenic sources) meant legislation and associated rules and regulations had to be crafted with these sources in mind.

27.2 U.S. REGULATION OF LEAD IN AMBIENT AIR

U.S. regulations for lead in ambient air are of quite recent vintage when compared to controls for this element in other environmental media. Ambient air lead regulations for the general population and vulnerable subsets of the general population in the United States were being implemented at about the same time workplace air Pb levels were being controlled through occupational hygiene standards. The U.S. Occupational Safety and Health Administration's body of occupational lead controls came into being in 1970 through various parts of the Occupational Safety and Health Act [§1910; 29 C.F.R.] at approximately the same time that the U.S. EPA began early attempts at lead control in ambient air in response to the requirements of the 1970 CAA. Consequently, there were few data from the world of occupational epidemiology to guide or inform the relative effectiveness of ambient air lead rulemaking efforts.

A second general aspect of air lead regulation in the United States is that its rulemaking was superimposed on existing general pollution control legislation. Specifically, it was preceded by air control initiatives implemented by certain states (California) negatively impacted by air pollution problems. A third aspect of U.S. air lead regulation, one not typically shared in nations elsewhere in the world, was vulnerability to legal scrutiny and subsequent challenge by various affected stakeholders. In two instances, one in the early 1970s and one in 1980, various industrial interests economically impacted by U.S. EPA rulemaking for air Pb sued unsuccessfully in the Federal appellate court of jurisdiction to reverse the regulatory rulings. In 2004, a Missouri environmental interest group successfully sued the U.S. EPA to force review and implementation of a new, lower primary and secondary NAAQS standard than the 1.5 $\mu g/m^3$ value in place since 1978.

The U.S. Clean Air Act (CAA) was first enacted by Congress in 1963 [77 Stat. 392; 42 U.S.C. §7401]. The statute was intended for air pollution in general, and from its beginning attempted to address general pollution and its components as emitted from point sources—steel mills, factories, foundries, and fossil-fueled electric power plants, among others—and mobile sources, particularly autos and commercial vehicles. Several years later, the Motor Vehicle Air Pollution Control Act of 1966 [PL 90−148; 81 Stat. 485] codified the distinctions in source emissions adding to overall air pollution, stationary versus mobile.

The first of two major amendments to the Act occurred in 1970 [PL 91−604; 84 Stat. 1676; 42 U.S.C. §§7401−7671q; 40 C.F.R. Part 50]. This amendment led to a number of major changes in the original regulatory machinery, changes implemented by the U.S. EPA. Notable for the current topic, lead was first identified as a pollutant to be regulated under §112 of the 1970 Amendments (CAAAs). This section dealt with point source Hazardous Air Pollutants (HAPs). Other pollutants were to be regulated by health-based primary and secondary NAAQS when identified as "criteria" pollutants. Primary NAAQS are intended to prevent threats to human health. Secondary standards are concerned with public welfare impacts, such as damage to physical infrastructure, crops, and domestic animals.

The Agency identified six criteria pollutants for which standards were first issued in 1971: carbon monoxide, hydrocarbons, nitrogen oxides, ozone, particulate matter, and sulfur oxides. Hydrocarbons were later dropped from the list. Particulate matter was later stratified in regulatory rulemaking into two sets of NAAQS, depending on particle size: the coarse fraction, $\geq 10\ \mu m$ and the fine portion, $\leq 2.5\ \mu m$. Citizen suits were permitted against the U.S. EPA or other agencies whose regulatory mandate encompassed ambient air pollutants and where the intent was to force actions mandated within the enabling legislation.

Third, sections of the 1970 Amendments (§§108, 109) had EPA prepare the scientific rationale for the health criteria for regulating Pb (§108) and then issue Pb NAAQS (§109) sufficiently protective of sensitive segments of the U.S. population with an adequate margin of safety. These two sections also barred consideration of economic costs of the regulations. They were to be implemented only for health protection considerations. Sensitive populations included children, the elderly, and those with respiratory diseases.

The amendments also allowed the U.S. EPA to delegate enforcement of the NAAQS to the states via State Implementation Plans (SIPs) that required approval of the agency. Not all states served a subordinate enforcement role in implementing these SIPs. The amendments identified a special role for California, which acquired the statutory and regulatory status of a source of air quality regulations co-equal to the Federal structures. Furthermore, other states could choose to adopt the California regulations where they are more stringent than Federal rules. California functions as an air quality regulatory entity through the California Air Resources Board (CARB). As a practical matter, the size of the California population's vehicular fleet was and remains large enough that CARB regulations for in-state vehicles essentially govern vehicles for sale elsewhere in the U.S. market. For example, California limited lead in gasoline in 1976 and in 1988 enacted the California CAA, setting forth a 20-year regulatory plan.

CARB functions through Air Quality Districts. One of these, the Bay Area Air Quality Management District—currently consisting of nine counties—in 1971 issued the first ambient air lead regulation anywhere in the United States. It also banned hydrogen sulfide emissions from point sources and banned open burning at trash dumps. Texas also was early in implementing air quality regulations through the Texas Air Control Board, later to be merged into the Texas Natural Resource Conservation Commission, effective in 1993.

An equally important mandate relevant to lead and other pollutant regulations was passage of the 1969 NEPA [42 U.S.C. 4321–4347], establishing the Council on Environmental Quality, and mandating environmental impact statements for all major actions.

27.2.1 U.S. Concurrent Regulation of Lead in Ambient Air and Leaded Gasoline

The history of the introduction of lead additives to gasoline and public awareness of that action offers both similarities to and differences with the early lead paint poisoning history in America's older, larger cities as illustrated by Baltimore, MD. In terms of similarities, the addition of lead to gasoline was nationwide and entailed potential exposure of much of the nation's risk populations. Second, like the major lead paint producers, those interests promoting the use of lead additives were economically powerful. The latter included the major auto makers and the major oil companies. In both cases, there were myriad corporate enterprises dependent on these two groups. With respect to differences, more economic strata of the population would be affected by leaded gasoline exposures, while deteriorating lead paint was seen as a problem for poor families in deteriorating neighborhoods.

Finally, the accountability and potential liability for health consequences of lead exposures from these two sources would take different paths.

The regulatory history of ambient air Pb is intimately tied to the parallel history of leaded fuel consumption in the United States and other industrialized nations in Europe and Asia. Leaded gasoline entered the environmental scene in the 1920s, with the development of the automobile culture and the engineering and technology requirements of less costly internal combustion engine power and efficiency. Ambient air Pb began to reflect its increasingly principal component, lead from gasoline combustion, with the passage of the twentieth century, particularly in the post-WWII decades. Until then, major contributors to air were a mélange of point source emissions, e.g., lead smelters and refineries, fossil-fueled boilers, metallurgical operations for other metals, and classes of incinerators.

The principal component of leaded gasoline has always been one or several organolead antiknock additives provisionally accepted for production and use by the Federal government. While TEL was the principal agent, TML has also been used. In 1927, the U.S. Surgeon General selected a voluntary level of TEL added to gasoline of 3 cc/gal, a value not health- or environmentally based, but corresponding to the level then being blended into gasoline at the oil refineries. In addition, given that TEL in gasoline in the 1920s and later was a provisional stance by the Federal health apparatus, the Surgeon General needed further research on public health and safety. This aspect of the TEL introduction history also notes that research into the human toxicity of TEL passed by default into the hands of additive industry researchers in academia and largely remained there for the next 40–60 years (Rosner and Markowitz, 1985).

This Federal voluntary level remained intact for more than 30 years. In 1958, the permissible additive level was actually raised, not lowered, to 4 cc/gal (4.4 g Pb/gal). With the establishment of the U.S. EPA in 1970, its first Administrator, William Ruckelshaus, publicly noted concerns about lead from leaded fuel use. In November 1973, the agency published its view that air Pb from auto exhaust was a threat to American public health, based on its own study and assessment of leaded auto exhaust and human health. This cleared the way for regulation of TEL through sections of the 1970 CAAAs dealing with fuel additives concluded to be a threat to human health.

EPA first dealt with lead in gasoline as a regulatory issue under §211(c)(1) and other 211 subsections of the CAA. Under this section, the EPA Administrator has broad authority to regulate or prohibit manufacture and sale of any fuel additive if its emission products would produce pollution that endangers public health or welfare.

The U.S. EPA promulgated regulations for lead in gasoline in December 1973, setting out a phase-down schedule starting in 1975 and continuing for 5 years (38 FR 33741; December 6, 1973). Beginning with a refinery average of 2.0 g/gal in 1973, the content was to be reduced to a maximum of 0.5 g/gal in 1979. In 1982, new rules (47 FR 49331; October 29, 1982) changed the basis of the standard to only leaded fuel versus all fuel (the latter permitting refiners to actually increase lead content as sales of leaded fuel declined and older vehicles were removed from the national fleet), set a limit of 1.1 g Pb/gal and directed that the limit be reduced to 0.5 g Pb/gal, January 1, 1985, and 0.10 g Pb/gal, 1 year later.

Regulation of lead in gasoline, versus control of ambient air Pb in the 1970 CAAAs, combined purposes that encompassed direct adverse health concerns about lead and lead-induced damage to catalytic converters, which affected emissions of gaseous criteria pollutants. Regulation of lead in gasoline via provisions of the CAA dealing with fuel additives did not avoid the need for a regulatory impact analysis (RIA) because controls produced added costs exceeding a $100 million threshold. This led to a detailed econometric analysis using various cost–benefit approaches (U.S. EPA/Schwartz and coworkers, 1985). The Schwartz et al. analysis showed that there were economic benefits to the phasedown of greater than $5 billion annually when blood pressure-related and other health-based benefits are included and over $1 billion without the blood pressure adverse impacts. The U.S. EPA, following this report, soon issued regulations lowering Pb content as per its schedule.

The phasedown of Pb in gasoline was replaced by a phase-out schedule in the 1990 CAAAs, with phaseout to be completed in 1995. Phaseout was subsequently achieved.

It was fully expected that regulating lead in gasoline because of its adverse health impacts would then direct attention to the need for reframing ambient air Pb as a criteria pollutant, for which primary and secondary NAAQS would be required. One compelling argument in this reframing was the fact that the overwhelming fraction of Pb in U.S. urban ambient air by the early 1970s was auto exhaust Pb. Second, this huge contribution was from a mobile source dispersed throughout the United States, that is, the U.S. vehicular fleet consuming leaded gasoline. Third, recasting air Pb controls via the criteria pollutant mechanism confined regulation to health-based considerations with little opportunity to slow regulation by econometric arguments.

Such logic induced an environmental group, the NRDC, to bring suit against the U.S. EPA in Federal Court (Southern District of New York) in 1975 requiring the Administrator to designate ambient air Pb as a criteria pollutant, requiring the issuance of primary and secondary NAAQS. Pb was designated a criteria pollutant on March 31, 1976.

27.2.2 Lead in Ambient Air: the Regulatory Process, the Debates, the Litigation

The early 1970s saw certain segments of the lead industry, particularly manufacturers and marketers of gasoline lead additives, resisting lead emissions regulation as it related to regulatory definitions of TEL as a fuel additive and the authority of EPA under the fuel additives section of the CAA to regulate

TEL. This particular phase of EPA's attempts to control TEL was related to both health hazard issues and issues of fuel economy and to the technological requirements of a growing national auto fleet equipped with catalytic converters. Optimal function in these converters was required to maximize efficiency of controls on the gaseous criteria pollutants such as nitrogen oxides emissions. Public health and safety was therefore not the sole or most critical explicit component of direct gasoline additive control.

Listing lead as a criteria, ambient air pollutant within §108 of the CAA by the Administrator in 1976 changed the calculus for the lead industry considerably. Health concerns were not only to be taken into account, they would be dominant. Health risk, furthermore, was to drive the process without regard to associated costs to lower health risks per §108 requirements for all regulated criteria pollutants. For the lead industry that meant the gasoline lead antiknock additive market now faced further economic pressures from both continued growth in the catalytic converter-equipped auto fleet and controls on public health and ecological impacts of its activity. As a result, the industry gave considerable attention to lead exposure and public health issues in its interactions with regulatory agencies and the scientific community.

The 1978 air lead standard issued by EPA Administrator Douglas Costle had three discrete chronological components: a pre-air lead criteria document period; criteria document preparation by EPA scientists and a group of outside scientific and public health consultants; and use of the criteria document information by EPA's OAQPS in preparing its required material and associated recommendations to the administrator. A Federal court in late 1975 directed EPA to prepare the relevant technical documentation as part of the process for issuing an air lead standard. The burden of this document effort fell to EPA's Office of Criteria and Special Studies (OCSS), under the Health Effects Research Laboratory (HERL) at EPA's National Environmental Research Center in Research Triangle Park, NC. Section 108 of the 1970 CAAAs, as amended in June 1974, defined criteria documents to be used by the agency in setting NAAQS values as documents that:

> ... accurately reflect the latest scientific knowledge useful in indicating the kind and extent of all identifiable effects on public health or welfare which may be expected from the presence of such pollutant in the ambient air ...

A lead criteria document, the scientific basis for subsequent regulatory action, was issued in December 1977 (U.S. EPA, 1977). An ambient air Pb standard was issued in October 1978 (43 FR 46246; October 5, 1978), with a NAAQS value of 1.5 $\mu g/m^3$ Pb, determined as a quarterly maximum arithmetic average. This value was the same for the primary and secondary NAAQS for Pb. This NAAQS for Pb was to be reviewed at 5-year intervals to determine if changes were necessary, with recommendations for changes to come from an independent Congressionally specified advisory body called the Clean Air Scientific Advisory Committee (CASAC).

The administrator's selection of a numerical value for the Pb NAAQS was determined by a number of health-based factors: (1) an adverse health effect, hemoglobin reduction; (2) the protective statistic, virtually all children not exposed to lead with blood levels ≥15 μg/dl; (3) a background blood Pb level not based on air Pb, 12 μg/dl; (4) a blood lead: air lead ratio of 2 units; and (5) selection as the standard which provided an air-based PbB increment of no more than 3 PbB units.

Affected industry stakeholders again brought suit against the agency in 1978, after issuance of the 1978 air Pb NAAQS. Various components of the lead industry, via the industry's main trade group, the LIA, and a trade group for the fuels industry, the American Petroleum Institute (API), brought suit in the U.S. Court of Appeals for the DC Circuit, the court of jurisdiction for Federal regulatory matters. They argued, among other things, that EPA's Administrator Costle had acted arbitrarily and capriciously in the choice of NAAQS and its scientific basis.

The LIA suit was rejected [647 F.2d 1130, 1154, D.C. Cir., 1980] and certiori denied [449 U.S. 1042, 1980] as was the action brought by the API [665 F. 2nd 1176, D.C. Cir., 1981] and certiori denied [455 U.S. 1034, 1982]. These legal actions against the agency to bar the issuance and enforcement of the 1.5 μg/m^3 standard took the form of challenges at various levels and in various ways. Details of these interactions are principally captured in documents filed by the industry and contained in EPA's air office docket on the matter.

For example, the lead industry's technical criticisms during the pending rulemaking phase correlated closely with the quantitative significance of EPA's strategy for air lead levels. Industry's statisticians particularly challenged EPA's choice of the ratio relating PbB to air lead. The EPA value of 2 reduced the permissible air lead level to 1.5 units, while a value of 1 or 0.75 would increase the allowed air lead concentration to 3 and 4 μg/m^3, respectively. EPA's selection of the ratio value of 2 for children was based on the detailed analysis of this topic in Chapter 12 of the final draft of the air lead criteria document. Industry analyses of this ratio were relatively numerous both at the time of document preparation and when the proposed standard for lead in air was offered.

Two additional areas of controversy based on industry criticisms were the choice of the PbB GSD and the selection of anemia as the adverse effect in children to be guarded against for purposes of developing the air lead standard. The size of the GSD, or the degree of PbB data variability, in the PbB log-normal distribution in child populations determined the size of the fraction of lead-exposed children in the upper tail of that distribution. The larger the value, the larger this fraction of children at segments of the upper tail, and the farther into the upper tail extended some fraction of highly exposed children. For some cut point in the distribution, the PbB values theoretically would be higher, the higher the GSD. EPA selected a GSD in its air lead standard staff deliberations that was within the range of values described in the lead criteria document, sparking lively discussions with various outside interests. It was clear to all parties that certain sites and sources were associated with lead exposures producing high heterogeneity and associated higher GSDs in the impacted child populations. This would be typical of stationary emission sources.

The CAA of 1970 and its early amendments addressed national air quality issues. After it passed, a number of references to air lead and air mercury were incorporated in the CAA.

As noted earlier, §108(a)(2) of the CAA as amended in 1974 and administrative action by the EPA Administrator in early 1976 led to designation of lead as a criteria pollutant. A criteria pollutant requires air quality criteria developed through an air quality criteria document that becomes the basis for setting primary and secondary NAAQS. Criteria pollutants at the time of the CAA mostly covered significant emissions from mobile sources—cars, trucks, buses, etc.—but with phaseout of leaded gasoline, there was more focus on stationary sources and those postdepositional pathways compartments for lead such as roadside dusts.

NAAQS promulgated for ambient air criteria pollutants have four components: the indicator to be compliance-tested, the averaging time, the form of the standard's averaging time in terms of statistical expression, and the level of the criteria pollutant. While most interested parties focus on the level of an air standard, the other components are also of critical importance. For example, the averaging time bears directly on the relative frequency of permissible exceedences of regulated emissions such as lead particulate. In brief, the shorter the averaging time, the fewer the number of permitted exceedences. The statistical form of the measurements is also relevant to the issue of compliance by emitting entities. The 1978 standard was based on a maximum quarterly average (arithmetic mean). Other pollutant characteristics impinge on aspects such as the particle size distributions within ambient air, e.g., sub-2.5 μm respirable particulate.

Sections 112 of the CAA and 122(a) of the 1977 amendments deal with HAPs, that is, those air pollutants that are typically emitted by industrial processes and fixed sources. While lead is still on the original list in the CAA as a HAP (§112(b)(7)), it was also redesignated in 1976 as a criteria, ambient air pollutant under §108 (see above). Mercury and other toxic metals are on the HAPs list.

27.2.3 The 1990 CAAAs

The 1990 CAAAs comprised the most sweeping legislative changes in the basic 1970 CAA with the host of amendments enacted in that year after considerable debate and maneuvering by affected parties. Broad initiatives mandating action by the U.S. EPA included the systematic and quantitative assessment of the acid rain problem, particularly as this issue was linked to contributory pollutants in U.S. ambient air. A number of other amendments that affected lead regulation were part of the changes. An important CAAA provision mandated the phaseout, rather than just the phasedown of lead in gasoline by 1995. Section 106 of the amendments [Title I, Part D, Subtitle 5 of the original CAA] deals with nonattainment provisions for lead. Solid waste combustion emissions of lead are covered in §305(a) of the amendments [§129(a)(4)] in the original CAA.

Section 112(n)(1)(A) of the amended CAA directed EPA to evaluate the health hazards associated with emissions of HAPs from electric utility power generating units. HAPS identified for further evaluation included lead, out of a screening list of 67 substances.

Interim Air Pb Rulemaking: 1978–2008

In 1982, by a 5-year timetable mandated for all criteria pollutants, lead was revisited with a second criteria review and OAQPS assessment process. This resulted in a criteria document preparation process that was more extensive and complex, and a product much more exhaustive than was the 1976–1977 effort. Four volumes of the revised criteria document were prepared by 1986 (U.S. EPA, 1986) after several rounds of peer review provided by the agency's independent scientific peer review mechanism, the CASAC. As noted earlier, this committee was created by Congress to assure its independence from the Executive Branch's agencies in serving its advisory functions. The chapters having to do with sources of human lead exposure, the range and extent of lead poisoning in human and ecological populations, the environmental epidemiology of lead exposures around the world, and synthesizing the data were much more comprehensive than in the earlier report.

In 1989, the U.S. EPA's OAQPS employed this large database to derive a range of acceptable alternative lead NAAQS (versus the 1978 air lead standard of 1.5 $\mu g/m^3$) covering the range of 0.50–0.75 μg Pb/m^3 air (U.S. EPA—OAQPS Staff Paper, 1989). The range was applied to a modeling approach using three U.S. case studies representing Dallas, TX; Tampa, FL; and East Helena, MT. In these case studies, the epidemiological metrics employed were mean PbB values and percent exceedence rates at 10 and 15 $\mu g/dl$ for either children or adults. While the 1978 air Pb standard of 1.5 $\mu g/m^3$ would allow 1.2–1.9% of children in the case areas to have their blood lead levels exceed 10 $\mu g/dl$, the 0.5 $\mu g/m^3$ alternative standard would be considerably lower, about 1.0%. On a national population basis, these differences are significant in total child numbers.

The CASAC panel (as created and charged by §109(d)(2), the 1977 CAAAs) reviewed the range of alternative air Pb standard values in the 1989 OAQPS Staff Paper. In its January 1990 report to the EPA Administrator, the panel recommended consideration and promulgation of the 0.5 $\mu g/m^3$

alternative to the existing 1.5 value. In addition, the members recommended further evaluation and inclusion of 0.25 $\mu g/m^3$ as one of the alternative options (CASAC, 1990).

The agency elected not to pursue further technical analysis or RIA of the alternative OAQPS-recommended and CASAC-reviewed NAAQS. Instead, it elected to adopt a multimedia approach to regulating lead (U.S. EPA, 2006). The stated rationale for this decision lay in the observation that, formerly, leaded gasoline combustion was the dominant source of lead in ambient air and required control for air input reasons. The phasedown and the later-mandated phaseout of Pb in gasoline (1990 CAAAs) reduced the need to focus on mobile source-based air lead and amplified attention to inputs from other sources.

In the 1990s, however, the U.S. EPA did issue air Pb regulations under Section 112 of the CAA dealing with HAPs [42 U.S.C. §7412], to limit emissions of Pb from primary and secondary Pb smelters. Section 112 requires that the Agency establish NESHAPS emitted in the amount of 10 tons of pollutant or more or any HAP combination of 25 tons or more (termed major sources). There are also area sources, the emitters that are not major emitters, e.g., smelters. Thirty of the latter, called "urban HAPs," came under regulation in July 1999 (64 FR 38715; July 19, 1999). Primary Pb smelters were regulated under § 112(d) provisions in 1999 (64 FR 30194; June 4, 1999), while revised standards for secondary facilities were issued in 1997 (62 FR 32209; June 13, 1997). Unlike ambient air controls applied for criteria pollutant Pb as primary and secondary NAAQS, smelter regulation hewed to a requirement that the maximum emission reductions attainable be sought.

More recently, provisions of §112 were applied to court-ordered HAPS standards for 11 source categories, nine of which were promulgated in July 2008 (73 FR 42978; July 23, 2008; 40 C.F.R. Part 63). The categories included nine metal fabrication and finishing sources, with the statutory requirement for a 90% reduction of emission levels for lead, cadmium, chromium, manganese, and nickel.

The 2008 U.S. Lead NAAQS

Air Pb levels in the 20-year interval from 1983 to 2002 declined by 94% in U.S. urban and suburban areas, with emission tonnages corresponding to these figures declining from 220,000 tons in 1970 to <4,000 tons in 1999 (U.S. EPA, 2006). However, the previous decades of airborne Pb emissions created an enormous terrestrial reservoir of the element that could be reentrained as fugitive dusts at smelter sites and along countless U.S. roadways. Furthermore, stationary sources such as smelters, foundries, and incinerators noted in earlier chapters were emitting discernible levels.

While air Pb levels declined considerably in recent years, so did the threshold values for exposures and toxicity markers identified in sensitive subsets of human populations such as infants and toddlers. In fact, the scientific and health policy consensus growing in multiple deliberative settings is that no clearly safe level of lead exposure has been identified. Rather, research has relied on empirical thresholds in terms of measured or modeled PbB concentrations that continue to correlate with sensitive toxic endpoints at blood values below 5 μg/dl. The upshot of these parallel movements in exposures and toxic endpoints downward is that even modest Pb exposures can be of toxicological and epidemiological significance.

In 2004, the Missouri Coalition for the Environment sued the U.S. EPA in U.S. Federal court [Civil Action No. 4:04-CV-00660, Eastern District, MO] to produce a new Pb criteria document and issue a revised NAAQS for lead. The new criteria document was finalized in October 2006 (U.S. EPA, 2006).

Subsequently, as described in an earlier chapter, the agency's OAQPS prepared several risk assessment exercises for three case studies, one of which was a general U.S. urban Pb exposure case and two of which addressed point sources in the form of a primary and a secondary smelter community. From this risk assessment methodology, employing combinations of modeled and empirical data sets, OAQPS evolved a range of proposed alternative NAAQS to substitute for the long-intact 1978 value of 1.5 $\mu g/m^3$. The 2006 Pb criteria document, OAQPS's full risk assessment modeling exercises, and the administrator's proposed NAAQS promulgation in May 2008 were all reviewed by the CASAC air Pb panel in the 2006–early 2008 period.

In November 2008, the EPA Administrator promulgated a significantly revised air Pb primary and secondary NAAQS (73 FR 66964; November 12, 2008), effective January 12, 2009. The air Pb level was revised downward 10-fold, from 1.5 to 0.15 $\mu g/m^3$. The indicator metric remained Pb in air, to be measured as total suspended particles, but the averaging time was revised from a quarterly maximum average to a rolling 3-month average with a maximum form.

One consequence of the marked revision downward of the air Pb NAAQS was reestablishment of the need for a broader urban air Pb monitoring network. The mandated reporting requirements for emitters of more than 0.5 tons annually of Pb (revised from the 1.0 tons annually in the original rulemaking), added to the need to expand the air Pb monitoring network.

27.3 INTERNATIONAL REGULATION OF LEAD IN AMBIENT AIR

27.3.1 Introductory History and Perspective

This portion of the chapter discusses the evolution and implementation of various ambient air Pb controls in countries outside the United States. The historical and regulatory aspects of air Pb controls in other countries and world areas differ in various ways from those in the United States. On the other hand, there are elements guiding the rationales for air Pb regulation that are common to both.

International controls on air Pb can be broadly divided into actions proposed or implemented for industrialized nations and those for developing countries. Typically, but not invariably, the latter adopt the lead regulations set by the former or their political organizations, e.g., the European Union.

The EEC is a postwar creation, formed in 1956. Managing European air quality control actions through the EEC began in the early 1970s. In this period, air quality actions varied considerably from country to country. These 1970s-era actions, e.g., Directive 70/220/EEC, can be seen as having a more economically tilted rationale than an environmental health-protective one. For example, there was a concerted push to harmonize nation-specific standards so as to ease the economics of vehicular marketing. EEC's role was constrained to some extent by questions of constitutional power issues in that a constitutional basis for European Parliament actions in the environmental protection sector did not become settled until the 1987 Single Act.

The involvement of the European Union in continental air quality matters began in 1980, with Directive 80/779/EEC, which set air quality limit values (AQLVs) for sulfur oxides (SO_2) and (total) suspended particulates. AQLVs can be viewed as the regulatory equivalent of the United States NAAQS for criteria pollutants described earlier in this chapter. Lead, nitrogen dioxide, and ozone limit values were promulgated by later directives. The EU's work in air quality actions focused on not only harmonizing levels of air pollutant limit values across EU members but also unifying the monitoring and compliance processes. Much as was the case for U.S. air quality actions by the U.S. EPA, the European Union and member nations faced such lead regulation problems as what indicator to measure, the level of the pollutant, averaging time, etc.

Structurally, air Pb regulation in the European Union is a binary process. The scientific and health risk rationales for controls on lead in various environmental media are first articulated by multistate bodies like the WHO, the UNEP, and the ILO. Subsequently, ministerial actions either collectively by EU Directives via the European Commission or through national ministries give consideration to and may often adopt WHO guidelines as regulations. In a number of cases, various nations had air quality controls under way before the evolution of concerted collective actions through the European Union. Industrialized countries outside Europe, e.g., Japan, will often adopt or harmonize actions with WHO guideline recommendations.

27.3.2 The WHO Environmental Criteria and Guideline Process

The WHO currently provides scientific and public health advice and articulates environmental health guidelines through two mechanisms. The first is preparation and dissemination of Environmental Health Criteria (EHC) documents. WHO shares responsibilities for preparation of EHC reports on lead and other environmental pollutants through the International Programme on

Chemical Safety, based in Geneva, Switzerland, with other programs of WHO. The second mechanism is more specifically directed to pollutants in ambient air. In both cases, documents are prepared through a multidraft process where authors first prepare working draft sections. These undergo peer evaluation by Working Group members meeting in a plenary and subgroup gathering attended by WHO and other staff and followed by preparation of drafts which then traverse the WHO administrative structure prior to publication.

The more global report on inorganic lead as a human health hazard is EHC 165: Inorganic Lead, published in 1995 (WHO, 1995). This dealt principally with lead and public health and updated an earlier document published in 1977 (WHO, 1977). An earlier EHC report, EHC 85: Lead—Environmental Aspects, appeared in 1989 and addressed ecological effects (WHO, 1989). The 1995 EHC document for Pb issued a number of recommendations for elimination or marked reduction of lead in sources impacting humans. In terms of air Pb, the 1995 report called for elimination of lead additives in gasoline and other media-specific lead elimination measures. Added recommendations included systematic lead screenings of children using measurement of PbB and monitoring of Pb levels in media relevant to human exposures.

A second mechanism providing guidance for air quality standards for European and non-European states, principally through the European Union, has been the Air Quality Guidelines for Europe (AQGE) reports. Two reports were issued by WHO's Regional Office for Europe, Copenhagen, Denmark. The first appeared in 1987 and the updated document, the Second Edition of the Guidelines, in 2000. The reports covered four classes of pollutants, mainly in terms of human health risks: Organic pollutants, inorganic pollutants, classical pollutants, and indoor air pollutants. Additional discussions addressed ecotoxic effects of the classic contaminants: nitrogen dioxide, ozone, particulate matter, and sulfur dioxide. The inorganic contaminants group included 8 metals and 1 metalloid (arsenic). Lead was included among the metals.

The 1987 report considered various aspects of airborne lead impacts on human health, then-current air Pb levels in different regions of Europe, the translation of airborne Pb levels to risk of Pb toxicity in risk groups such as young children, and it offered recommendations for air Pb guidelines to various member states. The recommended air Pb guideline, averaged on an annual basis, was in the range of $0.5-1.0\ \mu g/m^3$. This narrow range took into account both direct inhalation exposures to air Pb and ingestion, by children, of post-fallout Pb as residential leaded dusts. It included protection of 98% of affected children exposed to air Pb.

The 2000 air guidelines report refined and made more quantitative the health risk calculations for air Pb exposures of young children, which led to a lowered recommendation of $0.5\ \mu g/m^3$ Pb in its draft 1995 interim report, a value retained in the final draft document.

General Aspects of European Air Lead Regulatory Directives

The 1996 Air Quality Framework Directive (AQFD) provided, along with "daughter" directives, the wherewithal for some level of harmonizing management of air quality within the European Union, although individual members were provided considerable latitude in articulating the means and methods by which they implemented the broad EU's AQ standards' language. States were enjoined to issue controls for both mobile and stationary sources, the latter to include incinerators, boilers, and other large combustion operations. Analogous to regulatory machinery in the United States, the European Union required controls on both air emissions and fuel quality criteria. Specific air emissions limits were established as of 2004 for the gaseous pollutants ammonia, nitrogen oxides, sulfur dioxide and volatile organic compounds, carbon monoxide, particulates (at PM_{10}), ozone, the volatile organic benzene, and lead. Singled out for regulatory consideration were: polycyclic aromatic hydrocarbons, arsenic, cadmium, mercury, and nickel.

EU member states were also given latitude in such processes as defining the boundaries of geographical areas, including those with populations $\geq$250,000 residents (called agglomerations), within the overall jurisdictions, for purposes of air quality monitoring and management planning. Member states' autonomy in defining these zones and agglomerations was assured by the fact that review by the European Union was barred. However, states must provide the AQFD tabulations of those zones and agglomerations where at least one pollutant exceeds the sum of limit value and tolerance margins, those that exceed the limit but are within the margins, and those with all air contaminants within the limit values.

27.3.3 Regulatory Initiatives in the European Union for Lead in Ambient Air

Air Pb in the EEC was first considered as a category 1 pollutant (by EEC classification) in the early 1980s, and the European Community via Directive 82/884/EEC established an air Pb value of 2.0 $\mu g/m^3$ as an annual average. Compliance with this initial value was not deemed problematic within member states because phase-down actions for reducing and removing lead in gasoline had begun.

Air Pb was one of the air pollutants scheduled for inclusion for revision in the AQLVs in September 1996 (EU Directive 96/62/EC) and for which new AQLVs were proposed in October 1997. The lead quality limit value, formulated as a human health-based standard, was proposed at 0.5 $\mu g/m^3$, with an annual averaging time, and was to be achieved by January 1, 2005. This proposed value was identical to that recommended as an air Pb quality guideline within the WHO AQGE's interim 1995 draft document. The proposed set of air quality revisions was accompanied by a Position Paper on

Lead appearing in November 1997, setting out the public health status and the case for lowering the regulatory level of air Pb. As was the case with air Pb—gasoline Pb regulatory relationships in the United States, attention was paid to scheduling reductions for lead in both gasoline and ambient air.

On April 22, 1999, the Council of the European Union issued Directive 1999/30/EC, which updated the AQLVs for the four pollutants/pollutant classes: lead, nitrogen dioxide and oxides of nitrogen, particulate matter, and sulfur dioxide. Annex IV of the directive describes the reduction of the AQLV for Pb to 0.5 $\mu g/m^3$, expressed as an annual mean, to be fully implemented by January 1, 2005. For industrial point sources with historical high Pb contamination, the applicable deadline was January 1, 2010. In the time between 2005 and 2010, the point source air values were permitted to be higher, 1.0 $\mu g/m^3$ averaged annually. The directive detailed the placement of air quality monitoring and sampling points as a function of factors such as population density and spelled out measurement methodology.

Directive 2008/50/EC of the Council of the European Union was issued on May 21, 2008 and was an array of provisions by which one evaluated ongoing status of air quality within member states in a periodic fashion after issuance of the AQLVs for the indicated pollutants. The Directive, for example, articulated the assessment methodologies for exceedences and compliance in 5-year reviews. With lead and other ambient air contaminants in the European Union, exceedences are ascertained in terms of upper and lower assessment thresholds (Annex II). Air Pb upper and lower assessment thresholds were set at 70% or 0.35 $\mu g/m^3$ and 50% or 0.25 $\mu g/m^3$, respectively of the 0.5 AQLV.

Within the EU's AQFD, each member country is responsible for controlling the level of lead and other named pollutants within their territories and is simultaneously responsible for effecting reductions of lead emissions within its borders. Attainment plans are required for zones and agglomerations for those pollutants not within AQLV attainment; plans are not required if there is attainment. Transboundary considerations for ambient air Pb, as with other EU-relevant air pollutants, require that multiple countries pursue cooperation and consultation where one member state is negatively impacted by lead emissions from a neighboring state. Emission control plans for air Pb and other air pollutants require member states to list main emission sources, total emissions (metric tons), any imported pollutant problems, and planned implementation of methods for improving air quality.

A 2002 European Commission report (EC, 2002) presented some progress on the attainment of the AQLVs for lead and other EU pollutants in terms of graded performance. The assessments noted achievement of the limit value+margin of tolerance, the margin of tolerance being defined for each pollutant in preceding Directives. For air Pb, nine countries reported on exceedences in air Pb as the sum of the limit value and margin of tolerance. Of these, all but one member state reported no geographic zones in the countries exceeding this value. One country had some zones exceeding this combined level but no states had most of their zones exceeding the limit value+margin of tolerance figures.

27.3.4 Illustrative Air Pb Control Activities in Individual Countries

Different EU member countries have employed various regulatory approaches in dealing with ambient air Pb concentrations. The origins of these differences are multiple, involving factors such as the intensity of industrial activities producing point source Pb emissions, the volumes of national leaded gasoline consumption rates, and their economic aspects (see next section). For example, those 15 member countries comprising the European Union in 1994—Austria, Belgium, Denmark, Finland, France, Germany, Greece, Ireland, Italy, Luxembourg, the Netherlands, Portugal, Spain, Sweden, the United Kingdom—varied greatly in their percentage contributions. The relative percent contributions by the above-indicated nations were, in 1990: 1.6%, 3.1%, 0.9%, 1.2%, 16.2%, 15.5%, 2.7%, 0.7%, 15.0%, 0.5%, 2.0%, 3.4%, 17.5%, 2.4%, 17.2%, respectively (ESQUAD, 1994). Five countries collectively accounted for the majority of the EU Pb total emissions, 80% (rounding).

The CEC-Directorate-General XI (1997) in late 1997 issued summary air Pb regulatory particulars for the 15 EU member states then in the Union and two additional countries. The countries identified above as having the highest relative Pb emission rates to ambient air in or before 1997 had established the air Pb regulatory value of 2.0 μg/m^3, promulgated at that time by EU Directive 82/884/EEC. Many of the member states making smaller contributions of Pb to ambient air similarly adopted the earlier EU limit value from the 1980s. In late 1997, there was no Austrian air Pb limit value but the figure of 0.5 μg Pb/m^3 as an annual mean was being advanced. Belgium, functioning since 1993 as three regions for purposes of air quality regulatory initiatives, employed the old 2.0 μg/m^3 EU Directive value. Sweden, by contrast, adopted the 1978 U.S. air Pb NAAQS of 1.5 μg/m^3 as a quarterly maximum average. Poland and Switzerland, the nonmember states, had AQLV figures below 2.0 μg/m^3. By the late 1990s, Poland implemented two air standard maxima: 1.0 μg/m^3 Pb averaged for 24 hours and an annual mean of 0.2 μg/m^3. Switzerland implemented an annual mean regulatory value of 1.0 μg/m^3.

With the adoption of the annual mean 0.5 μg/m^3 air Pb AQLV by the European Union in 1999, many of the individual member nations adopted at least this level and some pursued an even more protective regulatory objective. For example, the United Kingdom in 2003 stated its intent to implement the WHO and EU annual mean figure of 0.5 μg/m^3 by December 31, 2004, and half-that, 0.25 μg/m^3, by December 31, 2008.

Air Pb Regulation in Canada

The Canadian regulation of ambient air Pb both resembles and differs from air Pb regulation in the United States with respect to primacy for control and air Pb control methodology. The principal resemblance lies in both nations relying on their respective CAAs. The principal difference is in the division of regulatory labor between the respective Federal governments and the state or provincial jurisdictions. In the United States, the Federal CAA and its various amendments spell out what and how regulation is to occur, but mandates to the 50 states the implementation of the regulations via the SIPs noted earlier. States, however, are offered limited latitude in deviating from the Federal language and intent with the notable exception of opting for California's parallel regulatory initiatives. The 1977 U.S. CAAAs in fact offered states the option to adopt either Federal or California air standards in their SIPs.

The Canadian structure is opposite, in that regulatory power largely resides with the various provincial ministries to establish and enforce ambient air pollutant standards. Canada currently has only three country-wide standards, those for particulate matter, ozone, and mercury. Ambient air Pb is regulated by the provinces.

The Canadian Federal government enacted two legislative initiatives for controlling ambient air pollutants like lead. The first, the CAA of 1970, regulated four specific air contaminants: lead, mercury, asbestos, and vinyl chloride. This was followed by a more comprehensive package of legislation in Ottawa called the Canadian Environmental Protection Act, first enacted in 1990 with amendments following in 1999 (effective in 2000). To date, a nationwide air lead standard is not in force. The provincial air Pb standard implemented by Ontario and in force through August 2008 consisted of two air Pb standards [R.R.O. 1990, Regulation 337, last amended O. Reg. 518/07] that depended on two averaging times: 2.0 $\mu g/m^3$ averaged over 24 hours, and an arithmetic mean of 0.7 $\mu g\ Pb/m^3$ as the long-term value.

These initiatives addressed ambient air as encountered by Canadian populations. Other lead regulations have also been in force. With respect to stationary source Pb emissions, Canada has regulated emissions exiting on-site points from secondary lead smelters, such as stacks. Secondary lead smelter regulations were first issued in 1976 under the 1970 Canadian CAA and later replaced by the requirements [SOR/91-155] of the 1990 Canadian Environmental Protection Act (§34(1)) in early 1991.

27.3.5 Controls on Leaded Gasoline Use Outside the United States

Controls on Leaded Gasoline in the European Union

The major focus on leaded gasoline consumption in controlling air Pb levels in the European Union paralleled that in the United States in the course of the latter's phase-out actions. These efforts were largely summarized in several reports by various international bodies and academic groups. The genesis of this interest, as in the American experience, was that leaded gasoline consumption was the overwhelming source of ambient air Pb levels adversely affecting general populations in member countries. As reported in the Commission of the European Communities' Position Paper on Lead (CEC-Directorate-General XI, 1997: Table 1.3), the relative percentage contribution of vehicular traffic to ambient air Pb levels in 15 member states ranged from 50.0% (Belgium) to 96.6% (Portugal), with 12 countries showing over 75% of air Pb due to mobile sources (vehicular traffic) and 8 showing over 80%.

The leaded gasoline phase-down and phase-out timelines in the European Union lagged behind comparable actions in both the United States (phaseout, 1995) and Canada (phaseout, 1990). However, some member states addressed the leaded gasoline matter much earlier than others. Germany, in 1972, announced a two-stage phasedown of lead content in gasoline: 0.4 g Pb/l in 1972, declining to 0.15 g/l in 1976. In the European Union, the fuel Pb content was set at 0.4 g/l in 1981, but by 1985 availability of lead-free gasoline (<0.013 g/l) was mandated, so catalytic converters in vehicles could achieve air quality standards for other pollutants.

In the years from the mid-1980s to the late 1990s, European countries hewed to a highly diverse timetable for the phaseout of leaded gasoline. A chronological history of these activities is contained in the estimates and tabulations of the World Bank Technical Paper No. 397 (Lovei, 1997). Table 3 of this Technical Paper included European Union member states in its stratified reporting of phaseout of leaded gasoline and phasing in of unleaded gasoline in nations around the world, based on data mainly from 1996. Low-lead-fuel nations (gasoline Pb <0.15 g Pb/l) had a high percentage of unleaded gasoline as a share of all fuel stocks, 70–100%, including in Europe: Austria, Belgium, Denmark, Finland, Germany, Luxembourg, the Netherlands, Norway, Slovak Republic, Sweden, and Switzerland. Non-European states with high unleaded fuel percentages included the following: Argentina, Brazil, Costa Rica, El Salvador, Guatemala, Honduras, Hong Kong, Japan, Thailand, and the United States.

As of February 2009, only 11 countries around the world used some amount of leaded gasoline and only 3 used it exclusively (Tables 27.1 and 27.2).

Control of Leaded Gasoline Use in Canada

Canada mandated the complete phaseout of lead additives from gasoline in 1990.

Control of Leaded Gasoline in Developing Regions

The UNEP reported in late 2005 that Sub-Saharan Africa would be free of leaded gasoline by January 2006. This achievement was a result first cited for

TABLE 27.1 Significant U.S. Regulations or Guidelines for Lead in Ambient Air

Statutory/Regulatory Initiatives	Codified Law or Regulation	Year(s)	Key Features
A. Federal actions			
U.S. CAA of 1963	77 Stat. 392; 42 U. S.C. §7401	1963–1966	General air pollution controls on point sources (steel mills, foundries, power plants) and mobile sources (vehicles)
Motor Vehicle Air Pollution Control Act of 1966 [PL 90–148]	81 Stat. 485	1966–1970	Regulatory distinctions in source emissions, mobile versus stationary
1970 CAAAs [PL 91–604]	84 Stat. 1676; U.S. C. §§7401–7681q; 40 C.F.R. Part 50	1970–1978	Statutory basis for beginning controls on leaded gasoline; issued mandates for "hazardous" and "criteria" pollutants emitted to air
1975–1976 Federal Court action in suit against EPA under CAA	EPA declares air Pb to be a "criteria" pollutant; §§108, 109 rulemaking [U.S.C. 7408, 7409]	1976–present	EPA action allowed controls on ambient air in the United States versus emissions from fixed sources
1970 CAA mandate for Pb standard as criteria pollutant	EPA issues air Pb primary and secondary NAAQS of 1.5 $\mu g/m^3$	1978–2008	Air Pb standard mostly focused on mobile U.S. sources such as the vehicular fleet
Federal Court action in 2005 under the CAA to review 1978 air Pb standard	EPA issues air Pb primary and secondary NAAQS of 0.15 $\mu g/m^3$	2008–present	New air Pb standard intended for both direct and postdeposition reentrained Pb
CAAAs of 1990	Changes to §106, dealing with nonattainment of air Pb standards; §129, dealing with waste combustion, §305(a)	1990–present	CAA 1990 Amendments concern Pb and other HAPs emitted from fossil-fueled power plants
B. U.S. state controls			
California air quality statutes, e.g., the 1988 California CAA	Air Quality Districts established the CARB	1971–present	1970 CAA spelled out CA regulations for air quality as co-equal with Federal requirements; states can use CA controls in certain cases; in 1971, CA issued first air Pb standard in the United States.

TABLE 27.2 Significant International Regulations or Guidelines for Lead in Ambient Air

A. WHO guidelines for air Pb		
EHC documents for Pb	EHC 3: Lead; EHC 85: Lead—Environmental Aspects; EHC 165: Inorganic Lead	EHC documents present scientific data for Pb in all environmental media (see text)
AQGE, Lead	1987 AQGE; 2000 AQGE update	Guidelines are for Europe, but are typically adopted as directives and regulations in Europe and elsewhere (see text)
B. Air Pb controls in the European Union		
EU Council Directives	82/884/EEC 1982–1996	Established an air Pb of 2.0 μg/m^3, annual average
EU Council Directives	96/62/EC 1996–1999; 99/30/EC 1999–2008	Established AQLVs for Pb and other pollutants, as an annual mean, to be in place in member states by January 1, 2005; Pb level set at 0.5 μg/m^3
EU Council Directives	2008/50/EC 2008–	Various measures for monitoring ongoing air quality status for Pb and certain other pollutants; retains ambient air Pb level of 0.5 μg/m^3
C. Gasoline Pb controls in the European Union		
EU Council Directives to member states on fuel Pb content	78/611/EEC; 85/210/EEC	1981–1985 Maximum Pb content is 0.40 g Pb/l gasoline; unleaded gasoline defined as having ≤0.013 g/l; leaded fuel permitted at 0.15–0.40 g/l; lowered to 0.15 g/l in 1985
EU Council Directive on banning leaded gasoline sales	98/70/EC	Banned sale of leaded gasoline in member states after January 1, 2000; unleaded gasoline to have 0.005 g Pb/l or less
C. Air Pb regulations in some individual countries outside the United States		
Canada Federal Government, Provinces	Gasoline Pb phaseout 1990 under Canada's CAA; air Pb now controlled via the Environmental Protection Act	Complete phaseout in 1990

(*Continued*)

TABLE 27.2 Significant International Regulations or Guidelines for Lead in Ambient Air—(cont.)

EU member and nonmember states, Western Europe: various national limits on gasoline Pb, 1997	0 g Pb/l permitted in Sweden, Finland, Denmark, Austria; 0.15 g/l permitted in Belgium, France, Germany, Italy, the United Kingdom, Switzerland, the Netherlands, Norway, Ireland, Iceland; Portugal, Spain, Greece, Turkey retained 0.40 g/l through late 1990s	Western Europe nations differ in maximum Pb levels permitted in gasoline; majority adopted 0.15 g Pb/l prior to ban on sale of leaded gasoline in the European Union

pursuit in the Dakar Declaration of 2001 calling for leaded fuel phaseout by 2006. A partnership of UNEP and NGOs indicated in 2006 intent to pursue a phasing out of leaded gasoline globally by 2008. At the end of 2005, 30 nations globally, many of them in developing regions, were still using leaded gasoline.

REFERENCES

Clean Air Scientific Advisory Committee, 1990. Report of the Clean Air Scientific Advisory Committee on its Review of The National Ambient Air Quality Standards for Lead. CASAC Closure of the 1989 Air Quality Criteria Document Addendum and the 1989 Staff Position Paper, January 3. U.S. Environmental Protection Agency Science Advisory Board, Washington, DC.

ESQUAD, 1994. van den Hout, K.D. (Ed.). The impact of atmospheric deposition of non-acidifying pollutants on the quality of European forest soils and the North Sea. Main report of the ESQUAD project. RIVM- National Institute of Public Health and Environmental Protection, Bilthoven, The Netherlands.

European Commission, 2002. Overview of Methods and Results of the Preliminary Assessment of Air Quality in Europe under Directives 96/62/EC and 1999/30/EC. Directorate-General, Environment. Brussels, Belgium.

Lovei, M., 1997. Phasing Out Lead From Gasoline. Worldwide Experience and Policy Implications. World Bank Technical Paper No. 397. Pollution Management Series. The World Bank, Washington, DC.

Rosner, D., Markowitz, G.A., 1985. 'Gift of God?' The public health controversy over leaded gasoline during the 1920s. Am. J. Publ. Health 75, 344–352.

Schwartz, J., Pitcher, H., Levin, R., Ostro, B., Nichols, A., Authors. Report No. EPA-230/05-85-006. Office of Policy Planning and Evaluation, Washington, DC.

U.S. Environmental Protection Agency, 1977. Air Quality Criteria for Lead. Report No. EPA-600/8-77-017. Office of Research and Development, Washington, DC.

U.S. Environmental Protection Agency, 1985. Costs and Benefits of Reducing Lead in Gasoline: Final Regulatory Impact Analysis.

U.S. Environmental Protection Agency, 1986. Air Quality Criteria for Lead, 4 vols. EPA 600/8-83/028bF. Environmental Criteria and Assessment Office, Research Triangle Park, NC.

U.S. Environmental Protection Agency, 1989. OAQPS Staff Paper: Review of the National Ambient Air Quality Standards for Lead: Assessment of Scientific and Technical Information. March. Office of Air Quality Planning and Standards, Washington, DC.

U.S. Environmental Protection Agency, 2006. Air Quality Criteria for Lead, vol. 1. Report No. EPA/600/R-05/144aF. National Center for Environmental Assessment, Washington, DC.

World Health Organization, 1977. Environmental Health Criteria 3: Lead. World Health Organization, Geneva, Switzerland.

World Health Organization, 1985. Environmental Health Criteria 85: Lead—Environmental Aspects. International Programme on Chemical Safety, Geneva, Switzerland.

World Health Organization, 1995. Environmental Health Criteria 165: Inorganic Lead. International Programme on Chemical Safety, Geneva, Switzerland.

Chapter 28

Regulation and Regulatory Policies for Lead in Water

28.1 A BRIEF PERSPECTIVE ON AMBIENT AND DRINKING WATER Pb REGULATION

The systematic regulation of lead in drinking and ambient waters began in the last 50–60 years. The absence of controls on the element in water until more recent decades contrasts with the much longer public health history of water Pb exposures in human populations. Water lead's toxic impacts trace in at least qualitative terms to Greco-Roman times or earlier and have been described in such writings as those of Nriagu (1983, 1985). By the nineteenth century, isolated studies of water Pb toxicity in humans comprised an established database in the United States, the United Kingdom, and continental Europe.

This belated regulatory attention was arguably attributable to several factors. First, no historical quantitative toxicity record was available to view dose–toxic response relationships for water Pb versus toxic hazards attributable to other lead-containing media. Second, water Pb was not a readily visible hazard, unlike suspended leaded dusts released to workplace air during dry sanding and scraping of old lead paint by painters. Third, no effective regulatory apparatus existed in industrialized countries for regulating toxic elements in drinking water carried by pipes, aqueducts, and other means. Fourth, water Pb levels, even at highly hazardous levels of pollution, were not readily measurable given the relatively crude analytical methods in use until more recent times. While Pb concentrations might be low and difficult to measure, volumes of the media consumed in typical populations were such that total daily intakes would be significant. Finally, unlike the case for lead in paint and lead in ambient air, human health hazard information was not available for potable water Pb exposures first arising in the workplace and then tracking into the common environments of general human populations.

28.1.1 A Brief Historical Summary of Water Pb and Human Health

Lead leaches into drinking water via lead pipes, household plumbing solders, kettle solders, lead-lined cisterns collecting rainwater or storage tanks, with

Trace Metals and other Contaminants in the Environment, Volume 10
ISSN: 1875-1121 DOI: 10.1016/B978-0-444-51554-4.00028-6

lead piping historically being the most common leaching source (Nriagu, 1983, 1985). A number of references have been made to water transport in lead lines since the Greco-Roman period (Nriagu, 1985), although we also know from archaeological pipe fragments that lead pipe was used in dynastic Egypt and Asia Minor (Aitchison, 1960). It is clear that drinking water was viewed as a hazardous source of lead exposures. In the Imperial Roman period, the architect Vitruvius counseled caution in using lead for water transport, although this caveat probably drew more upon known occupational lead poisoning in pipe manufacture than from direct water lead toxicity (Nriagu, 1985).

Although interest in the role of water Pb in human poisonings grew in post-Roman times, poisoning episodes were sporadic and idiosyncratic as to source. Tronchin (1764) described a widespread poisoning episode, characterized by clinical GI effects, in Amsterdam, the Netherlands, that arose from drinking rainwater collected in lead-lined cisterns and/or gathered from runoff over leaded roofing.

Reports of rainwater plumbosolvency linked to human toxicity populated the later public health record in the nineteenth century following the advent of the Industrial Revolution in Europe. This was particularly evident after central surface water supplies were established for growing cities. For example, in the British Isles where soft water supplies with low pH and a high propensity for plumbosolvency were piped into lead-lined storage tanks in areas such as Glasgow and Ayr in Scotland (Moore et al., 1981; Richards and Moore, 1984), in North Wales (Elwood et al., 1983), and elsewhere in Britain (Pocock et al., 1983), the public health history clearly shows water supply lead toxicity hazards. The potential for toxic lead exposures in Great Britain with lead plumbing and soft water was actually recognized relatively early in the nineteenth century, as in the report of Christison in Edinburgh (Christison, 1844).

Four years later, Thomson (1848) described clinical lead toxicity in English residents poisoned from lead-piped and cistern-stored domestic water for general consumption in two close-by villages, Weybridge and Chertsey (Addlestone). He further described some chemical analytical testing of lead in domestic waters. Thomson's case reports were shortly followed by those of Robertson (1851), who described in *Lancet* a mass lead poisoning of children at a girl's school, traced to a leaded cistern.

Similarly, areas of the United States with soft water supplies and industrializing areas using lead piping had long been known to provide a risk of lead toxicity from plumbosolvent water consumption. This was particularly so for New England. Dana (1848), in an Appendix to his translation of Tanquerel des Planches' treatise on lead poisoning among French lead workers and the general population, discussed lead toxicity accompanying installation of a water supply system that included lead piping and plumbosolvent water supplies in the mid-nineteenth century in the growing textile center of Lowell, MA. That system was installed in 1839.

Four years later, Bigelow (1852) described his simplistic arguments minimizing the corrosive action of a Boston surface water source, Cochituate Lake, on leaden pipes used to distribute potable water to the city's residents. He did so while not ruling out that, under certain conditions of seasonality and water chemistries, some risk to susceptible individuals might occur. Bigelow also drew attention to the findings of a commission that lead piping conducting soft waters to the city of London was held not to be a source of consumer lead exposures or associated lead poisoning.

In marked contrast to these early views and conclusions, epidemiological studies conducted in the twentieth century using more sophisticated analytical measurements of drinking water lead and lead in blood showed that, in fact, various natural conditions favored lead leaching from lead piping and other lead components, particularly in New England and Pacific Northwest communities using soft, corrosive waters (Ohanian, 1986). The collective impact of these findings has been to prompt regulatory controls on both drinking water Pb and on Pb use in community water distribution systems.

Lead levels in ambient water serving as actual or potential drinking water sources were reported in several earlier U.S. studies. Those findings predated later investigations that combined the inputs of both distribution and residential plumbing lead sources. Hem and Durum (1973) reported that $<0.5\%$ of surface water samples from >700 sites in the United States and Puerto Rico exceeded the then-current 50 μg Pb/l regulatory level. Finished water Pb levels at 969 U.S. treatment systems prior to distribution showed levels from undetectable to 640 μg/l, with 1.4% exceeding the then-current interim standard of 50 μg/l. Durfur and Becker (1964), reporting levels for systems of the largest 100 U.S. cities, noted that only 5% of these systems exceeded 10 μg/l, had a median value of 3.7 μg/l, and had a maximum level of 61 μg/l. The U.S. EPA (1975) survey of 592 interstate carrier water supplies reported only two of these had water >50 μg Pb/l.

Ohanian (1986) and Levin (U.S. EPA/Levin, 1986) summarized the components of U.S. water distribution and household plumbing systems likely to yield elevated water Pb levels under varying conditions. Lead pipes in household service connections, estimated to comprise about half (51%) of 539 major U.S. systems in the 1920s (Donaldson, 1924), would make up the most intractable source of water Pb, in that partial household line flushing, useful for reducing standing water lead in residences or commercial buildings due to joint soldering with lead and lead in brass fixtures, would still present some water Pb from outside segments to the street main. While the total interior surface area of outside yard water lines in contact with household water would dwarf that of even multiple lead–tin soldered plumbing using copper lines, areas of corrosivity, e.g., copper pipe with lead–tin soldered joints, produce electrochemical leaching, i.e., galvanic releases, at higher rates than passive releases.

This coexistence of large surface area and localized high leaching rate sources meant cumulative exposures to human populations could lead to significant intakes of Pb. By contrast, raw waters entering potable water treatment plants and treated water leaving the treatment plant and entering community water mains were relatively much lower in Pb content than water volumes leaving street-level main connections and stagnant water sitting for extended periods in household plumbing. The latter is the principal point at which the main culprit, interior pipe surface corrosion, occurs.

Water systems are lead sources where lead mobility is always a problematic factor and cannot be assumed to always be stable, whatever the amounts of lead components in community and household water supply segments. While the phenomenon of inert scaling and mineral depositions even on lead pipe distribution systems has been argued as a factor in reducing lead leaching, disruption of such slowly created interfaces via chemical reactions can occur rather rapidly. Edwards et al. (2009) reported that the introduction of drinking water disinfection via chloramination versus traditional chlorination can chemically mobilize pipe wall deposits and rapidly increase lead leaching from existing lead pipe, and in the case of several communities, increases in children's PbB levels. This speed of change is linked to changes occurring in a centralized fashion. That is, an abrupt change in a large community's central raw water treatment alters the water chemistry for all households connected to that water source.

Prior to the use of tap water Pb surveys in the United States and elsewhere, there was little quantitative appreciation of either the nature or the extent of Pb contamination of U.S. tap water being consumed by human populations. This was especially the case for any regulatory needs. Identifying water Pb contamination via distribution and plumbing on a specific community basis, however, occurred. Against a guideline of 20 μg Pb/l of tap water, standing mean Pb levels in tap water often exceeded this figure in U.S. soft water areas such as Boston (Karalekas et al., 1976; Worth et al., 1981) and Seattle (Sharett et al., 1982).

U.S. studies of plumbosolvency in the 1970s and 1980s clearly indicated that there was a high correlation between soft, acidic water and associated lead surface corrosivity and that the overall U.S. picture for contamination frequencies of water by Pb was a variable one. U.S. national household tap water lead surveys (Patterson and O'Brien, 1979; U.S. EPA/Levin, 1986; McCabe et al., 1970; Craun and McCabe, 1975) showed (1) one-sixth (16%) of U.S. tap water lead levels were ≥20 μg/l; (2) 100% of residences with new plumbing with recently soldered connections in copper systems showed Pb ≥20 μg/l; and that (3) for the sampling year 1988, a total of 42 million U.S. residents had some exposure to water Pb ≥20 μg/l, about 34 million from older housing exposures and 8 million from newly constructed residences (U.S. ATSDR, 1988).

A broader U.S. survey of inorganic contaminants, lead plus other regulated inorganics including primary and secondary substances, was carried out by the American Water Works Association (AWWA, 1985) with eventual combining of certain data sets with complementary information from the U.S. EPA on frequencies of existing Maximum Contaminant Level (MCL) violations for the substances from nonreporting states. Thirty-seven states and three territories reported with data. The mean Pb level in reported water samples was 250 μg/l with a concentration range of 50–5,000 μg/l.

The potable water supply Pb contamination picture for the United Kingdom in terms of need for regulatory intervention indicated that, as in the United States, water corrosivity, plumbing lead leaching, and associated Pb exposures varied considerably. During 1975–1976, the UK Department of the Environment carried out a countrywide survey of drinking water lead levels in residents' water supplies (Department of the Environment, 1977). This national survey showed that sampling points in Scotland had 21% ≥100 μg Pb/l versus 2.6% and 2.3% for England and Wales, respectively. The Scottish figure for water Pb 50–100 μg/l was 13.4% versus 5.2% and 6.5% for England and Wales, respectively.

The tap water Pb problem was especially acute for certain communities in Scotland with lead plumbing components in their residential units and soft, corrosive water in highlands impoundments. Of particular note was the finding that parts of Scotland and Wales with particularly corrosive water supplies also had housing stock with old, lead plumbing-based water systems. This included lead pipe for street connections and lead-lined storage tanks in the cities of Glasgow and Ayr (Richards and Moore, 1984; Sherlock et al., 1982).

28.1.2 Dose–Response Relationships in Humans for Drinking Water Pb Relevant to Regulation

A core question for effective lead regulation in drinking water and water supplies is how much of the human daily lead intake and total PbB are reliably attributable to drinking water Pb concentrations. Anecdotal and clinical case reports in Europe and the United States made it clear that lead pipe-based water distribution could and did produce frank lead poisoning. Quantitative linkages for dose–responses in these instances largely remained unknown in that early era of limited environmental and clinical epidemiology. In comparative terms, we would expect that areas with significant amounts of lead in plumbing systems and corrosive, soft potable water supplies would produce higher water lead levels and would make a much larger contribution to total body lead and PbB than areas with low corrosivity waters.

More current methods for estimating fractional contributions of drinking water Pb to total blood lead and body lead burdens and thereby the relative need for regulatory controls have been described and include such approaches as statistical associations of elevated drinking water Pb with elevations in PbB and/or exceedences of health advisories, the use of empirical statistical modeling, and the reported empirical results of changes in population PbB levels with interventions to reduce the amounts of water Pb ingested by exposed populations.

Two epidemiological assessments that are particularly useful here are those of Moore et al. (1985) and Sherlock et al. (1984), who reported reductions in both mean tap water Pb and population mean PbB levels for Scottish cities when their corrosive water supplies were treated to increase the pH and decrease corrosivity. In both cases, tap water Pb and blood lead levels declined relatively rapidly and dramatically with such remedial interventions. The data of Moore et al. (1985) showed that using a closed-loop water liming system for Glasgow resulted in a mean decline in Pb from 244 to 13 μg/l, while figures for Ayr, Scotland showed a decline in mean water Pb from 474 to 51 μg/l (Sherlock et al., 1984).

Blood Pb mean reductions in the above studies were similarly significant with treatment. Postpartum mothers in Glasgow prior to treatment showed a geometric mean PbB of 14.5 μg/dl versus a figure of 8.1 μg/dl postintervention. Furthermore, both minimum and maximum values for the PbB range were reduced, and the rate of PbB exceedences for PbB >35 μg/dl declined from 6% to 0%. Mothers of young children (<5 years old) in the Ayr study showed a reduction in geometric mean PbB from 20.5 to 9.5 μg/dl, reductions in the range, and decline in the percent >35 μg/dl from 13% to 0%.

Statistical regression modeling of the PbB relationship to tap water Pb in these studies showed that PbB varied with the cube root of the water level and that this relationship persisted to water Pb values <50 μg/l. The curvilinear relationship also indicated that the relative decline rate in PbB rose markedly with continued decline in water Pb. Moore et al. (1985) noted that reduction in water Pb at low (~10 μg/l) levels produced a sevenfold greater reduction in PbB than at water concentrations >50 μg/l. The two studies here showed that reducing corrosivity produced a corresponding mean PbB decline of 40–50% for women of childbearing age. Other risk groups, such as infants and toddlers and particularly bottle-fed infants, might well show a higher relative reduction.

The Scottish findings arguably represent a worst-case scenario, and areas with lower tap water Pb levels would presumably show a lower fractional contribution to body lead, but Bois et al. (1989) reported that the use of clearance concepts for tap water Pb consumption in children supported a U.S. MCL of 10 μg/l since the then-current value of 50 μg/l made water Pb a major Pb intake source. Elwood et al. (1984) found that Welsh women randomly sampled with respect to PbB and various sources of environmental Pb could sometimes have a higher exposure to water Pb than to the element in ambient air.

28.2 REGULATION OF Pb IN U.S. DRINKING WATER AND WATER SUPPLIES

Regulation of lead in drinking water can be classified into actions covering water supply quality *per se* and water levels at the tap. Water at the tap can derive from private wells or community sources serving a specified number of individuals. Projecting the U.S. EPA's regulatory jurisdiction into U.S. residences meeting certain legal definitions of their water source, using tap water Pb controls first implemented in the early 1990s, required a relatively more complex approach than merely promulgating a standard for water Pb as an MCL in isolation.

28.2.1 Historical Aspects of Regulation of Lead Levels in U.S. Public Drinking Water

The environmental health history of regulation of Pb in drinking and ambient waters predates to some extent that for ambient air Pb. As noted above, it is the level of Pb in drinking water that is of principal concern for human population Pb exposures and toxicity. Consequently, the bulk of this section tracks regulatory initiatives dealing with drinking water. Ambient U.S. waters generally are relatively low in Pb content, but this does not always foretell low human Pb exposures. Dissolved or suspended Pb in the water column is in contact with sediment compartments continuously and terrestrial compartments episodically, e.g., Pb mobilized to flooded surfaces during times of high precipitation and runoff. Furthermore, Pb can enter ambient waters from both point and nonpoint sources. Point source discharges of effluents high in lead content, such as those from lead or lead–zinc smelters, can produce locally high lead concentrations in receiving waters or in downstream sediments (NAS/NRC, 2005), while airborne Pb deposition into large bodies of freshwater serving as drinking water supplies, e.g., the Great Lakes, is an example of nonpoint source Pb reaching ambient waters.

In the case of U.S. drinking water, a relatively loose set of Federal guidelines issued by the U.S. PHS in 1914 first served as regulations in terms of any "standards." The 1914 regulatory initiative was narrow, confined to standards for bacteriological quality of water systems for interstate carriers, i.e., ships, trains, and buses (Borchardt and Walton, 1971; Pontius and Clark, 1999), and was more a set of guidelines than regulations. This was in keeping with the universal practice among public health systems of conflating drinking water safety with controls on pathogens and their propensity for inducing widespread disease. Concerns with waterborne toxic metal pollutants such as lead evolved subsequently. These U.S. PHS standards were expanded in 1925, 1942, and 1962.

The 1962 U.S. PHS guidelines/standards (U.S. DHEW, 1962) embraced 28 substances and were the immediate predecessor to the regulatory requirements spelled out in the 1974 SDWA. These requirements had the enforcement power of actual standards and were housed in the water office of the U.S. PHS. Over time, all 50 states implemented these U.S. PHS values as either standards or guidelines. The 1962 U.S. PHS standards included 11 metals or metalloids, of which the nonessential toxicants were arsenic, barium, cadmium, silver, and lead. The others were chromium, copper, iron, manganese, selenium, and zinc. Radionuclides were also included in the total of 28 substances: beta and photon emitters; gross alpha emitters; radium-226; strontium-90.

28.2.2 The 1974 SDWA and Drinking Water Lead

Post-WWII U.S. industrial and large-scale agribusiness growth led to both widespread increases in loadings of the 28 U.S. PHS contaminants to drinking water supplies and to the introduction of added substances with human toxicity risk. These developments called for a more systematic and effective regulatory approach in terms of legislation, regulatory initiatives, and enforcement. This led to the two principal legislative and regulatory mechanisms by which U.S. drinking water was controlled with respect to lead contamination: the 1974 SDWA and the 1972 CWA. The former is discussed here, the latter in a following section.

The 1974 SDWA [P.L. 93–523; 88 Stat. 1660 (1974); 42 U.S.C. §300f–399j-9] directed the U.S. EPA to carry out several groups of regulatory initiatives for lead and other contaminants, applied to all public water systems operating in the United States. The only drinking water category exempt was private wells serving fewer than 25 individuals. Regulatory coverage specifically applied to several categories of "public" systems. The broadest were the 170,000 U.S. systems providing water to 270 million residents, where service is provided to at least 25 individuals or via 15 connections for at least 60 days annually. This tally partitioned into community water systems—54,000 in number, where individuals receive water year-round and where all federal regulations apply—and noncommunity systems, such as campgrounds and gas stations.

Drinking water Pb and other contaminants are principally regulated via an NPDWR since this element is associated with adverse health effects [40 C.F.R. Part 141]. The 1974 SDWA first required the U.S. EPA to establish standards for each identified toxic contaminant, to include lead, in the form of an MCL and an MCL Goal (MCLG). The former are enforceable, mandatory levels, while the latter are in the way of nonenforceable recommended guideline values. Conceptually and in practice, the MCLs are to be set as close to the MCLGs as permitted by the available health risk science. In addition to health risk-based NPDWRs, the agency also issued nonenforceable, guideline values for National Secondary Drinking Water Regulations (NSDWRs). Fifteen contaminants were the focus of NSDWRs and they are those that had undesirable aesthetic and related properties for consuming populations but were not considered toxicologically adverse. The U.S. EPA not only set interim NPDWRs for lead and other toxic contaminants mandated by the 1974 SDWA but also required that states or other jurisdictional bodies with delegated authority monitor and analyze public drinking water for levels of contaminants and make public any violations. Second, the act directed the U.S. EPA to revise these interim values following recommendations from the National Academy of Sciences.

In late 1975, the U.S. EPA promulgated a set of national interim NPDWRs in the form of MCLs. The value for lead was specified at 0.05 mg/l (50 μg/l, 50 ppb), the same value established in the 1962 U.S. PHS regulations (40 FR 59570; December 24, 1975). The NPDWR MCL of 50 ppb for lead referred to the processed drinking water coming out of the treatment facility. Because of water corrosivity and leaded plumbing systems, the agency in 1980 amended its interim regulation language to include assessing water corrosivity postdistribution. This value continued to serve as the U.S. drinking water standard until the EPA's implementation of the 1991 Lead–Copper Rule (see discussion below).

The 1974 Act also mandated that there be testing and monitoring of system-treated water lead levels at the point of distribution.

28.2.3 Later SDWA Amendments and Water Pb Regulations

The first major amendments to the 1974 Act occurred with the SDWA Amendments of 1986 [P.L. 99–339; 100 Stat. 642]. Several sections of the SDWA Amendments applied to lead regulation by the U.S. EPA. Section 1412 mandated NPDWRs for contaminants that cause adverse human health effects, and this action converted the interim NPDWRs issued by the U.S. EPA in 1975 to NPDWRs. Furthermore, beginning January 1, 1988, and on a 3-year cycle afterward, the EPA Administrator was directed to publish a list of contaminants which do or may occur in public water systems and require regulatory actions.

It is important to recognize that an NPDWR is not necessarily the same as an MCL, in that protective and adequate regulation of a toxic substance may require technical modalities beyond reductions below an MCL. The amendments implicitly acknowledged this in §1412, allowing the EPA Administrator to regulate lead or other contaminants by means other than an MCL. This part of the Act provided the statutory means for U.S. EPA to regulate drinking water lead via the 1991 Lead and Copper Rule (LCR).

Additional important lead-specific sections of the amendments are in Sec. 1417, banning lead solder, flux, and lead pipe for new residential water plumbing systems or for the repair of existing ones. Lead-free solders and fluxes are defined as having 0.2% Pb or less. Lead-free pipe, fixtures, and pumps are those having 8.0% or less Pb. The basis of the 8% permissible lead content for the latter's "lead-free" designation is not apparent. Household brass plumbing fixtures and related materials such as water meters and well pumps will often have Pb at levels of 5–7%, providing measurable rates of Pb leaching under corrosive water conditions (Gardels and Sorg, 1989; Maas and Patch, 2004; Maas et al., 2005). Furthermore, some U.S. jurisdictions with responsibility for enforcing the amendments such as California have seen lawsuits successfully challenge the high Pb levels in these fixtures, resulting in low-lead brass fixtures becoming available in the market with little apparent effect on fixture installation or performance.

A second set of changes in the SDWA were the 1996 amendments [P.L. 104–182; 110 Stat. 1613]. The 1996 amendments addressed health-based risk as the basis for standard setting, providing flexibility and technical assistance for small water systems, public right-to-know and financial assistance to water systems through state revolving loan funds.

The 1991 U.S. EPA LCR

One result of the 1986 SDWA Amendments was the promulgation in 1991 of the U.S. EPA's LCR, a two-part NPDWR for drinking water Pb (56 FR 26460–26564; June 7, 1991). This regulatory regimen became, and remains, the principal Federal/State mechanism for controlling tap water Pb to minimize exposures in a major fraction of the U.S. population. Identified by the agency as the second and principal element of the regulatory rule was the imposition of an action level, not an MCL, whereby a community's tap water Pb levels were not to exceed a level of 15 μg/l Pb (15 ppb Pb) in a 1 liter sample in more than 10% of all test samples (90th percentile outcome). Such an approach is permitted by the SDWA [§1412(b)(7)(a)]. The computation of this percentile for a community system's results entails all results placed in ascending order and each result assigned a number, ascending from number 1. The highest number equals the total taken. The total is governed by the EPA tally for different-sized systems.

A community's exceedence of this U.S. EPA/state action level is not legally a violation but serves as a statistical trigger for a relatively involved series of sampling requirements, water treatment techniques, and, with regard to both community and household tap water Pb exceedences, a formulized schedule for replacing lead service lines (LSLs), the typical source of persistent elevations in tap water Pb levels after treatment methods have been tried.

Tap water sampling in any community is dictated by size of the water system. The number of samplings for those serving >100,000 is 100 and for a system serving 50,001–100,000 people is 60. Monitoring is done semiannually unless systems qualify for reduced monitoring. In cases where treatment techniques or other means have shown no exceedences for several rounds of testings, these large system sampling numbers are 50 and 30, respectively. Simultaneously with contaminant sampling rates, water quality parameter (WQP) testings are done but at a lower frequency than the contaminant samplings. WQP testings include those for pH, alkalinity, Ca level, conductivity, orthophosphate or silicate, depending on the species of corrosion inhibitor used and temperature. These measurements guide the water utilities for system-wide interventions.

Other provisions of the LCR are triggered when the action level is exceeded during periodic monitoring. Water departing the water system may require corrosion control treatment if the action level is exceeded and the WQP results identify corrosive water per requirements of the rule [40 C.F.R. 141.81(b)(2) and 141.81(b)(3)]. Public notification and public education provisions are dictated in the LCR, as is the need to evaluate source water monitoring and treatment. Finally, the rule directs LSLs be replaced on a timetable (7%/year) if a community's water Pb levels show LSLs contribute more than the action level value of 15 ppb. Schools, child-occupied facilities, and related public facilities are subject to the LCR if they have their own water supplies as defined in the SDWA. Otherwise, regulations for the community systems apply. In all cases, schools come under the regulatory provisions of the 1988 offshoot of the SDWA, the 1988 Lead Contamination Control Act (see below).

Various technical amendments, narrow in scope, were added to the rule by U.S. EPA in the 1990s and in the years 2000 and 2001. The greatest number of revisions in the LCR occurred with U.S. EPA's promulgation of seven sets of changes in October 2007 (72 FR 57782–57820; October 10, 2007); [40 C.F.R. Parts 141, 142]. Three of the seven groups of changes had to do with monitoring: minimum number of required samples; compliance and monitoring period definitions; reduced monitoring criteria. One had to do with treatment processes, specifically advanced notification and approval requirements for systems intending to change treatment method or add a new water source. Changes in the LCR having to do with customer awareness dealt with two of the topics: notification to the public of sampling results and public education requirements. Finally, lead service line replacement concerns prompted changes having to do with systems required to reevaluate lead pipelines classified as "replaced." (Table 28.1)

28.2.4 The 1988 Lead Contamination Control Act

A major amendment to the SDWA was addition of the Lead Contamination Control Act in October 1988 [PL 100–572; 42 U.S.C. Part F: §300]. This act addressed lead in school drinking water coolers and funding childhood lead screening programs through the U.S. CDC. Regulatory activities in connection with this act, such as school lead exposures from water coolers and related water sources, were assigned to the U.S. EPA and the CPSC. The U.S. EPA was directed to provide a list identifying the brands and models of water coolers that were not defined as "lead-free" (54 FR 14320; April 10, 1989), while the CPSC was directed to issue an order mandating that

TABLE 28.1 Significant U.S. Regulations or Guidelines for Lead in Drinking Water

Statutory/Regulatory Initiatives	Codified Law or Regulation	Years	Key Features
Public Health Service Act and Amendments	42 U.S.C. 300	1914, 1925, 1962	U.S. PHS issued guidelines/standards beginning with pathogenic controls in 1914; established a level of 100 μg Pb/l in 1925, reducing it to 50 μg/l in 1962
SDWA of 1974 [Title XIV of the Public Health Service Act; PL 93–523; December, 16, 1974]	88 Stat. 1600 (1974); 42 U.S.C. 300f–300j-9	1974	Mandated EPA to set water Pb primary and secondary regulations for drinking water, among other requirements
SDWA Amendments of 1986 [PL 99–339; June 19, 1986]	100 Stat. 642: §1412 and other sections	1986 and later	Mandated ban on Pb solder for plumbing, lead pipes, and other Pb components
1988 Lead Contamination Control Act [PL 100–572; October 31, 1988]	42 U.S.C. Part F: §300	1988 and later	Banned Pb in school drinking water coolers and restored central U.S. CDC funding for national PbB screening programs; existing school coolers to be repaired, replaced
U.S. EPA 1975 Interim Primary Drinking Water Regulations	40 C.F.R. Part 141; 40 FR 59566	1975–1991	Set interim contaminant standards for lead and other contaminants; an interim MCL of 50 μg/l (50 ppb) was established
U.S. EPA Lead/Copper Rule, 1991 (LCR)	SDWA authorizations (56 FR 26460–26564, June 7, 1991); use of action level, authorized under the act [§1412 (b)(7)(A)]	1991–present	Set an action level of 15 ppb at the 90th percentile of community sampling results; set MCLG of 0; required initial monitoring by December 1992
U.S. EPA changes to the LCR in 1990s, 2000, 2001, 2007	Major changes to the rule in 2007 (October 10, 2007; 40 C.F.R. Parts 141, 112)	1990s 2000 2001 2007	2007 changes deal with seven sets of items (see text)

makers and importers of water coolers with lead-lined tanks repair or replace existing coolers, or order a recall with refunds to the owners. Manufacturing new coolers with Pb fittings to be sold through U.S. interstate commerce was prohibited. The U.S. EPA was also directed to provide grants to states to assist local school agencies in remedying lead contamination via water in schools.

As a follow-on to mandates to U.S. EPA in the act, the agency issued a 1994 report detailing guidance and recommendations to school systems for dealing with lead-containing water coolers and other sources of contamination (U.S. EPA, 1994). The 1994 report contained a recommendation of 20 μg/l (20 ppb) as the guideline value not to be exceeded for school drinking water Pb in standing (stagnant) samples.

The provisions of the Act dealing with lead screening had three components: (1) screening infants and children for elevated PbB levels; (2) ensuring referral for intervention for lead-poisoned infants and children; and (3) providing public education about childhood lead poisoning.

28.2.5 The 1972 CWA and Later Amendments

The CWA is a shorter statutory label for the FWPCA of 1948 [P.L. 80−845] and its subsequent amendments enacted through 1996, with changes via revisions or amendments in 1956, 1961, 1965, 1966, 1970, 1972, 1977, 1981, and 1987, 1988, 1990−1992, 1994, 1995, and 1996. The label "Clean Water Act" was made official in 1977 with "The Clean Water Act of 1977" [P.L. 95−217]. The name principally attaches to the major 1972 reorganization of the FWPCA [P.L. 92−500] and is codified under Title 33 [33 U.S.C. §§1251−1387]. The 1972 changes were extensive: (1) required all municipal and industrial effluents be treated before discharge to public waters; (2) provided financial assistance to build public treatment plants; and (3) delegated to states the routine implementation of the act and simultaneously increased the Federal role.

From a health science and public health perspective, the CWA works in concert with the SDWA to provide added regulatory control in the area of surface ambient public water supplies such as lakes and rivers. These surface water bodies often serve or may serve eventually as drinking water sources. Ground water quality, by contrast, is regulated by the U.S. EPA through the SDWA. However, the CWA independently provided an exposure control function to attenuate human population lead exposures via additional pathways that otherwise occur through lead and other pollutant discharges and thereby produce exposures in both ecological and human populations. For example, lead in extractive industry effluents historically discharged from lead mills and smelters in the Western United States to adjacent waterways increased the levels of dissolved and particulate Pb and other inorganic toxicants in receiving sediments that ultimately resulted in increased lead burdens in fish and other fresh aquatic life serving as food sources for local populations (NAS/NRC, 2005).

The core role of the CWA is to control pollution of the country's ambient surface waters. The Act mandated implementation of its requirements by principally the U.S. EPA and, in specific cases set forth in the act, the U.S. Army Corps of Engineers. Two activities are carried out by the U.S. EPA under the CWA. One funds construction, modification, or expansion of sewage treatment facilities that produce processed wastewater effluents meeting Federal and state criteria. The other is the more direct water quality regulatory function of issuing discharge permits and enforcing regulations directed at priority and toxic pollutants from industrial and municipal discharges. Such discharges are from both point sources, e.g., an effluent discharge pipe, or from nonpoint sources such as storm water runoff from agricultural or urban impervious surfaces.

The U.S. EPA's water quality regulatory responsibilities take two forms: issuing discharge permits [The National Pollutant Discharge Elimination System (NPDES) permits] and enforcing controls on an array of priority pollutants that include lead.

NPDES Permit Process

One of the principal regulatory tools available to the U.S. EPA in enforcing the CWA is the NPDES permitting function set forth in §402 [33 U.S.C. §1342 and 40 C.F.R. Parts 123, 124, 125]. NPDES permits are issued by either the Agency or any of the delegated states. Currently, around 65,000 dischargers are required to obtain permits from the designated states or from U.S. EPA. For agency approval, an NPDES permit application is required to include monitoring and reporting schedules for discharges from industrial operations directly or indirectly to a publicly owned treatment works (POTW). Where the NPDES permit for a POTW is to take cognizance of industrial discharges to it, pretreatment of industrial effluents via pretreatment standards [§307(b), 33 U.S.C. §1317(b), §402(b)(8)] is required for discharge to a POTW where discharge pollutants are not amenable to a treatment process or where such pollutants adversely impact the treatment process or subsequent disposal of sludges.

U.S. EPA mandates an NPDES permit, issued by the 45 states that are empowered to manage the permit program, for all pollutant discharges from point sources linked to industrial or municipal dischargers (e.g., pipe or outfall). Permitting is predicated on the regulatory tenet enunciated by a recent Congressional Research Service summary of the CWA that "... all discharges into the nation's waters are unlawful, unless specifically authorized by a permit ... the act's principal enforcement tool."

Earlier incarnations of the act centered on regulating discharges of conventional, i.e., nontoxic or nonchemical, pollutants such as suspended solids and bacteria, using a best practicable control technology. Subsequent changes in the CWA required the agency to use a more stringent technology directed to toxic substances, best available technology. Toxic or priority pollutants, particularly inorganic contaminants, do not biodegrade.

Water quality is preserved by the U.S. EPA through two mechanisms: water quality standards and technology-based effluent limitations. Standards are employed for control of overall water quality, while effluent limitations via technological approaches are for certain contaminants from certain sources and function through numerical limitations identified in the NPDES. Standards are promulgated within a context of intended uses (e.g., recreation, water supply) and are numerical or narrative in form. Every state is required by the agency to promulgate water quality standards for all water bodies within its jurisdiction.

A total maximum daily load (TMDL) may be resorted to in situations where the implementation of technology-based effluent limits still allow exceedences in water quality contaminant limits. The concept of a TMDL hinges on the concept of allocation of pollution contributions for subsequent regulation among the relevant array of industrial and/or municipal dischargers.

For obvious reasons attending effective enforcement, the CWA does not regulate nonpoint sources such as storm water runoff through the permitting mechanism. Rather, §319 mandates state programs for runoff management.

The U.S. Army Corps of Engineers, via §404, is responsible for issuing permits for disposing of dredged or fill materials in waters and wetlands. Only two states have delegated authority for this form of discharge permitting compared to the NPDES process: Michigan and New Jersey. While the Corps administers the permitting, the U.S. EPA is charged with bringing enforcement and penalty actions for violations.

Water Quality Criteria and Standards for Lead and Other Pollutants

Water quality criteria are drawn from the available technical database for a particular ambient water pollutant. These databases provide the wherewithal to assess levels of toxicants which will not produce harm to aquatic life or human health. Standards are typically numerical values indicating the maximum permissible level of a metal or other toxicant in a body of water.

Much of the focus of CWA implementation agencies and offices in states and within the U.S. EPA has been on water quality criteria and standards for toxic pollutants, with toxics subsuming a second category of regulated substances, the priority pollutants. Distinctions between the two categories have engendered some confusion. Toxic pollutants refer to a group of 65 substances or substance groups which were the subject of extensive litigation among various parties in the 1970s and early 1980s. This tally of 65 toxics/toxic groupings is also known as the Toxic Consent Decree toxics. Each item on this list of 65 that refers to a chemical group rather than a chemically specific substance, however, can potentially embrace numerous individual contaminants. Furthermore, groups of chemically complex substances may be difficult to analyze quantitatively, a prerequisite for compliance and compliance enforcement. Consequently, the specific substances and substance groups within the 65 Decree toxics were subdivided into an original tally of 129 "Priority Pollutants," then reduced to 126. These are individual substances for which analytical methodology is available for compliance monitoring by the designated enforcers. Lead was in the decree list and is in the Priority Pollutant list.

The national recommended water quality criteria for lead (65 FR 31682) derived by the U.S. EPA [per §304(a)(1)], as with all the pollutant criteria, included freshwater and saltwater acute and chronic criterion concentrations to protect aquatic life. The corresponding values (μg Pb/l) are 65, 2.5, 210, and 8.1, respectively. As of June 1, 2010, there were no human health criteria for lead in terms of fish consumption or fish plus water consumption, previous values having been withdrawn (Table 28.2).

TABLE 28.2 Significant U.S. Regulations or Guidelines for Lead in Ambient Waters

Statutory/Regulatory Initiative	Codified Law or Regulation	Key Features
Water Pollution Control Act ("Clean Water Act", CWA) of 1948 [PL 80–845] and Amendments in 1956, 1961, 1965, 1966, 1970, 1972, 1977, 1981, 1987, 1988, 1990–92, 1994, 1995, 1996	Officially labeled the CWA in 1977 [PL 95–217]; codified under Title 33, 33 U.S.C. §§1251–1387	CWA works in concert with SDWA on regulatory controls on surface ambient water supplies for Pb and other contaminants; Act administratively empowers agencies to issue discharge permits and control priority pollutants
U.S. EPA role in issuing NPDES discharge permits	U.S. EPA controls NPDES permits; §402, 33 U.S.C. §1342, 40 C.F.R. Parts 123 et seq.	U.S. EPA works in concert with the U.S. Army Corps of Engineers in granting discharge permits
U.S. EPA role in establishing water quality criteria and standards	Water criteria promulgated under the CWA (§304(a)(1); Pb water quality values are expressed as acute or chronic criteria for freshwater or salt water; for freshwater, 65 and 2.5 μg/l; for salt water, 210 and 8.1 μg/l	No human health criteria for lead in fish or fish + water consumption

28.3 INTERNATIONAL REGULATION OF Pb IN DRINKING WATER AND WATER SUPPLIES

As is the case with European controls on lead in paint and ambient air, much of the regulatory effort for lead in drinking water and surface ambient waters lies within the purview of the European Union's regulatory apparatus. As is also the case with ambient air Pb, EU regulations for drinking water Pb and ambient water pollution rely on existing guidelines issued by the WHO.

28.3.1 WHO Guidelines for Lead in Drinking Water

The WHO first addressed the quality of international drinking water supplies in member states in 1958, publishing the document *International Standards for Drinking Water*. The document was revised in 1963 and 1971. In 1984, the first WHO Guidelines for Drinking Water Quality (GDWQ) (WHO, 1984) was published. The 1984 Guidelines appeared as three volumes: (1) recommendations; (2) supporting information such as health criteria; and (3) surveillance and control. The major conceptual and structural change in the WHO drinking water monographs, going from international standard for contaminants to a guideline approach, i.e., an enforcement versus a recommendation, was rationalized (WHO, 1984) on the basis of the preferred approach where individual member states or member state communities could use a risk–benefit approach. The elements of both risk and benefit as they vary across specific nations could be tailored to the individual states. This revised approach has been employed in tandem with the European Union issuing drinking water regulatory directives with accompanying statutory language providing how states can individually regulate contaminants in their drinking water or in their ambient water supplies.

The GDWQ for lead was first derived in 1984 as a value of 50 μg/l, recommended for measurement at the consumer's tap. The guideline was based on some rudimentary lead toxicokinetics in human populations and the PTWI recommendation of 3 mg Pb from the 1972 Joint FAO/WHO Expert Committee on Food Additives (JECFA-FAO/WHO, 1972) for adults, adjusted somewhat for child intakes.

WHO recognized and considered lead in its GDWQ figure as a contaminant whose level is mainly determined by water line and plumbing parameters such as corrosivity in recommending a value at the tap and not at the treatment facility's predistribution points. In WHO's terminology (WHO, 1984), lead typified a Type 2 substance, one that is influenced by distribution-stage parameters, and typified further a Type 2b contaminant, where distribution was the principal level determinant. Volume 1 of WHO (1984) included statistical and analytical methodological discussions to address lead's Type 2b characterization.

The WHO GDWQ value for lead and other contaminants was revisited in the late 1980s and the 1990s to reflect the findings of international research showing that adverse health effects in children and other human risk populations were occurring at lower PbB levels and at the lower intakes producing those PbB values. In 1993, the guideline value for Pb was revised downward to 10 μg/l.

In 2004, WHO issued its third edition of GDWQ to address in part the need to take a systematized approach to water safety plans, beyond simply suggesting GDWQ values in isolation. Other aspects of the report had to do with revisions of some existing guidelines and addition of some new guidelines. The 2004 edition of the WHO Guidelines reflects the interest of the agency in adopting a more holistic approach to drinking water protection by addressing the entire framework for safe drinking water: health-based targets via risk assessment and risk management; system assessment and design; operational monitoring; management plans, documentation, and communication; and surveillance of drinking water quality. That is, water intended for human consumption or other uses was to be managed in a fashion analogous to a utilitarian life cycle approach.

Lead as an issue for monitoring and preserving water quality guidelines is discussed in the 2004 Guidelines with respect to those matters also encountered in the United States and requiring attention via statutory remedy and U.S. EPA regulatory practices. This included dealing with water treatment and lead water distribution component removal or banning options in member countries. The guideline value of 10 μg/l measured at the tap was retained in the 2004 Guidelines. A support document (WHO–Europe, 2007) lays out the elements of a framework for implementing water safety plans in the European Union.

28.3.2 Drinking Water Lead Regulations in the European Union

Public waters and those waters serving as raw and processed drinking water in member states of the European Union comprise one of the most extensively regulated environmental media in terms of contaminant levels. The European Union regulates drinking water lead and lead in other environmental media via quality standards promulgated within Union-wide Directives. Directives are issued by the governing body of the European Union and do not take on the force of law until the various member states enact legislation and enforcement methodologies. EU Regulations, by contrast, are issued by the governing body and hold the force of law within all member states without the requirement for state-by-state enactment and enforcement actions. Regulations typically attend the distribution of products within the European Union. An example of the latter is the recently enacted regulation REACH (Registration, Evaluation, Authorization, and Restriction of Chemicals). REACH will eventually cover many thousands of chemical products entering EU member states.

EU drinking water quality directives have as their stated purpose protecting drinking water in terms of microbiological, chemical, and organoleptic (e.g., taste, smell) parameters. Directives have generally mirrored the WHO water Pb guidelines with respect to setting specific guideline values but also stated the time frame for implementation as deadlines. Member states have typically codified the latest WHO guideline numerical value, but there is variability in individual time frames for implementation.

Over the years, the European Commission through its Directorate for the Environment has issued a number of directives on lead in drinking water. These date to the 1970s. In 1977, the Council of the European Communities (CEC) issued Directive 77/321/EEC (CEC, 1977) for the biological screening of the EU population for Pb (PbB). It was the results of this action that helped identify the particular drinking water lead exposures in Scotland versus other parts of the United Kingdom.

Subsequently, the 1980 CEC Directive 80/778/EEC became the first directive dealing with water quality in the European Union, including the Pb content in drinking water. It came into force on July 15, 1980. There were 66 contaminants identified for regulation via stated limit values by July 15, 1987; these were enacted into law by July 15, 1982.

The issuance of WHO's GDWQ limits for lead and other contaminants in 1993 was followed by a new drinking water directive, Council Directive 98/83/EC, formally articulated on December 25, 1998 (Council of the European Union, 1998). The new directive was issued to the 15 member states then comprising the European Union for enacting into law by December 25, 2000, to meet the drinking water standards (1) by December 25, 2013, for an interim level of 25 μg/l and (2) at the final 10 μg/l value on or after December 25, 2013. The timetable was necessitated by the need to deal with the lead pipe problem. Twelve new EU members, entering the Union between 2004 and 2007, had to meet these standards by dates negotiated by each new member. This directive set mandatory limit values for 28 microbiological and chemical parameters evaluated at the tap although the option for testing at the treatment point where no additions are made to the water during distribution was offered (Table 28.3).

The interim 25 ppb Pb standard within the 1998 directive applied from December 25, 2003 to December 24, 2013. States such as England, Scotland, and Wales, via their respective statutory instruments, were able to promulgate the EC's two-phase time frame for reducing levels from 50 to 10 μg/l via an interim reduction to 25 μg/l on December 25, 2003, lasting until December 24, 2013, and followed then by full adoption of the guideline.

The deadline for member state enactment of and compliance with the 25 μg/l interim Pb level by the end of 2003 was deemed to be premature by other states, based on a collective evaluation (Drinking Water Seminar) done in 2003 with member states still having to comply with requirements and the upcoming issuance of new WHO guidelines (WHO, 2004).

TABLE 28.3 Significant International Regulations or Guidelines for Lead in Drinking or Ambient Waters

Statutory/ Regulatory Initiative	Codified Law or Regulation	Years	Key Features
A. WHO Guidelines			
GDWQ	1984; Recommended GDWQ=50 μg Pb/l	1984–1993	A recommended guideline approach was chosen in tandem with national or multinational (European Union) entities adopting regulations
WHO GDWQ	1993; Recommended GDWQ=10 μg Pb/l	1993–Present	Revision downward based on growing evidence of low-level-Pb toxicity in children
WHO GDWQ	2004; Retained value of 10 μg Pb/l; added framework for Pb control in drinking water	2004–Present	Expansion or adoption of holistic approach: risk assessments, system design; adequate monitoring and surveillance; approach presented in a 2007 WHO–Europe document
B. European Union (EU) water regulations			
Council Directive	77/321/CEC	–	Provided for beginning of EU-wide PbB screening in children
Council Directive	80/778/EEC (adopted July 15, 1980)	–	Provided for a water standard of 50 μg Pb/l
Council Directive	98/83/EC (adopted November 3, 1998)	–	Lowered the drinking water Pb standard in the European Union to 10 μg Pb/l on an extended reduction schedule, 15 years, to 2013. From 2003 to 2013, an interim level of 25 μg/l was implemented

During 2007, the Commission decided to pursue revision of the 1998 Directive to accommodate conceptual and other requirements for effective implementation of EU drinking water protections. Toward that end, it sought WHO's preparation of WHO–Europe's support document, issued in October 2007 (see above; WHO, 2007). In addition, it commissioned a detailed study on a list of chemical parameters for directive revision, with a final report released in September 2008 (DG Env/DHI, 2008). This report recommended retaining the 10 μg Pb/l standard, measured at the tap, but noted that four member states were having compliance problems, and there were problems in terms of lead connections and monitoring.

28.3.3 Drinking Water Lead Regulations in Canada

Lead in Canadian drinking water is regulated by guideline values promulgated by a Federal agency, Health Canada's Water Quality and Health Bureau. Canadian drinking water guideline values are stratified into one of three categories. Those levels derived from scientific data showing documented human or ecological health risk, i.e., are health-based, are termed maximum allowable concentrations (MACs), and this category includes lead. Other categories are aesthetic objectives and operational guidance values.

Health Canada's role in drinking water guidelines functions through a multilevel drinking water committee, the Federal–Provincial–Territorial Committee on Drinking Water. The regulatory division of labor is such that provinces and territories are responsible for implementing the guidelines, but municipalities are principally responsible for guideline monitoring and compliance. Use of guidelines in Canada at all jurisdictional levels, rather than statutorily established and enforceable standards, distinguishes this approach from those in the United States and the European Union. A second difference is the nature of the guidelines in terms of sampling and determination of compliance. While current EU directives and the U.S. water action level for Pb prescribe tap water Pb measures where distribution system Pb inputs are accommodated, the Canadian Guideline applies to fully flushed water line samples. The drinking water guideline for Pb in all provinces and territories is a MAC of 10 μg/l (10 ppb), established in 1992.

In addition, guidelines and objectives in Canada have been promulgated for protection of aquatic life in freshwater, and these have relevance to human populations. For aquatic life, the Federal Guideline in terms of Pb concentration is governed by water hardness, indexed as $CaCO_3$, ranging from 1 to 7 μg Pb/l, for carbonate levels of 0–120 mg/l, 120–180 mg/l, and >180 mg/l.

28.4 REGULATION OF LEAD IN BOTTLED WATER

In the United States, bottled water is regulated as a food by the U.S. FDA under provisions of the Federal FDCA. The U.S. FDA regulations for bottled water are codified in 21 C.F.R. 301, Parts 101, 110, and 129, §§165.110 a,b,c. Much of the regulating authority of the agency for bottled water quality and safety attaches to bottled water in interstate commerce.

Bottled water, although regulated by a different governmental agency than regular drinking water, is required by the U.S. FDA under §410 of the FDC Act to set the same NPDWR for contaminants such as lead as does the EPA, that is, use of an MCL or a treatment technique. The U.S. FDA opted to use the treatment facility MCL rather than the U.S. EPA action level of 15 ppb since The MCL for Pb promulgated by the U.S. EPA for water exiting the treatment plant is 5 μg/l (5 ppb) and would more closely resemble bottled waters in lead content. This value of 5 ppb is currently the Pb bottled water standard.

REFERENCES

Aitchison, L., 1960. A History of Metals, vol. 1. Interscience, New York.

American Water Works Association, 1985. Internal Corrosion of Water Distribution Systems. Cooperative Research Report: Research Foundation and the DVGW. AWWA Research Foundation.

Bigelow, J., 1852. Report on the action of Cochituate water on leaden pipes, and the influence of the same on health. Am. J. Med. Sci. 98–100.

Bois, F.Y., Tozer, T.N., Zeise, L., Benet, L.Z., 1989. Application of clearance concepts to the assessment of exposure to lead in drinking water. Am. J. Publ. Health 79, 827–831.

Borchardt, J.A., Walton, G., 1971. Water Quality. Water Quality and Treatment: A Handbook of Public Water Supply. American Water Works Association 1–52.

Christison, R., 1844. On the action of water upon lead. Trans. R. Soc. Edinburgh 15, 265–276.

Commission of the European Communities, 1977. Council directive of 29 March 1977 on biological screening of the population for lead (77/321/EEC). Off. J. Eur. Commun. 20, 10–17.

Council of the European Communities, 1980. Council directive of 15 July 1980 relating to the quality of water for human consumption (80/778/EEC). Off. J. Eur. Commun. L229, 11–29.

Council of the European Union, 1998. Council directive of 3 November 1998 on the quality of water intended for human consumption (98/83/EC). In force 25 December 1998. Off. J. Eur. Commun. L330, 32–54, 05/12/1998.

Craun, G.F., McCabe, L.G., 1975. Problems associated with metals in drinking water. J. Am. Water Works Assoc. 67, 593–599.

Dana, S.L., 1848. Lead Diseases. (Editor and Translator) L. Tanquerel des Planches. 1839. Traite des Maladies de Plomb, ou Saturnines. A Treatise from the French. With Notes and Additions of the Use of Lead Pipe and its Substitutes. Daniel Bixby & Co., Lowell, MA.

Department of the Environment, 1977. Lead in Drinking Water: A Survey in Great Britain. HMSO, London, United Kingdom, Pollution Paper 12.

Directorate General Environment, 2008. Final Report on Establishment of a List of Chemical Parameters for the Revision of the Drinking Water Directive. Report ENV.D2/ETU/2007/0077r. September. Prepared by DHI Harsholm Denmark.

Donaldson, W., 1924. The action of water on service pipes. J. Am. Water Works Assoc. 11, 649.

Durfur, C.N., Becker, E., 1964. Public Water Supplies of the 100 Largest Cities in the U.S., 1962. Water Supply Paper 1812. U.S. Geological Survey.

Edwards, M., Triantafyllidou, S., Best, D., 2009. Elevated blood lead in young children due to lead-contaminated drinking water: Washington, DC, 2001–2004. Environ. Sci. Technol. 43, 1618–1623.

Elwood, P.C., Phillips, K.M., Lowe, N., Phillips, J.K., Toothill, C., 1983. Hardness of domestic water and blood lead levels. Hum. Toxicol. 2, 645–648.

Elwood, P.C., Gallacher, J.E.J., Phillips, K.M., Davies, B.E., Toothill, C., 1984. Greater contribution to blood lead from water than from air. Nature 310, 138–140.

Food and Agricultural Organization/World Health Organization, 1972. Sixteenth Report of the Joint FAO/WHO Expert Committee on Food Additives, Technical Report Series: No. 505. World Health Organization, Geneva, Switzerland.

Gardels, M.C., Sorg, T.J., 1989. A laboratory study of the leaching of lead from water faucets. J. Am. Water Works Assoc. 81, 101–113.

Hem, J.D., Durum, W.H., 1973. Solubility and occurrence of lead in surface water. J. Am. Water Works Assoc. 65, 562–568.

Karalekas, P.C., Craun, G.F., Hammonds, A.F., Ryan, C.R., Darth, D.J., 1976. Lead and other trace metals in drinking water in the Boston metropolitan area. J. N. Engl. Water Works Assoc. 90, 150–172.

Maas, R.P., Patch, S.C., 2004. Get the lead out human exposure from brass water service components. J. Ontario Water Works Assoc. XVIII.

Maas, R.P., Patch, S.C., Morgan, D.M., Pandolfo, T.J., 2005. Reducing lead exposure from drinking water: Recent history and current status. Publ. Health Rep. 120, 316–321.

McCabe, L.J., Symons, J.M., Lee, R.D., Gordon, G.R., 1970. Survey of community water supply systems. J. Am. Water Works Assoc. 62, 670–677.

Moore, M.R., Goldberg, A., Fyfe, W.M., Low, R.A., Richards, W.N., 1981. Lead in water in Glasgow—A story of success. Scott. Med. J. 26, 354–355.

Moore, M.R., Richards, W.N., Sherlock, J.G., 1985. Successful abatement of lead exposure from water supplies in the West of Scotland. Environ. Res. 38, 67–76.

National Academy of Sciences: National Research Council, 2005. Superfund and Mining Megasites. Lessons from the Coeur d' Alene River Basin. National Academy Press, Washington, DC.

Nriagu, J.O., 1983. Lead and Lead Poisoning in Antiquity. John Wiley & Sons, New York.

Nriagu, J.O., 1985. Historical perspective on the contamination of food and beverages with lead. In: Mahaffey, K.R. (Ed.), Dietary and Environmental Lead: Human Health Effects. Elsevier, New York, pp. 1–41.

Ohanian, E.V., 1986. Health effects of corrosion products in drinking water. Trace Subst. Environ. Health 20, 122–138.

Patterson, J.W., O'Brien, J.E., 1979. Control of lead corrosion. J. Am. Water Works Assoc. 71, 264–271.

Pocock, S.J., Shaper, A.G., Walker, M., Wale, C.J., Clayton, B., Delves, T., et al., 1983. Effects of tap water lead, water hardness, alcohol, and cigarettes on blood lead concentrations. J. Epidemiol. Commun. Health 37, 1–7.

Pontius, F.W., Clark, S., 1999. Drinking Water Quality Standards, Regulations, and Goals. In: Letterman, R.D. (Ed.), Water Quality and Treatment, fifth ed. McGraw-Hill Book Co., New York (Chapter 1).

Richards, W.N., Moore, M.R., 1984. Lead hazard controlled in Scottish water systems. J. Am. Water Works Assoc. 75, 60–167.

Robertson, J., 1851. An account of cases of chronic lead poisoning, caused by drinking water kept in a leaden cistern. Lancet 202.

Sharett, A.R., Carter, A.P., Orheim, R.M., Feinleib, M., 1982. Daily intake of lead, cadmium, copper, and zinc from drinking water: the Seattle study of trace metal exposure. Environ. Res. 28, 456–475.

Sherlock, J., Smart, G., Forbes, G.I., Moore, M.R., Patterson, W.J., Richards, W.N., et al., 1982. Assessment of lead intakes and dose–response for a population in Ayr exposed to plumbosolvent water supply. Hum. Toxicol. 1, 115–122.

Sherlock, J.C., Ashby, D., Delves, H.T., Forbes, G.I., Moore, M.R., Patterson, W.J., et al., 1984. Reduction in exposure to lead from drinking water and its effect on blood lead concentrations. Hum. Toxicol. 3, 383–392.

Thomson, T., 1848. Observations on poisoning by water passed through leaden tubes and retained in leaden cisterns. Med. Times, pp. 193–196.

Tronchin, T., 1764. A treatise on the colica Pictonum; or the dry-belly-ache (R. Schomberg, Ed. and Trans.). W. Johnston, London.

U.S. Agency for Toxic Substances and Disease Registry, 1988. The Nature and Extent of Lead Poisoning in Children in the United States: A Report to Congress. U.S. Department of Health and Human Services, Washington, DC.

U.S. Department of Health, Education and Welfare, 1962. Public Health Service Drinking Water Standards—1962. Public Health Service Publication No. 956, U.S. Public Health Service, Washington, DC.

U.S. Environmental Protection Agency, 1975. National Interim Drinking Water Regulations. 40 FR 59570, December 24, 1975.

U.S. Environmental Protection Agency, 1986. (Levin, R., principal author). Reducing Lead in Drinking Water: A Benefit Analysis. Report No. EPA-230-09-86-019. Office of Policy Planning and Assessment, Washington, DC.

U.S. Environmental Protection Agency, 1994. Lead in Drinking Water in Schools and Non-Residential Buildings. Report No. 812-8-94-002. Office of Water, Washington, DC.

World Health Organization, 1984. Guidelines for Drinking Water Quality. Vol. 1: Recommendations. WHO, Geneva, Switzerland, pp. 55–56.

World Health Organization, 1993. Guidelines for Drinking Water Quality, second ed. Vol. 1: Recommendations. WHO, Geneva, Switzerland.

World Health Organization, 2004. Guidelines for Drinking Water Quality, third ed. Vol. 1: Recommendations. WHO, Geneva, Switzerland.

World Health Organization, 2007. Support for the Development of a Framework for the Implementation of Water Safety Plans in the European Union. EC Grant Agreement 07-0201-200546174/sub/D2. World Health Organization–Europe, Copenhagen, Denmark.

Worth, D., Matranga, A., Lieberman, M., Devos, E., Karalekas, P., Ryan, C., et al., 1981. Lead in drinking water: the contribution of household tap water to blood lead levels. In: Lynam, D.R., Piantanida, L.G., Cole, J.F. (Eds.), Environmental Lead. Academic Press, New York, pp. 199–225.

Chapter 29

Regulation and Regulatory Policies for Lead in Food

29.1 A BRIEF PERSPECTIVE ON LEAD IN HUMAN DIETS

The environmental health history of regulation of Pb in human diets dates to erratic and typically limited efforts to control or eliminate periodic outbreaks of overt lead poisonings associated with contaminated foods or beverages other than water. As chronicled by writers such as Nriagu (1983, 1985), contamination episodes were more likely than not purposeful adulteration, occasioned by an intent to cloak poor quality or other problems with foods or beverages in the marketplace. Typically, controls on such practices were highly localized, short-lived, rarely disseminated more broadly, and often ineffective.

The earliest efforts at alimentary lead controls were seen in Europe, with some efforts at reining in food contamination episodes recorded as early as the medieval period. Consumption of lead-laced beverages in Great Britain during the eighteenth and nineteenth centuries by a growing consumer class and the attending effects of Pb-adulterated comestibles such as saturnine gout was the stuff of satirical grist if not a catalyst for public health activism. In colonial America and other New World locales, some recognition of the need for controlling ingestible items in commerce dates to lead contamination plaguing rum production in Massachusetts distilleries, a hazard publicized by Benjamin Franklin (McCord, 1953).

Lead levels in food and beverages, in regulatory terms, comprise a complex and problem-laden pathway for reducing exposures of various risk populations. Lead enters the diet through various means: food production, food processing, food transport and marketing, food handling postmarket, etc., and Pb contamination can occur at one or more of these steps. This multiplicity of Pb source-derived mechanisms for contamination made early attempts at lead entry controls much worse given the paucity of data for lead's behavior in the human environment. The almost total lack of awareness of, or the nature of, input from sources of lead contamination in food supplies meant that any existing entities charged with health protection could

Trace Metals and other Contaminants in the Environment, Volume 10
ISSN: 1875-1121 DOI: 10.1016/B978-0-444-51554-4.00029-8

only speculate as to sources of lead poisonings, speculation which may or may not have been correct.

Complicating matters further, measurement methods with the requisite sensitivity and specificity for quantifying Pb content in the diet were largely absent, and even now reliable measurement of Pb content in dietary staples lies mainly in the hands of specialized laboratories and trained staff. Furthermore, the regulation of Pb or any other substance in human diets is complicated by highly variable behaviors and practices, parameters which control total time-specific intakes across populations. Again, stability of Pb intakes in terms of Pb regulation hinges on stability of both levels of lead and amounts of dietary components ingested daily.

Effective regulation of the lead content of foods, beverages, and related commodities such as dietary supplements had to await the establishment and stable operation of public agencies with the jurisdictional means to intervene through bans, proscriptions, and interventions. For the United States and other industrialized nations, this basically meant the twentieth century.

29.2 REGULATION OF DIETARY LEAD IN THE UNITED STATES

Regulating Pb in the diets of U.S. residents lies principally in the hands of the U.S. FDA with considerable responsibilities also assigned to its sister agencies: (1) the USDA principally through the Food Safety and Inspection Service (FSIS) for imported and domestic meat, poultry, frozen/dried/liquid eggs; (2) the U.S. EPA for regulating Pb-containing pesticides and water sources used in beverages; (3) labeling regulation for alcoholic beverages (>7% alcohol) through the U.S. Treasury Department's Alcohol and Tobacco Tax and Trade Bureau; and (4) the U.S. Commerce Department's National Marine Fisheries Service. However, the relative role of Federal agencies in terms of the regulatory chain for the U.S. food supply has traditionally been greater for the USDA than the U.S. FDA (see the U.S. Government Accountability Office report (U.S. GAO, 2005)). The former exercises its statutory mandates for food supply protection mainly through the FSIS but, with respect to lead contamination, also the Agricultural Research Service. However, two more recent laws have expanded the responsibilities of the U.S. FDA to more closely approach those of the USDA in connection with protection from terrorism threats (see below).

U.S. GAO (2005) found that, in the aggregate, food safety and quality is mandated through 30 statutory initiatives implemented by 15 different Federal agencies. Of particular concern are fragmented responsibilities for food safety. For example, different agencies are responsible for different food categories. The U.S. FDA is responsible for seafood processors and monitoring action levels of contaminants such as lead in seafood and certain other categories such as whole eggs, milk, grain products, fruits, and vegetables under several core statutes (FFDCA, the Public Health Service

Act, and the Egg Products Inspection Act), but USDA inspects and tests meat- and poultry-processing facilities as well as those for egg products [via the Federal Meat Inspection Act, the Poultry Products Inspection Act, the Egg Products Inspection Act]. These Federal divisions of labor are supplemented across the country by various agencies within the various states.

The U.S. FDA exercises regulatory functions via various mechanisms such as: (1) inspecting imported foods, (2) inspecting domestic and imported ceramic hollowware and dinnerware for compliance with action levels (guidelines), (3) developing Pb action levels and tolerable Pb intake levels, (4) testing food-related items for compliance, (5) issuing public warnings when violations occur, and (6) systematic and periodic surveys of lead in the U.S. diet. There is more reliance on use of guidelines, advisories, industry letters, and the like and less reliance on triggered compulsory enforcement actions for controlling public access to contaminated foods and beverages. Where excess levels of Pb in a U.S. FDA-regulated product are identified, products can be seized or detained (imports) and injunctions sought against distribution and sale.

Intra-agency divisions of lead control in foods and related materials, in turn, assign much of the U.S. FDA's regulatory mandates to the Center for Food Safety and Nutrition (CFSAN). Responsibility in the food and beverage sectors includes the following: safety of food and color additives; safety of foods and components produced through biotechnology; seafood and juice Hazard Analysis and Critical Control Point (HACCP) regulations; programs for addressing health risks from foodborne chemical and biological contaminants; correct labeling; safety of dietary supplements, infant foods; industry surveillance and compliance; and food standard and safety harmonization efforts with other national and international jurisdictions.

The general basis of U.S. FDA food lead control regulation and how it performs this function is spelled out in two basic statutes. These are, to a limited extent, the Food and Drug Act (F&D Act) of 1906 and the 1938 FFDCA. A number of important amendments to the 1938 Act have been enacted over the years, and those of particular relevance to the topic at issue are as follows: (1) the Infant Formula Act of 1980 [PL 96–359]; (2) Dietary Supplement Health and Education Act of 1994 [PL 103–417]; (3) Food Quality Protection Act of 1996 [PL 104–170]; (4) Pediatric Research Equity Act of 2003 [PL 108–155]; and (5) the Dietary Supplement and Nonprescription Drug Consumer Protection Act of 2006 [PL 109–462].

The U.S. FDA more recently had a major expansion in its surveillance and contaminant enforcement authority with respect to potential threats of terrorist acts against the U.S. food supply: (1) the Homeland Security Act of 2002 (PL 107–296; 116 Stat. 2135) and (2) the Public Health Security and Bioterrorism Preparedness and Response Act of 2002 [PL 107–188; 116 Stat. 594]. These two statutes prompted the development of the HACCP program within the agency's CFSAN. For example, an episode of deliberate contamination with hazardous substances within the food production, distribution, and consumption system requires the Agency to narrow an effective response to the point of contamination rapidly. CFSAN applies mandatory rules and regulations via the HACCP program for seafood- and juice-processing facilities. Similarly, USDA employs mandatory HACCP rules in meat- and poultry-processing facilities.

The regulatory mission of the U.S. FDA with respect to lead has been said to be a binary one (Kashtock, 2009): (1) prevent the avoidable entry of lead into food, while (2) controlling the amounts of unavoidable lead entering food. In that context, the historical use of the inorganic pesticide lead arsenate in the early decades of the twentieth century prior to widespread use of organic chemical pesticides exemplified an administratively defined unavoidable use and was controlled via use of a tolerance limit for commercially produced apple crops in tandem with surface contamination removal steps before marketing. Kashtock (2009) noted, however, that the first legal seizure of a commodity by the agency was in 1908, for Pb-contaminated green coffee beans that had been adulterated (colored) with lead chromate.

Lead arsenate was principally used as a larvicide in apple orchards to control the larval stage of the codling moth (*Carpocapsa pomonella*). As a consequence, considerable acreage in apple-growing areas of the United States that employed the lead arsenate remains impacted by elevated soil Pb and As concentrations. Inspecting and testing for Pb in apple crops comprised a significant amount of the U.S. FDA's regulatory efforts for food Pb right through WWII (Kashtock, 2009). Testing for pesticide use in the United States was mandated by the 1910 Federal Insecticides Act. This early phase was mainly directed to concerns of the USDA about pesticide quality in commercial channels. These provisions were superseded by the 1947 Federal Insecticide, Fungicide, and Rodenticide Act (FIFRA) and later amendments to FIFRA.

29.2.1 General Regulatory Activities of the U.S. FDA for Food Pb

The 1938 FDCA mandated that limits or tolerances for chemicals in food be set for human health protection. Limits and tolerances were food category-specific and differed from the later use of provisional total tolerable intakes (PTTIs), total intake levels for various U.S. consumer groups based on health-risk considerations (see Carrington and Bolger, 1992).

The U.S. FDA's regulatory authority and activity for regulated categories is found in §402(a)(1) [21 U.S.C. §342] of the FFDCA, which holds any food category subject to regulation because of being adulterated if it contains any poisonous or deleterious matter which may or ordinarily does render that food harmful to human health. Categories of adulterants include but are not limited to trace substances such as lead, natural toxins, pesticides, organic chemicals, adulterants formed as an artifact of food preparation (e.g., acrylamide), and radionuclides. The act also implicates foods as adulterated if they lack ingredients rather than are contaminated by some substance.

For cases where substances are present in foods by virtue of being unavoidably present, U.S. FDA acts to set tolerances for these toxicants under Section 406 [21 U.S.C. § 346] of the FFDCA.

29.2.2 Pb Provisional Total Tolerable Intakes (PTTIs) and U.S. Total Diet Pb Surveys (TDSs)

The U.S. FDA has established Pb PTTIs in the diets of four subsets of the U.S. population (Carrington and Bolger, 1992). Young children, those <7 years of age, have the most protective PTTIs, 6 μg Pb/day. PTTI levels for the remaining three groups are as follows: older children ≥7 years of age, 15 μg/day; pregnant women, 25 μg/day; other adults, 75 μg/day. Calculations used in arriving at these values were a mix of Pb intakes, age- and state-specific host factors and Pb biokinetics, culminating in a raw estimate of host-specific intakes adjusted by a safety factor of 10-fold. Children <7 years of age ingesting 6 μg Pb/day will typically absorb 3 μg/day. Of this amount, a biokinetic factor (~0.40) can be used to link absorbed Pb to a resulting PbB value of 1.2 μg/dl.

The PTTI is a cumulative metric for daily Pb intakes and does not guide apportionments by regulatory limits of Pb contents in individual foods or food groups (Carrington and Bolger, 1992). It was derived by estimating the amounts of dietary lead intake in various age segments of the population that produce a PbB of 10 μg/dl (young children) or the higher threshold of 30 μg/dl (adults). The PTTI for age-scaled U.S. diets provides no direct measure of the extent of Pb intakes from other sources but would be assumed to be accommodated in the 10-fold safety factor. One can, however, compare dietary intakes temporally to ascertain both time trends in intakes and relationship of intakes to the PTTIs.

The TDS is a periodic survey of the contaminant contents of multiple food groups arrayed by subsets of the U.S. consuming population (see Chapter 6 for detailed survey results and time trends in Pb content).

The trend in Pb content of the U.S. diet has been downward over the years, for reasons noted earlier. As a percentage of the daily U.S. FDA PTTI, daily U.S. intakes of Pb are in the range of 1–5% and 6–23%. The uppermost value in the upper range indicated that almost one-quarter of the PTTI is achieved through the actual dietary intake levels.

29.2.3 U.S. FDA Guidance or Action Levels for Food in U.S. Diets

The U.S. FDA does not promulgate, in policy or practice, a comprehensive schedule of regulatory limits for lead in the many scores of dietary components comprising the complex makeup of U.S. diets. The U.S. FDA case differs considerably from the parallel functions of regulations in the European Union, as noted in the next section. There are several reasons for this. First, statutory authority for U.S. FDA actions differ from those of other U.S. agencies, such as the U.S. EPA. Second, there are complex divisions of labor among U.S. FDA and sister agencies. As noted above, the USDA sees to food safety aspects of major food categories. Second, pesticide residue tolerance limits are established by the U.S. EPA and the U.S. FDA sees to enforcement of those. The U.S. EPA sees to regulation of lead in tap water, and thereby is the determinant agency for any Pb entering food during preparation using tap water. For example, this applies to cooked rice, other cereals, and vegetables.

Examples of segments of the food supply where U.S. FDA has issued guidance or regulatory actions specifically for lead are certain seafoods, wine, candy, and fruit juices. The agency, as noted earlier, does not promulgate action levels or guidance for lead and other toxic metal contaminants in each and every dietary item in the U.S. food supply. The seafood items regulated in terms of lead levels are shellfish and crustacea. The corresponding safety levels are 1.7 ppm Pb for shellfish and 1.5 ppm for crustacea. These are regulatory safety levels but not food Pb tolerances, and the U.S. FDA can take legal action to remove these products from public commerce if the limits are exceeded.

Juice lead levels are regulated when packed in cans with lead-soldered seams through an emergency action level established in 1993 of 0.08 μg/g (80 ppb) (58 FR 17233; April 1, 1993). Juice Pb levels are also to be controlled as a recognizable or anticipated hazard within the regulatory and security meanings of the Juice HACCP Regulation (see above). If lead is expected to be a likely hazard due to introduction during its production and packaging, producers are required to carry out a hazard analysis by experts (U.S. FDA: Guidance for Industry, 2001).

A particularly involved sequence of regulatory assessment was that for lead in candy popular with young children, particularly brands produced in Mexico and marketed in the United States and elsewhere. Sources of the lead contamination have been said to be either wrappers or a contaminant in sugar, the major ingredient. The agency has acted in this particular case via use of a notification instrument: *Guidance for Industry Communications: Manufacturers, Importers, and Distributors*. In its Recommendations of June 13, 1995, the U.S. FDA established a regulatory trigger for lead content of 0.5 ppm, on the assumption that Pb arose from wrapper printing ink or from sugar. The 0.5 ppm level was equivalent to the NAS/NRC Institute of Medicine's specification for sugar Pb in its Food Chemicals Codex (FCC).

In November 2006, the 0.5 ppm guidance was rescinded by the U.S. FDA and replaced by the much more protective value of 0.1 ppm, fivefold lower.

This change was concurrent with the same downward revision in the FCC for sugar Pb.

Guidance for Pb in alcoholic beverages such as wine with alcohol content >7% has also been issued. The agency recommended in March 2007 as guidance, not as regulation, a lead contaminant limit of 0.3 ppm (300 ppb) for wines (U.S. FDA: Guidance for Industry and FDA, 2007), using a 1991 recommended guidance value.

29.2.4 U.S. FDA Regulations and Activities for Food Contact Surfaces

Two areas of U.S. FDA activity related to food Pb levels spelled out in statute concern lead-releasing containers and tableware that come into contact with food. The first major action within this category was the effort to reduce the use of and ultimately ban lead-soldered containers for canned food in general and infant dietary components in particular. This activity was discussed in the earlier chapter on dietary Pb exposures. The agency in 1979 brokered a voluntary agreement by the food and beverage container industry to carry out a scheduled phase-out of production of lead-sealed containers (see NAS/NRC, 1993; U.S. ATSDR, 1988; U.S. EPA, 1986, Ch. 7).

By the late 1980s to early 1990s, such lead-seamed containers had largely disappeared from the American food and beverage scene, and by 1991 the industry had completely converted to alternative containers. In 1996, the U.S. FDA took formal action [§§201, 402, 409, 701 of the FFDCA (21 U.S.C. 189.240)] to ban lead-seamed cans (60 FR 33106–33109; June 27, 1995). The formal ban served at that point to limit canned food imports in lead-soldered containers for domestic canned food inventories.

A number of regulations address the lead in ceramic ware capable of leaching and contaminating foods prepared or stored in these containers. The lead comes from the glazes used in their manufacture, particularly improperly fired glazes. Several sets of rulemaking were promulgated over the years and the most recent action levels are presented here. Leaching limits of Pb (57 FR 29734) into the leaching fluid by different categories of tableware (level, container ware type) are as follows: 3.0 μg/ml, ceramic flatware; 2.0 μg/ml, small hollowware (but not cups and mugs); 0.5 μg/ml, cups and mugs; 0.5 μg/ml, pitchers; 1.0 μg/ml, large hollowware (excluding pitchers). U.S. FDA regulates these extractible Pb levels via the Compliance Policy Guides (CPGs), Subchapter 545: CPG 7117.07. These levels have been extensively employed in regulating imported ware from areas with problematic glazes.

Wine bottles formerly capped with tin–lead foil capsules over the corks were regulated subsequent to data showing lead leaching when the agency issued a final rule in February 1996, banning lead capsules (caps) for use as bottle covering over the corks (61 FR 4816–4820; February 8, 1996).

29.2.5 U.S. FDA Regulatory Controls over Food Supplements

The term "supplement" has various regulatory meanings in various jurisdictions. The U.S. FDA, acting under the 1994 Dietary Supplement Health and Education Act, defines a supplement as a product for supplementing the diet, and that can consist of vitamins, minerals, herbs, botanicals, or amino acids. Other countries define the term within the drugs category. Various classes of supplements for the U.S. diet are subject to U.S. FDA control actions. These include vitamins and energy supplements.

The U.S. FDA, as a practical matter, is limited under the 1994 Act and the FFDCA [§§402(a)(1), (a)(2), (a)(3), (a)(4)] as to how extensively it can regulate the introduction and sale of these products to the U.S. public. Because of their categorization under "food," rather than "drugs," no prior approval is required for market entry. Rather, manufacturers can simply inform the agency and include the ingredients, followed by a waiting period. The U.S. FDA posts nonproprietary parts of the new product but cannot certify it as "safe." Given the statutory strictures, the agency is largely limited to providing announcements of safety records.

For example, the agency on May 24, 2010, warned consumers to avoid a supplement heavily contaminated with lead, "Vita Breath," reported as 500 ppm Pb in the U.S. FDA's own testing and reported elsewhere with much higher amounts. The product was recalled.

In a recent survey by the U.S. FDA for lead content of 324 vitamin and mineral products marketed for use by women and children, test results indicated median and maximum Pb contents of 0.04 μg/g (rounding) and 0.40 μg/g (rounding) Pb across all products for the highest risk group, young children of 0–6 years of age. Daily intakes in this risk group in terms of median and maximum values were 0.12 and 2.9 μg/day. The survey (Mindak et al., 2008) indicated overall Pb exposures from these products were below the PTTI values for each of the four exposure groups used by the U.S. FDA. However, Miodovnik and Landrigan (2009) argued that the U.S. FDA employed a now-obsolete exposure yardstick in calculating the PTTI values, the biological reference level of 10 μg/dl. While this value has been retained by the U.S. CDC in its statements on childhood lead poisoning for reasons of practicality and logistics, that agency has also accepted that considerable evidence exists to show adverse effects in young children at levels below this threshold (see earlier chapters). Equally of concern to those authors, the relative inputs to the overall daily intakes from all sources combined needed to be considered. The range of U.S. Federal advisory and regulatory activities for Pb in the U.S. food supply are summarized in Table 29.1.

TABLE 29.1 Significant U.S. Regulations or Guidelines for Lead in Diet

Statutory/Regulatory Initiative	Codified Law/ Regulation	Year(s)	Key Features
A. Federal Statutes			
1906 Federal F&D Act, superceded by 1938 U.S. FDCA	§402(a)(1): 21 U.S.C. §342 §406: 21 U.S.C. §346 Other sections: 408(a), 512, 409	1938–forward	Principal statutory basis for U.S. FDA activities in Pb control in the U.S. diet. Agency judges toxic adulterants such as Pb if they make food harmful to health; enables use of tolerance limits for Pb
Some significant amendments to the FDCA: Infant Formula Act of 1980 [PL 96–359]; Dietary Supplement Health and Education Act of 1994 [PL 103–417]; Food Quality Protection Act of 1996 [104–170]; Pediatric Research Equity Act of 2003 [PL 108–155]; Dietary Supplement and Nonprescription Drug Consumer Protection Act of 1996 [PL 109–426]			Various amendments focus on specific problems dealing with the U.S. food supply, such as contaminants in infant formula, levels of contaminants in supplements such as vitamins; protection of U.S. children from food contaminants
1910 Federal Insecticide Act, superceded by 1947 Federal Insecticide, Fungicide, Rodenticide Act, with 1959, 1964, 1975, 1978, 1980, 1981, 1988 amendments		1910–forward	Statutory basis for earlier regulation by the U.S. FDA of the larvicide lead arsenate on apple orchards
B. Federal Regulatory Actions			
Promulgation of U.S. FDA action levels for Pb in various substances via CPGs	Leaching of Pb from ceramic ware, hollowware, pottery: CPG 454.450; 21 U.S. C., Parts 109, 509 (57 FR 55485)		Leaching of Pb from various ware is controlled by action levels (ppm) enumerated in the text. Action levels trigger U.S. FDA legal actions if exceeded

(*Continued*)

TABLE 29.1 Significant U.S. Regulations or Guidelines for Lead in Diet—(cont.)

Statutory/Regulatory Initiative	Codified Law/ Regulation	Year(s)	Key Features
U.S. FDA promulgated a ban on lead foil caps for wines, taking effect in 1996	21 C.F.R. Part 189; 61 FR 4816–4820, February 8, 1996	1996–forward	Ban on lead foil caps reduces a significant source of wine Pb contamination
U.S. FDA banned use of lead-seamed cans in domestic or imported canned food production in 1995	21 C.F.R. Part 189; 60 FR 33106–33110, June 27, 1995	1995–forward	Main regulation for an historically significant contributor to Pb in canned foods and evaporated milk. This action followed an emergency action in 1993 setting action levels for Pb in foods packed in lead-seamed cans. The fruit juice Pb level limit was 80 ppb (58 FR 17233)
U.S. FDA issued a November 2006 guidance letter publicly releasing its maximum candy Pb allowance at a level of 0.1 ppm Pb	58 FR 33860, June 21, 1993	2006–forward	New recommendation rescinded earlier guidance of 0.5 ppm Pb; candies favored by children were the focus
U.S. FDA issued guidance documents for permitted levels of Pb in shellfish	Appendix 5, Table 5-A: FDA and EPA Safety Levels in Regulations and Guidance: June 2001		U.S. FDA regulates through these guidance levels Pb in certain shellfish: 1.5 ppm, crustacea; 1.7 ppm Pb, moluscan bivalves
U.S. FDA monitors Pb in components of the U.S. food supply through the TDS; USDA also examines lead in the food components			These surveys document major declines over the last 20–25 years in dietary Pb intakes, children and adults

29.2.6 Illustrative Food Pb Regulations in States of the United States

The most prominent and influential U.S. state actions in terms of regulatory controls on food Pb above and beyond any promulgated by and within the Federal government are those issued by California. The state regulates Pb in both children's and adults' products, including food, through the CA SDWA and Toxic Enforcement Act of 1986, with the regulatory limit of 0.5 μg Pb/day based on avoiding Pb exposures associated with toxic reproductive effects in humans. The Act, popularly known as Proposition 65 or Prop. 65, is codified as CA Health and Safety Code, §§25249.6 et seq. Diet components and dietary supplements are included.

Eighteen states within the Toxics in Packaging Coalition regulate Pb in packaging materials: CA, CT, FL, GA, IL, IA, MD, ME, MO, NH, NJ, NY, PA, RI, VT, VA, WA, and WI.

29.3 REGULATION OF DIETARY Pb IN THE EUROPEAN UNION

The European Union and its member states regulate lead and other contaminants in diet and diet-related materials through a complex set of interlocking processes engaging different organizations. In a number of ways, the EU regulatory model differs from that of the U.S. approach via the U.S. FDA because of their differing statutory frameworks, divisions of governmental labor, and control mechanisms and associated resulting strictures or liberties. As noted earlier, the European Union can regulate lead in foods and related commodities through either directives or regulations. EU Directives wend a more complex path to ultimate controls on lead in that the ultimate food lead regulation is left to the member states within some broad guidelines.

A major technical factor in the subsequent workings of food Pb controls in the European Union are the recommendations and guidelines generated within the organizational expertise of the advisory bodies of the WHO, in like fashion to guidelines provided for lead in air, drinking water, etc.

29.3.1 The FAO/WHO International Guidance System for Lead in Foods

EU directives and regulations concerning food lead typically arise downstream of the various guidelines and standards set forth by the joint program of the FAO and WHO working through such sister organizations as the Codex Alimentarius (Food Code) Commission (CAC). The latter, in turn, works via its various committees, such as the Executive Committee and the Codex Committee on Contaminants in Foods (CCCF) and its technical expertise arm, the JECFA-FAO/WHO (FAO/WHO, 2010), which addresses and provides recommendations for control of both additives and contaminants in food.

The function of the Codex Alimentarius (Food Code) in its current incarnation is to establish food health standards that provide a harmonized reference for international trade in foodstuffs. Legislation developed by the European Union and others is assisted in the case of toxic elements such as lead through the Codex General Standard for Contaminants and Toxins in Foods (GSCTF). The Code's legal importance is significant, inasmuch as it has 171 member countries, including all states of the European Union and, since 2003, the European Community.

The CAC was created in the early 1960s by the FAO and WHO to implement their joint standards program, among other functions. The FAO and WHO periodically evaluate the work of the CAC through the World Health Assembly's resolutions process. The range of the CAC's activities and responsibilities include defining and setting international standards, codes of practice, guidelines and recommendations governing agricultural and fishery products, foodstuffs, food additives, food contaminants, animal feeds, labeling, inspection and certification, analysis methods, and ethical farming and food hygiene practices. A principal operational form of these duties is the continuous updating of them in the interest of the Codex, working through its committees.

The CAC standards are codified in one of the Codex's 13 volumes and the Commission convenes in annual meetings working through a formal agenda. Three expert committees advise the CAC in its functions, one of which, JECFA, includes lead as a food contaminant in its assessments. As noted by Berg and Licht (2002), the apparatus for controls on lead and other contaminants is centered on the development and presentation of Maximum Limits (MLs), which are generally similar to those promulgated via the directives and regulations of the European Union and joint with the EC from 2003.

The principal role of the JECFA, dating to the 1970s, has been the development of provisional tolerance values scaled by some time unit reflecting its propensity for body accumulation and per unit of body weight, derived using current scientific and toxicological information for human populations. Lead limits in food are presented as Provisional Tolerable Weekly Intakes (PTWIs). The Pb PTWI has been periodically revised downward over time as to both Pb level and the segment of the exposure population exposed to lead. By 1993, the PTWI was lowered to a figure of 25 μg Pb/kg body weight/week (3.5 μg Pb/kg body weight/day), first for children and then for all age groups (WHO, 1993).

A recent deliberation of the JECFA for lead and other metal contaminants occurred June 8–17, 2010, in Geneva, Switzerland (FAO/WHO, 2010). In a strong action, JEFCA concluded that the present PTWI for lead, 25 μg/kg body weight/week, was not sufficiently protective for young children; the group withdrew this PTWI for children. Withdrawal of the PTWI necessitates international evaluation of the potential health risks from ongoing food lead exposures by means other than calculating how close dietary Pb intakes are to total intake ceilings. The JECFA action was not totally unexpected since a committee of the regulatory body, the European Food Safety Authority (EFSA) Panel on Contaminants in the Food Chain (CONTAM, 2010), took the same action earlier in 2010.

Another function of the FAO/WHO organization has been the use of surveys of food lead levels and overall food Pb exposures in European and other international venues. Galal-Gorchev (1991) presented the findings of an earlier survey effort, the joint UNEP/FAO/WHO Food Contamination Monitoring Programme (GEMS/Food), which is an effort within the GEMS created by the UNEP. Adult intakes of Pb were presented for 25 countries over the years 1980–1988, with the Pb intakes measured as μg/kg body weight/week, ranging from 1 to 63 units. The then-extant PTWI value of 50 μg/kg body weight/week was exceeded by three countries (Cuba, India, Italy) and nearly exceeded by one (Thailand). With infants and children, three countries (Federal Republic of Germany, 1980; Hungary, 1983; and Poland, 1985) exceeded the childhood PTWI of 25 μg/kg body weight.

Surveys for EU member states in the 1990s were largely individual national surveys (summarized in Nasreddine and Parent-Massin, 2002). Survey data in the first decade of the twenty-first century have been gathered under the aegis of the European Community (see below) as part of its regulatory apparatus. Table 29.2 provides summary data for these various guidelines and regulatory highlights for lead in international diets.

29.3.2 Regulations and Directives on Food Lead in the European Union

Regulatory activity concerning lead in food of EU member states evolved and expanded in the last two decades. Evolution is likely to continue, meaning the status of regulations for food Pb described in this chapter may be outdated in coming years. The extent of change is not likely to be major in the case of food lead in developed European states, but new member states may witness changes versus their prior lagging status in overall lead controls.

Some key directives and regulations were promulgated during this period and those are given the most attention in this section. Earlier actions encompassed the original 15 members of the European Union. With recent admissions, the EU member count is 27: Austria, Belgium, Bulgaria, Czech Republic, Cyprus, Denmark, Estonia, Finland, France, Germany, Greece, Hungary,

TABLE 29.2 Significant International Statutory/Regulatory Initiatives for Lead in Human Diets

Statutory/Regulatory Initiatives	Codified Law or Regulation	Year(s)	Key Features
A. Guidelines of the FAO/WHO and CAC			
Codex GSCTF	See text	Early 1960s–forward	Has 171 member countries involved in its activities; work of the CAC in 13 volumes
JECFA-FAO/WHO	Periodic reports of recommendations to WHO, FAO, and CAC	1972–forward	JECFA has issued several intake limits for diet Pb expressed as weekly provisional tolerance values; in June 2010, withdrew Pb tolerance value for young children (see text for more details)
Panel on CONTAM	Advisory arm to the EFSA; issued 2004 and 2010 reports on Pb in foods	–	CONTAM group also withdrew its earlier views on a safe Pb level in children's diets (in Spring 2010)
UNEP/FAO/WHO Food Contamination Monitoring Programme; two surveys done later	Monitoring of Pb in food groups in different segments of global populations; later surveys, in 2004 under Directive 93/5/EEC, and 2003–2009 by EFSA CONTAM panel	GEMS/Food 1980 1988, later years, under directives	Major early survey of the status of the global food contamination problem with lead
B. Regulations and directives for EU food lead			
EU food Pb	EC Regulation: 315/93/EEC, February 8, 1993	–	Issued rulemaking banning toxic levels of Pb in foods, setting maximum limits (MLs) for Pb and other contaminants; diet Pb controlled by regulations, not directives
EFSA develops assessments of and MLs for Pb in EU member food supplies	EC's Directorate-General, Health and Consumer Protection	Early 2000s	EFSA works through the CONTAM expert panel to do health-risk assessments for Pb

(*Continued*)

TABLE 29.2 Significant International Statutory/Regulatory Initiatives for Lead in Human Diets—(cont.)

Statutory/Regulatory Initiatives	Codified Law or Regulation	Year(s)	Key Features
Setting of MLs for Pb by the European Union	EC Regulation No. 466/2001 (March 8, 2001), in force 2002; Previous Regulation, EEC No. 194/1997 repealed	2002	See text for Pb MLs in various food groups
	EC Regulation No. 1881/2006, issued December 19, 2006, in force March 1, 2007; supercedes the previous one	2007	Repealed but largely retained the MLs for Pb
	EC Regulation No. 629/2008 issued July 2, 2008 sets MLs in certain foods	–	–

Ireland, Italy, Latvia, Lithuania, Luxembourg, Malta, the Netherlands, Poland, Portugal, Romania, Slovakia, Slovenia, Spain, Sweden, United Kingdom.

The EU legislative initiatives on food safety relevant to lead and other toxicants are set forth in the February 8, 1993 EC Regulation 315/93/EEC: (1) food items having toxic levels of contaminants as to be a hazard to public health shall not be allowed on the market; (2) contaminant levels in all cases shall be kept as low as reasonably achievable; (3) maximum limits (MLs) must be set for certain contaminants in order to protect public health. Unlike the case with regulating lead in ambient air and drinking water, proceeding through directives and individual member state regulation, most of the control activities for toxicants in food are in the form of regulations. These regulations can be viewed as being both horizontal, i.e., lead in all 14–15 constituent food groups within the regulated dietary profiles for member states, or vertical, for all contaminants within a food group. The regulation mechanism, binding upon all member states, assures a harmonization of regulation, particularly with regard to toxic contaminants such as lead.

Administratively, the EU regulations are developed and communicated through the European Commission's Directorate-General, Health and Consumer Protections. Since the early 2000s, the scientific basis and development of MLs in various food categories has been the responsibility of the

EFSA. EFSA functions through various mechanisms to provide input to food safety in the European Union and to the quantitative assessment of food toxicant risks through MLs. That is, a major function of EFSA is risk assessment. EFSA empanels two types of collective expertise in fulfilling its duties to the European Commission. These are Scientific Opinion and Scientific Report committees. The March 18, 2010 Scientific Opinion on Lead in Food was prepared by the EFSA Panel on CONTAM (CONTAM, 2010). This is a standing committee to which technical questions with a risk component may be directed. The CONTAM, in its withdrawal action for the PTWI for lead in tandem with that of the FAO/WHO JECFA committee (see above), noted that current exposure data and associated margins of exposure (MOE) estimates do not rule out that some children 1–7 years of age may be having potentially toxic Pb exposures through food.

A second Scientific Opinion effort of the CONTAM Panel was the report on lead as an undesirable substance in animal feed (CONTAM, 2004). This is of public health concern since lead accumulation in food animals eventually becomes human dietary Pb exposure.

EFSA scientific reports are *ad hoc* efforts that involve different authors across the various reports. Two recent EFSA Scientific Reports concerning lead are those of Jorgensen (2010) and Boon et al. (2010). Jorgensen employed a BMD modeling approach to the pooled analysis data of Lanphear et al. (2005) for dose–response relationships in longitudinal studies for children's blood lead and IQ decrements. He calculated a BMD and its 95%-confidence lower bound for a 1% (one IQ point) change in IQ as the BM response. Boon et al. (2010) employed two exposure models to estimate the long-term dietary Pb exposure risks for young children in the European Union based on current survey data. The models were the beta-binomial-normal and the observed individual means approaches. Lower bound exposures were estimated as similar (0.4–1.7 μg/kg body weight/day) for both models at the median intakes, but there was divergence at the 95th-percentile intake.

29.3.3 EU Maximum Limits on Lead Levels (MLs) in European Diets

The EU picture for controls on lead in the diet differ markedly from that of the U.S. FDA, with the latter's mix of specific foods not otherwise restricted from regulation by either some statutory constraint or shared regulatory authority with sister agencies. Several core regulations underpin the setting of MLs for lead in various dietary items in the EU member states. EC Regulation No. 466/2001 was promulgated on March 8, 2001, to be in force in 2002. This regulation repealed its predecessor: EEC No. 194/1997.

Seven categories of substances, including the heavy metals group that includes lead, were specified. The remaining contaminant groups were nitrates, mycotoxins, 3-monochloropropane-1,2-diol, dioxin and dioxin-like PCBs, inorganic tin, and polycyclic aromatic hydrocarbons. The food products consumed by various segments of the member state populations and for which lead MLs were specified were as follows: vegetables, including certain leafy vegetables, fruit, fruit juices, berries, cereals, milk and milk products, baby foods via infant formulae and milk, wine, edible animal offal, meat from bovines, sheep, pig and poultry, all classes of fish products, fats and oils.

The food groups and their corresponding ML values (ppm Pb, wet weight) are as follows: milk, 0.02; fats and oils, 0.1; fruit, 0.1; berries and small fruits, 0.2; vegetables, 0.1; potatoes, 0.1; cereals, 0.2; meat, 0.1; edible offal, 0.5; fish muscle meat, 0.2/0.4; crustaceans, excluding meat of crab, 0.5; bivalve mollusks, 1.5; cephalopods, 1.0; fruit juices, 0.05; wine, 0.2.

The above EC regulation was subsequently repealed with respect to lead and other contaminants by EC Regulation No. 1881/2006, promulgated December 19, 2006 and effective March 1, 2007. The food groups and their ML values (ppm Pb, wet weight, values with virtually no change from the previous regulation) are as follows: raw, pasteurized milk and milk for processing, 0.02; infant formulae and follow-on formulae, 0.02; meat, 0.10; offal, 0.50; fish muscle meat, 0.30; crustaceans, with some exceptions, 0.50; bivalve mollusks, 1.5; cephalopods, without viscera, 1.0; cereals, legumes, and pulses, 0.20; vegetables with some exceptions, 0.10; other vegetables—Brassica, fungi, leafy vegetables, 0.30; fruit, not berries or small fruit, 0.10; berries and small fruit, 0.20; fats and oils, not milk fat, 0.10; fruit juices, 0.05; wine and related, 0.20; aromatized wine, 0.20. Accompanying this regulation was one covering required reliable sampling and analytical methodologies, EC No. 333/2007, promulgated March 28, 2007. This regulation repeals the Directives: 2001/22/EC, 2004/16/EC, and 2005/10/EC.

The various EU regulations imposed on member states the obligation of monitoring lead in food groups and transmitting their findings, of providing the wherewithal to evaluate such public health issues as trends in lead levels in European diets and also calculations of the Margins of Exposure (MOEs), the relative proximity of total daily Pb intakes through diets to total daily allowable Pb amounts. Two comprehensive lead exposure surveys have been carried out in the European Union by various components of the EC rule- and regulation-making system rather than through the FAO/WHO mechanism. Both surveys entailed food Pb data gathering from reporting member states and assessment of their significance.

The recent popularity of herbal, i.e., alternative or complementary medicine preparations and the finding in the United States and Europe of high amounts of lead in some of them, e.g., the use of herbal or mineral compounds, has led to the European Commission promulgating a regulation (EC Regulation No. 629/2008) on July 2, 2008, setting limits to such contaminants in foodstuffs under the category of food supplements. The MLs in Regulation No. 1881/2006 for Pb in foods are amended to include a food supplement ML of 3.0 ppm wet weight. Some additional amendments for lead and other contaminants were included in this action as well.

An expert committee sponsored by the European Commission under existing Directive 93/5/EEC within the Directorate-General, Health and Consumer Protection, carried out a survey of arsenic, cadmium, and lead levels in member state diets. The results were published in a March 2004 report (EC Scientific Cooperation (SCOOP) Task 3.2.11, 2004). With lead, 11 member states plus Norway submitted Pb occurrence and intake data for the adult population, but only two states—Denmark and the United Kingdom—provided adequate data for total intake estimates. This resulted in marked underestimates of daily Pb intakes. The mean daily dietary Pb intakes, μg/day, for the 12 states were as follows: Belgium, 38; Denmark, 18; Finland, 6; France, 57; Germany, 47; Greece, 25; Ireland, 1.1; Italy, 30; Norway, 21; Portugal, 133; Sweden, 5; United Kingdom, 27. Mean intake for the 12 states was 42 μg/day. Across the 12 states, the mean percentage of the PTWI value of 25 μg/kg body weight attained by the mean daily Pb intakes was 14%. Some states markedly exceeded others, with Portugal having intakes 53% of the PTWI. The major contributor to this country's intake was fruit and vegetable Pb.

The second survey was based on member state food Pb intake and analysis data over the period 2003–2009 gathered by the EFSA CONTAM expert committee and presented in its 2010 report. Fourteen member states and Norway submitted 140,000 measurements of food and tap water Pb, of which 94,126 data points were usable. Two-thirds of the measurements were below detection or quantification limits. Mean and 95th-percentile values were calculated for each country. Using both point-estimate and probabilistic methods, similar values were estimated. Lower and upper bounds for groups designated as either low- or high-end consumers were determined. From the measurements, daily Pb exposures for the states ranged from 0.36 to 1.24 μg/kg body weight/day for the low-end consumer group and 0.73–2.43 μg/kg body weight/day for the high-end consumers. Children aged 1–3 years who were high-end consumers had lower and upper bound values of 1.71 and 5.51 μg/kg body weight/day.

29.3.4 EU Regulations for Pb Contamination of Food from Surfaces

Two EU directives govern the contamination of foods by Pb leaching from contact surfaces, such as ceramic ware, metal containers, and metal utensils. Ceramic ware leaching of Pb was closely controlled first through a long-existing Directive 84/500/EEC that set strict limits on Pb releases from ceramic ware. Subsequently, ceramic ware controls were through 2005/311/

EEC. As was and remains the case in the United States, imported ceramic ware from certain countries appears to be the main persisting problem.

Lead leaching from metal containers and utensils is governed by Directive 89/109/EEC, Article 2.2. This was followed by a March 9, 2001 guideline report on "Guidelines on Metals and Alloys Used as Food Contact Materials" (Council of Europe, 2001).

29.4 REGULATION OF DIETARY Pb IN SOME OTHER COUNTRIES

29.4.1 Food Lead Regulations in Canada

Lead in the Canadian diet is regulated under two Federal statutes: The F&D Act for Pb as a contaminant in various categories of foods, including infant formulae in the Canadian diet, and leachable Pb levels from ceramic ware and food utensils through the Hazardous Products Act. Food health and safety standards are established by Health Canada and enforced by agencies such as the Ministry of National Health and Welfare.

Some tasks are now assumed by the Canadian Food Inspection Agency (CFIA). The CFIA was established in 1997 and broadly does the function of the USDA's FSIS (see above). For example, compliance with existing limits on lead in food is CFIA's responsibility. The Ministry of National Health and Welfare sees to establishing policies and standards for food safety and quality in Canada and accompanying tasks.

Lead leaching limits from containers and utensils into food were established within the Hazardous Products Act through R.S., c. H-3, Part 1 of Schedule 1, Item 2. The limits are similar to those of the U.S. FDA. Lead leaching provisions of the Hazardous Products Act are enforced through Health Canada's Product Safety Bureau. Any glass or cup, including crystal ware, that has an exterior decorative pattern around the rim also comes under the Canadian Product Safety Act. The F&D Act was amended by Chapter 42, Statutes of Canada, through Bill C-28, and formally codified in November 25, 2005.

Canadian food Pb regulations under the F&D Act are implemented through the Food and Drug Regulations C.R.C., c. 870. Part B covers foods, and Division 15 addresses lead as a food adulterant. Part B, Division 15 (B.15.003) of the regulations identifies in its Table 29.2 a number of food categories having tolerable Pb limits (ppm wet weight): infant formula, 0.08 ppm; evaporated milk and related products, 0.15 ppm; fruit juice, fruit nectar, ready-to-serve beverages, 0.20 ppm; fish protein and whole tomatoes, 0.50 ppm; tomato paste, tomato sauce, 1.5 ppm; edible bone meal, 10 ppm. The fruit juice figure of 0.2 ppm is fourfold higher than that promulgated by the international CAC of 0.05 ppm (see previous section). There is some indication that some of the tolerance values for Pb in different food categories will be revised, based on a petition (Petition No. 187, December 27, 2006).

29.4.2 Food Lead Regulations in the United Kingdom

England has the historical distinction of having had one of the earliest examples of regulatory intervention in the food system (Canadian Encyclopedia, 2010) via the English Assize of Bread Act, enacted ca. 1200 during the reign of King John. The act originally dealt with the volume of bread bakers were required to furnish for sale and consumption but slowly grew to encompass other food categories such as fish, meat, and ale during the thirteenth century. Both quantity and quality, i.e., freedom from adulteration, became subsumed within the act. In 1860, Britain enacted a broad body of food statutes dealing with barring adulteration of foods. Amendments were added in 1872 and 1875.

In England, Scotland, Wales, and Northern Ireland, the replacement regulations governing contamination of foods by lead migration from ceramic ware noted above for the EU regulatory process were brought into place in 2006. Also, adoption of the MLs for lead and other metal contaminants via the EC 2001 and 2006 regulations described in the previous section have proceeded. As part of the EU food lead surveys, the U.K. Food Standards Agency in October 2006 noted that baby food lead and other metal levels were well within the EC contaminant MLs.

REFERENCES

Berg, T., Licht, D., 2002. International legislation on trace elements as contaminants in food: a review. Food Addit. Contam. 19, 916–927.

Boon, P.E., Sioen, I., van der Voet, H., Huybrechts, I., De Neve, M., Aminao, P., et al., 2010. Long-term dietary exposure to lead in young children living in different European countries. Scientific Report to EFSA: EFSA-Q-2009-00837. European Food Safety Agency.

Carrington, C.D., Bolger, P.M., 1992. An assessment of the hazards of lead in food. Regul. Toxicol. Pharmacol. 16, 265–277.

CONTAM, 2004. Opinion of the Scientific Panel on Contaminants in the Food Chain (CONTAM) Related to Lead as Undesirable Substance in Animal Feed. <http://www.efsa.europa.eu/en/scdocs/scdoc/71htm> (accessed 4.08.10.).

CONTAM, 2010. EFSA Panel on Contaminants in the Food Chain (CONTAM): Scientific Opinion on Lead in Food. EFSA J. 8, 1570, doi:10.2903/jefsa.2010.1570.

Council of Europe, 2001. Technical Document: Guidelines on Metals and Alloys Used as Food Contact Materials. Policy Statement Concerning Metals and Alloys. March 9.

Food and Agriculture Organization/World Health Organization, 2010. Seventy-Third Meeting, Joint FAO/WHO Expert Committee on Food Additives (JECFA/73/SC), Geneva, June 8–17. World Health Organization, Geneva, Switzerland.

Galal-Gorchev, H., 1991. Global overview of dietary lead exposure. Chem. Speciation Bioavail. 3, 5–11.

Jorgensen, E.B., 2010. An international pooled analysis for obtaining a benchmark dose for environmental lead exposure in children. A Scientific/Technical Report Submitted to EFSA: CT/EFSA/CONTAM/2009/03. University of Copenhagen, Copenhagen, Denmark.

Kashtock, M., 2009. Reducing lead exposure from food. *Food Safety Magazine*, October/November, pp. 22–24.

Lanphear, B.P., Hornung, R., Khoury, J., Yolton, K., Baghurst, P., Bellinger, D.C., et al., 2005. Low-level environmental lead exposure and children's intellectual function: an international pooled analysis. Environ. Health Perspect. 113, 894–899.

McCord, C.P., 1953. Lead and lead poisoning in early America: Benjamin Franklin and lead poisoning. Ind. Med. Surg. 22, 393–399.

Mindak, W.R., Cheng, J., Canas, B.J., Bolger, P.M., 2008. Lead in women's and children's vitamins. J. Agric. Food Chem. 56, 6892–6896.

Miodovnik, A., Landrigan, P.J., 2009. The U.S. Food and Drug Administration risk assessment in women's and children's vitamins is based on outmoded assumptions. Environ. Health Perspect. 117, 1021–1022.

Nasreddine, L., Parent-Massin, D., 2002. Food contamination by metals and pesticides in the European Union. Should we worry? Toxicol. Lett. 127, 29–41.

National Academy of Sciences: National Research Council, 1993. Measuring Lead Exposure in Infants, Children, and Other Sensitive Populations. National Academy Press, Washington, DC.

Nriagu, J.O., 1983. Lead and Lead Poisoning in Antiquity. John Wiley & Sons, New York.

Nriagu, J.O., 1985. Historical perspective on the contamination of food and beverages with lead. In: Mahaffey, K.R. (Ed.), Dietary and Environmental Lead: Human Health Effects. Elsevier, New York, pp. 1–41.

U.S. Agency for Toxic Substances and Disease Registry, 1988. The Nature and Extent of Lead Poisoning in Children in the United States. Report to Congress. U.S. Department of Health and Human Services, Atlanta, GA.

U.S. Environmental Protection Agency, 1986. Air Quality Criteria for Lead, 4 vols. Report No. EPA/600/8-83/028bF. Ch. 7. Environmental Criteria and Assessment Office, Washington, DC.

U.S. Food and Drug Administration, 2001. Guidance for Industry: The Juice HACCP Regulation: August 31. <http://www.fda.gov/Food/GuidanceComplianceRegulatoryInformation/GuidanceDocuments/Juice/ucm072602.html> (accessed 2.08.10.).

U.S. Food and Drug Administration, 2006. Supporting Document for Recommended Maximum Level for Lead in Candy Likely to be Consumed Frequently by Small Children. Docket No. 2005D-0481. November. Washington, DC.

U.S. Food and Drug Administration, 2007. Guidance for Industry and FDA: 1991 Letter to Bureau of Alcohol, Tobacco and Firearms Regarding Lead in Wine, March. Washington, DC. <http://www.cfsan.fda.gov/guidance.html> (accessed 31.07.10.).

U.S. Government Accountability Office, 2005. Overseeing the U.S. Food Supply. Steps Should Be Taken to Reduce Overlapping Inspections and Related Activities. Report No. GAO-05-549T, May 17. Washington, DC.

World Health Organization, 1993. Guidelines for Drinking Water Quality. 2nd Ed. Vol. 1: Recommendations. World Health Organization, Geneva, Switzerland.

Index

D

E

F

G

H

J

K

O

R

S

T

U

V

W

X

Y

Z

Kolibri 360 GmbH
Ferdinand-Jühlke-Straße 7
99095 Erfurt, Deutschland
produktsicherheit@kolibri360.de